Operations Research: An Introduction

Eighth Edition

Hamdy A. Taha

University of Arkansas, Fayetteville

Upper Saddle River, New Jersey 07458

Library of Congress Cataloging-in-Publication Data

Taha, Hamdy A.
Operations research: an introduction / Hamdy A. Taha.—8th ed.
p. cm.
Includes bibliographical references and index.
ISBN 0-13-188923-0
1. Operations research. 2. Programming (Mathematics) 1. Title.

T57.6.T3 1997 96-37160
003 -dc21

96-37160

Vice President and Editorial Director, ECS: *Marcia J. Horton*
Senior Editor: *Holly Stark*
Executive Managing Editor: *Vince O'Brien*
Managing Editor: *David A. George*
Production Editor: *Craig Little*
Director of Creative Services: *Paul Belfanti*
Art Director: *Jayne Conte*
Cover Designer: *Bruce Kenselaar*
Art Editor: *Greg Dulles*
Manufacturing Manager: *Alexis Heydt-Long*
Manufacturing Buyer: *Lisa McDowell*

Pearson Prentice Hall
Pearson Education, Inc.
Upper Saddle River, NJ 07458

Printed in the United States of America
10 9 8 7 6 5 4 3 2

ISBN 0-13-188923-0

Pearson Education Ltd., *London*
Pearson Education Australia Pty. Ltd., *Sydney*
Pearson Education Singapore, Pte. Ltd.
Pearson Education North Asia Ltd., *Hong Kong*
Pearson Education Canada, Inc., *Toronto*
Pearson Educación de Mexico, S.A. de C.V.
Pearson Education—Japan, *Tokyo*
Pearson Education Malaysia, Pte. Ltd.
Pearson Education, Inc., *Upper Saddle River, New Jersey*

To Karen

Los ríos no llevan agua,
el sol las fuentes secó...
¡Yo sé donde hay una fuente
que no ha de secar el sol!
La fuente que no se agota
es mi propio corazón...

—V. Ruiz Aguilera (1862)

Contents

On the CD-ROM

Preface

The eighth edition is a major revision that streamlines the presentation of the text material with emphasis on the applications and computations in operations research:

- Chapter 2 is dedicated to linear programming modeling, with applications in the areas of urban renewal, currency arbitrage, investment, production planning, blending, scheduling, and trim loss. New end-of-section problems deal with topics ranging from water quality management and traffic control to warfare.
- Chapter 3 presents the general LP sensitivity analysis, including dual prices and reduced costs, in a simple and straightforward manner as a direct extension of the simplex tableau computations.
- Chapter 4 is now dedicated to LP post-optimal analysis based on duality.
- An Excel-based combined nearest neighbor-reversal heuristic is presented for the traveling salesperson problem.
- Markov chains treatment has been expanded into new Chapter 17.
- The totally new Chapter 24* presents 15 fully developed real-life applications. The analysis, which often cuts across more than one OR technique (e.g., heuristics and LP, or ILP and queuing), deals with the modeling, data collection, and computational aspects of solving the problem. These applications are cross-referenced in pertinent chapters to provide an appreciation of the use of OR techniques in practice.
- The new Appendix E* includes approximately 50 mini cases of real-life situations categorized by chapters.
- More than 1000 end-of-section problem are included in the book.
- Each chapter starts with a *study guide* that facilitates the understanding of the material and the effective use of the accompanying software.
- The integration of software in the text allows testing concepts that otherwise could not be presented effectively:
    1. Excel spreadsheet implementations are used throughout the book, including dynamic programming, traveling salesperson, inventory, AHP, Bayes' probabilities, "electronic" statistical tables, queuing, simulation, Markov chains, and nonlinear programming. The interactive user input in some spreadsheets promotes a better understanding of the underlying techniques.
    2. The use of Excel Solver has been expanded throughout the book, particularly in the areas of linear, network, integer, and nonlinear programming.
    3. The powerful commercial modeling language, AMPL®, has been integrated in the book using numerous examples ranging from linear and network to

*Contained on the CD-ROM.

integer and nonlinear programming. The syntax of AMPL is given in Appendix A and its material cross-referenced within the examples in the book.
4. TORA continue to play the key role of tutorial software.
- All computer-related material has been deliberately compartmentalized either in separate sections or as subsection titled *AMPL/Excel/Solver/TORA moment* to minimize disruptions in the main presentation in the book.

To keep the page count at a manageable level, some sections, complete chapters, and two appendixes have been moved to the accompanying CD. The selection of the excised material is based on the author's judgment regarding frequency of use in introductory OR classes.

ACKNOWLEDGMENTS

I wish to acknowledge the importance of the seventh edition reviews provided by Layek L. Abdel-Malek, New Jersey Institute of Technology, Evangelos Triantaphyllou, Louisiana State University, Michael Branson, Oklahoma State University, Charles H. Reilly, University of Central Florida, and Mazen Arafeh, Virginia Polytechnic Institute and State University. In particular, I owe special thanks to two individuals who have influenced my thinking during the preparation of the eighth edition: R. Michael Harnett (Kansas State University), who over the years has provided me with valuable feedback regarding the organization and the contents of the book, and Richard H. Bernhard (North Carolina State University), whose detailed critique of the seventh edition prompted a reorganization of the opening chapters in this edition.

Robert Fourer (Northwestern University) patiently provided me with valuable feedback about the AMPL material presented in this edition. I appreciate his help in editing the material and for suggesting changes that made the presentation more readable. I also wish to acknowledge his help in securing permissions to include the AMPL student version and the solvers CPLEX, KNITRO, LPSOLVE, LOQO, and MINOS on the accompanying CD.

As always, I remain indebted to my colleagues and to hundreds of students for their comments and their encouragement. In particular, I wish to acknowledge the support I receive from Yuh-Wen Chen (Da-Yeh University, Taiwan), Miguel Crispin (University of Texas, El Paso), David Elizandro (Tennessee Tech University), Rafael Gutiérrez (University of Texas, El Paso), Yasser Hosni (University of Central Florida), Erhan Kutanoglu (University of Texas, Austin), Robert E. Lewis (United States Army Management Engineering College), Gino Lim (University of Houston), Scott Mason (University of Arkansas), Heather Nachtman (University of Arkanas), Manuel Rossetti (University of Arkansas), Tarek Taha (JB Hunt, Inc.), and Nabeel Yousef (University of Central Florida).

I wish to express my sincere appreciation to the Pearson Prentice Hall editorial and production teams for their superb assistance during the production of the book: Dee Bernhard (Associate Editor), David George (Production Manager - Engineering), Bob Lentz (Copy Editor), Craig Little (Production Editor), and Holly Stark (Senior Acquisitions Editor).

HAMDY A. TAHA
hat@uark.edu
http://ineg.uark.edu/TahaORbook/

About the Author

Hamdy A. Taha is a University Professor Emeritus of Industrial Engineering with the University of Arkansas, where he taught and conducted research in operations research and simulation. He is the author of three other books on integer programming and simulation, and his works have been translated into Malay, Chinese, Korean, Spanish, Japanese, Russian, Turkish, and Indonesian. He is also the author of several book chapters, and his technical articles have appeared in *European Journal of Operations Research, IEEE Transactions on Reliability, IIE Transactions, Interfaces, Management Science, Naval Research Logistics Quarterly, Operations Research,* and *Simulation*.

Professor Taha was the recipient of the Alumni Award for excellence in research and the university-wide Nadine Baum Award for excellence in teaching, both from the University of Arkansas, and numerous other research and teaching awards from the College of Engineering, University of Arkansas. He was also named a Senior Fulbright Scholar to Carlos III University, Madrid, Spain. He is fluent in three languages and has held teaching and consulting positions in Europe, Mexico, and the Middle East.

Trademarks

AMPL is a registered trademark of AMPL Optimization, LLC, 900 Sierra Pl. SE, Albuquerque, NM 87108-3379.

CPLEX is a registered trademark of ILOG, Inc., 1080 Linda Vista Ave., Mountain View, CA 94043.

KNITRO is a registered trademark of Ziena Optimization Inc., 1801 Maple Ave, Evanston, IL 60201.

LOQO is a trademark of Princeton University, Princeton, NJ 08544.

Microsoft, Windows, and Excel registered trademarks of Microsoft Corporation in the United States and/or other countries.

MINOS is a trademark of Stanford University, Stanford, CA 94305.

Solver is a trademark of Frontline Systems, Inc., 7617 Little River Turnpike, Suite 960, Annandale, VA 22003.

TORA is a trademark of SimTec, Inc., P.O. Box 3492, Fayetteville, AR 72702

Note: Other product and company names that are mentioned herein may be trademarks or registered trademarks of their respective owners in the United States and/or other countries.

CHAPTER 1

What Is Operations Research?

Chapter Guide. The first formal activities of Operations Research (OR) were initiated in England during World War II, when a team of British scientists set out to make scientifically based decisions regarding the best utilization of war materiel. After the war, the ideas advanced in military operations were adapted to improve efficiency and productivity in the civilian sector.

This chapter will familiarize you with the basic terminology of operations research, including mathematical modeling, feasible solutions, optimization, and iterative computations. You will learn that defining the problem correctly is the most important (and most difficult) phase of practicing OR. The chapter also emphasizes that, while mathematical modeling is a cornerstone of OR, intangible (unquantifiable) factors (such as human behavior) must be accounted for in the final decision. As you proceed through the book, you will be presented with a variety of applications through solved examples and chapter problems. In particular, Chapter 24 (on the CD) is entirely devoted to the presentation of fully developed case analyses. Chapter materials are cross-referenced with the cases to provide an appreciation of the use of OR in practice.

1.1 OPERATIONS RESEARCH MODELS

Imagine that you have a 5-week business commitment between Fayetteville (FYV) and Denver (DEN). You fly out of Fayetteville on Mondays and return on Wednesdays. A regular round-trip ticket costs $400, but a 20% discount is granted if the dates of the ticket span a weekend. A one-way ticket in either direction costs 75% of the regular price. How should you buy the tickets for the 5-week period?

We can look at the situation as a decision-making problem whose solution requires answering three questions:

1. What are the decision **alternatives?**
2. Under what **restrictions** is the decision made?
3. What is an appropriate **objective criterion** for evaluating the alternatives?

Three alternatives are considered:

1. **1.** Buy five regular FYV-DEN-FYV for departure on Monday and return on Wednesday of the same week.
2. **2.** Buy one FYV-DEN, four DEN-FYV-DEN that span weekends, and one DEN-FYV.
3. **3.** Buy one FYV-DEN-FYV to cover Monday of the first week and Wednesday of the last week and four DEN-FYV-DEN to cover the remaining legs. All tickets in this alternative span at least one weekend.

The restriction on these options is that you should be able to leave FYV on Monday and return on Wednesday of the same week.

An obvious objective criterion for evaluating the proposed alternative is the price of the tickets. The alternative that yields the smallest cost is the best. Specifically, we have

$$\text{Alternative 1 cost} = 5 \times 400 = \$2000$$

$$\text{Alternative 2 cost} = .75 \times 400 + 4 \times (.8 \times 400) + .75 \times 400 = \$1880$$

$$\text{Alternative 3 cost} = 5 \times (.8 \times 400) = \mathbf{\$1600}$$

Thus, you should choose alternative 3.

Though the preceding example illustrates the three main components of an OR model—alternatives, objective criterion, and constraints—situations differ in the details of how each component is developed and constructed. To illustrate this point, consider forming a maximum-area rectangle out of a piece of wire of length L inches. What should be the width and height of the rectangle?

In contrast with the tickets example, the number of alternatives in the present example is not finite; namely, the width and height of the rectangle can assume an infinite number of values. To formalize this observation, the alternatives of the problem are identified by defining the width and height as continuous (algebraic) variables.

Let

$$w = \text{width of the rectangle in inches}$$

$$h = \text{height of the rectangle in inches}$$

Based on these definitions, the restrictions of the situation can be expressed verbally as

1. **1.** Width of rectangle + Height of rectangle = Half the length of the wire
2. **2.** Width and height cannot be negative

These restrictions are translated algebraically as

1. **1.** $2(w + h) = L$
2. **2.** $w \geq 0, h \geq 0$

The only remaining component now is the objective of the problem; name maximization of the area of the rectangle. Let z be the area of the rectangle, then the complete model becomes

$$\text{Maximize } z = wh$$

subject to

$$2(w + h) = L$$
$$w, h \geq 0$$

The optimal solution of this model is $w = h = \frac{L}{4}$, which calls for constructing a square shape.

Based on the preceding two examples, the general OR model can be organized in the following general format:

Maximize or minimize **Objective Function**

subject to

Constraints

A solution of the mode is **feasible** if it satisfies all the constraints. It is **optimal** if, in addition to being feasible, it yields the best (maximum or minimum) value of the objective function. In the tickets example, the problem presents three feasible alternatives, with the third alternative yielding the optimal solution. In the rectangle problem, a feasible alternative must satisfy the condition $w + h = \frac{L}{2}$ with w and h assuming nonnegative values. This leads to an infinite number of feasible solutions and, unlike the tickets problem, the optimum solution is determined by an appropriate mathematical tool (in this case, differential calculus).

Though OR models are designed to "optimize" a specific objective criterion subject to a set of constraints, the quality of the resulting solution depends on the completeness of the model in representing the real system. Take, for example, the tickets model. If one is not able to identify all the dominant alternatives for purchasing the tickets, then the resulting solution is optimum only relative to the choices represented in the model. To be specific, if alternative 3 is left out of the model, then the resulting "optimum" solution would call for purchasing the tickets for \$1880, which is a **suboptimal** solution. The conclusion is that "the" optimum solution of a model is best only for *that* model. If the model happens to represent the real system reasonably well, then its solution is optimum also for the real situation.

PROBLEM SET 1.1A

1. In the tickets example, identify a fourth feasible alternative.
2. In the rectangle problem, identify two feasible solutions and determine which one is better.
3. Determine the optimal solution of the rectangle problem. (*Hint*: Use the constraint to express the objective function in terms of one variable, then use differential calculus.)

, Jim, John, and Kelly are standing on the east bank of a river and wish to cross to est side using a canoe. The canoe can hold at most two people at a time. Amy, being ost athletic, can row across the river in 1 minute. Jim, John, and Kelly would take 2, 10 minutes, respectively. If two people are in the canoe, the slower person dictates the crossing time. The objective is for all four people to be on the other side of the river in the shortest time possible.

(a) Identify at least two feasible plans for crossing the river (remember, the canoe is the only mode of transportation and it cannot be shuttled empty).

(b) Define the criterion for evaluating the alternatives.

***(c)**[1] What is the smallest time for moving all four people to the other side of the river?

***5.** In a baseball game, Jim is the pitcher and Joe is the batter. Suppose that Jim can throw either a fast or a curve ball at random. If Joe correctly predicts a curve ball, he can maintain a .500 batting average, else if Jim throws a curve ball and Joe prepares for a fast ball, his batting average is kept down to .200. On the other hand, if Joe correctly predicts a fast ball, he gets a .300 batting average, else his batting average is only .100.

(a) Define the alternatives for this situation.

(b) Define the objective function for the problem and discuss how it differs from the familiar optimization (maximization or minimization) of a criterion.

6. During the construction of a house, six joists of 24 feet each must be trimmed to the correct length of 23 feet. The operations for cutting a joist involve the following sequence:

Operation	Time (seconds)
1. Place joist on saw horses	15
2. Measure correct length (23 feet)	5
3. Mark cutting line for circular saw	5
4. Trim joist to correct length	20
5. Stack trimmed joist in a designated area	20

Three persons are involved: Two loaders must work simultaneously on operations 1, 2, and 5, and one cutter handles operations 3 and 4. There are two pairs of saw horses on which untrimmed joists are placed in preparation for cutting, and each pair can hold up to three side-by-side joists. Suggest a good schedule for trimming the six joists.

1.2 SOLVING THE OR MODEL

In OR, we do not have a single general technique to solve all mathematical models that can arise in practice. Instead, the type and complexity of the mathematical model dictate the nature of the solution method. For example, in Section 1.1 the solution of the tickets problem requires simple ranking of alternatives based on the total purchasing price, whereas the solution of the rectangle problem utilizes differential calculus to determine the maximum area.

The most prominent OR technique is **linear programming**. It is designed for models with linear objective and constraint functions. Other techniques include **integer programming** (in which the variables assume integer values), **dynamic programming**

[1]An asterisk (*) designates problems whose solution is provided in Appendix C.

(in which the original model can be decomposed into more manageable subproblems), **network programming** (in which the problem can be modeled as a network), and **nonlinear programming** (in which functions of the model are nonlinear). These are only a few among many available OR tools.

A peculiarity of most OR techniques is that solutions are not generally obtained in (formulalike) closed forms. Instead, they are determined by **algorithms.** An algorithm provides fixed computational rules that are applied repetitively to the problem, with each repetition (called **iteration**) moving the solution closer to the optimum. Because the computations associated with each iteration are typically tedious and voluminous, it is imperative that these algorithms be executed on the computer.

Some mathematical models may be so complex that it is impossible to solve them by any of the available optimization algorithms. In such cases, it may be necessary to abandon the search for the *optimal* solution and simply seek a *good* solution using **heuristics** or *rules of thumb.*

1.3 QUEUING AND SIMULATION MODELS

Queuing and simulation deal with the study of waiting lines. They are not optimization techniques; rather, they determine measures of performance of the waiting lines, such as average waiting time in queue, average waiting time for service, and utilization of service facilities.

Queuing models utilize probability and stochastic models to analyze waiting lines, and simulation estimates the measures of performance by imitating the behavior of the real system. In a way, simulation may be regarded as the next best thing to observing a real system. The main difference between queuing and simulation is that queuing models are purely mathematical, and hence are subject to specific assumptions that limit their scope of application. Simulation, on the other hand, is flexible and can be used to analyze practically any queuing situation.

The use of simulation is not without drawbacks. The process of developing simulation models is costly in both time and resources. Moreover, the execution of simulation models, even on the fastest computer, is usually slow.

1.4 ART OF MODELING

The illustrative models developed in Section 1.1 are true representations of real situations. This is a rare occurrence in OR, as the majority of applications usually involve (varying degrees of) approximations. Figure 1.1 depicts the levels of abstraction that characterize the development of an OR model. We abstract the assumed real world from the real situation by concentrating on the dominant variables that control the behavior of the real system. The model expresses in an amenable manner the mathematical functions that represent the behavior of the assumed real world.

To illustrate levels of abstraction in modeling, consider the Tyko Manufacturing Company, where a variety of plastic containers are produced. When a production order is issued to the production department, necessary raw materials are acquired from the company's stocks or purchased from outside sources. Once the production batch is completed, the sales department takes charge of distributing the product to customers.

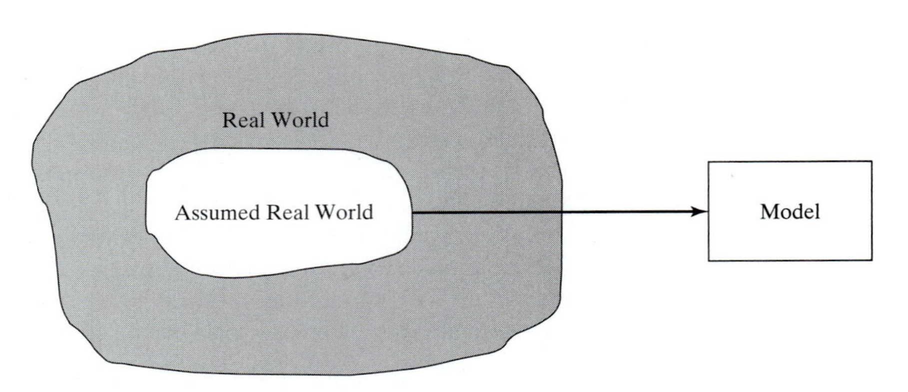

FIGURE 1.1
Levels of abstraction in model development

A logical question in the analysis of Tyko's situation is the determination of the size of a production batch. How can this situation be represented by a model?

Looking at the overall system, a number of variables can bear directly on the level of production, including the following (partial) list categorized by departments.

1. *Production Department:* Production capacity expressed in terms of available machine and labor hours, in-process inventory, and quality control standards.
2. *Materials Department:* Available stock of raw materials, delivery schedules from outside sources, and storage limitations.
3. *Sales Department:* Sales forecast, capacity of distribution facilities, effectiveness of the advertising campaign, and effect of competition.

Each of these variables affects the level of production at Tyko. Trying to establish explicit functional relationships between them and the level of production is a difficult task indeed.

A first level of abstraction requires defining the boundaries of the assumed real world. With some reflection, we can approximate the real system by two dominant variables:

1. Production rate.
2. Consumption rate.

Determination of the production rate involves such variables as production capacity, quality control standards, and availability of raw materials. The consumption rate is determined from the variables associated with the sales department. In essence, simplification from the real world to the assumed real world is achieved by "lumping" several real-world variables into a single assumed-real-world variable.

It is easier now to abstract a model from the assumed real world. From the production and consumption rates, measures of excess or shortage inventory can be established. The abstracted model may then be constructed to balance the conflicting costs of excess and shortage inventory—i.e., to minimize the total cost of inventory.

1.5 MORE THAN JUST MATHEMATICS

Because of the mathematical nature of OR models, one tends to think that an OR study is *always* rooted in mathematical analysis. Though mathematical modeling is a cornerstone of OR, simpler approaches should be explored first. In some cases, a "common sense" solution may be reached through simple observations. Indeed, since the human element invariably affects most decision problems, a study of the psychology of people may be key to solving the problem. Three illustrations are presented here to support this argument.

1. Responding to complaints of slow elevator service in a large office building, the OR team initially perceived the situation as a waiting-line problem that might require the use of mathematical queuing analysis or simulation. After studying the behavior of the people voicing the complaint, the psychologist on the team suggested installing full-length mirrors at the entrance to the elevators. Miraculously the complaints disappeared, as people were kept occupied watching themselves and others while waiting for the elevator.

2. In a study of the check-in facilities at a large British airport, a United States-Canadian consulting team used queuing theory to investigate and analyze the situation. Part of the solution recommended the use of well-placed signs to urge passengers who were within 20 minutes from departure time to advance to the head of the queue and request immediate service. The solution was not successful, because the passengers, being mostly British, were "conditioned to very strict queuing behavior" and hence were reluctant to move ahead of others waiting in the queue.

3. In a steel mill, ingots were first produced from iron ore and then used in the manufacture of steel bars and beams. The manager noticed a long delay between the ingots production and their transfer to the next manufacturing phase (where end products were manufactured). Ideally, to reduce the reheating cost, manufacturing should start soon after the ingots left the furnaces. Initially the problem was perceived as a line-balancing situation, which could be resolved either by reducing the output of ingots or by increasing the capacity of the manufacturing process. The OR team used simple charts to summarize the output of the furnaces during the three shifts of the day. They discovered that, even though the third shift started at 11:00 P.M., most of the ingots were produced between 2:00 and 7:00 A.M. Further investigation revealed that third-shift operators preferred to get long periods of rest at the start of the shift and then make up for lost production during morning hours. The problem was solved by "leveling out" the production of ingots throughout the shift.

Three conclusions can be drawn from these illustrations:

1. Before embarking on sophisticated mathematical modeling, the OR team should explore the possibility of using "aggressive" ideas to resolve the situation. The solution of the elevator problem by installing mirrors is rooted in human psychology rather than in mathematical modeling. It is also simpler and less costly than any recommendation a mathematical model might have produced. Perhaps this is the reason OR teams usually include the expertise of "outsiders" from nonmathematical fields (psychology in the case of the elevator problem). This point was recognized and implemented by the first OR team in Britain during World War II.

2. Solutions are rooted in people and not in technology. Any solution that does not take human behavior into account is apt to fail. Even though the mathematical solution of the British airport problem may have been sound, the fact that the consulting team was not aware of the cultural differences between the United States and Britain (Americans and Canadians tend to be less formal) resulted in an unimplementable recommendation.

3. An OR study should never start with a bias toward using a specific mathematical tool before its use can be justified. For example, because linear programming is a successful technique, there is a tendency to use it as the tool of choice for modeling "any" situation. Such an approach usually leads to a mathematical model that is far removed from the real situation. It is thus imperative that we first analyze available data, using the simplest techniques where possible (e.g., averages, charts, and histograms), with the objective of pinpointing the source of the problem. Once the problem is defined, a decision can be made regarding the most appropriate tool for the solution.[2] In the steel mill problem, simple charting of the ingots production was all that was needed to clarify the situation.

1.6 PHASES OF AN OR STUDY

An OR study is rooted in *teamwork*, where the OR analysts and the client work side by side. The OR analysts' expertise in modeling must be complemented by the experience and cooperation of the client for whom the study is being carried out.

As a decision-making tool, OR is both a science and an art. It is a science by virtue of the mathematical techniques it embodies, and it is an art because the success of the phases leading to the solution of the mathematical model depends largely on the creativity and experience of the operations research team. Willemain (1994) advises that "effective [OR] practice requires more than analytical competence: It also requires, among other attributes, technical judgement (e.g., when and how to use a given technique) and skills in communication and organizational survival."

It is difficult to prescribe specific courses of action (similar to those dictated by the precise theory of mathematical models) for these intangible factors. We can, however, offer general guidelines for the implementation of OR in practice.

The principal phases for implementing OR in practice include

1. Definition of the problem.
2. Construction of the model.

[2]Deciding on a specific mathematical model before justifying its use is like "putting the cart before the horse," and it reminds me of the story of a frequent air traveler who was paranoid about the possibility of a terrorist bomb on board the plane. He calculated the probability that such an event could occur, and though quite small, it wasn't small enough to calm his anxieties. From then on, he always carried a bomb in his briefcase on the plane because, according to his calculations, the probability of having *two* bombs aboard the plane was practically zero!

3. Solution of the model.
4. Validation of the model.
5. Implementation of the solution.

Phase 3, dealing with *model solution*, is the best defined and generally
plement in an OR study, because it deals mostly with precise mathematical models. Implementation of the remaining phases is more an art than a theory.

Problem definition involves defining the scope of the problem under investigation. This function should be carried out by the entire OR team. The aim is to identify three principal elements of the decision problem: (1) description of the decision alternatives, (2) determination of the objective of the study, and (3) specification of the limitations under which the modeled system operates.

Model construction entails an attempt to translate the problem definition into mathematical relationships. If the resulting model fits one of the standard mathematical models, such as linear programming, we can usually reach a solution by using available algorithms. Alternatively, if the mathematical relationships are too complex to allow the determination of an analytic solution, the OR team may opt to simplify the model and use a heuristic approach, or they may consider the use of simulation, if appropriate. In some cases, mathematical, simulation, and heuristic models may be combined to solve the decision problem, as the case analyses in Chapter 24 demonstrate.

Model solution is by far the simplest of all OR phases because it entails the use of well-defined optimization algorithms. An important aspect of the model solution phase is *sensitivity analysis*. It deals with obtaining additional information about the behavior of the optimum solution when the model undergoes some parameter changes. Sensitivity analysis is particularly needed when the parameters of the model cannot be estimated accurately. In these cases, it is important to study the behavior of the optimum solution in the neighborhood of the estimated parameters.

Model validity checks whether or not the proposed model does what it purports to do—that is, does it predict adequately the behavior of the system under study? Initially, the OR team should be convinced that the model's output does not include "surprises." In other words, does the solution make sense? Are the results intuitively acceptable? On the formal side, a common method for checking the validity of a model is to compare its output with historical output data. The model is valid if, under similar input conditions, it reasonably duplicates past performance. Generally, however, there is no assurance that future performance will continue to duplicate past behavior. Also, because the model is usually based on careful examination of past data, the proposed comparison is usually favorable. If the proposed model represents a new (nonexisting) system, no historical data would be available. In such cases, we may use simulation as an independent tool for verifying the output of the mathematical model.

Implementation of the solution of a validated model involves the translation of the results into understandable operating instructions to be issued to the people who will administer the recommended system. The burden of this task lies primarily with the OR team.

ABOUT THIS BOOK

Morris (1967) states that "the teaching of models is not equivalent to the teaching of modeling." I have taken note of this important statement during the preparation of the eighth edition, making an effort to introduce the art of modeling in OR by including realistic models throughout the book. Because of the importance of computations in OR, the book presents extensive tools for carrying out this task, ranging from the tutorial aid TORA to the commercial packages Excel, Excel Solver, and AMPL.

A first course in OR should give the student a good foundation in the mathematics of OR as well as an appreciation of its potential applications. This will provide OR users with the kind of confidence that normally would be missing if training were concentrated only on the philosophical and artistic aspects of OR. Once the mathematical foundation has been established, you can increase your capabilities in the artistic side of OR modeling by studying published practical cases. To assist you in this regard, Chapter 24 includes 15 fully developed and analyzed cases that cover most of the OR models presented in this book. There are also some 50 cases that are based on real-life applications in Appendix E on the CD. Additional case studies are available in journals and publications. In particular, *Interfaces* (published by INFORMS) is a rich source of diverse OR applications.

REFERENCES

Altier, W. J., *The Thinking Manager's Toolbox: Effective Processes for Problem Solving and Decision Making*, Oxford University Press, New York, 1999.

Checkland, P., *Systems Thinking, System Practice*, Wiley, New York, 1999.

Evans, J., *Creative Thinking in the Decision and Management Sciences*, South-Western Publishing, Cincinnati, 1991.

Gass, S., "Model World: Danger, Beware the User as a Modeler," *Interfaces*, Vol. 20, No. 3, pp. 60–64, 1990.

Morris, W., "On the Art of Modeling," *Management Science*, Vol. 13, pp. B707–B717, 1967.

Paulos, J.A., *Innumeracy: Mathematical Illiteracy and its Consequences*, Hill and Wang, New York, 1988.

Singh, Simon, *Fermat's Enigma*, Walker, New York, 1997.

Willemain, T. R., "Insights on Modeling from a Dozen Experts," *Operations Research*, Vol. 42, No. 2, pp. 213–222, 1994.

CHAPTER 2

Modeling with Linear Programming

Chapter Guide. This chapter concentrates on model formulation and computations in linear programming (LP). It starts with the modeling and graphical solution of a two-variable problem which, though highly simplified, provides a concrete understanding of the basic concepts of LP and lays the foundation for the development of the general *simplex* algorithm in Chapter 3. To illustrate the use of LP in the real world, applications are formulated and solved in the areas of urban planning, currency arbitrage, investment, production planning and inventory control, gasoline blending, manpower planning, and scheduling. On the computational side, two distinct types of software are used in this chapter. (1) TORA, a totally menu-driven and self-documenting tutorial program, is designed to help you understand the basics of LP through interactive feedback. (2) Spreadsheet-based Excel Solver and the AMPL modeling language are commercial packages designed for practical problems.

The material in Sections 2.1 and 2.2 is crucial for understanding later LP developments in the book. You will find TORA's interactive graphical module especially helpful in conjunction with Section 2.2. Section 2.3 presents diverse LP applications, each followed by targeted problems.

Section 2.4 introduces the commercial packages Excel Solver and AMPL. Models in Section 2.3 are solved with AMPL and Solver, and all the codes are included in folder ch2Files. Additional Solver and AMPL models are included opportunely in the succeeding chapters, and a detailed presentation of AMPL syntax is given in Appendix A. A good way to learn AMPL and Solver is to experiment with the numerous models presented throughout the book and to try to adapt them to the end-of-section problems. The AMPL codes are cross-referenced with the material in Appendix A to facilitate the learning process.

The TORA, Solver, and AMPL materials have been deliberately compartmentalized either in separate sections or under the subheadings *TORA/Solver/AMPL moment* to minimize disruptions in the main text. Nevertheless, you are encouraged to work end-of-section problems on the computer. The reason is that, at times, a model may look "correct" until you try to obtain a solution, and only then will you discover that the formulation needs modifications.

This chapter includes summaries of 2 real-life applications, 12 solved examples, 2 Solver models, 4 AMPL models, 94 end-of-section problems, and 4 cases. The cases are in Appendix E on the CD. The AMPL/Excel/Solver/TORA programs are in folder ch2Files.

Real-Life Application–Frontier Airlines Purchases Fuel Economically

The fueling of an aircraft can take place at any of the stopovers along the flight route. Fuel price varies among the stopovers, and potential savings can be realized by loading extra fuel (called *tankering*) at a cheaper location for use on subsequent flight legs. The disadvantage of tankering is the excess burn of gasoline resulting from the extra weight. LP (and heuristics) is used to determine the optimum amount of tankering that balances the cost of excess burn against the savings in fuel cost. The study, carried out in 1981, resulted in net savings of about $350,000 per year. Case 1 in Chapter 24 on the CD provides the details of the study. Interestingly, with the recent rise in the cost of fuel, many airlines are now using LP-based tankering software to purchase fuel.

2.1 TWO-VARIABLE LP MODEL

This section deals with the graphical solution of a two-variable LP. Though two-variable problems hardly exist in practice, the treatment provides concrete foundations for the development of the general simplex algorithm presented in Chapter 3.

Example 2.1-1 (The Reddy Mikks Company)

Reddy Mikks produces both interior and exterior paints from two raw materials, $M1$ and $M2$. The following table provides the basic data of the problem:

	Tons of raw material per ton of		Maximum daily availability (tons)
	Exterior paint	*Interior paint*	
Raw material, $M1$	6	4	24
Raw material, $M2$	1	2	6
Profit per ton ($1000)	5	4	

A market survey indicates that the daily demand for interior paint cannot exceed that for exterior paint by more than 1 ton. Also, the maximum daily demand for interior paint is 2 tons.

Reddy Mikks wants to determine the optimum (best) product mix of interior and exterior paints that maximizes the total daily profit.

The LP model, as in any OR model, has three basic components.

1. Decision **variables** that we seek to determine.
2. **Objective** (goal) that we need to optimize (maximize or minimize).
3. **Constraints** that the solution must satisfy.

The proper definition of the decision variables is an essential first step in the development of the model. Once done, the task of constructing the objective function and the constraints becomes more straightforward.

For the Reddy Mikks problem, we need to determine the daily amounts to be produced of exterior and interior paints. Thus the variables of the model are defined as

$$x_1 = \text{Tons produced daily of exterior paint}$$
$$x_2 = \text{Tons produced daily of interior paint}$$

To construct the objective function, note that the company wants to *maximize* (i.e., increase as much as possible) the total daily profit of both paints. Given that the profits per ton of exterior and interior paints are 5 and 4 (thousand) dollars, respectively, it follows that

$$\text{Total profit from exterior paint} = 5x_1 \text{ (thousand) dollars}$$
$$\text{Total profit from interior paint} = 4x_2 \text{ (thousand) dollars}$$

Letting z represent the total daily profit (in thousands of dollars), the objective of the company is

$$\text{Maximize } z = 5x_1 + 4x_2$$

Next, we construct the constraints that restrict raw material usage and product demand. The raw material restrictions are expressed verbally as

$$\begin{pmatrix}\text{Usage of a raw material}\\ \text{by both paints}\end{pmatrix} \leq \begin{pmatrix}\text{Maximum raw material}\\ \text{availability}\end{pmatrix}$$

The daily usage of raw material $M1$ is 6 tons per ton of exterior paint and 4 tons per ton of interior paint. Thus

$$\text{Usage of raw material } M1 \text{ by exterior paint} = 6x_1 \text{ tons/day}$$
$$\text{Usage of raw material } M1 \text{ by interior paint} = 4x_2 \text{ tons/day}$$

Hence

$$\text{Usage of raw material } M1 \text{ by both paints} = 6x_1 + 4x_2 \text{ tons/day}$$

In a similar manner,

$$\text{Usage of raw material } M2 \text{ by both paints} = 1x_1 + 2x_2 \text{ tons/day}$$

Because the daily availabilities of raw materials $M1$ and $M2$ are limited to 24 and 6 tons, respectively, the associated restrictions are given as

$$6x_1 + 4x_2 \leq 24 \quad \text{(Raw material } M1\text{)}$$
$$x_1 + 2x_2 \leq 6 \quad \text{(Raw material } M2\text{)}$$

The first demand restriction stipulates that the excess of the daily production of interior over exterior paint, $x_2 - x_1$, should not exceed 1 ton, which translates to

$$x_2 - x_1 \leq 1 \quad \text{(Market limit)}$$

The second demand restriction stipulates that the maximum daily demand of interior paint is limited to 2 tons, which translates to

$$x_2 \leq 2 \text{ (Demand limit)}$$

An implicit (or "understood-to-be") restriction is that variables x_1 and x_2 cannot assume negative values. The **nonnegativity restrictions,** $x_1 \geq 0, x_2 \geq 0$, account for this requirement.

The complete Reddy Mikks model is

$$\text{Maximize } z = 5x_1 + 4x_2$$

subject to

$$6x_1 + 4x_2 \leq 24 \quad (1)$$
$$x_1 + 2x_2 \leq 6 \quad (2)$$
$$-x_1 + x_2 \leq 1 \quad (3)$$
$$x_2 \leq 2 \quad (4)$$
$$x_1, x_2 \geq 0 \quad (5)$$

Any values of x_1 and x_2 that satisfy *all* five constraints constitute a **feasible solution**. Otherwise, the solution is **infeasible**. For example, the solution, $x_1 = 3$ tons per day and $x_2 = 1$ ton per day, is feasible because it does not violate *any* of the constraints, including the nonnegativity restrictions. To verify this result, substitute $(x_1 = 3, x_2 = 1)$ in the left-hand side of each constraint. In constraint (1) we have $6x_1 + 4x_2 = 6 \times 3 + 4 \times 1 = 22$, which is less than the right-hand side of the constraint (= 24). Constraints 2 through 5 will yield similar conclusions (verify!). On the other hand, the solution $x_1 = 4$ and $x_2 = 1$ is infeasible because it does not satisfy constraint (1)—namely, $6 \times 4 + 4 \times 1 = 28$, which is larger than the right-hand side (= 24).

The goal of the problem is to find the best *feasible* solution, or the **optimum**, that maximizes the total profit. Before we can do that, we need to know how many *feasible* solutions the Reddy Mikks problem has. The answer, as we will see from the graphical solution in Section 2.2, is "an infinite number," which makes it impossible to solve the problem by enumeration. Instead, we need a systematic procedure that will locate the optimum solution in a finite number of steps. The graphical method in Section 2.2 and its algebraic generalization in Chapter 3 will explain how this can be accomplished.

Properties of the LP Model. In Example 2.1-1, the objective and the constraints are all linear functions. **Linearity** implies that the LP must satisfy three basic properties:

1. Proportionality: This property requires the contribution of each decision variable in both the objective function and the constraints to be *directly proportional* to the value of the variable. For example, in the Reddy Mikks model, the quantities $5x_1$ and $4x_2$ give the profits for producing x_1 and x_2 tons of exterior and interior paint, respectively, with the unit profits per ton, 5 and 4, providing the constants of proportionality. If, on the other hand, Reddy Mikks grants some sort of quantity discounts when sales exceed certain amounts, then the profit will no longer be proportional to the production amounts, x_1 and x_2, and the profit function becomes nonlinear.

2. Additivity: This property requires the total contribution of all the variables in the objective function and in the constraints to be the direct sum of the individual contributions of each variable. In the Reddy Mikks model, the total profit equals the sum of the two individual profit components. If, however, the two products *compete* for market share in such a way that an increase in sales of one adversely affects the other, then the additivity property is not satisfied and the model is no longer linear.

3. Certainty: All the objective and constraint coefficients of the LP model are deterministic. This means that they are known constants—a rare occurrence in real life, where data are more likely to be represented by probabilistic distributions. In essence, LP coefficients are average-value approximations of the probabilistic distributions. If the standard deviations of these distributions are sufficiently small, then the approximation is acceptable. Large standard deviations can be accounted for directly by using stochastic LP algorithms (Section 19.2.3) or indirectly by applying sensitivity analysis to the optimum solution (Section 3.6).

PROBLEM SET 2.1A

1. For the Reddy Mikks model, construct each of the following constraints and express it with a linear left-hand side and a constant right-hand side:

*__(a)__ The daily demand for interior paint exceeds that of exterior paint by *at least* 1 ton.

(b) The daily usage of raw material $M2$ in tons is *at most* 6 and *at least* 3.

*__(c)__ The demand for interior paint cannot be less than the demand for exterior paint.

(d) The minimum quantity that should be produced of both the interior and the exterior paint is 3 tons.

*__(e)__ The proportion of interior paint to the total production of both interior and exterior paints must not exceed .5.

2. Determine the best *feasible* solution among the following (feasible and infeasible) solutions of the Reddy Mikks model:

(a) $x_1 = 1, x_2 = 4$.

(b) $x_1 = 2, x_2 = 2$.

(c) $x_1 = 3, x_2 = 1.5$.

(d) $x_1 = 2, x_2 = 1$.

(e) $x_1 = 2, x_2 = -1$.

*__3.__ For the feasible solution $x_1 = 2, x_2 = 2$ of the Reddy Mikks model, determine the unused amounts of raw materials $M1$ and $M2$.

4. Suppose that Reddy Mikks sells its exterior paint to a single wholesaler at a quantity discount. The profit per ton is $5000 if the contractor buys no more than 2 tons daily and $4500 otherwise. Express the objective function mathematically. Is the resulting function linear?

2.2 GRAPHICAL LP SOLUTION

The graphical procedure includes two steps:

1. Determination of the feasible solution space.
2. Determination of the optimum solution from among all the feasible points in the solution space.

The procedure uses two examples to show how maximization and minimization objective functions are handled.

2.2.1 Solution of a Maximization Model

Example 2.2-1

This example solves the Reddy Mikks model of Example 2.1-1.

Step 1. *Determination of the Feasible Solution Space:*
First, we account for the nonnegativity constraints $x_1 \geq 0$ and $x_2 \geq 0$. In Figure 2.1, the horizontal axis x_1 and the vertical axis x_2 represent the exterior- and interior-paint variables, respectively. Thus, the nonnegativity of the variables restricts the solution-space area to the first quadrant that lies above the x_1-axis and to the right of the x_2-axis.

To account for the remaining four constraints, first replace each inequality with an equation and then graph the resulting straight line by locating two distinct points on it. For example, after replacing $6x_1 + 4x_2 \leq 24$ with the straight line $6x_1 + 4x_2 = 24$, we can determine two distinct points by first setting $x_1 = 0$ to obtain $x_2 = \frac{24}{4} = 6$ and then setting $x_2 = 0$ to obtain $x_1 = \frac{24}{6} = 4$. Thus, the line passes through the two points (0, 6) and (4, 0), as shown by line (1) in Figure 2.1.

Next, consider the effect of the inequality. All it does is divide the (x_1, x_2)-plane into two half-spaces, one on each side of the graphed line. Only one of these two halves satisfies the inequality. To determine the correct side, choose (0, 0) as a *reference point*. If it satisfies the inequality, then the side in which it lies is the

FIGURE 2.1
Feasible space of the Reddy Mikks model

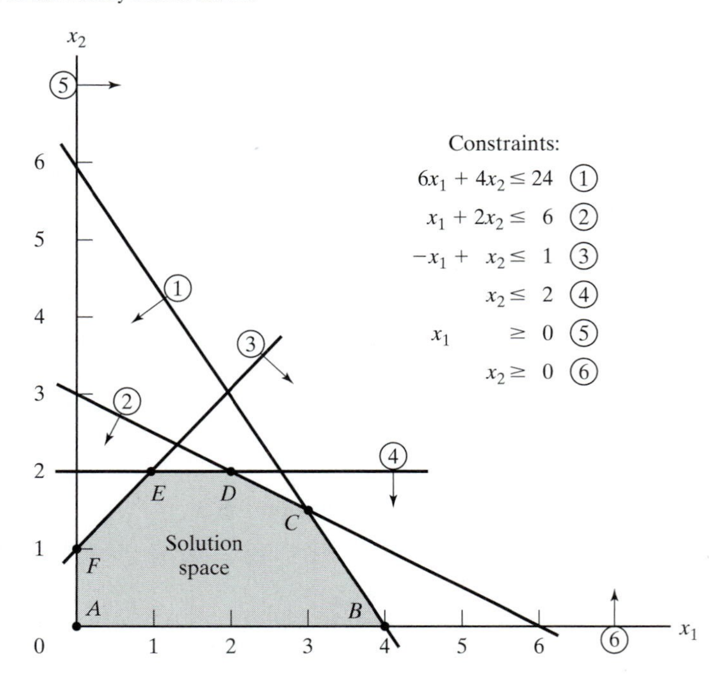

feasible half-space, otherwise the other side is. The use of the reference point $(0, 0)$ is illustrated with the constraint $6x_1 + 4x_2 \leq 24$. Because $6 \times 0 + 4 \times 0 = 0$ is less than 24, the half-space representing the inequality includes the origin (as shown by the arrow in Figure 2.1).

It is convenient computationally to select $(0, 0)$ as the reference point, unless the line happens to pass through the origin, in which case any other point can be used. For example, if we use the reference point $(6, 0)$, the left-hand side of the first constraint is $6 \times 6 + 4 \times 0 = 36$, which is larger than its right-hand side $(= 24)$, which means that the side in which $(6, 0)$ lies is not feasible for the inequality $6x_1 + 4x_2 \leq 24$. The conclusion is consistent with the one based on the reference point $(0, 0)$.

Application of the reference-point procedure to all the constraints of the model produces the constraints shown in Figure 2.1 (verify!). The **feasible solution space** of the problem represents the area in the first quadrant in which all the constraints are satisfied simultaneously. In Figure 2.1, any point in or on the boundary of the area *ABCDEF* is part of the feasible solution space. All points outside this area are infeasible.

TORA Moment.

The menu-driven TORA graphical LP module should prove helpful in reinforcing your understanding of how the LP constraints are graphed. Select Linear Programming from the MAIN menu. After inputting the model, select Solve ⇒ Graphical from the SOLVE/MODIFY menu. In the output screen, you will be able to experiment interactively with graphing the constraints one at a time, so you can see how each constraint affects the solution space.

Step 2. *Determination of the Optimum Solution:*

The feasible space in Figure 2.1 is delineated by the line segments joining the points *A, B, C, D, E*, and *F*. Any point within or on the boundary of the space *ABCDEF* is feasible. Because the feasible space *ABCDEF* consists of an *infinite* number of points, we need a systematic procedure to identify the optimum solution.

The determination of the optimum solution requires identifying the direction in which the profit function $z = 5x_1 + 4x_2$ increases (recall that we are *maximizing* z). We can do so by assigning *arbitrary* increasing values to z. For example, using $z = 10$ and $z = 15$ would be equivalent to graphing the two lines $5x_1 + 4x_2 = 10$ and $5x_1 + 4x_2 = 15$. Thus, the direction of increase in z is as shown Figure 2.2. The optimum solution occurs at C, which is the point in the solution space beyond which any further increase will put z outside the boundaries of *ABCDEF*.

The values of x_1 and x_2 associated with the optimum point C are determined by solving the equations associated with lines (1) and (2)—that is,

$$6x_1 + 4x_2 = 24$$
$$x_1 + 2x_2 = 6$$

The solution is $x_1 = 3$ and $x_2 = 1.5$ with $z = 5 \times 3 + 4 \times 1.5 = 21$. This calls for a daily product mix of 3 tons of exterior paint and 1.5 tons of interior paint. The associated daily profit is \$21,000.

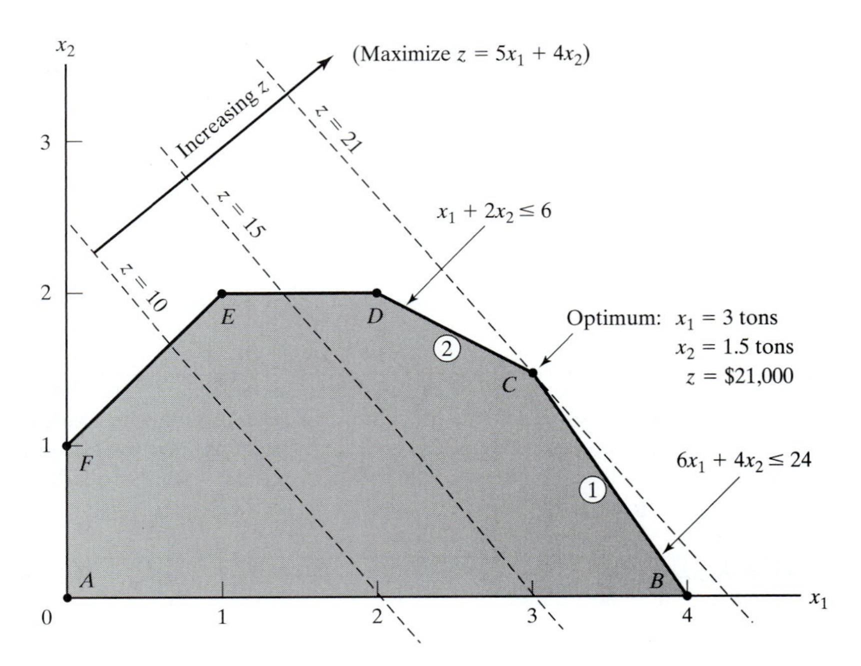

FIGURE 2.2
Optimum solution of the Reddy Mikks model

An important characteristic of the optimum LP solution is that it is *always* associated with a **corner point** of the solution space (where two lines intersect). This is true even if the objective function happens to be parallel to a constraint. For example, if the objective function is $z = 6x_1 + 4x_2$, which is parallel to constraint 1, we can always say that the optimum occurs at either corner point B or corner point C. Actually any point on the line segment BC will be an *alternative* optimum (see also Example 3.5-2), but the important observation here is that the line segment BC is totally defined by the *corner points* B and C.

TORA Moment.

You can use TORA interactively to see that the optimum is always associated with a corner point. From the output screen, you can click View/Modify Input Data to modify the objective coefficients and re-solve the problem graphically. You may use the following objective functions to test the proposed idea:

(a) $z = 5x_1 + x_2$

(b) $z = 5x_1 + 4x_2$

(c) $z = x_1 + 3x_2$

(d) $z = -x_1 + 2x_2$

(e) $z = -2x_1 + x_2$

(f) $z = -x_1 - x_2$

The observation that the LP optimum is always associated with a corner point means that the optimum solution can be found simply by enumerating all the corner points as the following table shows:

Corner point	(x_1, x_2)	z
A	$(0, 0)$	0
B	$(4, 0)$	20
C	**(3, 1.5)**	**21 (OPTIMUM)**
D	$(2, 2)$	18
E	$(1, 2)$	13
F	$(0, 1)$	4

As the number of constraints and variables increases, the number of corner points also increases, and the proposed enumeration procedure becomes less tractable computationally. Nevertheless, the idea shows that, from the standpoint of determining the LP optimum, the solution space $ABCDEF$ with its *infinite* number of solutions can, in fact, be replaced with a *finite* number of promising solution points—namely, the corner points, A, B, C, D, E, and F. This result is key for the development of the general algebraic algorithm, called the *simplex method*, which we will study in Chapter 3.

PROBLEM SET 2.2A

1. Determine the feasible space for each of the following independent constraints, given that $x_1, x_2 \geq 0$.

*(a) $-3x_1 + x_2 \leq 6$.

(b) $x_1 - 2x_2 \geq 5$.

(c) $2x_1 - 3x_2 \leq 12$.

*(d) $x_1 - x_2 \leq 0$.

(e) $-x_1 + x_2 \geq 0$.

2. Identify the direction of increase in z in each of the following cases:

*(a) Maximize $z = x_1 - x_2$.

(b) Maximize $z = -5x_1 - 6x_2$.

(c) Maximize $z = -x_1 + 2x_2$.

*(d) Maximize $z = -3x_1 + x_2$.

3. Determine the solution space and the optimum solution of the Reddy Mikks model for each of the following independent changes:

(a) The maximum daily demand for exterior paint is at most 2.5 tons.

(b) The daily demand for interior paint is at least 2 tons.

(c) The daily demand for interior paint is exactly 1 ton higher than that for exterior paint.

(d) The daily availability of raw material $M1$ is at least 24 tons.

(e) The daily availability of raw material $M1$ is at least 24 tons, and the daily demand for interior paint exceeds that for exterior paint by at least 1 ton.

4. A company that operates 10 hours a day manufactures two products on three sequential processes. The following table summarizes the data of the problem:

	Minutes per unit			
Product	*Process 1*	*Process 2*	*Process 3*	Unit profit
1	10	6	8	\$2
2	5	20	10	\$3

Determine the optimal mix of the two products.

***5.** A company produces two products, A and B. The sales volume for A is at least 80% of the total sales of both A and B. However, the company cannot sell more than 100 units of A per day. Both products use one raw material, of which the maximum daily availability is 240 lb. The usage rates of the raw material are 2 lb per unit of A and 4 lb per unit of B. The profit units for A and B are \$20 and \$50, respectively. Determine the optimal product mix for the company.

6. Alumco manufactures aluminum sheets and aluminum bars. The maximum production capacity is estimated at either 800 sheets or 600 bars per day. The maximum daily demand is 550 sheets and 580 bars. The profit per ton is \$40 per sheet and \$35 per bar. Determine the optimal daily production mix.

***7.** An individual wishes to invest \$5000 over the next year in two types of investment: Investment A yields 5% and investment B yields 8%. Market research recommends an allocation of at least 25% in A and at most 50% in B. Moreover, investment in A should be at least half the investment in B. How should the fund be allocated to the two investments?

8. The Continuing Education Division at the Ozark Community College offers a total of 30 courses each semester. The courses offered are usually of two types: practical, such as woodworking, word processing, and car maintenance; and humanistic, such as history, music, and fine arts. To satisfy the demands of the community, at least 10 courses of each type must be offered each semester. The division estimates that the revenues of offering practical and humanistic courses are approximately \$1500 and \$1000 per course, respectively.

(a) Devise an optimal course offering for the college.

(b) Show that the worth per additional course is \$1500, which is the same as the revenue per practical course. What does this result mean in terms of offering additional courses?

9. ChemLabs uses raw materials I and II to produce two domestic cleaning solutions, A and B. The daily availabilities of raw materials I and II are 150 and 145 units, respectively. One unit of solution A consumes .5 unit of raw material I and .6 unit of raw material II, and one unit of solution B uses .5 unit of raw material I and .4 unit of raw material II. The profits per unit of solutions A and B are \$8 and \$10, respectively. The daily demand for solution A lies between 30 and 150 units, and that for solution B between 40 and 200 units. Find the optimal production amounts of A and B.

10. In the Ma-and-Pa grocery store, shelf space is limited and must be used effectively to increase profit. Two cereal items, Grano and Wheatie, compete for a total shelf space of 60 ft^2. A box of Grano occupies .2 ft^2 and a box of Wheatie needs .4 ft^2. The maximum daily demands of Grano and Wheatie are 200 and 120 boxes, respectively. A box of Grano nets \$1.00 in profit and a box of Wheatie \$1.35. Ma-and-Pa thinks that because the unit profit of Wheatie is 35% higher than that of Grano, Wheatie should be allocated

35% more space than Grano, which amounts to allocating about 57% to Wheatie and 43% to Grano. What do you think?

11. Jack is an aspiring freshman at Ulern University. He realizes that "all work and no play make Jack a dull boy." As a result, Jack wants to apportion his available time of about 10 hours a day between work and play. He estimates that play is twice as much fun as work. He also wants to study at least as much as he plays. However, Jack realizes that if he is going to get all his homework assignments done, he cannot play more than 4 hours a day. How should Jack allocate his time to maximize his pleasure from both work and play?

12. Wild West produces two types of cowboy hats. A type 1 hat requires twice as much labor time as a type 2. If the all available labor time is dedicated to Type 2 alone, the company can produce a total of 400 Type 2 hats a day. The respective market limits for the two types are 150 and 200 hats per day. The profit is $8 per Type 1 hat and $5 per Type 2 hat. Determine the number of hats of each type that would maximize profit.

13. Show & Sell can advertise its products on local radio and television (TV). The advertising budget is limited to $10,000 a month. Each minute of radio advertising costs $15 and each minute of TV commercials $300. Show & Sell likes to advertise on radio at least twice as much as on TV. In the meantime, it is not practical to use more than 400 minutes of radio advertising a month. From past experience, advertising on TV is estimated to be 25 times as effective as on radio. Determine the optimum allocation of the budget to radio and TV advertising.

***14.** Wyoming Electric Coop owns a steam-turbine power-generating plant. Because Wyoming is rich in coal deposits, the plant generates its steam from coal. This, however, may result in emission that does not meet the Environmental Protection Agency standards. EPA regulations limit sulfur dioxide discharge to 2000 parts per million per ton of coal burned and smoke discharge from the plant stacks to 20 lb per hour. The Coop receives two grades of pulverized coal, $C1$ and $C2$, for use in the steam plant. The two grades are usually mixed together before burning. For simplicity, it can be assumed that the amount of sulfur pollutant discharged (in parts per million) is a weighted average of the proportion of each grade used in the mixture. The following data are based on consumption of 1 ton per hour of each of the two coal grades.

Coal grade	Sulfur discharge in parts per million	Smoke discharge in lb per hour	Steam generated in lb per hour
$C1$	1800	2.1	12,000
$C2$	2100	.9	9,000

(a) Determine the optimal ratio for mixing the two coal grades.

(b) Determine the effect of relaxing the smoke discharge limit by 1 lb on the amount of generated steam per hour.

15. Top Toys is planning a new radio and TV advertising campaign. A radio commercial costs $300 and a TV ad costs $2000. A total budget of $20,000 is allocated to the campaign. However, to ensure that each medium will have at least one radio commercial and one TV ad, the most that can be allocated to either medium cannot exceed 80% of the total budget. It is estimated that the first radio commercial will reach 5000 people, with each additional commercial reaching only 2000 new ones. For TV, the first ad will reach 4500 people and each additional ad an additional 3000. How should the budgeted amount be allocated between radio and TV?

16. The Burroughs Garment Company manufactures men's shirts and women's blouses for Walmark Discount Stores. Walmark will accept all the production supplied by Burroughs. The production process includes cutting, sewing, and packaging. Burroughs employs 25 workers in the cutting department, 35 in the sewing department, and 5 in the packaging department. The factory works one 8-hour shift, 5 days a week. The following table gives the time requirements and profits per unit for the two garments:

	Minutes per unit			
Garment	*Cutting*	*Sewing*	*Packaging*	Unit profit ($)
Shirts	20	70	12	8
Blouses	60	60	4	12

Determine the optimal weekly production schedule for Burroughs.

17. A furniture company manufactures desks and chairs. The sawing department cuts the lumber for both products, which is then sent to separate assembly departments. Assembled items are sent for finishing to the painting department. The daily capacity of the sawing department is 200 chairs or 80 desks. The chair assembly department can produce 120 chairs daily and the desk assembly department 60 desks daily. The paint department has a daily capacity of either 150 chairs or 110 desks. Given that the profit per chair is $50 and that of a desk is $100, determine the optimal production mix for the company.

***18.** An assembly line consisting of three consecutive stations produces two radio models: HiFi-1 and HiFi-2. The following table provides the assembly times for the three workstations.

	Minutes per unit	
Workstation	*HiFi-1*	*HiFi-2*
1	6	4
2	5	5
3	4	6

The daily maintenance for stations 1, 2, and 3 consumes 10%, 14%, and 12%, respectively, of the maximum 480 minutes available for each station each day. Determine the optimal product mix that will minimize the idle (or unused) times in the three workstations.

19. *TORA Experiment.* Enter the following LP into TORA and select the graphic solution mode to reveal the LP graphic screen.

$$\text{Minimize } z = 3x_1 + 8x_2$$

subject to

$$\begin{aligned} x_1 + x_2 &\geq 8 \\ 2x_1 - 3x_2 &\leq 0 \\ x_1 + 2x_2 &\leq 30 \\ 3x_1 - x_2 &\geq 0 \\ x_1 &\leq 10 \\ x_2 &\geq 9 \\ x_1, x_2 &\geq 0 \end{aligned}$$

Next, on a sheet of paper, graph and scale the x_1- and x_2-axes for the problem (you may also click Print Graph on the top of the right window to obtain a ready-to-use scaled sheet). Now, graph a constraint manually on the prepared sheet, then click it on the left window of the screen to check your answer. Repeat the same for each constraint and then terminate the procedure with a graph of the objective function. The suggested process is designed to test and reinforce your understanding of the graphical LP solution through immediate feedback from TORA.

20. *TORA Experiment.* Consider the following LP model:

$$\text{Maximize } z = 5x_1 + 4x_2$$

subject to

$$\begin{aligned} 6x_1 + 4x_2 &\leq 24 \\ 6x_1 + 3x_2 &\leq 22.5 \\ x_1 + x_2 &\leq 5 \\ x_1 + 2x_2 &\leq 6 \\ -x_1 + x_2 &\leq 1 \\ x_2 &\leq 2 \\ x_1, x_2 &\geq 0 \end{aligned}$$

In LP, a constraint is said to be *redundant* if its removal from the model leaves the feasible solution space unchanged. Use the graphical facility of TORA to identify the redundant constraints, then show that their removal (simply by not graphing them) does not affect the solution space or the optimal solution.

21. *TORA Experiment.* In the Reddy Mikks model, use TORA to show that the removal of the raw material constraints (constraints 1 and 2) would result in an *unbounded solution space*. What can be said in this case about the optimal solution of the model?

22. *TORA Experiment.* In the Reddy Mikks model, suppose that the following constraint is added to the problem.

$$x_2 \geq 3$$

Use TORA to show that the resulting model has conflicting constraints that cannot be satisfied simultaneously and hence it has *no feasible solution*.

2.2.2 Solution of a Minimization Model

Example 2.2-2 (Diet Problem)

Ozark Farms uses at least 800 lb of special feed daily. The special feed is a mixture of corn and soybean meal with the following compositions:

	lb per lb of feedstuff		
Feedstuff	*Protein*	*Fiber*	Cost ($/lb)
Corn	.09	.02	.30
Soybean meal	.60	.06	.90

The dietary requirements of the special feed are at least 30% protein and at most 5% fiber. Ozark Farms wishes to determine the daily minimum-cost feed mix.

Because the feed mix consists of corn and soybean meal, the decision variables of the model are defined as

$$x_1 = \text{lb of corn in the daily mix}$$
$$x_2 = \text{lb of soybean meal in the daily mix}$$

The objective function seeks to minimize the total daily cost (in dollars) of the feed mix and is thus expressed as

$$\text{Minimize } z = .3x_1 + .9x_2$$

The constraints of the model reflect the daily amount needed and the dietary requirements. Because Ozark Farms needs at least 800 lb of feed a day, the associated constraint can be expressed as

$$x_1 + x_2 \geq 800$$

As for the protein dietary requirement constraint, the amount of protein included in x_1 lb of corn and x_2 lb of soybean meal is $(.09x_1 + .6x_2)$ lb. This quantity should equal at least 30% of the total feed mix $(x_1 + x_2)$ lb—that is,

$$.09x_1 + .6x_2 \geq .3(x_1 + x_2)$$

In a similar manner, the fiber requirement of at most 5% is constructed as

$$.02x_1 + .06x_2 \leq .05(x_1 + x_2)$$

The constraints are simplified by moving the terms in x_1 and x_2 to the left-hand side of each inequality, leaving only a constant on the right-hand side. The complete model thus becomes

$$\text{minimize } z = .3x_1 + .9x_2$$

subject to

$$\begin{aligned} x_1 + \quad x_2 &\geq 800 \\ .21x_1 - .30x_2 &\leq 0 \\ .03x_1 - .01x_2 &\geq 0 \\ x_1, x_2 &\geq 0 \end{aligned}$$

Figure 2.3 provides the graphical solution of the model. Unlike those of the Reddy Mikks model (Example 2.2-1), the second and third constraints pass through the origin. To plot the associated straight lines, we need one additional point, which can be obtained by assigning a value to one of the variables and then solving for the other variable. For example, in the second constraint, $x_1 = 200$ will yield $.21 \times 200 - .3x_2 = 0$, or $x_2 = 140$. This means that the straight line $.21x_1 - .3x_2 = 0$ passes through $(0, 0)$ and $(200, 140)$. Note also that $(0, 0)$ cannot be used as a reference point for constraints 2 and 3, because both lines pass through the origin. Instead, any other point [e.g., $(100, 0)$ or $(0, 100)$] can be used for that purpose.

Solution:

Because the present model seeks the minimization of the objective function, we need to reduce the value of z as much as possible in the direction shown in Figure 2.3. The optimum solution is the intersection of the two lines $x_1 + x_2 = 800$ and $.21x_1 - .3x_2 = 0$, which yields $x_1 = 470.59$ lb and $x_2 = 329.41$ lb. The associated minimum cost of the feed mix is $z = .3 \times 470.59 + .9 \times 329.42 =$ \$437.65 per day.

Remarks. We need to take note of the way the constraints of the problem are constructed. Because the model is minimizing the total cost, one may argue that the solution will seek exactly 800 tons of feed. Indeed, this is what the optimum solution given above does. Does this mean then that the first constraint can be deleted altogether simply by including the amount 800 tons

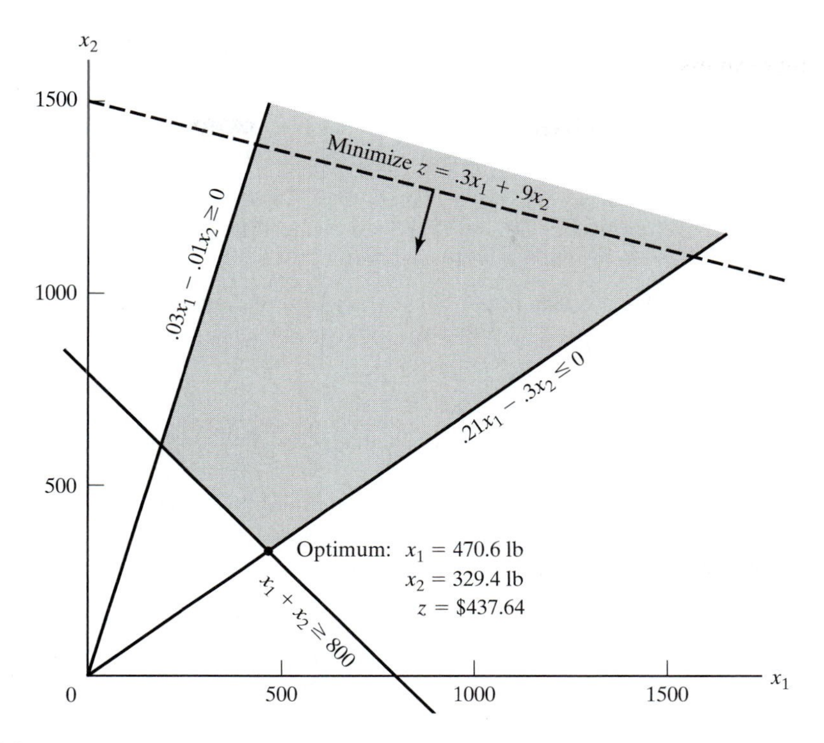

FIGURE 2.3
Graphical solution of the diet model

in the remaining constraints? To find the answer, we state the new protein and fiber constraints as

$$.09x_1 + .6x_2 \geq .3 \times 800$$
$$.02x_1 + .06x_2 \leq .05 \times 800$$

or

$$.09x_1 + .6x_2 \geq 240$$
$$.02x_1 + .06x_2 \leq 40$$

The new formulation yields the solution $x_1 = 0$, and $x_2 = 400$ lb (verify with TORA!), which does not satisfy the *implied* requirement for 800 lb of feed. This means that the constraint $x_1 + x_2 \geq 800$ must be used explicitly and that the protein and fiber constraints must remain exactly as given originally.

Along the same line of reasoning, one may be tempted to replace $x_1 + x_2 \geq 800$ with $x_1 + x_2 = 800$. In the present example, the two constraints yield the same answer. But in general this may not be the case. For example, suppose that the daily mix must include at least 500 lb of corn. In this case, the optimum solution will call for using 500 lb of corn and 350 lb of soybean (verify with TORA!), which is equivalent to a daily feed mix of $500 + 350 = 850$ lb. Imposing the equality constraint a priori will lead to the conclusion that the problem has no feasible solution (verify with TORA!). On the other hand, the use of the inequality is inclusive of the equality case, and hence its use does not prevent the model from producing exactly 800 lb of feed mix, should the remaining constraints allow it. The conclusion is that we should not "pre-guess" the solution by imposing the additional equality restriction, and we should always use inequalities unless the situation explicitly stipulates the use of equalities.

PROBLEM SET 2.2B

1. Identify the direction of decrease in z in each of the following cases:

(a) Minimize $z = 4x_1 - 2x_2$.

(b) Minimize $z = -3x_1 + x_2$.

(c) Minimize $z = -x_1 - 2x_2$.

2. For the diet model, suppose that the daily availability of corn is limited to 450 lb. Identify the new solution space, and determine the new optimum solution.

3. For the diet model, what type of optimum solution would the model yield if the feed mix should not exceed 800 lb a day? Does the solution make sense?

4. John must work at least 20 hours a week to supplement his income while attending school. He has the opportunity to work in two retail stores. In store 1, he can work between 5 and 12 hours a week, and in store 2 he is allowed between 6 and 10 hours. Both stores pay the same hourly wage. In deciding how many hours to work in each store, John wants to base his decision on work stress. Based on interviews with present employees, John estimates that, on an ascending scale of 1 to 10, the stress factors are 8 and 6 at stores 1 and 2, respectively. Because stress mounts by the hour, he assumes that the total stress for each store at the end of the week is proportional to the number of hours he works in the store. How many hours should John work in each store?

***5.** OilCo is building a refinery to produce four products: diesel, gasoline, lubricants, and jet fuel. The minimum demand (in bbl/day) for each of these products is 14,000, 30,000, 10,000, and 8,000, respectively. Iran and Dubai are under contract to ship crude to OilCo. Because of the production quotas specified by OPEC (Organization of Petroleum Exporting Countries) the new refinery can receive at least 40% of its crude from Iran and the remaining amount from Dubai. OilCo predicts that the demand and crude oil quotas will remain steady over the next ten years.

The specifications of the two crude oils lead to different product mixes: One barrel of Iran crude yields .2 bbl of diesel, .25 bbl of gasoline, .1 bbl of lubricant, and .15 bbl of jet fuel. The corresponding yields from Dubai crude are .1, .6, .15, and .1, respectively. OilCo needs to determine the minimum capacity of the refinery (in bbl/ day).

6. Day Trader wants to invest a sum of money that would generate an annual yield of at least $10,000. Two stock groups are available: blue chips and high tech, with average annual yields of 10% and 25%, respectively. Though high-tech stocks provide higher yield, they are more risky, and Trader wants to limit the amount invested in these stocks to no more than 60% of the total investment. What is the minimum amount Trader should invest in each stock group to accomplish the investment goal?

***7.** An industrial recycling center uses two scrap aluminum metals, *A* and *B*, to produce a special alloy. Scrap *A* contains 6% aluminum, 3% silicon, and 4% carbon. Scrap *B* has 3% aluminum, 6% silicon, and 3% carbon. The costs per ton for scraps *A* and *B* are $100 and $80, respectively. The specifications of the special alloy require that (1) the aluminum content must be at least 3% and at most 6%, (2) the silicon content must lie between 3%

and 5%, and (3) the carbon content must be between 3% and 7%. Determine the optimum mix of the scraps that should be used in producing 1000 tons of the alloy.

8. *TORA Experiment.* Consider the Diet Model and let the objective function be given as

$$\text{Minimize } z = .8x_1 + .8x_2$$

Use TORA to show that the optimum solution is associated with *two* distinct corner points and that both points yield the same objective value. In this case, the problem is said to have *alternative optima.* Explain the conditions leading to this situation and show that, in effect, the problem has an infinite number of alternative optima, then provide a formula for determining all such solutions.

2.3 SELECTED LP APPLICATIONS

This section presents realistic LP models in which the definition of the variables and the construction of the objective function and constraints are not as straightforward as in the case of the two-variable model. The areas covered by these applications include the following:

1. Urban planning.
2. Currency arbitrage.
3. Investment.
4. Production planning and inventory control.
5. Blending and oil refining.
6. Manpower planning.

Each model is fully developed and its optimum solution is analyzed and interpreted.

2.3.1 Urban Planning[1]

Urban planning deals with three general areas: (1) building new housing developments, (2) upgrading inner-city deteriorating housing and recreational areas, and (3) planning public facilities (such as schools and airports). The constraints associated with these projects are both economic (land, construction, financing) and social (schools, parks, income level). The objectives in urban planning vary. In new housing developments, profit is usually the motive for undertaking the project. In the remaining two categories, the goals involve social, political, economic, and cultural considerations. Indeed, in a publicized case in 2004, the mayor of a city in Ohio wanted to condemn an old area of the city to make way for a luxury housing development. The motive was to increase tax collection to help alleviate budget shortages. The example presented in this section is fashioned after the Ohio case.

---

[1]This section is based on Laidlaw (1972).

Example 2.3-1 (Urban Renewal Model)

The city of Erstville is faced with a severe budget shortage. Seeking a long-term solution, the city council votes to improve the tax base by condemning an inner-city housing area and replacing it with a modern development.

The project involves two phases: (1) demolishing substandard houses to provide land for the new development, and (2) building the new development. The following is a summary of the situation.

1. As many as 300 substandard houses can be demolished. Each house occupies a .25-acre lot. The cost of demolishing a condemned house is $2000.
2. Lot sizes for new single-, double-, triple-, and quadruple-family homes (units) are .18, .28, .4, and .5 acre, respectively. Streets, open space, and utility easements account for 15% of available acreage.
3. In the new development the triple and quadruple units account for at least 25% of the total. Single units must be at least 20% of all units and double units at least 10%.
4. The tax levied per unit for single, double, triple, and quadruple units is $1,000, $1,900, $2,700, and $3,400, respectively.
5. The construction cost per unit for single-, double-, triple-, and quadruple- family homes is $50,000, $70,000, $130,000, and $160,000, respectively. Financing through a local bank can amount to a maximum of $15 million.

How many units of each type should be constructed to maximize tax collection?

Mathematical Model: Besides determining the number of units to be constructed of each type of housing, we also need to decide how many houses must be demolished to make room for the new development. Thus, the variables of the problem can be defined as follows:

$$x_1 = \text{Number of units of single-family homes}$$

$$x_2 = \text{Number of units of double-family homes}$$

$$x_3 = \text{Number of units of triple-family homes}$$

$$x_4 = \text{Number of units of quadruple-family homes}$$

$$x_5 = \text{Number of old homes to be demolished}$$

The objective is to maximize total tax collection from all four types of homes—that is,

$$\text{Maximize } z = 1000x_1 + 1900x_2 + 2700x_3 + 3400x_4$$

The first constraint of the problem deals with land availability.

$$\begin{pmatrix}\text{Acreage used for new}\\ \text{home construction}\end{pmatrix} \leq \begin{pmatrix}\text{Net available}\\ \text{acreage}\end{pmatrix}$$

From the data of the problem we have

$$\text{Acreage needed for new homes} = .18x_1 + .28x_2 + .4x_3 + .5x_4$$

To determine the available acreage, each demolished home occupies a .25-acre lot, thus netting $.25x_5$ acres. Allowing for 15% open space, streets, and easements, the net acreage available is $.85(.25x_5) = .2125x_5$. The resulting constraint is

$$.18x_1 + .28x_2 + .4x_3 + .5x_4 \leq .2125x_5$$

or

$$.18x_1 + .28x_2 + .4x_3 + .5x_4 - .2125x_5 \leq 0$$

The number of demolished homes cannot exceed 300, which translates to

$$x_5 \leq 300$$

Next we add the constraints limiting the number of units of each home type.

$$(\text{Number of single units}) \geq (20\% \text{ of all units})$$

$$(\text{Number of double units}) \geq (10\% \text{ of all units})$$

$$(\text{Number of triple and quadruple units}) \geq (25\% \text{ of all units})$$

These constraints translate mathematically to

$$x_1 \geq .2(x_1 + x_2 + x_3 + x_4)$$

$$x_2 \geq .1(x_1 + x_2 + x_3 + x_4)$$

$$x_3 + x_4 \geq .25(x_1 + x_2 + x_3 + x_4)$$

The only remaining constraint deals with keeping the demolishition/construction cost within the allowable budget—that is,

$$(\text{Construction and demolition cost}) \leq (\text{Available budget})$$

Expressing all the costs in thousands of dollars, we get

$$(50x_1 + 70x_2 + 130x_3 + 160x_4) + 2x_5 \leq 15000$$

The complete model thus becomes

$$\text{Maximize } z = 1000x_1 + 1900x_2 + 2700x_3 + 3400x_4$$

subject to

$$.18x_1 + .28x_2 + .4x_3 + .5x_4 - .2125x_5 \leq 0$$

$$x_5 \leq 300$$

$$-.8x_1 + .2x_2 + .2x_3 + .2x_4 \leq 0$$

$$.1x_1 - .9x_2 + .1x_3 + .1x_4 \leq 0$$

$$.25x_1 + .25x_2 - .75x_3 - .75x_4 \leq 0$$

$$50x_1 + 70x_2 + 130x_3 + 160x_4 + 2x_5 \leq 15000$$

$$x_1, x_2, x_3, x_4, x_5 \geq 0$$

Solution:

The optimum solution (using file amplEX2.3-1.txt or solverEx2.3-1.xls) is:

Total tax collection $= z =$ \$343,965
Number of single homes $= x_1 = 35.83 \simeq 36$ units
Number of double homes $= x_2 = 98.53 \simeq 99$ units
Number of triple homes $= x_3 = 44.79 \simeq 45$ units
Number of quadruple homes $= x_4 = 0$ units
Number of homes demolished $= x_5 = 244.49 \simeq 245$ units

Remarks. Linear programming does not guarantee an integer solution automatically, and this is the reason for rounding the continuous values to the closest integer. The rounded solution calls for constructing $180\ (= 36 + 99 + 45)$ units and demolishing 245 old homes, which yields \$345,600 in taxes. Keep in mind, however, that, in general, the rounded solution may not be feasible. In fact, the current rounded solution violates the budget constraint by \$70,000 (verify!). Interestingly, the true optimum integer solution (using the algorithms in Chapter 9) is $x_1 = 36, x_2 = 98, x_3 = 45, x_4 = 0$, and $x_5 = 245$ with $z = \$343{,}700$. Carefully note that the rounded solution yields a better objective value, which appears contradictory. The reason is that the rounded solution calls for producing an extra double home, which is feasible only if the budget is increased by \$70,000.

PROBLEM SET 2.3A

1. A realtor is developing a rental housing and retail area. The housing area consists of efficiency apartments, duplexes, and single-family homes. Maximum demand by potential renters is estimated to be 500 efficiency apartments, 300 duplexes, and 250 single-family homes, but the number of duplexes must equal at least 50% of the number of efficiency apartments and single homes. Retail space is proportionate to the number of home units at the rates of at least 10 ft^2, 15 ft^2, and 18 ft^2 for efficiency, duplex, and single family units, respectively. However, land availability limits retail space to no more than 10,000 ft^2. The monthly rental income is estimated at \$600, \$750, and \$1200 for efficiency-, duplex-, and single-family units, respectively. The retail space rents for \$100/ft^2. Determine the optimal retail space area and the number of family residences.

2. The city council of Fayetteville is in the process of approving the construction of a new 200,000-ft^2 convention center. Two sites have been proposed, and both require exercising the "eminent domain" law to acquire the property. The following table provides data about proposed (contiguous) properties in both sites together with the acquisition cost.

	Site 1		Site 2	
Property	*Area* (1000 ft^2)	*Cost* (1000 \$)	*Area* (1000 ft^2)	*Cost* (1000 \$)
1	20	1,000	80	2,800
2	50	2,100	60	1,900
3	50	2,350	50	2,800
4	30	1,850	70	2,500
5	60	2,950		

Partial acquisition of property is allowed. At least 75% of property 4 must be acquired if site 1 is selected, and at least 50% of property 3 must be acquired if site 2 is selected.

Although site 1 property is more expensive (on a per ft^2 basis), the construction cost is less than at site 2, because the infrastructure at site 1 is in a much better shape. Construction cost is $25 million at site 1 and $27 million at site 2. Which site should be selected, and what properties should be acquired?

***3.** A city will undertake five urban renewal housing projects over the next five years. Each project has a different starting year and a different duration. The following table provides the basic data of the situation:

	Year 1	Year 2	Year 3	Year 4	Year 5	Cost (million $)	Annual income (million $)
Project 1	Start		End			5.0	.05
Project 2		Start			End	8.0	.07
Project 3	Start				End	15.0	.15
Project 4			Start	End		1.2	.02
Budget (million $)	3.0	6.0	7.0	7.0	7.0		

Projects 1 and 4 must be finished completely within their durations. The remaining two projects can be finished partially within budget limitations, if necessary. However, each project must be at least 25% completed within its duration. At the end of each year, the completed section of a project is immediately occupied by tenants and a proportional amount of income is realized. For example, if 40% of project 1 is completed in year 1 and 60% in year 3, the associated income over the five-year planning horizon is $.4 \times \$50,000$ (for year 2) $+ .4 \times \$50,000$ (for year 3) $+ (.4 + .6) \times \$50,000$ (for year 4) $+ (.4 + .6) \times \$50,000$ (for year 5) $= (4 \times .4 + 2 \times .6) \times \$50,000$. Determine the optimal schedule for the projects that will maximize the total income over the five-year horizon. For simplicity, disregard the time value of money.

4. The city of Fayetteville is embarking on an urban renewal project that will include lower- and middle-income row housing, upper-income luxury apartments, and public housing. The project also includes a public elementary school and retail facilities. The size of the elementary school (number of classrooms) is proportional to the number of pupils, and the retail space is proportional to the number of housing units. The following table provides the pertinent data of the situation:

	Lower income	Middle income	Upper income	Public housing	School room	Retail unit
Minimum number of units	100	125	75	300		0
Maximum number of units	200	190	260	600		25
Lot size per unit (acre)	.05	.07	.03	.025	.045	.1
Average number of pupils per unit	1.3	1.2	.5	1.4		
Retail demand per unit (acre)	.023	.034	.046	.023	.034	
Annual income per unit($)	7000	12,000	20,000	5000	—	15,000

The new school can occupy a maximum space of 2 acres at the rate of at most 25 pupils per room. The operating annual cost per school room is $10,000. The project will be located on a 50-acre vacant property owned by the city. Additionally, the project can make use of an adjacent property occupied by 200 condemned slum homes. Each condemned home occupies .25 acre. The cost of buying and demolishing a slum unit is $7000. Open space, streets, and parking lots consume 15% of total available land.

Develop a linear program to determine the optimum plan for the project.

5. Realco owns 800 acres of undeveloped land on a scenic lake in the heart of the Ozark Mountains. In the past, little or no regulation was imposed upon new developments around the lake. The lake shores are now dotted with vacation homes, and septic tanks, most of them improperly installed, are in extensive use. Over the years, seepage from the septic tanks led to severe water pollution. To curb further degradation of the lake, county officials have approved stringent ordinances applicable to all future developments: (1) Only single-, double-, and triple-family homes can be constructed, with single-family homes accounting for at least 50% of the total. (2) To limit the number of septic tanks, minimum lot sizes of 2, 3, and 4 acres are required for single-, double-, and triple-family homes, respectively. (3) Recreation areas of 1 acre each must be established at the rate of one area per 200 families. (4) To preserve the ecology of the lake, underground water may not be pumped out for house or garden use. The president of Realco is studying the possibility of developing the 800-acre property. The new development will include single-, double-, and triple-family homes. It is estimated that 15% of the acreage will be allocated to streets and utility easements. Realco estimates the returns from the different housing units as follows:

Housing unit	Single	Double	Triple
Net return per unit ($)	10,000	12,000	15,000

The cost of connecting water service to the area is proportionate to the number of units constructed. However, the county charges a minimum of $100,000 for the project. Additionally, the expansion of the water system beyond its present capacity is limited to 200,000 gallons per day during peak periods. The following data summarize the water service connection cost as well as the water consumption, assuming an average size family:

Housing unit	Single	Double	Triple	Recreation
Water service connection cost per unit ($)	1000	1200	1400	800
Water consumption per unit (gal/day)	400	600	840	450

Develop an optimal plan for Realco.

6. Consider the Realco model of Problem 5. Suppose that an additional 100 acres of land can be purchased for $450,000, which will increase the total acreage to 900 acres. Is this a profitable deal for Realco?

2.3.2 Currency Arbitrage[2]

In today's global economy, a multinational company must deal with currencies of the countries in which it operates. Currency arbitrage, or simultaneous purchase and sale of currencies in different markets, offers opportunities for advantageous movement of money from one currency to another. For example, converting £1000 to U.S. dollars in 2001 with an exchange rate of $1.60 to £1 will yield $1600. Another way of making the conversion is to first change the British pound to Japanese yen and then convert the yen to U.S. dollars using the 2001 exchange rates of £1 = ¥175 and $1 = ¥105. The

[2]This section is based on J. Kornbluth and G. Salkin (1987, Chapter 6).

resulting dollar amount is $\frac{(£1{,}000 \times ¥175)}{¥105} = \$1{,}666.67$. This example demonstrates the advantage of converting the British money first to Japanese yen and then to dollars. This section shows how the arbitrage problem involving many currencies can be formulated and solved as a linear program.

Example 2.3-2 (Currency Arbitrage Model)

Suppose that a company has a total of 5 million dollars that can be exchanged for euros (€), British pounds (£), yen (¥), and Kuwaiti dinars (KD). Currency dealers set the following limits on the amount of any single transaction: 5 million dollars, 3 million euros, 3.5 million pounds, 100 million yen, and 2.8 million KDs. The table below provides typical spot exchange rates. The bottom diagonal rates are the reciprocal of the top diagonal rates. For example, rate(€ → $) = 1/rate($ → €) = 1/.769 = 1.30.

	$	€	£	¥	KD
$	1	.769	.625	105	.342
€	$\frac{1}{.769}$	1	.813	137	.445
£	$\frac{1}{.625}$	$\frac{1}{.813}$	1	169	.543
¥	$\frac{1}{105}$	$\frac{1}{137}$	$\frac{1}{169}$	1	.0032
KD	$\frac{1}{.342}$	$\frac{1}{.445}$	$\frac{1}{.543}$	$\frac{1}{.0032}$	1

Is it possible to increase the dollar holdings (above the initial $5 million) by circulating currencies through the currency market?

Mathematical Model: The situation starts with $5 million. This amount goes through a number of conversions to other currencies before ultimately being reconverted to dollars. The problem thus seeks determining the amount of each conversion that will maximize the total dollar holdings.

For the purpose of developing the model and simplifying the notation, the following numeric code is used to represent the currencies.

Currency	$	€	£	¥	KD
Code	1	2	3	4	5

Define

$$x_{ij} = \text{Amount in currency } i \text{ converted to currency } j,\ i \text{ and } j = 1, 2, \ldots, 5$$

For example, x_{12} is the dollar amount converted to euros and x_{51} is the KD amount converted to dollars. We further define two additional variables representing the input and the output of the arbitrage problem:

$$I = \text{Initial dollar amount } (= \$5 \text{ million})$$
$$y = \text{Final dollar holdings (to be determined from the solution)}$$

Our goal is to determine the maximum final dollar holdings, y, subject to the currency flow restrictions and the maximum limits allowed for the different transactions.

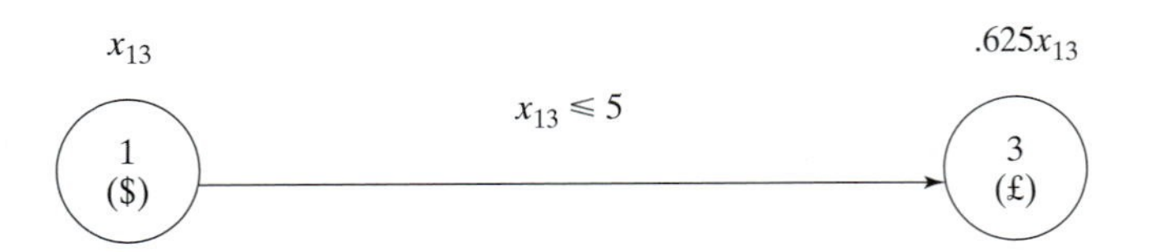

FIGURE 2.4
Definition of the input/output variable, x_{13}, between $ and £

We start by developing the constraints of the model. Figure 2.4 demonstrates the idea of converting dollars to pounds. The dollar amount x_{13} at originating currency 1 is converted to $.625x_{13}$ pounds at end currency 3. At the same time, the transacted dollar amount cannot exceed the limit set by the dealer, $x_{13} \leq 5$.

To conserve the flow of money from one currency to another, each currency must satisfy the following input-output equation:

$$\begin{pmatrix}\text{Total sum available}\\ \text{of a currency (input)}\end{pmatrix} = \begin{pmatrix}\text{Total sum converted to}\\ \text{other currencies (output)}\end{pmatrix}$$

1. *Dollar* ($i = 1$):

$$\begin{aligned}\text{Total available dollars} &= \text{Initial dollar amount} + \\ &\quad\ \text{dollar amount from other currencies} \\ &= I + (€ \rightarrow \$) + (£ \rightarrow \$) + (¥ \rightarrow \$) + (\text{KD} \rightarrow \$) \\ &= I + \tfrac{1}{.769}x_{21} + \tfrac{1}{.625}x_{31} + \tfrac{1}{105}x_{41} + \tfrac{1}{.342}x_{51}\end{aligned}$$

$$\begin{aligned}\text{Total distributed dollars} &= \text{Final dollar holdings} + \\ &\quad\ \text{dollar amount to other currencies} \\ &= y + (\$ \rightarrow €) + (\$ \rightarrow £) + (\$ \rightarrow ¥) + (\$ \rightarrow \text{KD}) \\ &= y + x_{12} + x_{13} + x_{14} + x_{15}\end{aligned}$$

Given $I = 5$, the dollar constraint thus becomes

$$y + x_{12} + x_{13} + x_{14} + x_{15} - \left(\tfrac{1}{.769}x_{21} + \tfrac{1}{.625}x_{31} + \tfrac{1}{105}x_{41} + \tfrac{1}{.342}x_{51}\right) = 5$$

2. *Euro* ($i = 2$):

$$\begin{aligned}\text{Total available euros} &= (\$ \rightarrow €) + (£ \rightarrow €) + (¥ \rightarrow €) + (\text{KD} \rightarrow €) \\ &= .769x_{12} + \tfrac{1}{.813}x_{32} + \tfrac{1}{137}x_{42} + \tfrac{1}{.445}x_{52} \\ \text{Total distributed euros} &= (€ \rightarrow \$) + (€ \rightarrow £) + (€ \rightarrow ¥) + (€ \rightarrow \text{KD}) \\ &= x_{21} + x_{23} + x_{24} + x_{25}\end{aligned}$$

Thus, the constraint is

$$x_{21} + x_{23} + x_{24} + x_{25} - \left(.769x_{12} + \tfrac{1}{.813}x_{32} + \tfrac{1}{137}x_{42} + \tfrac{1}{.445}x_{52}\right) = 0$$

3. *Pound* ($i = 3$):

$$\text{Total available pounds} = (\$ \rightarrow £) + (€ \rightarrow £) + (¥ \rightarrow £) + (\text{KD} \rightarrow £)$$
$$= .625x_{13} + .813x_{23} + \tfrac{1}{169}x_{43} + \tfrac{1}{.543}x_{53}$$

$$\text{Total distributed pounds} = (£ \rightarrow \$) + (£ \rightarrow €) + (£ \rightarrow ¥) + (£ \rightarrow \text{KD})$$
$$= x_{31} + x_{32} + x_{34} + x_{35}$$

Thus, the constraint is

$$x_{31} + x_{32} + x_{34} + x_{35} - .625x_{13} + .813x_{23} + \tfrac{1}{169}x_{43} + \tfrac{1}{.543}x_{53} = 0$$

4. *Yen* ($i = 4$):

$$\text{Total available yen} = (\$ \rightarrow ¥) + (€ \rightarrow ¥) + (£ \rightarrow ¥) + (\text{KD} \rightarrow ¥)$$
$$= 105x_{14} + 137x_{24} + 169x_{34} + \tfrac{1}{.0032}x_{54}$$

$$\text{Total distributed yen} = (¥ \rightarrow \$) + (¥ \rightarrow €) + (¥ \rightarrow £) + (¥ \rightarrow \text{KD})$$
$$= x_{41} + x_{42} + x_{43} + x_{45}$$

Thus, the constraint is

$$x_{41} + x_{42} + x_{43} + x_{45} - \left(105x_{14} + 137x_{24} + 169x_{34} + \tfrac{1}{.0032}x_{54}\right) = 0$$

5. *KD* ($i = 5$):

$$\text{Total available KDs} = (\text{KD} \rightarrow \$) + (\text{KD} \rightarrow €) + (\text{KD} \rightarrow £) + (\text{KD} \rightarrow ¥)$$
$$= .342x_{15} + .445x_{25} + .543x_{35} + .0032x_{45}$$

$$\text{Total distributed KDs} = (\$ \rightarrow \text{KD}) + (€ \rightarrow \text{KD}) + (£ \rightarrow \text{KD}) + (¥ \rightarrow \text{KD})$$
$$= x_{51} + x_{52} + x_{53} + x_{54}$$

Thus, the constraint is

$$x_{51} + x_{52} + x_{53} + x_{54} - (.342x_{15} + .445x_{25} + .543x_{35} + .0032x_{45}) = 0$$

The only remaining constraints are the transaction limits, which are 5 million dollars, 3 million euros, 3.5 million pounds, 100 million yen, and 2.8 million KDs. These can be translated as

$$x_{1j} \leq 5, j = 2, 3, 4, 5$$
$$x_{2j} \leq 3, j = 1, 3, 4, 5$$
$$x_{3j} \leq 3.5, j = 1, 2, 4, 5$$
$$x_{4j} \leq 100, j = 1, 2, 3, 5$$
$$x_{5j} \leq 2.8, j = 1, 2, 3, 4$$

The complete model is now given as

$$\text{Maximize } z = y$$

subject to

$$y + x_{12} + x_{13} + x_{14} + x_{15} - \left(\tfrac{1}{.769}x_{21} + \tfrac{1}{.625}x_{31} + \tfrac{1}{105}x_{41} + \tfrac{1}{.342}x_{51}\right) = 5$$

$$x_{21} + x_{23} + x_{24} + x_{25} - \left(.769x_{12} + \tfrac{1}{.813}x_{32} + \tfrac{1}{137}x_{42} + \tfrac{1}{.445}x_{52}\right) = 0$$

$$x_{31} + x_{32} + x_{34} + x_{35} - \left(.625x_{13} + .813x_{23} + \tfrac{1}{169}x_{43} + \tfrac{1}{.543}x_{53}\right) = 0$$

$$x_{41} + x_{42} + x_{43} + x_{45} - \left(105x_{14} + 137x_{24} + 169x_{34} + \tfrac{1}{.0032}x_{54}\right) = 0$$

$$x_{51} + x_{52} + x_{53} + x_{54} - (.342x_{15} + .445x_{25} + .543x_{35} + .0032x_{45}) = 0$$

$$x_{1j} \le 5,\ j = 2, 3, 4, 5$$

$$x_{2j} \le 3,\ j = 1, 3, 4, 5$$

$$x_{3j} \le 3.5,\ j = 1, 2, 4, 5$$

$$x_{4j} \le 100,\ j = 1, 2, 3, 5$$

$$x_{5j} \le 2.8,\ j = 1, 2, 3, 4$$

$$x_{ij} \ge 0, \text{ for all } i \text{ and } j$$

Solution:

The optimum solution (using file amplEx2.3-2.txt or solverEx2.3-2.xls) is:

Solution	Interpretation
$y = 5.09032$	Final holdings = \$5,090,320. Net dollar gain = \$90,320, which represents a 1.8064% rate of return
$x_{12} = 1.46206$	Buy \$1,462,060 worth of euros
$x_{15} = 5$	Buy \$5,000,000 worth of KD
$x_{25} = 3$	Buy €3,000,000 worth of KD
$x_{31} = 3.5$	Buy £3,500,000 worth of dollars
$x_{32} = 0.931495$	Buy £931,495 worth of euros
$x_{41} = 100$	Buy ¥100,000,000 worth of dollars
$x_{42} = 100$	Buy ¥100,000,000 worth of euros
$x_{43} = 100$	Buy ¥100,000,000 worth of pounds
$x_{53} = 2.085$	Buy KD2,085,000 worth of pounds
$x_{54} = .96$	Buy KD960,000 worth of yen

Remarks. At first it may appear that the solution is nonsensical because it calls for using $x_{12} + x_{15} = 1.46206 + 5 = 6.46206$, or \$6,462,060 to buy euros and KDs when the initial dollar amount is only \$5,000,000. Where do the extra dollars come from? What happens in practice is that the given solution is submitted to the currency dealer as *one* order, meaning we do not wait until we accumulate enough currency of a certain type before making a buy. In the end, the net result of all these transactions is a net cost of \$5,000,000 to the investor. This can be seen by summing up all the dollar transactions in the solution:

$$I = y + x_{12} + x_{13} + x_{14} + x_{15} - \left(\tfrac{1}{.769}x_{21} + \tfrac{1}{.625}x_{31} + \tfrac{1}{105}x_{41} + \tfrac{1}{.342}x_{51}\right)$$

$$= 5.09032 + 1.46206 + 5 - \left(\tfrac{3.5}{.625} + \tfrac{100}{105}\right) = 5$$

Notice that x_{21}, x_{31}, x_{41} and x_{51} are in euro, pound, yen, and KD, respectively, and hence must be converted to dollars.

PROBLEM SET 2.3B

1. Modify the arbitrage model to account for a commission that amounts to .1% of any currency buy. Assume that the commission does not affect the circulating funds and that it is collected after the entire order is executed. How does the solution compare with that of the original model?
*2. Suppose that the company is willing to convert the initial \$5 million to any other currency that will provide the highest rate of return. Modify the original model to determine which currency is the best.
3. Suppose the initial amount I = \$7 million and that the company wants to convert it optimally to a combination of euros, pounds, and yen. The final mix may not include more than €2 million, £3 million, and ¥200 million. Modify the original model to determine the optimal buying mix of the three currencies.
4. Suppose that the company wishes to buy \$6 million. The transaction limits for different currencies are the same as in the original problem. Devise a buying schedule for this transaction, given that mix may not include more than €3 million, £2 million, and KD2 million.
5. Suppose that the company has \$2 million, €5 million, £4 million. Devise a buy-sell order that will improve the overall holdings converted to yen.

2.3.3 Investment

Today's investors are presented with multitudes of investment opportunities. Examples of investment problems are capital budgeting for projects, bond investment strategy, stock portfolio selection, and establishment of bank loan policy. In many of these situations, linear programming can be used to select the optimal mix of opportunities that will maximize return while meeting the investment conditions set by the investor.

Example 2.3-3 (Loan Policy Model)

Thriftem Bank is in the process of devising a loan policy that involves a maximum of \$12 million. The following table provides the pertinent data about available types of loans.

Type of loan	Interest rate	Bad-debt ratio
Personal	.140	.10
Car	.130	.07
Home	.120	.03
Farm	.125	.05
Commercial	.100	.02

Bad debts are unrecoverable and produce no interest revenue.

Competition with other financial institutions requires that the bank allocate at least 40% of the funds to farm and commercial loans. To assist the housing industry in the region, home loans must equal at least 50% of the personal, car, and home loans. The bank also has a stated policy of not allowing the overall ratio of bad debts on all loans to exceed 4%.

Mathematical Model: The situation seeks to determine the amount of loan in each category, thus leading to the following definitions of the variables:

$$x_1 = \text{personal loans (in millions of dollars)}$$

$$x_2 = \text{car loans}$$

$$x_3 = \text{home loans}$$

$$x_4 = \text{farm loans}$$

$$x_5 = \text{commercial loans}$$

The objective of the Thriftem Bank is to maximize its net return, the difference between interest revenue and lost bad debts. The interest revenue is accrued only on loans in good standing. Thus, because 10% of personal loans are lost to bad debt, the bank will receive interest on only 90% of the loan—that is, it will receive 14% interest on $.9x_1$ of the original loan x_1. The same reasoning applies to the remaining four types of loans. Thus,

$$\begin{aligned}\text{Total interest} &= .14(.9x_1) + .13(.93x_2) + .12(.97x_3) + .125(.95x_4) + .1(.98x_5)\\ &= .126x_1 + .1209x_2 + .1164x_3 + .11875x_4 + .098x_5\end{aligned}$$

We also have

$$\text{Bad debt} = .1x_1 + .07x_2 + .03x_3 + .05x_4 + .02x_5$$

The objective function is thus expressed as

$$\begin{aligned}\text{Maximize } z &= \text{Total interest—Bad debt}\\ &= (.126x_1 + .1209x_2 + .1164x_3 + .11875x_4 + .098x_5)\\ &\quad - (.1x_1 + .07x_2 + .03x_3 + .05x_4 + .02x_5)\\ &= .026x_1 + .0509x_2 + .0864x_3 + .06875x_4 + .078x_5\end{aligned}$$

The problem has five constraints:

1. *Total funds should not exceed \$12 (million):*

$$x_1 + x_2 + x_3 + x_4 + x_5 \leq 12$$

2. *Farm and commercial loans equal at least 40% of all loans:*

$$x_4 + x_5 \geq .4(x_1 + x_2 + x_3 + x_4 + x_5)$$

or

$$.4x_1 + .4x_2 + .4x_3 - .6x_4 - .6x_5 \leq 0$$

3. *Home loans should equal at least 50% of personal, car, and home loans:*

$$x_3 \geq .5(x_1 + x_2 + x_3)$$

or

$$.5x_1 + .5x_2 - .5x_3 \leq 0$$

4. *Bad debts should not exceed 4% of all loans:*

$$.1x_1 + .07x_2 + .03x_3 + .05x_4 + .02x_5 \leq .04(x_1 + x_2 + x_3 + x_4 + x_5)$$

or

$$.06x_1 + .03x_2 - .01x_3 + .01x_4 - .02x_5 \leq 0$$

5. *Nonnegativity*:

$$x_1 \geq 0, x_2 \geq 0, x_3 \geq 0, x_4 \geq 0, x_5 \geq 0$$

A subtle assumption in the preceding formulation is that all loans are issued at approximately the same time. This assumption allows us to ignore differences in the time value of the funds allocated to the different loans.

Solution:

The optimal solution is

$$z = .99648, x_1 = 0, x_2 = 0, x_3 = 7.2, x_4 = 0, x_5 = 4.8$$

Remarks.

1. You may be wondering why we did not define the right-hand side of the second constraint as $.4 \times 12$ instead of $.4(x_1 + x_2 + x_3 + x_4 + x_5)$. After all, it seems logical that the bank would want to loan out all \$12 (million). The answer is that the second usage does not "rob" the model of this possibility. If the optimum solution needs all \$12 (million), the given constraint will allow it. But there are two important reasons why you should not use $.4 \times 12$: (1) If other constraints in the model are such that all \$12 (million) *cannot* be used (for example, the bank may set caps on the different loans), then the choice $.4 \times 12$ could lead to an infeasible or incorrect solution. (2) If you want to experiment with the effect of changing available funds (say from \$12 to \$13 million) on the optimum solution, there is a real chance that you may forget to change $.4 \times 12$ to $.4 \times 13$, in which case the solution you get will not be correct. A similar reasoning applies to the left-hand side of the fourth constraint.
2. The optimal solution calls for allocating all \$12 million: \$7.2 million to home loans and \$4.8 million to commercial loans. The remaining categories receive none. The return on the investment is computed as

$$\text{Rate of return} = \frac{z}{12} = \frac{.99648}{12} = .08034$$

This shows that the combined annual rate of return is 8.034%, which is less than the best *net* interest rate (=.0864 for home loans), and one wonders why the optimum does not take advantage of this opportunity. The answer is that the restriction stipulating that farm and commercial loans account for at least 40% of all loans (constraint 2) forces the solution to allocate \$4.8 million to commercial loans at the lower *net* rate of .078, hence lowering the overall interest rate to $\frac{.0864 \times 7.2 + .078 \times 4.8}{12} = .08034$. In fact, if we remove constraint 2, the optimum will allocate all the funds to home loans at the higher 8.64% rate.

PROBLEM SET 2.3C

1. Fox Enterprises is considering six projects for possible construction over the next four years. The expected (present value) returns and cash outlays for the projects are given below. Fox can undertake any of the projects partially or completely. A partial undertaking of a project will prorate both the return and cash outlays proportionately.

	Cash outlay ($1000)				
Project	*Year 1*	*Year 2*	*Year 3*	*Year 4*	Return ($1000)
1	10.5	14.4	2.2	2.4	32.40
2	8.3	12.6	9.5	3.1	35.80
3	10.2	14.2	5.6	4.2	17.75
4	7.2	10.5	7.5	5.0	14.80
5	12.3	10.1	8.3	6.3	18.20
6	9.2	7.8	6.9	5.1	12.35
Available funds ($1000)	60.0	70.0	35.0	20.0	

(a) Formulate the problem as a linear program, and determine the optimal project mix that maximizes the total return. Ignore the time value of money.

(b) Suppose that if a portion of project 2 is undertaken then at least an equal portion of project 6 must undertaken. Modify the formulation of the model and find the new optimal solution.

(c) In the original model, suppose that any funds left at the end of a year are used in the next year. Find the new optimal solution, and determine how much each year "borrows" from the preceding year. For simplicity, ignore the time value of money.

(d) Suppose in the original model that the yearly funds available for any year can be exceeded, if necessary, by borrowing from other financial activities within the company. Ignoring the time value of money, reformulate the LP model, and find the optimum solution. Would the new solution require borrowing in any year? If so, what is the rate of return on borrowed money?

***2.** Investor Doe has $10,000 to invest in four projects. The following table gives the cash flow for the four investments.

	Cash flow ($1000) at the start of				
Project	*Year 1*	*Year 2*	*Year 3*	*Year 4*	*Year 5*
1	−1.00	0.50	0.30	1.80	1.20
2	−1.00	0.60	0.20	1.50	1.30
3	0.00	−1.00	0.80	1.90	0.80
4	−1.00	0.40	0.60	1.80	0.95

The information in the table can be interpreted as follows: For project 1, $1.00 invested at the start of year 1 will yield $.50 at the start of year 2, $.30 at the start of year 3, $1.80 at the start of year 4, and $1.20 at the start of year 5. The remaining entries can be interpreted similarly. The entry 0.00 indicates that no transaction is taking place. Doe has the additional option of investing in a bank account that earns 6.5% annually. All funds accumulated at the end of one year can be reinvested in the following year. Formulate the problem as a linear program to determine the optimal allocation of funds to investment opportunities.

3. HiRise Construction can bid on two 1-year projects. The following table provides the quarterly cash flow (in millions of dollars) for the two projects.

	Cash flow (in millions of $) at				
Project	*1/1/08*	*4/1/08*	*7/1/08*	*10/1/08*	*12/31/08*
I	−1.0	−3.1	−1.5	1.8	5.0
II	−3.0	−2.5	1.5	1.8	2.8

HiRise has cash funds of $1 million at the beginning of each quarter and may borrow at most $1 million at a 10% nominal annual interest rate. Any borrowed money must be returned at the end of the quarter. Surplus cash can earn quarterly interest at an 8% nominal annual rate. Net accumulation at the end of one quarter is invested in the next quarter.

(a) Assume that HiRise is allowed partial or full participation in the two projects. Determine the level of participation that will maximize the net cash accumulated on 12/31/2008.

(b) Is it possible in any quarter to borrow money and simultaneously end up with surplus funds? Explain.

4. In anticipation of the immense college expenses, a couple have started an annual investment program on their child's eighth birthday that will last until the eighteenth birthday. The couple estimate that they will be able to invest the following amounts at the beginning of each year:

Year	1	2	3	4	5	6	7	8	9	10
Amount ($)	2000	2000	2500	2500	3000	3500	3500	4000	4000	5000

To avoid unpleasant surprises, they want to invest the money safely in the following options: Insured savings with 7.5% annual yield, six-year government bonds that yield 7.9% and have a current market price equal to 98% of face value, and nine-year municipal bonds yielding 8.5% and having a current market price of 1.02 of face value. How should the couple invest the money?

*5. A business executive has the option to invest money in two plans: Plan A guarantees that each dollar invested will earn $.70 a year later, and plan B guarantees that each dollar invested will earn $2 after 2 years. In plan A, investments can be made annually, and in plan B, investments are allowed for periods that are multiples of two years only. How should the executive invest $100,000 to maximize the earnings at the end of 3 years?

6. A gambler plays a game that requires dividing bet money among four choices. The game has three outcomes. The following table gives the corresponding gain or loss per dollar for the different options of the game.

	Return per dollar deposited in choice			
Outcome	*1*	*2*	*3*	*4*
1	−3	4	−7	15
2	5	−3	9	4
3	3	−9	10	−8

The gambler has a total of $500, which may be played only once. The exact outcome of the game is not known a priori. Because of this uncertainty, the gambler's strategy is to maximize the *minimum* return produced by the three outcomes. How should the gambler allocate the $500 among the four choices? (*Hint:* The gambler's net return may be positive, zero, or negative.)

7. (Lewis, 1996) Monthly bills in a household are received monthly (e.g., utilities and home mortgage), quarterly (e.g., estimated tax payment), semiannually (e.g., insurance) , or annually (e.g., subscription renewals and dues). The following table provides the monthly bills for next year.

Month	Jan.	Feb.	Mar.	Apr.	May	Jun.	Jul.	Aug.	Sep.	Oct.	Nov.	Dec.	Total
$	800	1200	400	700	600	900	1500	1000	900	1100	1300	1600	12000

To account for these expenses, the family sets aside $1000 per month, which is the average of the total divided by 12 months. If the money is deposited in a regular savings account, it can earn 4% annual interest, provided it stays in the account at least one month. The bank also offers 3-month and 6-month certificates of deposit that can earn 5.5% and 7% annual interest, respectively. Develop a 12-month investment schedule that will maximize the family's total return for the year. State any assumptions or requirements needed to reach a feasible solution.

2.3.4 Production Planning and Inventory Control

There is a wealth of LP applications to production and inventory control, ranging from simple allocation of machining capacity to meet demand to the more complex case of using inventory to "dampen" the effect of erratic change in demand over a given planning horizon and of using hiring and firing to respond to changes in workforce needs. This section presents three examples. The first deals with the scheduling of products using common production facilities to meet demand during a single period, the second deals with the use of inventory in a multiperiod production system to fill future demand, and the third deals with the use of a combined inventory and worker hiring/firing to "smooth" production over a multiperiod planning horizon with fluctuating demand.

Example 2.3-4 (Single-Period Production Model)

In preparation for the winter season, a clothing company is manufacturing parka and goose overcoats, insulated pants, and gloves. All products are manufactured in four different departments: cutting, insulating, sewing, and packaging. The company has received firm orders for its products. The contract stipulates a penalty for undelivered items. The following table provides the pertinent data of the situation.

	Time per units (hr)				
Department	*Parka*	*Goose*	*Pants*	*Gloves*	Capacity (hr)
Cutting	.30	.30	.25	.15	1000
Insulating	.25	.35	.30	.10	1000
Sewing	.45	.50	.40	.22	1000
Packaging	.15	.15	.1	.05	1000
Demand	800	750	600	500	
Unit profit	$30	$40	$20	$10	
Unit penalty	$15	$20	$10	$8	

Devise an optimal production plan for the company.

Mathematical Model: The definition of the variables is straightforward. Let

$$x_1 = \text{number of parka jackets}$$
$$x_2 = \text{number of goose jackets}$$
$$x_3 = \text{number of pairs of pants}$$
$$x_4 = \text{number of pairs of gloves}$$

The company is penalized for not meeting demand. This means that the objective of the problem is to maximize the net receipts, defined as

$$\text{Net receipts} = \text{Total profit} - \text{Total penalty}$$

The total profit is readily expressed as $30x_1 + 40x_2 + 20x_3 + 10x_4$. The total penalty is a function of the shortage quantities (= demand − units supplied of each product). These quantities can be determined from the following demand limits:

$$x_1 \leq 800, x_2 \leq 750, x_3 \leq 600, x_4 \leq 500$$

A demand is not fulfilled if its constraint is satisfied as a strict inequality. For example, if 650 parka jackets are produced, then $x_1 = 650$, which leads to a shortage of $800 - 650 = 150$ parka jackets. We can express the shortage of any product algebraically by defining a new nonnegative variable—namely,

$$s_j = \text{Number of shortage units of product } j, j = 1, 2, 3, 4$$

In this case, the demand constraints can be written as

$$x_1 + s_1 = 800, x_2 + s_2 = 750, x_3 + s_3 = 600, x_4 + s_4 = 500$$
$$x_j \geq 0, s_j \geq 0, j = 1, 2, 3, 4$$

We can now compute the shortage penalty as $15s_1 + 20s_2 + 10s_3 + 8s_4$. Thus, the objective function can be written as

$$\text{Maximize } z = 30x_1 + 40x_2 + 20x_3 + 10x_4 - (15s_1 + 20s_2 + 10s_3 + 8s_4)$$

To complete the model, the remaining constraints deal with the production capacity restrictions; namely

$$.30x_1 + .30x_2 + .25x_3 + .15x_4 \leq 1000 \quad \text{(Cutting)}$$
$$.25x_1 + .35x_2 + .30x_3 + .10x_4 \leq 1000 \quad \text{(Insulating)}$$
$$.45x_1 + .50x_2 + .40x_3 + .22x_4 \leq 1000 \quad \text{(Sewing)}$$
$$.15x_1 + .15x_2 + .10x_3 + .05x_4 \leq 1000 \quad \text{(Packaging)}$$

The complete model thus becomes

$$\text{Maximize } z = 30x_1 + 40x_2 + 20x_3 + 10x_4 - (15s_1 + 20s_2 + 10s_3 + 8s_4)$$

subject to

$$.30x_1 + .30x_2 + .25x_3 + .15x_4 \leq 1000$$
$$.25x_1 + .35x_2 + .30x_3 + .10x_4 \leq 1000$$
$$.45x_1 + .50x_2 + .40x_3 + .22x_4 \leq 1000$$
$$.15x_1 + .15x_2 + .10x_3 + .05x_4 \leq 1000$$
$$x_1 + s_1 = 800, x_2 + s_2 = 750, x_3 + s_3 = 600, x_4 + s_4 = 500$$
$$x_j \geq 0, s_j \geq 0, j = 1, 2, 3, 4$$

Solution:

The optimum solution is $z = \$64{,}625$, $x_1 = 850$, $x_2 = 750$, $x_3 = 387.5$, $x_4 = 500$, $s_1 = s_2 = s_4 = 0$, $s_3 = 212.5$. The solution satisfies all the demand for both types of jackets and the gloves. A shortage of 213 (rounded up from 212.5) pairs of pants will result in a penalty cost of $213 \times \$10 = \$2130$.

Example 2.3-5 (Multiple Period Production-Inventory Model)

Acme Manufacturing Company has contracted to deliver home windows over the next 6 months. The demands for each month are 100, 250, 190, 140, 220, and 110 units, respectively. Production cost per window varies from month to month depending on the cost of labor, material, and utilities. Acme estimates the production cost per window over the next 6 months to be \$50, \$45, \$55, \$48, \$52, and \$50, respectively. To take advantage of the fluctuations in manufacturing cost, Acme may elect to produce more than is needed in a given month and hold the excess units for delivery in later months. This, however, will incur storage costs at the rate of \$8 per window per month assessed on end-of-month inventory. Develop a linear program to determine the optimum production schedule.

Mathematical Model: The variables of the problem include the monthly production amount and the end-of-month inventory. For $i = 1, 2, \ldots, 6$, let

$$x_i = \text{Number of units produced in month } i$$
$$I_i = \text{Inventory units left at the end of month } i$$

The relationship between these variables and the monthly demand over the six-month horizon is represented by the schematic diagram in Figure 2.5. The system starts empty, which means that $I_0 = 0$.

The objective function seeks to minimize the sum of the production and end-of-month inventory costs. Here we have,

$$\text{Total production cost} = 50x_1 + 45x_2 + 55x_3 + 48x_4 + 52x_5 + 50x_6$$
$$\text{Total inventory cost} = 8(I_1 + I_2 + I_3 + I_4 + I_5 + I_6)$$

Thus the objective function is

$$\text{Minimize } z = 50x_1 + 45x_2 + 55x_3 + 48x_4 + 52x_5 + 50x_6$$
$$+ 8(I_1 + I_2 + I_3 + I_4 + I_5 + I_6)$$

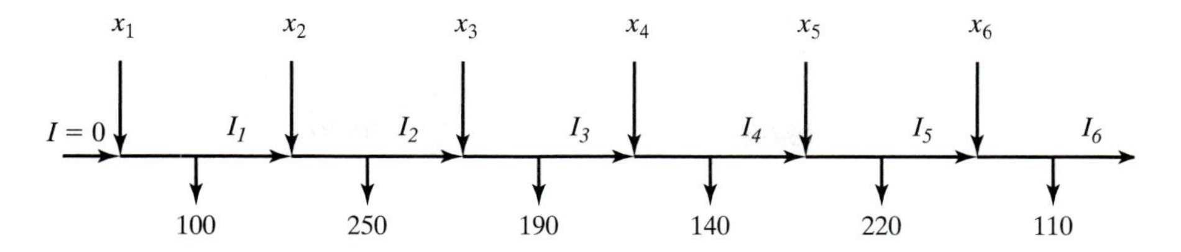

FIGURE 2.5
Schematic representation of the production-inventory system

The constraints of the problem can be determined directly from the representation in Figure 2.5. For each period we have the following balance equation:

$$\text{Beginning inventory} + \text{Production amount} - \text{Ending inventory} = \text{Demand}$$

This is translated mathematically for the individual months as

$$\begin{aligned}
I_0 + x_1 - I_1 &= 100 \quad \text{(Month 1)}\\
I_1 + x_2 - I_2 &= 250 \quad \text{(Month 2)}\\
I_2 + x_3 - I_3 &= 190 \quad \text{(Month 3)}\\
I_3 + x_4 - I_4 &= 140 \quad \text{(Month 4)}\\
I_4 + x_5 - I_5 &= 220 \quad \text{(Month 5)}\\
I_5 + x_6 - I_6 &= 110 \quad \text{(Month 6)}\\
x_i, I_i &\geq 0, \text{ for all } i = 1, 2, \ldots, 6\\
I_0 &= 0
\end{aligned}$$

For the problem, $I_0 = 0$ because the situation starts with no initial inventory. Also, in any optimal solution, the ending inventory I_6 will be zero, because it is not logical to end the horizon with positive inventory, which can only incur additional inventory cost without serving any purpose.

The complete model is now given as

$$\begin{aligned}
\text{Minimize } z = {} & 50x_1 + 45x_2 + 55x_3 + 48x_4 + 52x_5 + 50x_6\\
& + 8(I_1 + I_2 + I_3 + Ix_4 + I_5 + I_6)
\end{aligned}$$

subject to

$$\begin{aligned}
x_1 - I_1 &= 100 \quad \text{(Month 1)}\\
I_1 + x_2 - I_2 &= 250 \quad \text{(Month 2)}\\
I_2 + x_3 - I_3 &= 190 \quad \text{(Month 3)}\\
I_3 + x_4 - I_4 &= 140 \quad \text{(Month 4)}\\
I_4 + x_5 - I_5 &= 220 \quad \text{(Month 5)}\\
I_5 + x_6 - I_6 &= 110 \quad \text{(Month 6)}\\
x_i, I_i &\geq 0, \text{ for all } i = 1, 2, \ldots, 6
\end{aligned}$$

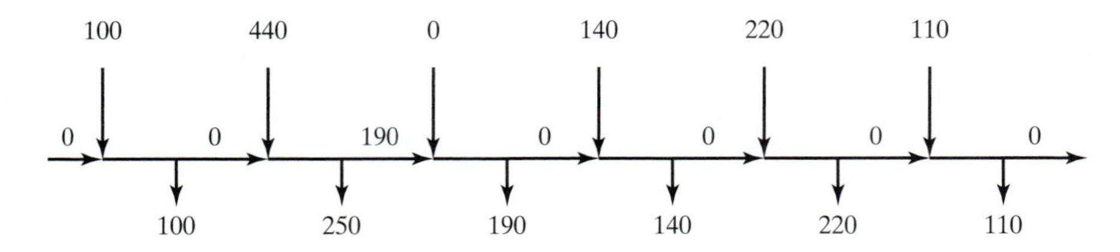

FIGURE 2.6

Optimum solution of the production-inventory problem

Solution:

The optimum solution is summarized in Figure 2.6. It shows that each month's demand is satisfied directly from the month's production, except for month 2 whose production quantity of 440 units covers the demand for both months 2 and 3. The total associated cost is $z = \$49{,}980$.

Example 2.3-6 (Multiperiod Production Smoothing Model)

A company will manufacture a product for the next four months: March, April, May, and June. The demands for each month are 520, 720, 520, and 620 units, respectively. The company has a steady workforce of 10 employees but can meet fluctuating production needs by hiring and firing temporary workers, if necessary. The extra costs of hiring and firing in any month are \$200 and \$400 per worker, respectively. A permanent worker can produce 12 units per month, and a temporary worker, lacking comparable experience, only produce 10 units per month. The company can produce more than needed in any month and carry the surplus over to a succeeding month at a holding cost of \$50 per unit per month. Develop an optimal hiring/firing policy for the company over the four-month planning horizon.

Mathematical Model: This model is similar to that of Example 2.3-5 in the general sense that each month has its production, demand, and ending inventory. There are two exceptions: (1) accounting for the permanent versus the temporary workforce, and (2) accounting for the cost of hiring and firing in each month.

Because the permanent 10 workers cannot be fired, their impact can be accounted for by subtracting the units they produce from the respective monthly demand. The remaining demand, if any, is satisfied through hiring and firing of temps. From the standpoint of the model, the net demand for each month is

$$\text{Demand for March} = 520 - 12 \times 10 = 400 \text{ units}$$

$$\text{Demand for April} = 720 - 12 \times 10 = 600 \text{ units}$$

$$\text{Demand for May} = 520 - 12 \times 10 = 400 \text{ units}$$

$$\text{Demand for June} = 620 - 12 \times 10 = 500 \text{ units}$$

For $i = 1, 2, 3, 4$, the variables of the model can be defined as

x_i = Net number of temps at the start of month i *after* any hiring or firing

S_i = Number of temps hired or fired at the start of month i

I_i = Units of ending inventory for month i

The variables x_i and I_i, by definition, must assume nonnegative values. On the other hand, the variable S_i can be positive when new temps are hired, negative when workers are fired, and zero if no hiring or firing occurs. As a result, the variable must be *unrestricted in sign*. This is the first instance in this chapter of using an unrestricted variable. As we will see shortly, special substitution is needed to allow the implementation of hiring and firing in the model.

The objective is to minimize the sum of the cost of hiring and firing plus the cost of holding inventory from one month to the next. The treatment of the inventory cost is similar to the one given in Example 2.3-5—namely,

$$\text{Inventory holding cost} = 50(I_1 + I_2 + I_3)$$

(Note that $I_4 = 0$ in the optimum solution.) The cost of hiring and firing is a bit more involved. We know that in any optimum solution, at least 40 temps $\left(= \frac{400}{10}\right)$ must be hired at the start of March to meet the month's demand. However, rather than treating this situation as a special case, we can let the optimization process take care of it automatically. Thus, given that the costs of hiring and firing a temp are \$200 and \$400, respectively, we have

$$\begin{pmatrix}\text{Cost of hiring}\\ \text{and firing}\end{pmatrix} = 200\begin{pmatrix}\text{Number of hired temps}\\ \text{at the start of}\\ \text{March, April, May, and June}\end{pmatrix}$$
$$+ 400\begin{pmatrix}\text{Number of fired temps}\\ \text{at the start of}\\ \text{March, April, May, and June}\end{pmatrix}$$

To translate this equation mathematically, we will need to develop the constraints first.

The constraints of the model deal with inventory and hiring and firing. First we develop the inventory constraints. Defining x_i as the number of temps available in month i and given that the productivity of a temp is 10 units per month, the number of units produced in the same month is $10x_i$. Thus the inventory constraints are

$$\begin{aligned}
10x_1 &= 400 + I_1 && \text{(March)}\\
I_1 + 10x_2 &= 600 + I_2 && \text{(April)}\\
I_2 + 10x_3 &= 400 + I_3 && \text{(May)}\\
I_3 + 10x_4 &= 500 && \text{(June)}\\
x_1, x_2, x_3, x_4 &\geq 0, I_1, I_2, I_3 \geq 0
\end{aligned}$$

Next, we develop the constraints dealing with hiring and firing. First, note that the temp workforce starts with x_1 workers at the beginning of March. At the start of April, x_1 will be adjusted (up or down) by S_2 to generate x_2. The same idea applies to x_3 and x_4. These observations lead to the following equations

$$\begin{aligned}
x_1 &= S_1\\
x_2 &= x_1 + S_2\\
x_3 &= x_2 + S_3
\end{aligned}$$

$$x_4 = x_3 + S_4$$
$$S_1, S_2, S_3, S_4 \text{ unrestricted in sign}$$
$$x_1, x_2, x_3, x_4 \geq 0$$

The variables S_1, S_2, S_3, and S_4 represent hiring when they are strictly positive and firing when they are strictly negative. However, this "qualitative" information cannot be used in a mathematical expression. Instead, we use the following substitution:

$$S_i = S_i^- - S_i^+, \text{ where } S_i^-, S_i^+ \geq 0$$

The unrestricted variable S_i is now the difference between two nonnegative variables S_i^- and S_i^+. We can think of S_i^- as the number of temps hired and S_i^+ as the number of temps fired. For example, if $S_i^- = 5$ and $S_i^+ = 0$ then $S_i = 5 - 0 = +5$, which represents hiring. If $S_i^- = 0$ and $S_i^+ = 7$ then $S_i = 0 - 7 = -7$, which represents firing. In the first case, the corresponding cost of hiring is $200S_i^- = 200 \times 5 = \$1000$ and in the second case the corresponding cost of firing is $400S_i^+ = 400 \times 7 = \$2800$. This idea is the basis for the development of the objective function.

First we need to address an important point: What if both S_i^- and S_i^+ are positive? The answer is that this cannot happen because it implies that the solution calls for both hiring and firing in the same month. Interestingly, the theory of linear programming (see Chapter 7) tells us that S_i^- and S_i^+ can never be positive simultaneously, a result that confirms intuition.

We can now write the cost of hiring and firing as follows:

$$\text{Cost of hiring} = 200(S_1^- + S_2^- + S_3^- + S_4^-)$$
$$\text{Cost of firing} = 400(S_1^+ + S_2^+ + S_3^+ + S_4^+)$$

The complete model is

$$\text{Minimize } z = 50(I_1 + I_2 + I_3 + I_4) + 200(S_1^- + S_2^- + S_3^- + S_4^-) + 400(S_1^+ + S_2^+ + S_3^+ + S_4^+)$$

subject to

$$10x_1 = 400 + I_1$$
$$I_1 + 10x_2 = 600 + I_2$$
$$I_2 + 10x_3 = 400 + I_3$$
$$I_3 + 10x_4 = 500$$
$$x_1 = S_1^- - S_1^+$$
$$x_2 = x_1 + S_2^- - S_2^+$$
$$x_3 = x_2 + S_3^- - S_3^+$$
$$x_4 = x_3 + S_4^- - S_4^+$$
$$S_1^-, S_1^+, S_2^-, S_2^+, S_3^-, S_3^+, S_4^-, S_4^+ \geq 0$$
$$x_1, x_2, x_3, x_4 \geq 0$$
$$I_1, I_2, I_3 \geq 0$$

Solution:

The optimum solution is $z = \$19{,}500$, $x_1 = 50$, $x_2 = 50$, $x_3 = 45$, $x_4 = 45$, $S_1^- = 50$, $S_3^+ = 5$, $I_1 = 100$, $I_3 = 50$. All the remaining variables are zero. The solution calls for hiring 50 temps in March ($S_1^- = 50$) and holding the workforce steady till May, when 5 temps are fired ($S_3^+ = 5$). No further hiring or firing is recommended until the end of June, when, presumably, all temps are terminated. This solution requires 100 units of inventory to be carried into May and 50 units to be carried into June.

PROBLEM SET 2.3D

1. Toolco has contracted with AutoMate to supply their automotive discount stores with wrenches and chisels. AutoMate's weekly demand consists of at least 1500 wrenches and 1200 chisels. Toolco cannot produce all the requested units with its present one-shift capacity and must use overtime and possibly subcontract with other tool shops. The result is an increase in the production cost per unit, as shown in the following table. Market demand restricts the ratio of chisels to wrenches to at least 2:1.

Tool	Production type	Weekly production range (units)	Unit cost ($)
Wrenches	Regular	0–550	2.00
	Overtime	551–800	2.80
	Subcontracting	$801-\infty$	3.00
Chisel	Regular	0–620	2.10
	Overtime	621–900	3.20
	Subcontracting	$901-\infty$	4.20

   (a) Formulate the problem as a linear program, and determine the optimum production schedule for each tool.

   (b) Relate the fact that the production cost function has increasing unit costs to the validity of the model.

2. Four products are processed sequentially on three machines. The following table gives the pertinent data of the problem.

Machine	Cost per hr ($)	Manufacturing time (hr) per unit				Capacity (hr)
		Product 1	*Product 2*	*Product 3*	*Product 4*	
1	10	2	3	4	2	500
2	5	3	2	1	2	380
3	4	7	3	2	1	450
Unit selling price ($)		75	70	55	45	

Formulate the problem as an LP model, and find the optimum solution.

***3.** A manufacturer produces three models, I, II, and III, of a certain product using raw materials A and B. The following table gives the data for the problem:

	Requirements per unit			
Raw material	*I*	*II*	*III*	Availability
A	2	3	5	4000
B	4	2	7	6000
Minimum demand	200	200	150	
Profit per unit($)	30	20	50	

The labor time per unit of model I is twice that of II and three times that of III. The entire labor force of the factory can produce the equivalent of 1500 units of model I. Market requirements specify the ratios 3:2:5 for the production of the three respective models. Formulate the problem as a linear program, and find the optimum solution.

4. The demand for ice cream during the three summer months (June, July, and August) at All-Flavors Parlor is estimated at 500, 600, and 400 20-gallon cartons, respectively. Two wholesalers, 1 and 2, supply All-Flavors with its ice cream. Although the flavors from the two suppliers are different, they are interchangeable. The maximum number of cartons either supplier can provide is 400 per month. Also, the prices the two suppliers charge change from one month to the next according to the following schedule:

	Price per carton in month		
	June	*July*	*August*
Supplier 1	$100	$110	$120
Supplier 2	$115	$108	$125

To take advantage of price fluctuation, All-Flavors can purchase more than is needed for a month and store the surplus to satisfy the demand in a later month. The cost of refrigerating an ice cream carton is $5 per month. It is realistic in the present situation to assume that the refrigeration cost is a function of the average number of cartons on hand during the month. Develop an optimum schedule for buying ice cream from the two suppliers.

5. The demand for an item over the next four quarters is 300, 400, 450, and 250 units, respectively. The price per unit starts at $20 in the first quarter and increases by $2 each quarter thereafter. The supplier can provide no more than 400 units in any one quarter. Although we can take advantage of lower prices in early quarters, a storage cost of $3.50 is incurred per unit per quarter. In addition, the maximum number of units that can be held over from one quarter to the next cannot exceed 100. Develop an optimum schedule for purchasing the item to meet the demand.

6. A company has contracted to produce two products, A and B, over the months of June, July, and August. The total production capacity (expressed in hours) varies monthly. The following table provides the basic data of the situation:

	June	July	August
Demand for A (units)	500	5000	750
Demand for B (units)	1000	1200	1200
Capacity (hours)	3000	3500	3000

The production rates in units per hour are 1.25 and 1 for products *A* and *B*, respectively. All demand must be met. However, demand for a later month may be filled from the production in an earlier one. For any carryover from one month to the next, holding costs of $.90 and $.75 per unit per month are charged for products *A* and *B*, respectively. The unit production costs for the two products are $30 and $28 for *A* and *B*, respectively. Determine the optimum production schedule for the two products.

***7.** The manufacturing process of a product consists of two successive operations, I and II. The following table provides the pertinent data over the months of June, July, and August:

	June	July	August
Finished product demand (units)	500	450	600
Capacity of operation I (hr)	800	700	550
Capacity of operation II (hr)	1000	850	700

Producing a unit of the product takes .6 hour on operation I plus .8 hour on operation II. Overproduction of either the semifinished product (operation I) or the finished product (operation II) in any month is allowed for use in a later month. The corresponding holding costs are $.20 and $.40 per unit per month. The production cost varies by operation and by month. For operation 1, the unit production cost is $10, $12, and $11 for June, July, and August. For operation 2, the corresponding unit production cost is $15, $18, and $16. Determine the optimal production schedule for the two operations over the 3-month horizon.

8. Two products are manufactured sequentially on two machines. The time available on each machine is 8 hours per day and may be increased by up to 4 hours of overtime, if necessary, at an additional cost of $100 per hour. The table below gives the production rate on the two machines as well as the price per unit of the two products. Determine the optimum production schedule and the recommended use of overtime, if any.

	Production rate (units/hr)	
	Product 1	*Product 2*
Machine 1	5	5
Machine 2	8	4
Price per unit ($)	110	118

2.3.5 Blending and Refining

A number of LP applications deal with blending different input materials to produce products that meet certain specifications while minimizing cost or maximizing profit. The input materials could be ores, metal scraps, chemicals, or crude oils and the output products could be metal ingots, paints, or gasoline of various grades. This section presents a (simplified) model for oil refining. The process starts with distilling crude oil to produce intermediate gasoline stocks and then blending these stocks to produce final gasolines. The final products must satisfy certain quality specifications (such as octane rating). In addition, distillation capacities and demand limits can directly affect the level of production of the different grades of gasoline. One goal of the model is determine the optimal mix of final products that will maximize an appropriate profit function. In some cases, the goal may be to minimize a cost function.

Example 2.3-7 (Crude Oil Refining and Gasoline Blending)

Shale Oil, located on the island of Aruba, has a capacity of 1,500,000 bbl of crude oil per day. The final products from the refinery include three types of unleaded gasoline with different octane numbers (ON): regular with ON = 87, premium with ON = 89, and super with ON = 92. The refining process encompasses three stages: (1) a distillation tower that produces feedstock (ON = 82) at the rate of .2 bbl per bbl of crude oil, (2) a cracker unit that produces gasoline stock (ON = 98) by using a portion of the feedstock produced from the distillation tower at the rate of .5 bbl per bbl of feedstock, and (3) a blender unit that blends the gasoline stock from the cracker unit and the feedstock from the distillation tower. The company estimates the net profit per barrel of the three types of gasoline to be $6.70, $7.20, and $8.10, respectively. The input capacity of the cracker unit is 200,000 barrels of feedstock a day. The demand limits for regular, premium, and super gasoline are 50,000, 30,000, and 40,000 barrels per day. Develop a model for determining the optimum production schedule for the refinery.

Mathematical Model: Figure 2.7 summarizes the elements of the model. The variables can be defined in terms of two input streams to the blender (feedstock and cracker gasoline) and the three final products. Let

$$x_{ij} = \text{bbl/day of input stream } i \text{ used to blend final product } j,\ i = 1, 2;\ j = 1, 2, 3$$

Using this definition, we have

$$\begin{aligned}
\text{Daily production of regular gasoline} &= x_{11} + x_{21} \text{ bbl/day} \\
\text{Daily production of premium gasoline} &= x_{12} + x_{22} \text{ bbl/day} \\
\text{Daily production of super gasoline} &= x_{13} + x_{23} \text{ bbl/day}
\end{aligned}$$

$$\begin{aligned}
\begin{pmatrix}\text{Daily output}\\ \text{of blender unit}\end{pmatrix} &= \begin{pmatrix}\text{Daily production}\\ \text{of regular gas}\end{pmatrix} + \begin{pmatrix}\text{Daily production}\\ \text{of premium gas}\end{pmatrix} \\
&\quad + \begin{pmatrix}\text{Daily production}\\ \text{of super gas}\end{pmatrix} \\
&= (x_{11} + x_{21}) + (x_{12} + x_{22}) + (x_{13} + x_{23})\text{bbl/day}
\end{aligned}$$

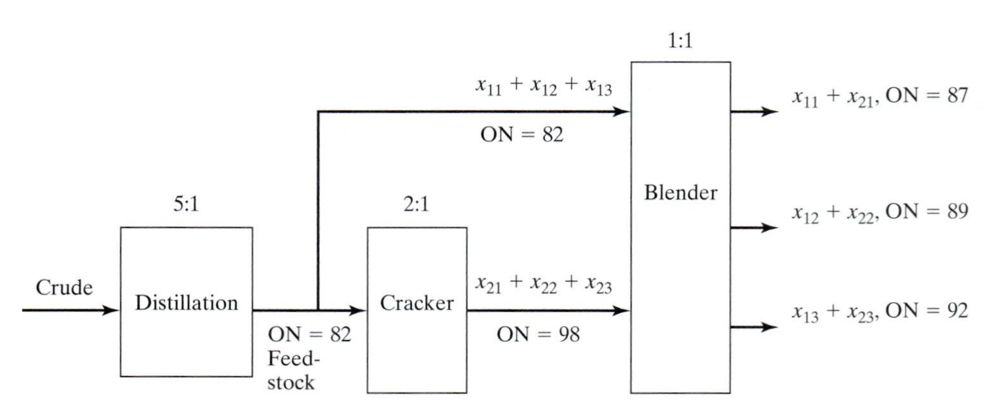

FIGURE 2.7

Product flow in the refinery problem

$$\begin{pmatrix}\text{Daily feedstock}\\\text{to blender}\end{pmatrix} = x_{11} + x_{12} + x_{13} \text{ bbl/day}$$

$$\begin{pmatrix}\text{Daily cracker unit}\\\text{feed to blender}\end{pmatrix} = x_{21} + x_{22} + x_{23} \text{ bbl/day}$$

$$\begin{pmatrix}\text{Daily feedstock}\\\text{to cracker}\end{pmatrix} = 2(x_{21} + x_{22} + x_{23})\text{bbl/day}$$

$$\begin{pmatrix}\text{Daily crude oil used}\\\text{in the refinery}\end{pmatrix} = 5(x_{11} + x_{12} + x_{13}) + 10(x_{21} + x_{22} + x_{23})\text{bbl/day}$$

The objective of the model is to maximize the total profit resulting from the sale of all three grades of gasoline. From the definitions given above, we get

$$\text{Maximize } z = 6.70(x_{11} + x_{21}) + 7.20(x_{12} + x_{22}) + 8.10(x_{13} + x_{23})$$

The constraints of the problem are developed as follows:

1. *Daily crude oil supply does not exceed 1,500,000 bbl/day:*

$$5(x_{11} + x_{12} + x_{13}) + 10(x_{21} + x_{22} + x_{23}) \leq 1{,}500{,}000$$

2. *Cracker unit input capacity does not exceed 200,000 bbl/day:*

$$2(x_{21} + x_{22} + x_{23}) \leq 200{,}000$$

3. *Daily demand for regular does not exceed 50,000 bbl:*

$$x_{11} + x_{21} \leq 50{,}000$$

4. *Daily demand for premium does not exceed 30,000:*

$$x_{12} + x_{22} \leq 30{,}000$$

5. *Daily demand for super does not exceed 40,000 bbl:*

$$x_{13} + x_{23} \leq 40{,}000$$

6. *Octane number (ON) for regular is at least 87:*

The octane number of a gasoline product is the weighted average of the octane numbers of the input streams used in the blending process and can be computed as

$$\begin{pmatrix}\text{Average ON of}\\\text{regular gasoline}\end{pmatrix} = \frac{\text{Feedstock ON} \times \text{feedstock bbl/day} + \text{Cracker unit ON} \times \text{Cracker unit bbl/day}}{\text{Total bbl/day of regular gasoline}}$$

$$= \frac{82x_{11} + 98x_{21}}{x_{11} + x_{21}}$$

Thus, octane number constraint for regular gasoline becomes

$$\frac{82x_{11} + 98x_{21}}{x_{11} + x_{21}} \geq 87$$

The constraint is linearized as

$$82x_{11} + 98x_{21} \geq 87(x_{11} + x_{21})$$

7. *Octane number (ON) for premium is at least 89:*

$$\frac{82x_{12} + 98x_{22}}{x_{12} + x_{22}} \geq 89$$

which is linearized as

$$82x_{12} + 98x_{22} \geq 89(x_{12} + x_{22})$$

8. *Octane number (ON) for super is at least 92:*

$$\frac{82x_{13} + 98x_{23}}{x_{13} + x_{23}} \geq 92$$

or

$$82x_{13} + 98x_{23} \geq 92(x_{13} + x_{23})$$

The complete model is thus summarized as

$$\text{Maximize } z = 6.70(x_{11} + x_{21}) + 7.20(x_{12} + x_{22}) + 8.10(x_{13} + x_{23})$$

subject to

$$\begin{aligned}
5(x_{11} + x_{12} + x_{13}) + 10(x_{21} + x_{22} + x_{23}) &\leq 1{,}500{,}000 \\
2(x_{21} + x_{22} + x_{23}) &\leq 200{,}000 \\
x_{11} + x_{21} &\leq 50{,}000 \\
x_{12} + x_{22} &\leq 30{,}000 \\
x_{13} + x_{23} &\leq 40{,}000 \\
82x_{11} + 98x_{21} &\geq 87(x_{11} + x_{21}) \\
82x_{12} + 98x_{22} &\geq 89(x_{12} + x_{22}) \\
82x_{13} + 98x_{23} &\geq 92(x_{13} + x_{23})
\end{aligned}$$

$$x_{11}, x_{12}, x_{13}, x_{21}, x_{22}, x_{23} \geq 0$$

The last three constraints can be simplified to produce a constant right-hand side.

Solution:

The optimum solution (using file amplEx2.3-7.txt) is $z = 1{,}482{,}000$, $x_{11} = 20{,}625$, $x_{21} = 9375$, $x_{12} = 16{,}875$, $x_{22} = 13{,}125$, $x_{13} = 15{,}000$, $x_{23} = 25{,}000$. This translates to

Daily profit = \$1,482,000

Daily amount of regular gasoline = $x_{11} + x_{21} = 20{,}625 + 9375 = 30{,}000$ bbl/day

Daily amount of premium gasoline = $x_{12} + x_{22} = 16{,}875 + 13{,}125 = 30{,}000$ bbl/day

Daily amount of regular gasoline = $x_{13} + x_{23} = 15{,}000 + 25{,}000 = 40{,}000$ bbl/day

The solution shows that regular gasoline production is 20,000 bbl/day short of satisfying the maximum demand. The demand for the remaining two grades is satisfied.

PROBLEM SET 2.3E

1. Hi-V produces three types of canned juice drinks, *A*, *B*, and *C*, using fresh strawberries, grapes, and apples. The daily supply is limited to 200 tons of strawberries, 100 tons of grapes, and 150 tons of apples. The cost per ton of strawberries, grapes, and apples is $200, $100, and $90, respectively. Each ton makes 1500 lb of strawberry juice, 1200 lb of grape juice, and 1000 lb of apple juice. Drink *A* is a 1:1 mix of strawberry and apple juice. Drink *B* is 1:1:2 mix of strawberry, grape, and apple juice. Drink *C* is a 2:3 mix of grape and apple juice. All drinks are canned in 16-oz (1 lb) cans. The price per can is $1.15, $1.25, and $1.20 for drinks *A*, *B*, and *C*. Determine the optimal production mix of the three drinks.

*2. A hardware store packages handyman bags of screws, bolts, nuts, and washers. Screws come in 100-lb boxes and cost $110 each, bolts come in 100-lb boxes and cost $150 each, nuts come in 80-lb boxes and cost $70 each, and washers come in 30-lb boxes and cost $20 each. The handyman package weighs at least 1 lb and must include, by weight, at least 10% screws and 25% bolts, and at most 15% nuts and 10% washers. To balance the package, the number of bolts cannot exceed the number of nuts or the number of washers. A bolt weighs 10 times as much as a nut and 50 times as much as a washer. Determine the optimal mix of the package.

3. All-Natural Coop makes three breakfast cereals, *A*, *B*, and *C*, from four ingredients: rolled oats, raisins, shredded coconuts, and slivered almonds. The daily availabilities of the ingredients are 5 tons, 2 tons, 1 ton, and 1 ton, respectively. The corresponding costs per ton are $100, $120, $110, and $200. Cereal *A* is a 50:5:2 mix of oats, raisins, and almond. Cereal *B* is a 60:2:3 mix of oats, coconut, and almond. Cereal *C* is a 60:3:4:2 mix of oats, raisins, coconut, and almond. The cereals are produced in jumbo 5-lb sizes. All-Natural sells *A*, *B*, and *C* at $2, $2.50, and $3.00 per box, respectively. The minimum daily demand for cereals *A*, *B*, and *C* is 500, 600, and 500 boxes. Determine the optimal production mix of the cereals and the associated amounts of ingredients.

4. A refinery manufactures two grades of jet fuel, *F*1 and *F*2, by blending four types of gasoline, *A*, *B*, *C*, and *D*. Fuel *F*1 uses gasolines *A*, *B*, *C*, and *D* in the ratio 1:1:2:4, and fuel *F*2 uses the ratio 2:2:1:3. The supply limits for *A*, *B*, *C*, and *D* are 1000, 1200, 900, and 1500 bbl/day, respectively. The costs per bbl for gasolines *A*, *B*, *C*, and *D* are $120, $90, $100, and $150, respectively. Fuels *F*1 and *F*2 sell for $200 and $250 per bbl. The minimum demand for *F*1 and *F*2 is 200 and 400 bbl/day. Determine the optimal production mix for *F*1 and *F*2.

5. An oil company distills two types of crude oil, *A* and *B*, to produce regular and premium gasoline and jet fuel. There are limits on the daily availability of crude oil and the minimum demand for the final products. If the production is not sufficient to cover demand, the shortage must be made up from outside sources at a penalty. Surplus production will not be sold immediately and will incur storage cost. The following table provides the data of the situation:

	Fraction yield per bbl				
Crude	*Regular*	*Premium*	*Jet*	Price/bbl ($)	bbl/day
Crude A	.20	.1	.25	30	2500
Crude B	.25	.3	.10	40	3000
Demand (bbl/day)	500	700	400		
Revenue ($/bbl)	50	70	120		
Storage cost for surplus production ($/bbl)	2	3	4		
Penalty for unfilled demand ($/bbl)	10	15	20		

Determine the optimal product mix for the refinery.

6. In the refinery situation of Problem 5, suppose that the distillation unit actually produces the intermediate products naphtha and light oil. One bbl of crude A produces .35 bbl of naphtha and .6 bbl of light oil, and one bbl of crude B produces .45 bbl of naphtha and .5 bbl of light oil. Naphtha and light oil are blended to produce the three final gasoline products: One bbl of regular gasoline has a blend ratio of 2:1 (naphtha to light oil), one bbl of premium gasoline has a blend ratio of ratio of 1:1, and one bbl of jet fuel has a blend ratio of 1:2. Determine the optimal production mix.
7. Hawaii Sugar Company produces brown sugar, processed (white) sugar, powdered sugar, and molasses from sugar cane syrup. The company purchases 4000 tons of syrup weekly and is contracted to deliver at least 25 tons weekly of each type of sugar. The production process starts by manufacturing brown sugar and molasses from the syrup. A ton of syrup produces .3 ton of brown sugar and .1 ton of molasses. White sugar is produced by processing brown sugar. It takes 1 ton of brown sugar to produce .8 ton of white sugar. Powdered sugar is produced from white sugar through a special grinding process that has a 95% conversion efficiency (1 ton of white sugar produces .95 ton of powdered sugar). The profits per ton for brown sugar, white sugar, powdered sugar, and molasses are $150, $200, $230, and $35, respectively. Formulate the problem as a linear program, and determine the weekly production schedule.
8. Shale Oil refinery blends two petroleum stocks, A and B, to produce two high-octane gasoline products, I and II. Stocks A and B are produced at the maximum rates of 450 and 700 bbl/hour, respectively. The corresponding octane numbers are 98 and 89, and the vapor pressures are 10 and 8 lb/in^2. Gasoline I and gasoline II must have octane numbers of at least 91 and 93, respectively. The vapor pressure associated with both products should not exceed 12 lb/in^2. The profits per bbl of I and II are $7 and $10, respectively. Determine the optimum production rate for I and II and their blend ratios from stocks A and B. (*Hint*: Vapor pressure, like the octane number, is the weighted average of the vapor pressures of the blended stocks.)
9. A foundry smelts steel, aluminum, and cast iron scraps to produce two types of metal ingots, I and II, with specific limits on the aluminum, graphite and silicon contents. Aluminum and silicon briquettes may be used in the smelting process to meet the desired specifications. The following tables set the specifications of the problem:

	Contents (%)				
Input item	*Aluminum*	*Graphite*	*Silicon*	Cost/ton ($)	Available tons/day
Steel scrap	10	5	4	100	1000
Aluminum scrap	95	1	2	150	500
Cast iron scrap	0	15	8	75	2500
Aluminum briquette	100	0	0	900	Any amount
Silicon briquette	0	0	100	380	Any amount

	Ingot I		Ingot II	
Ingredient	*Minimum*	*Maximum*	*Minimum*	*Maximum*
Aluminum	8.1%	10.8%	6.2%	8.9%
Graphite	1.5%	3.0%	4.1%	∞
Silicon	2.5%	∞	2.8%	4.1%
Demand (tons/day)	130		250	

Determine the optimal input mix the foundry should smelt.

10. Two alloys, A and B, are made from four metals, I, II, III, and IV, according to the following specifications:

Alloy	Specifications	Selling price ($)
A	At most 80% of I At most 30% of II At least 50% of IV	200
B	Between 40% and 60% of II At least 30% of III At most 70% of IV	300

The four metals, in turn, are extracted from three ores according to the following data:

		Constituents (%)					
Ore	Maximum quantity (tons)	*I*	*II*	*III*	*IV*	*Others*	Price/ton ($)
1	1000	20	10	30	30	10	30
2	2000	10	20	30	30	10	40
3	3000	5	5	70	20	0	50

How much of each type of alloy should be produced? (*Hint:* Let x_{kj} be tons of ore i allocated to alloy k, and define w_k as tons of alloy k produced.)

2.3.6 Manpower Planning

Fluctuations in a labor force to meet variable demand over time can be achieved through the process of hiring and firing, as demonstrated in Example 2.3-6. There are situations in which the effect of fluctuations in demand can be "absorbed" by adjusting the start and end times of a work shift. For example, instead of following the traditional three 8-hour-shift start times at 8:00 A.M., 3:00 P.M., and 11:00 P.M., we can use overlapping 8-hour shifts in which the start time of each is made in response to increase or decrease in demand.

The idea of redefining the start of a shift to accommodate fluctuation in demand can be extended to other operating environments as well. Example 2.3-8 deals with the determination of the minimum number of buses needed to meet rush-hour and off-hour transportation needs.

Real-Life Application—Telephone Sales Manpower Planning at Qantas Airways

Australian airline Qantas operates its main reservation offices from 7:00 till 22:00 using 6 shifts that start at different times of the day. Qantas used linear programming (with imbedded queuing analysis) to staff its main telephone sales reservation office efficiently while providing convenient service to its customers. The study, carried out in the late 1970s, resulted in annual savings of over 200,000 Australian dollars per year. The study is detailed in Case 15, Chapter 24 on the CD.

Example 2.3-8 (Bus Scheduling)

Progress City is studying the feasibility of introducing a mass-transit bus system that will alleviate the smog problem by reducing in-city driving. The study seeks the minimum number of buses that can handle the transportation needs. After gathering necessary information, the city engineer noticed that the minimum number of buses needed fluctuated with the time of the day and that the required number of buses could be approximated by constant values over successive 4-hour intervals. Figure 2.8 summarizes the engineer's findings. To carry out the required daily maintenance, each bus can operate 8 successive hours a day only.

Mathematical Model: Determine the number of operating buses in each shift (variables) that will meet the minimum demand (constraints) while minimizing the total number of buses in operation (objective).

You may already have noticed that the definition of the variables is ambiguous. We know that each bus will run for 8 consecutive hours, but we do not know when a shift should start. If we follow a normal three-shift schedule (8:01 A.M.-4:00 P.M., 4:01 P.M.-12:00 midnight, and 12:01 A.M.-8:00 A.M.) and assume that x_1, x_2, and x_3 are the number of buses starting in the first, second, and third shifts, we can see from Figure 2.8 that $x_1 \geq 10$, $x_2 \geq 12$, and $x_3 \geq 8$. The corresponding minimum number of daily buses is $x_1 + x_2 + x_3 = 10 + 12 + 8 = 30$.

The given solution is acceptable only if the shifts *must* coincide with the normal three-shift schedule. It may be advantageous, however, to allow the optimization process to choose the "best" starting time for a shift. A reasonable way to accomplish this is to allow a shift to start every 4 hours. The bottom of Figure 2.8 illustrates this idea where overlapping 8-hour shifts

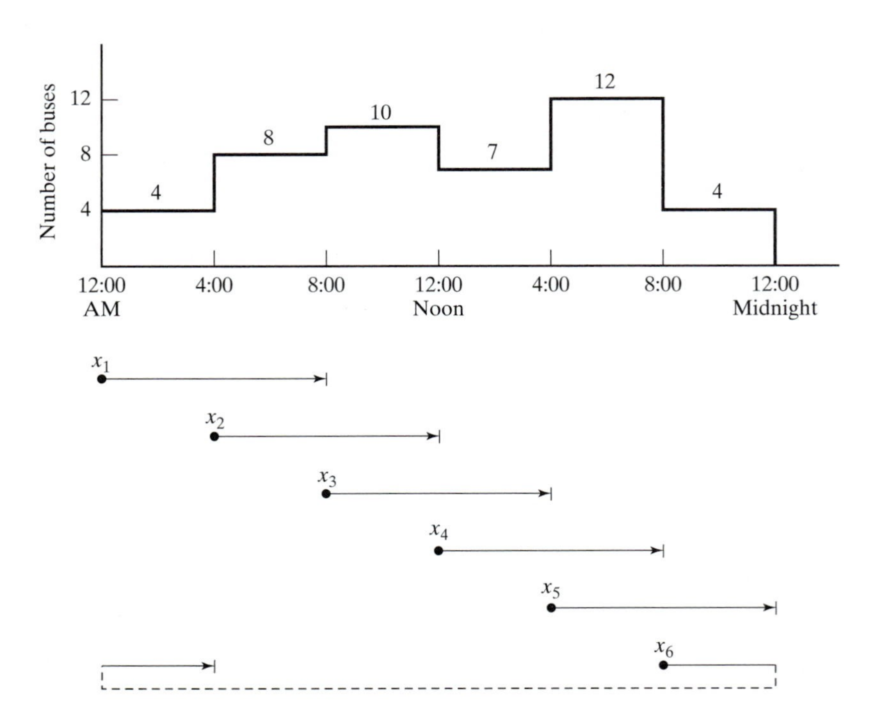

FIGURE 2.8

Number of buses as a function of the time of the day

may start at 12:01 A.M., 4:01 A.M., 8:01 A.M., 12:01 P.M., 4:01 P.M., and 8:01 P.M. Thus, the variables may be defined as

$$x_1 = \text{number of buses starting at 12: 01 A.M.}$$
$$x_2 = \text{number of buses starting at 4:01 A.M.}$$
$$x_3 = \text{number of buses starting at 8:01 A.M.}$$
$$x_4 = \text{number of buses starting at 12:01 P.M.}$$
$$x_5 = \text{number of buses starting at 4:01 P.M.}$$
$$x_6 = \text{number of buses starting at 8:01 P.M.}$$

We can see from Figure 2.8 that because of the overlapping of the shifts, the number of buses for the successive 4-hour periods is given as

Time period	Number of buses in operation
12:01 A.M. – 4:00 A.M.	$x_1 + x_6$
4:01 A.M. – 8:00 A.M.	$x_1 + x_2$
8:01 A.M. – 12:00 noon	$x_2 + x_3$
12:01 P.M. – 4:00 P.M.	$x_3 + x_4$
4:01 P.M. – 8:00 P.M.	$x_4 + x_5$
8:01 A.M. – 12:00 A.M.	$x_5 + x_6$

The complete model is thus written as

$$\text{Minimize } z = x_1 + x_2 + x_3 + x_4 + x_5 + x_6$$

subject to

$$\begin{aligned}
x_1 \qquad\qquad\qquad\qquad + x_6 &\geq 4 \text{ (12:01 A.M.-4:00 A.M.)}\\
x_1 + x_2 \qquad\qquad\qquad\qquad &\geq 8 \text{ (4:01 A.M.-8:00 A.M.)}\\
x_2 + x_3 \qquad\qquad\qquad &\geq 10 \text{ (8:01 A.M.-12:00 noon)}\\
x_3 + x_4 \qquad\qquad &\geq 7 \text{ (12:01 P.M.-4:00 P.M.)}\\
x_4 + x_5 \qquad &\geq 12 \text{ (4:01 P.M.-8:00 P.M.)}\\
x_5 + x_6 &\geq 4 \text{ (8:01 P.M.-12:00 P.M.)}\\
x_j \geq 0, j &= 1, 2, \ldots, 6
\end{aligned}$$

Solution:

The optimal solution calls for using 26 buses to satisfy the demand with $x_1 = 4$ buses to start at 12:01 A.M., $x_2 = 10$ at 4:01 A.M., $x_4 = 8$ at 12:01 P.M., and $x_5 = 4$ at 4:01 P.M.

PROBLEM SET 2.3F

***1.** In the bus scheduling example suppose that buses can run either 8- or 12-hour shifts. If a bus runs for 12 hours, the driver must be paid for the extra hours at 150% of the regular hourly pay. Do you recommend the use of 12-hour shifts?

2. A hospital employs volunteers to staff the reception desk between 8:00 A.M. and 10:00 P.M. Each volunteer works three consecutive hours except for those starting at 8:00 P.M. who work for two hours only. The minimum need for volunteers is approximated by a step function over 2-hour intervals starting at 8:00 A.M. as 4, 6, 8, 6, 4, 6, 8. Because most volunteers are retired individuals, they are willing to offer their services at any hour of the day (8:00 A.M. to 10:00 P.M.). However, because of the large number of charities competing for their service, the number needed must be kept as low as possible. Determine an optimal schedule for the start time of the volunteers

3. In Problem 2, suppose that no volunteers will start at noon or 6:00 P.M. to allow for lunch and dinner. Determine the optimal schedule.

4. In an LTL (less-than-truckload) trucking company, terminal docks include *casual* workers who are hired temporarily to account for peak loads. At the Omaha, Nebraska, dock, the minimum demand for casual workers during the seven days of the week (starting on Monday) is 20, 14, 10, 15, 18, 10, 12 workers. Each worker is contracted to work five consecutive days. Determine an optimal weekly hiring practice of casual workers for the company.

***5.** On most university campuses students are contracted by academic departments to do errands, such as answering the phone and typing. The need for such service fluctuates during work hours (8:00 A.M. to 5:00 P.M.). In the IE department, the minimum number of students needed is 2 between 8:00 A.M. and 10:00 A.M., 3 between 10:01 A.M. and 11:00 A.M., 4 between 11:01 A.M. and 1:00 P.M., and 3 between 1:01 P.M. and 5:00 P.M. Each student is allotted 3 consecutive hours (except for those starting at 3:01, who work for 2 hours and those who start at 4:01, who work for one hour). Because of their flexible schedule, students can usually report to work at any hour during the work day, except that no student wants to start working at lunch time (12:00 noon). Determine the minimum number of students the IE department should employ and specify the time of the day at which they should report to work.

6. A large department store operates 7 days a week. The manager estimates that the minimum number of salespersons required to provide prompt service is 12 for Monday, 18 for Tuesday, 20 for Wednesday, 28 for Thursday, 32 for Friday, and 40 for each of Saturday and Sunday. Each salesperson works 5 days a week, with the two consecutive off-days staggered throughout the week. For example, if 10 salespersons start on Monday, two can take their off-days on Tuesday and Wednesday, five on Wednesday and Thursday, and three on Saturday and Sunday. How many salespersons should be contracted and how should their off-days be allocated?

2.3.7 Additional Applications

The preceding sections have demonstrated the application of LP to six representative areas. The fact is that LP enjoys diverse applications in an enormous number of areas. The problems at the end of this section demonstrate some of these areas, ranging from agriculture to military applications. This section also presents an interesting application that deals with cutting standard stocks of paper rolls to sizes specified by customers.

Example 2.3-9 (Trim Loss or Stock Slitting)

The Pacific Paper Company produces paper rolls with a standard width of 20 feet each. Special customer orders with different widths are produced by slitting the standard rolls. Typical orders (which may vary daily) are summarized in the following table:

Order	Desired width (ft)	Desired number of rolls
1	5	150
2	7	200
3	9	300

In practice, an order is filled by setting the knives to the desired widths. Usually, there are a number of ways in which a standard roll may be slit to fill a given order. Figure 2.9 shows three feasible knife settings for the 20-foot roll. Although there are other feasible settings, we limit the discussion for the moment to settings 1, 2, and 3 in Figure 2.9. We can combine the given settings in a number of ways to fill orders for widths 5, 7, and 9 feet. The following are examples of feasible combinations:

1. Slit 300 (standard) rolls using setting 1 and 75 rolls using setting 2.
2. Slit 200 rolls using setting 1 and 100 rolls using setting 3.

Which combination is better? We can answer this question by considering the "waste" each combination generates. In Figure 2.9 the shaded portion represents surplus rolls not wide enough to fill the required orders. These surplus rolls are referred to as *trim loss*. We can evaluate the "goodness" of each combination by computing its trim loss. However, because the surplus rolls may have different widths, we should base the evaluation on the trim loss *area* rather than on the *number* of surplus rolls. Assuming that the standard roll is of length L feet, we can compute the trim-loss area as follows:

$$\text{Combination 1: } 300\,(4 \times L) + 75\,(3 \times L) = 1425L \text{ ft}^2$$

$$\text{Combination 2: } 200\,(4 \times L) + 100\,(1 \times L) = 900L \text{ ft}^2$$

These areas account only for the shaded portions in Figure 2.9. Any surplus production of the 5-, 7- and 9-foot rolls must be considered also in the computation of the trim-loss area. In

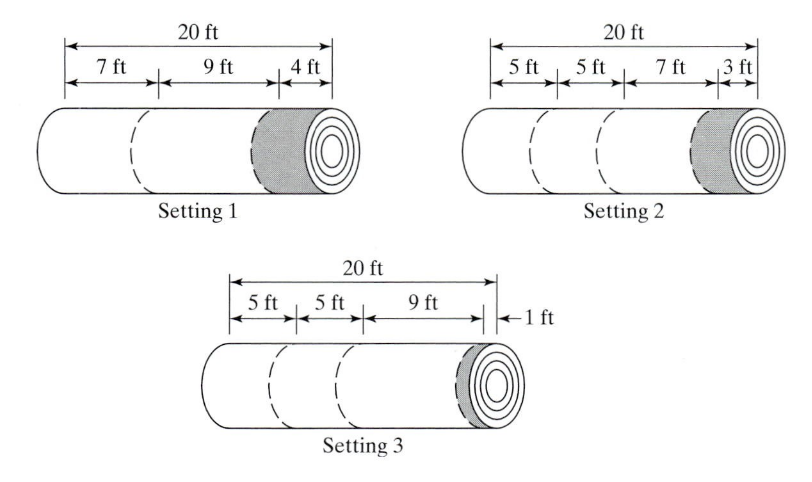

FIGURE 2.9
Trim loss (shaded) for knife settings 1, 2, and 3

combination 1, setting 1 produces a surplus of $300 - 200 = 100$ extra 7-foot rolls and setting 2 produces 75 extra 7-foot rolls. Thus the additional waste area is $175\ (7 \times L) = 1225L\ \text{ft}^2$. Combination 2 does not produce surplus rolls of the 7- and 9-foot rolls but setting 3 does produce $200 - 150 = 50$ extra 5-foot rolls, with an added waste area of $50\ (5 \times L) = 250L\ \text{ft}^2$. As a result we have

$$\text{Total trim-loss area for combination 1} = 1425L + 1225L = 2650L\ \text{ft}^2$$

$$\text{Total trim-loss area for combination 2} = 900L + 250L = 1150L\ \text{ft}^2$$

Combination 2 is better, because it yields a smaller trim-loss area.

Mathematical Model: The problem can be summarized verbally as determining the *knife-setting combinations* (variables) that will *fill the required orders* (constraints) with the *least trim-loss area* (objective).

The definition of the variables as given must be translated in a way that the mill operator can use. Specifically, the variables are defined as *the number of standard rolls to be slit according to a given knife setting*. This definition requires identifying all possible knife settings as summarized in the following table (settings 1, 2, and 3 are given in Figure 2.9). You should convince yourself that settings 4, 5, and 6 are valid and that no "promising" settings have been excluded. Remember that a promising setting cannot yield a trim-loss roll of width 5 feet or larger.

	Knife setting						
Required width (ft)	*1*	*2*	*3*	*4*	*5*	*6*	Minimum number of rolls
5	0	2	2	4	1	0	150
7	1	1	0	0	2	0	200
9	1	0	1	0	0	2	300
Trim loss per foot of length	4	3	1	0	1	2	

To express the model mathematically, we define the variables as

$$x_j = \text{number of standard rolls to be slit according to setting } j,\ j = 1, 2, \ldots, 6$$

The constraints of the model deal directly with satisfying the demand for rolls.

$$\text{Number of 5-ft rolls produced} = 2x_2 + 2x_3 + 4x_4 + x_5 \geq 150$$

$$\text{Number of 7-ft rolls produced} = x_1 + x_2 + 2x_5 \geq 200$$

$$\text{Number of 9-ft rolls produced} = x_1 \quad x_2 + x_3 + 2x_5 + 2x_6 \geq 300$$

To construct the objective function, we observe that the total trim loss area is the difference between the total area of the standard rolls used and the total area representing all the orders. Thus

$$\text{Total area of standard rolls} = 20L(x_1 + x_2 + x_3 + x_4 + x_5 + x_6)$$

$$\text{Total area of orders} = L(150 \times 5 + 200 \times 7 + 300 \times 9) = 4850L$$

The objective function then becomes

$$\text{Minimize } z = 20L(x_1 + x_2 + x_3 + x_4 + x_5 + x_6) - 4850L$$

Because the length L of the standard roll is a constant, the objective function equivalently reduces to

$$\text{Minimize } z = x_1 + x_2 + x_3 + x_4 + x_5 + x_6$$

The model may thus be written as

$$\text{Minimize } z = x_1 + x_2 + x_3 + x_4 + x_5 + x_6$$

subject to

$$\begin{aligned}
2x_2 + 2x_3 + 4x_4 + x_5 &\geq 150 \text{ (5-ft rolls)} \\
x_1 + x_2 + 2x_5 &\geq 200 \text{ (7-ft rolls)} \\
x_1 + x_3 + 2x_6 &\geq 300 \text{ (9-ft rolls)} \\
x_j \geq 0, j = 1,2, \ldots, 6 &
\end{aligned}$$

Solution:

The optimum solution calls for cutting 12.5 standard rolls according to setting 4, 100 according to setting 5, and 150 according to setting 6. The solution is not implementable because x_4 is noninteger. We can either use an integer algorithm to solve the problem (see Chapter 9) or round x_4 conservatively to 13 rolls.

Remarks. The trim-loss model as presented here assumes that all the feasible knife settings can be determined in advance. This task may be difficult for large problems, and viable feasible combinations may be missed. The problem can be remedied by using an LP model with imbedded integer programs designed to generate promising knife settings on demand until the optimum solution is found. This algorithm, sometimes referred to as **column generation**, is detailed in Comprehensive Problem 7-3, Appendix E on the CD. The method is rooted in the use of (reasonably advanced) linear programming *theory*, and may serve to refute the argument that, in practice, it is unnecessary to learn LP theory.

PROBLEM SET 2.3G

*__1.__ Consider the trim-loss model of Example 2.3-9.

(a) If we slit 200 rolls using setting 1 and 100 rolls using setting 3, compute the associated trim-loss area.

(b) Suppose that the only available standard roll is 15 feet wide. Generate all possible knife settings for producing 5-, 7-, and 9-foot rolls, and compute the associated trim loss per foot length.

(c) In the original model, if the demand for 7-foot rolls is decreased by 80, what is the minimum number of standard 20-foot rolls that will be needed to fill the demand for of all three types of rolls?

(d) In the original model, if the demand for 9-foot rolls is changed to 400, how many additional standard 20-foot rolls will be needed to satisfy the new demand?

2. *Shelf Space Allocation.* A grocery store must decide on the shelf space to be allocated to each of five types of breakfast cereals. The maximum daily demand is 100, 85, 140, 80, and 90 boxes, respectively. The shelf space in square inches for the respective boxes is 16, 24, 18, 22, and 20. The total available shelf space is 5000 in^2. The profit per unit is \$1.10, \$1.30, \$1.08, \$1.25, and \$1.20, respectively. Determine the optimal space allocation for the five cereals.
3. *Voting on Issues.* In a particular county in the State of Arkansas, four election issues are on the ballot: Build new highways, increase gun control, increase farm subsidies, and increase gasoline tax. The county includes 100,000 urban voters, 250,000 suburban voters, and 50,000 rural voters, all with varying degrees of support for and opposition to election issues. For example, rural voters are opposed to gun control and gas tax and in favor of road building and farm subsidies. The county is planning a TV advertising campaign with a budget of \$100,000 at the cost of \$1500 per ad. The following table summarizes the impact of a single ad in terms of the number of pro and con votes as a function of the different issues:

	Expected number of pro (+) and con (−) votes per ad		
Issue	*Urban*	*Suburban*	*Rural*
New highways	−30,000	+60,000	+30,000
Gun control	+80,000	+30,000	−45,000
Smog control	+40,000	+10,000	0
Gas tax	+90,000	0	−25,000

An issue will be adopted if it garners at least 51% of the votes. Which issues will be approved by voters, and how many ads should be allocated to these issues?

4. *Assembly-Line Balancing.* A product is assembled from three different parts. The parts are manufactured by two departments at different production rates as given in the following table:

		Production rate (units/hr)		
Department	Capacity (hr/wk)	*Part 1*	*Part 2*	*Part 3*
1	100	8	5	10
2	80	6	12	4

Determine the maximum number of final assembly units that can be produced weekly. (*Hint*: Assembly units = min {units of part 1, units of part 2, units of part 3}. Maximize $z = \min\{x_1, x_2\}$ is equivalent to max z subject to $z \leq x_1$ and $z \leq x_2$.)

5. *Pollution Control.* Three types of coal, C1, C2, and C3, are pulverized and mixed together to produce 50 tons per hour needed to power a plant for generating electricity. The burning of coal emits sulfur oxide (in parts per million) which must meet the Environmental Protection Agency (EPA) specifications of at most 2000 parts per million. The following table summarizes the data of the situation:

	C1	C2	C3
Sulfur (parts per million)	2500	1500	1600
Pulverizer capacity (ton/hr)	30	30	30
Cost per ton	\$30	\$35	\$33

Determine the optimal mix of the coals.

*6. *Traffic Light Control.* (Stark and Nicholes, 1972) Automobile traffic from three highways, H1, H2, and H3, must stop and wait for a green light before exiting to a toll road. The tolls are \$3, \$4, and \$5 for cars exiting from H1, H2, and H3, respectively. The flow rates from H1, H2, and H3 are 500, 600, and 400 cars per hour. The traffic light cycle may not exceed 2.2 minutes, and the green light on any highway must be at least 25 seconds. The yellow light is on for 10 seconds. The toll gate can handle a maximum of 510 cars per hour. Assuming that no cars move on yellow, determine the optimal green time interval for the three highways that will maximize toll gate revenue per traffic cycle.

7. *Fitting a Straight Line into Empirical Data (Regression).* In a 10-week typing class for beginners, the average speed per student (in words per minute) as a function of the number of weeks in class is given in the following table.

Week, x	1	2	3	4	5	6	7	8	9	10
Words per minute, y	5	9	15	19	21	24	26	30	31	35

Determine the coefficients a and b in the straight-line relationship, $\hat{y} = ax + b$, that best fit the given data. (*Hint*: Minimize the sum of the *absolute value* of the deviations between theoretical $\hat{y}$ and empirical y. Min $|x|$ is equivalent to min z subject to $z \leq x$ and $z \geq -x$.)

8. *Leveling the Terrain for a New Highway.* (Stark and Nicholes, 1972) The Arkansas Highway Department is planning a new 10-mile highway on uneven terrain as shown by the profile in Figure 2.10. The width of the construction terrain is approximately 50 yards. To simplify the situation, the terrain profile can be replaced by a step function as shown in the figure. Using heavy machinery, earth removed from high terrain is hauled to fill low areas. There are also two burrow pits, I and II, located at the ends of the 10-mile stretch from which additional earth can be hauled, if needed. Pit I has a capacity of 20,000 cubic yards and pit II a capacity of 15,000 cubic yards. The costs of removing earth from pits I and II are, respectively, \$1.50 and \$1.90 per cubic yard. The transportation cost per cubic yard per mile is \$.15 and the cost of using heavy machinery to load hauling trucks is \$.20 per cubic yard. This means that a

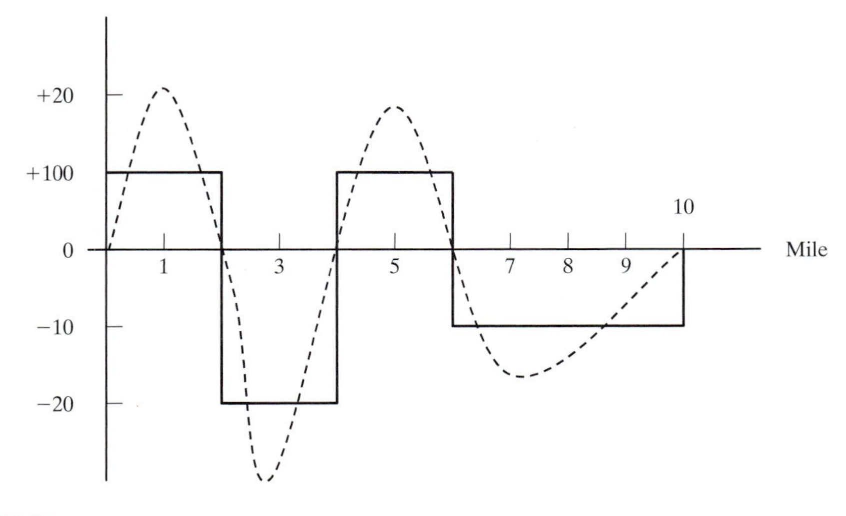

FIGURE 2.10
Terrain profile for Problem 8

cubic yard from pit I hauled one mile will cost a total of $(1.5 + .20) + 1 \times .15 = \$1.85$ and a cubic yard hauled one mile from a hill to a fill area will cost $.20 + 1 \times .15 = \$.35$. Develop a minimum cost plan for leveling the 10-mile stretch.

9. *Military Planning*. (Shepard and Associates, 1988) The Red Army (R) is trying to invade the territory defended by the Blue Army (B). Blue has three defense lines and 200 regular combat units and can draw also on a reserve pool of 200 units. Red plans to attack on two fronts, north and south, and Blue has set up three east-west defense lines, I, II, and III. The purpose of defense lines I and II is to delay the Red Army attack by at least 4 days in each line and to maximize the total duration of the battle. The advance time of the Red Army is estimated by the following empirical formula:

$$\text{Battle duration in days} = a + b\left(\frac{\text{Blue units}}{\text{Red units}}\right)$$

The constants a and b are a function of the defense line and the north/south front as the following table shows:

	a			*b*		
	I	*II*	*III*	*I*	*II*	*III*
North front	.5	.75	.55	8.8	7.9	10.2
South front	1.1	1.3	1.5	10.5	8.1	9.2

The Blue Army reserve units can be used in defense lines II and II only. The allocation of units by the Red Army to the three defense lines is given in the following table.

	Number of Red Army attack units		
	Defense Line I	*Defense Line II*	*Defense Line III*
North front	30	60	20
South front	30	40	20

How should Blue allocate its resources among the three defense lines and the north/south fronts?

10. *Water Quality Management.* (Stark and Nicholes, 1972) Four cities discharge waste water into the same stream. City 1 is upstream, followed downstream by city 2, then city 3, then city 4. Measured alongside the stream, the cities are approximately 15 miles apart. A measure of the amount of pollutants in waste water is the BOD (biochemical oxygen demand), which is the weight of oxygen required to stabilize the waste constituent in water. A higher BOD indicates worse water quality. The Environmental Protection Agency (EPA) sets a maximum allowable BOD loading, expressed in lb BOD per gallon. The removal of pollutants from waste water takes place in two forms: (1) natural decomposition activity stimulated by the oxygen in the air, and (2) treatment plants at the points of discharge before the waste reaches the stream. The objective is to determine the most economical efficiency of each of the four plants that will reduce BOD to acceptable levels. The maximum possible plant efficiency is 99%.

To demonstrate the computations involved in the process, consider the following definitions for plant 1:

Q_1 = Stream flow (gal/hour) on the 15-mile reach 1-2 leading to city 2
p_1 = BOD discharge rate (in lb/hr)

x_1 = efficiency of plant 1 ($\leq .99$)
b_1 = maximum allowable BOD loading in reach 1-2 (in lb BOD/gal)

To satisfy the BOD loading requirement in reach 1-2, we must have

$$p_1(1 - x_1) \leq b_1 Q_1$$

In a similar manner, the BOD loading constraint for reach 2-3 takes the form

$$(1 - r_{12})\begin{pmatrix}\text{BOD discharge} \\ \text{rate in reach 1-2}\end{pmatrix} + \begin{pmatrix}\text{BOD discharge} \\ \text{rate in reach 2-3}\end{pmatrix} \leq b_2 Q_2$$

or

$$(1 - r_{12})p_1(1 - x_1) + p_2(1 - x_2) \leq b_2 Q_2$$

The coefficient r_{12} (<1) represents the fraction of waste removed in reach 1-2 by decomposition. For reach 2-3, the constraint is

$$(1 - r_{23})[(1 - r_{12})p_1(1 - x_1) + p_2(1 - x_2)] + p_3(1 - x_3) \leq b_3 Q_3$$

Determine the most economical efficiency for the four plants using the following data (the fraction of BOD removed by decomposition is 6% for all four reaches):

	Reach 1-2 ($i = 1$)	Reach 2-3 ($i = 2$)	Reach 2-3 ($i = 3$)	Reach 3-4 ($i = 4$)
Q_i (gal/hr)	215,000	220,000	200,000	210,000
p_i (lb/hr)	500	3,000	6,000	1,000
b_i (lb BOD/gal)	.00085	.0009	.0008	.0008
Treatment cost ($/lb BOD removed)	.20	.25	.15	.18

11. *Loading Structure.* (Stark and Nicholes, 1972) The overhead crane with two lifting yokes in Figure 2.11 is used to transport mixed concrete to a yard for casting concrete barriers. The concrete bucket hangs at midpoint from the yoke. The crane end rails can support a maximum of 25 kip each and the yoke cables have a 20-kip capacity each. Determine the maximum load capacity, W_1 and W_2. (*Hint*: At equilibrium, the sum of moments about any point on the girder or yoke is zero.)

12. *Allocation of Aircraft to Routes.* Consider the problem of assigning aircraft to four routes according to the following data:

Aircraft type	Capacity (passengers)	Number of aircraft	Number of daily trips on route			
			1	*2*	*3*	*4*
1	50	5	3	2	2	1
2	30	8	4	3	3	2
3	20	10	5	5	4	2
Daily number of customers			1000	2000	900	1200

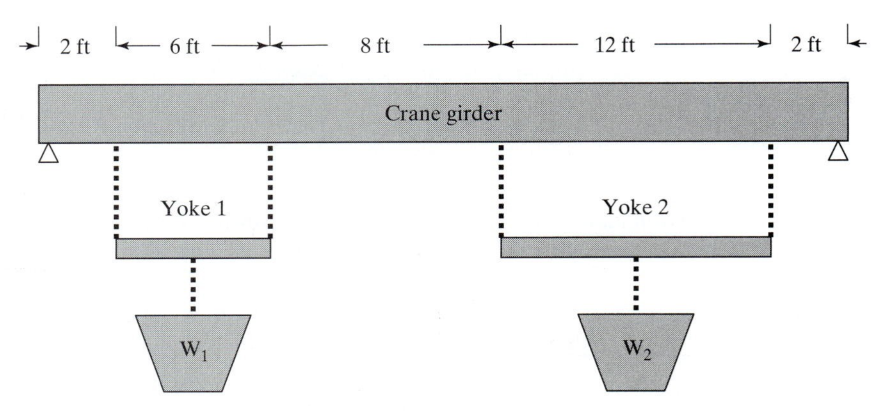

FIGURE 2.11
Overhead crane with two yokes (Problem 11)

The associated costs, including the penalties for losing customers because of space unavailability, are

	Operating cost ($) per trip on route			
Aircraft type	*1*	*2*	*3*	*4*
1	1000	1100	1200	1500
2	800	900	1000	1000
3	600	800	800	900
Penalty ($) per lost customer	40	50	45	70

Determine the optimum allocation of aircraft to routes and determine the associated number of trips.

2.4 COMPUTER SOLUTION WITH SOLVER AND AMPL

In practice, where typical linear programming models may involve thousands of variables and constraints, the only feasible way to solve such models is to use the computer. This section presents two distinct types of popular software: Excel Solver and AMPL. Solver is particularly appealing to spreadsheet users. AMPL is an algebraic modeling language that, like any other programming language, requires more expertise. Nevertheless, AMPL, and other similar languages,[3] offer great flexibility in modeling and executing large and complex LP models. Although the presentation in this section concentrates on LPs, both AMPL and Solver can be used with integer and nonlinear programs, as will be shown later in the book.

[3]Other known commercial packages include AIMMS, GAMS, LINGO, MPL, OPL Studio, and Xpress-Mosel.

2.4.1 LP Solution with Excel Solver

In Excel Solver, the spreadsheet is the input and output medium for the LP. Figure 2.12 shows the layout of the data for the Reddy Mikks model (file solverRM1.xls). The top of the figure includes four types of information: (1) input data cells (shaded areas,

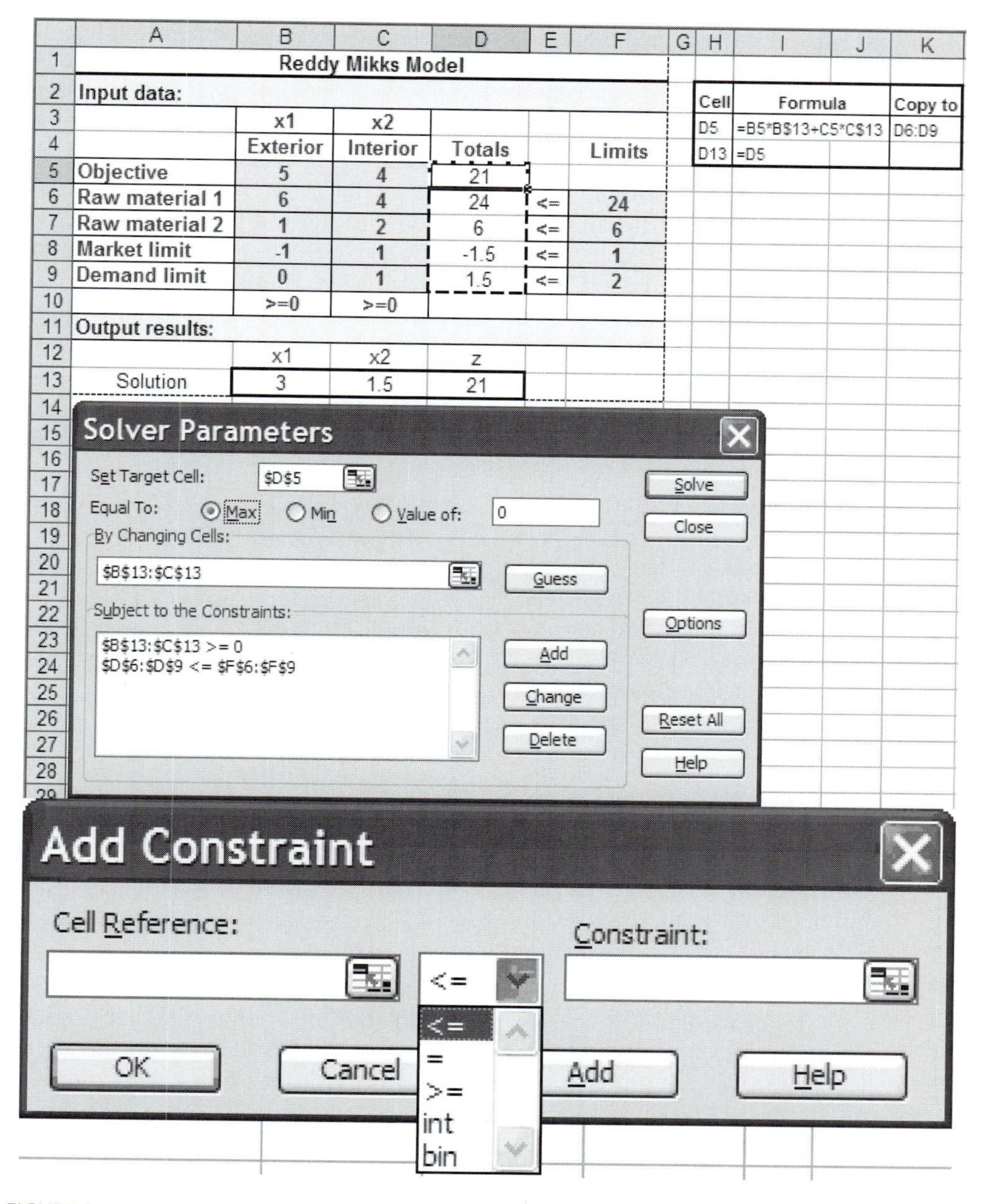

FIGURE 2.12

Defining the Reddy Mikks model with Excel Solver (file solverRM1.xls)

B5:C9 and F6:F9), (2) cells representing the variables and the objective function we seek to evaluate (solid rectangle cells, B13:D13), (3) algebraic definitions of the objective function and the left-hand side of the constraints (dashed rectangle cells, D5:D9), and (4) cells that provides explanatory names or symbols. Solver requires the first three types only. The fourth type enhances the readability of the model and serves no other purpose. The relative positioning of the four types of information on the spreadsheet need not follow the layout shown in Figure 2.12. For example, the cells defining the objective function and the variables need not be contiguous, nor do they have to be placed below the problem. What is important is that we know where they are so they can be referenced by Solver. Nonetheless, it is a good idea to use a format similar to the one suggested in Figure 2.12, because it makes the model more readable.

How does Solver link to the spreadsheet data? First we provide equivalent "algebraic" definitions of the objective function and the left-hand side of the constraints using the input data (shaded cells B5:C9 and F6:F9) and the objective function and variables (solid rectangle cells B13:D13), and then we place the resulting formulas in the appropriate cells of the dashed rectangle D5:D9. The following table shows the original LP functions and their placement in the appropriate cells:

	Algebraic expression	Spreadsheet formula	Entered in cell
Objective, z	$5x_1 + 4x_2$	=B5*$B$13+C5*$C$13	D5
Constraint 1	$6x_1 + 4x_2$	=B6*$B$13+C6*$C$13	D6
Constraint 2	$x_1 + 2x_2$	=B7*$B$13+C7*$C$13	D7
Constraint 3	$-x_1 + x_2$	=B8*$B$13+C8*$C$13	D8
Constraint 4	$0x_1 + x_2$	=B9*$B$13+C9*$C$13	D9

Actually, you only need to enter the formula for cell D5 and then copy it into cells D6:D9. To do so correctly, the fixed references $B$13 and $C$13 representing x_1 and x_2 must be used. For larger linear programs, it is more efficient to enter

=SUMPRODUCT(B5:C5,$B$13:$C$13)

in cell D5 and copy it into cells D6:D9.

All the elements of the LP model are now ready to be linked with Solver. From Excel's Tools menu, select Solver[4] to open the **Solver Parameters** dialogue box shown in the middle of Figure 2.12. First, you define the objective function, z, and the sense of optimization by entering the following data:

Set Target Cell: $D$5
Equal To: ⊙ Max
By Changing Cells: $B$13:$C$13

This information tells Solver that the variables defined by cells $B$13 and $C$13 are determined by maximizing the objective function in cell $D$5.

[4]If Solver does not appear under Tools, click Add-ins in the same menu and check Solver Add-in, then click OK.

The next step is to set up the constraints of the problems by clicking Add in the **Solver Parameters** dialogue box. The **Add Constraint** dialogue box will be displayed (see the bottom of Figure 2.12) to facilitate entering the elements of the constraints (left-hand side, inequality type, and right-hand side) as[5]

$D$6:$D$9<=$F$6:$F$9

A convenient substitute to typing in the cell ranges is to highlight cells D6:D9 to enter the left-hand sides and then cells F6:F9 to enter the right-hand sides. The same procedure can be used with Target Cell.

The only remaining constraints are the nonnegativity restrictions, which are added to the model by clicking Add in the **Add Constraint** dialogue box to enter

$B$13:$C$13>=0

Another way to enter the nonnegative constraints is to click Options on the **Solver Parameters** dialogue box to access the **Solver Options** dialogue box (see Figure 2.13) and then check ☑ Assume Non-Negative. While you are in the **Solver Options** box, you also need to check ☑ Assume Linear Model.

In general, the remaining default settings in **Solver Options** need not be changed. However, the default precision of .000001 may be set too "high" for some problems, and Solver may return the message "Solver could not find a feasible solution" when in fact the problem does have a feasible solution. In such cases, the precision needs to be adjusted to reflect less precision. If the same message persists, then the problem may be infeasible.

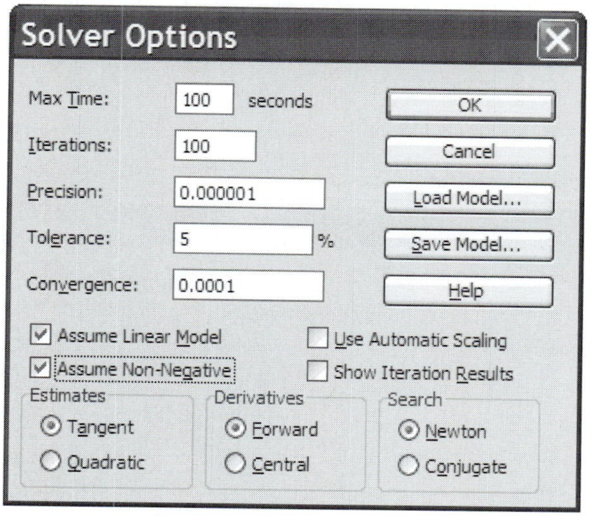

FIGURE 2.13

Solver options dialogue box

[5]You will notice that in the **Add Constraint** dialogue box (Figure 2.12), the middle box specifying the type of inequalities (<= and >=) has two additional options, int and bin, which stand for **int**eger and **bin**ary and can be used with integer programs to restrict variables to integer or binary values (see Chapter 9).

For readability, you can use descriptive Excel range names instead of cell names. A range is created by highlighting the desired cells, typing the range name in the top left box of the sheet, and then pressing Return. Figure 2.14 (file solverRM2.xls) provides the details with a summary of the range names used in the model. You should contrast file solverRM2.xls with file solverRM1.xls to see how ranges are used in the formulas.

To solve the problem, click Solve on **Solver Parameters** (Figure 2.14). A new dialogue box, **Solver Results**, will then give the status of the solution. If the model setup is correct, the optimum value of z will appear in cell D5 and the values of x_1 and x_2 will go to cells B13 and C13, respectively. For convenience, we use cell D13 to exhibit the optimum value of z by entering the formula =D5 in cell D13 to display the entire optimum solution in contiguous cells.

If a problem has no feasible solution, Solver will issue the explicit message "Solver could not find a feasible solution." If the optimal objective value is unbounded, Solver will issue the somewhat ambiguous message "The Set Cell values do not converge." In either case, the message indicates that there is something wrong with the formulation of the model, as will be discussed in Section 3.5.

The **Solver Results** dialogue box will give you the opportunity to request further details about the solution, including the important sensitivity analysis report. We will discuss these additional results in Section 3.6.4.

The solution of the Reddy Mikks by Solver is straightforward. Other models may require a "bit of ingenuity" before they can be defined in a convenient manner. A class

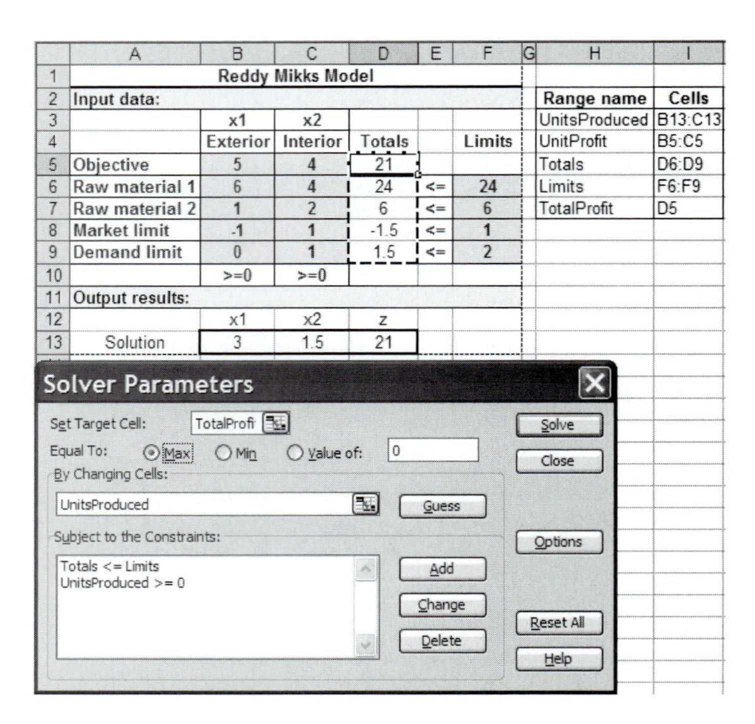

FIGURE 2.14

Use of range names in Excel Solver (file solverRM2.xls)

of LP models that falls in this category deals with network optimization, as will be demonstrated in Chapter 6.

PROBLEM SET 2.4A

1. Modify the Reddy Mikks Solver model of Figure 2.12 to account for a third type of paint named "marine." Requirements per ton of raw materials 1 and 2 are .5 and .75 ton, respectively. The daily demand for the new paint lies between .5 ton and 1.5 tons. The profit per ton is $3.5 (thousand).
2. Develop the Excel Solver model for the following problems:
   (a) The diet model of Example 2.2-2.
   (b) Problem 16, Set 2.2a
   (c) The urban renewal model of Example 2.3-1.
   ***(d)** The currency arbitrage model of Example 2.3-2. (Hint: You will find it convenient to use the entire currency conversion matrix rather than the top diagonal elements only. Of course, you generate the bottom diagonal elements by using appropriate Excel formulas.)
   (e) The multi-period production-inventory model of Example 2.3-5.

2.4.2 LP Solution with AMPL[6]

This section provides a brief introduction to AMPL. The material in Appendix A provides detailed coverage of AMPL syntax and will be cross-referenced opportunely with the presentation in this section as well as with other AMPL presentations throughout the book.

Four examples are presented here: The first two deal with the basics of AMPL, and the remaining two demonstrate more advanced usages to make a case for the advantages of AMPL.

Reddy Mikks Problem—a Rudimentary Model. AMPL provides a facility for modeling an LP in a rudimentary long-hand format. Figure 2.15 gives the self-explanatory code

```
var x1 >=0;
var x2 >=0;
maximize z: 5*x1+4*x2;
subject to
  c1: 6*x1+4*x2<=24;
  c2: x1+2*x2<=6;
  c3: -x1+x2<=1;
  c4: x2<=2;
solve;
display z,x1,x2;
```

Figure 2.15

Rudimentary AMPL model for the Reddy Mikks problem (file amplRM1.txt)

[6]For convenience, the AMPL student version, provided by AMPL Optimization LLC with instructions, is on the accompanying CD. Future updates may be downloaded from *www.ampl.com*. AMPL uses line commands and operates in a DOS (rather than Windows) environment. A recent beta version of a Windows interface can be found in *www.OptiRisk-Systems.com*.

for the Reddy Mikks model (file amplRM1.txt). All reserved keywords are in bold. All other names are user generated. The objective function and each of the constraints must be given a distinct user-generated name followed by a colon. Each statement closes with a semi-colon.

This rudimentary AMPL model is too specific in the sense that it requires developing a new code each time the data of the problem are changed. For practical problems with hundreds (even thousands) of variables and constraints, this long-hand format is cumbersome. AMPL alleviates this difficulty by dividing the problem into two components: (1) A general model that expresses the problem algebraically for any desired number of variables and constraints, and (2) specific data that drive the algebraic model. We will use the Reddy Mikks model to demonstrate the basic ideas of AMPL.

Reddy Mikks Problem—an Algebraic Model. Figure 2.16 lists the statements of the model (file amplRM2.txt). The file must be strictly text (ASCII). Comments are preceded with # and may appear anywhere in the model. The language is case sensitive and all its keywords (with few exceptions) must be in lower case. (Section A.2 provides more details.)

```
#-------------------------------------------algebraic model
param m;
param n;
param c{1..n};
param b{1..m};
param a{1..m,1..n};

var x{1..n}>=0;

maximize z: sum{j in 1..n}c[j]*x[j];
subject to restr{i in 1..m}:
           sum{j in 1..n}a[i,j]*x[j]<=b[i];
#------------------------------------------specify model data
data;
param n:=2;
param m:=4;
param c:=1 5 2 4;
param b:=1 24  2 6  3 1  4 2;
param a:     1   2 :=
         1   6   4
         2   1   2
         3  -1   1
         4   0   1;
#------------------------------------------solve the problem
solve;
display z, x;
```

FIGURE 2.16

AMPL model of the Reddy Mikks problem with input data (file amplRM2.txt)

The algebraic model in AMPL views the general Reddy Mikks problem in the following generic format

$$\textbf{Maximize } z\text{: } \sum_{j=1}^{n} c_j x_j$$

$$\textbf{subject to } restr_i\text{: } \sum_{j=1}^{n} a_{ij} x_j \le b_i, i = 1, 2, \ldots, m$$

$$x_j \ge 0, j = 1, 2, \ldots, n$$

It assumes that the problem has n variables and m constraints. It gives the objective function and constraint i the (arbitrary) names z and $restr_i$. The rest of the parameters c_j, b_i, and a_{ij} are self-explanatory.

The model starts with the **param** statements that declare m, n, c, b, and a_{ij} as parameters (or constants) whose specific values are given in the input data section of the model. It translates $c_j (j = 1, 2, \ldots, n)$ as `c{1..n}`, $b_i (i = 1, 2, \ldots, m)$ as `b{1..m}`, and $a_{ij} (i = 1, 2, \ldots, m, j = 1, 2, \ldots, n)$ as `a{1..m,1..n}`. Next, the variables $x_j (j = 1, 2, \ldots, n)$ together with the nonnegativity restriction are defined by the **var** statement

```
var x{1..n}>=0;
```

If `>=0` is removed from the definition of x_j, then the variable is assumed unrestricted. The notation in `{}` represents the *set* of subscripts over which a `param` or a `var` is defined.

After defining all the parameters and the variables, we can develop the model itself. The objective function and constraints must each carry a distinct user-defined name followed by a colon (`:`). In the Reddy Mikks model the objective is given the name `z:` preceded by `maximize`, as the following AMPL statement states:

```
maximize z: sum{j in 1..n}c[j]*x[j];
```

The statement is a direct translation of maximize $z = \sum_{j=1}^{n} c_j x_j$ (with = replaced by :). Note the use of the brackets `[]` for representing the subscripts.

Constraint i is given the *root* name `restr` indexed over the set `{1..m}`:

```
restr{i in 1..m}:sum{j in 1..n}a[i,j]*x[j]<=b[i];
```

The statement is a direct translation of $\sum_{j=1}^{n} a_{ij} x_j \le b_i$. The keywords `subject to` are optional. This general model may now be used to solve any problem with any set of input data representing any number of constraints `m` and any number of variables `n`.

The `data;` section allows tailoring the model to the specific Reddy Mikks problem. Thus, `param n:=2;` and `param m:=4;` tell AMPL that the problem has 2 variables and 4 constraints. Note that the compound operator `:=` must be used and that the statement must start with the keyword param. For the single-subscripted parameter c, each element is represented by the subscript j followed by c_j separated by a blank space. Thus, the two values $c_1 = 5$ and $c_2 = 4$ translate to

```
param c:= 1 5  2  4;
```

The data for parameter b are entered in a similar manner.

For the double-subscripted parameter a, the top line defines the subscript j, and the subscript i is entered at the start of each row as

```
param a:    1   2 :=
        1   6   4
        2   1   2
        3  -1   1
        4   0   1;
```

In effect, the data a_{ij} read as a two-dimensional matrix with its rows designating i and its columns designating j. Note that a semicolon is needed only at the end of all a_{ij} data.

The model and its data are now ready. The command solve; invokes the solution and the command display z, x; provides the solution.

To execute the model, first invoke AMPL (by clicking ampl.exe in the AMPL directory). At the ampl prompt, enter the following **model** command, then press Return:

```
ampl: model AmplRM2.txt;
```

The output of the system will then appear on the screen as follows:

```
MINOS 5.5: Optimal solution found.
2 iterations, objective = 21

z = 21
x[*]:=

1 = 3
2 = 1.5
```

The bottom four lines are the result of executing display z,x;.

Actually, AMPL allows separating the algebraic model and the data into two independent files. This arrangement is advisable because once the model has been developed, only the data file needs to be changed. (See the end of Section A.2 for details.) In this book, we elect not to separate the model and data files, mainly for reasons of compactness.

The Arbitrage Problem. The simple Reddy Mikks model introduces some of the basic elements of AMPL. The more complex arbitrage model of Example 2.3-2 offers the opportunity to introduce additional AMPL capabilities that include: (1) imposing conditions on the elements of a set, (2) use of if then else to represent conditional values, (3) use of computed parameters, and (4) use of a simple print statement to retrieve output. These points are also discussed in more detail in Appendix A.

```
param inCurrency;                          #initial amount I
param outCurrency;                         #maximized holding y
param n;                                   #nbr of currencies
param r{i in 1..n,j in 1..n:i<=j};#above-diagonal rates
param I;                                   #initial amt of inCurrency
param maxTransaction{1..n};                #limit on transaction amt

var x{i in 1..n,j in 1..n}>=0;             #amt of i converted to j
var y>=0;                                  #max amt of outCurrency

maximize z: y;
subject to
 r1{i in 1..n,j in 1..n}:x[i,j]<=maxTransaction[i];
 r2{i in 1..n}:(if i=inCurrency then I else 0)+
   sum{k in 1..n}(if k<i then r[k,i] else 1/r[i,k])*x[k,i]=
      (if i=outCurrency then y else 0)+sum{j in 1..n}x[i,j];
#----------------------------------------input data
data;
param inCurrency=1;
param outCurrency=1;
param n:=5;
#                      $      euro  pound     yen    KD
param r:               1      2       3       4       5:=
                 1     1      .769    .625    105     .342   #$
                 2     .      1       .813    137     .445   #euro
                 3     .      .       1       169     .543   #pound
                 4     .      .       .       1       .0032 #yen
                 5     .      .       .       .       1;     #KD
param I:= 5;
param maxTransaction:=1 5 2 3 3 3.5 4 100 5 2.8;
#----------------------------------------Solution command
solve;
display z,y,x>file2.out;
print "rate of return =",trunc(100*(z-I)/I,4),"%">file2.out;
```

FIGURE 2.17

AMPL model of the Arbitrage problem (file amplEx2.3-2.txt)

Figure 2.17 (file amplEx2.3-2.txt) gives the AMPL code for the arbitrage problem. The model is general in the sense that it can be used to maximize the final holdings `y` of any currency, named `outCurrency`, starting with an initial amount `I` of another currency, named `inCurrency`. Additionally, any number of currencies, `n`, can be involved in the arbitrage process.

The exchange rates are defined as

```
param r{i in 1..n,j in 1..n:i<=j};
```

The definition gives only the diagonal and above-diagonal elements by imposing the condition `i<=j` (preceded by a colon) on the set `{i in 1..n,j in 1..n}`. With this definition, reciprocals are used to compute the below-diagonal rates, as will be shown shortly.

The variable x_{ij}, representing the amount of currency i converted to currency j, is defined as

```
var x{i in 1..n,j in 1..n}>=0;
```

The model has two sets of constraints: The first set with the root name `r1` sets the limits on the amounts of any currency conversion transaction by using the statement

```
r1{i in 1..n,j in 1..n}: x[i,j]<=maxTransaction[i];
```

The second set of constraints with the root name `r2` is a translation of the restriction

$$(\text{Input to currency } i) = (\text{Output from currency } i)$$

Its statement is given as

```
r2{i in 1..n}:
  (if i=inCurrency then I else 0)+
   sum{k in 1..n}(if k<i then r[k,i] else 1/r[i,k])*x[k,i]
     =(if i=outCurrency then y else 0)+sum{j in 1..n}x[i,j];
```

This type of constraints is ideal for the use of the special construct `if then else` to specify conditional values. In the left-hand side of the constraint, the expression

```
(if i=inCurrency then I else 0)
```

says that in the constraint for the input currency `(i=inCurrency)` there is an external input `I`, else the external input is zero. Next, the expression

```
sum{k in 1..n}(if k<i then r[k,i] else 1/r[i,k])*x[k,i]
```

computes the input funds from other currency converted to the input currency. If you review Example 2.3-2 you will notice that when `k<i`, the conversion uses the above-diagonal elements of the exchange rate `r`. Otherwise, the *row* reciprocal is used for the below-diagonal elements (diagonal elements are 1). This is precisely what `if then else` does. (See Section A.3 for details.)

The `if`-expression in the right-hand side of constraint `r2` can be explained in a similar manner—namely,

```
(if i=outCurrency then y else 0)
```

says that the external output is `y` for `outCurrency` and zero for all others.

We can enhance the readability of constraints `r2` by defining the following **computed parameter** (see Section A.3) that defines the entire exchange rate table:

```
Param rate{k in 1..n,i in 1..n}
          =(if k<i then r[k,i] else 1/r[i,k])
```

In this case, constraints `r2` become

```
r2{i in 1..n}:
  (if i=inCurrency then I else 0)+ sum{k in 1..n}rate[k,i]*x[k,i]
     =(if i=outCurrency then y else 0)+sum{j in 1..n}x[i,j];
```

In the `data;` section, `inCurrency` and `outCurrency` each equal 1, which means that the problem is seeking the maximum dollar output using an initial amount of $5 million. In general, `inCurrency` and `outCurrency` may designate any distinct currencies. For example, setting `inCurrency` equal to `2` and `outCurrency` equal to 4 maximizes the yen output given a 5 million euros initial investment.

The unspecified entries of `param r` are flagged in AMPL with dots (`.`). These values are then overridden either by using the reciprocal as shown in Figure 2.17 or through the use of the computed parameter `rate` as shown above. The alternative to using dots is to unnecessarily compute and enter the below-diagonal elements as data.

The `display` statement sends the output to file file2.out instead of defaulting it to the screen. The `print` statement computes and truncates the rate of return and sends the output to file file2.out. The `print` statement can also be formatted using `printf`, just as in any higher level programming language. (See Section A.5.2 for details.)

It is important to notice that input data in AMPL need not be hard-coded in the model, as they can be retrieved from external files, spreadsheets, and databases (see Section A.5 for details). This is crucial in the arbitrage model, where the volatile exchange rates must often be accepted within less than 10 seconds. By allowing the AMPL model to receive its data from a database that automatically updates the exchange rates, the model can provide timely optimal solutions.

The Bus Scheduling Problem. The bus scheduling problem of Example 2.3-8 provides an interesting modeling situation in AMPL. Of course, we can always use a two-subscripted parameter, similar to parameter `a` in the Reddy Mikks model in Section 2.4.2 (Figure 2.16), but this may be cumbersome in this case. Instead, we can take advantage of the special structure of the constraints and use conditional expressions to represent them implicitly.

The left-hand side of constraint 1 is $x_1 + x_m$, where m is the total number of periods in a 24-hour day (= 6 in the present example). For the remaining constraints, the left-hand side takes the form $x_{i-1} + x_i, i = 2, 3, \ldots, m$. Using `if then else` (as we did in the arbitrage problem), all m constraints can be represented compactly by one statement as shown in Figure 2.18 (file amplEx2.3-8.txt). This representation is superior to defining the left-hand side of the constraints as an explicit parameter.

AMPL offers a wide range of programming capabilities. For example, the input/output data can be secured from/sent to external files, spreadsheets, and databases and the model can be executed interactively for a wide variety of options that allow testing different scenarios. The details are given in Appendix A. Also, many AMPL models are presented throughout the book with cross references to the material in Appendix A to assist you in understanding these options.

```
param m;
param  min_nbr_buses{1..m};
var x_nbr_buses{1..m} >= 0;
minimize tot_nbr_buses: sum {i in 1..m} x_nbr_buses[i];
subject to constr_nbr{i in 1..m}:
        if i=1 then
              x_nbr_buses[i]+x_nbr_buses[m]
        else
              x_nbr_buses[i-1]+x_nbr_buses[i] >= min_nbr_buses[i];

data;
param m:=6;
param min_nbr_buses:= 1 4  2 8  3 10  4 7  5 12  6 4;

solve;
display tot_nbr_buses, x_nbr_buses;
```

FIGURE 2.18

AMPL model of the bus scheduling problem of Example 2.3-8 (file amplEx2.3-8.txt)

PROBLEM SET 2.4B

1. In the Reddy Mikks model, suppose that a third type of paint, named "marine," is produced. The requirements per ton of raw materials $M1$ and $M2$ are .5 and .75 ton, respectively. The daily demand for the new paint lies between .5 ton and 1.5 tons and the profit per ton is $3.5 (thousand). Modify the Excel Solver model solverRM2.xls and the AMPL model amplRM2.txt to account for the new situation and determine the optimum solution. Compare the additional effort associated with each modification.

2. Develop AMPL models for the following problems:

(a) The diet problem of Example 2.2-2 and find the optimum solution.

(b) Problem 4, Set 2.3b.

*__(c)__ Problem 7, Set 2.3d.

(d) Problem 7, Set 2.3g.

(e) Problem 9, Set 2.3g.

*__(f)__ Problem 10, Set 2.3g.

REFERENCES

Fourer, R., D. Gay, and B. Kernighan, *AMPL, A Modeling Language for Mathematical Programming*, 2nd ed., Brooks/Cole-Thomson, Pacific Grove, CA, 2003.

Kornbluth, J., and G. Salkin, *The Management for Corporate Financial Assets: Applications of Mathematical Programming Models*, Academic Press, London, 1987.

Shepard, R., D. Hartley, P. Hasman, L. Thorpe, and M. Bathe, *Applied Operations Research*, Plenum Press, New York, 1988.

Stark, R., and R. Nicholes, *Mathematical Programming Foundations for Design: Civil Engineering Systems*, McGraw-Hill, New York, 1972.

Laidlaw, C. *Linear Programming for Urban Development Plan Evaluation*, Praegers, London, 1972.

Lewis, T., "Personal Operations Research: Practicing OR on Ourselves," *Interfaces,* Vol. 26, No. 5, pp. 34–41, 1996.

William, H., *Model Building in Mathematical Programming*, 4th ed., Wiley, New York, 1994.

CHAPTER 3

The Simplex Method and Sensitivity Analysis

Chapter Guide. This chapter details the simplex method for solving the general LP problem. It also explains how simplex-based sensitivity analysis is used to provide important economic interpretations about the optimum solution, including the *dual prices* and the *reduced cost.*

The simplex method computations are particularly tedious, repetitive, and, above all, boring. As you do these computations, you should not lose track of the big picture; namely, the simplex method attempts to move from one corner point of the solution space to a better corner point until the optimum is found. To assist you in this regard, TORA's interactive *user-guided* module (with instant feedback) allows you to decide how the computations should proceed while relieving you of the burden of the tedious computations. In this manner, you get to understand the concepts without being overwhelmed by the computational details. Rest assured that once you have learned how the simplex method works (and it is important that you do understand the concepts), computers will carry out the tedious work and you will *never* again need to solve an LP manually.

Throughout my teaching experience, I have noticed that while students can easily carry out the tedious simplex method computations, in the end, some cannot tell why they are doing them or what the solution is. To assist in overcoming this potential difficulty, the material in the chapter stresses the interpretation of each iteration in terms of the solution to the original problem.

When you complete the material in this chapter, you will be in a position to read and interpret the output reports provided by commercial software. The last section describes how these reports are generated in AMPL, Excel Solver, and TORA.

This chapter includes a summary of 1 real-life application, 11 solved examples, 1 AMPL model, 1 Solver model, 1 TORA model, 107 end-of-section problems, and 3 cases. The cases are in Appendix E on the CD. The AMPL/Excel/Solver/TORA programs are in folder ch3Files.

Real Life Application—Optimization of Heart Valve Production

Biological heart valves in different sizes are bioprostheses manufactured from porcine hearts for human implantation. On the supply side, porcine hearts cannot be "produced" to specific sizes. Moreover, the exact size of a manufactured valve cannot be determined until the biological component of pig heart has been processed. As a result, some sizes may be overstocked and others understocked. A linear programming model was developed to reduce overstocked sizes and increase understocked sizes. The resulting savings exceeded $1,476,000 in 1981, the year the study was made. The details of this study are presented in Case 2, Chapter 24 on the CD.

3.1 LP MODEL IN EQUATION FORM

The development of the simplex method computations is facilitated by imposing two requirements on the constraints of the problem:

1. All the constraints (with the exception of the nonnegativity of the variables) are equations with nonnegative right-hand side.
2. All the variables are nonnegative.

These two requirements are imposed here primarily to standardize and streamline the simplex method calculations. It is important to know that all commercial packages (and TORA) directly accept inequality constraints, nonnegative right-hand side, and unrestricted variables. Any necessary preconditioning of the model is done internally in the software before the simplex method solves the problem.

3.1.1 Converting Inequalities into Equations with Nonnegative Right-Hand Side

In ($\leq$) constraints, the right-hand side can be thought of as representing the limit on the availability of a resource, in which case the left-hand side would represent the usage of this limited resource by the activities (variables) of the model. The difference between the right-hand side and the left-hand side of the ($\leq$) constraint thus yields the *unused* or *slack* amount of the resource.

To convert a ($\leq$)-inequality to an equation, a nonnegative **slack variable** is added to the left-hand side of the constraint. For example, in the Reddy Mikks model (Example 2.1-1), the constraint associated with the use of raw material $M1$ is given as

$$6x_1 + 4x_2 \leq 24$$

Defining s_1 as the slack or unused amount of $M1$, the constraint can be converted to the following equation:

$$6x_1 + 4x_2 + s_1 = 24, s_1 \geq 0$$

Next, a ($\geq$)-constraint sets a lower limit on the activities of the LP model, so that the amount by which the left-hand side exceeds the minimum limit represents a *surplus*. The conversion from ($\geq$) to ($=$) is achieved by subtracting a nonnegative

surplus variable from the left-hand side of the inequality. For example, in the diet model (Example 2.2-2), the constraint representing the minimum feed requirements is

$$x_1 + x_2 \geq 800$$

Defining S_1 as the surplus variable, the constraint can be converted to the following equation

$$x_1 + x_2 - S_1 = 800, S_1 \geq 0$$

The only remaining requirement is for the right-hand side of the resulting equation to be nonnegative. The condition can always be satisfied by multiplying both sides of the resulting equation by -1, where necessary. For example, the constraint

$$-x_1 + x_2 \leq -3$$

is equivalent to the equation

$$-x_1 + x_2 + s_1 = -3, s_1 \geq 0$$

Now, multiplying both sides by -1 will render a nonnegative right-hand side, as desired—that is,

$$x_1 - x_2 - s_1 = 3$$

PROBLEM SET 3.1A

*1. In the Reddy Mikks model (Example 2.2-1), consider the feasible solution $x_1 = 3$ tons and $x_2 = 1$ ton. Determine the value of the associated slacks for raw materials $M1$ and $M2$.

2. In the diet model (Example 2.2-2), determine the surplus amount of feed consisting of 500 lb of corn and 600 lb of soybean meal.

3. Consider the following inequality

$$10x_1 - 3x_2 \geq -5$$

Show that multiplying both sides of the inequality by -1 and then converting the resulting inequality into an equation is the same as converting it first to an equation and then multiplying both sides by -1.

*4. Two different products, $P1$ and $P2$, can be manufactured by one or both of two different machines, $M1$ and $M2$. The unit processing time of either product on either machine is the same. The daily capacity of machine $M1$ is 200 units (of either $P1$ or $P2$, or a mixture of both) and the daily capacity of machine $M2$ is 250 units. The shop supervisor wants to balance the production schedule of the two machines such that the total number of units produced on one machine is within 5 units of the number produced on the other. The profit per unit of $P1$ is \$10 and that of $P2$ is \$15. Set up the problem as an LP in equation form.

5. Show how the following objective function can be presented in equation form:

$$\text{Minimize } z = \max\{|x_1 - x_2 + 3x_3|, |-x_1 + 3x_2 - x_3|\}$$

$$x_1, x_2, x_3 \geq 0$$

(*Hint*: $|a| \leq b$ is equivalent to $a \leq b$ and $a \geq -b$.)

6. Show that the m equations:

$$\sum_{j=1}^{n} a_{ij}x_j = b_i, i = 1, 2, \ldots, m$$

are equivalent to the following $m + 1$ inequalities:

$$\sum_{j=1}^{n} a_{ij}x_j \leq b_i, i = 1, 2, \ldots, m$$

$$\sum_{j=1}^{n} \left(\sum_{i=1}^{m} a_{ij} \right) x_j \geq \sum_{i=1}^{m} b_i$$

3.1.2 Dealing with Unrestricted Variables

In Example 2.3-6 we presented a multiperiod production smoothing model in which the workforce at the start of each period is adjusted up or down depending on the demand for that period. Specifically, if x_i (≥ 0) is the workforce size in period i, then x_{i+1} (≥ 0) the workforce size in period $i + 1$ can be expressed as

$$x_{i+1} = x_i + y_{i+1}$$

The variable y_{i+1} must be unrestricted in sign to allow x_{i+1} to increase or decrease relative to x_i depending on whether workers are hired or fired, respectively.

As we will see shortly, the simplex method computations require all the variables be nonnegative. We can always account for this requirement by using the substitution

$$y_{i+1} = y_{i+1}^{-} - y_{i+1}^{+}, \text{ where } y_{i+1}^{-} \geq 0 \text{ and } y_{i+1}^{+} \geq 0$$

To show how this substitution works, suppose that in period 1 the workforce is $x_1 = 20$ workers and that the workforce in period 2 will be increased by 5 to reach 25 workers. In terms of the variables y_2^{-} and y_2^{+}, this will be equivalent to $y_2^{-} = 5$ and $y_2^{+} = 0$ or $y_2 = 5 - 0 = 5$. Similarly, if the workforce in period 2 is reduced to 16, then we have $y_2^{-} = 0$ and $y_2^{+} = 4$, or $y_2 = 0 - 4 = -4$. The substitution also allows for the possibility of no change in the workforce by letting both variables assume a zero value.

You probably are wondering about the possibility that both y_2^{-} and y_2^{+} may assume positive values simultaneously. Intuitively, as we explained in Example 2.3-6, this cannot happen, because it means that we can hire and fire a worker at the same time. This intuition is also supported by a mathematical proof that shows that, in any simplex method solution, it is impossible that both variables will assume positive values simultaneously.

PROBLEM SET 3.1B

1. McBurger fast-food restaurant sells quarter-pounders and cheeseburgers. A quarter-pounder uses a quarter of a pound of meat, and a cheeseburger uses only .2 lb. The restaurant starts the day with 200 lb of meat but may order more at an additional cost of 25 cents per pound to cover the delivery cost. Any surplus meat at the end of the day is donated to charity. McBurger's profits are 20 cents for a quarter-pounder and 15 cents for a cheeseburger. McBurger does not expect to sell more than 900 sandwiches in any one

day. How many of each type sandwich should McBurger plan for the day? Solve the problem using TORA, Solver, or AMPL.

2. Two products are manufactured in a machining center. The productions times per unit of products 1 and 2 are 10 and 12 minutes, respectively. The total regular machine time is 2500 minutes per day. In any one day, the manufacturer can produce between 150 and 200 units of product 1, but no more than 45 units of product 2. Overtime may be used to meet the demand at an additional cost of $.50 per minute. Assuming that the unit profits for products 1 and 2 are $6.00 and $7.50, respectively, formulate the problem as an LP model, then solve with TORA, Solver, or AMPL to determine the optimum production level for each product as well as any overtime needed in the center.

***3.** JoShop manufactures three products whose unit profits are $2, $5, and $3, respectively. The company has budgeted 80 hours of labor time and 65 hours of machine time for the production of three products. The labor requirements per unit of products 1, 2, and 3 are 2, 1, and 2 hours, respectively. The corresponding machine-time requirements per unit are 1, 1, and 2 hours. JoShop regards the budgeted labor and machine hours as goals that may be exceeded, if necessary, but at the additional cost of $15 per labor hour and $10 per machine hour. Formulate the problem as an LP, and determine its optimum solution using TORA, Solver, or AMPL.

4. In an LP in which there are several unrestricted variables, a transformation of the type $x_j = x_j^- - x_j^+, x_j^-, x_j^+ \geq 0$ will double the corresponding number of nonnegative variables. We can, instead, replace k unrestricted variables with exactly $k + 1$ nonnegative variables by using the substitution $x_j = x_j' - w, x_j', w \geq 0$. Use TORA, Solver, or AMPL to show that the two methods produce the same solution for the following LP:

$$\text{Maximize } z = -2x_1 + 3x_2 - 2x_3$$

subject to

$$4x_1 - x_2 - 5x_3 = 10$$

$$2x_1 + 3x_2 + 2x_3 = 12$$

$$x_1 \geq 0, x_2, x_3 \text{ unrestricted}$$

3.2 TRANSITION FROM GRAPHICAL TO ALGEBRAIC SOLUTION

The ideas conveyed by the graphical LP solution in Section 2.2 lay the foundation for the development of the algebraic simplex method. Figure 3.1 draws a parallel between the two methods. In the graphical method, the solution space is delineated by the half-spaces representing the constraints, and in the simplex method the solution space is represented by m simultaneous linear equations and n nonnegative variables.

We can see visually why the graphical solution space has an infinite number of solution points, but how can we draw a similar conclusion from the algebraic representation of the solution space? The answer is that in the algebraic representation the number of equations m is always *less than or equal to* the number of variables n.[1] If $m = n$, and the equations are consistent, the system has only one solution; but if $m < n$ (which

[1]If the number of equations m is larger than the number of variables n, then at least $m - n$ equations must be redundant.

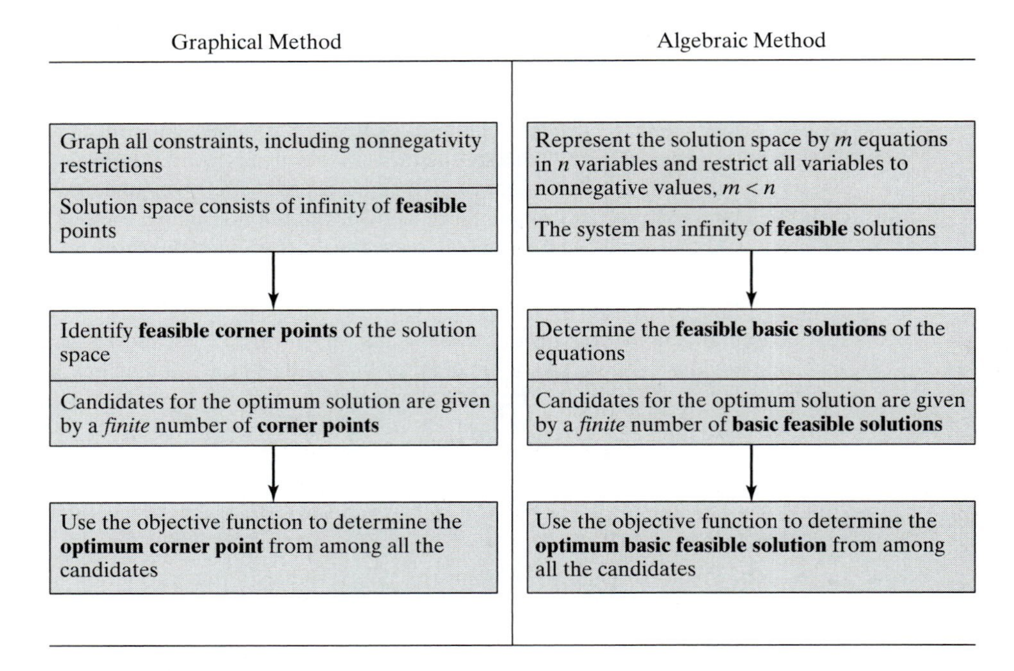

FIGURE 3.1

Transition from graphical to algebraic solution

represents the majority of LPs), then the system of equations, again if consistent, will yield an infinite number of solutions. To provide a simple illustration, the equation $x = 2$ has $m = n = 1$, and the solution is obviously unique. But, the equation $x + y = 1$ has $m = 1$ and $n = 2$, and it yields an infinite number of solutions (any point on the straight line $x + y = 1$ is a solution).

Having shown how the LP solution space is represented algebraically, the candidates for the optimum (i.e., corner points) are determined from the simultaneous linear equations in the following manner:

Algebraic Determination of Corner Points.

In a set of $m \times n$ equations ($m < n$), if we set $n - m$ variables equal to zero and then solve the m equations for the remaining m variables, the resulting solution, if unique, is called a **basic solution** and must correspond to a (feasible or infeasible) corner point of the solution space. This means that the *maximum* number of corner points is

$$C_m^n = \frac{n!}{m!(n-m)!}$$

The following example demonstrates the procedure.

Example 3.2-1

Consider the following LP with two variables:

$$\text{Maximize } z = 2x_1 + 3x_2$$

subject to

$$2x_1 + x_2 \leq 4$$
$$x_1 + 2x_2 \leq 5$$
$$x_1, x_2 \geq 0$$

Figure 3.2 provides the graphical solution space for the problem.

Algebraically, the solution space of the LP is represented as:

$$2x_1 + x_2 + s_1 = 4$$
$$x_1 + 2x_2 + s_2 = 5$$
$$x_1, x_2, s_1, s_2 \geq 0$$

The system has $m = 2$ equations and $n = 4$ variables. Thus, according to the given rule, the corner points can be determined algebraically by setting $n - m = 4 - 2 = 2$ variables equal to

FIGURE 3.2

LP solution space of Example 3.2-1

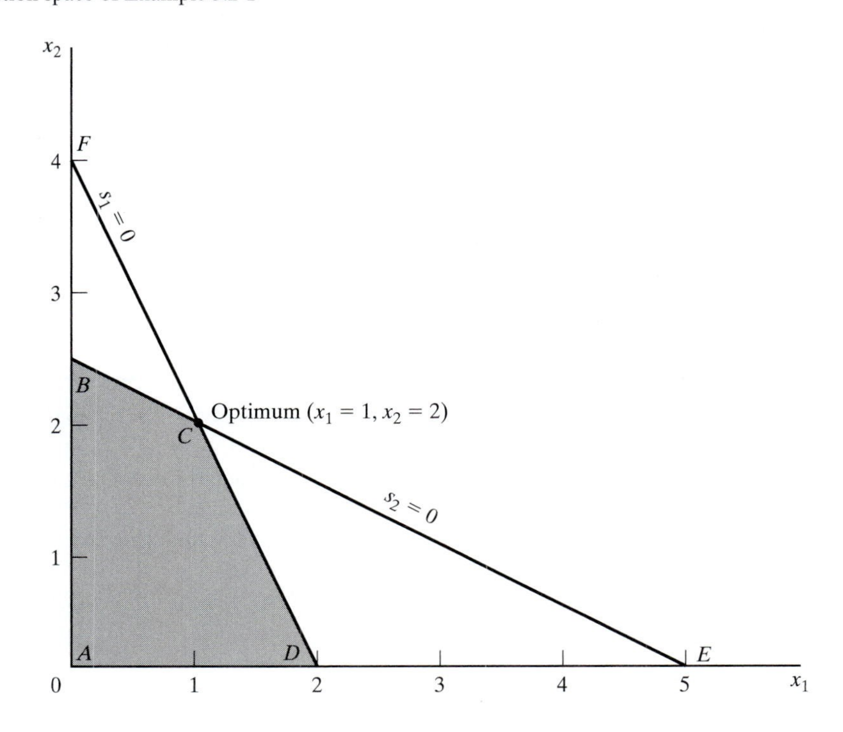

zero and then solving for the remaining $m = 2$ variables. For example, if we set $x_1 = 0$ and $x_2 = 0$, the equations provide the unique (basic) solution

$$s_1 = 4, s_2 = 5$$

This solution corresponds to point A in Figure 3.2 (convince yourself that $s_1 = 4$ and $s_2 = 5$ at point A). Another point can be determined by setting $s_1 = 0$ and $s_2 = 0$ and then solving the two equations

$$2x_1 + x_2 = 4$$

$$x_1 + 2x_2 = 5$$

This yields the basic solution ($x_1 = 1, x_2 = 2$), which is point C in Figure 3.2.

You probably are wondering how one can decide which $n - m$ variables should be set equal to zero to target a specific corner point. Without the benefit of the graphical solution (which is available only for two or three variables), we cannot say which $(n - m)$ zero variables are associated with which corner point. But that does not prevent us from enumerating *all* the corner points of the solution space. Simply consider *all* combinations in which $n - m$ variables are set to zero and solve the resulting equations. Once done, the optimum solution is the feasible basic solution (corner point) that yields the best objective value.

In the present example we have $C_2^4 = \frac{4!}{2!2!} = 6$ corner points. Looking at Figure 3.2, we can immediately spot the four corner points A, B, C, and D. Where, then, are the remaining two? In fact, points E and F also are corner points for the problem, but they are *infeasible* because they do not satisfy all the constraints. These infeasible corner points are not candidates for the optimum.

To summarize the transition from the graphical to the algebraic solution, the zero $n - m$ variables are known as **nonbasic variables**. The remaining m variables are called **basic variables** and their solution (obtained by solving the m equations) is referred to as **basic solution**. The following table provides all the basic and nonbasic solutions of the current example.

Nonbasic (zero) variables	Basic variables	Basic solution	Associated corner point	Feasible?	Objective value, z
(x_1, x_2)	(s_1, s_2)	(4, 5)	A	Yes	0
(x_1, s_1)	(x_2, s_2)	(4, −3)	F	No	—
(x_1, s_2)	(x_2, s_1)	(2.5, 1.5)	B	Yes	7.5
(x_2, s_1)	(x_1, s_2)	(2, 3)	D	Yes	4
(x_2, s_2)	(x_1, s_1)	(5, −6)	E	No	—
$\mathbf{(s_1, s_2)}$	$\mathbf{(x_1, x_2)}$	**(1, 2)**	$\mathbf{C}$	**Yes**	**8 (optimum)**

Remarks. We can see from the computations above that as the problem size increases (that is, m and n become large), the procedure of enumerating all the corner points involves prohibitive computations. For example, for $m = 10$ and $n = 20$, it is necessary to solve $C_{10}^{20} = 184{,}756$ sets of 10×10 equations, a staggering task indeed, particularly when we realize that a (10×20)-LP is a small size in most real-life situations, where hundreds or even thousands of variables and constraints are not unusual. The simplex method alleviates this computational burden dramatically by investigating only a fraction of all possible basic feasible solutions (corner points) of the solution space. In essence, the simplex method utilizes an intelligent search procedure that locates the optimum corner point in an efficient manner.

PROBLEM SET 3.2A

1. Consider the following LP:

$$\text{Maximize } z = 2x_1 + 3x_2$$

subject to

$$x_1 + 3x_2 \leq 6$$
$$3x_1 + 2x_2 \leq 6$$
$$x_1, x_2 \geq 0$$

(a) Express the problem in equation form.

(b) Determine all the basic solutions of the problem, and classify them as feasible and infeasible.

*__(c)__ Use direct substitution in the objective function to determine the optimum basic feasible solution.

(d) Verify graphically that the solution obtained in (c) is the optimum LP solution—hence, conclude that the optimum solution can be determined algebraically by considering the basic feasible solutions only.

*__(e)__ Show how the *infeasible* basic solutions are represented on the graphical solution space.

2. Determine the optimum solution for each of the following LPs by enumerating all the basic solutions.

(a) Maximize $z = 2x_1 - 4x_2 + 5x_3 - 6x_4$

subject to

$$x_1 + 4x_2 - 2x_3 + 8x_4 \leq 2$$
$$-x_1 + 2x_2 + 3x_3 + 4x_4 \leq 1$$
$$x_1, x_2, x_3, x_4 \geq 0$$

(b) Minimize $z = x_1 + 2x_2 - 3x_3 - 2x_4$

subject to

$$x_1 + 2x_2 - 3x_3 + x_4 = 4$$
$$x_1 + 2x_2 + x_3 + 2x_4 = 4$$
$$x_1, x_2, x_3, x_4 \geq 0$$

*__3.__ Show algebraically that all the basic solutions of the following LP are infeasible.

$$\text{Maximize } z = x_1 + x_2$$

subject to

$$x_1 + 2x_2 \leq 6$$
$$2x_1 + x_2 \leq 16$$
$$x_1, x_2 \geq 0$$

4. Consider the following LP:

$$\text{Maximize } z = 2x_1 + 3x_2 + 5x_3$$

subject to

$$-6x_1 + 7x_2 - 9x_3 \geq 4$$
$$x_1 + x_2 + 4x_3 = 10$$
$$x_1, x_3 \geq 0$$
$$x_2 \text{ unrestricted}$$

Conversion to the equation form involves using the substitution $x_2 = x_2^- - x_2^+$. Show that a basic solution cannot include both x_2^- and x_2^+ simultaneously.

5. Consider the following LP:

$$\text{Maximize } z = x_1 + 3x_2$$

subject to

$$x_1 + x_2 \leq 2$$
$$-x_1 + x_2 \leq 4$$
$$x_1 \text{ unrestricted}$$
$$x_2 \geq 0$$

(a) Determine all the basic feasible solutions of the problem.

(b) Use direct substitution in the objective function to determine the best basic solution.

(c) Solve the problem graphically, and verify that the solution obtained in (c) is the optimum.

3.3 THE SIMPLEX METHOD

Rather than enumerating *all* the basic solutions (corner points) of the LP problem (as we did in Section 3.2), the simplex method investigates only a "select few" of these solutions. Section 3.3.1 describes the *iterative* nature of the method, and Section 3.3.2 provides the computational details of the simplex algorithm.

3.3.1 Iterative Nature of the Simplex Method

Figure 3.3 provides the solution space of the LP of Example 3.2-1. Normally, the simplex method starts at the origin (point A) where $x_1 = x_2 = 0$. At this starting point, the value of the objective function, z, is zero, and the logical question is whether an increase in nonbasic x_1 and/or x_2 above their current zero values can improve (increase) the value of z. We answer this question by investigating the objective function:

$$\text{Maximize } z = 2x_1 + 3x_2$$

The function shows that an increase in either x_1 or x_2 (or both) above their current zero values will *improve* the value of z. The design of the simplex method calls for increasing *one variable at a time*, with the selected variable being the one with the *largest*

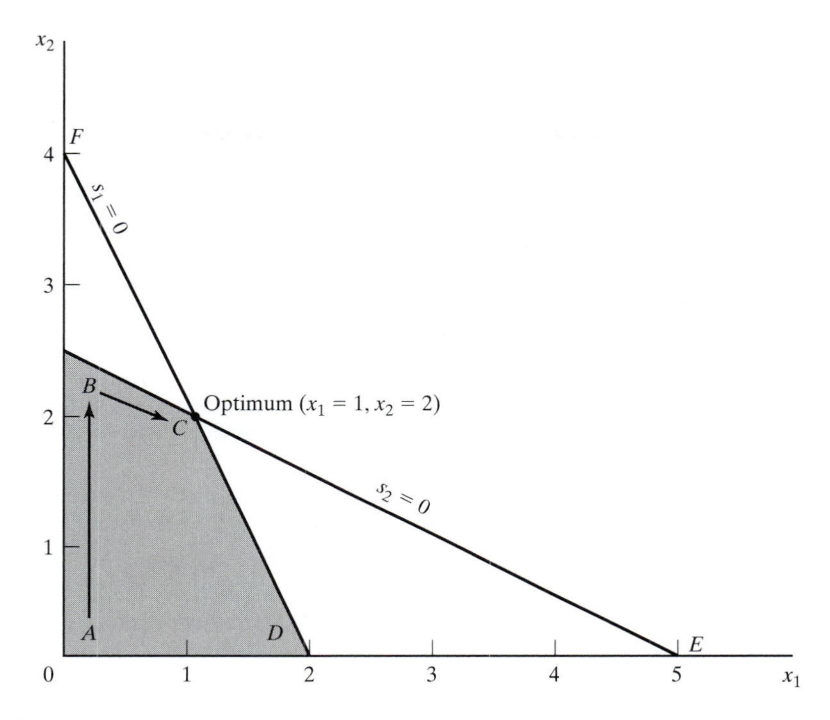

FIGURE 3.3
Iterative process of the simplex method

rate of improvement in z. In the present example, the value of z will increase by 2 for each unit increase in x_1 and by 3 for each unit increase in x_2. This means that the *rate* of improvement in the value of z is 2 for x_1 and 3 for x_2. We thus elect to increase x_2, the variable with the largest rate of improvement. Figure 3.3 shows that the value of x_2 must be increased until corner point B is reached (recall that stopping short of reaching corner point B is not optimal because a candidate for the optimum must be a corner point). At point B, the simplex method will then increase the value of x_1 to reach the improved corner point C, which is the optimum. The path of the simplex algorithm is thus defined as $A \rightarrow B \rightarrow C$. Each corner point along the path is associated with an **iteration**. It is important to note that the simplex method moves alongside the **edges** of the solution space, which means that the method cannot cut across the solution space, going from A to C directly.

We need to make the transition from the graphical solution to the algebraic solution by showing how the points A, B, and C are represented by their basic and nonbasic variables. The following table summarizes these representations:

Corner point	Basic variables	Nonbasic (zero) variables
A	s_1, s_2	x_1, x_2
B	s_1, x_2	x_1, s_2
C	x_1, x_2	s_1, s_2

Notice the change pattern in the basic and nonbasic variables as the solution moves along the path $A \rightarrow B \rightarrow C$. From A to B, nonbasic x_2 at A becomes basic at B and basic s_2 at A becomes nonbasic at B. In the terminology of the simplex method, we say that x_2 is the **entering variable** (because it enters the basic solution) and s_2 is the **leaving variable** (because it leaves the basic solution). In a similar manner, at point B, x_1 *enters* (the basic solution) and s_1 *leaves*, thus leading to point C.

PROBLEM SET 3.3A

1. In Figure 3.3, suppose that the objective function is changed to

$$\text{Maximize } z = 8x_1 + 4x_2$$

Identify the path of the simplex method and the basic and nonbasic variables that define this path.

2. Consider the graphical solution of the Reddy Mikks model given in Figure 2.2. Identify the path of the simplex method and the basic and nonbasic variables that define this path.

***3.** Consider the three-dimensional LP solution space in Figure 3.4, whose feasible extreme points are $A, B, \ldots,$ and J.

(a) Which of the following pairs of corner points cannot represent *successive* simplex iterations: (A, B), (B, D), (E, H), and (A, I)? Explain the reason.

(b) Suppose that the simplex iterations start at A and that the optimum occurs at H. Indicate whether any of the following paths are *not* legitimate for the simplex algorithm, and state the reason.

(i) $A \rightarrow B \rightarrow G \rightarrow H$.

(ii) $A \rightarrow E \rightarrow I \rightarrow H$.

(iii) $A \rightarrow C \rightarrow E \rightarrow B \rightarrow A \rightarrow D \rightarrow G \rightarrow H$.

4. For the solution space in Figure 3.4, all the constraints are of the type $\leq$ and all the variables x_1, x_2, and x_3 are nonnegative. Suppose that s_1, s_2, s_3, and s_4 (≥ 0) are the slacks associated with constraints represented by the planes $CEIJF$, $BEIHG$, $DFJHG$, and IJH, respectively. Identify the basic and nonbasic variables associated with each feasible extreme point of the solution space.

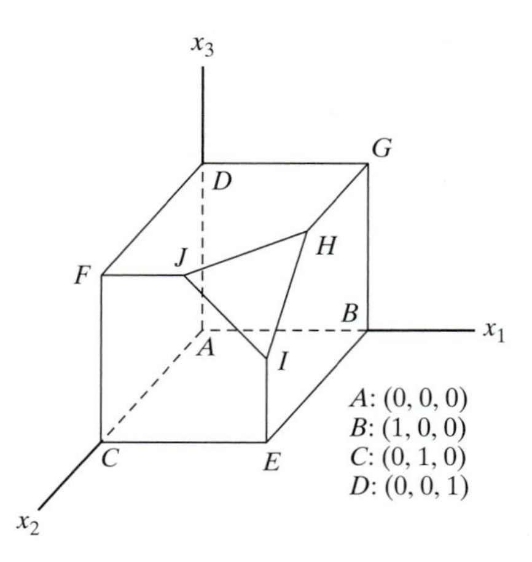

FIGURE 3.4

Solution space of Problem 3, Set 3.2b

5. Consider the solution space in Figure 3.4, where the simplex algorithm starts at point A. Determine the entering variable in the *first* iteration together with its value and the improvement in z for each of the following objective functions:
   ***(a)** Maximize $z = x_1 - 2x_2 + 3x_3$
   (b) Maximize $z = 5x_1 + 2x_2 + 4x_3$
   (c) Maximize $z = -2x_1 + 7x_2 + 2x_3$
   (d) Maximize $z = x_1 + x_2 + x_3$

3.3.2 Computational Details of the Simplex Algorithm

This section provides the computational details of a simplex iteration, including the rules for determining the entering and leaving variables as well as for stopping the computations when the optimum solution has been reached. The vehicle of explanation is a numerical example.

Example 3.3-1

We use the Reddy Mikks model (Example 2.1-1) to explain the details of the simplex method. The problem is expressed in equation form as

$$\text{Maximize } z = 5x_1 + 4x_2 + 0s_1 + 0s_2 + 0s_3 + 0s_4$$

subject to

$$\begin{aligned}
6x_1 + 4x_2 + s_1 \qquad\qquad\qquad &= 24 \quad \text{(Raw material } M1) \\
x_1 + 2x_2 \qquad + s_2 \qquad\qquad &= 6 \quad \text{(Raw material } M2) \\
-x_1 + x_2 \qquad\qquad + s_3 \qquad &= 1 \quad \text{(Market limit)} \\
x_2 \qquad\qquad\qquad + s_4 &= 2 \quad \text{(Demand limit)} \\
x_1, x_2, s_1, s_2, s_3, s_4 &\geq 0
\end{aligned}$$

The variables s_1, s_2, s_3, and s_4 are the slacks associated with the respective constraints.

Next, we write the objective equation as

$$z - 5x_1 - 4x_2 = 0$$

In this manner, the starting simplex tableau can be represented as follows:

Basic	z	x_1	x_2	s_1	s_2	s_3	s_4	Solution	
z	1	−5	−4	0	0	0	0	0	z-row
s_1	0	6	4	1	0	0	0	24	s_1-row
s_2	0	1	2	0	1	0	0	6	s_2-row
s_3	0	−1	1	0	0	1	0	1	s_3-row
s_4	0	0	1	0	0	0	1	2	s_4-row

The design of the tableau specifies the set of basic and nonbasic variables as well as provides the solution associated with the starting iteration. As explained in Section 3.3.1, the simplex iterations start at the origin $(x_1, x_2) = (0, 0)$ whose associated set of nonbasic and basic variables are defined as

$$\text{Nonbasic (zero) variables: } (x_1, x_2)$$

$$\text{Basic variables: } (s_1, s_2, s_3, s_4)$$

Substituting the nonbasic variables $(x_1, x_2) = (0, 0)$ and noting the special 0-1 arrangement of the coefficients of z and the basic variables (s_1, s_2, s_3, s_4) in the tableau, the following solution is immediately available (without any calculations):

$$\begin{aligned} z &= 0 \\ s_1 &= 24 \\ s_2 &= 6 \\ s_3 &= 1 \\ s_4 &= 2 \end{aligned}$$

This information is shown in the tableau by listing the basic variables in the leftmost *Basic* column and their values in the rightmost *Solution* column. In effect, the tableau defines the current corner point by specifying its basic variables and their values, as well as the corresponding value of the objective function, z. Remember that the nonbasic variables (those not listed in the *Basic* column) always equal zero.

Is the starting solution optimal? The objective function $z = 5x_1 + 4x_2$ shows that the solution can be improved by increasing x_1 or x_2. Using the argument in Section 3.3.1, x_1 with the *most positive* coefficient is selected as the *entering variable*. Equivalently, because the simplex tableau expresses the objective function as $z - 5x_1 - 4x_2 = 0$, the entering variable will correspond to the variable with the *most negative* coefficient in the objective equation. This rule is referred to as the **optimality condition.**

The mechanics of determining the leaving variable from the simplex tableau calls for computing the *nonnegative* **ratios** of the right-hand side of the equations (*Solution* column) to the corresponding constraint coefficients under the entering variable, x_1, as the following table shows.

Basic	Entering x_1	Solution	Ratio (or Intercept)
s_1	6	24	$x_1 = \frac{24}{6} = 4$ ← minimum
s_2	1	6	$x_1 = \frac{6}{1} = 6$
s_3	−1	1	$x_1 = \frac{1}{-1} = -1$ (ignore)
s_4	0	2	$x_1 = \frac{2}{0} = \infty$ (ignore)
Conclusion: x_1 enters and s_1 leaves			

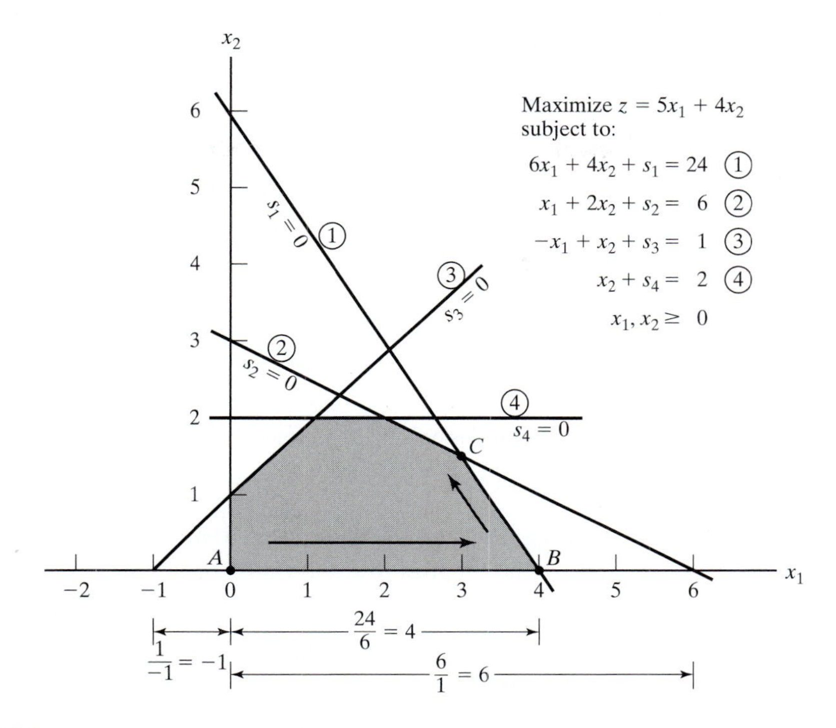

FIGURE 3.5
Graphical interpretation of the simplex method ratios in the Reddy Mikks model

The *minimum nonnegative* ratio automatically identifies the current basic variable s_1 as the leaving variable and assigns the entering variable x_1 the new value of 4.

How do the computed ratios determine the leaving variable and the value of the entering variable? Figure 3.5 shows that the computed ratios are actually the intercepts of the constraints with the entering variable (x_1) axis. We can see that the value of x_1 must be increased to 4 at corner point B, which is the smallest nonnegative intercept with the x_1-axis. An increase beyond B is infeasible. At point B, the current basic variable s_1 associated with constraint 1 assumes a zero value and becomes the *leaving variable*. The rule associated with the ratio computations is referred to as the **feasibility condition** because it guarantees the feasibility of the new solution.

The new solution point B is determined by "swapping" the entering variable x_1 and the leaving variable s_1 in the simplex tableau to produce the following sets of nonbasic and basic variables:

$$\text{Nonbasic (zero) variables at } B: (s_1, x_2)$$

$$\text{Basic variables at } B: (x_1, s_2, s_3, s_4)$$

The swapping process is based on the **Gauss-Jordan row operations**. It identifies the entering variable column as the **pivot column** and the leaving variable row as the **pivot row**. The intersection of the pivot column and the pivot row is called the **pivot element**. The following tableau is a restatement of the starting tableau with its pivot row and column highlighted.

	Basic	z	x_1 (Enter ↓)	x_2	s_1	s_2	s_3	s_4	Solution	
	z	1	−5	−4	0	0	0	0	0	
Leave ←	s_1	0	**6**	4	1	0	0	0	24	Pivot row
	s_2	0	1	2	0	1	0	0	6	
	s_3	0	−1	1	0	0	1	0	1	
	s_4	0	0	1	0	0	0	1	2	
			Pivot column							

The Gauss-Jordan computations needed to produce the new basic solution include two types.

1. *Pivot row*

a. Replace the leaving variable in the *Basic* column with the entering variable.
b. New pivot row = Current pivot row ÷ Pivot element

2. *All other rows, including z*
New Row = (Current row) − (Its pivot column coefficient) × (New pivot row)

These computations are applied to the preceding tableau in the following manner:

1. Replace s_1 in the *Basic* column with x_1:

New x_1-row = Current s_1-row ÷ 6

$$= \tfrac{1}{6}(0\ 6\ 4\ 1\ 0\ 0\ 0\ 24)$$
$$= \left(0\ 1\ \tfrac{2}{3}\ \tfrac{1}{6}\ 0\ 0\ 0\ 4\right)$$

2. New z-row = Current z-row − (−5) × New x_1-row

$$= \left(1\ -5\ -4\ 0\ 0\ 0\ 0\ 0\right) - (-5) \times \left(0\ 1\ \tfrac{2}{3}\ \tfrac{1}{6}\ 0\ 0\ 0\ 4\right)$$
$$= \left(1\ 0\ -\tfrac{2}{3}\ \tfrac{5}{6}\ 0\ 0\ 0\ 20\right)$$

3. New s_2-row = Current s_2-row − (1) × New x_1-row

$$= (0\ 1\ 2\ 0\ 1\ 0\ 0\ 6) - (1) \times \left(0\ 1\ \tfrac{2}{3}\ \tfrac{1}{6}\ 0\ 0\ 0\ 4\right)$$
$$= \left(0\ 0\ \tfrac{4}{3}\ -\tfrac{1}{6}\ 1\ 0\ 0\ 2\right)$$

4. New s_3-row = Current s_3-row − (−1) × New x_1-row

$$= (0\ -1\ 1\ 0\ 0\ 1\ 0\ 1) - (-1) \times \left(0\ 1\ \tfrac{2}{3}\ \tfrac{1}{6}\ 0\ 0\ 0\ 4\right)$$
$$= \left(0\ 0\ \tfrac{5}{3}\ \tfrac{1}{6}\ 0\ 1\ 0\ 5\right)$$

5. New s_4-row = Current s_4-row − (0) × New x_1-row

$$= (0\ 0\ 1\ 0\ 0\ 0\ 1\ 2) - (0)\left(0\ 1\ \tfrac{2}{3}\ \tfrac{1}{6}\ 0\ 0\ 0\ 4\right)$$
$$= (0\ 0\ 1\ 0\ 0\ 0\ 1\ 2)$$

The new basic solution is (x_1, s_2, s_3, s_4), and the new tableau becomes

	Basic	z	x_1	x_2 ↓	s_1	s_2	s_3	s_4	Solution
	z	1	0	$-\frac{2}{3}$	$\frac{5}{6}$	0	0	0	20
	x_1	0	1	$\frac{2}{3}$	$\frac{1}{6}$	0	0	0	4
←	s_2	0	0	$\frac{4}{3}$	$-\frac{1}{6}$	1	0	0	2
	s_3	0	0	$\frac{5}{3}$	$\frac{1}{6}$	0	1	0	5
	s_4	0	0	1	0	0	0	1	2

Observe that the new tableau has the same properties as the starting tableau. When we set the new nonbasic variables x_2 and s_1 to zero, the *Solution* column automatically yields the new basic solution ($x_1 = 4, s_2 = 2, s_3 = 5, s_4 = 2$). This "conditioning" of the tableau is the result of the application of the Gauss-Jordan row operations. The corresponding new objective value is $z = 20$, which is consistent with

$$\begin{aligned}\text{New z} &= \text{Old } z + \text{New } x_1\text{-value} \times \text{its objective coefficient} \\ &= 0 + 4 \times 5 = 20\end{aligned}$$

In the last tableau, the *optimality condition* shows that x_2 is the entering variable. The feasibility condition produces the following

Basic	Entering x_2	Solution	Ratio
x_1	$\frac{2}{3}$	4	$x_2 = 4 \div \frac{2}{3} = 6$
s_2	$\frac{4}{3}$	2	$x_2 = 2 \div \frac{4}{3} = 1.5$ (minimum)
s_3	$\frac{5}{3}$	5	$x_2 = 5 \div \frac{5}{3} = 3$
s_4	1	2	$x_2 = 2 \div 1 = 2$

Thus, s_2 leaves the basic solution and new value of x_2 is 1.5. The corresponding increase in z is $\frac{2}{3}x_2 = \frac{2}{3} \times 1.5 = 1$, which yields new $z = 20 + 1 = 21$.

Replacing s_2 in the *Basic* column with entering x_2, the following Gauss-Jordan row operations are applied:

1. New pivot x_2-row = Current s_2-row $\div \frac{4}{3}$
2. New z-row = Current z-row $- \left(-\frac{2}{3}\right) \times$ New x_2-row
3. New x_1-row = Current x_1-row $- \left(\frac{2}{3}\right) \times$ New x_2-row
4. New s_3-row = Current s_3-row $- \left(\frac{5}{3}\right) \times$ New x_2-row
5. New s_4-row = Current s_4-row $- (1) \times$ New x_2-row

These computations produce the following tableau:

Basic	z	x_1	x_2	s_1	s_2	s_3	s_4	Solution
z	1	0	0	$\frac{3}{4}$	$\frac{1}{2}$	0	0	21
x_1	0	1	0	$\frac{1}{4}$	$-\frac{1}{2}$	0	0	3
x_2	0	0	1	$-\frac{1}{8}$	$\frac{3}{4}$	0	0	$\frac{3}{2}$
s_3	0	0	0	$\frac{3}{8}$	$-\frac{5}{4}$	1	0	$\frac{5}{2}$
s_4	0	0	0	$\frac{1}{8}$	$-\frac{3}{4}$	0	1	$\frac{1}{2}$

Based on the optimality condition, *none* of the z-row coefficients associated with the nonbasic variables, s_1 and s_2, are negative. Hence, the last tableau is optimal.

The optimum solution can be read from the simplex tableau in the following manner. The optimal values of the variables in the *Basic* column are given in the right-hand-side *Solution* column and can be interpreted as

Decision variable	Optimum value	Recommendation
x_1	3	Produce 3 tons of exterior paint daily
x_2	$\frac{3}{2}$	Produce 1.5 tons of interior paint daily
z	21	Daily profit is $21,000

You can verify that the values $s_1 = s_2 = 0$, $s_3 = \frac{5}{2}$, $s_4 = \frac{1}{2}$ are consistent with the given values of x_1 and x_2 by substituting out the values of x_1 and x_2 in the constraints.

The solution also gives the status of the resources. A resource is designated as **scarce** if the activities (variables) of the model use the resource completely. Otherwise, the resource is **abundant**. This information is secured from the optimum tableau by checking the value of the slack variable associated with the constraint representing the resource. If the slack value is zero, the resource is used completely and, hence, is classified as scarce. Otherwise, a positive slack indicates that the resource is abundant. The following table classifies the constraints of the model:

Resource	Slack value	Status
Raw material, $M1$	$s_1 = 0$	Scarce
Raw material, $M2$	$s_2 = 0$	Scarce
Market limit	$s_3 = \frac{5}{2}$	Abundant
Demand limit	$s_4 = \frac{1}{2}$	Abundant

Remarks. The simplex tableau offers a wealth of additional information that includes:

1. *Sensitivity analysis,* which deals with determining the conditions that will keep the current solution unchanged.
2. *Post-optimal analysis,* which deals with finding a new optimal solution when the data of the model are changed.

Section 3.6 deals with sensitivity analysis. The more involved topic of post-optimal analysis is covered in Chapter 4.

TORA Moment.

The Gauss-Jordan computations are tedious, voluminous, and, above all, boring. Yet, they are the least important, because in practice these computations are carried out by the computer. What is important is that you understand *how* the simplex method works. TORA's interactive *user-guided* option (with instant feedback) can be of help in this regard because it allows you to decide the course of the computations in the simplex method without the burden of carrying out the Gauss-Jordan calculations. To use TORA with the Reddy Mikks problem, enter the model and then, from the SOLVE/MODIFY menu, select Solve ⇒ Algebraic ⇒ Iterations ⇒ All-Slack. (The All-Slack selection indicates that the starting basic solution consists of slack variables only. The remaining options will be presented in Sections 3.4, 4.3, and 7.4.2.) Next, click Go To Output Screen. You can generate one or all iterations by clicking Next Iteration or All Iterations. If you opt to generate the iterations one at a time, you can interactively specify the entering and leaving variables by clicking the headings of their corresponding column and row. If your selections are correct, the column turns green and the row turns red. Else, an error message will be posted.

3.3.3 Summary of the Simplex Method

So far we have dealt with the maximization case. In minimization problems, the *optimality condition* calls for selecting the entering variable as the nonbasic variable with the most *positive* objective coefficient in the objective equation, the exact opposite rule of the maximization case. This follows because max z is equivalent to min $(-z)$. As for the *feasibility condition* for selecting the leaving variable, the rule remains unchanged.

Optimality condition. The entering variable in a maximization (minimization) problem is the *nonbasic* variable having the most negative (positive) coefficient in the z-row. Ties are broken arbitrarily. The optimum is reached at the iteration where all the z-row coefficients of the nonbasic variables are nonnegative (nonpositive).

Feasibility condition. For both the maximization and the minimization problems, the leaving variable is the *basic* variable associated with the smallest nonnegative ratio (with *strictly positive* denominator). Ties are broken arbitrarily.

Gauss-Jordan row operations.

1. Pivot row

a. Replace the leaving variable in the *Basic* column with the entering variable.
b. New pivot row $=$ Current pivot row $\div$ Pivot element

2. All other rows, including z
New row $=$ (Current row) $-$ (pivot column coefficient) $\times$ (New pivot row)

The steps of the simplex method are

Step 1. Determine a starting basic feasible solution.

Step 2. Select an *entering variable* using the optimality condition. Stop if there is no entering variable; the last solution is optimal. Else, go to step 3.

Step 3. Select a *leaving variable* using the feasibility condition.

Step 4. Determine the new basic solution by using the appropriate Gauss-Jordan computations. Go to step 2.

PROBLEM SET 3.3B

1. This problem is designed to reinforce your understanding of the simplex feasibility condition. In the first tableau in Example 3.3-1, we used the minimum (nonnegative) ratio test to determine the leaving variable. Such a condition guarantees that none of the new values of the basic variables will become negative (as stipulated by the definition of the LP). To demonstrate this point, force s_2, instead of s_1, to leave the basic solution. Now, look at the resulting simplex tableau, and you will note that s_1 assumes a negative value $(= -12)$, meaning that the new solution is infeasible. This situation will never occur if we employ the minimum-ratio feasibility condition.

2. Consider the following set of constraints:

$$x_1 + 2x_2 + 2x_3 + 4x_4 \leq 40$$

$$2x_1 - x_2 + x_3 + 2x_4 \leq 8$$

$$4x_1 - 2x_2 + x_3 - x_4 \leq 10$$

$$x_1, x_2, x_3, x_4 \geq 0$$

Solve the problem for each of the following objective functions.

(a) Maximize $z = 2x_1 + x_2 - 3x_3 + 5x_4$.

(b) Maximize $z = 8x_1 + 6x_2 + 3x_3 - 2x_4$.

(c) Maximize $z = 3x_1 - x_2 + 3x_3 + 4x_4$.

(d) Minimize $z = 5x_1 - 4x_2 + 6x_3 - 8x_4$.

***3.** Consider the following system of equations:

$$x_1 + 2x_2 - 3x_3 + 5x_4 + x_5 = 4$$

$$5x_1 - 2x_2 + 6x_4 + x_6 = 8$$

$$2x_1 + 3x_2 - 2x_3 + 3x_4 + x_7 = 3$$

$$-x_1 + x_3 - 2x_4 + x_8 = 0$$

$$x_1, x_2, \ldots, x_8 \geq 0$$

Let $x_5, x_6, \ldots$, and x_8 be a given initial basic feasible solution. Suppose that x_1 becomes basic. Which of the given basic variables must become nonbasic at zero level to guarantee that all the variables remain nonnegative, and what is the value of x_1 in the new solution? Repeat this procedure for x_2, x_3, and x_4.

4. Consider the following LP:

$$\text{Maximize } z = x_1$$

subject to

$$\begin{aligned} 5x_1 + x_2 \quad\quad\quad &= 4 \\ 6x_1 \quad\quad + x_3 \quad &= 8 \\ 3x_1 \quad\quad\quad + x_4 &= 3 \\ x_1, x_2, x_3, x_4 &\geq 0 \end{aligned}$$

(a) Solve the problem *by inspection* (do not use the Gauss-Jordan row operations), and justify the answer in terms of the basic solutions of the simplex method.

(b) Repeat (a) assuming that the objective function calls for minimizing $z = x_1$.

5. Solve the following problem by *inspection*, and justify the method of solution in terms of the basic solutions of the simplex method.

$$\text{Maximize } z = 5x_1 - 6x_2 + 3x_3 - 5x_4 + 12x_5$$

subject to

$$x_1 + 3x_2 + 5x_3 + 6x_4 + 3x_5 \leq 90$$

$$x_1, x_2, x_3, x_4, x_5 \geq 0$$

(*Hint:* A basic solution consists of one variable only.)

6. The following tableau represents a specific simplex iteration. All variables are nonnegative. The tableau is not optimal for either a maximization or a minimization problem. Thus, when a nonbasic variable enters the solution it can either increase or decrease z or leave it unchanged, depending on the parameters of the entering nonbasic variable.

Basic	x_1	x_2	x_3	x_4	x_5	x_6	x_7	x_8	Solution
z	0	−5	0	4	−1	−10	0	0	620
x_8	0	3	0	−2	−3	−1	5	1	12
x_3	0	1	1	3	1	0	3	0	6
x_1	1	−1	0	0	6	−4	0	0	0

(a) Categorize the variables as basic and nonbasic and provide the current values of all the variables.

***(b)** Assuming that the problem is of the maximization type, identify the nonbasic variables that have the potential to improve the value of z. If each such variable enters the basic solution, determine the associated leaving variable, if any, and the associated change in z. Do not use the Gauss-Jordan row operations.

(c) Repeat part (b) assuming that the problem is of the minimization type.

(d) Which nonbasic variable(s) will not cause a change in the value of z when selected to enter the solution?

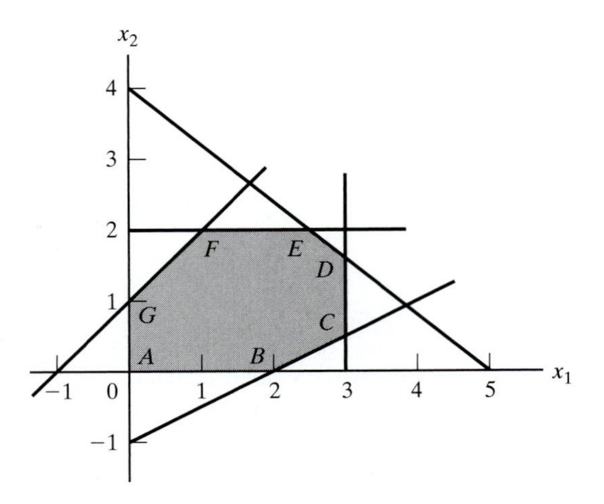

FIGURE 3.6
Solution space for Problem 7, Set 3.3b

7. Consider the two-dimensional solution space in Figure 3.6.
   (a) Suppose that the objective function is given as

$$\text{Maximize } z = 3x_1 + 6x_2$$

   If the simplex iterations start at point A, identify the path to the optimum point E.

   (b) Determine the entering variable, the corresponding ratios of the feasibility condition, and the change in the value of z, assuming that the starting iteration occurs at point A and that the objective function is given as

$$\text{Maximize } z = 4x_1 + x_2$$

   (c) Repeat (b), assuming that the objective function is

$$\text{Maximize } z = x_1 + 4x_2$$

8. Consider the following LP:

$$\text{Maximize } z = 16x_1 + 15x_2$$

subject to

$$\begin{aligned} 40x_1 + 31x_2 &\le 124 \\ -x_1 + x_2 &\le 1 \\ x_1 &\le 3 \\ x_1, x_2 &\ge 0 \end{aligned}$$

   (a) Solve the problem by the simplex method, where the entering variable is the nonbasic variable with the *most* negative z-row coefficient.

   (b) Resolve the problem by the simplex algorithm, always selecting the entering variable as the nonbasic variable with the *least* negative z-row coefficient.

   (c) Compare the number of iterations in (a) and (b). Does the selection of the entering variable as the nonbasic variable with the *most* negative z-row coefficient lead to a smaller number of iterations? What conclusion can be made regarding the optimality condition?

   (d) Suppose that the sense of optimization is changed to minimization by multiplying z by -1. How does this change affect the simplex iterations?

***9.** In Example 3.3-1, show how the second best optimal value of z can be determined from the optimal tableau.

10. Can you extend the procedure in Problem 9 to determine the third best optimal value of z?

11. The Gutchi Company manufactures purses, shaving bags, and backpacks. The construction includes leather and synthetics, leather being the scarce raw material. The production process requires two types of skilled labor: sewing and finishing. The following table gives the availability of the resources, their usage by the three products, and the profits per unit.

	Resource requirements per unit			
Resource	*Purse*	*Bag*	*Backpack*	Daily availability
Leather (ft^2)	2	1	3	42 ft^2
Sewing (hr)	2	1	2	40 hr
Finishing (hr)	1	.5	1	45 hr
Selling price ($)	24	22	45	

(a) Formulate the problem as a linear program and find the optimum solution (using TORA, Excel Solver, or AMPL).

(b) From the optimum solution determine the status of each resource.

12. *TORA experiment.* Consider the following LP:

$$\text{Maximize } z = x_1 + x_2 + 3x_3 + 2x_4$$

subject to

$$\begin{aligned} x_1 + 2x_2 - 3x_3 + 5x_4 &\leq 4 \\ 5x_1 - 2x_2 \qquad\quad + 6x_4 &\leq 8 \\ 2x_1 + 3x_2 - 2x_3 + 3x_4 &\leq 3 \\ -x_1 \qquad\quad + x_3 + 2x_4 &\leq 0 \\ x_1, x_2, x_3, x_4 &\geq 0 \end{aligned}$$

(a) Use TORA's iterations option to determine the optimum tableau.

(b) Select any nonbasic variable to "enter" the basic solution, and click Next Iteration to produce the associated iteration. How does the new objective value compare with the optimum in (a)? The idea is to show that the tableau in (a) is optimum because none of the nonbasic variables can improve the objective value.

13. *TORA experiment.* In Problem 12, use TORA to find the next-best optimal solution.

3.4 ARTIFICIAL STARTING SOLUTION

As demonstrated in Example 3.3-1, LPs in which all the constraints are ($\leq$) with non-negative right-hand sides offer a convenient all-slack starting basic feasible solution. Models involving ($=$) and/or ($\geq$) constraints do not.

The procedure for starting "ill-behaved" LPs with ($=$) and ($\geq$) constraints is to use **artificial variables** that play the role of slacks at the first iteration, and then dispose of them legitimately at a later iteration. Two closely related methods are introduced here: the M-method and the two-phase method.

3.4.1 *M*-Method

The M-method starts with the LP in equation form (Section 3.1). If equation i does not have a slack (or a variable that can play the role of a slack), an artificial variable, R_i, is added to form a starting solution similar to the convenient all-slack basic solution. However, because the artificial variables are not part of the original LP model, they are assigned a very high **penalty** in the objective function, thus forcing them (eventually) to equal zero in the optimum solution. This will always be the case if the problem has a feasible solution. The following rule shows how the penalty is assigned in the cases of maximization and minimization:

Penalty Rule for Artificial Variables.

Given M, a sufficiently large positive value (mathematically, $M \rightarrow \infty$), the objective coefficient of an artificial variable represents an appropriate **penalty** if:

$$\text{Artificial variable objective coefficient} = \begin{cases} -M, & \text{in maximization problems} \\ M, & \text{in minimization problems} \end{cases}$$

Example 3.4-1

$$\text{Minimize } z = 4x_1 + x_2$$

subject to

$$\begin{aligned} 3x_1 + x_2 &= 3 \\ 4x_1 + 3x_2 &\geq 6 \\ x_1 + 2x_2 &\leq 4 \\ x_1, x_2 &\geq 0 \end{aligned}$$

Using x_3 as a surplus in the second constraint and x_4 as a slack in the third constraint, the equation form of the problem is given as

$$\text{Minimize } z = 4x_1 + x_2$$

subject to

$$\begin{aligned} 3x_1 + x_2 \qquad\qquad &= 3 \\ 4x_1 + 3x_2 - x_3 \qquad &= 6 \\ x_1 + 2x_2 \qquad + x_4 &= 4 \\ x_1, x_2, x_3, x_4 &\geq 0 \end{aligned}$$

The third equation has its slack variable, x_4, but the first and second equations do not. Thus, we add the artificial variables R_1 and R_2 in the first two equations and penalize them in the objective function with $MR_1 + MR_2$ (because we are minimizing). The resulting LP is given as

$$\text{Minimize } z = 4x_1 + x_2 + MR_1 + MR_2$$

subject to

$$\begin{aligned} 3x_1 + x_2 \quad + R_1 \quad &= 3 \\ 4x_1 + 3x_2 - x_3 \quad + R_2 &= 6 \\ x_1 + 2x_2 \quad + x_4 \quad &= 4 \\ x_1, x_2, x_3, x_4, R_1, R_2 &\geq 0 \end{aligned}$$

The associated starting basic solution is now given by $(R_1, R_2, x_4) = (3, 6, 4)$.

From the standpoint of solving the problem on the computer, M must assume a numeric value. Yet, in practically all textbooks, including the first seven editions of this book, M is manipulated algebraically in all the simplex tableaus. The result is an added, and unnecessary, layer of difficulty which can be avoided simply by substituting an appropriate numeric value for M (which is what we do anyway when we use the computer). In this edition, we will break away from the long tradition of manipulating M algebraically and use a numerical substitution instead. The intent, of course, is to simplify the presentation without losing substance.

What value of M should we use? The answer depends on the data of the original LP. Recall that M must be sufficiently large *relative to the original objective coefficients* so it will act as a penalty that forces the artificial variables to zero level in the optimal solution. At the same time, since computers are the main tool for solving LPs, we do not want M to be too large (even though mathematically it should tend to infinity) because potential severe round-off error can result when very large values are manipulated with much smaller values. In the present example, the objective coefficients of x_1 and x_2 are 4 and 1, respectively. It thus appears reasonable to set $M = 100$.

Using $M = 100$, the starting simplex tableau is given as follows (for convenience, the z-column is eliminated because it does not change in all the iterations):

Basic	x_1	x_2	x_3	R_1	R_2	x_4	Solution
z	-4	-1	0	-100	-100	0	0
R_1	3	1	0	1	0	0	3
R_2	4	3	-1	0	1	0	6
x_4	1	2	0	0	0	1	4

Before proceeding with the simplex method computations, we need to make the z-row consistent with the rest of the tableau. Specifically, in the tableau, $x_1 = x_2 = x_3 = 0$, which yields the starting basic solution $R_1 = 3$, $R_2 = 6$, and $x_4 = 4$. This solution yields $z = 100 \times 3 + 100 \times 6 = 900$ (instead of 0, as the right-hand side of the z-row currently shows). This inconsistency stems from the fact that R_1 and R_2 have nonzero coefficients $(-100, -100)$ in the z-row (compare with the all-slack starting solution in Example 3.3-1, where the z-row coefficients of the slacks are zero).

We can eliminate this inconsistency by substituting out R_1 and R_2 in the z-row using the appropriate constraint equations. In particular, notice the highlighted elements $(= 1)$ in the R_1-row and the R_2-row. Multiplying *each* of R_1-row and R_2-row by 100 and adding the *sum* to the z-row will substitute out R_1 and R_2 in the objective row—that is,

$$\text{New } z\text{-row} = \text{Old } z\text{-row} + (100 \times R_1\text{-row} + 100 \times R_2\text{-row})$$

The modified tableau thus becomes (verify!)

Basic	x_1	x_2	x_3	R_1	R_2	x_4	Solution
z	696	399	−100	0	0	0	900
R_1	3	1	0	1	0	0	3
R_2	4	3	−1	0	1	0	6
x_4	1	2	0	0	0	1	4

Notice that $z = 900$, which is consistent now with the values of the starting basic feasible solution: $R_1 = 3$, $R_2 = 6$, and $x_4 = 4$.

The last tableau is ready for us to apply the simplex method using the simplex optimality and the feasibility conditions, exactly as we did in Section 3.3.2. Because we are minimizing the objective function, the variable x_1 having the most *positive* coefficient in the z-row (= 696) enters the solution. The minimum ratio of the feasibility condition specifies R_1 as the leaving variable (verify!).

Once the entering and the leaving variables have been determined, the new tableau can be computed by using the familiar Gauss-Jordan operations.

Basic	x_1	x_2	x_3	R_1	R_2	x_4	Solution
z	0	167	−100	−232	0	0	204
x_1	1	$\frac{1}{3}$	0	$\frac{1}{3}$	0	0	1
R_2	0	$\frac{5}{3}$	−1	$-\frac{4}{3}$	1	0	2
x_4	0	$\frac{5}{3}$	0	$-\frac{1}{3}$	0	1	3

The last tableau shows that x_2 and R_2 are the entering and leaving variables, respectively. Continuing with the simplex computations, two more iterations are needed to reach the optimum: $x_1 = \frac{2}{5}$, $x_2 = \frac{9}{5}$, $z = \frac{17}{5}$ (verify with TORA!).

Note that the artificial variables R_1 and R_2 leave the basic solution in the first and second iterations, a result that is consistent with the concept of penalizing them in the objective function.

Remarks. The use of the penalty M will not force an artificial variable to zero level in the final simplex iteration if the LP does not have a feasible solution (i.e., the constraints are not consistent). In this case, the final simplex iteration will include at least one artificial variable at a positive level. Section 3.5.4 explains this situation.

PROBLEM SET 3.4A

1. Use hand computations to complete the simplex iteration of Example 3.4-1 and obtain the optimum solution.
2. *TORA experiment.* Generate the simplex iterations of Example 3.4-1 using TORA's Iterations ⇒ M-method module (file toraEx3.4-1.txt). Compare the effect of using $M = 1$, $M = 10$, and $M = 1000$ on the solution. What conclusion can be drawn from this experiment?

3. In Example 3.4-1, identify the starting tableau for each of the following (independent) cases, and develop the associated z-row after substituting out all the artificial variables:

*__(a)__ The third constraint is $x_1 + 2x_2 \geq 4$.

*__(b)__ The second constraint is $4x_1 + 3x_2 \leq 6$.

(c) The second constraint is $4x_1 + 3x_2 = 6$.

(d) The objective function is to maximize $z = 4x_1 + x_2$.

4. Consider the following set of constraints:

$$
\begin{aligned}
-2x_1 + 3x_2 &= 3 \quad (1)\\
4x_1 + 5x_2 &\geq 10 \quad (2)\\
x_1 + 2x_2 &\leq 5 \quad (3)\\
6x_1 + 7x_2 &\leq 3 \quad (4)\\
4x_1 + 8x_2 &\geq 5 \quad (5)\\
x_1, x_2 &\geq 0
\end{aligned}
$$

For each of the following problems, develop the z-row after substituting out the artificial variables:

(a) Maximize $z = 5x_1 + 6x_2$ subject to (1), (3), and (4).

(b) Maximize $z = 2x_1 - 7x_2$ subject to (1), (2), (4), and (5).

(c) Minimize $z = 3x_1 + 6x_2$ subject to (3), (4), and (5).

(d) Minimize $z = 4x_1 + 6x_2$ subject to (1), (2), and (5).

(e) Minimize $z = 3x_1 + 2x_2$ subject to (1) and (5).

5. Consider the following set of constraints:

$$
\begin{aligned}
x_1 + x_2 + x_3 &= 7\\
2x_1 - 5x_2 + x_3 &\geq 10\\
x_1, x_2, x_3 &\geq 0
\end{aligned}
$$

Solve the problem for each of the following objective functions:

(a) Maximize $z = 2x_1 + 3x_2 - 5x_3$.

(b) Minimize $z = 2x_1 + 3x_2 - 5x_3$.

(c) Maximize $z = x_1 + 2x_2 + x_3$.

(d) Minimize $z = 4x_1 - 8x_2 + 3x_3$.

***6.** Consider the problem

$$\text{Maximize } z = 2x_1 + 4x_2 + 4x_3 - 3x_4$$

subject to

$$
\begin{aligned}
x_1 + x_2 + x_3 \qquad &= 4\\
x_1 + 4x_2 \qquad + x_4 &= 8\\
x_1, x_2, x_3, x_4, &\geq 0
\end{aligned}
$$

The problem shows that x_3 and x_4 can play the role of slacks for the two equations. They differ from slacks in that they have nonzero coefficients in the objective function. We can use x_3 and x_4 as starting variable, but, as in the case of artificial variables, they must be substituted out in the objective function before the simplex iterations are carried out. Solve the problem with x_3 and x_4 as the starting basic variables and without using any artificial variables.

7. Solve the following problem using x_3 and x_4 as starting basic feasible variables. As in Problem 6, do not use any artificial variables.

$$\text{Minimize } z = 3x_1 + 2x_2 + 3x_3$$

subject to

$$\begin{aligned} x_1 + 4x_2 + x_3 \quad\quad &\geq 7 \\ 2x_1 + x_2 \quad\quad + x_4 &\geq 10 \\ x_1, x_2, x_3, x_4 &\geq 0 \end{aligned}$$

8. Consider the problem

$$\text{Maximize } z = x_1 + 5x_2 + 3x_3$$

subject to

$$\begin{aligned} x_1 + 2x_2 + x_3 &= 3 \\ 2x_1 - x_2 \quad\quad &= 4 \\ x_1, x_2, x_3 &\geq 0 \end{aligned}$$

The variable x_3 plays the role of a slack. Thus, no artificial variable is needed in the first constraint. However, in the second constraint, an artificial variable is needed. Use this starting solution (i.e., x_3 in the first constraint and R_2 in the second constraint) to solve this problem.

9. Show how the M-method will indicate that the following problem has no feasible solution.

$$\text{Maximize } z = 2x_1 + 5x_2$$

subject to

$$\begin{aligned} 3x_1 + 2x_2 &\geq 6 \\ 2x_1 + x_2 &\leq 2 \\ x_1, x_2 &\geq 0 \end{aligned}$$

3.4.2 Two-Phase Method

In the M-method, the use of the penalty M, which by definition must be large relative to the actual objective coefficients of the model, can result in roundoff error that may impair the accuracy of the simplex calculations. The two-phase method alleviates this difficulty by eliminating the constant M altogether. As the name suggests, the method solves the LP in two phases: Phase I attempts to find a starting basic feasible solution, and, if one is found, Phase II is invoked to solve the original problem.

Summary of the Two-Phase Method

Phase I. Put the problem in equation form, and add the necessary artificial variables to the constraints (exactly as in the M-method) to secure a starting basic solution. Next, find a basic solution of the resulting equations that, regardless of whether the LP is maximization or minimization, *always* minimizes the sum of the artificial variables. If the minimum value of the sum is positive, the LP problem has no feasible solution, which ends the process (recall that a positive artificial variable signifies that an original constraint is not satisfied). Otherwise, proceed to Phase II.

Phase II. Use the feasible solution from Phase I as a starting basic feasible solution for the *original* problem.

Example 3.4-2

We use the same problem in Example 3.4-1.

Phase I

$$\text{Minimize } r = R_1 + R_2$$

subject to

$$\begin{aligned} 3x_1 + x_2 \quad + R_1 \quad &= 3 \\ 4x_1 + 3x_2 - x_3 \quad + R_2 \quad &= 6 \\ x_1 + 2x_2 \quad + x_4 &= 4 \\ x_1, x_2, x_3, x_4, R_1, R_2 &\geq 0 \end{aligned}$$

The associated tableau is given as

Basic	x_1	x_2	x_3	R_1	R_2	x_4	Solution
r	0	0	0	−1	−1	0	0
R_1	3	1	0	1	0	0	3
R_2	4	3	−1	0	1	0	6
x_4	1	2	0	0	0	1	4

As in the M-method, R_1 and R_2 are substituted out in the r-row by using the following computations:

$$\text{New } r\text{-row} = \text{Old } r\text{-row} + (1 \times R_1\text{-row} + 1 \times R_2\text{-row})$$

The new r-row is used to solve Phase I of the problem, which yields the following optimum tableau (verify with TORA's Iterations ⇒ Two-phase Method):

Basic	x_1	x_2	x_3	R_1	R_2	x_4	Solution
r	0	0	0	−1	−1	0	0
x_1	1	0	$\frac{1}{5}$	$\frac{3}{5}$	$-\frac{1}{5}$	0	$\frac{3}{5}$
x_2	0	1	$-\frac{3}{5}$	$-\frac{4}{5}$	$\frac{3}{5}$	0	$\frac{6}{5}$
x_4	0	0	1	1	−1	1	1

Because minimum $r = 0$, Phase I produces the basic feasible solution $x_1 = \frac{3}{5}$, $x_2 = \frac{6}{5}$, and $x_4 = 1$. At this point, the artificial variables have completed their mission, and we can eliminate their columns altogether from the tableau and move on to Phase II.

Phase II

After deleting the artificial columns, we write the *original* problem as

$$\text{Minimize } z = 4x_1 + x_2$$

subject to

$$\begin{aligned} x_1 \quad + \tfrac{1}{5}x_3 \qquad &= \tfrac{3}{5} \\ x_2 - \tfrac{3}{5}x_3 \qquad &= \tfrac{6}{5} \\ x_3 + x_4 &= 1 \\ x_1, x_2, x_3, x_4 &\geq 0 \end{aligned}$$

Essentially, Phase I is a procedure that transforms the original constraint equations in a manner that provides a starting basic feasible solution for the problem, if one exists. The tableau associated with Phase II problem is thus given as

Basic	x_1	x_2	x_3	x_4	Solution
z	-4	-1	0	0	0
x_1	1	0	$\frac{1}{5}$	0	$\frac{3}{5}$
x_2	0	1	$-\frac{3}{5}$	0	$\frac{6}{5}$
x_4	0	0	1	1	1

Again, because the basic variables x_1 and x_2 have nonzero coefficients in the z-row, they must be substituted out, using the following computations.

$$\text{New } z\text{-row} = \text{Old } z\text{-row} + (4 \times x_1\text{-row} + 1 \times x_2\text{-row})$$

The initial tableau of Phase II is thus given as

Basic	x_1	x_2	x_3	x_4	Solution
z	0	0	$\frac{1}{5}$	0	$\frac{18}{5}$
x_1	1	0	$\frac{1}{5}$	0	$\frac{3}{5}$
x_2	0	1	$-\frac{3}{5}$	0	$\frac{6}{5}$
x_4	0	0	1	1	1

Because we are minimizing, x_3 must enter the solution. Application of the simplex method will produce the optimum in one iteration (verify with TORA).

Remarks. Practically all commercial packages use the two-phase method to solve LP. The *M*-method with its potential adverse roundoff error is probably never used in practice. Its inclusion in this text is purely for historical reasons, because its development predates the development of the two-phase method.

The removal of the artificial variables and their columns at the end of Phase I can take place only when they are all *nonbasic* (as Example 3.4-2 illustrates). If one or more artificial variables are *basic* (at *zero* level) at the end of Phase I, then the following additional steps must be undertaken to remove them prior to the start of Phase II.

Step 1. Select a zero artificial variable to leave the basic solution and designate its row as the *pivot row*. The entering variable can be *any* nonbasic (nonartificial) variable with a *nonzero* (positive or negative) coefficient in the pivot row. Perform the associated simplex iteration.

Step 2. Remove the column of the (just-leaving) artificial variable from the tableau. If all the zero artificial variables have been removed, go to Phase II. Otherwise, go back to Step 1.

The logic behind Step 1 is that the feasibility of the remaining basic variables will not be affected when a zero artificial variable is made nonbasic regardless of whether the pivot element is positive or negative. Problems 5 and 6, Set 3.4b illustrate this situation. Problem 7 provides an additional detail about Phase I calculations.

PROBLEM SET 3.4B

*1. In Phase I, if the LP is of the maximization type, explain why we do not maximize the sum of the artificial variables in Phase I.

2. For each case in Problem 4, Set 3.4a, write the corresponding Phase I objective function.

3. Solve Problem 5, Set 3.4a, by the two-phase method.

4. Write Phase I for the following problem, and then solve (with TORA for convenience) to show that the problem has no feasible solution.

$$\text{Maximize } z = 2x_1 + 5x_2$$

subject to

$$3x_1 + 2x_2 \geq 6$$

$$2x_1 + x_2 \leq 2$$

$$x_1, x_2 \geq 0$$

5. Consider the following problem:

$$\text{Maximize } z = 2x_1 + 2x_2 + 4x_3$$

subject to

$$2x_1 + x_2 + x_3 \leq 2$$

$$3x_1 + 4x_2 + 2x_3 \geq 8$$

$$x_1, x_2, x_3 \geq 0$$

(a) Show that Phase I will terminate with an artificial *basic* variable at zero level (you may use TORA for convenience).

(b) Remove the zero artificial variable prior to the start of Phase II, then carry out Phase II iterations.

6. Consider the following problem:

$$\text{Maximize } z = 3x_1 + 2x_2 + 3x_3$$

subject to

$$\begin{aligned} 2x_1 + x_2 + x_3 &= 2 \\ x_1 + 3x_2 + x_3 &= 6 \\ 3x_1 + 4x_2 + 2x_3 &= 8 \\ x_1, x_2, x_3 &\geq 0 \end{aligned}$$

(a) Show that Phase I terminates with two zero artificial variables in the basic solution (use TORA for convenience).

(b) Show that when the procedure of Problem 5(b) is applied at the end of Phase I, only one of the two zero artificial variables can be made nonbasic.

(c) Show that the original constraint associated with the zero artificial variable that cannot be made nonbasic in (b) must be redundant—hence, its row and its column can be dropped altogether at the start of Phase II.

***7.** Consider the following LP:

$$\text{Maximize } z = 3x_1 + 2x_2 + 3x_3$$

subject to

$$\begin{aligned} 2x_1 + x_2 + x_3 &\leq 2 \\ 3x_1 + 4x_2 + 2x_3 &\geq 8 \\ x_1, x_2, x_3 &\geq 0 \end{aligned}$$

The optimal simplex tableau at the end of Phase I is given as

Basic	x_1	x_2	x_3	x_4	x_5	R	Solution
z	−5	0	−2	−1	−4	0	0
x_2	2	1	1	0	1	0	2
R	−5	0	−2	−1	−4	1	0

Explain why the nonbasic variables x_1, x_3, x_4, and x_5 can never assume positive values at the end of Phase II. Hence, conclude that their columns can dropped before we start Phase II. In essence, the removal of these variables reduces the constraint equations of the problem to $x_2 = 2$. This means that it will not be necessary to carry out Phase II at all, because the solution space is reduced to one point only.

8. Consider the LP model

$$\text{Minimize } z = 2x_1 - 4x_2 + 3x_3$$

subject to

$$\begin{aligned} 5x_1 - 6x_2 + 2x_3 &\geq 5 \\ -x_1 + 3x_2 + 5x_3 &\geq 8 \\ 2x_1 + 5x_2 - 4x_3 &\leq 4 \\ x_1, x_2, x_3 &\geq 0 \end{aligned}$$

Show how the inequalities can be modified to a set of equations that requires the use of a single artificial variable only (instead of two).

3.5 SPECIAL CASES IN THE SIMPLEX METHOD

This section considers four special cases that arise in the use of the simplex method.

1. Degeneracy
2. Alternative optima
3. Unbounded solutions
4. Nonexisting (or infeasible) solutions

Our interest in studying these special cases is twofold: (1) to present a *theoretical* explanation of these situations and (2) to provide a *practical* interpretation of what these special results could mean in a real-life problem.

3.5.1 Degeneracy

In the application of the feasibility condition of the simplex method, a tie for the minimum ratio may occur and can be broken arbitrarily. When this happens, at least one *basic* variable will be zero in the next iteration and the new solution is said to be **degenerate**.

There is nothing alarming about a degenerate solution, with the exception of a small theoretical inconvenience, called **cycling** or **circling**, which we shall discuss shortly. From the practical standpoint, the condition reveals that the model has at least one *redundant* constraint. To provide more insight into the practical and theoretical impacts of degeneracy, a numeric example is used.

Example 3.5-1 (Degenerate Optimal Solution)

$$\text{Maximize } z = 3x_1 + 9x_2$$

subject to

$$x_1 + 4x_2 \leq 8$$
$$x_1 + 2x_2 \leq 4$$
$$x_1, x_2 \geq 0$$

Given the slack variables x_3 and x_4, the following tableaus provide the simplex iterations of the problem:

Iteration	Basic	x_1	x_2	x_3	x_4	Solution
0	z	-3	-9	0	0	0
x_2 enters	x_3	1	4	1	0	8
x_3 leaves	x_4	1	2	0	1	4
1	z	$-\frac{3}{4}$	0	$\frac{9}{4}$	0	18
x_1 enters	x_2	$\frac{1}{4}$	1	$\frac{1}{4}$	0	2
x_4 leaves	x_4	$\frac{1}{2}$	0	$-\frac{1}{2}$	1	0
2	z	0	0	$\frac{3}{2}$	$\frac{3}{2}$	18
(optimum)	x_2	0	1	$\frac{1}{2}$	$-\frac{1}{2}$	2
	x_1	1	0	-1	2	0

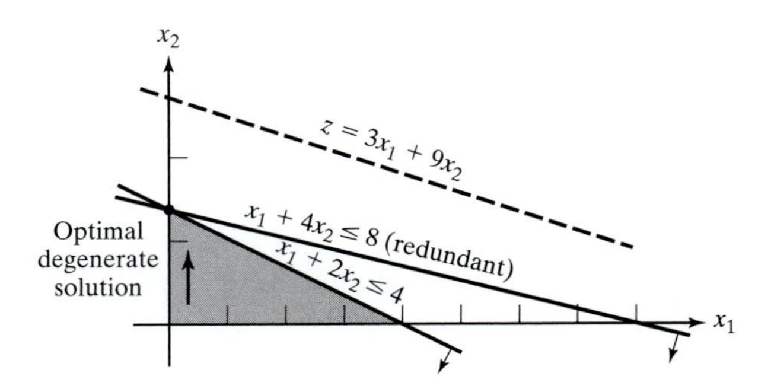

FIGURE 3.7
LP degeneracy in Example 3.5-1

In iteration 0, x_3 and x_4 tie for the leaving variable, leading to degeneracy in iteration 1 because the basic variable x_4 assumes a zero value. The optimum is reached in one additional iteration.

What is the practical implication of degeneracy? Look at the graphical solution in Figure 3.7. Three lines pass through the optimum point ($x_1 = 0, x_2 = 2$). Because this is a two-dimensional problem, the point is *overdetermined* and one of the constraints is redundant.[2] In practice, the mere knowledge that some resources are superfluous can be valuable during the implementation of the solution. The information may also lead to discovering irregularities in the construction of the model. Unfortunately, there are no efficient computational techniques for identifying the redundant constraints directly from the tableau.

From the theoretical standpoint, degeneracy has two implications. The first is the phenomenon of **cycling** or **circling**. Looking at simplex iterations 1 and 2, you will notice that the objective value does not improve ($z = 18$). It is thus possible for the simplex method to enter a repetitive sequence of iterations, never improving the objective value and never satisfying the optimality condition (see Problem 4, Set 3.5a). Although there are methods for eliminating cycling, these methods lead to drastic slowdown in computations. For this reason, most LP codes do not include provisions for cycling, relying on the fact that it is a rare occurrence in practice.

The second theoretical point arises in the examination of iterations 1 and 2. Both iterations, though differing in the basic-nonbasic categorization of the variables, yield identical values for all the variables and objective value—namely,

$$x_1 = 0, x_2 = 2, x_3 = 0, x_4 = 0, z = 18$$

Is it possible then to stop the computations at iteration 1 (when degeneracy first appears), even though it is not optimum? The answer is no, because the solution may be *temporarily* degenerate as Problem 2, Set 3.5a demonstrates.

---

[2]Redundancy generally implies that constraints can be removed without affecting the feasible solution space. A sometimes quoted counterexample is $x + y \leq 1, x \geq 1, y \geq 0$. Here, the removal of any one constraint will change the feasible space from a single point to a region. Suffice it to say, however, that this condition is true only if the solution space consists of a single feasible point, a highly unlikely occurrence in real-life LPs.

PROBLEM SET 3.5A

*1. Consider the graphical solution space in Figure 3.8. Suppose that the simplex iterations start at A and that the optimum solution occurs at D. Further, assume that the objective function is defined such that at A, x_1 enters the solution first.

(a) Identify (on the graph) the corner points that define the simplex method path to the optimum point.

(b) Determine the maximum possible number of simplex iterations needed to reach the optimum solution, assuming no cycling.

2. Consider the following LP:

$$\text{Maximize } z = 3x_1 + 2x_2$$

subject to

$$4x_1 - x_2 \leq 8$$
$$4x_1 + 3x_2 \leq 12$$
$$4x_1 + x_2 \leq 8$$
$$x_1, x_2 \geq 0$$

(a) Show that the associated simplex iterations are temporarily degenerate (you may use TORA for convenience).

(b) Verify the result by solving the problem graphically (TORA's Graphic module can be used here).

3. *TORA experiment.* Consider the LP in Problem 2.

(a) Use TORA to generate the simplex iterations. How many iterations are needed to reach the optimum?

(b) Interchange constraints (1) and (3) and re-solve the problem with TORA. How many iterations are needed to solve the problem?

(c) Explain why the numbers of iterations in (a) and (b) are different.

FIGURE 3.8

Solution space of Problem 1, Set 3.5a

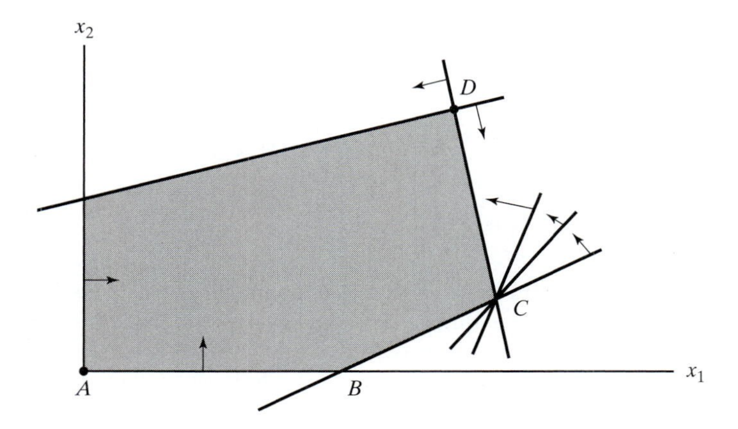

4. *TORA Experiment.* Consider the following LP (authored by E.M. Beale to demonstrate cycling):

$$\text{Maximize } z = \tfrac{3}{4}x_1 - 20x_2 + \tfrac{1}{2}x_3 - 6x_4$$

subject to

$$\begin{aligned}
\tfrac{1}{4}x_1 - 8x_2 - x_3 + 9x_4 &\le 0 \\
\tfrac{1}{2}x_1 - 12x_2 - \tfrac{1}{2}x_3 + 3x_4 &\le 0 \\
x_3 &\le 1 \\
x_1, x_2, x_3, x_4 &\ge 0
\end{aligned}$$

From TORA's SOLVE/MODIFY menu, select Solve ⇒ Algebraic ⇒ Iterations ⇒ All-slack. Next, "thumb" through the successive simplex iterations using the command Next iteration (do not use All iterations, because the simplex method will then cycle indefinitely). You will notice that the starting all-slack basic feasible solution at iteration 0 will reappear identically in iteration 6. This example illustrates the occurrence of cycling in the simplex iterations and the possibility that the algorithm may never converge to the optimum solution.

It is interesting that cycling will not occur in this example if all the coefficients in this LP are converted to integer values by using proper multiples (try it!).

3.5.2 Alternative Optima

When the objective function is parallel to a nonredundant **binding constraint** (i.e., a constraint that is satisfied as an equation at the optimal solution), the objective function can assume the same optimal value at more than one solution point, thus giving rise to alternative optima. The next example shows that there is an *infinite* number of such solutions. It also demonstrates the practical significance of encountering such solutions.

Example 3.5-2 (Infinite Number of Solutions)

$$\text{Maximize } z = 2x_1 + 4x_2$$

subject to

$$\begin{aligned}
x_1 + 2x_2 &\le 5 \\
x_1 + x_2 &\le 4 \\
x_1, x_2 &\ge 0
\end{aligned}$$

Figure 3.9 demonstrates how alternative optima can arise in the LP model when the objective function is parallel to a binding constraint. Any point on the *line segment BC* represents an alternative optimum with the same objective value $z = 10$.

The iterations of the model are given by the following tableaus.

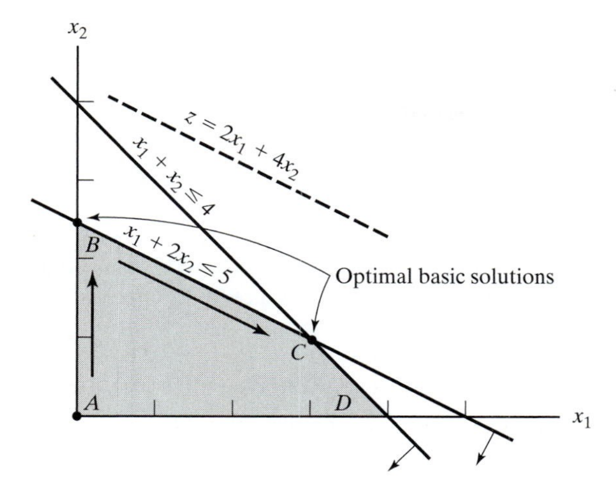

FIGURE 3.9

LP alternative optima in Example 3.5-2

Iteration	Basic	x_1	x_2	x_3	x_4	Solution
0	z	-2	-4	0	0	0
x_2 enters	x_3	1	2	1	0	5
x_3 leaves	x_4	1	1	0	1	4
1 (optimum)	z	0	0	2	0	10
x_1 enters	x_2	$\frac{1}{2}$	1	$\frac{1}{2}$	0	$\frac{5}{2}$
x_4 leaves	x_4	$\frac{1}{2}$	0	$-\frac{1}{2}$	1	$\frac{3}{2}$
2	z	0	0	2	0	10
(alternative optimum)	x_2	0	1	1	-1	1
	x_1	1	0	-1	2	3

Iteration 1 gives the optimum solution $x_1 = 0, x_2 = \frac{5}{2}$, and $z = 10$, which coincides with point B in Figure 3.9. How do we know from this tableau that alternative optima exist? Look at the z-equation coefficients of the *non*basic variables in iteration 1. The coefficient of nonbasic x_1 is zero, indicating that x_1 can enter the basic solution without changing the value of z, but causing a change in the values of the variables. Iteration 2 does just that—letting x_1 enter the basic solution and forcing x_4 to leave. The new solution point occurs at $C(x_1 = 3, x_2 = 1, z = 10)$. (TORA's Iterations option allows determining one alternative optimum at a time.)

The simplex method determines only the two corner points B and C. Mathematically, we can determine all the points (x_1, x_2) on the line segment BC as a nonnegative weighted average of points B and C. Thus, given

$$B: x_1 = 0, x_2 = \frac{5}{2}$$

$$C: x_1 = 3, x_2 = 1$$

then all the points on the line segment BC are given by

$$\left.\begin{aligned}\hat{x}_1 &= \alpha(0) + (1-\alpha)(3) = 3 - 3\alpha \\ \hat{x}_2 &= \alpha\left(\tfrac{5}{2}\right) + (1-\alpha)(1) = 1 + \tfrac{3}{2}\alpha\end{aligned}\right\}, 0 \leq \alpha \leq 1$$

When $\alpha = 0$, $(\hat{x}_1, \hat{x}_2) = (3,1)$, which is point C. When $\alpha = 1$, $(\hat{x}_1, \hat{x}_2) = \left(0, \frac{5}{2}\right)$, which is point B. For values of α between 0 and 1, $(\hat{x}_1, \hat{x}_2)$ lies between B and C.

Remarks. In practice, alternative optima are useful because we can choose from many solutions without experiencing deterioration in the objective value. For instance, in the present example, the solution at B shows that activity 2 only is at a positive level, whereas at C both activities are positive. If the example represents a product-mix situation, there may be advantages in producing two products rather than one to meet market competition. In this case, the solution at C may be more appealing.

PROBLEM SET 3.5B

***1.** For the following LP, identify three alternative optimal basic solutions, and then write a general expression for all the nonbasic alternative optima comprising these three basic solutions.

$$\text{Maximize } z = x_1 + 2x_2 + 3x_3$$

subject to

$$\begin{aligned} x_1 + 2x_2 + 3x_3 &\leq 10 \\ x_1 + x_2 \quad\quad &\leq 5 \\ x_1 \quad\quad\quad\quad &\leq 1 \\ x_1, x_2, x_3 &\geq 0 \end{aligned}$$

Note: Although the problem has more than three alternative basic solution optima, you are only required to identify three of them. You may use TORA for convenience.

2. Solve the following LP:

$$\text{Maximize } z = 2x_1 - x_2 + 3x_3$$

subject to

$$\begin{aligned} x_1 - x_2 + 5x_3 &\leq 10 \\ 2x_1 - x_2 + 3x_3 &\leq 40 \\ x_1, x_2, x_3 &\geq 0 \end{aligned}$$

From the optimal tableau, show that all the alternative optima are not corner points (i.e., nonbasic). Give a two-dimensional graphical demonstration of the type of solution space and objective function that will produce this result. (You may use TORA for convenience.)

3. For the following LP, show that the optimal solution is degenerate and that none of the alternative solutions are corner points (you may use TORA for convenience).

$$\text{Maximize } z = 3x_1 + x_2$$

subject to

$$x_1 + 2x_2 \leq 5$$
$$x_1 + x_2 - x_3 \leq 2$$
$$7x_1 + 3x_2 - 5x_3 \leq 20$$
$$x_1, x_2, x_3 \geq 0$$

3.5.3 Unbounded Solution

In some LP models, the values of the variables may be increased indefinitely without violating any of the constraints—meaning that the solution space is *unbounded* in at least one variable. As a result, the objective value may increase (maximization case) or decrease (minimization case) indefinitely. In this case, both the solution space and the optimum objective value are unbounded.

Unboundedness points to the possibility that the model is poorly constructed. The most likely irregularity in such models is that one or more nonredundant constraints have not been accounted for, and the parameters (constants) of some constraints may not have been estimated correctly.

The following examples show how unboundedness, in both the solution space and the objective value, can be recognized in the simplex tableau.

Example 3.5-3 (Unbounded Objective Value)

$$\text{Maximize } z = 2x_1 + x_2$$

subject to

$$x_1 - x_2 \leq 10$$
$$2x_1 \leq 40$$
$$x_1, x_2 \geq 0$$

Starting Iteration

Basic	x_1	x_2	x_3	x_4	Solution
z	-2	-1	0	0	0
x_3	1	-1	1	0	10
x_4	2	0	0	1	40

In the starting tableau, both x_1 and x_2 have negative z-equation coefficients. Hence either one can improve the solution. Because x_1 has the most negative coefficient, it is normally selected as the entering variable. However, *all* the *constraint* coefficients under x_2 (i.e., the denominators of the ratios of the feasibility condition) are *negative* or *zero*. This means that there is no leaving variable and that x_2 can be increased indefinitely without violating any of the constraints (compare with the graphical interpretation of the minimum ratio in Figure 3.5). Because each unit increase in x_2 will increase z by 1, an infinite increase in x_2 leads to an infinite increase in z. Thus, the problem has no bounded solution. This result can be seen in Figure 3.10. The solution space is unbounded in the direction of x_2, and the value of z can be increased indefinitely.

Remarks. What would have happened if we had applied the strict optimality condition that calls for x_1 to enter the solution? The answer is that a succeeding tableau would eventually have led to an entering variable with the same characteristics as x_2. See Problem 1, Set3.5c.

PROBLEM SET 3.5C

1. *TORA Experiment.* Solve Example 3.5-3 using TORA's Iterations option and show that even though the solution starts with x_1 as the entering variable (per the optimality condition), the simplex algorithm will point eventually to an unbounded solution.

*2. Consider the LP:

$$\text{Maximize } z = 20x_1 + 10x_2 + x_3$$

subject to

$$\begin{aligned} 3x_1 - 3x_2 + 5x_3 &\leq 50 \\ x_1 \qquad + x_3 &\leq 10 \\ x_1 - x_2 + 4x_3 &\leq 20 \\ x_1, x_2, x_3 &\geq 0 \end{aligned}$$

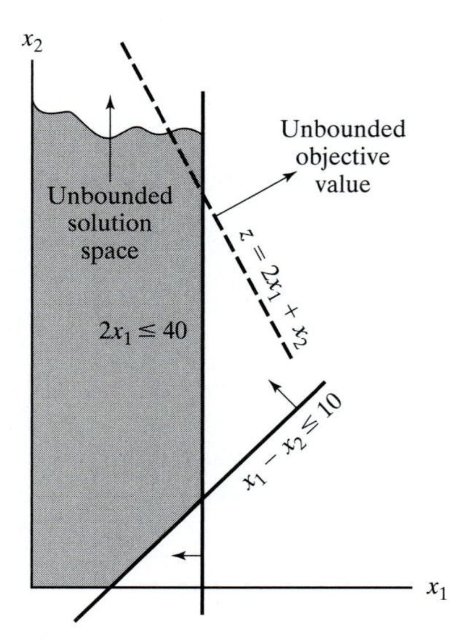

FIGURE 3.10
LP unbounded solution in Example 3.5-3

(a) By inspecting the constraints, determine the direction (x_1, x_2, or x_3) in which the solution space is unbounded.

(b) Without further computations, what can you conclude regarding the optimum objective value?

3. In some ill-constructed LP models, the solution space may be unbounded even though the problem may have a bounded objective value. Such an occurrence can point only to irregularities in the construction of the model. In large problems, it may be difficult to detect unboundedness by inspection. Devise a procedure for determining whether or not a solution space is unbounded.

3.5.4 Infeasible Solution

LP models with inconsistent constraints have no feasible solution. This situation can never occur if *all* the constraints are of the type $\leq$ with nonnegative right-hand sides because the slacks provide a feasible solution. For other types of constraints, we use artificial variables. Although the artificial variables are penalized in the objective function to force them to zero at the optimum, this can occur only if the model has a feasible space. Otherwise, at least one artificial variable will be *positive* in the optimum iteration. From the practical standpoint, an infeasible space points to the possibility that the model is not formulated correctly.

Example 3.5-4 (Infeasible Solution Space)

Consider the following LP:

$$\text{Maximize } z = 3x_1 + 2x_2$$

subject to

$$2x_1 + x_2 \leq 2$$

$$3x_1 + 4x_2 \geq 12$$

$$x_1, x_2 \geq 0$$

Using the penalty $M = 100$ for the artificial variable R, the following tableaux provide the simplex iterations of the model.

Iteration	Basic	x_1	x_2	x_4	x_3	R	Solution
0	z	−303	−402	100	0	0	−1200
x_2 enters	x_3	2	1	0	1	0	2
x_3 leaves	R	3	4	−1	0	1	12
1	z	501	0	100	402	0	−396
(pseudo-optimum)	x_2	2	1	0	1	0	2
	R	−5	0	−1	−4	1	4

Optimum iteration 1 shows that the artificial variable R is *positive* $(= 4)$, which indicates that the problem is infeasible. Figure 3.11 demonstrates the infeasible solution space. By allowing

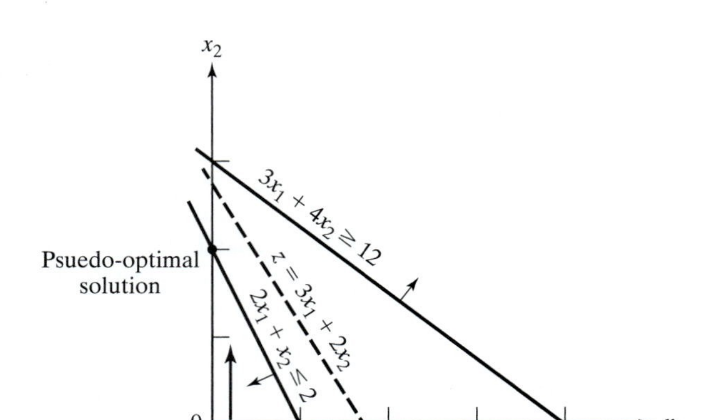

FIGURE 3.11

Infeasible solution of Example 3.5-4

the artificial variable to be positive, the simplex method, in essence, has reversed the direction of the inequality from $3x_1 + 4x_2 \geq 12$ to $3x_1 + 4x_2 \leq 12$ (can you explain how?). The result is what we may call a **pseudo-optimal** solution.

PROBLEM SET 3.5D

***1.** Toolco produces three types of tools, *T1*, *T2*, and *T3*. The tools use two raw materials, *M1* and *M2*, according to the data in the following table:

	Number of units of raw materials per tool		
Raw material	*T1*	*T2*	*T3*
M1	3	5	6
M2	5	3	4

The available daily quantities of raw materials *M1* and *M2* are 1000 units and 1200 units, respectively. The marketing department informed the production manager that according to their research, the daily demand for all three tools must be at least 500 units. Will the manufacturing department be able to satisfy the demand? If not, what is the most Toolco can provide of the three tools?

2. *TORA Experiment.* Consider the LP model

$$\text{Maximize } z = 3x_1 + 2x_1 + 3x_3$$

subject to

$$2x_1 + x_2 + x_3 \leq 2$$
$$3x_1 + 4x_2 + 2x_3 \geq 8$$
$$x_1, x_2, x_3 \geq 0$$

Use TORA's Iterations ⇒ M-Method to show that the optimal solution includes an artificial basic variable, but at zero level. Does the problem have a *feasible* optimal solution?

3.6 SENSITIVITY ANALYSIS

In LP, the parameters (input data) of the model can change within certain limits without causing the optimum solution to change. This is referred to as *sensitivity analysis*, and will be the subject matter of this section. Later, in Chapter 4, we will study *post-optimal analysis* which deals with determining the new optimum solution resulting from making targeted changes in the input data.

In LP models, the parameters are usually not exact. With sensitivity analysis, we can ascertain the impact of this uncertainty on the quality of the optimum solution. For example, for an estimated unit profit of a product, if sensitivity analysis reveals that the optimum remains the same for a $\pm 10\%$ change in the unit profit, we can conclude that the solution is more robust than in the case where the indifference range is only $\pm 1\%$.

We will start with the more concrete graphical solution to explain the basics of sensitivity analysis. These basics will then be extended to the general LP problem using the simplex tableau results.

3.6.1 Graphical Sensitivity Analysis

This section demonstrates the general idea of sensitivity analysis. Two cases will be considered:

1. Sensitivity of the optimum solution to changes in the availability of the resources (right-hand side of the constraints).
2. Sensitivity of the optimum solution to changes in unit profit or unit cost (coefficients of the objective function).

We will consider the two cases separately, using examples of two-variable graphical LPs.

Example 3.6-1 (Changes in the Right-Hand Side)

JOBCO produces two products on two machines. A unit of product 1 requires 2 hours on machine 1 and 1 hour on machine 2. For product 2, a unit requires 1 hour on machine 1 and 3 hours on machine 2. The revenues per unit of products 1 and 2 are \$30 and \$20, respectively. The total daily processing time available for each machine is 8 hours.

Letting x_1 and x_2 represent the daily number of units of products 1 and 2, respectively, the LP model is given as

$$\text{Maximize } z = 30x_1 + 20x_2$$

subject to

$$\begin{aligned} 2x_1 + x_2 &\leq 8 \quad \text{(Machine 1)} \\ x_1 + 3x_2 &\leq 8 \quad \text{(Machine 2)} \\ x_1, x_2 &\geq 0 \end{aligned}$$

Figure 3.12 illustrates the change in the optimum solution when changes are made in the capacity of machine 1. If the daily capacity is increased from 8 hours to 9 hours, the new optimum will occur at point G. The rate of change in optimum z resulting from changing machine 1 capacity from 8 hours to 9 hours can be computed as follows:

$$\left(\begin{array}{c}\text{Rate of revenue change} \\ \text{resulting from increasing} \\ \text{machine 1 capacity by 1 hr} \\ (\text{point } C \text{ to point } G)\end{array}\right) = \frac{z_G - z_C}{(\text{Capacity change})} = \frac{142 - 128}{9 - 8} = \$14.00/\text{hr}$$

The computed rate provides a *direct link* between the model input (resources) and its output (total revenue) that represents the **unit worth of a resource** (in \$/hr)—that is, the change in the optimal objective value per unit change in the availability of the resource (machine capacity). This means that a unit increase (decrease) in machine 1 capacity will increase (decrease) revenue by \$14.00. Although *unit worth of a resource* is an apt description of the rate of change of the objective function, the technical name **dual** or **shadow price** is now standard in the LP literature and all software packages and, hence, will be used throughout the book.

FIGURE 3.12

Graphical sensitivity of optimal solution to changes in the availability of resources (right-hand side of the constraints)

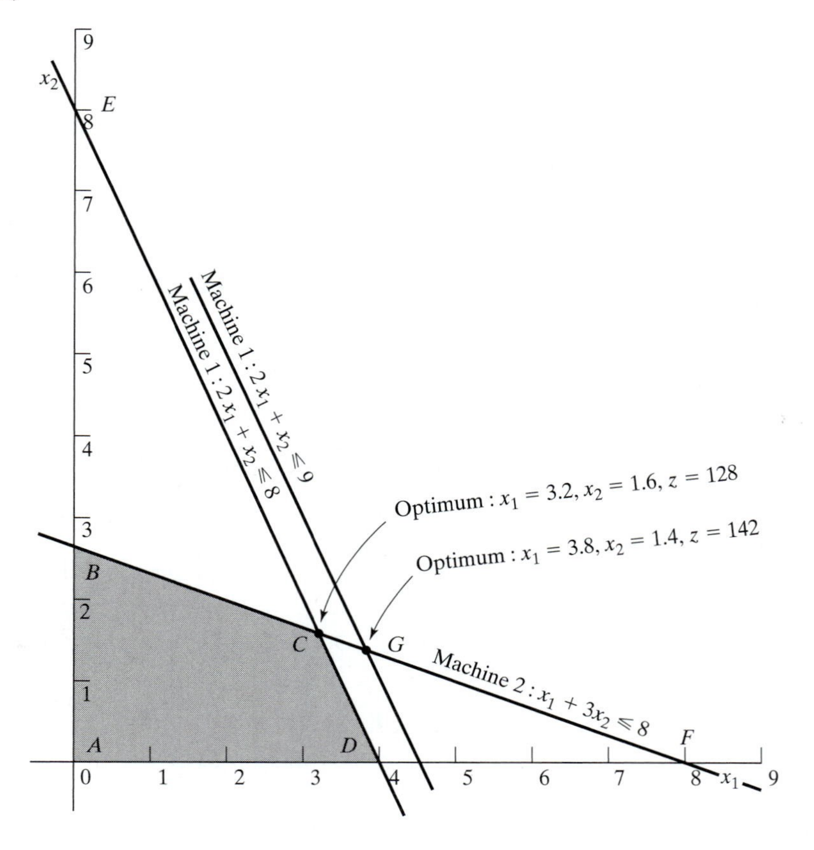

Looking at Figure 3.12, we can see that the dual price of \$14.00/hr remains valid for changes (increases or decreases) in machine 1 capacity that move its constraint parallel to itself to any point on the line segment BF. This means that the range of applicability of the given dual price can be computed as follows:

$$\text{Minimum machine 1 capacity [at } B = (0, 2.67)] = 2 \times 0 + 1 \times 2.67 = 2.67 \text{ hr}$$

$$\text{Maximum machine 1 capacity [at } F = (8, 0)] = 2 \times 8 + 1 \times 0 = 16 \text{ hr}$$

We can thus conclude that the dual price of \$14.00/hr will remain valid for the range

$$2.67 \text{ hrs} \leq \text{Machine 1 capacity} \leq 16 \text{ hrs}$$

Changes outside this range will produce a different dual price (worth per unit).

Using similar computations, you can verify that the dual price for machine 2 capacity is \$2.00/hr and it remains valid for changes (increases or decreases) that move its constraint parallel to itself to any point on the line segment DE, which yields the following limits:

$$\text{Minimum machine 2 capacity [at } D = (4, 0)] = 1 \times 4 + 3 \times 0 = 4 \text{ hr}$$

$$\text{Maximum machine 2 capacity [at } E = (8, 0)] = 1 \times 0 + 3 \times 8 = 24 \text{ hr}$$

The conclusion is that the dual price of \$2.00/hr for machine 2 will remain applicable for the range

$$4 \text{ hr} \leq \text{Machine 2 capacity} \leq 24 \text{ hr}$$

The computed limits for machine 1 and 2 are referred to as the **feasibility ranges**. All software packages provide information about the dual prices and their feasibility ranges. Section 3.6.4 shows how AMPL, Solver, and TORA generate this information.

The dual prices allow making economic decisions about the LP problem, as the following questions demonstrate:

Question 1. If JOBCO can increase the capacity of both machines, which machine should receive higher priority?

The dual prices for machines 1 and 2 are \$14.00/hr and \$2.00/hr. This means that each additional hour of machine 1 will increase revenue by \$14.00, as opposed to only \$2.00 for machine 2. Thus, priority should be given to machine 1.

Question 2. A suggestion is made to increase the capacities of machines 1 and 2 at the additional cost of \$10/hr. Is this advisable?

For machine 1, the additional net revenue per hour is $14.00 - 10.00 = \$4.00$ and for machine 2, the net is $\$2.00 - \$10.00 = -\$8.00$. Hence, only the capacity of machine 1 should be increased.

Question 3. If the capacity of machine 1 is increased from the present 8 hours to 13 hours, how will this increase impact the optimum revenue?

The dual price for machine 1 is \$14.00 and is applicable in the range (2.67, 16) hr. The proposed increase to 13 hours falls within the feasibility range. Hence, the increase in revenue is $\$14.00(13 - 8) = \$70.00$, which means that the total revenue will be increased to (current revenue + change in revenue) $= 128 + 70 = \$198.00$.

Question 4. Suppose that the capacity of machine 1 is increased to 20 hours, how will this increase impact the optimum revenue?

The proposed change is outside the range (2.67, 16) hr for which the dual price of \$14.00 remains applicable. Thus, we can only make an immediate conclusion regarding an increase up to 16 hours. Beyond that, further calculations are needed to find the answer (see Chapter 4). Remember that falling outside the feasibility range does *not* mean that the problem has no solution. It only means that we do not have sufficient information to make an *immediate* decision.

Question 5. We know that the change in the optimum objective value equals (dual price $\times$ change in resource) so long as the change in the resource is within the feasibility range. What about the associated optimum values of the variables?

The optimum values of the variables will definitely change. However, the level of information we have from the graphical solution is not sufficient to determine the new values. Section 3.6.2, which treats the sensitivity problem algebraically, provides this detail.

PROBLEM SET 3.6A

1. A company produces two products, A and B. The unit revenues are \$2 and \$3, respectively. Two raw materials, $M1$ and $M2$, used in the manufacture of the two products have respective daily availabilities of 8 and 18 units. One unit of A uses 2 units of $M1$ and 2 units of $M2$, and 1 unit of B uses 3 units of $M1$ and 6 units of $M2$.

- **(a)** Determine the dual prices of $M1$ and $M2$ and their feasibility ranges.
- **(b)** Suppose that 4 additional units of $M1$ can be acquired at the cost of 30 cents per unit. Would you recommend the additional purchase?
- **(c)** What is the most the company should pay per unit of $M2$?
- **(d)** If $M2$ availability is increased by 5 units, determine the associated optimum revenue.

***2.** Wild West produces two types of cowboy hats. A Type 1 hat requires twice as much labor time as a Type 2. If all the available labor time is dedicated to Type 2 alone, the company can produce a total of 400 Type 2 hats a day. The respective market limits for the two types are 150 and 200 hats per day. The revenue is \$8 per Type 1 hat and \$5 per Type 2 hat.

- **(a)** Use the graphical solution to determine the number of hats of each type that maximizes revenue.
- **(b)** Determine the dual price of the production capacity (in terms of the Type 2 hat) and the range for which it is applicable.
- **(c)** If the daily demand limit on the Type 1 hat is decreased to 120, use the dual price to determine the corresponding effect on the optimal revenue.
- **(d)** What is the dual price of the market share of the Type 2 hat? By how much can the market share be increased while yielding the computed worth per unit?

Example 3.6-2 (Changes in the Objective Coefficients)

Figure 3.13 shows the graphical solution space of the JOBCO problem presented in Example 3.6-1. The optimum occurs at point C ($x_1 = 3.2$, $x_2 = 1.6$, $z = 128$). Changes in revenue units (i.e., objective-function coefficients) will change the slope of z. However, as can be seen from the figure, the optimum solution will remain at point C so long as the objective function lies between lines BF and DE, the two constraints that define the optimum point. This means that there is a range for the coefficients of the objective function that will keep the optimum solution unchanged at C.

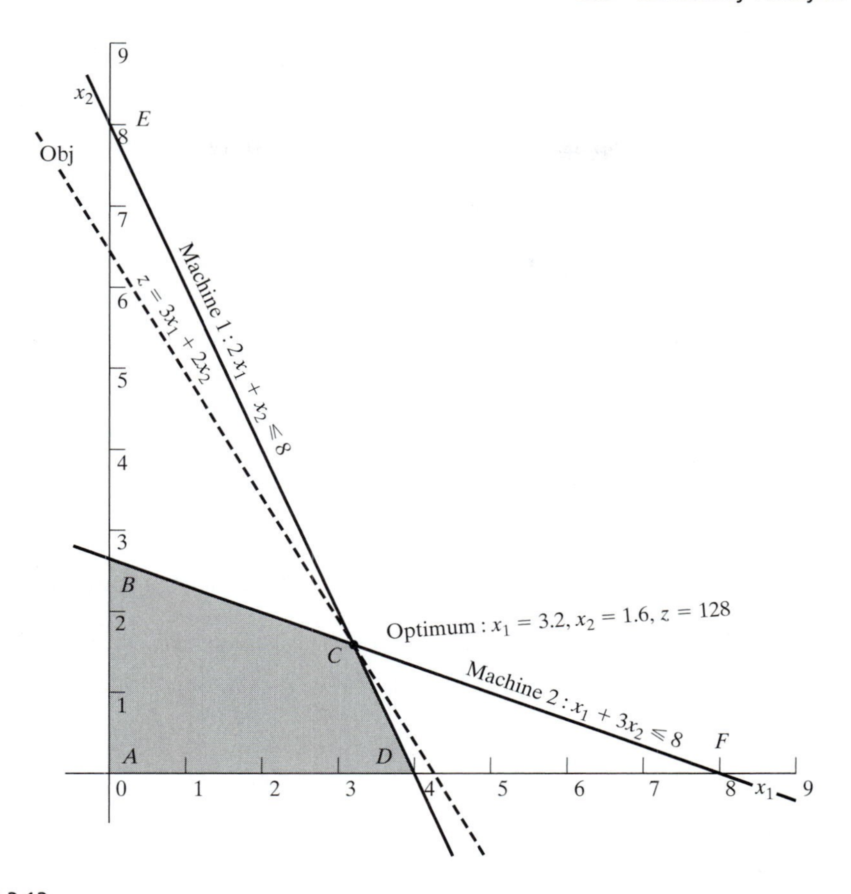

FIGURE 3.13

Graphical sensitivity of optimal solution to changes in the revenue units (coefficients of the objective function)

We can write the objective function in the general format

$$\text{Maximize } z = c_1x_1 + c_2x_2$$

Imagine now that the line z is pivoted at C and that it can rotate clockwise and counterclockwise. The optimum solution will remain at point C so long as $z = c_1x_1 + c_2x_2$ lies between the two lines $x_1 + 3x_2 = 8$ and $2x_1 + x_2 = 8$. This means that the ratio $\frac{c_1}{c_2}$ can vary between $\frac{1}{3}$ and $\frac{2}{1}$, which yields the following condition:

$$\frac{1}{3} \leq \frac{c_1}{c_2} \leq \frac{2}{1} \quad \text{or} \quad .333 \leq \frac{c_1}{c_2} \leq 2$$

This information can provide immediate answers regarding the optimum solution as the following questions demonstrate:

Question 1. Suppose that the unit revenues for products 1 and 2 are changed to \$35 and \$25, respectively. Will the current optimum remain the same?

The new objective function is

$$\text{Maximize } z = 35x_1 + 25x_2$$

The solution at C will remain optimal because $\frac{c_1}{c_2} = \frac{35}{25} = 1.4$ remains within the optimality range (.333, 2). When the ratio falls outside this range, additional calculations are needed to find the new optimum (see Chapter 4). Notice that although the values of the variables at the optimum point C remain unchanged, the optimum value of z changes to $35 \times (3.2) + 25 \times (1.6) = \$152.00$.

Question 2. Suppose that the unit revenue of product 2 is fixed at its current value of $c_2 = \$20.00$. What is the associated range for c_1, the unit revenue for product 1 that will keep the optimum unchanged?

Substituting $c_2 = 20$ in the condition $\frac{1}{3} \leq \frac{c_1}{c_2} \leq 2$, we get

$$\tfrac{1}{3} \times 20 \leq c_1 \leq 2 \times 20$$

Or

$$6.67 \leq c_1 \leq 40$$

This range is referred to as the **optimality range** for c_1, and it implicitly assumes that c_2 is fixed at $20.00.

We can similarly determine the *optimality range* for c_2 by fixing the value of c_1 at $30.00. Thus,

$$c_2 \leq 30 \times 3 \text{ and } c_2 \geq \tfrac{30}{2}$$

Or

$$15 \leq c_2 \leq 90$$

As in the case of the right-hand side, all software packages provide the optimality ranges. Section 3.6.4 shows how AMPL, Solver, and TORA generate these results.

Remark. Although the material in this section has dealt only with two variables, the results lay the foundation for the development of sensitivity analysis for the general LP problem in Sections 3.6.2 and 3.6.3.

PROBLEM SET 3.6B

1. Consider Problem 1, Set 3.6a.
   - **(a)** Determine the optimality condition for $\frac{c_A}{c_B}$ that will keep the optimum unchanged.
   - **(b)** Determine the optimality ranges for c_A and c_B, assuming that the other coefficient is kept constant at its present value.
   - **(c)** If the unit revenues c_A and c_B are changed simultaneously to $5 and $4, respectively, determine the new optimum solution.
   - **(d)** If the changes in (c) are made one at a time, what can be said about the optimum solution?
2. In the Reddy Mikks model of Example 2.2-1;
   - **(a)** Determine the range for the ratio of the unit revenue of exterior paint to the unit revenue of interior paint.

(b) If the revenue per ton of exterior paint remains constant at \$5000 per ton, determine the maximum unit revenue of interior paint that will keep the present optimum solution unchanged.

(c) If for marketing reasons the unit revenue of interior paint must be reduced to \$3000, will the current optimum production mix change?

***3.** In Problem 2, Set 3.6a:

(a) Determine the optimality range for the unit revenue ratio of the two types of hats that will keep the current optimum unchanged.

(b) Using the information in (b), will the optimal solution change if the revenue per unit is the same for both types?

3.6.2 Algebraic Sensitivity Analysis—Changes in the Right-Hand Side

In Section 3.6.1, we used the graphical solution to determine the *dual prices* (the unit worths of resources) and their feasibility ranges. This section extends the analysis to the general LP model. A numeric example (the TOYCO model) will be used to facilitate the presentation.

Example 3.6-2 (TOYCO Model)

TOYCO assembles three types of toys—trains, trucks, and cars—using three operations. The daily limits on the available times for the three operations are 430, 460, and 420 minutes, respectively, and the revenues per unit of toy train, truck, and car are \$3, \$2, and \$5, respectively. The assembly times per train at the three operations are 1, 3, and 1 minutes, respectively. The corresponding times per train and per car are (2, 0, 4) and (1, 2, 0) minutes (a zero time indicates that the operation is not used).

Letting x_1, x_2, and x_3 represent the daily number of units assembled of trains, trucks, and cars, respectively, the associated LP model is given as:

$$\text{Maximize } z = 3x_1 + 2x_2 + 5x_3$$

subject to

$$\begin{aligned} x_1 + 2x_2 + x_3 &\leq 430 \text{ (Operation 1)} \\ 3x_1 + 2x_3 &\leq 460 \text{ (Operation 2)} \\ x_1 + 4x_2 &\leq 420 \text{ (Operation 3)} \\ x_1, x_2, x_3 &\geq 0 \end{aligned}$$

Using x_4, x_5, and x_6 as the slack variables for the constraints of operations 1, 2, and 3, respectively, the optimum tableau is

Basic	x_1	x_2	x_3	x_4	x_5	x_6	Solution
z	4	0	0	1	2	0	1350
x_2	$-\frac{1}{4}$	1	0	$\frac{1}{2}$	$-\frac{1}{4}$	0	100
x_3	$\frac{3}{2}$	0	1	0	$\frac{1}{2}$	0	230
x_6	2	0	0	-2	1	1	20

The solution recommends manufacturing 100 trucks and 230 cars but no trains. The associated revenue is $1350.

Determination of Dual Prices. The constraints of the model after adding the slack variables x_4, x_5, and x_6 can be written as follows:

$$\begin{aligned} x_1 + 2x_2 + x_3 + x_4 &= 430 \quad \text{(Operation 1)} \\ 3x_1 + 2x_3 + x_5 &= 460 \quad \text{(Operation 2)} \\ x_1 + 4x_2 + x_6 &= 420 \quad \text{(Operation 3)} \end{aligned}$$

or

$$\begin{aligned} x_1 + 2x_2 + x_3 &= 430 - x_4 \quad \text{(Operation 1)} \\ 3x_1 + 2x_3 &= 460 - x_5 \quad \text{(Operation 2)} \\ x_1 + 4x_2 &= 420 - x_6 \quad \text{(Operation 3)} \end{aligned}$$

With this representation, the slack variables have the same units (minutes) as the operation times. Thus, we can say that a one-minute *decrease* in the slack variable is equivalent to a one-minute *increase* in the operation time.

We can use the information above to determine the *dual prices* from the z-equation in the optimal tableau:

$$z + 4x_1 + x_4 + 2x_5 + 0x_6 = 1350$$

This equation can be written as

$$\begin{aligned} z &= 1350 - 4x_1 - x_4 - 2x_5 - 0x_6 \\ &= 1350 - 4x_1 + 1(-x_4) + 2(-x_5) + 0(-x_6) \end{aligned}$$

Given that a *decrease* in the value of a slack variable is equivalent to an *increase* in its operation time, we get

$$\begin{aligned} z = 1350 - 4x_1 &+ 1 \times \text{(increase in operation 1 time)} \\ &+ 2 \times \text{(increase in operation 2 time)} \\ &+ 0 \times \text{(increase in operation 3 time)} \end{aligned}$$

This equation reveals that (1) a one-minute increase in operation 1 time increases z by \$1, (2) a one-minute increase in operation 2 time increases z by \$2, and (3) a one-minute increase in operation 3 time does not change z.

To summarize, the z-row in the optimal tableau:

Basic	x_1	x_2	x_3	x_4	x_5	x_6	Solution
z	4	0	0	1	2	0	1350

yields directly the dual prices, as the following table shows:

Resource	Slack variable	Optimal z-equation coefficient of slack variable	Dual price
Operation 1	x_4	1	\$1/min
Operation 2	x_5	2	\$2/min
Operation 3	x_6	0	\$0/min

The zero dual price for operation 3 means that there is no economic advantage in allocating more production time to this operation. The result makes sense because the resource is already abundant, as is evident by the fact that the slack variable associated with Operation 3 is positive ($= 20$) in the optimum solution. As for each of Operations 1 and 2, a one minute increase will improve revenue by \$1 and \$2, respectively. The dual prices also indicate that, when allocating additional resources, Operation 2 may be given higher priority because its dual price is twice as much as that of Operation 1.

The computations above show how the dual prices are determined from the optimal tableau for $\leq$ constraints. For $\geq$ constraints, the same idea remains applicable except that the dual price will assume the opposite sign of that associated with the $\leq$ constraint. As for the case where the constraint is an equation, the determination of the dual price from the optimal simplex tableau requires somewhat "involved" calculations as will be shown in Chapter 4.

Determination of the Feasibility Ranges. Having determined the dual prices, we show next how the *feasibility ranges* in which they remain valid are determined. Let D_1, D_2, and D_3 be the changes (positive or negative) in the daily manufacturing time allocated to operations 1, 2, and 3, respectively. The model can be written as follows:

$$\text{Maximize } z = 3x_1 + 2x_2 + 5x_3$$

subject to

$$\begin{aligned}
x_1 + 2x_2 + x_3 &\leq 430 + D_1 \quad \text{(Operation 1)} \\
3x_1 \qquad + 2x_3 &\leq 460 + D_2 \quad \text{(Operation 2)} \\
x_1 + 4x_2 \qquad &\leq 420 + D_3 \quad \text{(Operation 3)} \\
x_1, x_2, x_3 &\geq 0
\end{aligned}$$

We will consider the general case of making the changes simultaneously. The special cases of making change one at a time are derived from these results.

The procedure is based on recomputing the optimum simplex tableau with the modified right-hand side and then deriving the conditions that will keep the solution feasible—that is, the right-hand side of the optimum tableau remains nonnegative. To show how the right-hand side is recomputed, we start by modifying the *Solution* column of the starting tableau using the new right-hand sides: $430 + D_1$, $460 + D_2$, and $420 + D_3$. The starting tableau will thus appear as

							Solution			
Basic	x_1	x_2	x_3	x_4	x_5	x_6	RHS	D_1	D_2	D_3
z	-3	-2	-5	0	0	0	0	0	0	0
x_4	1	2	1	1	0	0	430	1	0	0
x_5	3	0	2	0	1	0	460	0	1	0
x_6	1	4	0	0	0	1	420	0	0	1

The columns under D_1, D_2, and D_3 are identical to those under the starting basic columns x_4, x_5, and x_6. This means that when we carry out the *same* simplex iterations as in the *original* model, the columns in the two groups must come out identical as well. Effectively, the new optimal tableau will become

							Solution			
Basic	x_1	x_2	x_3	x_4	x_5	x_6	RHS	D_1	D_2	D_3
z	4	0	0	1	2	0	1350	1	2	0
x_2	$-\frac{1}{4}$	1	0	$\frac{1}{2}$	$-\frac{1}{4}$	0	100	$\frac{1}{2}$	$-\frac{1}{4}$	0
x_3	$\frac{3}{2}$	0	1	0	$\frac{1}{2}$	0	230	0	$\frac{1}{2}$	0
x_6	2	0	0	-2	1	1	20	-2	1	1

The new optimum tableau provides the following optimal solution:

$$
\begin{aligned}
z &= 1350 + D_1 + 2D_2 \\
x_2 &= 100 + \tfrac{1}{2}D_1 - \tfrac{1}{4}D_2 \\
x_3 &= 230 + \tfrac{1}{2}D_2 \\
x_6 &= 20 - 2D_1 + D_2 + D_3
\end{aligned}
$$

Interestingly, as shown earlier, the new z-value confirms that the dual prices for operations 1, 2, and 3 are 1, 2, and 0, respectively.

The current solution remains feasible so long as all the variables are nonnegative, which leads to the following **feasibility conditions**:

$$
\begin{aligned}
x_2 &= 100 + \tfrac{1}{2}D_1 - \tfrac{1}{4}D_2 \geq 0 \\
x_3 &= 230 + \tfrac{1}{2}D_2 \geq 0 \\
x_6 &= 20 - 2D_1 + D_2 + D_3 \geq 0
\end{aligned}
$$

Any simultaneous changes D_1, D_2, and D_3 that satisfy these inequalities will keep the solution feasible. If all the conditions are satisfied, then the new optimum solution can be found through direct substitution of D_1, D_2, and D_3 in the equations given above.

To illustrate the use of these conditions, suppose that the manufacturing time available for operations 1, 2, and 3 are 480, 440, and 410 minutes respectively. Then, $D_1 = 480 - 430 = 50$, $D_2 = 440 - 460 = -20$, and $D_3 = 410 - 420 = -10$. Substituting in the feasibility conditions, we get

$$\begin{aligned} x_2 &= 100 + \tfrac{1}{2}(50) - \tfrac{1}{4}(-20) = 130 > 0 && \text{(feasible)} \\ x_3 &= 230 + \tfrac{1}{2}(-20) = 220 > 0 && \text{(feasible)} \\ x_6 &= 20 - 2(50) + (-20) + (-10) = -110 < 0 && \text{(infeasible)} \end{aligned}$$

The calculations show that $x_6 < 0$, hence the current solution does not remain feasible. Additional calculations will be needed to find the new solution. These calculations are discussed in Chapter 4 as part of the post-optimal analysis.

Alternatively, if the changes in the resources are such that $D_1 = -30$, $D_2 = -12$, and $D_3 = 10$, then

$$\begin{aligned} x_2 &= 100 + \tfrac{1}{2}(-30) - \tfrac{1}{4}(-12) = 88 > 0 && \text{(feasible)} \\ x_3 &= 230 + \tfrac{1}{2}(-12) = 224 > 0 && \text{(feasible)} \\ x_6 &= 20 - 2(-30) + (-12) + (10) = 78 > 0 && \text{(feasible)} \end{aligned}$$

The new feasible solution is $x_1 = 88$, $x_3 = 224$, and $x_6 = 68$ with $z = 3(0) + 2(88) + 5(224) = \$1296$. Notice that the optimum objective value can also be computed as $z = 1350 + 1(-30) + 2(-12) = \$1296$.

The given conditions can be specialized to produce the individual *feasibility ranges* that result from changing the resources *one at a time* (as defined in Section 3.6.1).

Case 1. Change in operation 1 time from 460 *to* 460 + D_1 *minutes.* This change is equivalent to setting $D_2 = D_3 = 0$ in the simultaneous conditions, which yields

$$\left.\begin{aligned} x_2 &= 100 + \tfrac{1}{2}D_1 \geq 0 \Rightarrow D_1 \geq -200 \\ x_3 &= 230 > 0 \\ x_6 &= 20 - 2D_1 \geq 0 \Rightarrow D_1 \leq 10 \end{aligned}\right\} \Rightarrow -200 \leq D_1 \leq 10$$

Case 2. Change in operation 2 time from 430 *to* 430 + D_2 *minutes.* This change is equivalent to setting $D_1 = D_3 = 0$ in the simultaneous conditions, which yields

$$\left.\begin{aligned} x_2 &= 100 - \tfrac{1}{4}D_2 \geq 0 \Rightarrow D_2 \leq 400 \\ x_3 &= 230 + \tfrac{1}{2}D_2 \geq 0 \Rightarrow D_2 \geq -460 \\ x_6 &= 20 + D_2 \geq 0 \quad \Rightarrow D_2 \geq -20 \end{aligned}\right\} \Rightarrow -20 \leq D_2 \leq 400$$

Case 3. Change in operation 3 time from 420 *to* 420 + D_3 *minutes.* This change is equivalent to setting $D_1 = D_2 = 0$ in the simultaneous conditions, which yields

$$\left.\begin{aligned} x_2 &= 100 > 0 \\ x_3 &= 230 > 0 \\ x_6 &= 20 + D_3 \geq 0 \end{aligned}\right\} \Rightarrow -20 \leq D_3 < \infty$$

We can now summarize the dual prices and their feasibility ranges for the TOYCO model as follows:[3]

Resource	Dual price	Feasibility range	Resource amount (minutes)		
			Minimum	*Current*	*Maximum*
Operation 1	1	$-200 \leq D_1 \leq 10$	230	430	440
Operation 2	2	$-20 \leq D_2 \leq 400$	440	440	860
Operation 3	0	$-20 \leq D_3 < \infty$	400	420	∞

It is important to notice that the dual prices will remain applicable for any *simultaneous* changes that keep the solution feasible, even if the changes violate the individual ranges. For example, the changes $D_1 = 30$, $D_2 = -12$, and $D_3 = 100$, will keep the solution feasible even though $D_1 = 30$ violates the feasibility range $-200 \leq D_1 \leq 10$, as the following computations show:

$$x_2 = 100 + \tfrac{1}{2}(30) - \tfrac{1}{4}(-12) = 118 > 0 \quad \text{(feasible)}$$

$$x_3 = 230 + \tfrac{1}{2}(-12) = 224 > 0 \quad \text{(feasible)}$$

$$x_6 = 20 - 2(30) + (-12) + (100) = 48 > 0 \quad \text{(feasible)}$$

This means that the dual prices will remain applicable, and we can compute the new optimum objective value from the dual prices as $z = 1350 + 1(30) + 2(-12) + 0(100) = \$1356$

The results above can be summarized as follows:

1. The dual prices remain valid so long as the changes D_i, $i = 1, 2, \ldots, m$, in the right-hand sides of the constraints satisfy all the feasibility conditions when the changes are simultaneous or fall within the feasibility ranges when the changes are made individually.

2. For other situations where the dual prices are not valid because the simultaneous feasibility conditions are not satisfied or because the individual feasibility ranges are violated, the recourse is to either re-solve the problem with the new values of D_i or apply the post-optimal analysis presented in Chapter 4.

PROBLEM SET 3.6C[4]

1. In the TOYCO model, suppose that the changes D_1, D_2, and D_3 are made *simultaneously* in the three operations.
   (a) If the availabilities of operations 1, 2, and 3 are changed to 438, 500, and 410 minutes, respectively, use the simultaneous conditions to show that the current basic solution

[3] Available LP packages usually present this information as standard output. Practically none provide the case of simultaneous conditions, presumably because its display is cumbersome, particularly for large LPs.

[4] In this problem set, you may find it convenient to generate the optimal simplex tableau with TORA.

remains feasible, and determine the change in the optimal revenue by using the optimal dual prices.

(b) If the availabilities of the three operations are changed to 460, 440, and 380 minutes, respectively, use the simultaneous conditions to show that the current basic solution becomes infeasible.

***2.** Consider the TOYCO model.

(a) Suppose that any additional time for operation 1 beyond its current capacity of 430 minutes per day must be done on an overtime basis at $50 an hour. The hourly cost includes both labor and the operation of the machine. Is it economically advantageous to use overtime with operation 1?

(b) Suppose that the operator of operation 2 has agreed to work 2 hours of overtime daily at $45 an hour. Additionally, the cost of the operation itself is $10 an hour. What is the net effect of this activity on the daily revenue?

(c) Is overtime needed for operation 3?

(d) Suppose that the daily availability of operation 1 is increased to 440 minutes. Any overtime used beyond the current maximum capacity will cost $40 an hour. Determine the new optimum solution, including the associated net revenue.

(e) Suppose that the availability of operation 2 is decreased by 15 minutes a day and that the hourly cost of the operation during regular time is $30. Is it advantageous to decrease the availability of operation 2?

3. A company produces three products, A, B, and C. The sales volume for A is at least 50% of the total sales of all three products. However, the company cannot sell more than 75 units of A per day. The three products use one raw material, of which the maximum daily availability is 240 lb. The usage rates of the raw material are 2 lb per unit of A, 4 lb per unit of B, and 3 lb per unit of C. The unit prices for A, B, and C are $20, $50, and $35, respectively.

(a) Determine the optimal product mix for the company.

(b) Determine the dual price of the raw material resource and its allowable range. If available raw material is increased by 120 lb, determine the optimal solution and the change in total revenue using the dual price.

(c) Use the dual price to determine the effect of changing the maximum demand for product A by ± 10 units.

4. A company that operates 10 hours a day manufactures three products on three sequential processes. The following table summarizes the data of the problem:

	Minutes per unit			
Product	*Process 1*	*Process 2*	*Process 3*	Unit price
1	10	6	8	$4.50
2	5	8	10	$5.00
3	6	9	12	$4.00

(a) Determine the optimal product mix.

(b) Use the dual prices to prioritize the three processes for possible expansion.

(c) If additional production hours can be allocated, what would be a fair cost per additional hour for each process?

5. The Continuing Education Division at the Ozark Community College offers a total of 30 courses each semester. The courses offered are usually of two types: practical, such as woodworking, word processing, and car maintenance; and humanistic, such as history, music, and fine arts. To satisfy the demands of the community, at least 10 courses of each type must be offered each semester. The division estimates that the revenues of offering practical and humanistic courses are approximately \$1500 and \$1000 per course, respectively.

(a) Devise an optimal course offering for the college.

(b) Show that the dual price of an additional course is \$1500, which is the same as the revenue per practical course. What does this result mean in terms of offering additional courses?

(c) How many more courses can be offered while guaranteeing that each will contribute \$1500 to the total revenue?

(d) Determine the change in revenue resulting from increasing the minimum requirement of humanistics by one course.

***6.** Show & Sell can advertise its products on local radio and television (TV), or in newspapers. The advertising budget is limited to \$10,000 a month. Each minute of advertising on radio costs \$15 and each minute on TV costs \$300. A newspaper ad costs \$50. Show & Sell likes to advertise on radio at least twice as much as on TV. In the meantime, the use of at least 5 newspaper ads and no more than 400 minutes of radio advertising a month is recommended. Past experience shows that advertising on TV is 50 times more effective than on radio and 10 times more effective than in newspapers.

(a) Determine the optimum allocation of the budget to the three media.

(b) Are the limits set on radio and newspaper advertising justifiable economically?

(c) If the monthly budget is increased by 50%, would this result in a proportionate increase in the overall effectiveness of advertising?

7. The Burroughs Garment Company manufactures men's shirts and women's blouses for Walmark Discount Stores. Walmark will accept all the production supplied by Burroughs. The production process includes cutting, sewing, and packaging. Burroughs employs 25 workers in the cutting department, 35 in the sewing department, and 5 in the packaging department. The factory works one 8-hour shift, 5 days a week. The following table gives the time requirements and prices per unit for the two garments:

	Minutes per unit			
Garment	*Cutting*	*Sewing*	*Packaging*	Unit price (\$)
Shirts	20	70	12	8.00
Blouses	60	60	4	12.00

(a) Determine the optimal weekly production schedule for Burroughs.

(b) Determine the worth of one hour of cutting, sewing, and packaging in terms of the total revenue.

(c) If overtime can be used in cutting and sewing, what is the maximum hourly rate Burroughs should pay for overtime?

8. ChemLabs uses raw materials *I* and *II* to produce two domestic cleaning solutions, *A* and *B*. The daily availabilities of raw materials *I* and *II* are 150 and 145 units, respectively. One unit of solution *A* consumes .5 unit of raw material *I* and .6 unit of raw material *II*, and one unit of solution *B* uses .5 unit of raw material *I* and .4 unit of raw material *II*. The prices per unit of solutions A and B are \$8 and \$10, respectively. The daily demand for solution A lies between 30 and 150 units, and that for solution B between 40 and 200 units.

(a) Find the optimal amounts of A and B that ChemLab should produce.

(b) Use the dual prices to determine which demand limits on products A and B should be relaxed to improve profitability.

(c) If additional units of raw material can be acquired at \$20 per unit, is this advisable? Explain.

(d) A suggestion is made to increase raw material II by 25% to remove a bottleneck in production. Is this advisable? Explain.

9. An assembly line consisting of three consecutive workstations produces two radio models: DiGi-1 and DiGi-2. The following table provides the assembly times for the three workstations.

	Minutes per unit	
Workstation	*DiGi-1*	*DiGi-2*
1	6	4
2	5	4
3	4	6

The daily maintenance for workstations 1, 2, and 3 consumes 10%, 14%, and 12%, respectively, of the maximum 480 minutes available for each workstation each day.

(a) The company wishes to determine the optimal product mix that will minimize the idle (or unused) times in the three workstations. Determine the optimum utilization of the workstations. [*Hint*: Express the sum of the idle times (slacks) for the three operations in terms of the original variables.]

(b) Determine the worth of decreasing the daily maintenance time for each workstation by 1 percentage point.

(c) It is proposed that the operation time for all three workstations be increased to 600 minutes per day at the additional cost of \$1.50 per minute. Can this proposal be improved?

10. The Gutchi Company manufactures purses, shaving bags, and backpacks. The construction of the three products requires leather and synthetics, with leather being the limiting raw material. The production process uses two types of skilled labor: sewing and finishing. The following table gives the availability of the resources, their usage by the three products, and the prices per unit.

	Resource requirements per unit			
Resource	*Purse*	*Bag*	*Backpack*	Daily availability
Leather (ft^2)	2	1	3	42
Sewing (hr)	2	1	2	40
Finishing (hr)	1	.5	1	45
Price (\$)	24	22	45	

Formulate the problem as a linear program and find the optimum solution. Next, indicate whether the following changes in the resources will keep the current solution feasible.

For the cases where feasibility is maintained, determine the new optimum solution (values of the variables and the objective function).

(a) Available leather is increased to 45 ft^2.

(b) Available leather is decreased by 1 ft^2.

(c) Available sewing hours are changed to 38 hours.

(d) Available sewing hours are changed to 46 hours.

(e) Available finishing hours are decreased to 15 hours.

(f) Available finishing hours are increased to 50 hours.

(g) Would you recommend hiring an additional sewing worker at $15 an hour?

11. HiDec produces two models of electronic gadgets that use resistors, capacitors, and chips. The following table summarizes the data of the situation:

	Unit resource requirements		
Resource	*Model 1 (units)*	*Model 2 (units)*	Maximum availability (units)
Resistor	2	3	1200
Capacitor	2	1	1000
Chips	0	4	800
Unit price ($)	3	4	

Let x_1 and x_2 be the amounts produced of Models 1 and 2, respectively. Following are the LP model and its associated optimal simplex tableau.

$$\text{Maximize } z = 3x_1 + 4x_2$$

subject to

$$\begin{aligned} 2x_1 + 3x_2 &\leq 1200 \quad \text{(Resistors)} \\ 2x_1 + x_2 &\leq 1000 \quad \text{(Capacitors)} \\ 4x_2 &\leq 800 \quad \text{(Chips)} \\ x_1, x_2 &\geq 0 \end{aligned}$$

Basic	x_1	x_2	s_1	s_2	s_3	Solution
z	0	0	$\frac{5}{4}$	$\frac{1}{4}$	0	1750
x_1	1	0	$-\frac{1}{4}$	$\frac{3}{4}$	0	450
s_3	0	0	-2	2	1	400
x_2	0	1	$\frac{1}{2}$	$-\frac{1}{2}$	0	100

***(a)** Determine the status of each resource.

***(b)** In terms of the optimal revenue, determine the dual prices for the resistors, capacitors, and chips.

(c) Determine the feasibility ranges for the dual prices obtained in (b).

(d) If the available number of resistors is increased to 1300 units, find the new optimum solution.

*(e) If the available number of chips is reduced to 350 units, will you be able to determine the new optimum solution directly from the given information? Explain.

(f) If the availability of capacitors is limited by the feasibility range computed in (c), determine the corresponding range of the optimal revenue and the corresponding ranges for the numbers of units to be produced of Models 1 and 2.

(g) A new contractor is offering to sell HiDec additional resistors at 40 cents each, but only if HiDec would purchase at least 500 units. Should HiDec accept the offer?

12. *The 100% feasibility rule.* A simplified rule based on the *individual* changes $D_1, D_2, \ldots,$ and D_m in the right-hand side of the constraints can be used to test whether or not *simultaneous* changes will maintain the feasibility of the current solution. Assume that the right-hand side b_i of constraint i is changed to $b_i + D_i$ *one at a time*, and that $p_i \leq D_i \leq q_i$ is the corresponding feasibility range obtained by using the procedure in Section 3.6.2. By definition, we have $p_i \leq 0$ $(q_i \geq 0)$ because it represents the maximum allowable decrease (increase) in b_i. Next, define r_i to equal $\frac{D_i}{p_i}$ if D_i is negative and $\frac{D_i}{q_i}$ if D_i is positive. By definition, we have $0 \leq r_i \leq 1$. The 100% rule thus says that, given the changes $D_1, D_2, \ldots,$ and D_m, then a *sufficient* (but not necessary) condition for the current solution to remain feasible is that $r_1 + r_2 + \cdots + r_m \leq 1$. If the condition is not satisfied, then the current solution may or may not remain feasible. The rule is not applicable if D_i falls outside the range (p_i, q_i).

In reality, the 100% rule is too weak to be consistently useful. Even in the cases where feasibility can be confirmed, we still need to obtain the new solution using the regular simplex feasibility conditions. Besides, the direct calculations associated with simultaneous changes given in Section 3.6.2 are straightforward and manageable.

To demonstrate the weakness of the rule, apply it to parts (a) and (b) of Problem 1 in this set. The rule fails to confirm the feasibility of the solution in (a) and does not apply in (b) because the changes in D_i are outside the admissible ranges. Problem 13 further demonstrates this point.

13. Consider the problem

$$\text{Maximize } z = x_1 + x_2$$

subject to

$$2x_1 + x_2 \leq 6$$
$$x_1 + 2x_2 \leq 6$$
$$x_1 + x_2 \geq 0$$

(a) Show that the optimal basic solution includes both x_1 and x_2 and that the feasibility ranges for the two constraints, considered one at a time, are $-3 \leq D_1 \leq 6$ and $-3 \leq D_2 \leq 6$.

*(b) Suppose that the two resources are increased simultaneously by $\Delta > 0$ each. First, show that the basic solution remains feasible for all $\Delta > 0$. Next, show that the 100% rule will confirm feasibility only if the increase is in the range $0 < \Delta \leq 3$ units. Otherwise, the rule fails for $3 < \Delta \leq 6$ and does not apply for $\Delta > 6$.

3.6.3 Algebraic Sensitivity Analysis—Objective Function

In Section 3.6.1, we used graphical sensitivity analysis to determine the conditions that will maintain the optimality of a two-variable LP solution. In this section, we extend these ideas to the general LP problem.

Definition of Reduced Cost. To facilitate the explanation of the objective function sensitivity analysis, first we need to define *reduced costs*. In the TOYCO model (Example 3.6-2), the objective z-equation in the optimal tableau is

$$z + 4x_1 + x_4 + 2x_5 = 1350$$

or

$$z = 1350 - 4x_1 - x_4 - 2x_5$$

The optimal solution does not recommend the production of toy trains ($x_1 = 0$). This recommendation is confirmed by the information in the z-equation because each unit increase in x_1 above its current zero level will decrease the value of z by \$4 — namely, $z = 1350 - 4 \times (1) - 1 \times (0) - 2 \times (0) = \$1346$.

We can think of the coefficient of x_1 in the z-equation ($= 4$) as a unit *cost* because it causes a reduction in the revenue z. But where does this "cost" come from? We know that x_1 has a unit revenue of \$3 in the original model. We also know that each toy train consumes resources (operations time), which in turn incur cost. Thus, the "attractiveness" of x_1 from the standpoint of optimization depends on the relative values of the revenue per unit and the cost of the resources consumed by one unit. This relationship is formalized in the LP literature by defining the reduced cost as

$$\begin{pmatrix}\text{Reduced cost} \\ \text{per unit}\end{pmatrix} = \begin{pmatrix}\text{Cost of consumed} \\ \text{resources per unit}\end{pmatrix} - (\text{Revenue per unit})$$

To appreciate the significance of this definition, in the original TOYCO model the revenue per unit for toy trucks ($= \$2$) is less than that for toy trains ($= \$3$). Yet the optimal solution elects to manufacture toy trucks ($x_2 = 100$ units) and no toy trains ($x_1 = 0$). The reason for this (seemingly nonintuitive) result is that the unit cost of the resources used by toy trucks (i.e., operations time) is smaller than its unit price. The opposite applies in the case of toy trains.

With the given definition of *reduced cost* we can now see that an unprofitable variable (such as x_1) can be made profitable in two ways:

1. By increasing the unit revenue.
2. By decreasing the unit cost of consumed resources.

In most real-life situations, the price per unit may not be a viable option because its value is dictated by market conditions. The real option then is to reduce the consumption of resources, perhaps by making the production process more efficient, as will be shown in Chapter 4.

Determination of the Optimality Ranges. We now turn our attention to determining the conditions that will keep an optimal solution unchanged. The presentation is based on the definition of *reduced cost*.

In the TOYCO model, let d_1, d_2, and d_3 represent the change in unit revenues for toy trucks, trains, and cars, respectively. The objective function then becomes

$$\text{Maximize } z = (3 + d_1)x_1 + (2 + d_2)x_2 + (5 + d_3)x_3$$

As we did for the right-hand side sensitivity analysis in Section 3.6.2, we will first deal with the general situation in which all the coefficients of the objective function are changed *simultaneously* and then specialize the results to the one-at-a-time case.

With the simultaneous changes, the z-row in the starting tableau appears as:

Basic	x_1	x_2	x_3	x_4	x_5	x_6	Solution
z	$-3 - d_1$	$-2 - d_2$	$-5 - d_3$	0	0	0	0

When we generate the simplex tableaus using the same sequence of entering and leaving variables in the original model (before the changes d_j are introduced), the optimal iteration will appear as follows (convince yourself that this is indeed the case by carrying out the simplex row operations):

Basic	x_1	x_2	x_3	x_4	x_5	x_6	Solution
z	$4 - \frac{1}{4}d_2 + \frac{3}{2}d_3 - d_1$	0	0	$1 + \frac{1}{2}d_2$	$2 - \frac{1}{4}d_2 + \frac{1}{2}d_3$	0	$1350 + 100d_2 + 230d_3$
x_2	$-\frac{1}{4}$	1	0	$\frac{1}{2}$	$-\frac{1}{4}$	0	100
x_3	$\frac{3}{2}$	0	1	0	$\frac{1}{2}$	0	230
x_6	$-\frac{1}{4}$	0	0	-2	1	1	20

The new optimal tableau is exactly the same as in the *original* optimal tableau except that the *reduced costs* (z-equation coefficients) have changed. This means that *changes in the objective-function coefficients can affect the optimality of the problem only.*

You really do not need to carry out the row operation to compute the new reduced costs. An examination of the new z-row shows that the coefficients of d_j are taken directly from the constraint coefficients of the optimum tableau. A convenient way for computing the new reduced cost is to add a new top row and a new leftmost column to the optimum tableau, as shown by the shaded areas below. The entries in the top row are the change d_j associated with each variable. For the leftmost column, the entries are 1 in the z-row and the associated d_j in the row of each basic variable. Keep in mind that $d_j = 0$ for the slack variables.

		d_1	d_2	d_3	0	0	0	
	Basic	x_1	x_2	x_3	x_4	x_5	x_6	Solution
1	z	4	0	0	1	2	0	1350
d_2	x_2	$-\frac{1}{4}$	1	0	$\frac{1}{2}$	$-\frac{1}{4}$	0	100
d_3	x_3	$\frac{3}{2}$	0	1	0	$\frac{1}{2}$	0	230
0	x_6	2	0	0	-2	1	1	20

Now, to compute the new reduced cost for any variable (or the value of z), multiply the elements of its column by the corresponding elements in the leftmost column, add them up, and subtract the top-row element from the sum. For example, for x_1, we have

	d_1	
Left column	x_1	(x_1-column × left-column)
1	4	4×1
d_2	$-\frac{1}{4}$	$-\frac{1}{4}d_2$
d_3	$\frac{3}{2}$	$\frac{3}{2}d_3$
0	2	2×0
Reduced cost for $x_1 = 4 - \frac{1}{4}d_2 + \frac{3}{2}d_3 - d_1$		

Note that the application of these computations to the *basic* variables will always produce a zero reduced cost, a proven theoretical result. Also, applying the same rule to the *Solution* column produces $z = 1350 + 100d_2 + 230d_3$.

Because we are dealing with a maximization problem, the current solution remains optimal so long as the new reduced costs (z-equation coefficients) remain nonnegative for all the nonbasic variables. We thus have the following **optimality conditions** corresponding to nonbasic x_1, x_4, and x_5:

$$4 - \tfrac{1}{4}d_2 + \tfrac{3}{2}d_3 - d_1 \geq 0$$
$$1 + \tfrac{1}{2}d_2 \geq 0$$
$$2 - \tfrac{1}{4}d_2 + \tfrac{1}{2}d_3 \geq 0$$

These conditions must be satisfied *simultaneously* to maintain the optimality of the current optimum.

To illustrate the use of these conditions, suppose that the objective function of TOYCO is changed from

$$\text{Maximize } z = 3x_1 + 2x_2 + 5x_3$$

to

$$\text{Maximize } z = 2x_1 + x_2 + 6x_3$$

Then, $d_1 = 2 - 3 = -\$1$, $d_2 = 1 - 2 = -\$1$, and $d_3 = 6 - 5 = \$1$. Substitution in the given conditions yields

$$4 - \tfrac{1}{4}d_2 + \tfrac{3}{2}d_3 - d_1 = 4 - \tfrac{1}{4}(-1) + \tfrac{3}{2}(1) - (-1) = 6.75 > 0 \text{ (satisfied)}$$
$$1 + \tfrac{1}{2}d_2 = 1 + \tfrac{1}{2}(-1) = .5 > 0 \qquad \text{(satisfied)}$$
$$2 - \tfrac{1}{4}d_2 + \tfrac{1}{2}d_3 = 2 - \tfrac{1}{4}(-1) + \tfrac{1}{2}(1) = 2.75 > 0 \qquad \text{(satisfied)}$$

The results show that the proposed changes will keep the current solution ($x_1 = 0$, $x_2 = 100$, $x_3 = 230$) optimal. Hence no further calculations are needed, except that the objective value will change to $z = 1350 + 100d_2 + 230d_3 = 1350 + 100 \times -1 + 230 \times 1 = \$1480$. If any of the conditions is not satisfied, a new solution must be determined (see Chapter 4).

The discussion so far has dealt with the maximization case. The only difference in the minimization case is that the reduced costs (z-equations coefficients) must be ≤ 0 to maintain optimality.

The general optimality conditions can be used to determine the special case where the changes d_j occur *one at a time* instead of simultaneously. This analysis is equivalent to considering the following three cases:

1. Maximize $z = (3 + d_1)x_1 + 2x_2 + 5x_3$
2. Maximize $z = 3x_1 + (2 + d_2)x_2 + 5x_3$
3. Maximize $z = 3x_1 + 2x_2 + (5 + d_3)x_3$

The individual conditions can be accounted for as special cases of the simultaneous case.[5]

Case 1. Set $d_2 = d_3 = 0$ in the simultaneous conditions, which gives

$$4 - d_1 \geq 0 \Rightarrow -\infty < d_1 \leq 4$$

Case 2. Set $d_1 = d_3 = 0$ in the simultaneous conditions, which gives

$$\left.\begin{array}{l} 4 - \frac{1}{4}d_2 \geq 0 \Rightarrow d_2 \leq 16 \\ 1 + \frac{1}{2}d_2 \geq 0 \Rightarrow d_2 \geq -2 \\ 2 - \frac{1}{4}d_2 \geq 0 \Rightarrow d_2 \leq 8 \end{array}\right\} \Rightarrow -2 \leq d_2 \leq 8$$

Case 3. Set $d_1 = d_2 = 0$ in the simultaneous conditions, which gives

$$\left.\begin{array}{l} 4 + \frac{3}{2}d_3 \geq 0 \Rightarrow d_3 \geq -\frac{8}{3} \\ 2 + \frac{1}{2}d_3 \geq 0 \Rightarrow d_3 \geq -4 \end{array}\right\} \Rightarrow -\frac{8}{3} \leq d_3 < \infty$$

The given individual conditions can be translated in terms of the total unit revenue. For example, for toy trucks (variable x_2), the total unit revenue is $2 + d_2$ and the associated condition $-2 \leq d_2 \leq 8$ translates to

$$2 + (-2) \leq 2 + d_2 \leq 2 + 8$$

or

$$\$0 \leq (\text{Unit revenue of toy truck}) \leq \$10$$

This condition assumes that the unit revenues for toy trains and toy cars remain fixed at \$3 and \$5, respectively.

The allowable range (\$0, \$10) indicates that the unit revenue of toy trucks (variable x_2) can be as low as \$0 or as high as \$10 without changing the current optimum, $x_1 = 0, x_2 = 100, x_3 = 230$. The total revenue will change to $1350 + 100d_2$, however.

[5]The individual ranges are standard outputs in all LP software. Simultaneous conditions usually are not part of the output, presumably because they are cumbersome for large problems.

It is important to notice that the changes d_1, d_2, and d_3 may be within their allowable individual ranges without satisfying the simultaneous conditions, and vice versa. For example, consider

$$\text{Maximize } z = 6x_1 + 8x_2 + 3x_3$$

Here $d_1 = 6 - 3 = \$3$, $d_2 = 8 - 2 = \$6$, and $d_3 = 3 - 5 = -\$2$, which are all within the permissible individual ranges ($-\infty < d_1 \leq 4$, $-2 \leq d_2 \leq 8$, and $-\frac{8}{3} \leq d_3 < \infty$). However, the corresponding simultaneous conditions yield

$$4 - \tfrac{1}{4}d_2 + \tfrac{3}{2}d_3 - d_1 = 4 - \tfrac{1}{4}(6) + \tfrac{3}{2}(-2) - 3 = -3.5 < 0 \text{ (not satisfied)}$$

$$1 + \tfrac{1}{2}d_2 = 1 + \tfrac{1}{2}(6) = 4 > 0 \qquad \text{(satisfied)}$$

$$2 - \tfrac{1}{4}d_2 + \tfrac{1}{2}d_3 = 2 - \tfrac{1}{4}(6) + \tfrac{1}{2}(-2) = -.5 < 0 \qquad \text{(not satisfied)}$$

The results above can be summarized as follows:

---

1. The optimal values of the variables remain unchanged so long as the changes d_j, $j = 1, 2, \ldots, n$, in the objective function coefficients satisfy all the optimality conditions when the changes are simultaneous or fall within the optimality ranges when a change is made individually.
2. For other situations where the simultaneous optimality conditions are not satisfied or the individual feasibility ranges are violated, the recourse is to either resolve the problem with the new values of d_j or apply the post-optimal analysis presented in Chapter 4.

PROBLEM SET 3.6D[6]

1. In the TOYCO model, determine if the current solution will change in each of the following cases:
   (i) $z = 2x_1 + x_2 + 4x_3$
   (ii) $z = 3x_1 + 6x_2 + x_3$
   (iii) $z = 8x_1 + 3x_2 + 9x_3$

*2. B&K grocery store sells three types of soft drinks: the brand names A1 Cola and A2 Cola and the cheaper store brand BK Cola. The price per can for A1, A2, and BK are 80, 70, and 60 cents, respectively. On the average, the store sells no more than 500 cans of all colas a day. Although A1 is a recognized brand name, customers tend to buy more A2 and BK because they are cheaper. It is estimated that at least 100 cans of A1 are sold daily and that A2 and BK combined outsell A1 by a margin of at least 4:2.
   (a) Show that the optimum solution does not call for selling the A3 brand.
   (b) By how much should the price per can of A3 be increased to be sold by B&K?
   (c) To be competitive with other stores, B&K decided to lower the price on all three types of cola by 5 cents per can. Recompute the reduced costs to determine if this promotion will change the current optimum solution.

---

[6]In this problem set, you may find it convenient to generate the optimal simplex tableau with TORA.

3. Baba Furniture Company employs four carpenters for 10 days to assemble tables and chairs. It takes 2 person-hours to assemble a table and .5 person-hour to assemble a chair. Customers usually buy one table and four to six chairs. The prices are $135 per table and $50 per chair. The company operates one 8-hour shift a day.
   (a) Determine the 10-day optimal production mix.
   (b) If the present unit prices per table and chair are each reduced by 10%, use sensitivity analysis to determine if the optimum solution obtained in (a) will change.
   (c) If the present unit prices per table and chair are changed to $120 and $25, will the solution in (a) change?
4. The Bank of Elkins is allocating a maximum of $200,000 for personal and car loans during the next month. The bank charges 14% for personal loans and 12% for car loans. Both types of loans are repaid at the end of a 1-year period. Experience shows that about 3% of personal loans and 2% of car loans are not repaid. The bank usually allocates at least twice as much to car loans as to personal loans.
   (a) Determine the optimal allocation of funds between the two loans and the net rate of return on all the loans.
   (b) If the percentages of personal and car loans are changed to 4% and 3%, respectively, use sensitivity analysis to determine if the optimum solution in (a) will change.

*5. Electra produces four types of electric motors, each on a separate assembly line. The respective capacities of the lines are 500, 500, 800, and 750 motors per day. Type 1 motor uses 8 units of a certain electronic component, type 2 motor uses 5 units, type 3 motor uses 4 units, and type 4 motor uses 6 units. The supplier of the component can provide 8000 pieces a day. The prices per motor for the respective types are $60, $40, $25, $30.
   (a) Determine the optimum daily production mix.
   (b) The present production schedule meets Electra's needs. However, because of competition, Electra may need to lower the price of type 2 motor. What is the most reduction that can be effected without changing the present production schedule?
   (c) Electra has decided to slash the price of all motor types by 25%. Use sensitivity analysis to determine if the optimum solution remains unchanged.
   (d) Currently, type 4 motor is not produced. By how much should its price be increased to be included in the production schedule?

6. Popeye Canning is contracted to receive daily 60,000 lb of ripe tomatoes at 7 cents per pound from which it produces canned tomato juice, tomato sauce, and tomato paste. The canned products are packaged in 24-can cases. A can of juice uses 1 lb of fresh tomatoes, a can of sauce uses $\frac{1}{2}$ lb, and a can of paste uses $\frac{3}{4}$ lb. The company's daily share of the market is limited to 2000 cases of juice, 5000 cases of sauce, and 6000 cases of paste. The wholesale prices per case of juice, sauce, and paste are $21, $9, and $12, respectively.
   (a) Develop an optimum daily production program for Popeye.
   (b) If the price per case for juice and paste remains fixed as given in the problem, use sensitivity analysis to determine the unit price range Popeye should charge for a case of sauce to keep the optimum product mix unchanged.
7. Dean's Furniture Company assembles regular and deluxe kitchen cabinets from precut lumber. The regular cabinets are painted white, and the deluxe are varnished. Both painting and varnishing are carried out in one department. The daily capacity of the assembly department is 200 regular cabinets and 150 deluxe. Varnishing a deluxe unit takes twice as much time as painting a regular one. If the painting/varnishing department is dedicated to the deluxe units only, it can complete 180 units daily. The company estimates that the revenues per unit for the regular and deluxe cabinets are $100 and $140, respectively.

(a) Formulate the problem as a linear program and find the optimal production schedule per day.

(b) Suppose that competition dictates that the price per unit of each of regular and deluxe cabinets be reduced to $80. Use sensitivity analysis to determine whether or not the optimum solution in (a) remains unchanged.

8. *The 100% Optimality Rule.* A rule similar to the *100% feasibility rule* outlined in Problem 12, Set 3.6c, can also be developed for testing the effect of simultaneously changing all c_j to $c_j + d_j, j = 1, 2, \ldots, n$, on the optimality of the current solution. Suppose that $u_j \leq d_j \leq v_j$ is the optimality range obtained as a result of changing each c_j to $c_j + d_j$ one at a time, using the procedure in Section 3.6.3. In this case, $u_j \leq 0$ ($v_j \geq 0$), because it represents the maximum allowable decrease (increase) in c_j that will keep the current solution optimal. For the cases where $u_j \leq d_j \leq v_j$, define r_j equal to $\frac{d_j}{v_j}$ if d_j is positive and $\frac{d_j}{u_j}$ if d_j is negative. By definition, $0 \leq r_j \leq 1$. The 100% rule says that a sufficient (but not necessary) condition for the current solution to remain optimal is that $r_1 + r_2 + \cdots + r_n \leq 1$. If the condition is not satisfied, the current solution may or may not remain optimal. The rule does not apply if d_j falls outside the specified ranges.

Demonstrate that the 100% optimality rule is too weak to be consistently reliable as a decision-making tool by applying it to the following cases:

(a) Parts (ii) and (iii) of Problem 1.

(b) Part (b) of Problem 7.

3.6.4 Sensitivity Analysis with TORA, Solver, and AMPL

We now have all the tools needed to decipher the output provided by LP software, particularly with regard to sensitivity analysis. We will use the TOYCO example to demonstrate the TORA, Solver, and AMPL output.

TORA's LP output report provides the sensitivity analysis data automatically as shown in Figure 3.14 (file toraTOYCO.txt). The output includes the reduced costs and the dual prices as well as their allowable optimality and feasibility ranges.

FIGURE 3.14

TORA sensitivity analysis for the TOYCO model

Sensitivity Analysis

Variable	CurrObjCoeff	MinObjCoeff	MaxObjCoeff	Reduced Cost
x1:	3.00	-infinity	7.00	4.00
x2:	2.00	0.00	10.00	0.00
x3:	5.00	2.33	infinity	0.00
Constraint	**Curr RHS**	**Min RHS**	**Max RHS**	**Dual Price**
1(<):	430.00	230.00	440.00	1.00
2(<):	460.00	440.00	860.00	2.00
3(<):	420.00	400.00	infinity	0.00

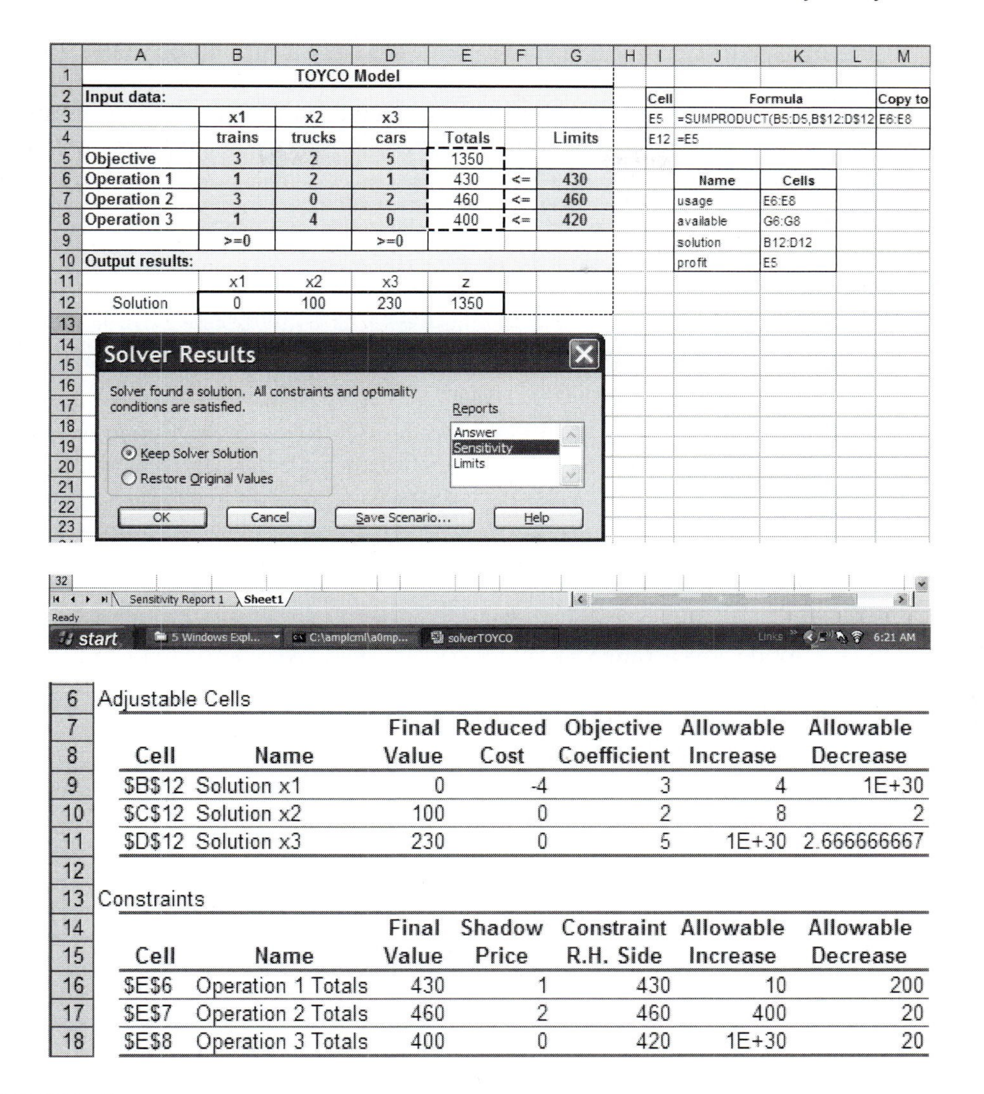

Adjustable Cells

Cell	Name	Final Value	Reduced Cost	Objective Coefficient	Allowable Increase	Allowable Decrease
$B$12	Solution x1	0	-4	3	4	1E+30
$C$12	Solution x2	100	0	2	8	2
$D$12	Solution x3	230	0	5	1E+30	2.666666667

Constraints

Cell	Name	Final Value	Shadow Price	Constraint R.H. Side	Allowable Increase	Allowable Decrease
$E$6	Operation 1 Totals	430	1	430	10	200
$E$7	Operation 2 Totals	460	2	460	400	20
$E$8	Operation 3 Totals	400	0	420	1E+30	20

FIGURE 3.15

Excel Solver sensitivity analysis report for the TOYCO model

Figure 3.15 provides the Solver TOYCO model (file solverTOYCO.xls) and its sensitivity analysis report. After you click Solve in the **Solver Parameters** dialogue box, the new dialogue box **Solver Results** will give you the opportunity to request further details about the solution, including the important sensitivity analysis report. The report will be stored in a separate Excel sheet, as shown by the choices on the bottom of the screen. You can then click **Sensitivity Report 1** to view the results. The report is similar to TORA's with three exceptions: (1) The reduced cost carries an opposite sign. (2) The name *shadow price* replaces the name *dual price*. (3) The optimality ranges are for the changes d_j and D_i rather than for the total objective coefficients and constraints on the right-hand side. The differences are minor and the interpretation of the results remains the same.

In AMPL, the sensitivity analysis report is readily available. File amplTOYCO.txt provides the code necessary to determine the sensitivity analysis output. It requires the following additional statements:

```
option solver cplex;
option cplex_options 'sensitivity';
solve;
#-------------------------------sensitivity analysis
display oper.down,oper.current,oper.up,oper.dual>a.out;
display x.down,x.current,x.up,x.rc>a.out;
```

The CPLEX `option` statements are needed to be able to obtain the standard sensitivity analysis report. In the TOYCO model, the indexed variables and constraints use the root names `x` and `oper`, respectively. Using these names, the suggestive suffixes `.down`, `.current`, and `.up` in the `display` statements automatically generate the formatted sensitivity analysis report in Figure 3.16. The suffixes `.dual` and `.rc` provide the dual price and the reduced cost.

An alternative to AMPL's standard sensitivity analysis report is to actually solve the LP model for a range of values for the objective coefficients and the right-hand side of the constraints. AMPL automates this process through the use of `commands` (see Section A.7). Suppose in the TOYCO model, file amplTOYCO.txt, that we want to investigate the effect of making changes in `b[1]`, the total available time for operation 1. We can do so by moving `solve` and `display` from amplTOYCO.txt to a new file, which we arbitrarily name analysis.txt:

```
repeat while b[1]<=500
{
solve;
display z, x;
let b[1]:=b[1]+1;
};
```

Next, enter the following lines at the ampl prompt:

```
ampl: model amplTOYCO.txt;
ampl: commands analysis.txt;
```

```
: oper.down oper.current oper.up oper.dual    :=
1     230        430         440       1
2     440        460         860       2
3     400        420       1e+20       0
;

:      x.down      x.current     x.up  x.rc     :=
1    -1e+20            3            7    -4
2         0            2           10     0
3         2.33333      5        1e+20     0
;
```

FIGURE 3.16

AMPL sensitivity analysis report for the TOYCO model

The first line will provide the model and its data and the second line will provide the optimum solutions starting with `b[1]` at 430 (the initial value given in amplTOYCO.txt) and continuing in increments of 1 until `b[1]` reaches 500. An examination of the output will then allow us to study the sensitivity of the optimum solution to changes in `b[1]`. Similar procedures can be followed with other coefficients including the case of making simultaneous changes.

PROBLEM SET 3.6E[7]

1. Consider Problem 1, Set 2.3c (Chapter 2). Use the dual price to decide if it is worthwhile to increase the funding for year 4.

2. Consider Problem 2, Set 2.3c (Chapter 2).

(a) Use the dual prices to determine the overall return on investment.

(b) If you wish to spend $1000 on pleasure at the end of year 1, how would this affect the accumulated amount at the start of year 5?

3. Consider Problem 3, Set 2.3c (Chapter 2).

(a) Give an economic interpretation of the dual prices of the model.

(b) Show how the dual price associated with the upper bound on borrowed money at the beginning of the third quarter can be derived from the dual prices associated with the balance equations representing the in-out cash flow at the five designated dates of the year.

4. Consider Problem 4, Set 2.3c (Chapter 2). Use the dual prices to determine the rate of return associated with each year.

***5.** Consider Problem 5, Set 2.3c (Chapter 2). Use the dual price to determine if it is worthwhile for the executive to invest more money in the plans.

6. Consider Problem 6, Set 2.3c (Chapter 2). Use the dual price to decide if it is advisable for the gambler to bet additional money.

7. Consider Problem 1, Set 2.3d (Chapter 2). Relate the dual prices to the unit production costs of the model.

8. Consider Problem 2, Set 2.3d (Chapter 2). Suppose that any additional capacity of machines 1 and 2 can be acquired only by using overtime. What is the maximum cost per hour the company should be willing to incur for either machine?

***9.** Consider Problem 3, Set 2.3d (Chapter 2).

(a) Suppose that the manufacturer can purchase additional units of raw material A at $12 per unit. Would it be advisable to do so?

(b) Would you recommend that the manufacturer purchase additional units of raw material B at $5 per unit?

10. Consider Problem 10, Set 2.3e (Chapter 2).

(a) Which of the specification constraints impacts the optimum solution adversely?

(b) What is the most the company should pay per ton of each ore?

---

[7]Before answering the problems in this set, you are expected to generate the sensitivity analysis report using AMPL, Solver, or TORA.

REFERENCES

Bazaraa, M., J. Jarvis, and H. Sherali, *Linear Programming and Network Flows*, 2nd ed., Wiley, New York, 1990.

Dantzig, G., *Linear Programming and Extensions*, Princeton University Press, Princeton, NJ, 1963.

Dantzig, G., and M. Thapa, *Linear Programming 1: Introduction,* Springer, New York, 1997.

Fourer, R., D. Gay, and B. Kernighan, *AMPL, A Modeling Language for Mathematical Programming*, 2nd ed., Brooks/Cole-Thompson, Pacific Grove, CA, 2003.

Nering, E., and A. Tucker, *Linear Programming and Related Problems*, Academic Press, Boston, 1992.

Schrage, L., *Optimization Modeling with LINGO,* LINDO Systems, Inc., Chicago, 1999.

Taha, H., "Linear Programming," Chapter II-1 in *Handbook of Operations Research*, J. Moder and S. Elmaghraby (eds.), Van Nostrand Reinhold, New York, 1987.

CHAPTER 4

Duality and Post-Optimal Analysis

Chapter Guide. Chapter 3 dealt with the sensitivity of the optimal solution by determining the ranges for the model parameters that will keep the optimum basic solution unchanged. A natural sequel to sensitivity analysis is *post-optimal analysis*, where the goal is to determine the new optimum that results from making targeted changes in the model parameters. Although post-optimal analysis can be carried out using the simplex tableau computations in Section 3.6, this chapter is based entirely on the dual problem.

At a minimum, you will need to study the dual problem and its economic interpretation (Sections 4.1, 4.2, and 4.3). The mathematical definition of the dual problem in Section 4.1 is purely abstract. Yet, when you study Section 4.3, you will see that the dual problem leads to intriguing economic interpretations of the LP model, including *dual prices* and *reduced costs*. It also provides the foundation for the development of the new *dual simplex algorithm*, a prerequisite for post-optimal analysis. The dual simplex algorithm is also needed for integer programming in Chapter 9.

The *generalized simplex algorithm* in Section 4.4.2 is intended to show that the simplex method is not rigid, in the sense that you can modify the rules to handle problems that start both infeasible and nonoptimal. However, this material may be skipped without loss of continuity.

You may use TORA's interactive mode to reinforce your understanding of the computational details of the dual simplex method.

This chapter includes 14 solved examples, 56 end-of-section problems, and 2 cases. The cases are in Appendix E on the CD.

4.1 DEFINITION OF THE DUAL PROBLEM

The **dual** problem is an LP defined directly and systematically from the **primal** (or original) LP model. The two problems are so closely related that the optimal solution of one problem automatically provides the optimal solution to the other.

In most LP treatments, the dual is defined for various forms of the primal depending on the sense of optimization (maximization or minimization), types of constraints

($\leq$, $\geq$, or $=$), and orientation of the variables (nonnegative or unrestricted). This type of treatment is somewhat confusing, and for this reason we offer a *single* definition that automatically subsumes *all* forms of the primal.

Our definition of the dual problem requires expressing the primal problem in the *equation form* presented in Section 3.1 (all the constraints are equations with nonnegative right-hand side and all the variables are nonnegative). This requirement is consistent with the format of the simplex starting tableau. Hence, any results obtained from the primal optimal solution will apply directly to the associated dual problem.

To show how the dual problem is constructed, define the primal in *equation form* as follows:

$$\text{Maximize or minimize } z = \sum_{j=1}^{n} c_j x_j$$

subject to

$$\sum_{j=1}^{n} a_{ij} x_j = b_i, i = 1, 2, \ldots, m$$

$$x_j \geq 0, j = 1, 2, \ldots, n$$

The variables x_j, $j = 1, 2, \ldots, n$, include the surplus, slack, and artificial variables, if any.

Table 4.1 shows how the dual problem is constructed from the primal. Effectively, we have

1. A dual variable is defined for each primal (constraint) equation.
2. A dual constraint is defined for each primal variable.
3. The constraint (column) coefficients of a primal variable define the left-hand-side coefficients of the dual constraint and its objective coefficient define the right-hand side.
4. The objective coefficients of the dual equal the right-hand side of the primal constraint equations.

TABLE 4.1 Construction of the Dual from the Primal

	Primal variables						
	x_1	x_2	$\cdots$	x_j	$\cdots$	x_n	
Dual variables	c_1	c_2	$\cdots$	c_j	$\cdots$	c_n	Right-hand side
y_1	a_{11}	a_{12}	$\cdots$	a_{1j}	$\cdots$	a_{1n}	b_1
y_2	a_{21}	a_{22}	$\cdots$	a_{2j}	$\cdots$	a_{2n}	b_2
$\vdots$	$\vdots$	$\vdots$	$\vdots$	$\vdots$	$\vdots$	$\vdots$	$\vdots$
y_m	a_{m1}	a_{m2}	$\cdots$	a_{mj}	$\cdots$	a_{mn}	b_m
				↑ *j*th dual constraint			↑ Dual objective coefficients

TABLE 4.2 Rules for Constructing the Dual Problem

Primal problem objective[a]	Dual problem		
	Objective	*Constraints type*	*Variables sign*
Maximization	Minimization	$\geq$	Unrestricted
Minimization	Maximization	$\leq$	Unrestricted

[a] All primal constraints are equations with nonnegative right-hand side and all the variables are nonnegative.

The rules for determining the sense of optimization (maximization or minimization), the type of the constraint ($\leq, \geq$, or $=$), and the sign of the dual variables are summarized in Table 4.2. Note that the sense of optimization in the dual is always opposite to that of the primal. An easy way to remember the constraint type in the dual (i.e., $\leq$ or $\geq$) is that if the dual objective is *minimization* (i.e., pointing *down*), then the constraints are all of the type $\geq$ (i.e., pointing *up*). The opposite is true when the dual objective is maximization.

The following examples demonstrate the use of the rules in Table 4.2 and also show that our definition incorporates all forms of the primal automatically.

Example 4.1-1

Primal	Primal in equation form	Dual variables
Maximize $z = 5x_1 + 12x_2 + 4x_3$	Maximize $z = 5x_1 + 12x_2 + 4x_3 + 0x_4$	
subject to	subject to	
$x_1 + 2x_2 + x_3 \leq 10$	$x_1 + 2x_2 + x_3 + x_4 = 10$	y_1
$2x_1 - x_2 + 3x_3 = 8$	$2x_1 - x_2 + 3x_3 + 0x_4 = 8$	y_2
$x_1, x_2, x_3 \geq 0$	$x_1, x_2, x_3, x_4 \geq 0$	

Dual Problem

$$\text{Minimize } w = 10y_1 + 8y_2$$

subject to

$$\begin{aligned} y_1 + 2y_2 &\geq 5 \\ 2y_1 - y_2 &\geq 12 \\ y_1 + 3y_2 &\geq 4 \\ \left.\begin{aligned} y_1 + 0y_2 &\geq 0 \\ y_1, y_2 &\text{ unrestricted} \end{aligned}\right\} &\Rightarrow (y_1 \geq 0,\ y_2 \text{ unrestricted}) \end{aligned}$$

Example 4.1-2

Primal	Primal in equation form	Dual variables
Minimize $z = 15x_1 + 12x_2$	Minimize $z = 15x_1 + 12x_2 + 0x_3 + 0x_4$	
subject to	subject to	
$x_1 + 2x_2 \geq 3$	$x_1 + 2x_2 - x_3 + 0x_4 = 3$	y_1
$2x_1 - 4x_2 \leq 5$	$2x_1 - 4x_2 + 0x_3 + x_4 = 5$	y_2
$x_1, x_2 \geq 0$	$x_1, x_2, x_3, x_4 \geq 0$	

Dual Problem

$$\text{Maximize } w = 3y_1 + 5y_2$$

subject to

$$\begin{aligned} y_1 + 2y_2 &\leq 15 \\ 2y_1 - 4y_2 &\leq 12 \\ \left.\begin{aligned} -y_1 \quad &\leq 0 \\ y_2 &\leq 0 \\ y_1, y_2 \text{ unrestricted} \end{aligned}\right\} &\Rightarrow (y_1 \geq 0, y_2 \leq 0) \end{aligned}$$

Example 4.1-3

Primal	Primal in equation form	Dual variables
	Substitute $x_1 = x_1^+ - x_1^-$	
Maximize $z = 5x_1 + 6x_2$	Maximize $z = 5x_1^+ - 5x_1^- + 6x_2$	
subject to	subject to	
$x_1 + 2x_2 = 5$	$x_1^- - x_1^+ + 2x_2 = 5$	y_1
$-x_1 + 5x_2 \geq 3$	$-x_1^- + x_1^+ + 5x_2 - x_3 = 3$	y_2
$4x_1 + 7x_2 \leq 8$	$4x_1^- - 4x_1^+ + 7x_2 + x_4 = 8$	y_3
x_1 unrestricted, $x_2 \geq 0$	$x_1^-, x_1^+, x_2, x_3, x_4 \geq 0$	

Dual Problem

$$\text{Minimize } z = 5y_1 + 3y_2 + 8y_3$$

subject to

$$\left.\begin{aligned} y_1 - y_2 + 4y_3 &\geq 5 \\ -y_1 + y_2 - 4y_3 &\geq -5 \end{aligned}\right\} \Rightarrow (y_1 - y_2 + 4y_3 = 5)$$

$$2y_1 + 5y_2 + 7y_3 \geq 6$$

$$\left.\begin{aligned} -y_2 \quad &\geq 0 \\ y_3 &\geq 0 \\ y_1, y_2, y_3 \text{ unrestricted} \end{aligned}\right\} \Rightarrow (y_1 \text{ unrestricted}, y_2 \leq 0, y_3 \geq 0)$$

The first and second constraints are replaced by an equation. The general rule in this case is that an unrestricted primal variable always corresponds to an equality dual constraint. Conversely, a primal equation produces an unrestricted dual variable, as the first primal constraint demonstrates.

Summary of the Rules for Constructing the Dual. The general conclusion from the preceding examples is that the variables and constraints in the primal and dual problems are defined by the rules in Table 4.3. It is a good exercise to verify that these explicit rules are subsumed by the general rules in Table 4.2.

TABLE 4.3 Rules for Constructing the Dual Problem

Maximization problem		Minimization problem
Constraints		*Variables*
$\geq$	$\Leftrightarrow$	≤ 0
$\leq$	$\Leftrightarrow$	≥ 0
$=$	$\Leftrightarrow$	Unrestricted
Variables		*Constraints*
≥ 0	$\Leftrightarrow$	$\geq$
≤ 0	$\Leftrightarrow$	$\leq$
Unrestricted	$\Leftrightarrow$	$=$

Note that the table does not use the designation primal and dual. What matters here is the sense of optimization. If the primal is maximization, then the dual is minimization, and vice versa.

PROBLEM SET 4.1A

1. In Example 4.1-1, derive the associated dual problem if the sense of optimization in the primal problem is changed to minimization.

***2.** In Example 4.1-2, derive the associated dual problem given that the primal problem is augmented with a third constraint, $3x_1 + x_2 = 4$.

3. In Example 4.1-3, show that even if the sense of optimization in the primal is changed to minimization, an unrestricted primal variable always corresponds to an equality dual constraint.

4. Write the dual for each of the following primal problems:

(a) Maximize $z = -5x_1 + 2x_2$
subject to

$$-x_1 + x_2 \leq -2$$
$$2x_1 + 3x_2 \leq 5$$
$$x_1, x_2 \geq 0$$

(b) Minimize $z = 6x_1 + 3x_2$
subject to

$$6x_1 - 3x_2 + x_3 \geq 2$$
$$3x_1 + 4x_2 + x_3 \geq 5$$
$$x_1, x_2, x_3 \geq 0$$

***(c)** Maximize $z = x_1 + x_2$
subject to

$$2x_1 + x_2 = 5$$
$$3x_1 - x_2 = 6$$
$$x_1, x_2 \text{ unrestricted}$$

*5. Consider Example 4.1-1. The application of the simplex method to the primal requires the use of an artificial variable in the second constraint of the standard primal to secure a starting basic solution. Show that the presence of an artificial primal in equation form variable does not affect the definition of the dual because it leads to a redundant dual constraint.

6. True or False?
   (a) The dual of the dual problem yields the original primal.
   (b) If the primal constraint is originally in equation form, the corresponding dual variable is necessarily unrestricted.
   (c) If the primal constraint is of the type $\leq$, the corresponding dual variable will be nonnegative (nonpositive) if the primal objective is maximization (minimization).
   (d) If the primal constraint is of the type $\geq$, the corresponding dual variable will be nonnegative (nonpositive) if the primal objective is minimization (maximization).
   (e) An unrestricted primal variable will result in an equality dual constraint.

4.2 PRIMAL-DUAL RELATIONSHIPS

Changes made in the original LP model will change the elements of the current optimal tableau, which in turn may affect the optimality and/or the feasibility of the current solution. This section introduces a number of primal-dual relationships that can be used to recompute the elements of the optimal simplex tableau. These relationships will form the basis for the economic interpretation of the LP model as well as for post-optimality analysis.

This section starts with a brief review of matrices, a convenient tool for carrying out the simplex tableau computations.

4.2.1 Review of Simple Matrix Operations

The simplex tableau computations use only three elementary matrix operations: (row vector) $\times$ (matrix), (matrix) $\times$ (column vector), and (scalar) $\times$ (matrix). These operations are summarized here for convenience. First, we introduce some matrix definitions:[1]

1. A *matrix*, **A**, of size $(m \times n)$ is a rectangular array of elements with m rows and n columns.
2. A *row vector*, **V**, of size m is a $(1 \times m)$ matrix.
3. A *column vector*, **P**, of size n is an $(n \times 1)$ matrix.

These definitions can be represented mathematically as

$$\mathbf{V} = (v_1, v_2, \ldots, v_m), \mathbf{A} = \begin{pmatrix} a_{11} & a_{12} & \vdots & a_{1n} \\ a_{21} & a_{22} & \vdots & a_{2n} \\ \cdots & \cdots & \cdots & \cdots \\ a_{m1} & a_{m2} & \vdots & a_{mn} \end{pmatrix}, \mathbf{P} = \begin{pmatrix} p_1 \\ p_2 \\ \cdots \\ p_n \end{pmatrix}$$

[1]Appendix D on the CD provides a more complete review of matrices.

1. (Row vector $\times$ matrix, VA). The operation is defined only if the size of the row vector **V** equals the number of rows of **A**. In this case,

$$\mathbf{VA} = \left(\sum_{i=1}^{m} v_i a_{i1}, \sum_{i=1}^{m} v_i a_{i2}, \ldots, \sum_{i=1}^{m} v_i a_{in} \right)$$

For example,

$$(11, 22, 33)\begin{pmatrix} 1 & 2 \\ 3 & 4 \\ 5 & 6 \end{pmatrix} = (1 \times 11 + 3 \times 22 + 5 \times 33, 2 \times 11 + 4 \times 22 + 6 \times 33)$$

$$= (242, 308)$$

2. (Matrix $\times$ column vector, AP). The operation is defined only if the number of columns of **A** equals the size of column vector **P**. In this case,

$$\mathbf{AP} = \begin{pmatrix} \sum_{j=1}^{n} a_{1j} p_j \\ \sum_{j=1}^{n} a_{2j} p_j \\ \vdots \\ \sum_{j=1}^{n} a_{mj} p_j \end{pmatrix}$$

As an illustration, we have

$$\begin{pmatrix} 1 & 3 & 5 \\ 2 & 4 & 6 \end{pmatrix}\begin{pmatrix} 11 \\ 22 \\ 33 \end{pmatrix} = \begin{pmatrix} 1 \times 11 + 3 \times 22 + 5 \times 33 \\ 2 \times 11 + 4 \times 22 + 6 \times 33 \end{pmatrix} = \begin{pmatrix} 242 \\ 308 \end{pmatrix}$$

3. (Scalar $\times$ matrix, αA). Given the scalar (or constant) quantity α, the multiplication operation $\alpha\mathbf{A}$ will result in a matrix of the same size as **A** whose (i, j)th element equals αa_{ij}. For example, given $\alpha = 10$,

$$(10)\begin{pmatrix} 1 & 2 & 3 \\ 4 & 5 & 6 \end{pmatrix} = \begin{pmatrix} 10 & 20 & 30 \\ 40 & 50 & 60 \end{pmatrix}$$

In general, $\alpha\mathbf{A} = \mathbf{A}\alpha$. The same operation is extended equally to the multiplication of vectors by scalars. For example, $\alpha\mathbf{V} = \mathbf{V}\alpha$ and $\alpha\mathbf{P} = \mathbf{P}\alpha$.

PROBLEM SET 4.2A

1. Consider the following matrices:

$$\mathbf{A} = \begin{pmatrix} 1 & 4 \\ 2 & 5 \\ 3 & 6 \end{pmatrix}, \mathbf{P}_1 = \begin{pmatrix} 1 \\ 2 \end{pmatrix}, \mathbf{P}_2 = \begin{pmatrix} 1 \\ 2 \\ 3 \end{pmatrix}$$

$$\mathbf{V}_1 = (11, 22), \mathbf{V}_2 = (-1, -2, -3)$$

In each of the following cases, indicate whether the given matrix operation is legitimate, and, if so, calculate the result.

*(a) $\mathbf{AV}_1$

(b) $\mathbf{AP}_1$

(c) $\mathbf{AP}_2$

(d) $\mathbf{V}_1\mathbf{A}$

*(e) $\mathbf{V}_2\mathbf{A}$

(f) $\mathbf{P}_1\mathbf{P}_2$

(g) $\mathbf{V}_1\mathbf{P}_1$

4.2.2 Simplex Tableau Layout

In Chapter 3, we followed a specific format for setting up the simplex tableau. This format is the basis for the development in this chapter.

Figure 4.1 gives a schematic representation of the *starting* and *general* simplex tableaus. In the starting tableau, the constraint coefficients under the starting variables form an **identity matrix** (all main-diagonal elements equal 1 and all off-diagonal elements equal zero). With this arrangement, subsequent iterations of the simplex tableau generated by the Gauss-Jordan row operations (see Chapter 3) will modify the elements of the identity matrix to produce what is known as the **inverse matrix**. As we will see in the remainder of this chapter, the inverse matrix is key to computing all the elements of the associated simplex tableau.

FIGURE 4.1

Schematic representation of the starting and general simplex tableaus

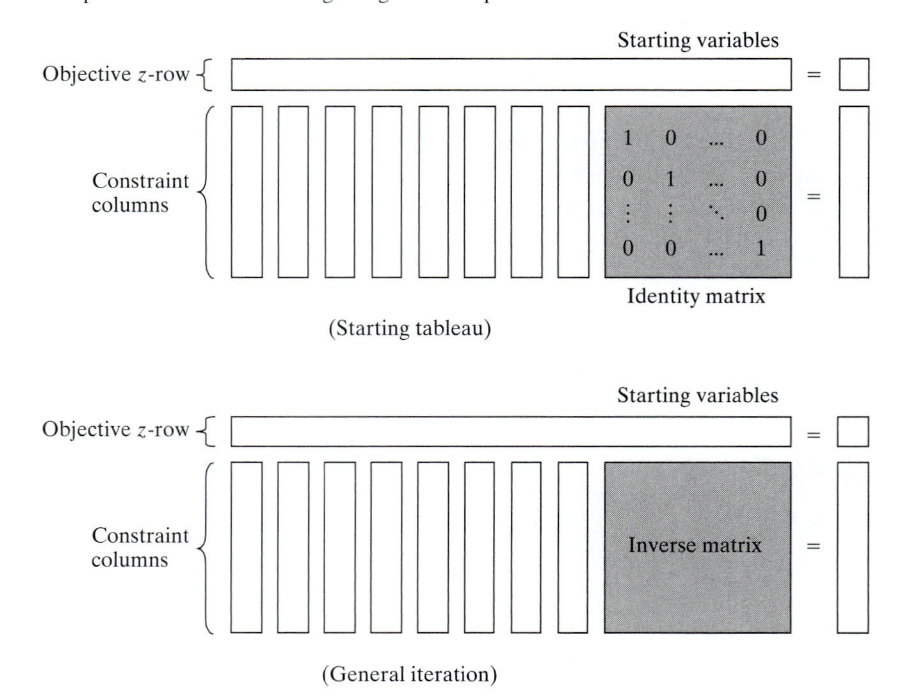

PROBLEM SET 4.2B

1. Consider the optimal tableau of Example 3.3-1.
   *(a) Identify the optimal inverse matrix.
   (b) Show that the right-hand side equals the inverse multiplied by the original right-hand side vector of the original constraints.
2. Repeat Problem 1 for the last tableau of Example 3.4-1.

4.2.3 Optimal Dual Solution

The primal and dual solutions are so closely related that the optimal solution of either problem directly yields (with little additional computation) the optimal solution to the other. Thus, in an LP model in which the number of variables is considerably smaller than the number of constraints, computational savings may be realized by solving the dual, from which the primal solution is determined automatically. This result follows because the amount of simplex computation depends largely (though not totally) on the number of constraints (see Problem 2, Set 4.2c).

This section provides two methods for determining the dual values. Note that the dual of the dual is itself the primal, which means that the dual solution can also be used to yield the optimal primal solution automatically.

Method 1.

$$\begin{pmatrix}\text{Optimal value of} \\ \textit{dual} \text{ variable } y_i\end{pmatrix} = \begin{pmatrix}\text{Optimal primal } z\text{-coefficient of } \textit{starting} \text{ variable } x_i \\ + \\ \textit{Original} \text{ objective coefficient of } x_i\end{pmatrix}$$

Method 2.

$$\begin{pmatrix}\text{Optimal values} \\ \text{of } \textit{dual} \text{ variables}\end{pmatrix} = \begin{pmatrix}\text{Row vector of} \\ \text{original objective coefficients} \\ \text{of optimal } \textit{primal} \text{ basic variables}\end{pmatrix} \times \begin{pmatrix}\text{Optimal } \textit{primal} \\ \text{inverse}\end{pmatrix}$$

The elements of the row vector must appear in the same order in which the basic variables are listed in the *Basic* column of the simplex tableau.

Example 4.2-1

Consider the following LP:

$$\text{Maximize } z = 5x_1 + 12x_2 + 4x_3$$

subject to

$$\begin{aligned} x_1 + 2x_2 + x_3 &\leq 10 \\ 2x_1 - x_2 + 3x_3 &= 8 \\ x_1, x_2, x_3 &\geq 0 \end{aligned}$$

To prepare the problem for solution by the simplex method, we add a slack x_4 in the first constraint and an artificial R in the second. The resulting primal and the associated dual problems are thus defined as follows:

Primal	Dual
Maximize $z = 5x_1 + 12x_2 + 4x_3 - MR$ subject to $x_1 + 2x_2 + x_3 + x_4 = 10$ $2x_1 - x_2 + 3x_3 + R = 8$ $x_1, x_2, x_3, x_4, R \geq 0$	Minimize $w = 10y_1 + 8y_2$ subject to $y_1 + 2y_2 \geq 5$ $2y_1 - y_2 \geq 12$ $y_1 + 3y_2 \geq 4$ $y_1 \geq 0$ $y_2 \geq -M$ ($\Rightarrow y_2$ unrestricted)

Table 4.4 provides the optimal primal tableau.

We now show how the optimal dual values are determined using the two methods described at the start of this section.

Method 1. In Table 4.4, the starting primal variables x_4 and R uniquely correspond to the dual variables y_1 and y_2, respectively. Thus, we determine the optimum dual solution as follows:

Starting primal basic variables	x_4	R
z-equation coefficients	$\frac{29}{5}$	$-\frac{2}{5} + M$
Original objective coefficient	0	$-M$
Dual variables	y_1	y_2
Optimal dual values	$\frac{29}{5} + 0 = \frac{29}{5}$	$-\frac{2}{5} + M + (-M) = -\frac{2}{5}$

Method 2. The optimal inverse matrix, highlighted under the starting variables x_4 and R, is given in Table 4.4 as

$$\text{Optimal inverse} = \begin{pmatrix} \frac{2}{5} & -\frac{1}{5} \\ \frac{1}{5} & \frac{2}{5} \end{pmatrix}$$

First, we note that the optimal primal variables are listed in the tableau in *row order* as x_2 and then x_1. This means that the elements of the original objective coefficients for the two variables must appear in the same order—namely,

$$\begin{aligned}(\text{Original objective coefficients}) &= (\text{Coefficient of } x_2, \text{coefficient of } x_1)\\ &= (12, 5)\end{aligned}$$

TABLE 4.4 Optimal Tableau of the Primal of Example 4.2-1

Basic	x_1	x_2	x_3	x_4	R	Solution
z	0	0	$\frac{3}{5}$	$\frac{29}{5}$	$-\frac{2}{5} + M$	$54\frac{4}{5}$
x_2	0	1	$-\frac{1}{5}$	$\frac{2}{5}$	$-\frac{1}{5}$	$\frac{12}{5}$
x_1	1	0	$\frac{7}{5}$	$\frac{1}{5}$	$\frac{2}{5}$	$\frac{26}{5}$

Thus, the optimal dual values are computed as

$$(y_1, y_2) = \begin{pmatrix}\text{Original objective} \\ \text{coefficients of } x_2, x_1\end{pmatrix} \times (\text{Optimal inverse})$$

$$= (12, 5)\begin{pmatrix}\frac{2}{5} & -\frac{1}{5} \\ \frac{1}{5} & \frac{2}{5}\end{pmatrix}$$

$$= \left(\tfrac{29}{5}, -\tfrac{2}{5}\right)$$

Primal-dual objective values. Having shown how the optimal dual values are determined, next we present the relationship between the primal and dual objective values. For any pair of *feasible* primal and dual solutions,

$$\begin{pmatrix}\text{Objective value in the} \\ \textit{maximization} \text{ problem}\end{pmatrix} \leq \begin{pmatrix}\text{Objective value in the} \\ \textit{minimization} \text{ problem}\end{pmatrix}$$

At the optimum, the relationship holds as a strict equation. The relationship does not specify which problem is primal and which is dual. Only the sense of optimization (maximization or minimization) is important in this case.

The optimum cannot occur with z strictly less than w (i.e., $z < w$) because, no matter how close z and w are, there is always room for improvement, which contradicts optimality as Figure 4.2 demonstrates.

Example 4.2-2

In Example 4.2-1, $\left(x_1 = 0, x_2 = 0, x_3 = \frac{8}{3}\right)$ and $(y_1 = 6, y_2 = 0)$ are feasible primal and dual solutions. The associated values of the objective functions are

$$z = 5x_1 + 12x_2 + 4x_3 = 5(0) + 12(0) + 4\left(\tfrac{8}{3}\right) = 10\tfrac{2}{3}$$

$$w = 10y_1 + 8y_2 = 10(6) + 8(0) = 60$$

Thus, $z\left(= 10\frac{2}{3}\right)$ for the maximization problem (primal) is less than $w\ (= 60)$ for the minimization problem (dual). The optimum value of $z\left(= 54\frac{4}{5}\right)$ falls within the range $\left(10\frac{2}{3}, 60\right)$.

FIGURE 4.2

Relationship between maximum z and minimum w

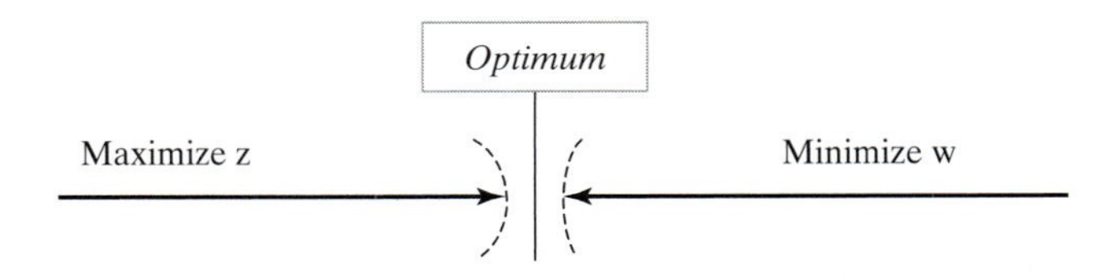

PROBLEM SET 4.2C

1. Find the optimal value of the objective function for the following problem by inspecting only its dual. (Do not solve the dual by the simplex method.)

$$\text{Minimize } z = 10x_1 + 4x_2 + 5x_3$$

subject to

$$5x_1 - 7x_2 + 3x_3 \geq 50$$
$$x_1, x_2, x_3 \geq 0$$

2. Solve the dual of the following problem, then find its optimal solution from the solution of the dual. Does the solution of the dual offer computational advantages over solving the primal directly?

$$\text{Minimize } z = 5x_1 + 6x_2 + 3x_3$$

subject to

$$\begin{aligned}
5x_1 + 5x_2 + 3x_3 &\geq 50 \\
x_1 + x_2 - x_3 &\geq 20 \\
7x_1 + 6x_2 - 9x_3 &\geq 30 \\
5x_1 + 5x_2 + 5x_3 &\geq 35 \\
2x_1 + 4x_2 - 15x_3 &\geq 10 \\
12x_1 + 10x_2 &\geq 90 \\
x_2 - 10x_3 &\geq 20 \\
x_1, x_2, x_3 &\geq 0
\end{aligned}$$

***3.** Consider the following LP:

$$\text{Maximize } z = 5x_1 + 2x_2 + 3x_3$$

subject to

$$\begin{aligned}
x_1 + 5x_2 + 2x_3 &= 30 \\
x_1 - 5x_2 - 6x_3 &\leq 40 \\
x_1, x_2, x_3 &\geq 0
\end{aligned}$$

Given that the artificial variable x_4 and the slack variable x_5 form the starting basic variables and that M was set equal to 100 when solving the problem, the *optimal* tableau is given as

Basic	x_1	x_2	x_3	x_4	x_5	Solution
z	0	23	7	105	0	150
x_1	1	5	2	1	0	30
x_5	0	−10	−8	−1	1	10

Write the associated dual problem and determine its optimal solution in two ways.

4. Consider the following LP:

$$\text{Minimize } z = 4x_1 + x_2$$

subject to

$$3x_1 + x_2 = 3$$
$$4x_1 + 3x_2 \geq 6$$
$$x_1 + 2x_2 \leq 4$$
$$x_1, x_2 \geq 0$$

The starting solution consists of artificial x_4 and x_5 for the first and second constraints and slack x_6 for the third constraint. Using $M = 100$ for the artificial variables, the optimal tableau is given as

Basic	x_1	x_2	x_3	x_4	x_5	x_6	Solution
z	0	0	0	−98.6	−100	−.2	3.4
x_1	1	0	0	.4	0	−.2	.4
x_2	0	1	0	.2	0	.6	1.8
x_3	0	0	1	1	−1	1	1.0

Write the associated dual problem and determine its optimal solution in two ways.

5. Consider the following LP:

$$\text{Maximize } z = 2x_1 + 4x_2 + 4x_3 - 3x_4$$

subject to

$$x_1 + x_2 + x_3 = 4$$
$$x_1 + 4x_2 + x_4 = 8$$
$$x_1, x_2, x_3, x_4 \geq 0$$

Using x_3 and x_4 as starting variables, the optimal tableau is given as

Basic	x_1	x_2	x_3	x_4	Solution
z	2	0	0	3	16
x_3	.75	0	1	−.25	2
x_2	.25	1	0	.25	2

Write the associated dual problem and determine its optimal solution in two ways.

***6.** Consider the following LP:

$$\text{Maximize } z = x_1 + 5x_2 + 3x_3$$

subject to

$$x_1 + 2x_2 + x_3 = 3$$
$$2x_1 - x_2 = 4$$
$$x_1, x_2, x_3 \geq 0$$

The starting solution consists of x_3 in the first constraint and an artificial x_4 in the second constraint with $M = 100$. The optimal tableau is given as

Basic	x_1	x_2	x_3	x_4	Solution
z	0	2	0	99	5
x_3	1	2.5	1	$-.5$	1
x_1	0	$-.5$	0	.5	2

Write the associated dual problem and determine its optimal solution in two ways.

7. Consider the following set of inequalities:

$$\begin{aligned} 2x_1 + 3x_2 &\leq 12 \\ -3x_1 + 2x_2 &\leq -4 \\ 3x_1 - 5x_2 &\leq 2 \\ x_1 &\text{ unrestricted} \\ x_2 &\geq 0 \end{aligned}$$

A feasible solution can be found by augmenting the trivial objective function Maximize $z = x_1 + x_2$ and then solving the problem. Another way is to solve the dual; from which a solution for the set of inequalities can be found. Apply the two methods.

8. Estimate a range for the optimal objective value for the following LPs:

***(a)** Minimize $z = 5x_1 + 2x_2$
subject to

$$\begin{aligned} x_1 - x_2 &\geq 3 \\ 2x_1 + 3x_2 &\geq 5 \\ x_1, x_2 &\geq 0 \end{aligned}$$

(b) Maximize $z = x_1 + 5x_2 + 3x_3$
subject to

$$\begin{aligned} x_1 + 2x_2 + x_3 &= 3 \\ 2x_1 - x_2 &= 4 \\ x_1, x_2, x_3 &\geq 0 \end{aligned}$$

(c) Maximize $z = 2x_1 + x_2$
subject to

$$\begin{aligned} x_1 - x_2 &\leq 10 \\ 2x_1 &\leq 40 \\ x_1, x_2 &\geq 0 \end{aligned}$$

(d) Maximize $z = 3x_1 + 2x_2$
subject to

$$\begin{aligned} 2x_1 + x_2 &\leq 3 \\ 3x_1 + 4x_2 &\leq 12 \\ x_1, x_2 &\geq 0 \end{aligned}$$

9. In Problem 7(a), let y_1 and y_2 be the dual variables. Determine whether the following pairs of primal-dual solutions are optimal:

***(a)** $(x_1 = 3, x_2 = 1; y_1 = 4, y_2 = 1)$

(b) $(x_1 = 4, x_2 = 1; y_1 = 1, y_2 = 0)$

(c) $(x_1 = 3, x_2 = 0; y_1 = 5, y_2 = 0)$

4.2.4 Simplex Tableau Computations

This section shows how *any iteration* of the entire simplex tableau can be generated from the *original* data of the problem, the *inverse* associated with the iteration, and the dual problem. Using the layout of the simplex tableau in Figure 4.1, we can divide the computations into two types:

1. Constraint columns (left- and right-hand sides).
2. Objective z-row.

Formula 1: Constraint Column Computations. In any simplex iteration, a left-hand or a right-hand side column is computed as follows:

$$\begin{pmatrix}\text{Constraint column} \\ \text{in iteration } i\end{pmatrix} = \begin{pmatrix}\text{Inverse in} \\ \text{iteration } i\end{pmatrix} \times \begin{pmatrix}\text{Original} \\ \text{constraint column}\end{pmatrix}$$

Formula 2: Objective z-row Computations. In any simplex iteration, the objective equation coefficient (reduced cost) of x_j is computed as follows:

$$\begin{pmatrix}\textit{Primal } z\text{-equation} \\ \text{coefficient of variable } x_j\end{pmatrix} = \begin{pmatrix}\text{Left-hand side of} \\ j\text{th } \textit{dual} \text{ constraint}\end{pmatrix} - \begin{pmatrix}\text{Right-hand side of} \\ j\text{th } \textit{dual} \text{ constraint}\end{pmatrix}$$

Example 4.2-3

We use the LP in Example 4.2-1 to illustrate the application of Formulas 1 and 2. From the optimal tableau in Table 4.4, we have

$$\text{Optimal inverse} = \begin{pmatrix}\frac{2}{5} & -\frac{1}{5} \\ \frac{1}{5} & \frac{2}{5}\end{pmatrix}$$

The use of Formula 1 is illustrated by computing all the left- and right-hand side columns of the optimal tableau:

$$\begin{pmatrix}x_1\text{-column in} \\ \text{optimal iteration}\end{pmatrix} = \begin{pmatrix}\text{Inverse in} \\ \text{optimal iteration}\end{pmatrix} \times \begin{pmatrix}\text{original} \\ x_1\text{-column}\end{pmatrix}$$

$$= \begin{pmatrix}\frac{2}{5} & -\frac{1}{5} \\ \frac{1}{5} & \frac{2}{5}\end{pmatrix} \times \begin{pmatrix}1 \\ 2\end{pmatrix} = \begin{pmatrix}0 \\ 1\end{pmatrix}$$

In a similar manner, we compute the remaining constraint columns; namely,

$$\begin{pmatrix} x_2\text{-column in} \\ \text{optimal iteration} \end{pmatrix} = \begin{pmatrix} \frac{2}{5} & -\frac{1}{5} \\ \frac{1}{5} & \frac{2}{5} \end{pmatrix} \times \begin{pmatrix} 2 \\ -1 \end{pmatrix} = \begin{pmatrix} 1 \\ 0 \end{pmatrix}$$

$$\begin{pmatrix} x_3\text{-column in} \\ \text{optimal iteration} \end{pmatrix} = \begin{pmatrix} \frac{2}{5} & -\frac{1}{5} \\ \frac{1}{5} & \frac{2}{5} \end{pmatrix} \times \begin{pmatrix} 1 \\ 3 \end{pmatrix} = \begin{pmatrix} -\frac{1}{5} \\ \frac{7}{5} \end{pmatrix}$$

$$\begin{pmatrix} x_4\text{-column in} \\ \text{optimal iteration} \end{pmatrix} = \begin{pmatrix} \frac{2}{5} & -\frac{1}{5} \\ \frac{1}{5} & \frac{2}{5} \end{pmatrix} \times \begin{pmatrix} 1 \\ 0 \end{pmatrix} = \begin{pmatrix} \frac{2}{5} \\ \frac{1}{5} \end{pmatrix}$$

$$\begin{pmatrix} R\text{-column in} \\ \text{optimal iteration} \end{pmatrix} = \begin{pmatrix} \frac{2}{5} & -\frac{1}{5} \\ \frac{1}{5} & \frac{2}{5} \end{pmatrix} \times \begin{pmatrix} 0 \\ 1 \end{pmatrix} = \begin{pmatrix} -\frac{1}{5} \\ \frac{2}{5} \end{pmatrix}$$

$$\begin{pmatrix} \text{Right-hand side} \\ \text{column in} \\ \text{optimal iteration} \end{pmatrix} = \begin{pmatrix} x_2 \\ x_1 \end{pmatrix} = \begin{pmatrix} \frac{2}{5} & -\frac{1}{5} \\ \frac{1}{5} & \frac{2}{5} \end{pmatrix} \times \begin{pmatrix} 10 \\ 8 \end{pmatrix} = \begin{pmatrix} \frac{12}{5} \\ \frac{26}{5} \end{pmatrix}$$

Next, we demonstrate how the objective row computations are carried out using Formula 2. The optimal values of the dual variables, $(y_1, y_2) = \left(\frac{29}{5}, -\frac{2}{5}\right)$, were computed in Example 4.2-1 using two different methods. These values are used in Formula 2 to determine the associated z-coefficients; namely,

$$\begin{aligned}
z\text{-cofficient of } x_1 &= y_1 + 2y_2 - 5 &&= \tfrac{29}{5} + 2 \times -\tfrac{2}{5} - 5 &&= 0 \\
z\text{-cofficient of } x_2 &= 2y_1 - y_2 - 12 &&= 2 \times \tfrac{29}{5} - \left(-\tfrac{2}{5}\right) - 12 &&= 0 \\
z\text{-cofficient of } x_3 &= y_1 + 3y_2 - 4 &&= \tfrac{29}{5} + 3 \times -\tfrac{2}{5} - 4 &&= \tfrac{3}{5} \\
z\text{-cofficient of } x_4 &= y_1 - 0 &&= \tfrac{29}{5} - 0 &&= \tfrac{29}{5} \\
z\text{-cofficient of } R &= y_2 - (-M) &&= -\tfrac{2}{5} - (-M) &&= -\tfrac{2}{5} + M
\end{aligned}$$

Notice that Formula 1 and Formula 2 calculations can be applied at any iteration of either the primal or the dual problems. All we need is the inverse associated with the (primal or dual) iteration and the original LP data.

PROBLEM SET 4.2D

1. Generate the first simplex iteration of Example 4.2-1 (you may use TORA's Iterations ⇒ M-method for convenience), then use Formulas 1 and 2 to verify all the elements of the resulting tableau.
2. Consider the following LP model:

$$\text{Maximize } z = 4x_1 + 14x_2$$

subject to

$$\begin{aligned}
2x_1 + 7x_2 + x_3 \quad &= 21 \\
7x_1 + 2x_2 \quad + x_4 &= 21 \\
x_1, x_2, x_3, x_4 &\geq 0
\end{aligned}$$

Check the optimality and feasibility of each of the following basic solutions.

***(a)** Basic variables $= (x_2, x_4)$, Inverse $= \begin{pmatrix} \frac{1}{7} & 0 \\ -\frac{2}{7} & 1 \end{pmatrix}$

(b) Basic variables $= (x_2, x_3)$, Inverse $= \begin{pmatrix} 0 & \frac{1}{2} \\ 1 & -\frac{7}{2} \end{pmatrix}$

(c) Basic variables $= (x_2, x_1)$, Inverse $= \begin{pmatrix} \frac{7}{45} & -\frac{2}{45} \\ -\frac{2}{45} & \frac{7}{45} \end{pmatrix}$

(d) Basic variables $= (x_1, x_4)$, Inverse $= \begin{pmatrix} \frac{1}{2} & 0 \\ -\frac{7}{2} & 1 \end{pmatrix}$

3. Consider the following LP model:

$$\text{Maximize } z = 3x_1 + 2x_2 + 5x_3$$

subject to

$$\begin{aligned} x_1 + 2x_2 + x_3 + x_4 \quad\quad &= 30 \\ 3x_1 \quad + 2x_3 \quad + x_5 \quad &= 60 \\ x_1 + 4x_2 \quad\quad + x_6 &= 20 \\ x_1, x_2, x_3, x_4, x_5, x_6 &\geq 0 \end{aligned}$$

Check the optimality and feasibility of the following basic solutions:

(a) Basic variables $= (x_4, x_3, x_6)$, Inverse $= \begin{pmatrix} 1 & -\frac{1}{2} & 0 \\ 0 & \frac{1}{2} & 0 \\ 0 & 0 & 1 \end{pmatrix}$

(b) Basic variables $= (x_2, x_3, x_1)$, Inverse $= \begin{pmatrix} \frac{1}{4} & -\frac{1}{8} & \frac{1}{8} \\ \frac{3}{2} & -\frac{1}{4} & -\frac{3}{4} \\ -1 & \frac{1}{2} & \frac{1}{2} \end{pmatrix}$

(c) Basic variables $= (x_2, x_3, x_6)$, Inverse $= \begin{pmatrix} \frac{1}{2} & -\frac{1}{4} & 0 \\ 0 & \frac{1}{2} & 0 \\ -2 & 1 & 1 \end{pmatrix}$

***4.** Consider the following LP model:

$$\text{Minimize } z = 2x_1 + x_2$$

subject to

$$\begin{aligned} 3x_1 + x_2 - x_3 \quad\quad &= 3 \\ 4x_1 + 3x_2 \quad - x_4 \quad &= 6 \\ x_1 + 2x_2 \quad\quad + x_5 &= 3 \\ x_1, x_2, x_3, x_4, x_5 &\geq 0 \end{aligned}$$

Compute the entire simplex tableau associated with the following basic solution and check it for optimality and feasibility.

$$\text{Basic variables} = (x_1, x_2, x_5), \text{Inverse} = \begin{pmatrix} \frac{3}{5} & -\frac{1}{5} & 0 \\ -\frac{4}{5} & \frac{3}{5} & 0 \\ 1 & -1 & 1 \end{pmatrix}$$

5. Consider the following LP model:

$$\text{Maximize } z = 5x_1 + 12x_2 + 4x_3$$

subject to

$$\begin{aligned} x_1 + 2x_2 + x_3 + x_4 &= 10 \\ 2x_1 - x_2 + 3x_3 &= 2 \\ x_1, x_2, x_3, x_4 &\geq 0 \end{aligned}$$

(a) Identify the best solution from among the following basic feasible solutions:

(i) $\text{Basic variables} = (x_4, x_3), \text{Inverse} = \begin{pmatrix} 1 & -\frac{1}{3} \\ 0 & \frac{1}{3} \end{pmatrix}$

(ii) $\text{Basic variables} = (x_2, x_1), \text{Inverse} = \begin{pmatrix} \frac{2}{5} & -\frac{1}{5} \\ \frac{1}{5} & \frac{2}{5} \end{pmatrix}$

(iii) $\text{Basic variables} = (x_2, x_3), \text{Inverse} = \begin{pmatrix} \frac{3}{7} & -\frac{1}{7} \\ \frac{1}{7} & \frac{2}{7} \end{pmatrix}$

(b) Is the solution obtained in (a) optimum for the LP model?

6. Consider the following LP model:

$$\text{Maximize } z = 5x_1 + 2x_2 + 3x_3$$

subject to

$$\begin{aligned} x_1 + 5x_2 + 2x_3 &\leq b_1 \\ x_1 - 5x_2 - 6x_3 &\leq b_2 \\ x_1, x_2, x_3 &\geq 0 \end{aligned}$$

The following optimal tableau corresponds to specific values of b_1 and b_2:

Basic	x_1	x_2	x_3	x_4	x_5	Solution
z	0	a	7	d	e	150
x_1	1	b	2	1	0	30
x_5	0	c	−8	−1	1	10

Determine the following:

(a) The right-hand-side values, b_1 and b_2.

(b) The optimal dual solution.

(c) The elements a, b, c, d, e.

***7.** The following is the optimal tableau for a maximization LP model with three ($\leq$) constraints and all nonnegative variables. The variables x_3, x_4, and x_5 are the slacks associated with the three constraints. Determine the associated optimal objective value in two different ways by using the primal and dual objective functions.

Basic	x_1	x_2	x_3	x_4	x_5	Solution
z	0	0	0	3	2	?
x_3	0	0	1	1	-1	2
x_2	0	1	0	1	0	6
x_1	1	0	0	-1	1	2

8. Consider the following LP:

$$\text{Maximize } z = 2x_1 + 4x_2 + 4x_3 - 3x_4$$

subject to

$$\begin{aligned} x_1 + x_2 + x_3 \quad\quad &= 4 \\ x_1 + 4x_2 \quad\quad + x_4 &= 8 \\ x_1, x_2, x_3, x_4 &\geq 0 \end{aligned}$$

Use the dual problem to show that the basic solution (x_1, x_2) is not optimal.

9. Show that Method 1 in Section 4.2.3 for determining the optimal dual values is actually based on the Formula 2 in Section 4.2.4.

4.3 ECONOMIC INTERPRETATION OF DUALITY

The linear programming problem can be viewed as a resource allocation model in which the objective is to maximize revenue subject to the availability of limited resources. Looking at the problem from this standpoint, the associated dual problem offers interesting economic interpretations of the LP resource allocation model.

To formalize the discussion, we consider the following representation of the general primal and dual problems:

Primal	Dual
Maximize $z = \sum_{j=1}^{n} c_j x_j$	Minimize $w = \sum_{i=1}^{m} b_i y_i$
subject to	subject to
$\sum_{j=1}^{n} a_{ij} x_j \leq b_i,\ i = 1, 2, \ldots, m$	$\sum_{i=1}^{m} a_{ij} y_i \geq c_j,\ j = 1, 2, \ldots, n$
$x_j \geq 0,\ j=1, 2, \ldots, n$	$y_i \geq 0,\ i = 1, 2, \ldots, m$

Viewed as a resource allocation model, the primal problem has n economic activities and m resources. The coefficient c_j in the primal represents the revenue per unit of activity j. Resource i, whose maximum availability is b_i, is consumed at the rate a_{ij} units per unit of activity j.

4.3.1 Economic Interpretation of Dual Variables

Section 4.2.3 states that for any two primal and dual *feasible* solutions, the values of the objective functions, when finite, must satisfy the following inequality:

$$z = \sum_{j=1}^{n} c_j x_j \leq \sum_{i=1}^{m} b_i y_i = w$$

The strict equality, $z = w$, holds when both the primal and dual solutions are optimal.

Let us consider the optimal condition $z = w$ first. Given that the primal problem represents a resource allocation model, we can think of z as representing revenue dollars. Because b_i represents the number of units available of resource i, the equation $z = w$ can be expressed dimensionally as

$$\$ = \sum_{i} (\text{units of resource } i) \times (\$ \text{ per unit of resource } i)$$

This means that the dual variable, y_i, represents the **worth per unit** of resource i. As stated in Section 3.6, the standard name **dual** (or **shadow**) **price** of resource i replaces the name *worth per unit* in all LP literature and software packages.

Using the same logic, the inequality $z < w$ associated with any two feasible primal and dual solutions is interpreted as

$$(\text{Revenue}) < (\text{Worth of resources})$$

This relationship says that so long as the total revenue from all the activities is less than the worth of the resources, the corresponding primal and dual solutions are not optimal. Optimality (maximum revenue) is reached only when the resources have been exploited completely, which can happen only when the input (worth of the resources) equals the output (revenue dollars). In economic terms, the system is said to be *unstable* (nonoptimal) when the input (worth of the resources) exceeds the output (revenue). Stability occurs only when the two quantities are equal.

Example 4.3-1

The Reddy Mikks model (Example 2.1-1) and its dual are given as:

Reddy Mikks primal	Reddy Mikks dual
Maximize $z = 5x_1 + 4x_2$	Minimize $w = 24y_1 + 6y_2 + y_3 + 2y_4$
subject to	subject to
$6x_1 + 4x_2 \leq 24$ (resource 1, $M1$)	$6y_1 + y_2 - y_3 \geq 5$
$x_1 + 2x_2 \leq 6$ (resource 2, $M2$)	$4y_1 + 2y_2 + y_3 + y_4 \geq 4$
$-x_1 + x_2 \leq 1$ (resource 3, market)	$y_1, y_2, y_3, y_4 \geq 0$
$x_2 \leq 2$ (resource 4, demand)	
$x_1, x_2 \geq 0$	
Optimal solution:	Optimal solution:
$x_1 = 3, x_2 = 1.5, z = 21$	$y_1 = .75, y_2 = 0.5, y_3 = y_4 = 0, w = 21$

Briefly, the Reddy Mikks model deals with the production of two types of paint (interior and exterior) using two raw materials $M1$ and $M2$ (resources 1 and 2) and subject to market and demand limits represented by the third and fourth constraints. The model determines the amounts (in tons/day) of interior and exterior paints that maximize the daily revenue (expressed in thousands of dollars).

The optimal dual solution shows that the dual price (worth per unit) of raw material $M1$ (resource 1) is $y_1 = .75$ (or \$750 per ton), and that of raw material $M2$ (resource 2) is $y_2 = .5$ (or \$500 per ton). These results hold true for specific *feasibility ranges* as we showed in Section 3.6. For resources 3 and 4, representing the market and demand limits, the dual prices are both zero, which indicates that their associated resources are abundant. Hence, their worth per unit is zero.

PROBLEM SET 4.3A

1. In Example 4.3-1, compute the change in the optimal revenue in each of the following cases (use TORA output to obtain the *feasibility ranges*):

(a) The constraint for raw material $M1$ (resource 1) is $6x_1 + 4x_2 \leq 22$.

(b) The constraint for raw material $M2$ (resource 2) is $x_1 + 2x_2 \leq 4.5$.

(c) The market condition represented by resource 4 is $x_2 \leq 10$.

***2.** NWAC Electronics manufactures four types of simple cables for a defense contractor. Each cable must go through four sequential operations: splicing, soldering, sleeving, and inspection. The following table gives the pertinent data of the situation.

	Minutes per unit				
Cable	*Splicing*	*Soldering*	*Sleeving*	*Inspection*	Unit revenue (\$)
SC320	10.5	20.4	3.2	5.0	9.40
SC325	9.3	24.6	2.5	5.0	10.80
SC340	11.6	17.7	3.6	5.0	8.75
SC370	8.2	26.5	5.5	5.0	7.80
Daily capacity (minutes)	4800.0	9600.0	4700.0	4500.0	

The contractor guarantees a minimum production level of 100 units for each of the four cables.

(a) Formulate the problem as a linear programming model, and determine the optimum production schedule.

(b) Based on the dual prices, do you recommend making increases in the daily capacities of any of the four operations? Explain.

(c) Does the minimum production requirements for the four cables represent an advantage or a disadvantage for NWAC Electronics? Provide an explanation based on the dual prices.

(d) Can the present unit contribution to revenue as specified by the dual price be guaranteed if we increase the capacity of soldering by 10%?

3. BagCo produces leather jackets and handbags. A jacket requires 8 m^2 of leather, and a handbag only 2 m^2. The labor requirements for the two products are 12 and 5 hours, respectively. The current weekly supplies of leather and labor are limited to 1200 m^2 and 1850 hours. The company sells the jackets and handbags at \$350 and \$120, respectively. The objective is to determine the production schedule that maximizes the net revenue. BagCo is considering an expansion of production. What is the maximum purchase price the company should pay for additional leather? For additional labor?

4.3.2 Economic Interpretation of Dual Constraints

The dual constraints can be interpreted by using Formula 2 in Section 4.2.4, which states that at any primal iteration,

$$\text{Objective coefficient of } x_j = \begin{pmatrix}\text{Left-hand side of} \\ \text{dual constraint } j\end{pmatrix} - \begin{pmatrix}\text{Right-hand side of} \\ \text{dual constraint } j\end{pmatrix}$$

$$= \sum_{i=1}^{m} a_{ij}y_i - c_j$$

We use dimensional analysis once again to interpret this equation. The revenue per unit, c_j, of activity j is in dollars per unit. Hence, for consistency, the quantity $\sum_{i=1}^{m} a_{ij}y_i$ must also be in dollars per unit. Next, because c_j represents revenue, the quantity $\sum_{i=1}^{m} a_{ij}y_i$, which appears in the equation with an opposite sign, must represent cost. Thus we have

$$\$ \text{ cost} = \sum_{i=1}^{m} a_{ij}y_i = \sum_{i=1}^{m} \begin{pmatrix}\text{usage of resource } i \\ \text{per unit of activity } j\end{pmatrix} \times \begin{pmatrix}\text{cost per unit} \\ \text{of resource } i\end{pmatrix}$$

The conclusion here is that the dual variable y_i represents the **imputed cost** per unit of resource i, and we can think of the quantity $\sum_{i=1}^{m} a_{ij}y_i$ as the imputed cost of all the resources needed to produce one unit of activity j.

In Section 3.6, we referred to the quantity $\left(\sum_{i=1}^{m} a_{ij}y_i - c_j\right)$ as the **reduced cost** of activity j. The maximization optimality condition of the simplex method says that an increase in the level of an unused (nonbasic) activity j can improve revenue only if its reduced cost is negative. In terms of the preceding interpretation, this condition states that

$$\begin{pmatrix}\text{Imputed cost of} \\ \text{resources used by} \\ \text{one unit of activity } j\end{pmatrix} < \begin{pmatrix}\text{Revenue per unit} \\ \text{of activity } j\end{pmatrix}$$

The maximization optimality condition thus says that it is economically advantageous to increase an activity to a positive level if its unit revenue exceeds its unit imputed cost.

We will use the TOYCO model of Section 3.6 to demonstrate the computation. The details of the model are restated here for convenience.

Example 4.3-2

TOYCO assembles three types of toys: trains, trucks, and cars using three operations. Available assembly times for the three operations are 430, 460, and 420 minutes per day, respectively, and the revenues per toy train, truck, and car are \$3, \$2, and \$5, respectively. The assembly times per train for the three operations are 1, 3, and 1 minutes, respectively. The corresponding times per truck and per car are (2, 0, 4) and (1, 2, 0) minutes (a zero time indicates that the operation is not used).

Letting x_1, x_2, and x_3 represent the daily number of units assembled of trains, trucks and cars, the associated LP model and its dual are given as:

TOYCO primal	TOYCO dual
Maximize $z = 3x_1 + 2x_2 + 5x_3$	Minimize $w = 430y_1 + 460y_2 + 420y_3$
subject to	subject to
$x_1 + 2x_2 + x_3 \leq 430$ (Operation 1)	$y_1 + 3y_2 + y_3 \geq 3$
$3x_1 + 2x_3 \leq 460$ (Operation 2)	$2y_1 + 4y_3 \geq 2$
$x_1 + 4x_2 \leq 420$ (Operation 3)	$y_1 + 2y_2 \geq 5$
$x_1, x_2, x_3 \geq 0$	$y_1, y_2, y_3 \geq 0$
Optimal solution:	Optimal solution:
$x_1 = 0, x_2 = 100, x_3 = 230, z = \$1350$	$y_1 = 1, y_2 = 2, y_3 = 0, w = \$1350$

The optimal primal solution calls for producing no toy trains, 100 toy trucks, and 230 toy cars. Suppose that TOYCO is interested in producing toy trains as well. How can this be achieved? Looking at the problem from the standpoint of the interpretation of the *reduced cost* for x_1, toy trains will become attractive economically only if the imputed cost of the resources used to produce one toy train is strictly less than its unit revenue. TOYCO thus can either increase the unit revenue per unit by raising the unit price, or it can decrease the imputed cost of the used resources ($= y_1 + 3y_2 + y_3$). An increase in unit price may not be possible because of market competition. A decrease in the unit imputed cost is more plausible because it entails making improvements in the assembly operations. Letting r_1, r_2, and r_3 represent the proportions by which the unit times of the three operations are reduced, the problem requires determining r_1, r_2, and r_3 such that the new imputed cost per per toy train is less than its unit revenue—that is,

$$1(1 - r_1)y_1 + 3(1 - r_2)y_2 + 1(1 - r_3)y_3 < 3$$

For the given optimal values of $y_1 = 1$, $y_2 = 2$, and $y_3 = 0$, this inequality reduces to (verify!)

$$r_1 + 6r_2 > 4$$

Thus, any values of r_1 and r_2 between 0 and 1 that satisfy $r_1 + 6r_2 > 4$ should make toy trains profitable. However, this goal may not be achievable because it requires practically impossible reductions in the times of operations 1 and 2. For example, even reductions as high as 50% in these times (that is, $r_1 = r_2 = .5$) fail to satisfy the given condition. Thus, TOYCO should not produce toy trains unless an increase in its unit price is possible.

PROBLEM SET 4.3B

1. In Example 4.3-2, suppose that for toy trains the per-unit time of operation 2 can be reduced from 3 minutes to at most 1.25 minutes. By how much must the per-unit time of operation 1 be reduced to make toy trains just profitable?

***2.** In Example 4.3-2, suppose that TOYCO is studying the possibility of introducing a fourth toy: fire trucks. The assembly does not make use of operation 1. Its unit assembly times on operations 2 and 3 are 1 and 3 minutes, respectively. The revenue per unit is $4. Would you advise TOYCO to introduce the new product?

***3.** JoShop uses lathes and drill presses to produce four types of machine parts, *PP*1, *PP*2, *PP*3, and *PP*4. The table below summarizes the pertinent data.

	Machining time in minutes per unit of				
Machine	*PP1*	*PP2*	*PP3*	*PP4*	Capacity (minutes)
Lathes	2	5	3	4	5300
Drill presses	3	4	6	4	5300
Unit revenue ($)	3	6	5	4	

For the parts that are not produced by the present optimum solution, determine the rate of deterioration in the optimum revenue per unit increase of each of these products.

4. Consider the optimal solution of JoShop in Problem 3. The company estimates that for each part that is not produced (per the optimum solution), an across-the-board 20% reduction in machining time can be realized through process improvements. Would these improvements make these parts profitable? If not, what is the minimum percentage reduction needed to realize revenueability?

4.4 ADDITIONAL SIMPLEX ALGORITHMS

In the simplex algorithm presented in Chapter 3 the problem starts at a (basic) feasible solution. Successive iterations continue to be feasible until the optimal is reached at the last iteration. The algorithm is sometimes referred to as the **primal simplex** method.

This section presents two additional algorithms: The **dual simplex** and the **generalized simplex**. In the dual simplex, the LP starts at a better than optimal *infeasible* (basic) solution. Successive iterations remain infeasible and (better than) optimal until feasibility is restored at the last iteration. The generalized simplex combines both the primal and dual simplex methods in one algorithm. It deals with problems that start both nonoptimal and infeasible. In this algorithm, successive iterations are associated with basic feasible or infeasible (basic) solutions. At the final iteration, the solution becomes optimal and feasible (assuming that one exists).

All three algorithms, the primal, the dual, and the generalized, are used in the course of post-optimal analysis calculations, as will be shown in Section 4.5.

4.4.1 Dual Simplex Algorithm

The crux of the dual simplex method is to start with a better than optimal and infeasible basic solution. The optimality and feasibility conditions are designed to preserve the optimality of the basic solutions while moving the solution iterations toward feasibility.

Dual feasibility condition. The leaving variable, x_r, is the basic variable having the most negative value (ties are broken arbitrarily). If all the basic variables are nonnegative, the algorithm ends.

Dual optimality condition. Given that x_r is the leaving variable, let $\bar{c}_j$ be the reduced cost of nonbasic variable x_j and α_{rj} the constraint coefficient in the x_r-row and x_j-column of the tableau. The entering variable is the nonbasic variable with $\alpha_{rj} < 0$ that corresponds to

$$\min_{\text{Nonbasic } x_j} \left\{ \left|\frac{\bar{c}_j}{\alpha_{rj}}\right|, \alpha_{rj} < 0 \right\}$$

(Ties are broken arbitrarily.) If $\alpha_{rj} \geq 0$ for all nonbasic x_j, the problem has no feasible solution.

To start the LP optimal and infeasible, two requirements must be met:

1. The objective function must satisfy the optimality condition of the regular simplex method (Chapter 3).

2. All the constraints must be of the type $(\leq)$.

The second condition requires converting any $(\geq)$ to $(\leq)$ simply by multiplying both sides of the inequality $(\geq)$ by -1. If the LP includes $(=)$ constraints, the equation can be replaced by two inequalities. For example,

$$x_1 + x_2 = 1$$

is equivalent to

$$x_1 + x_2 \leq 1, x_1 + x_2 \geq 1$$

or

$$x_1 + x_2 \leq 1, -x_1 - x_2 \leq -1$$

After converting all the constraints to $(\leq)$, the starting solution is infeasible if at least one of the right-hand sides of the inequalities is strictly negative.

Example 4.4-1

$$\text{Minimize } z = 3x_1 + 2x_2 + x_3$$

subject to

$$\begin{aligned} 3x_1 + x_2 + x_3 &\geq 3 \\ -3x_1 + 3x_2 + x_3 &\geq 6 \\ x_1 + x_2 + x_3 &\leq 3 \\ x_1, x_2, x_3 &\geq 0 \end{aligned}$$

In the present example, the first two inequalities are multiplied by -1 to convert them to $(\leq)$ constraints. The starting tableau is thus given as:

Basic	x_1	x_2	x_3	x_4	x_5	x_6	Solution
z	-3	-2	-1	0	0	0	0
x_4	-3	-1	-1	1	0	0	-3
x_5	3	-3	-1	0	1	0	-6
x_6	1	1	1	0	0	1	3

The tableau is optimal because all the reduced costs in the z-row are ≤ 0 ($\bar{c}_1 = -3, \bar{c}_2 = -2, \bar{c}_3 = -1, \bar{c}_4 = 0, \bar{c}_5 = 0, \bar{c}_6 = 0$). It is also infeasible because at least one of the basic variables is negative ($x_4 = -3, x_5 = -6, x_6 = 3$).

According to the dual feasibility condition, x_5 $(= -6)$ is the leaving variable. The next table shows how the dual optimality condition is used to determine the entering variable.

	$j = 1$	$j = 2$	$j = 3$
Nonbasic variable	x_1	x_2	x_3
z-row $(\bar{c}_j)$	-3	-2	-1
x_5-row, α_{4j}	3	-3	-1
Ratio, $\lvert\frac{\bar{c}_j}{\alpha_{5j}}\rvert$, $\alpha_{5j} < 0$	—	$\frac{2}{3}$	1

The ratios show that x_2 is the entering variable. Notice that a nonbasic variable x_j is a candidate for entering the basic solution only if its α_{rj} is strictly negative. This is the reason x_1 is excluded in the table above.

The next tableau is obtained by using the familiar row operations, which give

Basic	x_1	x_2	x_3	x_4	x_5	x_6	Solution
z	-5	0	$-\frac{1}{3}$	0	$-\frac{2}{3}$	0	4
x_4	-4	0	$-\frac{2}{3}$	1	$-\frac{1}{3}$	0	-1
x_2	-1	1	$\frac{1}{3}$	0	$-\frac{1}{3}$	0	2
x_6	2	0	$\frac{2}{3}$	0	$\frac{1}{3}$	1	1
Ratio	$\frac{5}{4}$	—	$\frac{1}{2}$	—	2	—	

The preceding tableau shows that x_4 leaves and x_3 enters, thus yielding the following tableau, which is both optimal and feasible:

Basic	x_1	x_2	x_3	x_4	x_5	x_6	Solution
z	-3	0	0	$-\frac{1}{2}$	$-\frac{1}{2}$	0	$\frac{9}{2}$
x_3	6	0	1	$-\frac{3}{2}$	$\frac{1}{2}$	0	$\frac{3}{2}$
x_2	-3	1	0	$\frac{1}{2}$	$-\frac{1}{2}$	0	$\frac{3}{2}$
x_6	-2	0	0	1	0	1	0

Notice how the dual simplex works. In all the iterations, optimality is maintained (all reduced costs are ≤ 0). At the same time, each new iteration moves the solution toward feasibility. At iteration 3, feasibility is restored for the first time and the process ends with the optimal feasible solution given as $x_1 = 0, x_2 = \frac{3}{2}, x_2 = \frac{3}{2}$, and $z = \frac{9}{2}$.

TORA Moment.

TORA provides a tutorial module for the dual simplex method. From the SOLVE/MODIFY menu select Solve ⇒ Algebraic ⇒ Iterations ⇒ Dual Simplex. Remember that you need to convert $(=)$ constraints to inequalities. You do not need to convert $(\geq)$ constraints because TORA will do the conversion internally. If the LP does not satisfy the initial requirements of the dual simplex, a message will appear on the screen.

As in the regular simplex method, the tutorial module allows you to select the entering and the leaving variables beforehand. An appropriate feedback then tells you if your selection is correct.

PROBLEM SET 4.4A[2]

1. Consider the solution space in Figure 4.3, where it is desired to find the optimum extreme point that uses the *dual* simplex method to minimize $z = 2x_1 + x_2$. The optimal solution occurs at point $F = (0.5, 1.5)$ on the graph.
   (a) Can the dual simplex start at point A?
   ***(b)** If the starting basic (infeasible but better than optimum) solution is given by point G, would it be possible for the iterations of the dual simplex method to follow the path $G \rightarrow E \rightarrow F$? Explain.
   (c) If the starting basic (infeasible) solution starts at point L, identify a possible path of the dual simplex method that leads to the optimum feasible point at point F.
2. Generate the dual simplex iterations for the following problems (using TORA for convenience), and trace the path of the algorithm on the graphical solution space.
   (a) Minimize $z = 2x_1 + 3x_2$

FIGURE 4.3

Solution space for Problem 1, Set 4.4a

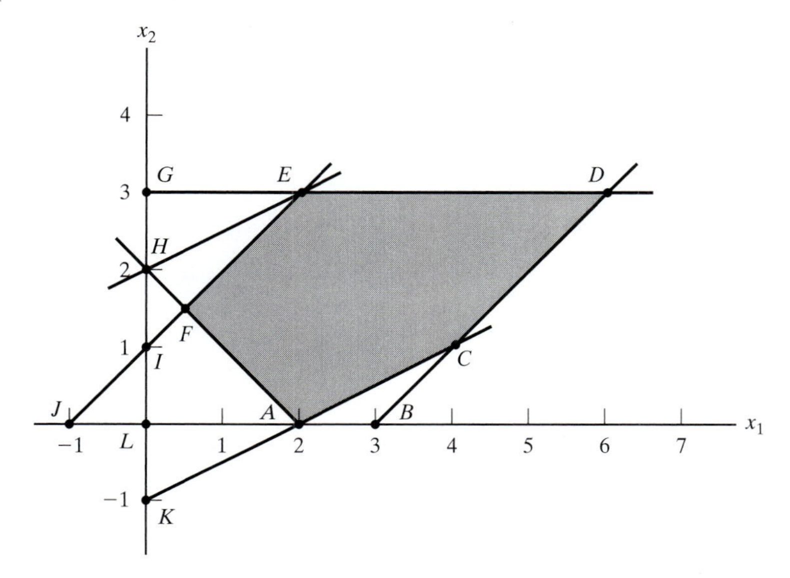

---

[2]You are encouraged to use TORA's tutorial mode where possible to avoid the tedious task of carrying out the Gauss-Jordan row operations. In this manner, you can concentrate on understanding the main ideas of the method.

subject to

$$2x_1 + 2x_2 \leq 30$$
$$x_1 + 2x_2 \geq 10$$
$$x_1, x_2 \geq 0$$

(b) Minimize $z = 5x_1 + 6x_2$
subject to

$$x_1 + x_2 \geq 2$$
$$4x_1 + x_2 \geq 4$$
$$x_1, x_2 \geq 0$$

(c) Minimize $z = 4x_1 + 2x_2$
subject to

$$x_1 + x_2 = 1$$
$$3x_1 - x_2 \geq 2$$
$$x_1, x_2 \geq 0$$

(d) Minimize $z = 2x_1 + 3x_2$
subject to

$$2x_1 + x_2 \geq 3$$
$$x_1 + x_2 = 2$$
$$x_1, x_2 \geq 0$$

3. *Dual Simplex with Artificial Constraints.* Consider the following problem:

$$\text{Maximize } z = 2x_1 - x_2 + x_3$$

subject to

$$2x_1 + 3x_2 - 5x_3 \geq 4$$
$$-x_1 + 9x_2 - x_3 \geq 3$$
$$4x_1 + 6x_2 + 3x_3 \leq 8$$
$$x_1, x_2, x_3 \geq 0$$

The starting basic solution consisting of surplus variables x_4 and x_5 and slack variable x_6 is infeasible because $x_4 = -4$ and $x_5 = -3$. However, the dual simplex is not applicable directly, because x_1 and x_3 do not satisfy the maximization optimality condition. Show that by adding the artificial constraint $x_1 + x_3 \leq M$ (where M is sufficiently large not to eliminate any feasible points in the original solution space), and then using the new constraint as a pivot row, the selection of x_1 as the entering variable (because it has the most negative objective coefficient) will render an all-optimal objective row. Next, carry out the regular dual simplex method on the modified problem.

4. Using the artificial constraint procedure introduced in Problem 3, solve the following problems by the dual simplex method. In each case, indicate whether the resulting solution is feasible, infeasible, or unbounded.

(a) Maximize $z = 2x_3$
subject to

$$\begin{aligned} -x_1 + 2x_2 - 2x_3 &\geq 8 \\ -x_1 + x_2 + x_3 &\leq 4 \\ 2x_1 - x_2 + 4x_3 &\leq 10 \\ x_1, x_2, x_3 &\geq 0 \end{aligned}$$

(b) Maximize $z = x_1 - 3x_2$
subject to

$$\begin{aligned} x_1 - x_2 &\leq 2 \\ x_1 + x_2 &\geq 4 \\ 2x_1 - 2x_2 &\geq 3 \\ x_1, x_2 &\geq 0 \end{aligned}$$

***(c)** Minimize $z = -x_1 + x_2$
subject to

$$\begin{aligned} x_1 - 4x_2 &\geq 5 \\ x_1 - 3x_2 &\leq 1 \\ 2x_1 - 5x_2 &\geq 1 \\ x_1, x_2 &\geq 0 \end{aligned}$$

(d) Maximize $z = 2x_3$
subject to

$$\begin{aligned} -x_1 + 3x_2 - 7x_3 &\geq 5 \\ -x_1 + x_2 - x_3 &\leq 1 \\ 3x_1 + x_2 - 10x_3 &\leq 8 \\ x_1, x_2, x_3 &\geq 0 \end{aligned}$$

5. Solve the following LP in three different ways (use TORA for convenience). Which method appears to be the most efficient computationally?

$$\text{Minimize } z = 6x_1 + 7x_2 + 3x_3 + 5x_4$$

subject to

$$\begin{aligned} 5x_1 + 6x_2 - 3x_3 + 4x_4 &\geq 12 \\ x_2 - 5x_3 - 6x_4 &\geq 10 \\ 2x_1 + 5x_2 + x_3 + x_4 &\geq 8 \\ x_1, x_2, x_3, x_4 &\geq 0 \end{aligned}$$

4.4.2 Generalized Simplex Algorithm

The (primal) simplex algorithm in Chapter 3 starts feasible but nonoptimal. The dual simplex in Section 4.4.1 starts (better than) optimal but infeasible. What if an LP model starts both nonoptimal and infeasible? We have seen that the primal simplex accounts for the infeasibility of the starting solution by using artificial variables. Similarly, the dual simplex accounts for the nonoptimality by using an artificial constraint (see Problem 3, Set 4.4a). Although these procedures are designed to enhance *automatic* computations, such details may cause one to lose sight of what the simplex algorithm truly entails—namely, the optimum solution of an LP is associated with a corner point (or basic) solution. Based on this observation, you should be able to "tailor" your own simplex algorithm for LP models that start both nonoptimal and infeasible. The following example illustrates what we call the generalized simplex algorithm.

Example 4.4-2

Consider the LP model of Problem 4(a), Set 4.4a. The model can be put in the following tableau form in which the starting basic solution (x_3, x_4, x_5) is both nonoptimal (because x_3 has a negative reduced cost) and infeasible (because $x_4 = -8$). (The first equation has been multiplied by -1 to reveal the infeasibility directly in the *Solution* column.)

Basic	x_1	x_2	x_3	x_4	x_5	x_6	Solution
z	0	0	-2	0	0	0	0
x_4	1	-2	2	1	0	0	-8
x_5	-1	1	1	0	1	0	4
x_6	2	-1	4	0	0	1	10

We can solve the problem without the use of any artificial variables or artificial constraints as follows: Remove infeasibility first by applying a version of the dual simplex feasibility condition that selects x_4 as the leaving variable. To determine the entering variable, all we need is a nonbasic variable whose constraint coefficient in the x_4-row is strictly negative. The selection can be done without regard to optimality, because it is nonexistent at this point anyway (compare with the dual optimality condition). In the present example, x_2 has a negative coefficient in the x_4-row and is selected as the entering variable. The result is the following tableau:

Basic	x_1	x_2	x_3	x_4	x_5	x_6	Solution
z	0	0	-2	0	0	0	0
x_2	$-\frac{1}{2}$	1	-1	$-\frac{1}{2}$	0	0	4
x_5	$-\frac{1}{2}$	0	2	$\frac{1}{2}$	1	0	0
x_6	$\frac{3}{2}$	0	3	$-\frac{1}{2}$	0	1	14

The solution in the preceding tableau is now feasible but nonoptimal, and we can use the primal simplex to determine the optimal solution. In general, had we not restored feasibility in the preceding tableau, we would repeat the procedure as necessary until feasibility is satisfied or there is evidence that the problem has no feasible solution (which happens if a basic variable is negative and all its constraint coefficients are nonnegative). Once feasibility is established, the next step is to pay attention to optimality by applying the proper optimality condition of the primal simplex method.

Remarks. The essence of Example 4.4-2 is that the simplex method is not rigid. The literature abounds with variations of the simplex method (e.g., the primal-dual method, the symmetrical method, the criss-cross method, and the multiplex method) that give the impression that each procedure is different, when, in effect, they all seek a corner point solution, with a slant toward automated computations and, perhaps, computational efficiency.

PROBLEM SET 4.4B

1. The LP model of Problem 4(c), Set 4.4a, has no feasible solution. Show how this condition is detected by the *generalized simplex procedure*.
2. The LP model of Problem 4(d), Set 4.4a, has no bounded solution. Show how this condition is detected by the *generalized simplex procedure*.

4.5 POST-OPTIMAL ANALYSIS

In Section 3.6, we dealt with the sensitivity of the optimum solution by determining the ranges for the different parameters that would keep the optimum basic solution unchanged. In this section, we deal with making changes in the parameters of the model and finding the new optimum solution. Take, for example, a case in the poultry industry where an LP model is commonly used to determine the optimal feed mix per broiler (see Example 2.2-2). The weekly consumption per broiler varies from .26 lb (120 grams) for a one-week-old bird to 2.1 lb (950 grams) for an eight-week-old bird. Additionally, the cost of the ingredients in the mix may change periodically. These changes require periodic recalculation of the optimum solution. *Post-optimal analysis* determines the new solution in an efficient way. The new computations are rooted in the use duality and the primal-dual relationships given in Section 4.2.

The following table lists the cases that can arise in post-optimal analysis and the actions needed to obtain the new solution (assuming one exists):

Condition after parameters change	Recommended action
Current solution remains optimal and feasible.	No further action is necessary.
Current solution becomes infeasible.	Use dual simplex to recover feasibility.
Current solution becomes nonoptimal.	Use primal simplex to recover optimality.
Current solution becomes both nonoptimal and infeasible.	Use the generalized simplex method to obtain new solution.

The first three cases are investigated in this section. The fourth case, being a combination of cases 2 and 3, is treated in Problem 6, Set 4.5a.

The TOYCO model of Example 4.3-2 will be used to explain the different procedures. Recall that the TOYCO model deals with the assembly of three types of toys: trains, trucks, and cars. Three operations are involved in the assembly. We wish to determine the number of units of each toy that will maximize revenue. The model and its dual are repeated here for convenience.

TOYCO primal	TOYCO dual
Maximize $z = 3x_1 + 2x_2 + 5x_3$	Minimize $z = 430y_1 + 460y_2 + 420y_3$
subject to	subject to
$x_1 + 2x_2 + x_3 \leq 430$ (Operation 1)	$y_1 + 3y_2 + y_3 \geq 3$
$3x_1 + 2x_3 \leq 460$ (Operation 2)	$2y_1 + 4y_3 \geq 2$
$x_1 + 4x_2 \leq 420$ (Operation 3)	$y_1 + 2y_2 \geq 5$
$x_1, x_2, x_3 \geq 0$	$y_1, y_2, y_3 \geq 0$
Optimal solution:	Optimal solution:
$x_1 = 0, x_2 = 100, x_3 = 230, z = \$1350$	$y_1 = 1, y_2 = 2, y_3 = 0, w = \$1350$

The associated optimum tableau for the primal is given as

Basic	x_1	x_2	x_3	x_4	x_5	x_6	Solution
z	4	0	0	1	2	0	1350
x_2	$-\frac{1}{4}$	1	0	$\frac{1}{2}$	$-\frac{1}{4}$	0	100
x_3	$\frac{3}{2}$	0	1	0	$\frac{1}{2}$	0	230
x_6	2	0	0	-2	1	1	20

4.5.1 Changes Affecting Feasibility

The feasibility of the current optimum solution may be affected only if (1) the right-hand side of the constraints is changed, or (2) a new constraint is added to the model. In both cases, infeasibility occurs when at least one element of the right-hand side of the optimal tableau becomes negative—that is, one or more of the current basic variables become negative.

Changes in the right-hand side. This change requires recomputing the right-hand side of the tableau using Formula 1 in Section 4.2.4:

$$\begin{pmatrix}\text{New right-hand side of}\\ \text{tableau in iteration } i\end{pmatrix} = \begin{pmatrix}\text{Inverse in}\\ \text{iteration } i\end{pmatrix} \times \begin{pmatrix}\text{New right-hand}\\ \text{side of constraints}\end{pmatrix}$$

Recall that the right-hand side of the tableau gives the values of the basic variables.

Example 4.5-1

Situation 1. Suppose that TOYCO wants to expand its assembly lines by increasing the daily capacity of operations 1, 2, and 3 by 40% to 602, 644, and 588 minutes, respectively. How would this change affect the total revenue?

With these increases, the only change that will take place in the optimum tableau is the right-hand side of the constraints (and the optimum objective value). Thus, the new basic solution is computed as follows:

$$\begin{pmatrix} x_2 \\ x_3 \\ x_6 \end{pmatrix} = \begin{pmatrix} \frac{1}{2} & -\frac{1}{4} & 0 \\ 0 & \frac{1}{2} & 0 \\ -2 & 1 & 1 \end{pmatrix} \begin{pmatrix} 602 \\ 644 \\ 588 \end{pmatrix} = \begin{pmatrix} 140 \\ 322 \\ 28 \end{pmatrix}$$

Thus, the current basic variables, x_2, x_3, and x_6, remain feasible at the new values 140, 322, and 28, respectively. The associated optimum revenue is \$1890, which is \$540 more than the current revenue of \$1350.

Situation 2. Although the new solution is appealing from the standpoint of increased revenue, TOYCO recognizes that its implementation may take time. Another proposal was thus made to shift the slack capacity of operation 3 ($x_6 = 20$ *minutes*) to the capacity of operation 1. How would this change impact the optimum solution?

The capacity mix of the three operations changes to 450, 460, and 400 minutes, respectively. The resulting solution is

$$\begin{pmatrix} x_2 \\ x_3 \\ x_6 \end{pmatrix} = \begin{pmatrix} \frac{1}{2} & -\frac{1}{4} & 0 \\ 0 & \frac{1}{2} & 0 \\ -2 & 1 & 1 \end{pmatrix} \begin{pmatrix} 450 \\ 460 \\ 400 \end{pmatrix} = \begin{pmatrix} 110 \\ 230 \\ -40 \end{pmatrix}$$

The resulting solution is infeasible because $x_6 = -40$, which requires applying the dual simplex method to recover feasibility. First, we modify the right-hand side of the tableau as shown by the shaded column. Notice that the associated value of $z = 3 \times 0 + 2 \times 110 + 5 \times 230 = \$1370$.

Basic	x_1	x_2	x_3	x_4	x_5	x_6	Solution
z	4	0	0	1	2	0	1370
x_2	$-\frac{1}{4}$	1	0	$\frac{1}{2}$	$-\frac{1}{4}$	0	110
x_3	$\frac{3}{2}$	0	1	0	$\frac{1}{2}$	0	230
x_6	2	0	0	-2	1	1	-40

From the dual simplex, x_6 leaves and x_4 enters, which yields the following optimal feasible tableau (in general, the dual simplex may take more than one iteration to recover feasibility).

Basic	x_1	x_2	x_3	x_4	x_5	x_6	Solution
z	5	0	0	0	$\frac{5}{2}$	$\frac{1}{2}$	1350
x_2	$\frac{1}{4}$	1	0	0	0	$\frac{1}{4}$	100
x_3	$\frac{3}{2}$	0	1	0	$\frac{1}{2}$	0	230
x_4	-1	0	0	1	$-\frac{1}{2}$	$-\frac{1}{2}$	20

The optimum solution (in terms of x_1, x_2, and x_3) remains the same as in the original model. This means that the proposed shift in capacity allocation is not advantageous in this case because all it does is shift the surplus capacity in operation 3 to a surplus capacity in operation 1. The conclusion is that operation 2 is the bottleneck and it may be advantageous to shift the surplus to operation 2 instead (see Problem 1, Set 4.5a). The selection of operation 2 over operation 1 is also reinforced by the fact that the dual price for operation 2 (\$2/min) is higher than that for operation 2 (= \$1/min).

PROBLEM SET 4.5A

1. In the TOYCO model listed at the start of Section 4.5, would it be more advantageous to assign the 20-minute excess capacity of operation 3 to operation 2 instead of operation 1?
2. Suppose that TOYCO wants to change the capacities of the three operations according to the following cases:

   (a) $\begin{pmatrix} 460 \\ 500 \\ 400 \end{pmatrix}$ **(b)** $\begin{pmatrix} 500 \\ 400 \\ 600 \end{pmatrix}$ **(c)** $\begin{pmatrix} 300 \\ 800 \\ 200 \end{pmatrix}$ **(d)** $\begin{pmatrix} 450 \\ 700 \\ 350 \end{pmatrix}$

   Use post-optimal analysis to determine the optimum solution in each case.
3. Consider the Reddy Mikks model of Example 2.1-1. Its optimal tableau is given in Example 3.3-1. If the daily availabilities of raw materials $M1$ and $M2$ are increased to 28 and 8 tons, respectively, use post-optimal analysis to determine the new optimal solution.

*4. The Ozark Farm has 20,000 broilers that are fed for 8 weeks before being marketed. The weekly feed per broiler varies according to the following schedule:

Week	1	2	3	4	5	6	7	8
lb/broiler	.26	.48	.75	1.00	1.30	1.60	1.90	2.10

For the broiler to reach a desired weight gain in 8 weeks, the feedstuffs must satisfy specific nutritional needs. Although a typical list of feedstuffs is large, for simplicity we will limit the model to three items only: limestone, corn, and soybean meal. The nutritional needs will also be limited to three types: calcium, protein, and fiber. The following table summarizes the nutritive content of the selected ingredients together with the cost data.

	Content (lb) per lb of			
Ingredient	*Calcium*	*Protein*	*Fiber*	\$ per lb
Limestone	.380	.00	.00	.12
Corn	.001	.09	.02	.45
Soybean meal	.002	.50	.08	1.60

The feed mix must contain

(a) At least .8% but not more than 1.2% calcium

(b) At least 22% protein

(c) At most 5% crude fiber

Solve the LP for week 1 and then use post-optimal analysis to develop an optimal schedule for the remaining 7 weeks.

5. Show that the 100% feasibility rule in Problem 12, Set 3.6c (Chapter 3) is based on the condition

$$\begin{pmatrix}\text{Optimum}\\ \text{inverse}\end{pmatrix}\begin{pmatrix}\text{Original right-hand}\\ \text{side vector}\end{pmatrix} \geq 0$$

6. *Post-optimal Analysis for Cases Affecting Both Optimality and Feasibility.* Suppose that you are given the following simultaneous changes in the Reddy Mikks model: The revenue per ton of exterior and interior paints are \$1000 and \$4000, respectively, and the maximum daily availabilities of raw materials, $M1$ and $M2$, are 28 and 8 tons, respectively.
   (a) Show that the proposed changes will render the current optimal solution both nonoptimal and infeasible.
   (b) Use the *generalized simplex algorithm* (Section 4.4.2) to determine the new optimal feasible solution.

Addition of New Constraints. The addition of a new constraint to an existing model can lead to one of two cases.

1. The new constraint is *redundant*, meaning that it is satisfied by the current optimum solution, and hence can be dropped from the model altogether.
2. The current solution violates the new constraint, in which case the dual simplex method is used to restore feasibility.

Notice that the addition of a new constraint can never improve the current optimum objective value.

Example 4.5-3

Situation 1. Suppose that TOYCO is changing the design of its toys, and that the change will require the addition of a fourth operation in the assembly lines. The daily capacity of the new operation is 500 minutes and the times per unit for the three products on this operation are 3, 1, and 1 minutes, respectively. Study the effect of the new operation on the optimum solution.

The constraint for operation 4 is

$$3x_1 + x_2 + x_3 \leq 500$$

This constraint is redundant because it is satisfied by the current optimum solution $x_1 = 0$, $x_2 = 100$, and $x_3 = 230$. Hence, the current optimum solution remains unchanged.

Situation 2. Suppose, instead, that TOYCO unit times on the fourth operation are changed to 3, 3, and 1 minutes, respectively. All the remaining data of the model remain the same. Will the optimum solution change?

The constraint for operation 4 is

$$3x_1 + 3x_2 + x_3 \leq 500$$

This constraint is not satisfied by the current optimum solution. Thus, the new constraint must be added to the current optimum tableau as follows (x_7 is a slack):

Basic	x_1	x_2	x_3	x_4	x_5	x_6	x_7	Solution
z	4	0	0	1	2	0	0	1350
x_2	$-\frac{1}{4}$	1	0	$\frac{1}{2}$	$-\frac{1}{4}$	0	0	100
x_3	$\frac{3}{2}$	0	1	0	$\frac{1}{2}$	0	0	230
x_6	2	0	0	-2	1	1	0	20
x_7	3	3	1	0	0	0	1	500

The tableau shows that $x_7 = 500$, which is not consistent with the values of x_2 and x_3 in the rest of the tableau. The reason is that the basic variables x_2 and x_3 have not been substituted out in the new constraint. This substitution is achieved by performing the following operation:

$$\text{New } x_7\text{-row} = \text{Old } x_7\text{-row} - \{3 \times (x_2\text{-row}) + 1 \times (x_3\text{-row})\}$$

This operation is exactly the same as substituting

$$x_2 = 100 - \left(-\tfrac{1}{4}x_1 + \tfrac{1}{2}x_4 - \tfrac{1}{4}x_5\right)$$
$$x_3 = 230 - \left(\tfrac{3}{2}x_1 + \tfrac{1}{2}x_5\right)$$

in the new constraint. The new tableau is thus given as

Basic	x_1	x_2	x_3	x_4	x_5	x_6	x_7	Solution
z	4	0	0	1	2	0	0	1350
x_2	$-\frac{1}{4}$	1	0	$\frac{1}{2}$	$-\frac{1}{4}$	0	0	100
x_3	$\frac{3}{2}$	0	1	0	$\frac{1}{2}$	0	0	230
x_6	2	0	0	-2	1	1	0	20
x_7	$\frac{9}{4}$	0	0	$-\frac{3}{2}$	$\frac{1}{4}$	0	1	-30

Application of the dual simplex method will produce the new optimum solution $x_1 = 0$, $x_2 = 90$, $x_3 = 230$, and $z = \$1330$ (verify!). The solution shows that the addition of operation 4 will worsen the revenues from \$1350 to \$1330.

PROBLEM SET 4.5B

1. In the TOYCO model, suppose the fourth operation has the following specifications: The maximum production rate based on 480 minutes a day is either 120 units of product 1, 480 units of product 2, or 240 units of product 3. Determine the optimal solution, assuming that the daily capacity is limited to
   *(a) 570 minutes.
   (b) 548 minutes.
2. *Secondary Constraints.* Instead of solving a problem using all of its constraints, we can start by identifying the so-called *secondary constraints*. These are the constraints that we

suspect are least restrictive in terms of the optimum solution. The model is solved using the remaining (primary) constraints. We may then add the secondary constraints one at a time. A secondary constraint is discarded if it satisfies the available optimum. The process is repeated until all the secondary constraints are accounted for.

Apply the proposed procedure to the following LP:

$$\text{Maximize } z = 5x_1 + 6x_2 + 3x_2$$

subject to

$$\begin{aligned} 5x_1 + 5x_2 + 3x_3 &\leq 50 \\ x_1 + x_2 - x_3 &\leq 20 \\ 7x_1 + 6x_2 - 9x_3 &\leq 30 \\ 5x_1 + 5x_2 + 5x_3 &\leq 35 \\ 12x_1 + 6x_2 &\leq 90 \\ x_2 - 9x_3 &\leq 20 \\ x_1, x_2, x_3 &\geq 0 \end{aligned}$$

4.5.2 Changes Affecting Optimality

This section considers two particular situations that could affect the optimality of the current solution:

1. Changes in the original objective coefficients.
2. Addition of a new economic activity (variable) to the model.

Changes in the Objective Function Coefficients. These changes affect only the optimality of the solution. Such changes thus require recomputing the z-row coefficients (reduced costs) according to the following procedure:

1. Compute the dual values using Method 2 in Section 4.2.3.
2. Use the new dual values in Formula 2, Section 4.2.4, to determine the new reduced costs (z-row coefficients).

Two cases will result:

1. New z-row satisfies the optimality condition. The solution remains unchanged (the optimum objective value may change, however).
2. The optimality condition is not satisfied. Apply the (primal) simplex method to recover optimality.

Example 4.5-4

Situation 1. In the TOYCO model, suppose that the company has a new pricing policy to meet the competition. The unit revenues under the new policy are \$2, \$3, and \$4 for train, truck, and car toys, respectively. How is the optimal solution affected?

The new objective function is

$$\text{Maximize } z = 2x_1 + 3x_2 + 4x_3$$

Thus,

$$(\text{New objective coefficients of basic } x_2, x_3, \text{ and } x_6) = (3, 4, 0)$$

Using Method 2, Section 4.2.3, the dual variables are computed as

$$(y_1, y_2, y_3) = (3, 4, 0)\begin{pmatrix} \frac{1}{2} & -\frac{1}{4} & 0 \\ 0 & \frac{1}{2} & 0 \\ -2 & 1 & 1 \end{pmatrix} = \left(\tfrac{3}{2}, \tfrac{5}{4}, 0\right)$$

The z-row coefficients are determined as the difference between the left- and right-hand sides of the dual constraints (Formula 2, Section 4.2.4). It is not necessary to recompute the objective-row coefficients of the basic variables x_2, x_3, and x_6 because they always equal zero regardless of any changes made in the objective coefficients (verify!).

$$\text{Reduced cost of } x_1 = y_1 + 3y_2 + y_3 - \mathbf{2} = \tfrac{3}{2} + 3\left(\tfrac{5}{4}\right) + 0 - \mathbf{2} = \tfrac{13}{4}$$

$$\text{Reduced cost of } x_4 = y_1 - 0 = \tfrac{3}{2}$$

$$\text{Reduced cost of } x_5 = y_2 - 0 = \tfrac{5}{4}$$

Note that the right-hand side of the first dual constraint is 2, the *new* coefficient in the modified objective function.

The computations show that the current solution, $x_1 = 0$ train, $x_2 = 100$ trucks, and $x_3 = 230$ cars, remains optimal. The corresponding new revenue is computed as $2 \times 0 + 3 \times 100 + 4 \times 230 = \$1220$. The new pricing policy is not advantageous because it leads to lower revenue.

Situation 2. Suppose now that the TOYCO objective function is changed to

$$\text{Maximize } z = 6x_1 + 3x_2 + 4x_3$$

Will the optimum solution change?

We have

$$(y_1, y_2, y_3) = (3, 4, 0)\begin{pmatrix} \frac{1}{2} & -\frac{1}{4} & 0 \\ 0 & \frac{1}{2} & 0 \\ -2 & 1 & 1 \end{pmatrix} = \left(\tfrac{3}{2}, \tfrac{5}{4}, 0\right)$$

$$\text{Reduced cost of } x_1 = y_1 + 3y_2 + y_3 - \mathbf{6} = \tfrac{3}{2} + 3\left(\tfrac{5}{4}\right) + 0 - \mathbf{6} = -\tfrac{3}{4}$$

$$\text{Reduced cost of } x_4 = y_1 - 0 = \tfrac{3}{2}$$

$$\text{Reduced cost of } x_5 = y_2 - 0 = \tfrac{5}{4}$$

The new reduced cost of x_1 shows that the current solution is not optimum.

To determine the new solution, the z-row is changed as highlighted in the following tableau:

Basic	x_1	x_2	x_3	x_4	x_5	x_6	Solution
z	$-\frac{3}{4}$	0	0	$\frac{3}{2}$	$\frac{5}{4}$	0	1220
x_2	$-\frac{1}{4}$	1	0	$\frac{1}{2}$	$-\frac{1}{4}$	0	100
x_3	$\frac{3}{2}$	0	1	0	$\frac{1}{2}$	0	230
x_6	2	0	0	-2	1	1	20

The elements shown in the shaded cells are the new *reduced cost* for the nonbasic variables x_1, x_4, and x_5. All the remaining elements are the same as in the original optimal tableau. The new optimum solution is then determined by letting x_1 enter and x_6 leave, which yields $x_1 = 10, x_2 = 102.5, x_3 = 215$, and $z = \$1227.50$ (verify!). Although the new solution recommends the production of all three toys, the optimum revenue is less than when two toys only are manufactured.

PROBLEM SET 4.5C

1. Investigate the optimality of the TOYCO solution for each of the following objective functions. If the solution changes, use post-optimal analysis to determine the new optimum. (The optimum tableau of TOYCO is given at the start of Section 4.5.)
   (a) $z = 2x_1 + x_2 + 4x_3$
   (b) $z = 3x_1 + 6x_2 + x_3$
   (c) $z = 8x_1 + 3x_2 + 9x_3$
2. Investigate the optimality of the Reddy Mikks solution (Example 4.3-1) for each of the following objective functions. If the solution changes, use post-optimal analysis to determine the new optimum. (The optimal tableau of the model is given in Example 3.3-1.)
   ***(a)** $z = 3x_1 + 2x_2$
   (b) $z = 8x_1 + 10x_2$
   ***(c)** $z = 2x_1 + 5x_2$
3. Show that the 100% optimality rule (Problem 8, Set 3.6d, Chapter 3) is derived from (reduced costs) ≥ 0 for maximization problems and (reduced costs) ≤ 0 for minimization problems.

Addition of a New Activity. The addition of a new activity in an LP model is equivalent to adding a new variable. Intuitively, the addition of a new activity is desirable only if it is profitable—that is, if it improves the optimal value of the objective function. This condition can be checked by computing the reduced cost of the new variable using Formula 2, Section 4.2.4. If the new activity satisfies the optimality condition, then the activity is not profitable. Else, it is advantageous to undertake the new activity.

Example 4.5-5

TOYCO recognizes that toy trains are not currently in production because they are not profitable. The company wants to replace toy trains with a new product, a toy fire engine, to be assembled on the existing facilities. TOYCO estimates the revenue per toy fire engine to be $4 and the assembly times per unit to be 1 minute on each of operations 1 and 2, and 2 minutes on operation 3. How would this change impact the solution?

Let x_7 represent the new fire engine product. Given that $(y_1, y_2, y_3) = (1, 2, 0)$ are the optimal dual values, we get

$$\text{Reduced cost of } x_7 = 1y_1 + 1y_2 + 2y_3 - 4 = 1 \times 1 + 1 \times 2 + 2 \times 0 - 4 = -1$$

The result shows that it is profitable to include x_7 in the optimal basic solution. To obtain the new optimum, we first compute its column constraint using Formula 1, Section 4.2.4, as

$$x_7\text{-constraint colum} = \begin{pmatrix} \frac{1}{2} & -\frac{1}{4} & 0 \\ 0 & \frac{1}{2} & 0 \\ -2 & 1 & 1 \end{pmatrix} \begin{pmatrix} 1 \\ 1 \\ 2 \end{pmatrix} = \begin{pmatrix} \frac{1}{4} \\ \frac{1}{2} \\ 1 \end{pmatrix}$$

Thus, the current simplex tableau must be modified as follows

Basic	x_1	x_2	x_3	x_7	x_4	x_5	x_6	Solution
z	4	0	0	-1	1	2	0	1350
x_2	$-\frac{1}{4}$	1	0	$\frac{1}{4}$	$\frac{1}{2}$	$-\frac{1}{4}$	0	100
x_3	$\frac{3}{2}$	0	1	$\frac{1}{2}$	0	$\frac{1}{2}$	0	230
x_6	2	0	0	1	-2	1	1	20

The new optimum is determined by letting x_7 enter the basic solution, in which case x_6 must leave. The new solution is $x_1 = 0, x_2 = 0, x_3 = 125, x_7 = 210$, and $z = \$1465$ (verify!), which improves the revenues by $115.

PROBLEM SET 4.5D

*1. In the original TOYCO model, toy trains are not part of the optimal product mix. The company recognizes that market competition will not allow raising the unit price of the toy. Instead, the company wants to concentrate on improving the assembly operation itself. This entails reducing the assembly time per unit in each of the three operations by a specified percentage, $p\%$. Determine the value of p that will make toy trains just profitable. (The optimum tableau of the TOYCO model is given at the start of Section 4.5.)

2. In the TOYCO model, suppose that the company can reduce the unit times on operations 1, 2, and 3 for toy trains from the current levels of 1, 3, and 1 minutes to .5, 1, and .5 minutes, respectively. The revenue per unit remains unchanged at $3. Determine the new optimum solution.

3. In the TOYCO model, suppose that a new toy (fire engine) requires 3, 2, 4 minutes, respectively, on operations 1, 2, and 3. Determine the optimal solution when the revenue per unit is given by

   *(a) $5.

   (b) $10.

4. In the Reddy Mikks model, the company is considering the production of a cheaper brand of exterior paint whose input requirements per ton include .75 ton of each of raw materials $M1$ and $M2$. Market conditions still dictate that the excess of interior paint over the production of *both* types of exterior paint be limited to 1 ton daily. The revenue per ton of the new exterior paint is \$3500. Determine the new optimal solution. (The model is explained in Example 4.5-1, and its optimum tableau is given in Example 3.3-1.)

REFERENCES

Bradley, S., A. Hax, and T. Magnanti, *Applied Mathematical Programming*, Addison-Wesley, Reading, MA, 1977.

Bazaraa, M., J. Jarvis, and H. Sherali, *Linear Programming and Network Flows*, 2nd ed., Wiley, New York, 1990.

Diwckar, U., *Introduction to Applied Optimization*, Kluwer Academic Publishers, Boston, 2003.

Nering, E., and A. Tucker, *Linear Programming and Related Problems*, Academic Press, Boston, 1992.

Vanderbei, R., *Linear Programming: Foundation and Optimization*, 2nd ed., Kluwer Academic Publishers, Boston MA, 2001.

CHAPTER 5

Transportation Model and Its Variants

Chapter Guide. The transportation model is a special class of linear programs that deals with shipping a commodity from *sources* (e.g., factories) to *destinations* (e.g., warehouses). The objective is to determine the shipping schedule that minimizes the total shipping cost while satisfying supply and demand limits. The application of the transportation model can be extended to other areas of operation, including inventory control, employment scheduling, and personnel assignment.

As you study the material in this chapter, keep in mind that the steps of the transportation algorithm are precisely those of the simplex method. Another point is that the transportation algorithm was developed in the early days of OR to enhance hand computations. Now, with the tremendous power of the computer, such shortcuts may not be warranted and, indeed, are never used in commercial codes in the strict manner presented in this chapter. Nevertheless, the presentation shows that the special transportation tableau is useful in modeling a class of problems in a concise manner (as opposed to the familiar LP model with explicit objective function and constraints). In particular, the transportation tableau format simplifies the solution of the problem by Excel Solver. The representation also provides interesting ideas about how the basic theory of linear programming is exploited to produce shortcuts in computations.

You will find TORA's tutorial module helpful in understanding the details of the transportation algorithm. The module allows you to make the decisions regarding the logic of the computations with immediate feedback.

This chapter includes a summary of 1 real-life application, 12 solved examples, 1 Solver model, 4 AMPL models, 46 end-of-section problems, and 5 cases. The cases are in Appendix E on the CD. The AMPL/Excel/Solver/TORA programs are in folder ch5Files.

Real-life Application—Scheduling Appointments at Australian Trade Events

The Australian Tourist Commission (ATC) organizes trade events around the world to provide a forum for Australian sellers to meet international buyers of tourism products, including accommodation, tours, and transport. During these events, sellers are

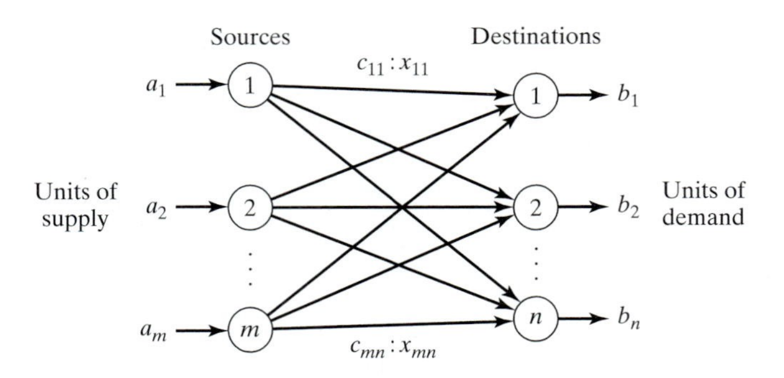

FIGURE 5.1
Representation of the transportation model with nodes and arcs

stationed in booths and are visited by buyers according to scheduled appointments. Because of the limited number of time slots available in each event and the fact that the number of buyers and sellers can be quite large (one such event held in Melbourne in 1997 attracted 620 sellers and 700 buyers), ATC attempts to schedule the seller-buyer appointments in advance of the event in a manner that maximizes preferences. The model has resulted in greater satisfaction for both the buyers and sellers. Case 3 in Chapter 24 on the CD provides the details of the study.

5.1 DEFINITION OF THE TRANSPORTATION MODEL

The general problem is represented by the network in Figure 5.1. There are m sources and n destinations, each represented by a **node**. The **arcs** represent the routes linking the sources and the destinations. Arc (i, j) joining source i to destination j carries two pieces of information: the transportation cost per unit, c_{ij}, and the amount shipped, x_{ij}. The amount of supply at source i is a_i and the amount of demand at destination j is b_j. The objective of the model is to determine the unknowns x_{ij} that will minimize the total transportation cost while satisfying all the supply and demand restrictions.

Example 5.1-1

MG Auto has three plants in Los Angeles, Detroit, and New Orleans, and two major distribution centers in Denver and Miami. The capacities of the three plants during the next quarter are 1000, 1500, and 1200 cars. The quarterly demands at the two distribution centers are 2300 and 1400 cars. The mileage chart between the plants and the distribution centers is given in Table 5.1.

The trucking company in charge of transporting the cars charges 8 cents per mile per car. The transportation costs per car on the different routes, rounded to the closest dollar, are given in Table 5.2.

The LP model of the problem is given as

$$\text{Minimize } z = 80x_{11} + 215x_{12} + 100x_{21} + 108x_{22} + 102x_{31} + 68x_{32}$$

TABLE 5.1 Mileage Chart

	Denver	Miami
Los Angeles	1000	2690
Detroit	1250	1350
New Orleans	1275	850

TABLE 5.2 Transportation Cost per Car

	Denver (1)	Miami (2)
Los Angeles (1)	\$80	\$215
Detroit (2)	\$100	\$108
New Orleans (3)	\$102	\$68

subject to

$$
\begin{array}{llll}
x_{11} + x_{12} & & = 1000 & \text{(Los Angeles)} \\
& x_{21} + x_{22} & = 1500 & \text{(Detroit)} \\
& + x_{31} + x_{32} & = 1200 & \text{(New Oreleans)} \\
x_{11} & + x_{21} + \quad x_{31} & = 2300 & \text{(Denver)} \\
x_{12} + & x_{22} \quad + x_{32} & = 1400 & \text{(Miami)}
\end{array}
$$

$$x_{ij} \geq 0, i = 1, 2, 3, j = 1, 2$$

These constraints are all equations because the total supply from the three sources (= 1000 + 1500 + 1200 = 3700 cars) equals the total demand at the two destinations (= 2300 + 1400 = 3700 cars).

The LP model can be solved by the simplex method. However, with the special structure of the constraints we can solve the problem more conveniently using the **transportation tableau** shown in Table 5.3.

TABLE 5.3 MG Transportation Model

	Denver	Miami	Supply
Los Angeles	80 x_{11}	215 x_{12}	**1000**
Detroit	100 x_{21}	108 x_{22}	**1500**
New Orleans	102 x_{31}	68 x_{32}	**1200**
Demand	**2300**	**1400**	

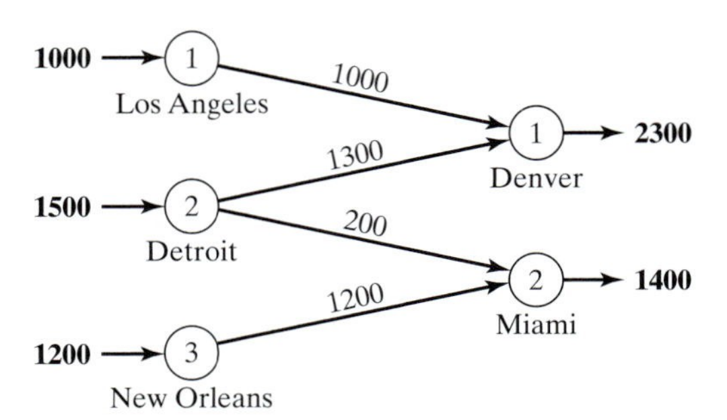

FIGURE 5.2
Optimal solution of MG Auto model

The optimal solution in Figure 5.2 (obtained by TORA[1]) calls for shipping 1000 cars from Los Angeles to Denver, 1300 from Detroit to Denver, 200 from Detroit to Miami, and 1200 from New Orleans to Miami. The associated minimum transportation cost is computed as $1000 \times \$80 + 1300 \times \$100 + 200 \times \$108 + 1200 \times \$68 = \$313{,}200$.

Balancing the Transportation Model. The transportation algorithm is based on the assumption that the model is balanced, meaning that the total demand equals the total supply. If the model is unbalanced, we can always add a dummy source or a dummy destination to restore balance.

Example 5.1-2

In the MG model, suppose that the Detroit plant capacity is 1300 cars (instead of 1500). The total supply (= 3500 cars) is less than the total demand (= 3700 cars), meaning that part of the demand at Denver and Miami will not be satisfied.

Because the demand exceeds the supply, a dummy source (plant) with a capacity of 200 cars (= 3700 − 3500) is added to balance the transportation model. The unit transportation costs from the dummy plant to the two destinations are zero because the plant does not exist.

Table 5.4 gives the balanced model together with its optimum solution. The solution shows that the dummy plant ships 200 cars to Miami, which means that Miami will be 200 cars short of satisfying its demand of 1400 cars.

We can make sure that a specific destination does not experience shortage by assigning a very high unit transportation cost from the dummy source to that destination. For example, a penalty of \$1000 in the dummy-Miami cell will prevent shortage at Miami. Of course, we cannot use this "trick" with all the destinations, because shortage must occur somewhere in the system.

The case where the supply exceeds the demand can be demonstrated by assuming that the demand at Denver is 1900 cars only. In this case, we need to add a dummy distribution center to "receive" the surplus supply. Again, the unit transportation costs to the dummy distribution center are zero, unless we require a factory to "ship out" completely. In this case, we must assign a high unit transportation cost from the designated factory to the dummy destination.

---

[1]To use TORA, from Main Menu select Transportation Model. From the SOLVE/MODIFY menu, select Solve ⇒ Final solution to obtain a summary of the optimum solution. A detailed description of the iterative solution of the transportation model is given in Section 5.3.3.

TABLE 5.4 MG Model with Dummy Plant

	Denver	Miami	Supply
Los Angeles	80 / **1000**	215	**1000**
Detroit	100 / **1300**	108	**1300**
New Orleans	102	68 / **1200**	**1200**
Dummy Plant	0	0 / **200**	**200**
Demand	**2300**	**1400**	

TABLE 5.5 MG Model with Dummy Destination

	Denver	Miami	Dummy	
Los Angeles	80 / **1000**	215	0	**1000**
Detroit	100 / **900**	108 / **200**	0 / **400**	**1500**
New Orleans	102	68 / **1200**	0	**1200**
Demand	**1900**	**1400**	**400**	

Table 5.5 gives the new model and its optimal solution (obtained by TORA). The solution shows that the Detroit plant will have a surplus of 400 cars.

PROBLEM SET 5.1A[2]

1. True or False?

(a) To balance a transportation model, it may be necessary to add both a dummy source and a dummy destination.

(b) The amounts shipped to a dummy destination represent surplus at the shipping source.

(c) The amounts shipped from a dummy source represent shortages at the receiving destinations.

[2]In this set, you may use TORA to find the optimum solution. AMPL and Solver models for the transportation problem will be introduced at the end of Section 5.3.2.

2. In each of the following cases, determine whether a dummy source or a dummy destination must be added to balance the model.

(a) Supply: $a_1 = 10, a_2 = 5, a_3 = 4, a_4 = 6$
Demand: $b_1 = 10, b_2 = 5, b_3 = 7, b_4 = 9$

(b) Supply: $a_1 = 30, a_2 = 44$
Demand: $b_1 = 25, b_2 = 30, b_3 = 10$

3. In Table 5.4 of Example 5.1-2, where a dummy plant is added, what does the solution mean when the dummy plant "ships" 150 cars to Denver and 50 cars to Miami?

***4.** In Table 5.5 of Example 5.1-2, where a dummy destination is added, suppose that the Detroit plant must ship out *all* its production. How can this restriction be implemented in the model?

5. In Example 5.1-2, suppose that for the case where the demand exceeds the supply (Table 5.4), a penalty is levied at the rate of $200 and $300 for each undelivered car at Denver and Miami, respectively. Additionally, no deliveries are made from the Los Angeles plant to the Miami distribution center. Set up the model, and determine the optimal shipping schedule for the problem.

***6.** Three electric power plants with capacities of 25, 40, and 30 million kWh supply electricity to three cities. The maximum demands at the three cities are estimated at 30, 35, and 25 million kWh. The price per million kWh at the three cities is given in Table 5.6.

During the month of August, there is a 20% increase in demand at each of the three cities, which can be met by purchasing electricity from another network at a premium rate of $1000 per million kWh. The network is not linked to city 3, however. The utility company wishes to determine the most economical plan for the distribution and purchase of additional energy.

(a) Formulate the problem as a transportation model.

(b) Determine an optimal distribution plan for the utility company.

(c) Determine the cost of the additional power purchased by each of the three cities.

7. Solve Problem 6, assuming that there is a 10% power transmission loss through the network.

8. Three refineries with daily capacities of 6, 5, and 8 million gallons, respectively, supply three distribution areas with daily demands of 4, 8, and 7 million gallons, respectively. Gasoline is transported to the three distribution areas through a network of pipelines. The transportation cost is 10 cents per 1000 gallons per pipeline mile. Table 5.7 gives the mileage between the refineries and the distribution areas. Refinery 1 is not connected to distribution area 3.

(a) Construct the associated transportation model.

(b) Determine the optimum shipping schedule in the network.

TABLE 5.6 Price/Million kWh for Problem 6

		City 1	City 2	City 3
Plant	1	$600	$700	$400
Plant	2	$320	$300	$350
Plant	3	$500	$480	$450

TABLE 5.7 Mileage Chart for Problem 8

	Distribution area 1	2	3
Refinery 1	120	180	—
Refinery 2	300	100	80
Refinery 3	200	250	120

*9. In Problem 8, suppose that the capacity of refinery 3 is 6 million gallons only and that distribution area 1 must receive all its demand. Additionally, any shortages at areas 2 and 3 will incur a penalty of 5 cents per gallon.

(a) Formulate the problem as a transportation model.

(b) Determine the optimum shipping schedule.

10. In Problem 8, suppose that the daily demand at area 3 drops to 4 million gallons. Surplus production at refineries 1 and 2 is diverted to other distribution areas by truck. The transportation cost per 100 gallons is $1.50 from refinery 1 and $2.20 from refinery 2. Refinery 3 can divert its surplus production to other chemical processes within the plant.

(a) Formulate the problem as a transportation model.

(b) Determine the optimum shipping schedule.

11. Three orchards supply crates of oranges to four retailers. The daily demand amounts at the four retailers are 150, 150, 400, and 100 crates, respectively. Supplies at the three orchards are dictated by available regular labor and are estimated at 150, 200, and 250 crates daily. However, both orchards 1 and 2 have indicated that they could supply more crates, if necessary, by using overtime labor. Orchard 3 does not offer this option. The transportation costs per crate from the orchards to the retailers are given in Table 5.8.

(a) Formulate the problem as a transportation model.

(b) Solve the problem.

(c) How many crates should orchards 1 and 2 supply using overtime labor?

12. Cars are shipped from three distribution centers to five dealers. The shipping cost is based on the mileage between the sources and the destinations, and is independent of whether the truck makes the trip with partial or full loads. Table 5.9 summarizes the mileage between the distribution centers and the dealers together with the monthly supply and demand figures given in *number* of cars. A full truckload includes 18 cars. The transportation cost per truck mile is $25.

(a) Formulate the associated transportation model.

(b) Determine the optimal shipping schedule.

TABLE 5.8 Transportation Cost/Crate for Problem 11

	Retailer 1	2	3	4
Orchard 1	$1	$2	$3	$2
Orchard 2	$2	$4	$1	$2
Orchard 3	$1	$3	$5	$3

TABLE 5.9 Mileage Chart and Supply and Demand for Problem 12

	Dealer 1	2	3	4	5	Supply
Center 1	100	150	200	140	35	**400**
Center 2	50	70	60	65	80	**200**
Center 3	40	90	100	150	130	**150**
Demand	**100**	**200**	**150**	**160**	**140**	

13. MG Auto, of Example 5.1-1, produces four car models: *M*1, *M*2, *M*3, and *M*4. The Detroit plant produces models *M*1, *M*2, and *M*4. Models *M*1 and *M*2 are also produced in New Orleans. The Los Angeles plant manufactures models *M*3 and *M*4. The capacities of the various plants and the demands at the distribution centers are given in Table 5.10.

The mileage chart is the same as given in Example 5.1-1, and the transportation rate remains at 8 cents per car mile for all models. Additionally, it is possible to satisfy a percentage of the demand for some models from the supply of others according to the specifications in Table 5.11.

(a) Formulate the corresponding transportation model.

(b) Determine the optimum shipping schedule.

(Hint: Add four new destinations corresponding to the new combinations [*M*1, *M*2], [*M*3, *M*4], [*M*1, *M*2], and [*M*2, *M*4]. The demands at the new destinations are determined from the given percentages.)

TABLE 5.10 Capacities and Demands for Problem 13

	Model *M*1	*M*2	*M*3	*M*4	Totals
Plant					
Los Angeles	—	—	700	300	1000
Detroit	500	600	—	400	1500
New Orleans	800	400	—	—	1200
Distribution center					
Denver	700	500	500	600	2300
Miami	600	500	200	100	1400

TABLE 5.11 Interchangeable Models in Problem 13

Distribution center	Percentage of demand	Interchangeable models
Denver	10	*M*1, *M*2
	20	*M*3, *M*4
Miami	10	*M*1, *M*2
	5	*M*2, *M*4

5.2 NONTRADITIONAL TRANSPORTATION MODELS

The application of the transportation model is not limited to *transporting* commodities between geographical sources and destinations. This section presents two applications in the areas of production-inventory control and tool sharpening service.

Example 5.2-1 (Production-Inventory Control)

Boralis manufactures backpacks for serious hikers. The demand for its product occurs during March to June of each year. Boralis estimates the demand for the four months to be 100, 200, 180, and 300 units, respectively. The company uses part-time labor to manufacture the backpacks and, accordingly, its production capacity varies monthly. It is estimated that Boralis can produce 50, 180, 280, and 270 units in March through June. Because the production capacity and demand for the different months do not match, a current month's demand may be satisfied in one of three ways.

1. Current month's production.
2. Surplus production in an earlier month.
3. Surplus production in a later month (backordering).

In the first case, the production cost per backpack is $40. The second case incurs an additional holding cost of $.50 per backpack per month. In the third case, an additional penalty cost of $2.00 per backpack is incurred for each month delay. Boralis wishes to determine the optimal production schedule for the four months.

The situation can be modeled as a transportation model by recognizing the following parallels between the elements of the production-inventory problem and the transportation model:

Transportation	Production-inventory
1. Source i	1. Production period i
2. Destination j	2. Demand period j
3. Supply amount at source i	3. Production capacity of period i
4. Demand at destination j	4. Demand for period j
5. Unit transportation cost from source i to destination j	5. Unit cost (production + inventory + penalty) in period i for period j

The resulting transportation model is given in Table 5.12.

TABLE 5.12 Transportation Model for Example 5.2-1

	1	2	3	4	Capacity
1	$40.00	$40.50	$41.00	$41.50	**50**
2	$42.00	$40.00	$40.50	$41.00	**180**
3	$44.00	$42.00	$40.00	$40.50	**280**
4	$46.00	$44.00	$42.00	$40.00	**270**
Demand	**100**	**200**	**180**	**300**	

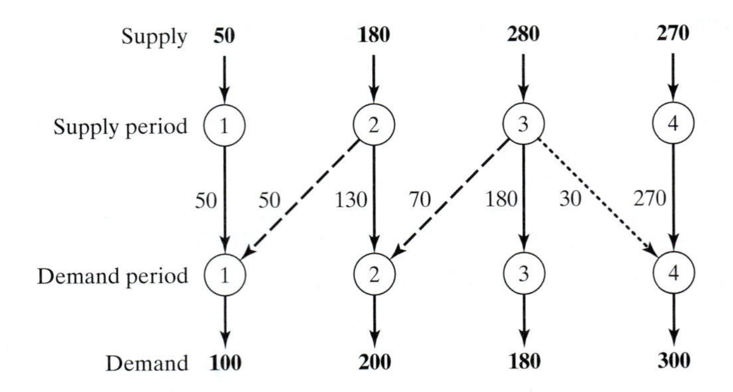

FIGURE 5.3
Optimal solution of the production–inventory model

The unit "transportation" cost from period i to period j is computed as

$$c_{ij} = \begin{cases} \text{Production cost in } i, i = j \\ \text{Production cost in } i + \text{holding cost from } i \text{ to } j, i < j \\ \text{Production cost in } i + \text{penaty cost from } i \text{ to } j, i > j \end{cases}$$

For example,

$$c_{11} = \$40.00$$

$$c_{24} = \$40.00 + (\$.50 + \$.50) = \$41.00$$

$$c_{41} = \$40.00 + (\$2.00 + \$2.00 + \$2.00) = \$46.00$$

The optimal solution is summarized in Figure 5.3. The dashed lines indicate back-ordering, the dotted lines indicate production for a future period, and the solid lines show production in a period for itself. The total cost is $31,455.

Example 5.2-2 (Tool Sharpening)

Arkansas Pacific operates a medium-sized saw mill. The mill prepares different types of wood that range from soft pine to hard oak according to a weekly schedule. Depending on the type of wood being milled, the demand for sharp blades varies from day to day according to the following 1-week (7-day) data:

Day	Mon.	Tue.	Wed.	Thu.	Fri.	Sat.	Sun.
Demand (blades)	24	12	14	20	18	14	22

The mill can satisfy the daily demand in the following manner:

1. Buy new blades at the cost of \$12 a blade.
2. Use an overnight sharpening service at the cost of \$6 a blade.
3. Use a slow 2-day sharpening service at the cost of \$3 a blade.

The situation can be represented as a transportation model with eight sources and seven destinations. The destinations represent the 7 days of the week. The sources of the model are defined as follows: Source 1 corresponds to buying new blades, which, in the extreme case, can provide sufficient supply to cover the demand for all 7 days ($= 24 + 12 + 14 + 20 + 18 + 14 + 22 = 124$). Sources 2 to 8 correspond to the 7 days of the week. The amount of supply for each of these sources equals the number of used blades at the end of the associated day. For example, source 2 (i.e., Monday) will have a supply of used blades equal to the demand for Monday. The unit "transportation cost" for the model is \$12, \$6, or \$3, depending on whether the blade is supplied from new blades, overnight sharpening, or 2-day sharpening. Notice that the overnight service means that used blades sent at the *end* of day i will be available for use at the *start* of day $i + 1$ or day $i + 2$, because the slow 2-day service will not be available until the *start* of day $i + 3$. The "disposal" column is a dummy destination needed to balance the model. The complete model and its solution are given in Table 5.13.

TABLE 5.13 Tool Sharpening Problem Expressed as a Transportation Model

	1 Mon.	2 Tue.	3 Wed.	4 Thu.	5 Fri.	6 Sat.	7 Sun.	8 Disposal	
1-New	\$12 **24**	\$12 **2**	\$12	\$12	\$12	\$12	\$12	\$0 **98**	**124**
2-Mon.	*M*	\$6 **10**	\$6 **8**	\$3 **6**	\$3	\$3	\$3	\$0	**24**
3-Tue.	*M*	*M*	\$6 **6**	\$6	\$3 **6**	\$3	\$3	\$0	**12**
4-Wed.	*M*	*M*	*M*	\$6 **14**	\$6	\$3	\$3	\$0	**14**
5-Thu.	*M*	*M*	*M*	*M*	\$6 **12**	\$6	\$3 **8**	\$0	**20**
6-Fri.	*M*	*M*	*M*	*M*	*M*	\$6 **14**	\$6	\$0 **4**	**18**
7-Sat.	*M*	*M*	*M*	*M*	*M*	*M*	\$6 **14**	\$0	**14**
8-Sun.	*M*	*M*	*M*	*M*	*M*	*M*	*M*	\$0 **22**	**22**
	24	**12**	**14**	**20**	**18**	**14**	**22**	**124**	

The problem has alternative optima at a cost of $840 (file toraEx5.2-2.txt). The following table summarizes one such solution.

	Number of sharp blades (Target day)			
Period	*New*	*Overnight*	*2-day*	Disposal
Mon.	24 (Mon.)	10 (Tue.) + 8 (Wed.)	6 (Thu.)	0
Tues.	2 (Tue.)	6 (Wed.)	6 (Fri.)	0
Wed.	0	14 (Thu.)	0	0
Thu.	0	12 (Fri.)	8 (Sun.)	0
Fri.	0	14 (Sat.)	0	4
Sat.	0	14 (Sun.)	0	0
Sun.	0	0	0	22

Remarks. The model in Table 5.13 is suitable only for the first week of operation because it does not take into account the *rotational* nature of the days of the week, in the sense that this week's days can act as sources for next week's demand. One way to handle this situation is to assume that the very first week of operation starts with all new blades for each day. From then on, we use a model consisting of exactly 7 sources and 7 destinations corresponding to the days of the week. The new model will be similar to Table 5.13 less source "New" and destination "Disposal." Also, only diagonal cells will be blocked (unit cost $= M$). The remaining cells will have a unit cost of either $3.00 or $6.00. For example, the unit cost for cell (Sat., Mon.) is $6.00 and that for cells (Sat., Tue.), (Sat., Wed.), (Sat., Thu.), and (Sat., Fri.) is $3.00. The table below gives the solution costing $372. As expected, the optimum solution will always use the 2-day service only. The problem has alternative optima (see file toraEx5.2-2a.txt).

	Week $i + 1$							
Week i	*Mon.*	*Tue.*	*Wed.*	*Thu.*	*Fri.*	*Sat.*	*Sun.*	Total
Mon.				6			18	24
Tue.					8		4	12
Wed.	12					2		14
Thu.	8	12						20
Fri.	4		14					18
Sat.				14				14
Sun.					10	12		22
Total	24	12	14	20	18	14	22	

PROBLEM SET 5.2A[3]

1. In Example 5.2-1, suppose that the holding cost per unit is period-dependent and is given by 40, 30, and 70 cents for periods 1, 2, and 3, respectively. The penalty and production costs remain as given in the example. Determine the optimum solution and interpret the results.

[3]In this set, you may use TORA to find the optimum solution. AMPL and Solver models for the transportation problem will be introduced at the end of Section 5.3.2.

*2. In Example 5.2-2, suppose that the sharpening service offers 3-day service for $1 a blade on Monday and Tuesday (days 1 and 2). Reformulate the problem, and interpret the optimum solution.

3. In Example 5.2-2, if a blade is not used the day it is sharpened, a holding cost of 50 cents per blade per day is incurred. Reformulate the model, and interpret the optimum solution.

4. JoShop wants to assign four different categories of machines to five types of tasks. The numbers of machines available in the four categories are 25, 30, 20, and 30. The numbers of jobs in the five tasks are 20, 20, 30, 10, and 25. Machine category 4 cannot be assigned to task type 4. Table 5.14 provides the unit cost (in dollars) of assigning a machine category to a task type. The objective of the problem is to determine the optimum number of machines in each category to be assigned to each task type. Solve the problem and interpret the solution.

*5. The demand for a perishable item over the next four months is 400, 300, 420, and 380 tons, respectively. The supply capacities for the same months are 500, 600, 200, and 300 tons. The purchase price per ton varies from month to month and is estimated at $100, $140, $120, and $150, respectively. Because the item is perishable, a current month's supply must be consumed within 3 months (starting with current month). The storage cost per ton per month is $3. The nature of the item does not allow back-ordering. Solve the problem as a transportation model and determine the optimum delivery schedule for the item over the next 4 months.

6. The demand for a special small engine over the next five quarters is 200, 150, 300, 250, and 400 units. The manufacturer supplying the engine has different production capacities estimated at 180, 230, 430, 300, and 300 for the five quarters. Back-ordering is not allowed, but the manufacturer may use overtime to fill the immediate demand, if necessary. The overtime capacity for each period is half the regular capacity. The production costs per unit for the five periods are $100, $96, $116, $102, and $106, respectively. The overtime production cost per engine is 50% higher than the regular production cost. If an engine is produced now for use in later periods, an additional storage cost of $4 per engine per period is incurred. Formulate the problem as a transportation model. Determine the optimum number of engines to be produced during regular time and overtime of each period.

7. Periodic preventive maintenance is carried out on aircraft engines, where an important component must be replaced. The numbers of aircraft scheduled for such maintenance over the next six months are estimated at 200, 180, 300, 198, 230, and 290, respectively. All maintenance work is done during the first day of the month, where a used component may be replaced with a new or an overhauled component. The overhauling of used components may be done in a local repair facility, where they will be ready for use at the beginning of next month, or they may be sent to a central repair shop, where a delay of

TABLE 5.14 Unit Costs for Problem 4

		Task type				
		1	2	3	4	5
Machine category	1	10	2	3	15	9
	2	5	10	15	2	4
	3	15	5	14	7	15
	4	20	15	13	—	8

TABLE 5.15 Bids per Acre for Problem 8

		Location		
		1	2	3
Bidder	1	\$520	\$210	\$570
	2	—	\$510	\$495
	3	\$650	—	\$240
	4	\$180	\$430	\$710

3 months (including the month in which maintenance occurs) is expected. The repair cost in the local shop is \$120 per component. At the central facility, the cost is only \$35 per component. An overhauled component used in a later month will incur an additional storage cost of \$1.50 per unit per month. New components may be purchased at \$200 each in month 1, with a 5% price increase every 2 months. Formulate the problem as a transportation model, and determine the optimal schedule for satisfying the demand for the component over the next six months.

8. The National Parks Service is receiving four bids for logging at three pine forests in Arkansas. The three locations include 10,000, 20,000, and 30,000 acres. A single bidder can bid for at most 50% of the total acreage available. The bids per acre at the three locations are given in Table 5.15. Bidder 2 does not wish to bid on location 1, and bidder 3 cannot bid on location 2.

(a) In the present situation, we need to *maximize* the total bidding revenue for the Parks Service. Show how the problem can be formulated as a transportation model.

(b) Determine the acreage that should be assigned to each of the four bidders.

5.3 THE TRANSPORTATION ALGORITHM

The transportation algorithm follows the *exact steps* of the simplex method (Chapter 3). However, instead of using the regular simplex tableau, we take advantage of the special structure of the transportation model to organize the computations in a more convenient form.

The special transportation algorithm was developed early on when hand computations were the norm and the shortcuts were warranted. Today, we have powerful computer codes that can solve a transportation model of any size as a regular LP.[4] Nevertheless, the transportation algorithm, aside from its historical significance, does provide insight into the use of the theoretical primal-dual relationships (introduced in Section 4.2) to achieve a practical end result, that of improving hand computations. The exercise is theoretically intriguing.

The details of the algorithm are explained using the following numeric example.

---

[4]In fact, TORA handles all necessary computations in the background using the regular simplex method and uses the transportation model format only as a screen "veneer."

TABLE 5.16 SunRay Transportation Model

	Mill 1	Mill 2	Mill 3	Mill 4	Supply
Silo 1	10 / x_{11}	2 / x_{12}	20 / x_{13}	11 / x_{14}	**15**
Silo 2	12 / x_{21}	7 / x_{22}	9 / x_{23}	20 / x_{24}	**25**
Silo 3	4 / x_{31}	14 / x_{32}	16 / x_{33}	18 / x_{34}	**10**
Demand	**5**	**15**	**15**	**15**	

Example 5.3-1 (SunRay Transport)

SunRay Transport Company ships truckloads of grain from three silos to four mills. The supply (in truckloads) and the demand (also in truckloads) together with the unit transportation costs per truckload on the different routes are summarized in the transportation model in Table 5.16. The unit transportation costs, c_{ij}, (shown in the northeast corner of each box) are in hundreds of dollars. The model seeks the minimum-cost shipping schedule x_{ij} between silo i and mill j $(i = 1, 2, 3; j = 1, 2, 3, 4)$.

Summary of the Transportation Algorithm. The steps of the transportation algorithm are exact parallels of the simplex algorithm.

Step 1. Determine a *starting* basic feasible solution, and go to step 2.

Step 2. Use the optimality condition of the simplex method to determine the *entering variable* from among all the nonbasic variables. If the optimality condition is satisfied, stop. Otherwise, go to step 3.

Step 3. Use the feasibility condition of the simplex method to determine the *leaving variable* from among all the current basic variables, and find the new basic solution. Return to step 2.

5.3.1 Determination of the Starting Solution

A general transportation model with m sources and n destinations has $m + n$ constraint equations, one for each source and each destination. However, because the transportation model is always balanced (sum of the supply = sum of the demand), one of these equations is redundant. Thus, the model has $m + n - 1$ independent constraint equations, which means that the starting basic solution consists of $m + n - 1$ basic variables. Thus, in Example 5.3-1, the starting solution has $3 + 4 - 1 = 6$ basic variables.

The special structure of the transportation problem allows securing a nonartificial starting basic solution using one of three methods:[5]

1. Northwest-corner method
2. Least-cost method
3. Vogel approximation method

The three methods differ in the "quality" of the starting basic solution they produce, in the sense that a better starting solution yields a smaller objective value. In general, though not always, the Vogel method yields the best starting basic solution, and the northwest-corner method yields the worst. The tradeoff is that the northwest-corner method involves the least amount of computations.

Northwest-Corner Method. The method starts at the northwest-corner cell (route) of the tableau (variable x_{11}).

Step 1. Allocate as much as possible to the selected cell, and adjust the associated amounts of supply and demand by subtracting the allocated amount.

Step 2. Cross out the row or column with zero supply or demand to indicate that no further assignments can be made in that row or column. If both a row and a column net to zero simultaneously, *cross out one only*, and leave a zero supply (demand) in the uncrossed-out row (column).

Step 3. If *exactly one* row or column is left uncrossed out, stop. Otherwise, move to the cell to the right if a column has just been crossed out or below if a row has been crossed out. Go to step 1.

Example 5.3-2

The application of the procedure to the model of Example 5.3-1 gives the starting basic solution in Table 5.17. The arrows show the order in which the allocated amounts are generated.

The starting basic solution is

$$x_{11} = 5, x_{12} = 10$$

$$x_{22} = 5, x_{23} = 15, x_{24} = 5$$

$$x_{34} = 10$$

The associated cost of the schedule is

$$z = 5 \times 10 + 10 \times 2 + 5 \times 7 + 15 \times 9 + 5 \times 20 + 10 \times 18 = \$520$$

Least-Cost Method. The least-cost method finds a better starting solution by concentrating on the cheapest routes. The method assigns as much as possible to the cell with the smallest unit cost (ties are broken arbitrarily). Next, the satisfied row or column is crossed out and the amounts of supply and demand are adjusted accordingly.

[5] All three methods are featured in TORA's tutorial module. See the end of Section 5.3.3.

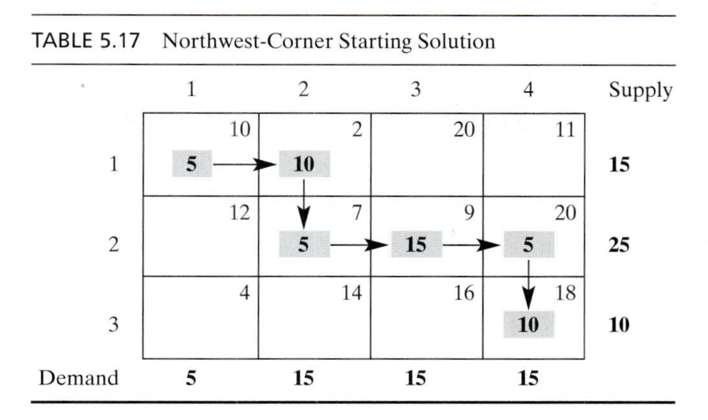

TABLE 5.17 Northwest-Corner Starting Solution

	1	2	3	4	Supply
1	(cost 10) **5**	(cost 2) **10**	(cost 20)	(cost 11)	**15**
2	(cost 12)	(cost 7) **5**	(cost 9) **15**	(cost 20) **5**	**25**
3	(cost 4)	(cost 14)	(cost 16)	(cost 18) **10**	**10**
Demand	**5**	**15**	**15**	**15**	

If both a row and a column are satisfied simultaneously, *only one is crossed out*, the same as in the northwest-corner method. Next, look for the uncrossed-out cell with the smallest unit cost and repeat the process until exactly one row or column is left uncrossed out.

Example 5.3-3

The least-cost method is applied to Example 5.3-1 in the following manner:

1. Cell (1, 2) has the least unit cost in the tableau (= \$2). The most that can be shipped through (1, 2) is $x_{12} = 15$ truckloads, which happens to satisfy both row 1 and column 2 simultaneously. We arbitrarily cross out column 2 and adjust the supply in row 1 to 0.
2. Cell (3, 1) has the smallest uncrossed-out unit cost (= \$4). Assign $x_{31} = 5$, and cross out column 1 because it is satisfied, and adjust the demand of row 3 to $10 - 5 = 5$ truckloads.
3. Continuing in the same manner, we successively assign 15 truckloads to cell (2, 3), 0 truckloads to cell (1, 4), 5 truckloads to cell (3, 4), and 10 truckloads to cell (2, 4) (verify!).

The resulting starting solution is summarized in Table 5.18. The arrows show the order in which the allocations are made. The starting solution (consisting of 6 basic variables) is $x_{12} = 15, x_{14} = 0, x_{23} = 15, x_{24} = 10, x_{31} = 5, x_{34} = 5$. The associated objective value is

$$z = 15 \times 2 + 0 \times 11 + 15 \times 9 + 10 \times 20 + 5 \times 4 + 5 \times 18 = \$475$$

The quality of the least-cost starting solution is better than that of the northwest-corner method (Example 5.3-2) because it yields a smaller value of z (\$475 versus \$520 in the northwest-corner method).

Vogel Approximation Method (VAM). VAM is an improved version of the least-cost method that generally, but not always, produces better starting solutions.

Step 1. For each row (column), determine a penalty measure by subtracting the *smallest* unit cost element in the row (column) from the *next smallest* unit cost element in the same row (column).

TABLE 5.18 Least-Cost Starting Solution

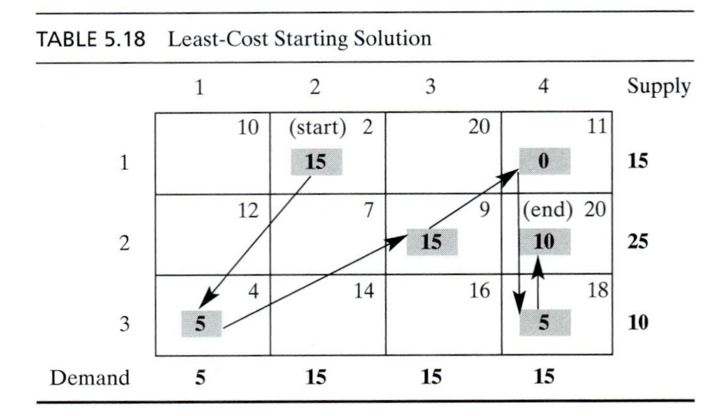

	1	2	3	4	Supply
1	10	(start) 2 **15**	20	11 **0**	**15**
2	12	7	9 **15**	(end) 20 **10**	**25**
3	4 **5**	14	16	18 **5**	**10**
Demand	**5**	**15**	**15**	**15**	

Step 2. Identify the row or column with the largest penalty. Break ties arbitrarily. Allocate as much as possible to the variable with the least unit cost in the selected row or column. Adjust the supply and demand, and cross out the satisfied row *or* column. If a row and a column are satisfied simultaneously, only one of the two is crossed out, and the remaining row (column) is assigned zero supply (demand).

Step 3. **(a)** If exactly one row or column with zero supply or demand remains uncrossed out, stop.

(b) If one row (column) with *positive* supply (demand) remains uncrossed out, determine the basic variables in the row (column) by the least-cost method. Stop.

(c) If all the uncrossed out rows and columns have (remaining) zero supply and demand, determine the *zero* basic variables by the least-cost method. Stop.

(d) Otherwise, go to step 1.

Example 5.3-4

VAM is applied to Example 5.3-1. Table 5.19 computes the first set of penalties.

Because row 3 has the largest penalty (= 10) and cell (3, 1) has the smallest unit cost in that row, the amount 5 is assigned to x_{31}. Column 1 is now satisfied and must be crossed out. Next, new penalties are recomputed as in Table 5.20.

Table 5.20 shows that row 1 has the highest penalty (= 9). Hence, we assign the maximum amount possible to cell (1, 2), which yields $x_{12} = 15$ and simultaneously satisfies both row 1 and column 2. We arbitrarily cross out column 2 and adjust the supply in row 1 to zero.

Continuing in the same manner, row 2 will produce the highest penalty (= 11), and we assign $x_{23} = 15$, which crosses out column 3 and leaves 10 units in row 2. Only column 4 is left, and it has a positive supply of 15 units. Applying the least-cost method to that column, we successively assign $x_{14} = 0$, $x_{34} = 5$, and $x_{24} = 10$ (verify!). The associated objective value for this solution is

$$z = 15 \times 2 + 0 \times 11 + 15 \times 9 + 10 \times 20 + 5 \times 4 + 5 \times 18 = \$475$$

This solution happens to have the same objective value as in the least-cost method.

TABLE 5.19 Row and Column Penalties in VAM

	1	2	3	4		Row penalty
1	10	2	20	11	**15**	10 − 2 = 8
2	12	7	9	20	**25**	9 − 7 = 2
3	4 **5**	14	16	18	**10**	14 − 4 = 10
	5	**15**	**15**	**15**		
Column penalty	10 − 4 = 6	7 − 2 = 5	16 − 9 = 7	18 − 11 = 7		

TABLE 5.20 First Assignment in VAM ($x_{31} = 5$)

	1	2	3	4		Row penalty
1	10	2	20	11	**15**	9
2	12	7	9	20	**25**	2
3	4 **5**	14	16	18	**10**	2
	5	**15**	**15**	**15**		
Column penalty	—	5	7	7		

PROBLEM SET 5.3A

1. Compare the starting solutions obtained by the northwest-corner, least-cost, and Vogel methods for each of the following models:

*(a)

0	2	1	**6**
2	1	5	**7**
2	4	3	**7**
5	**5**	**10**	

(b)

1	2	6	**7**
0	4	2	**12**
3	1	5	**11**
10	**10**	**10**	

(c)

5	1	8	**12**
2	4	0	**14**
3	6	7	**4**
9	**10**	**11**	

5.3.2 Iterative Computations of the Transportation Algorithm

After determining the starting solution (using any of the three methods in Section 5.3.1), we use the following algorithm to determine the optimum solution:

Step 1. Use the simplex *optimality condition* to determine the *entering variable* as the current nonbasic variable that can improve the solution. If the optimality condition is satisfied, stop. Otherwise, go to step 2.

Step 2. Determine the *leaving variable* using the simplex *feasibility condition*. Change the basis, and return to step 1.

The optimality and feasibility conditions do not involve the familiar row operations used in the simplex method. Instead, the special structure of the transportation model allows simpler computations.

Example 5.3-5

Solve the transportation model of Example 5.3-1, starting with the northwest-corner solution.

Table 5.21 gives the northwest-corner starting solution as determined in Table 5.17, Example 5.3-2.

The determination of the entering variable from among the current nonbasic variables (those that are not part of the starting basic solution) is done by computing the nonbasic coefficients in the z-row, using the **method of multipliers** (which, as we show in Section 5.3.4, is rooted in LP duality theory).

In the method of multipliers, we associate the multipliers u_i and v_j with row i and column j of the transportation tableau. For each current *basic* variable x_{ij}, these multipliers are shown in Section 5.3.4 to satisfy the following equations:

$$u_i + v_j = c_{ij}, \text{ for each } basic\ x_{ij}$$

As Table 5.21 shows, the starting solution has 6 basic variables, which leads to 6 equations in 7 unknowns. To solve these equations, the method of multipliers calls for arbitrarily setting any $u_i = 0$, and then solving for the remaining variables as shown below.

Basic variable	(u, v) Equation	Solution
x_{11}	$u_1 + v_1 = 10$	Set $u_1 = 0 \rightarrow v_1 = 10$
x_{12}	$u_1 + v_2 = 2$	$u_1 = 0 \rightarrow v_2 = 2$
x_{22}	$u_2 + v_2 = 7$	$v_2 = 2 \rightarrow u_2 = 5$
x_{23}	$u_2 + v_3 = 9$	$u_2 = 5 \rightarrow v_3 = 4$
x_{24}	$u_2 + v_4 = 20$	$u_2 = 5 \rightarrow v_4 = 15$
x_{34}	$u_3 + v_4 = 18$	$v_4 = 15 \rightarrow u_3 = 3$

To summarize, we have

$$u_1 = 0, u_2 = 5, u_3 = 3$$

$$v_1 = 10, v_2 = 2, v_3 = 4, v_4 = 15$$

Next, we use u_i and v_j to evaluate the nonbasic variables by computing

$$u_i + v_j - c_{ij}, \text{ for each } nonbasic\ x_{ij}$$

TABLE 5.21 Starting Iteration

	1	2	3	4	Supply
1	(10) **5**	(2) **10**	(20)	(11)	**15**
2	(12)	(7) **5**	(9) **15**	(20) **5**	**25**
3	(4)	(14)	(16)	(18) **10**	**10**
Demand	**5**	**15**	**15**	**15**	

The results of these evaluations are shown in the following table:

Nonbasic variable	$u_i + v_j - c_{ij}$
x_{13}	$u_1 + v_3 - c_{13} = 0 + 4 - 20 = -16$
x_{14}	$u_1 + v_4 - c_{14} = 0 + 15 - 11 = 4$
x_{21}	$u_2 + v_1 - c_{21} = 5 + 10 - 12 = 3$
x_{31}	$u_3 + v_1 - c_{31} = 3 + 10 - 4 = \mathbf{9}$
x_{32}	$u_3 + v_2 - c_{32} = 3 + 2 - 14 = -9$
x_{33}	$u_3 + v_3 - c_{33} = 3 + 4 - 16 = -9$

The preceding information, together with the fact that $u_i + v_j - c_{ij} = 0$ for each basic x_{ij}, is actually equivalent to computing the z-row of the simplex tableau, as the following summary shows.

Basic	x_{11}	x_{12}	x_{13}	x_{14}	x_{21}	x_{22}	x_{23}	x_{24}	↓ x_{31}	x_{32}	x_{33}	x_{34}
z	0	0	−16	4	3	0	0	0	9	−9	−9	0

Because the transportation model seeks to *minimize* cost, the entering variable is the one having the *most positive* coefficient in the z-row. Thus, x_{31} is the entering variable.

The preceding computations are usually done directly on the transportation tableau as shown in Table 5.22, meaning that it is not necessary really to write the (u, v)-equations explicitly. Instead, we start by setting $u_1 = 0$.[6] Then we can compute the v-values of all the columns that have *basic* variables in row 1—namely, v_1 and v_2. Next, we compute u_2 based on the (u, v)-equation of basic x_{22}. Now, given u_2, we can compute v_3 and v_4. Finally, we determine u_3 using the basic equation of x_{33}. Once all the u's and v's have been determined, we can evaluate the nonbasic variables by computing $u_i + v_j - c_{ij}$ for each nonbasic x_{ij}. These evaluations are shown in Table 5.22 in the boxed southeast corner of each cell.

Having identified x_{31} as the entering variable, we need to determine the leaving variable. Remember that if x_{31} enters the solution to become basic, one of the current basic variables must leave as nonbasic (at zero level).

TABLE 5.22 Iteration 1 Calculations

	$v_1 = 10$	$v_2 = 2$	$v_3 = 4$	$v_4 = 15$	Supply
$u_1 \equiv 0$	cost 10; **5**	cost 2; **10**	cost 20; [−16]	cost 11; [4]	**15**
$u_2 = 5$	cost 12; [3]	cost 7; **5**	cost 9; **15**	cost 20; **5**	**25**
$u_3 = 3$	cost 4; [9]	cost 14; [−9]	cost 16; [−9]	cost 18; **10**	**10**
Demand	**5**	**15**	**15**	**15**	

[6]The tutorial module of TORA is designed to demonstrate that assigning a zero initial value to any u or v does not affect the optimization results. See TORA Moment on page 216.

The selection of x_{31} as the entering variable means that we want to ship through this route because it reduces the total shipping cost. What is the most that we can ship through the new route? Observe in Table 5.22 that if route (3, 1) ships θ units (i.e., $x_{31} = \theta$), then the maximum value of θ is determined based on two conditions.

1. Supply limits and demand requirements remain satisfied.
2. Shipments through all routes remain nonnegative.

These two conditions determine the maximum value of θ and the leaving variable in the following manner. First, construct a *closed loop* that starts and ends at the entering variable cell, (3, 1). The loop consists of *connected horizontal* and *vertical* segments only (no diagonals are allowed).[7] Except for the entering variable cell, each corner of the closed loop must coincide with a basic variable. Table 5.23 shows the loop for x_{31}. Exactly one loop exists for a given entering variable.

Next, we assign the amount θ to the entering variable cell (3, 1). For the supply and demand limits to remain satisfied, we must alternate between subtracting and adding the amount θ at the successive *corners* of the loop as shown in Table 5.23 (it is immaterial whether the loop is traced in a clockwise or counterclockwise direction). For $\theta \geq 0$, the new values of the variables then remain nonnegative if

$$x_{11} = 5 - \theta \geq 0$$
$$x_{22} = 5 - \theta \geq 0$$
$$x_{34} = 10 - \theta \geq 0$$

The corresponding maximum value of θ is 5, which occurs when both x_{11} and x_{22} reach zero level. Because only one current basic variable must leave the basic solution, we can choose either x_{11} or x_{22} as the leaving variable. We arbitrarily choose x_{11} to leave the solution.

The selection of x_{31} $(= 5)$ as the entering variable and x_{11} as the leaving variable requires adjusting the values of the basic variables at the corners of the closed loop as Table 5.24 shows. Because each unit shipped through route (3, 1) reduces the shipping cost by \$9 $(= u_3 + v_1 - c_{31})$, the total cost associated with the new schedule is \$9 × 5 = \$45 less than in the previous schedule. Thus, the new cost is \$520 − \$45 = \$475.

TABLE 5.23 Determination of Closed Loop for x_{31}

	$v_1 = 10$	$v_2 = 2$	$v_3 = 4$	$v_4 = 15$	Supply
$u_1 = 0$	10 / **5 − Θ** (−)	2 / **10 + Θ** (+)	20 / −16	11 / 4	**15**
$u_2 = 5$	12 / 3	7 / **5 − Θ** (−)	9 / **15**	20 / **5 + Θ** (+)	**25**
$u_3 = 3$	4 / **Θ** (+) / 9	14 / −9	16 / −9	18 / **10 − Θ** (−)	**10**
Demand	**5**	**15**	**15**	**15**	

[7] TORA's tutorial module allows you to determine the cells of the *closed loop* interactively with immediate feedback regarding the validity of your selections. See TORA Moment on page 216.

TABLE 5.24 Iteration 2 Calculations

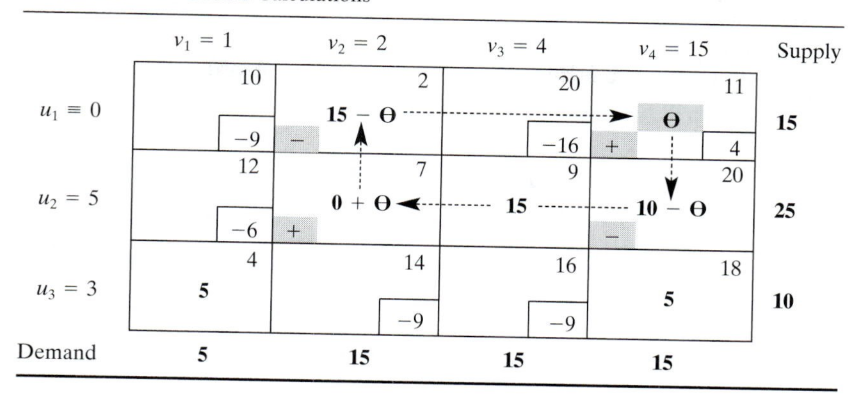

	$v_1 = 1$	$v_2 = 2$	$v_3 = 4$	$v_4 = 15$	Supply
$u_1 \equiv 0$	10 / −9	2 / **15 − θ** (−)	20 / −16	11 / **θ** (+) / 4	**15**
$u_2 = 5$	12 / −6	7 / **0 + θ** (+)	9 / **15**	20 / **10 − θ** (−)	**25**
$u_3 = 3$	4 / **5**	14 / −9	16 / −9	18 / **5**	**10**
Demand	**5**	**15**	**15**	**15**	

TABLE 5.25 Iteration 3 Calculations (Optimal)

	$v_1 = -3$	$v_2 = 2$	$v_3 = 4$	$v_4 = 11$	Supply
$u_1 \equiv 0$	10 / −13	2 / **5**	20 / −16	11 / **10**	**15**
$u_2 = 5$	12 / −10	7 / **10**	9 / **15**	20 / −4	**25**
$u_3 = 7$	4 / **5**	14 / −5	16 / −5	18 / **5**	**10**
Demand	**5**	**15**	**15**	**15**	

Given the new basic solution, we repeat the computation of the multipliers u and v, as Table 5.24 shows. The entering variable is x_{14}. The closed loop shows that $x_{14} = 10$ and that the leaving variable is x_{24}.

The new solution, shown in Table 5.25, costs $4 \times 10 = \$40$ less than the preceding one, thus yielding the new cost $\$475 - \$40 = \$435$. The new $u_i + v_j - c_{ij}$ are now negative for all nonbasic x_{ij}. Thus, the solution in Table 5.25 is optimal.

The following table summarizes the optimum solution.

From silo	To mill	Number of truckloads
1	2	5
1	4	10
2	2	10
2	3	15
3	1	5
3	4	5
	Optimal cost = $435	

TORA Moment.

From Solve/Modify Menu, select Solve ⇒ Iterations, and choose one of the three methods (northwest corner, least-cost, or Vogel) to start the transportation model iterations. The iterations module offers two useful interactive features:

1. You can set any u or v to zero before generating Iteration 2 (the default is $u_1 = 0$). Observe then that although the values of u_i and v_j change, the evaluation of the nonbasic cells ($= u_i + v_j - c_{ij}$) remains the same. This means that, initially, any u or v can be set to zero (in fact, any value) without affecting the optimality calculations.
2. You can test your understanding of the selection of the *closed loop* by clicking (in any order) the *corner* cells that comprise the path. If your selection is correct, the cell will change color (green for entering variable, red for leaving variable, and gray otherwise).

Solver Moment.

Entering the transportation model into Excel spreadsheet is straightforward. Figure 5.4 provides the Excel Solver template for Example 5.3-1 (file solverEx5.3-1.xls), together with all the formulas and the definition of range names.

In the input section, data include unit cost matrix (cells B4:E6), source names (cells A4:A6), destination names (cells B3:E3), supply (cells F4:F6), and demand (cells B7:E7). In the output section, cells B11:E13 provide the optimal solution in matrix form. The total cost formula is given in target cell A10.

AMPL Moment.

Figure 5.5 provides the AMPL model for the transportation model of Example 5.3-1 (file amplEx5.3-1a.txt). The names used in the model are self-explanatory. Both the constraints and the objective function follow the format of the LP model presented in Example 5.1-1.

The model uses the sets `sNodes` and `dNodes` to conveniently allow the use of the alphanumeric set members `{S1, S2, S3}` and `{D1, D2, D3, D4}` which are entered in the data section. All the input data are then entered in terms of these set members as shown in Figure 5.5.

Although the alphanumeric code for set members is more readable, generating them for large problems may not be convenient. File amplEx5.3-1b shows how the same sets can be defined as `{1..m}` and `{1..n}`, where `m` and `n` represent the number of sources and the number of destinations. By simply assigning numeric values for `m` and `n`, the sets are automatically defined for any size model.

The data of the transportation model can be retrieved from a spreadsheet (file TM.xls) using the AMPL `table` statement. File amplEx3.5-1c.txt provides the details. To study this model, you will need to review the material in Section A.5.5.

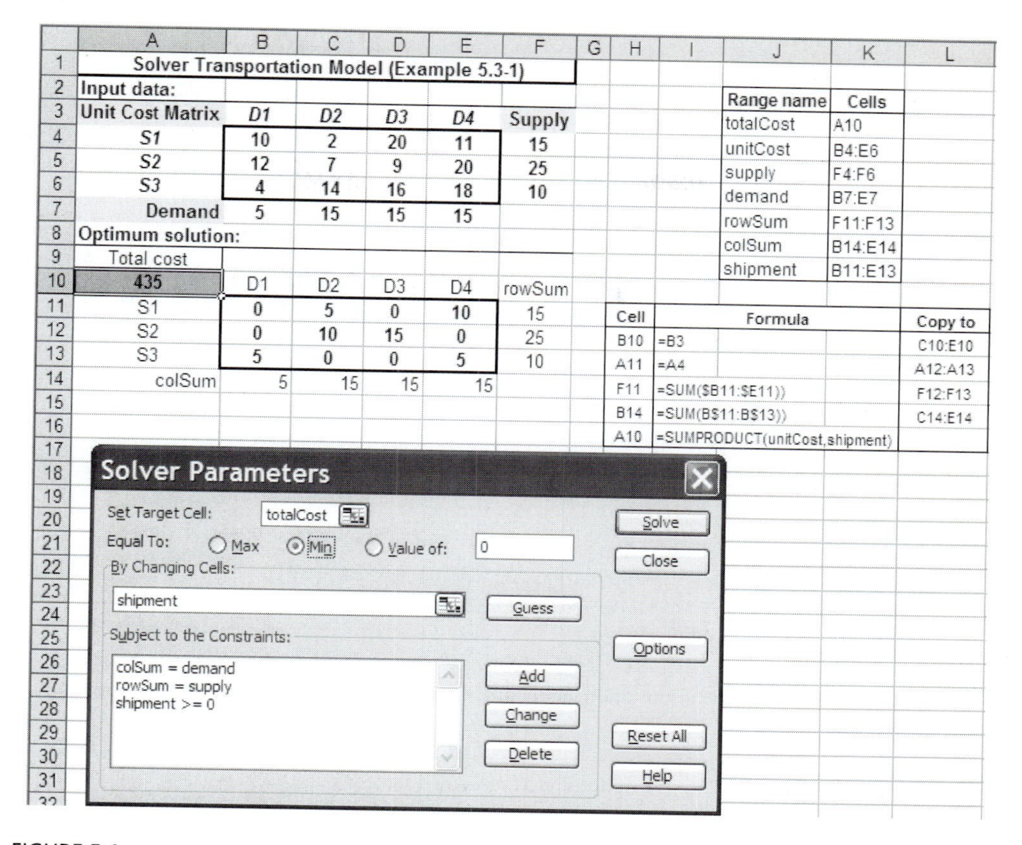

FIGURE 5.4

Excel Solver solution of the transportation model of Example 5.3-1 (File solverEx5.3-1.xls)

PROBLEM SET 5.3B

1. Consider the transportation models in Table 5.26.
   (a) Use the northwest-corner method to find the starting solution.
   (b) Develop the iterations that lead to the optimum solution.
   (c) *TORA Experiment.* Use TORA's Iterations module to compare the effect of using the northwest-corner rule, least-cost method, and Vogel method on the number of iterations leading to the optimum solution.
   (d) *Solver Experiment.* Solve the problem by modifying file solverEx5.3-1.xls.
   (e) *AMPL Experiment.* Solve the problem by modifying file amplEx5.3-1b.txt.
2. In the transportation problem in Table 5.27, the total demand exceeds the total supply. Suppose that the penalty costs per unit of unsatisfied demand are \$5, \$3, and \$2 for destinations 1, 2, and 3, respectively. Use the least-cost starting solution and compute the iterations leading to the optimum solution.

```
#----- Transporation model (Example 5.3-1)-----
set sNodes;
set dNodes;
param c{sNodes,dNodes};
param supply{sNodes};
param demand{dNodes};
var x{sNodes,dNodes}>=0;
minimize z:sum {i in sNodes,j in dNodes}c[i,j]*x[i,j];
subject to
source {i in sNodes}:sum{j in dNodes}x[i,j]=supply[i];
dest {j in dNodes}:sum{i in sNodes}x[i,j]=demand[j];
data;
set sNodes:=S1 S2 S3;
set dNodes:=D1 D2 D3 D4;
param c:
   D1  D2  D3  D4 :=
S1 10  2   20  11
S2 12  7   9   20
S3 4   14  16  18;
param supply:= S1 15 S2 25 S3 10;
param demand:=D1 5 D2 15 D3 15 D4 15;
solve;display z, x;
```

FIGURE 5.5

AMPL model of the transportation model of Example 5.3-1 (File amplEx5.3-1a.txt)

TABLE 5.26 Transportation Models for Problem 1

(i)

\$0	\$2	\$1	**6**
\$2	\$1	\$5	**9**
\$2	\$4	\$3	**5**
5	**5**	**10**	

(ii)

\$10	\$4	\$2	**8**
\$2	\$3	\$4	**5**
\$1	\$2	\$0	**6**
7	**6**	**6**	

(iii)

—	\$3	\$5	**4**
\$7	\$4	\$9	**7**
\$1	\$8	\$6	**19**
5	**6**	**19**	

TABLE 5.27 Data for Problem 2

\$5	\$1	\$7	**10**
\$6	\$4	\$6	**80**
\$3	\$2	\$5	**15**
75	**20**	**50**	

3. In Problem 2, suppose that there are no penalty costs, but that the demand at destination 3 must be satisfied completely.

(a) Find the optimal solution.

(b) *Solver Experiment.* Solve the problem by modifying file solverEx5.3-1.xls.

(c) *AMPL Experiment.* Solve the problem by modifying file amplEx5.3b-1.txt.

TABLE 5.28 Data for Problem 4

$1	$2	$1	**20**
$3	$4	$5	**40**
$2	$3	$3	**30**
30	**20**	**20**	

TABLE 5.29 Data for Problem 6

10			**10**
	20	**20**	**40**
10	**20**	**20**	

4. In the unbalanced transportation problem in Table 5.28, if a unit from a source is not shipped out (to any of the destinations), a storage cost is incurred at the rate of $5, $4, and $3 per unit for sources 1, 2, and 3, respectively. Additionally, all the supply at source 2 must be shipped out completely to make room for a new product. Use Vogel's starting solution and determine all the iterations leading to the optimum shipping schedule.

***5.** In a 3×3 transportation problem, let x_{ij} be the amount shipped from source i to destination j and let c_{ij} be the corresponding transportation cost per unit. The amounts of supply at sources 1, 2, and 3 are 15, 30, and 85 units, respectively, and the demands at destinations 1, 2, and 3 are 20, 30, and 80 units, respectively. Assume that the starting northwest-corner solution is optimal and that the associated values of the multipliers are given as $u_1 = -2, u_2 = 3, u_3 = 5, v_1 = 2, v_2 = 5$, and $v_3 = 10$.

(a) Find the associated optimal cost.

(b) Determine the smallest value of c_{ij} for each nonbasic variable that will maintain the optimality of the northwest-corner solution.

6. The transportation problem in Table 5.29 gives the indicated *degenerate* basic solution (i.e., at least one of the basic variables is zero). Suppose that the multipliers associated with this solution are $u_1 = 1, u_2 = -1, v_1 = 2, v_2 = 2$, and $v_3 = 5$ and that the unit cost for all (basic and nonbasic) *zero* x_{ij} variables is given by

$$c_{ij} = i + j\theta, -\infty < \theta < \infty$$

(a) If the given solution is optimal, determine the associated optimal value of the objective function.

(b) Determine the value of θ that will guarantee the optimality of the given solution. (*Hint:* Locate the zero basic variable.)

7. Consider the problem

$$\text{Minimize } z = \sum_{i=1}^{m}\sum_{j=1}^{n} c_{ij}x_{ij}$$

TABLE 5.30 Data for Problem 7

\$1	\$1	\$2	**5**
\$6	\$5	\$1	**6**
2	**7**	**1**	

subject to

$$\sum_{j=1}^{n} x_{ij} \geq a_i, i = 1, 2, \ldots, m$$

$$\sum_{i=1}^{m} x_{ij} \geq b_j, j = 1, 2, \ldots, n$$

$$x_{ij} \geq 0, \text{ all } i \text{ and } j$$

It may appear logical to assume that the optimum solution will require the first (second) set of inequalities to be replaced with equations if $\Sigma a_i \geq \Sigma b_j$ ($\Sigma a_i \leq \Sigma b_j$). The counterexample in Table 5.30 shows that this assumption is not correct.

Show that the application of the suggested procedure yields the solution $x_{11} = 2$, $x_{12} = 3$, $x_{22} = 4$, and $x_{23} = 2$, with $z = \$27$, which is worse than the feasible solution $x_{11} = 2$, $x_{12} = 7$, and $x_{23} = 6$, with $z = \$15$.

5.3.3 Simplex Method Explanation of the Method of Multipliers

The relationship between the method of multipliers and the simplex method can be explained based on the primal-dual relationships (Section 4.2). From the special structure of the LP representing the transportation model (see Example 5.1-1 for an illustration), the associated dual problem can be written as

$$\text{Maximize } z = \sum_{i=1}^{m} a_i u_i + \sum_{j=1}^{n} b_j v_j$$

subject to

$$u_i + v_j \leq c_{ij}, \text{ all } i \text{ and } j$$

$$u_i \text{ and } v_j \text{ unrestricted}$$

where

a_i = Supply amount at source i

b_j = Demand amount at destination j

c_{ij} = Unit transportation cost from source i to destination j

u_i = Dual variable of the constraint associated with source i

v_j = Dual variable of the constraint associated with destination j

From Formula 2, Section 4.2.4, the objective-function coefficients (reduced costs) of the variable x_{ij} equal the difference between the left- and right-hand sides of the corresponding dual constraint—that is, $u_i + v_j - c_{ij}$. However, we know that this quantity must equal zero for each *basic variable*, which then produces the following result:

$$u_i + v_j = c_{ij}, \text{ for each } \textit{basic} \text{ variable } x_{ij}.$$

There are $m + n - 1$ such equations whose solution (after assuming an arbitrary value $u_1 = 0$) yields the multipliers u_i and v_j. Once these multipliers are computed, the entering variable is determined from all the *nonbasic* variables as the one having the largest positive $u_i + v_j - c_{ij}$.

The assignment of an arbitrary value to one of the dual variables (i.e., $u_1 = 0$) may appear inconsistent with the way the dual variables are computed using Method 2 in Section 4.2.3. Namely, for a given basic solution (and, hence, inverse), the dual values must be unique. Problem 2, Set 5.3c, addresses this point.

PROBLEM SET 5.3C

1. Write the dual problem for the LP of the transportation problem in Example 5.3-5 (Table 5.21). Compute the associated optimum *dual* objective value using the optimal dual values given in Table 5.25, and show that it equals the optimal cost given in the example.
2. In the transportation model, one of the dual variables assumes an arbitrary value. This means that for the same basic solution, the values of the associated dual variables are not unique. The result appears to contradict the theory of linear programming, where the dual values are determined as the product of the vector of the objective coefficients for the basic variables and the associated inverse basic matrix (see Method 2, Section 4.2.3). Show that for the transportation model, although the inverse basis is unique, the vector of *basic* objective coefficients need not be so. Specifically, show that if c_{ij} is changed to $c_{ij} + k$ for all i and j, where k is a constant, then the optimal values of x_{ij} will remain the same. Hence, the use of an arbitrary value for a dual variable is implicitly equivalent to assuming that a specific constant k is added to all c_{ij}.

5.4 THE ASSIGNMENT MODEL

"The best person for the job" is an apt description of the assignment model. The situation can be illustrated by the assignment of workers with varying degrees of skill to jobs. A job that happens to match a worker's skill costs less than one in which the operator is not as skillful. The objective of the model is to determine the minimum-cost assignment of workers to jobs.

The general assignment model with n workers and n jobs is represented in Table 5.31.

The element c_{ij} represents the cost of assigning worker i to job j $(i, j = 1, 2, \ldots, n)$. There is no loss of generality in assuming that the number of workers always

TABLE 5.31 Assignment Model

		Jobs				
		1	2	...	n	
Worker	1	c_{11}	c_{12}	...	c_{1n}	**1**
	2	c_{21}	c_{22}	...	c_{2n}	**1**
	⋮	⋮	⋮	⋮	⋮	⋮
	n	c_{n1}	c_{n2}	...	c_{nn}	**1**
		1	**1**	...	**1**	

equals the number of jobs, because we can always add fictitious workers or fictitious jobs to satisfy this assumption.

The assignment model is actually a special case of the transportation model in which the workers represent the sources, and the jobs represent the destinations. The supply (demand) amount at each source (destination) exactly equals 1. The cost of "transporting" worker i to job j is c_{ij}. In effect, the assignment model can be solved directly as a regular transportation model. Nevertheless, the fact that all the supply and demand amounts equal 1 has led to the development of a simple solution algorithm called the **Hungarian method.** Although the new solution method appears totally unrelated to the transportation model, the algorithm is actually rooted in the simplex method, just as the transportation model is.

5.4.1 The Hungarian Method[8]

We will use two examples to present the mechanics of the new algorithm. The next section provides a simplex-based explanation of the procedure.

Example 5.4-1

Joe Klyne's three children, John, Karen, and Terri, want to earn some money to take care of personal expenses during a school trip to the local zoo. Mr. Klyne has chosen three chores for his children: mowing the lawn, painting the garage door, and washing the family cars. To avoid anticipated sibling competition, he asks them to submit (secret) bids for what they feel is fair pay for each of the three chores. The understanding is that all three children will abide by their father's decision as to who gets which chore. Table 5.32 summarizes the bids received. Based on this information, how should Mr. Klyne assign the chores?

The assignment problem will be solved by the Hungarian method.

Step 1. For the original cost matrix, identify each row's minimum, and subtract it from all the entries of the row.

[8]As with the transportation model, the classical Hungarian method, designed primarily for *hand* computations, is something of the past and is presented here purely for historical reasons. Today, the need for such computational shortcuts is not warranted as the problem can be solved as a regular LP using highly efficient computer codes.

TABLE 5.32 Klyne's Assignment Problem

	Mow	Paint	Wash
John	\$15	\$10	\$9
Karen	\$9	\$15	\$10
Terri	\$10	\$12	\$8

Step 2. For the matrix resulting from step 1, identify each column's minimum, and subtract it from all the entries of the column.

Step 3. Identify the optimal solution as the feasible assignment associated with the zero elements of the matrix obtained in step 2.

Let p_i and q_j be the minimum costs associated with row i and column j as defined in steps 1 and 2, respectively. The row minimums of step 1 are computed from the original cost matrix as shown in Table 5.33.

Next, subtract the row minimum from each respective row to obtain the reduced matrix in Table 5.34.

The application of step 2 yields the column minimums in Table 5.34. Subtracting these values from the respective columns, we get the reduced matrix in Table 5.35.

TABLE 5.33 Step 1 of the Hungarian Method

	Mow	Paint	Wash	Row minimum
John	15	10	9	$p_1 = 9$
Karen	9	15	10	$p_2 = 9$
Terri	10	12	8	$p_3 = 8$

TABLE 5.34 Step 2 of the Hungarian Method

	Mow	Paint	Wash
John	6	1	0
Karen	0	6	1
Terri	2	4	0
Column minimum	$q_1 = 0$	$q_2 = 1$	$q_3 = 0$

TABLE 5.35 Step 3 of the Hungarian Method

	Mow	Paint	Wash
John	6	**0**	0
Karen	**0**	5	1
Terri	2	3	**0**

The cells with underscored zero entries provide the optimum solution. This means that John gets to paint the garage door, Karen gets to mow the lawn, and Terri gets to wash the family cars. The total cost to Mr. Klyne is $9 + 10 + 8 = \$27$. This amount also will always equal $(p_1 + p_2 + p_3) + (q_1 + q_2 + q_3) = (9 + 9 + 8) + (0 + 1 + 0) = \$27$. (A justification of this result is given in the next section.)

The given steps of the Hungarian method work well in the preceding example because the zero entries in the final matrix happen to produce a *feasible* assignment (in the sense that each child is assigned a distinct chore). In some cases, the zeros created by steps 1 and 2 may not yield a feasible solution directly, and further steps are needed to find the optimal (feasible) assignment. The following example demonstrates this situation.

Example 5.4-2

Suppose that the situation discussed in Example 5.4-1 is extended to four children and four chores. Table 5.36 summarizes the cost elements of the problem.

The application of steps 1 and 2 to the matrix in Table 5.36 (using $p_1 = 1, p_2 = 7, p_3 = 4, p_4 = 5, q_1 = 0, q_2 = 0, q_3 = 3$, and $q_4 = 0$) yields the reduced matrix in Table 5.37 (verify!).

The locations of the zero entries do not allow assigning unique chores to all the children. For example, if we assign child 1 to chore 1, then column 1 will be eliminated, and child 3 will not have a zero entry in the remaining three columns. This obstacle can be accounted for by adding the following step to the procedure outlined in Example 5.4-1:

Step 2a. If no feasible assignment (with all zero entries) can be secured from steps 1 and 2,

(i) Draw the *minimum* number of horizontal and vertical lines in the last reduced matrix that will cover *all* the zero entries.

TABLE 5.36 Assignment Model

		Chore			
		1	2	3	4
Child	1	\$1	\$4	\$6	\$3
	2	\$9	\$7	\$10	\$9
	3	\$4	\$5	\$11	\$7
	4	\$8	\$7	\$8	\$5

TABLE 5.37 Reduced Assignment Matrix

		Chore			
		1	2	3	4
Child	1	**0**	3	2	2
	2	2	**0**	**0**	2
	3	**0**	1	4	3
	4	3	2	**0**	**0**

TABLE 5.38 Application of Step 2a

		Chore 1	2	3	4
Child	1	0	3	2	2
	2	**2**	0	0	2
	3	0	*1*	4	3
	4	**3**	2	0	0

TABLE 5.39 Optimal Assignment

		Chore 1	2	3	4
Child	1	**0**	2	1	1
	2	3	0	**0**	2
	3	0	**0**	3	2
	4	4	2	0	**0**

(ii) Select the *smallest uncovered* entry, subtract it from every uncovered entry, then add it to every entry at the intersection of two lines.

(iii) If no feasible assignment can be found among the resulting zero entries, repeat step 2a. Otherwise, go to step 3 to determine the optimal assignment.

The application of step 2a to the last matrix produces the shaded cells in Table 5.38. The smallest unshaded entry (shown in italics) equals 1. This entry is added to the bold intersection cells and subtracted from the remaining shaded cells to produce the matrix in Table 5.39.

The optimum solution (shown by the underscored zeros) calls for assigning child 1 to chore 1, child 2 to chore 3, child 3 to chore 2, and child 4 to chore 4. The associated optimal cost is $1 + 10 + 5 + 5 = \$21$. The same cost is also determined by summing the p_i's, the q_j's, and the entry that was subtracted after the shaded cells were determined—that is, $(1 + 7 + 4 + 5) + (0 + 0 + 3 + 0) + (1) = \$21$.

AMPL Moment.

File amplEx5.4-2.txt provides the AMPL model for the assignment model. The model is very similar to that of the transportation model.

PROBLEM SET 5.4A

1. Solve the assignment models in Table 5.40.
   - **(a)** Solve by the Hungarian method.
   - **(b)** *TORA Experiment.* Express the problem as an LP and solve it with TORA.
   - **(c)** *TORA Experiment.* Use TORA to solve the problem as a transportation model.

TABLE 5.40 Data for Problem 1

(i)

$3	$8	$2	$10	$3
$8	$7	$2	$9	$7
$6	$4	$2	$7	$5
$8	$4	$2	$3	$5
$9	$10	$6	$9	$10

(ii)

$3	$9	$2	$3	$7
$6	$1	$5	$6	$6
$9	$4	$7	$10	$3
$2	$5	$4	$2	$1
$9	$6	$2	$4	$5

(d) *Solver Experiment.* Modify Excel file solverEx5.3-1.xls to solve the problem.

(e) *AMPL Experiment.* Modify amplEx5.3-1b.txt to solve the problem.

2. JoShop needs to assign 4 jobs to 4 workers. The cost of performing a job is a function of the skills of the workers. Table 5.41 summarizes the cost of the assignments. Worker 1 cannot do job 3 and worker 3 cannot do job 4. Determine the optimal assignment using the Hungarian method.

3. In the JoShop model of Problem 2, suppose that an additional (fifth) worker becomes available for performing the four jobs at the respective costs of $60, $45, $30, and $80. Is it economical to replace one of the current four workers with the new one?

4. In the model of Problem 2, suppose that JoShop has just received a fifth job and that the respective costs of performing it by the four current workers are $20, $10, $20, and $80. Should the new job take priority over any of the four jobs JoShop already has?

***5.** A business executive must make the four round trips listed in Table 5.42 between the head office in Dallas and a branch office in Atlanta.

The price of a round-trip ticket from Dallas is $400. A discount of 25% is granted if the dates of arrival and departure of a ticket span a weekend (Saturday and Sunday). If the stay in Atlanta lasts more than 21 days, the discount is increased to 30%. A one-way

TABLE 5.41 Data for Problem 2

		Job 1	Job 2	Job 3	Job 4
Worker	1	$50	$50	—	$20
	2	$70	$40	$20	$30
	3	$90	$30	$50	—
	4	$70	$20	$60	$70

TABLE 5.42 Data for Problem 5

Departure date from Dallas	Return date to Dallas
Monday, June 3	Friday, June 7
Monday, June 10	Wednesday, June 12
Monday, June 17	Friday, June 21
Tuesday, June 25	Friday, June 28

ticket between Dallas and Atlanta (either direction) costs \$250. How should the executive purchase the tickets?

***6.** Figure 5.6 gives a schematic layout of a machine shop with its existing work centers designated by squares 1, 2, 3, and 4. Four new work centers, I, II, III, and IV, are to be added to the shop at the locations designated by circles *a, b, c,* and *d*. The objective is to assign the new centers to the proposed locations to minimize the total materials handling traffic between the existing centers and the proposed ones. Table 5.43 summarizes the frequency of trips between the new centers and the old ones. Materials handling equipment travels along the rectangular aisles intersecting at the locations of the centers. For example, the one-way travel distance (in meters) between center 1 and location *b* is 30 + 20 = 50 m.

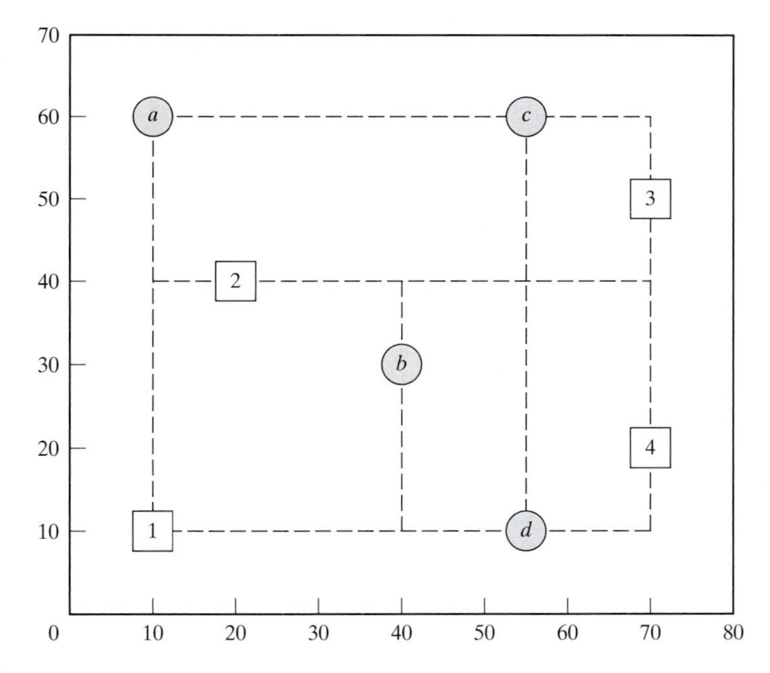

FIGURE 5.6
Machine shop layout for Problem 6, Set 5.4a

TABLE 5.43 Data for Problem 6

		New center			
		I	II	III	IV
Existing center	1	10	2	4	3
	2	7	1	9	5
	3	0	8	6	2
	4	11	4	0	7

7. In the Industrial Engineering Department at the University of Arkansas, INEG 4904 is a capstone design course intended to allow teams of students to apply the knowledge and skills learned in the undergraduate curriculum to a practical problem. The members of each team select a project manager, identify an appropriate scope for their project, write and present a proposal, perform necessary tasks for meeting the project objectives, and write and present a final report. The course instructor identifies potential projects and provides appropriate information sheets for each, including contact at the sponsoring organization, project summary, and potential skills needed to complete the project. Each design team is required to submit a report justifying the selection of team members and the team manager. The report also provides a ranking for each project in order of preference, including justification regarding proper matching of the team's skills with the project objectives. In a specific semester, the following projects were identified: Boeing F-15, Boeing F-18, Boeing Simulation, Cargil, Cobb-Vantress, ConAgra, Cooper, DaySpring (layout), DaySpring (material handling), J.B. Hunt, Raytheon, Tyson South, Tyson East, Wal-Mart, and Yellow Transportation. The projects for Boeing and Raytheon require U.S. citizenship of all team members. Of the eleven design teams available for this semester, four do not meet this requirement.

Devise a procedure for assigning projects to teams and justify the arguments you use to reach a decision.

5.4.2 Simplex Explanation of the Hungarian Method

The assignment problem in which n workers are assigned to n jobs can be represented as an LP model in the following manner: Let c_{ij} be the cost of assigning worker i to job j, and define

$$x_{ij} = \begin{cases} 1, & \text{if worker } i \text{ is assigned to job } j \\ 0, & \text{otherwise} \end{cases}$$

Then the LP model is given as

$$\text{Minimize } z = \sum_{i=1}^{n}\sum_{j=1}^{n} c_{ij}x_{ij}$$

subject to

$$\sum_{j=1}^{n} x_{ij} = 1, i = 1, 2, \ldots, n$$

$$\sum_{i=1}^{n} x_{ij} = 1, j = 1, 2, \ldots, n$$

$$x_{ij} = 0 \text{ or } 1$$

The optimal solution of the preceding LP model remains unchanged if a constant is added to or subtracted from any row or column of the cost matrix (c_{ij}). To prove this point, let p_i and q_j be constants subtracted from row i and column j. Thus, the cost element c_{ij} is changed to

$$c'_{ij} = c_{ij} - p_i - q_j$$

Now

$$\sum_i \sum_j c'_{ij} x_{ij} = \sum_i \sum_j (c_{ij} - p_i - q_j) x_{ij} = \sum_i \sum_j c_{ij} x_{ij} - \sum_i p_i \Big(\sum_j x_{ij} \Big) - \sum_j q_j \Big(\sum_i x_{ij} \Big)$$

$$= \sum_i \sum_j c_{ij} x_{ij} - \sum_i p_i(1) - \sum_j q_j(1)$$

$$= \sum_i \sum_j c_{ij} x_{ij} - \text{constant}$$

Because the new objective function differs from the original one by a constant, the optimum values of x_{ij} must be the same in both cases. The development thus shows that steps 1 and 2 of the Hungarian method, which call for subtracting p_i from row i and then subtracting q_j from column j, produce an equivalent assignment model. In this regard, if a feasible solution can be found among the zero entries of the cost matrix created by steps 1 and 2, then it must be optimum because the cost in the modified matrix cannot be less than zero.

If the created zero entries cannot yield a feasible solution (as Example 5.4-2 demonstrates), then step 2a (dealing with the covering of the zero entries) must be applied. The validity of this procedure is again rooted in the simplex method of linear programming and can be explained by duality theory (Chapter 4) and the complementary slackness theorem (Chapter 7). We will not present the details of the proof here because they are somewhat involved.

The reason $(p_1 + p_2 + \cdots + p_n) + (q_1 + q_2 + \cdots + q_n)$ gives the optimal objective value is that it represents the dual objective function of the assignment model. This result can be seen through comparison with the dual objective function of the transportation model given in Section 5.3.4. [See Bazaraa and Associates (1990, pp. 499–508) for the details.]

5.5 THE TRANSSHIPMENT MODEL

The transshipment model recognizes that it may be cheaper to ship through intermediate or *transient* nodes before reaching the final destination. This concept is more general than that of the regular transportation model, where direct shipments only are allowed between a source and a destination.

This section shows how a transshipment model can be converted to (and solved as) a regular transportation model using the idea of a **buffer**.

Example 5.5-1

Two automobile plants, $P1$ and $P2$, are linked to three dealers, $D1$, $D2$, and $D3$, by way of two transit centers, $T1$ and $T2$, according to the network shown in Figure 5.7. The supply amounts at plants $P1$ and $P2$ are 1000 and 1200 cars, and the demand amounts at dealers $D1$, $D2$, and $D3$, are 800, 900, and 500 cars. The shipping costs per car (in hundreds of dollars) between pairs of nodes are shown on the connecting links (or arcs) of the network.

Transshipment occurs in the network in Figure 5.7 because the entire supply amount of 2200 (= 1000 + 1200) cars at nodes $P1$ and $P2$ could conceivably pass through any node of the

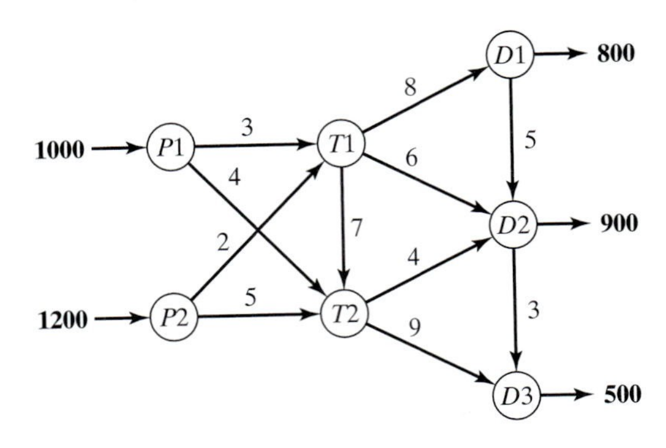

FIGURE 5.7
Transshipment network between plants and dealers

network before ultimately reaching their destinations at nodes $D1$, $D2$, and $D3$. In this regard, each node of the network with both input and output arcs ($T1$, $T2$, $D1$, and $D2$) acts as both a source and a destination and is referred to as a **transshipment node**. The remaining nodes are either **pure supply nodes** ($P1$ and $P2$) or **pure demand nodes** ($D3$).

The transshipment model can be converted into a regular transportation model with six sources ($P1, P2, T1, T2, D1$, and $D2$) and five destinations ($T1, T2, D1, D2$, and $D3$). The amounts of supply and demand at the different nodes are computed as

$$\begin{aligned} \text{Supply at a } \textit{pure supply node} &= \text{Original supply} \\ \text{Demand at a } \textit{pure demand node} &= \text{Original demand} \\ \text{Supply at a } \textit{transshipment node} &= \text{Original supply} + \text{Buffer amount} \\ \text{Demand at a } \textit{transshipment node} &= \text{Original demand} + \text{Buffer amount} \end{aligned}$$

The buffer amount should be sufficiently large to allow all of the *original* supply (or demand) units to pass through any of the *transshipment* nodes. Let B be the desired buffer amount; then

$$\begin{aligned} B &= \text{Total supply (or demand)} \\ &= 1000 + 1200 \text{ (or } 800 + 900 + 500) \\ &= 2200 \text{ cars} \end{aligned}$$

Using the buffer B and the unit shipping costs given in the network, we construct the equivalent regular transportation model as in Table 5.44.

The solution of the resulting transportation model (determined by TORA) is shown in Figure 5.8. Note the effect of transshipment: Dealer $D2$ receives 1400 cars, keeps 900 cars to satisfy its demand, and sends the remaining 500 cars to dealer $D3$.

PROBLEM SET 5.5A[9]

1. The network in Figure 5.9 gives the shipping routes from nodes 1 and 2 to nodes 5 and 6 by way of nodes 3 and 4. The unit shipping costs are shown on the respective arcs.
   (a) Develop the corresponding transshipment model.
   (b) Solve the problem, and show how the shipments are routed from the sources to the destinations.

[9]You are encouraged to use TORA, Excel Solver, or AMPL to solve the problems in this set.

TABLE 5.44 Transshipment Model

	T1	*T2*	*D1*	*D2*	*D3*	
*P*1	3	4	*M*	*M*	*M*	**1000**
*P*2	2	5	*M*	*M*	*M*	**1200**
*T*1	0	7	8	6	*M*	***B***
*T*2	*M*	0	*M*	4	9	***B***
*D*1	*M*	*M*	0	5	*M*	***B***
*D*2	*M*	*M*	*M*	0	3	***B***
	B	***B***	**800 + *B***	**900 + *B***	**500**	

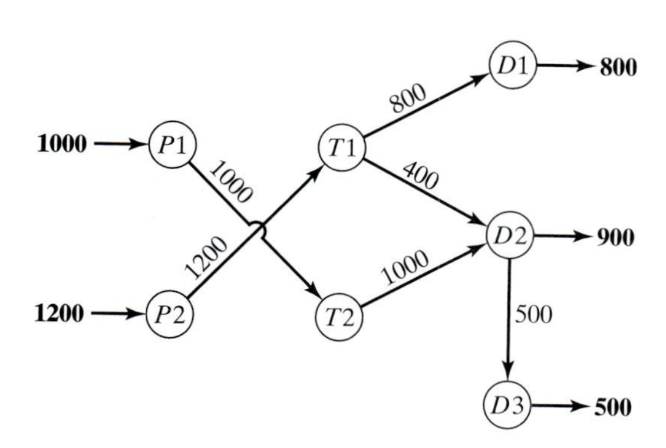

FIGURE 5.8
Solution of the transshipment model

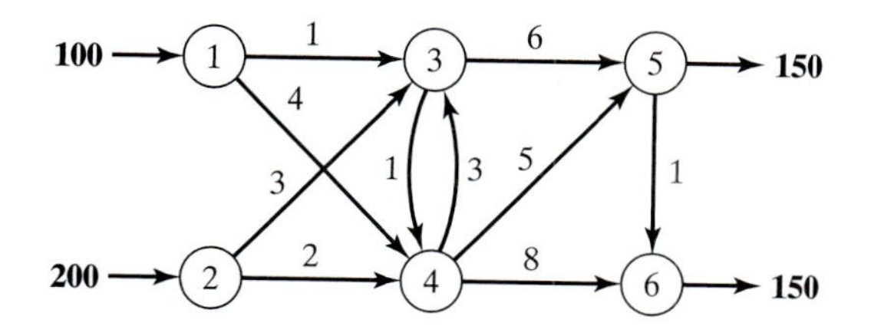

FIGURE 5.9
Network for Problem 1, Set 5.5a

2. In Problem 1, suppose that source node 1 can be linked to source node 2 with a unit shipping cost of \$1. The unit shipping cost from node 1 to node 3 is increased to \$5. Formulate the problem as a transshipment model, and find the optimum shipping schedule.

3. The network in Figure 5.10 shows the routes for shipping cars from three plants (nodes 1, 2, and 3) to three dealers (nodes 6 to 8) by way of two distribution centers (nodes 4 and 5). The shipping costs per car (in \$100) are shown on the arcs.

(a) Solve the problem as a transshipment model.

(b) Find the new optimum solution assuming that distribution center 4 can sell 240 cars directly to customers.

***4.** Consider the transportation problem in which two factories supply three stores with a commodity. The numbers of supply units available at sources 1 and 2 are 200 and 300; those demanded at stores 1, 2, and 3 are 100, 200, and 50, respectively. Units may be transshipped among the factories and the stores before reaching their final destination. Find the optimal shipping schedule based on the unit costs in Table 5.45.

5. Consider the oil pipeline network shown in Figure 5.11. The different nodes represent pumping and receiving stations. Distances in miles between the stations are shown on the network. The transportation cost per gallon between two nodes is directly proportional to the length of the pipeline. Develop the associated transshipment model, and find the optimum solution.

6. *Shortest-Route Problem.* Find the shortest route between nodes 1 and 7 of the network in Figure 5.12 by formulating the problem as a transshipment model. The distances between the different nodes are shown on the network. (*Hint*: Assume that node 1 has a net supply of 1 unit, and node 7 has a net demand also of 1 unit.)

FIGURE 5.10

Network for Problem 3, Set 5.5a

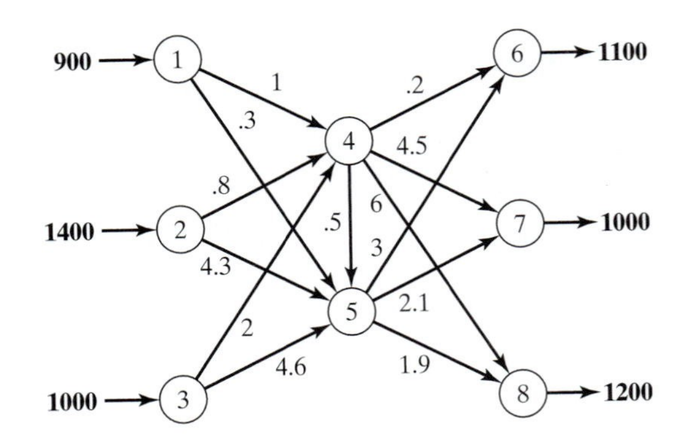

TABLE 5.45 Data for Problem 4

		Factory 1	Factory 2	Store 1	Store 2	Store 3
Factory	1	\$0	\$6	\$7	\$8	\$9
	2	\$6	\$0	\$5	\$4	\$3
Store	1	\$7	\$2	\$0	\$5	\$1
	2	\$1	\$5	\$1	\$0	\$4
	3	\$8	\$9	\$7	\$6	\$0

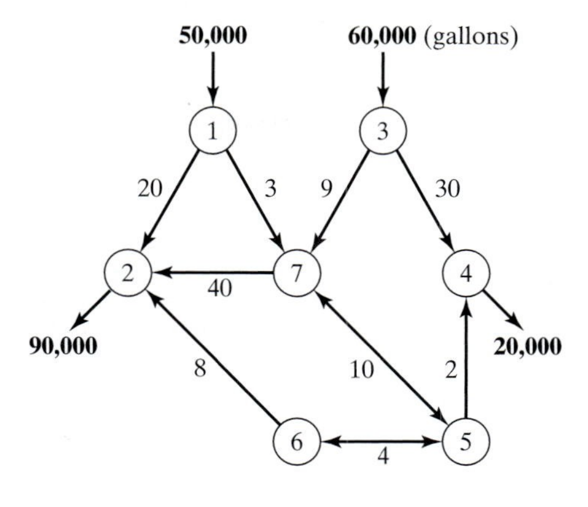

FIGURE 5.11
Network for Problem 5, Set 5.5a

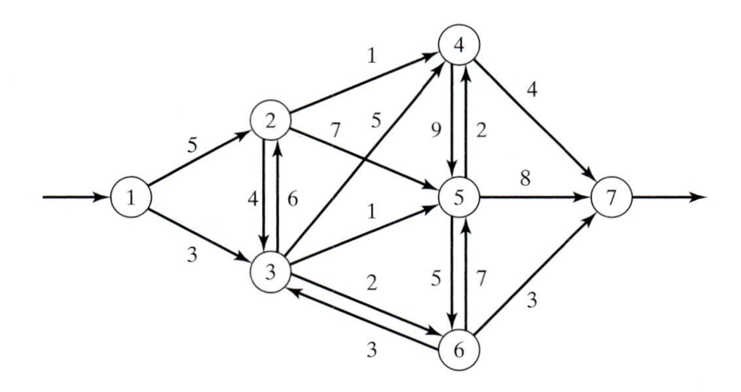

FIGURE 5.12
Network for Problem 6, Set 5.5a

7. In the transshipment model of Example 5.5-1, define x_{ij} as the amount shipped from node i to node j. The problem can be formulated as a linear program in which each node produces a constraint equation. Develop the linear program, and show that the resulting formulation has the characteristic that the constraint coefficients, a_{ij}, of the variable x_{ij} are

$$a_{ij} = \begin{cases} 1, \text{in constraint } i \\ -1, \text{in constraint } j \\ 0, \text{otherwise} \end{cases}$$

8. An employment agency must provide the following laborers over the next 5 months:

Month	1	2	3	4	5
No. of laborers	100	120	80	170	50

Because the cost of labor depends on the length of employment, it may be more economical to keep more laborers than needed during some months of the 5-month planning horizon. The following table estimates the labor cost as a function of the length of employment:

Months of employment	1	2	3	4	5
Cost per laborer ($)	100	130	180	220	250

Formulate the problem as a linear program. Then, using proper algebraic manipulations of the constraint equations, show that the model can be converted to a transshipment model, and find the optimum solution. (*Hint*: Use the transshipment characteristic in Problem 7 to convert the constraints of the scheduling problem into those of the transshipment model.)

REFERENCES

Bazaraa, M., J. Jarvis, and H. Sherali, *Linear Programming and Network Flows*, 2nd ed., Wiley, New York, 1990.

Dantzig, G., *Linear Programming and Extensions*, Princeton University Press, Princeton, N.J., 1963.

Hansen, P. and R. Wendell, "A Note on Airline Commuting," *Interfaces*, Vol. 12, No. 1, pp. 85–87, 1982.

Murty, K., *Network Programming*, Prentice Hall, Upper Saddle River, NJ, 1992.

CHAPTER 6

Network Models

Chapter Guide. The network models in this chapter include the traditional applications of finding the most efficient way to link a number of locations directly or indirectly, finding the shortest route between two cities, determining the maximum flow in a pipeline network, determining the minimum-cost flow in a network that satisfies supply and demand requirements at different locations, and scheduling the activities of a project.

The minimum-cost capacitated algorithm is a generalized network that subsumes the shortest-route and the maximal-flow models presented in this chapter. Its details can be found in Section 20.1 on the CD.

As you study the material in this chapter, you should pay special attention to the nontraditional applications of these models. For example, the shortest-route model can be used to determine the optimal equipment replacement policy and the maximum-flow model can be used to determine the optimum number of ships that meet a specific shipping schedule. These situations are included in the chapter as solved examples, problems, or cases.

Throughout the chapter, the formulation and solution of a network model as a linear program is emphasized. It is recommended that you study these relationships, because most commercial codes solve network problems as mere linear programs. Additionally, some formulations require imposing side constraints, which can be implemented only if the problem is solved as an LP.

To understand the computational details, you are encouraged to use TORA's interactive modules that create the steps of the solution in the exact manner presented in the book. For large-scale problems, the chapter offers both Excel Solver and AMPL models for the different algorithms.

This chapter includes a summary of 1 real-life application, 17 solved examples, 2 Solver models, 3 AMPL models, 69 end-of-section problems, and 5 cases. The cases are in Appendix E on the CD. The AMPL/Excel/Solver/TORA programs are in folder ch6Files.

Real-Life Application—Saving Federal Travel Dollars

U.S. Federal Government employees are required to attend development conferences and training courses in different locations around the country. Because the federal employees are located in offices scattered around the United States, the selection of the host city impacts travel cost. Currently, the selection of the city hosting conferences /training events is done without consideration of incurred travel cost. The problem seeks the determination of the optimal location of the host city. For Fiscal Year 1997, the developed model was estimated to save at least $400,000. Case 4 in Chapter 24 on the CD provides the details of the study.

6.1 SCOPE AND DEFINITION OF NETWORK MODELS

A multitude of operations research situations can be modeled and solved as networks (nodes connected by branches):

1. Design of an offshore natural-gas pipeline network connecting well heads in the Gulf of Mexico to an inshore delivery point. The objective of the model is to minimize the cost of constructing the pipeline.

2. Determination of the shortest route between two cities in an existing network of roads.

3. Determination of the maximum capacity (in tons per year) of a coal slurry pipeline network joining coal mines in Wyoming with power plants in Houston. (Slurry pipelines transport coal by pumping water through specially designed pipes.)

4. Determination of the time schedule (start and completion dates) for the activities of a construction project.

5. Determination of the minimum-cost flow schedule from oil fields to refineries through a pipeline network.

The solution of these situations, and others like it, is accomplished through a variety of network optimization algorithms. This chapter presents four of these algorithms.

1. Minimal spanning tree (situation 1)
2. Shortest-route algorithm (situation 2)
3. Maximal-flow algorithm (situation 3)
4. Critical path (CPM) algorithm (situation 4)

For the fifth situation, the minimum-cost capacitated network algorithm is presented in Section 20.1 on the CD.

Network Definitions. A network consists of a set of **nodes** linked by **arcs** (or **branches**). The notation for describing a network is (N, A), where N is the set of nodes and A is the set of arcs. As an illustration, the network in Figure 6.1 is described as

$$N = \{1, 2, 3, 4, 5\}$$

$$A = \{(1, 2), (1, 3), (2, 3), (2, 5), (3, 4), (3, 5), (4, 2), (4, 5)\}$$

Associated with each network is a **flow** (e.g., oil products flow in a pipeline and automobile traffic flows in highways). In general, the flow in a network is limited by the capacity of its arcs, which may be finite or infinite.

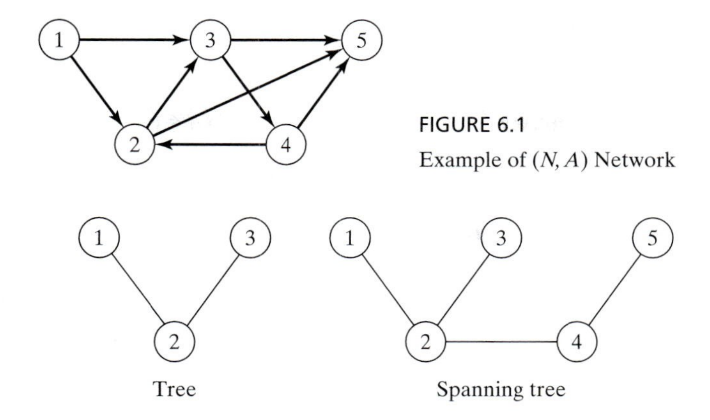

FIGURE 6.1

Example of (N, A) Network

FIGURE 6.2

Examples of a tree and a spanning tree

An arc is said to be **directed** or **oriented** if it allows positive flow in one direction and zero flow in the opposite direction. A **directed network** has all directed arcs.

A **path** is a sequence of distinct arcs that join two nodes through other nodes regardless of the direction of flow in each arc. A path forms a **cycle** or a **loop** if it connects a node to itself through other nodes. For example, in Figure 6.1, the arcs $(2, 3)$, $(3, 4)$, and $(4, 2)$ form a cycle.

A **connected network** is such that every two distinct nodes are linked by at least one path. The network in Figure 6.1 demonstrates this type of network. A **tree** is a *cycle-free* connected network comprised of a *subset* of all the nodes, and a **spanning tree** is a tree that links *all* the nodes of the network. Figure 6.2 provides examples of a tree and a spanning tree from the network in Figure 6.1.

Example 6.1-1 (Bridges of Königsberg)

The Prussian city of Königsberg (now Kalingrad in Russia) was founded in 1254 on the banks of river Pergel with seven bridges connecting its four sections (labeled A, B, C, and D) as shown in Figure 6.3. A problem circulating among the inhabitants of the city was to find out if a *round trip*

FIGURE 6.3

Bridges of Königsberg

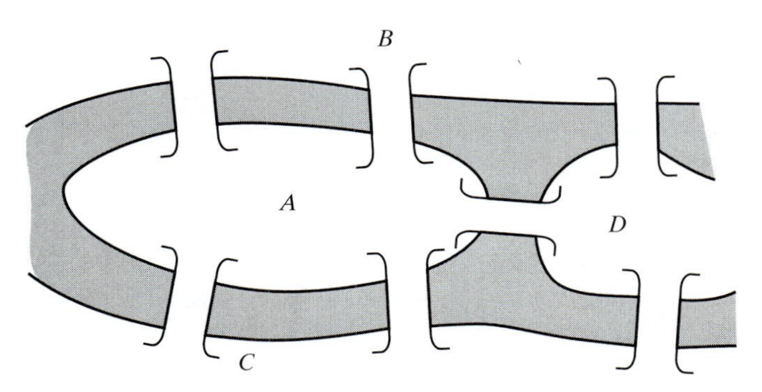

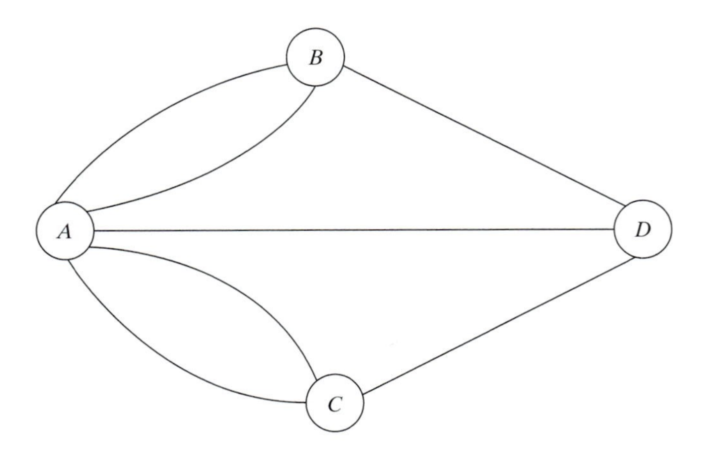

FIGURE 6.4
Network representation of Königsberg problem

of the four sections could be made with each bridge being crossed exactly once. No limits were set on the number of times any of the four sections could be visited.

In the mid-eighteenth century, the famed mathematician Leonhard Euler developed a special "path construction" argument to prove that it was impossible to make such a trip. Later, in the early nineteenth century the same problem was solved by representing the situation as a network in which each of the four sections (*A, B, C,* and *D*) is a node and each bridge is an arc joining applicable nodes, as shown in Figure 6.4.

The network-based solution is that the desired round trip (starting and ending in one section of the city) is impossible, because there are four nodes and each is associated with an *odd* number of arcs, which does not allow distinct entrance and exit (and hence distinct use of the bridges) to each section of the city.[1] The example demonstrates how the solution of the problem is facilitated by using network representation.

PROBLEM SET 6.1A

***1.** For each network in Figure 6.5 determine (a) a path, (b) a cycle, (c) a tree, and (d) a spanning tree.

2. Determine the sets N and A for the networks in Figure 6.5.

3. Draw the network defined by

$$N = \{1, 2, 3, 4, 5, 6\}$$
$$A = \{(1, 2), (1, 5), (2, 3), (2, 4), (3, 4), (3, 5), (4, 3), (4, 6), (5, 2), (5, 6)\}$$

***4.** Consider eight equal squares arranged in three rows, with two squares in the first row, four in the second, and two in the third. The squares of each row are arranged symmetrically about the vertical axis. It is desired to fill the squares with distinct numbers in the range 1 through 8

[1]General solution: A tour exists if all nodes have an even number of branches or if exactly two nodes have an odd number of branches. Else no tour exists. See B. Hopkins and R. Wilson, "The Truth about Königsberg," *College Math Journal*, Vol. 35, No. 3, pp. 198–207, 2004.

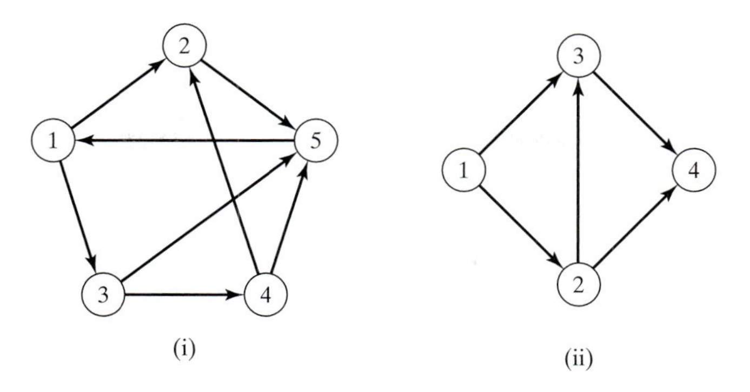

FIGURE 6.5
Networks for Problem 1, Set 6.1a

so that no two *adjacent* vertical, horizontal, or diagonal squares hold consecutive numbers. Use some form of a network representation to find the solution in a systematic way.

5. Three inmates escorted by 3 guards must be transported by boat from the mainland to a penitentiary island to serve their sentences. The boat cannot transfer more than two persons in either direction. The inmates are certain to overpower the guards if they outnumber them at any time. Develop a network model that designs the boat trips in a manner that ensures a smooth transfer of the inmates.

6.2 MINIMAL SPANNING TREE ALGORITHM

The minimal spanning tree algorithm deals with linking the nodes of a network, directly or indirectly, using the shortest total length of connecting branches. A typical application occurs in the construction of paved roads that link several rural towns. The road between two towns may pass through one or more other towns. The most economical design of the road system calls for minimizing the total miles of paved roads, a result that is achieved by implementing the minimal spanning tree algorithm.

The steps of the procedure are given as follows. Let $N = \{1, 2, \ldots, n\}$ be the set of nodes of the network and define

$$C_k = \text{Set of nodes that have been permanently connected at iteration } k$$
$$\overline{C}_k = \text{Set of nodes as yet to be connected permanently after iteration } k$$

Step 0. Set $C_0 = \emptyset$ and $\overline{C}_0 = N$.

Step 1. Start with *any* node i in the unconnected set $\overline{C}_0$ and set $C_1 = \{i\}$, which renders $\overline{C}_1 = N - \{i\}$. Set $k = 2$.

General step *k*. Select a node, j^*, in the unconnected set $\overline{C}_{k-1}$ that yields the shortest arc to a node in the connected set C_{k-1}. Link j^* permanently to C_{k-1} and remove it from $\overline{C}_{k-1}$; that is,

$$C_k = C_{k-1} + \{j^*\}, \overline{C}_k = \overline{C}_{k-1} - \{j^*\}$$

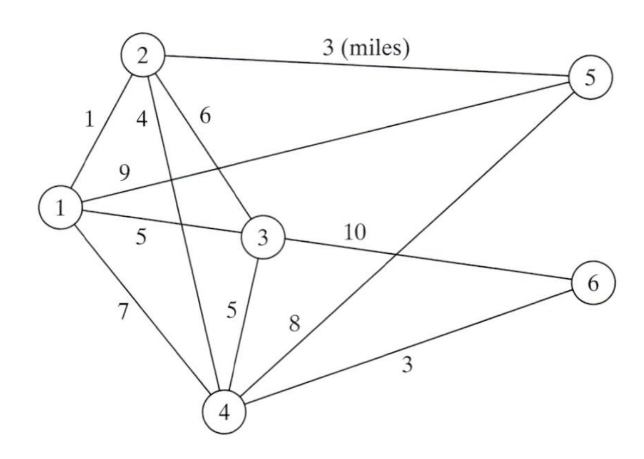

FIGURE 6.6
Cable connections for Midwest TV Company

If the set of unconnected nodes, $\overline{C}_k$, is empty, stop. Otherwise, set $k = k + 1$ and repeat the step.

Example 6.2-1

Midwest TV Cable Company is in the process of providing cable service to five new housing development areas. Figure 6.6 depicts possible TV linkages among the five areas. The cable miles are shown on each arc. Determine the most economical cable network.

The algorithm starts at node 1 (any other node will do as well), which gives

$$C_1 = \{1\}, \overline{C}_1 = \{2, 3, 4, 5, 6\}$$

The iterations of the algorithm are summarized in Figure 6.7. The thin arcs provide all the candidate links between C and $\overline{C}$. The thick branches represent the permanent links between the nodes of the connected set C, and the dashed branch represents the new (permanent) link added at each iteration. For example, in iteration 1, branch $(1, 2)$ is the shortest link ($= 1$ mile) among all the candidate branches from node 1 to nodes 2, 3, 4, 5, and 6 of the unconnected set $\overline{C}_1$. Hence, link $(1, 2)$ is made permanent and $j^* = 2$, which yields

$$C_2 = \{1, 2\}, \overline{C}_2 = \{3, 4, 5, 6\}$$

The solution is given by the minimal spanning tree shown in iteration 6 of Figure 6.7. The resulting minimum cable miles needed to provide the desired cable service are $1 + 3 + 4 + 3 + 5 = 16$ miles.

TORA Moment

You can use TORA to generate the iterations of the minimal spanning tree. From Main menu, select Network models ⇒ Minimal spanning tree. Next, from SOLVE/MODIFY menu, select Solve problem ⇒ Go to output screen. In the output screen, select a Starting node then use Next iteration or All iterations to generate the successive iterations. You can restart the iterations by selecting a new Starting Node. File toraEx6.2-1.txt gives TORA's data for Example 6.2-1.

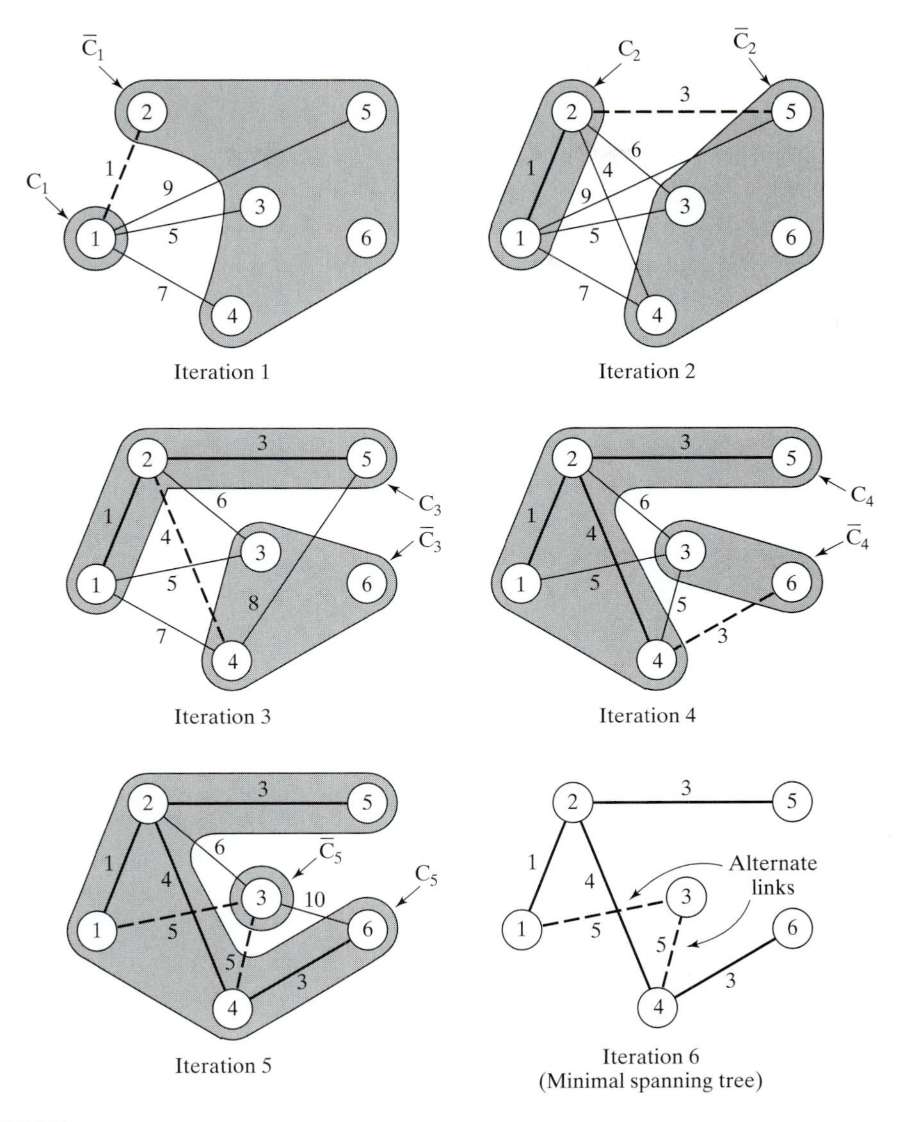

FIGURE 6.7

Solution iterations for Midwest TV Company

PROBLEM SET 6.2A

1. Solve Example 6.2-1 starting at node 5 (instead of node 1), and show that the algorithm produces the same solution.
2. Determine the minimal spanning tree of the network of Example 6.2-1 under each of the following separate conditions:
   *(a) Nodes 5 and 6 are linked by a 2-mile cable.
   (b) Nodes 2 and 5 cannot be linked.
   (c) Nodes 2 and 6 are linked by a 4-mile cable.

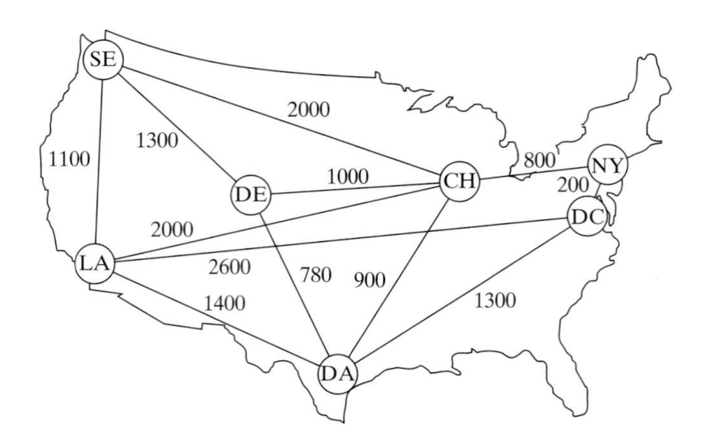

FIGURE 6.8
Network for Problem 3, Set 6.2a

(d) The cable between nodes 1 and 2 is 8 miles long.

(e) Nodes 3 and 5 are linked by a 2-mile cable.

(f) Node 2 cannot be linked directly to nodes 3 and 5.

3. In intermodal transportation, loaded truck trailers are shipped between railroad terminals on special flatbed carts. Figure 6.8 shows the location of the main railroad terminals in the United States and the existing railroad tracks. The objective is to decide which tracks should be "revitalized" to handle the intermodal traffic. In particular, the Los Angeles (LA) terminal must be linked directly to Chicago (CH) to accommodate expected heavy traffic. Other than that, all the remaining terminals can be linked, directly or indirectly, such that the total length (in miles) of the selected tracks is minimized. Determine the segments of the railroad tracks that must be included in the revitalization program.

4. Figure 6.9 gives the mileage of the feasible links connecting nine offshore natural gas wellheads with an inshore delivery point. Because wellhead 1 is the closest to shore, it is equipped with sufficient pumping and storage capacity to pump the output of the remaining eight wells to the delivery point. Determine the minimum pipeline network that links the wellheads to the delivery point.

***5.** In Figure 6.9 of Problem 4, suppose that the wellheads can be divided into two groups depending on gas pressure: a high-pressure group that includes wells 2, 3, 4, and 6, and a low-pressure group that includes wells 5, 7, 8, and 9. Because of pressure difference, it is not possible to link the wellheads from the two groups. At the same time, both groups must be connected to the delivery point through wellhead 1. Determine the minimum pipeline network for this situation.

6. Electro produces 15 electronic parts on 10 machines. The company wants to group the machines into cells designed to minimize the "dissimilarities" among the parts processed in each cell. A measure of "dissimilarity," d_{ij}, among the parts processed on machines i and j can be expressed as

$$d_{ij} = 1 - \frac{n_{ij}}{n_{ij} + m_{ij}}$$

where n_{ij} is the number of parts shared between machines i and j, and m_{ij} is the number of parts that are used by either machine i or machine j only.

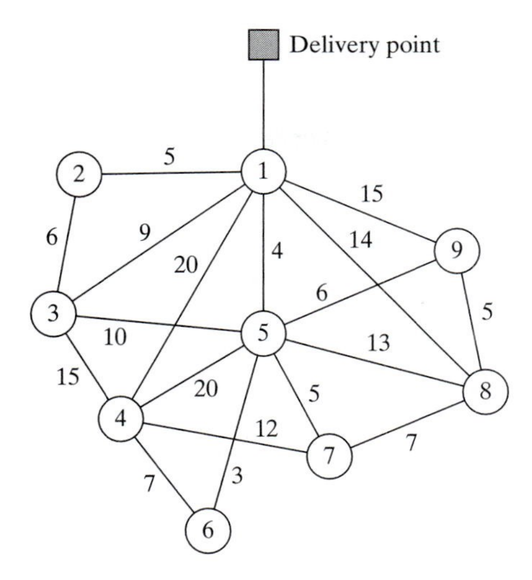

FIGURE 6.9

Network for Problem 4, Set 6.2a

The following table assigns the parts to machines:

Machine	Assigned parts
1	1, 6
2	2, 3, 7, 8, 9, 12, 13, 15
3	3, 5, 10, 14
4	2, 7, 8, 11, 12, 13
5	3, 5, 10, 11, 14
6	1, 4, 5, 9, 10
7	2, 5, 7, 8, 9, 10
8	3, 4, 15
9	4, 10
10	3, 8, 10, 14, 15

(a) Express the problem as a network model.

(b) Show that the determination of the cells can be based on the minimal spanning tree solution.

(c) For the data given in the preceding table, construct the two- and three-cell solutions.

6.3 SHORTEST-ROUTE PROBLEM

The shortest-route problem determines the shortest route between a source and destination in a transportation network. Other situations can be represented by the same model, as illustrated by the following examples.

6.3.1 Examples of the Shortest-Route Applications

Example 6.3-1 (Equipment Replacement)

RentCar is developing a replacement policy for its car fleet for a 4-year planning horizon. At the start of each year, a decision is made as to whether a car should be kept in operation or replaced.

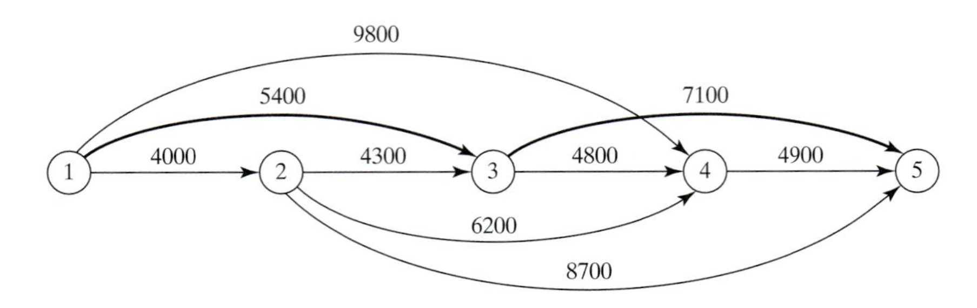

FIGURE 6.10
Equipment replacement problem as a shortest route model

A car must be in service a minimum of 1 year and a maximum of 3 years. The following table provides the replacement cost as a function of the year a car is acquired and the number of years in operation.

Equipment acquired at start of year	Replacement cost ($) for given years in operation		
	1	*2*	*3*
1	4000	5400	9800
2	4300	6200	8700
3	4800	7100	—
4	4900	—	—

The problem can be formulated as a network in which nodes 1 to 5 represent the start of years 1 to 5. Arcs from node 1 (year 1) can reach only nodes 2, 3, and 4 because a car must be in operation between 1 and 3 years. The arcs from the other nodes can be interpreted similarly. The length of each arc equals the replacement cost. The solution of the problem is equivalent to finding the shortest route between nodes 1 and 5.

Figure 6.10 shows the resulting network. Using TORA,[2] the shortest route (shown by the thick path) is $1 \rightarrow 3 \rightarrow 5$. The solution means that a car acquired at the start of year 1 (node 1) must be replaced after 2 years at the start of year 3 (node 3). The replacement car will then be kept in service until the end of year 4. The total cost of this replacement policy is \$12,500 (= \$5400 + \$7100).

Example 6.3-2 (Most Reliable Route)

I. Q. Smart drives daily to work. Having just completed a course in network analysis, Smart is able to determine the shortest route to work. Unfortunately, the selected route is heavily patrolled by police, and with all the fines paid for speeding, the shortest route may not be the best choice. Smart has thus decided to choose a route that maximizes the probability of *not* being stopped by police.

The network in Figure 6.11 shows the possible routes between home and work, and the associated probabilities of not being stopped on each segment. The probability of not being

[2]From Main menu, select Network models ⇒ Shortest route. From SOLVE/MODIFY menu, select Solve problem ⇒ Shortest routes.

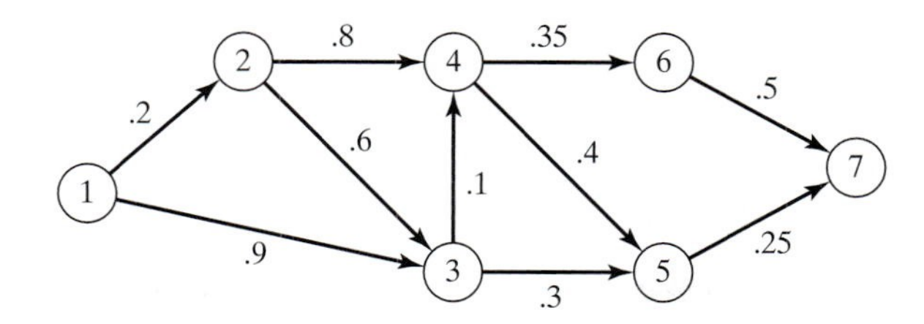

FIGURE 6.11

Most-reliable-route network model

stopped on a route is the product of the probabilities associated with its segments. For example, the probability of not receiving a fine on the route $1 \rightarrow 3 \rightarrow 5 \rightarrow 7$ is $.9 \times .3 \times .25 = .0675$. Smart's objective is to select the route that *maximizes* the probability of not being fined.

The problem can be formulated as a shortest-route model by using a logarithmic transformation that converts the product probability into the sum of the logarithms of probabilities—that is, if $p_{1k} = p_1 \times p_2 \times \cdots \times p_k$ is the probability of not being stopped, then $\log p_{1k} = \log p_1 + \log p_2 + \cdots + \log p_k$.

Mathematically, the maximization of $\log p_{1k}$ is equivalent to the maximization of $\log p_{1k}$. Because $\log p_{1k} \leq 0$, the maximization of $\log p_{1k}$ is equivalent to the minimization of $-\log p_{1k}$. Using this transformation, the individual probabilities p_j in Figure 6.11 are replaced with $-\log p_j$ for all j in the network, thus yielding the shortest-route network in Figure 6.12.

Using TORA, the shortest route in Figure 6.12 is defined by the nodes 1,3,5, and 7 with a corresponding "length" of 1.1707 ($= -\log p_{17}$). Thus, the maximum probability of not being stopped is $p_{17} = .0675$ only, not very encouraging news for Smart!

Example 6.3-3 (Three-Jug Puzzle)

An 8-gallon jug is filled with fluid. Given two empty 5- and 3-gallon jugs, we want to divide the 8 gallons of fluid into two equal parts using the three jugs. No other measuring devices are allowed. What is the smallest number of transfers (decantations) needed to achieve this result?

You probably can guess the solution to this puzzle. Nevertheless, the solution process can be systematized by representing the problem as a shortest-route problem.

A node is defined to represent the amount of fluid in the 8-, 5-, and 3-gallon jugs, respectively. This means that the network starts with node (8, 0, 0) and terminates with the desired

FIGURE 6.12

Most-reliable-route representation as a shortest-route model

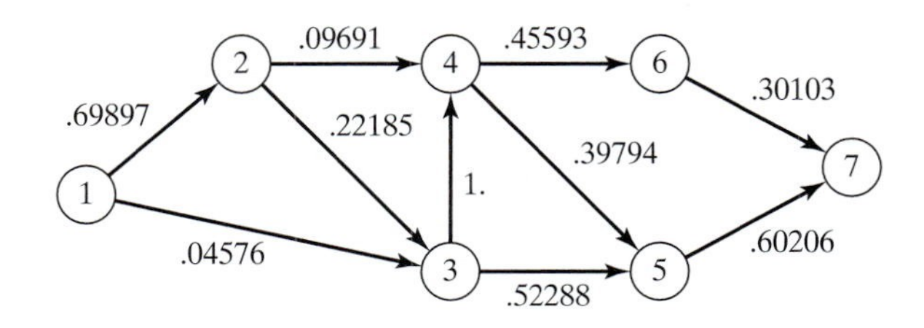

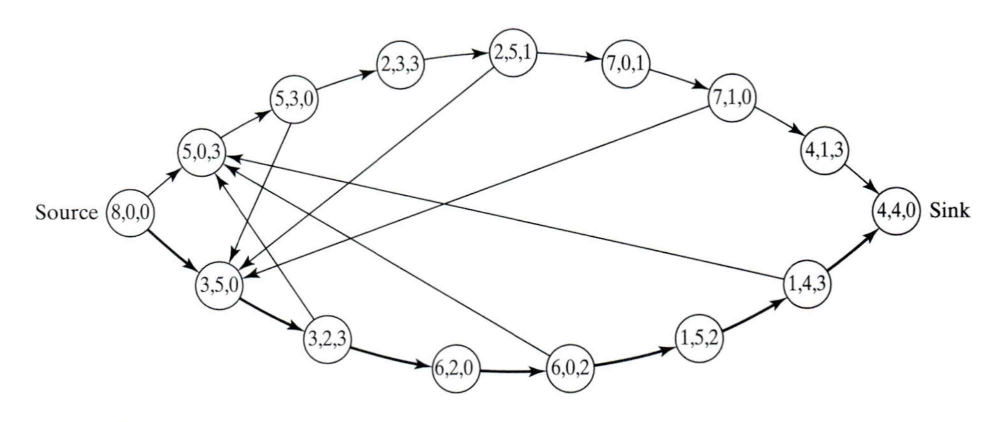

FIGURE 6.13

Three-jug puzzle representation as a shortest-route model

solution node (4, 4, 0). A new node is generated from the current node by decanting fluid from one jug into another.

Figure 6.13 shows different routes that lead from the start node (8, 0, 0) to the end node (4, 4, 0). The arc between two successive nodes represents a single transfer, and hence can be assumed to have a length of 1 unit. The problem thus reduces to determining the shortest route between node (8, 0, 0) and node (4, 4, 0).

The optimal solution, given by the bottom path in Figure 6.13, requires 7 transfers.

PROBLEM SET 6.3A

***1.** Reconstruct the equipment replacement model of Example 6.3-1, assuming that a car must be kept in service at least 2 years, with a maximum service life of 4 years. The planning horizon is from the start of year 1 to the end of year 5. The following table provides the necessary data.

	Replacement cost ($) for given years in operation		
Year acquired	*2*	*3*	*4*
1	3800	4100	6800
2	4000	4800	7000
3	4200	5300	7200
4	4800	5700	—
5	5300	—	—

2. Figure 6.14 provides the communication network between two stations, 1 and 7. The probability that a link in the network will operate without failure is shown on each arc. Messages are sent from station 1 to station 7, and the objective is to determine the route that will maximize the probability of a successful transmission. Formulate the situation as a shortest-route model and determine the optimum solution.

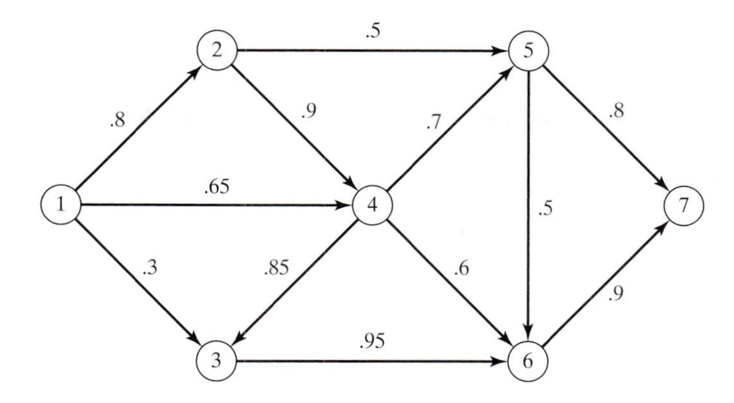

FIGURE 6.14
Network for Problem 2, Set 6.3a

3. *Production Planning.* DirectCo sells an item whose demands over the next 4 months are 100, 140, 210, and 180 units, respectively. The company can stock just enough supply to meet each month's demand, or it can overstock to meet the demand for two or more successive and consecutive months. In the latter case, a holding cost of $1.20 is charged per overstocked unit per month. DirectCo estimates the unit purchase prices for the next 4 months to be $15, $12, $10, and $14, respectively. A setup cost of $200 is incurred each time a purchase order is placed. The company wants to develop a purchasing plan that will minimize the total costs of ordering, purchasing, and holding the item in stock. Formulate the problem as a shortest-route model, and use TORA to find the optimum solution.

*4. *Knapsack Problem.* A hiker has a 5-ft^3 backpack and needs to decide on the most valuable items to take on the hiking trip. There are three items from which to choose. Their volumes are 2, 3, and 4 ft^3, and the hiker estimates their associated values on a scale from 0 to 100 as 30, 50, and 70, respectively. Express the problem as longest-route network, and find the optimal solution. (*Hint:* A node in the network may be defined as $[i, v]$, where i is the item number considered for packing, and v is the volume remaining immediately before a decision is made on i.)

5. An old-fashioned electric toaster has two spring-loaded base-hinged doors. The two doors open outward in opposite directions away from the heating element. A slice of bread is toasted one side at a time by pushing open one of the doors with one hand and placing the slice with the other hand. After one side is toasted, the slice is turned over to get the other side toasted. The goal is to determine the sequence of operations (placing, toasting, turning, and removing) needed to toast three slices of bread in the shortest possible time. Formulate the problem as a shortest-route model, using the following elemental times for the different operations:

Operation	Time (seconds)
Place one slice in either side	3
Toast one side	30
Turn slice already in toaster	1
Remove slice from either side	3

6.3.2 Shortest-Route Algorithms

This section presents two algorithms for solving both cyclic (i.e., containing loops) and acyclic networks:

1. Dijkstra's algorithm
2. Floyd's algorithm

Dijkstra's algorithm is designed to determine the shortest routes between the source node and every other node in the network. Floyd's algorithm is more general because it allows the determination of the shortest route between *any* two nodes in the network.

Dijkstra's Algorithm. Let u_i be the shortest distance from source node 1 to node i, and define d_{ij} (≥ 0) as the length of arc (i, j). Then the algorithm defines the label for an immediately succeeding node j as

$$[u_j, i] = [u_i + d_{ij}, i], d_{ij} \geq 0$$

The label for the starting node is [0, —], indicating that the node has no predecessor.

Node labels in Dijkstra's algorithm are of two types: *temporary* and *permanent*. A temporary label is modified if a shorter route to a node can be found. If no better route can be found, the status of the temporary label is changed to permanent.

Step 0. Label the source node (node 1) with the *permanent* label [0, —]. Set $i = 1$.

Step i. **(a)** Compute the *temporary* labels $[u_i + d_{ij}, i]$ for each node j that can be reached from node i, *provided j is not permanently labeled.* If node j is already labeled with $[u_j, k]$ through another node k and if $u_i + d_{ij} < u_j$, replace $[u_j, k]$ with $[u_i + d_{ij}, i]$.

(b) If *all* the nodes have *permanent* labels, stop. Otherwise, select the label $[u_r, s]$ having the shortest distance ($= u_r$) among all the *temporary* labels (break ties arbitrarily). Set $i = r$ and repeat step i.

Example 6.3-4

The network in Figure 6.15 gives the permissible routes and their lengths in miles between city 1 (node 1) and four other cities (nodes 2 to 5). Determine the shortest routes between city 1 and each of the remaining four cities.

Iteration 0. Assign the *permanent* label [0, —] to node 1.

Iteration 1. Nodes 2 and 3 can be reached from (the last permanently labeled) node 1. Thus, the list of labeled nodes (temporary and permanent) becomes

Node	Label	Status
1	**[0, —]**	**Permanent**
2	[0 + 100, 1] = [100, 1]	Temporary
3	[0 + 30, 1] = [30, 1]	Temporary

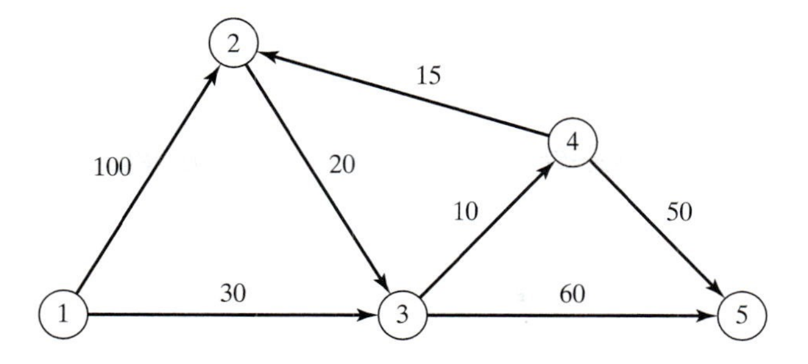

FIGURE 6.15
Network example for Dijkstra's shortest-route algorithm

For the two temporary labels [100, 1] and [30, 1], node 3 yields the smaller distance ($u_3 = 30$). Thus, the status of node 3 is changed to permanent.

Iteration 2. Nodes 4 and 5 can be reached from node 3, and the list of labeled nodes becomes

Node	Label	Status
1	[0, —]	Permanent
2	[100, 1]	Temporary
3	**[30, 1]**	**Permanent**
4	[30 + 10, 3] = [40, 3]	Temporary
5	[30 + 60, 3] = [90, 3]	Temporary

The status of the temporary label [40, 3] at node 4 is changed to permanent ($u_4 = 40$).

Iteration 3. Nodes 2 and 5 can be reached from node 4. Thus, the list of labeled nodes is updated as

Node	Label	Status
1	[0, —]	Permanent
2	[40 + 15, 4] = [55, 4]	Temporary
3	[30, 1]	Permanent
4	**[40, 3]**	**Permanent**
5	[90, 3] or [40 + 50, 4] = [90, 4]	Temporary

Node 2's temporary label [100, 1] obtained in iteration 1 is changed to [55, 4] in iteration 3 to indicate that a shorter route has been found through node 4. Also, in iteration 3, node 5 has two alternative labels with the same distance $u_5 = 90$.

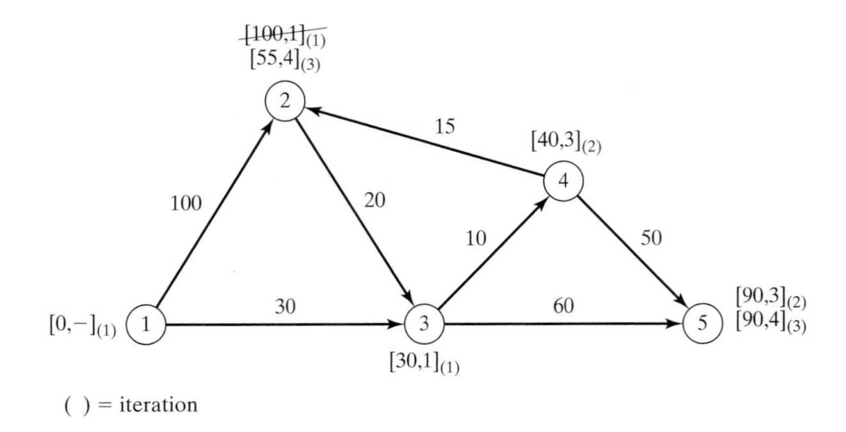

FIGURE 6.16
Dijkstra's labeling procedure

The list for iteration 3 shows that the label for node 2 is now permanent.

Iteration 4. Only node 3 can be reached from node 2. However, node 3 has a permanent label and cannot be relabeled. The new list of labels remains the same as in iteration 3 except that the label at node 2 is now permanent. This leaves node 5 as the only temporary label. Because node 5 does not lead to other nodes, its status is converted to permanent, and the process ends.

The computations of the algorithm can be carried out more easily on the network, as Figure 6.16 demonstrates.

The shortest route between nodes 1 and any other node in the network is determined by starting at the desired destination node and backtracking through the nodes using the information given by the permanent labels. For example, the following sequence determines the shortest route from node 1 to node 2:

$$(2) \rightarrow [55, 4] \rightarrow (4) \rightarrow [40, 3] \rightarrow (3) \rightarrow [30, 1] \rightarrow (1)$$

Thus, the desired route is $1 \rightarrow 3 \rightarrow 4 \rightarrow 2$ with a total length of 55 miles.

TORA Moment

TORA can be used to generate Dijkstra's iterations. From SOLVE/MODIFY menu, select Solve problem ⇒ Iterations ⇒ Dijkstra's algorithm. File toraEx6.3-4.txt provides TORA's data for Example 6.3-4.

PROBLEM SET 6.3B

1. The network in Figure 6.17 gives the distances in miles between pairs of cities 1, 2, ..., and 8. Use Dijkstra's algorithm to find the shortest route between the following cities:
   (a) Cities 1 and 8
   (b) Cities 1 and 6

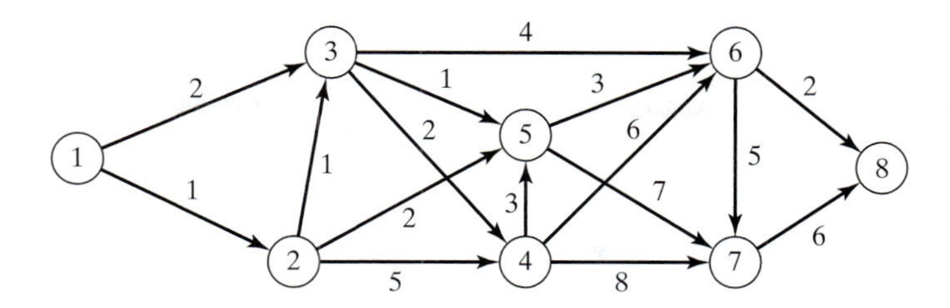

FIGURE 6.17
Network for Problem 1, Set 6.3b

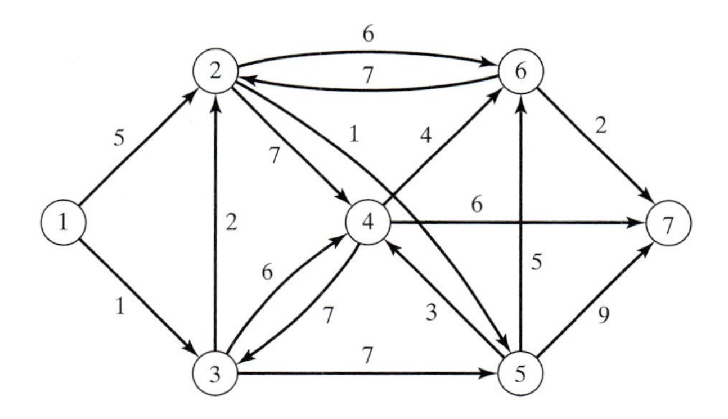

FIGURE 6.18
Network for Problem 2, Set 6.3b

*(c) Cities 4 and 8

(d) Cities 2 and 6

2. Use Dijkstra's algorithm to find the shortest route between node 1 and every other node in the network of Figure 6.18.

3. Use Dijkstr'a algorithm to determine the optimal solution of each of the following problems:

(a) Problem 1, Set 6.3a.

(b) Problem 2, Set 6.3a.

(c) Problem 4, Set 6.3a.

Floyd's Algorithm. Floyd's algorithm is more general than Dijkstra's because it determines the shortest route between *any* two nodes in the network. The algorithm represents an n-node network as a square matrix with n rows and n columns. Entry (i, j) of the matrix gives the distance d_{ij} from node i to node j, which is finite if i is linked directly to j, and infinite otherwise.

The idea of Floyd's algorithm is straightforward. Given three nodes i, j, and k in Figure 6.19 with the connecting distances shown on the three arcs, it is shorter to reach j from i passing through k if

$$d_{ik} + d_{kj} < d_{ij}$$

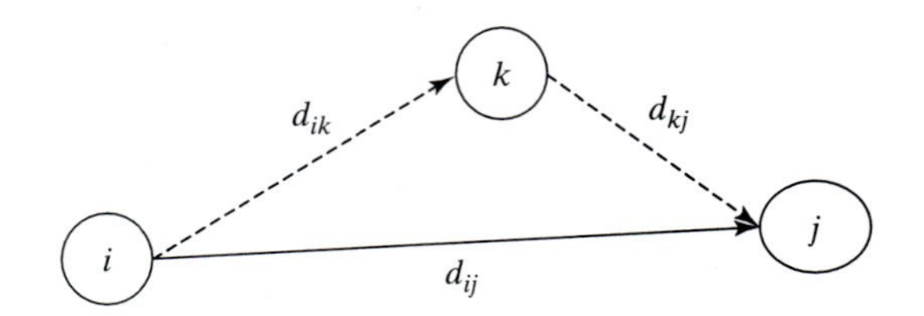

FIGURE 6.19
Floyd's triple operation

In this case, it is optimal to replace the direct route from $i \rightarrow j$ with the indirect route $i \rightarrow k \rightarrow j$. This **triple operation** exchange is applied systematically to the network using the following steps:

Step 0. Define the starting distance matrix D_0 and node sequence matrix S_0 as given below. The diagonal elements are marked with (—) to indicate that they are blocked. Set $k = 1$.

$D_0 =$

	1	2	...	j	...	n
1	—	d_{12}	...	d_{ij}	...	d_{1n}
2	d_{21}	—	...	d_{2j}	...	d_{2n}
⋮	⋮	⋮	⋮	⋮	⋮	⋮
i	d_{i1}	d_{i2}	...	d_{ij}	...	d_{in}
⋮	⋮	⋮	⋮	⋮	⋮	⋮
n	D_{n1}	d_{n2}	...	d_{nj}	...	—

$S_0 =$

	1	2	...	j	...	n
1	—	2	...	j	...	n
2	1	—	...	j	...	n
⋮	⋮	⋮	⋮	⋮	⋮	⋮
i	1	2	...	j	...	n
⋮	⋮	⋮	⋮	⋮	⋮	⋮
n	1	2	...	j	...	—

General step *k*. Define row k and column k as *pivot row* and *pivot column.* Apply the *triple operation* to each element d_{ij} in D_{k-1}, for all i and j. If the condition

$$d_{ik} + d_{kj} < d_{ij}, (i \neq k, j \neq k, \text{and } i \neq j)$$

is satisfied, make the following changes:

(a) Create D_k by replacing d_{ij} in D_{k-1} with $d_{ik} + d_{kj}$

(b) Create S_k by replacing s_{ij} in S_{k-1} with k. Set $k = k + 1$. If $k = n + 1$, stop; else repeat step k.

Step k of the algorithm can be explained by representing D_{k-1} as shown in Figure 6.20. Here, row k and column k define the current pivot row and column. Row i

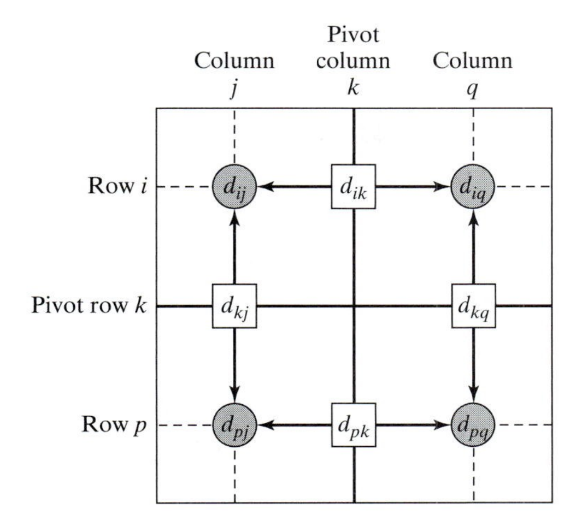

FIGURE 6.20
Implementation of triple operation in matrix form

represents any of the rows $1, 2, \ldots,$ and $k - 1$, and row p represents any of the rows $k + 1, k + 2, \ldots,$ and n. Similarly, column j represents any of the columns $1, 2, \ldots,$ and $k - 1$, and column q represents any of the columns $k + 1, k + 2, \ldots,$ and n. The *triple operation* can be applied as follows. If the sum of the elements on the pivot row and the pivot column (shown by squares) is smaller than the associated intersection element (shown by a circle), then it is optimal to replace the intersection distance by the sum of the pivot distances.

After n steps, we can determine the shortest route between nodes i and j from the matrices D_n and S_n using the following rules:

1. From D_n, d_{ij} gives the shortest distance between nodes i and j.
2. From S_n, determine the intermediate node $k = s_{ij}$ that yields the route $i \rightarrow k \rightarrow j$. If $s_{ik} = k$ and $s_{kj} = j$, stop; all the intermediate nodes of the route have been found. Otherwise, repeat the procedure between nodes i and k, and between nodes k and j.

Example 6.3-5

For the network in Figure 6.21, find the shortest routes between every two nodes. The distances (in miles) are given on the arcs. Arc (3,5) is directional, so that no traffic is allowed from node 5 to node 3. All the other arcs allow two-way traffic.

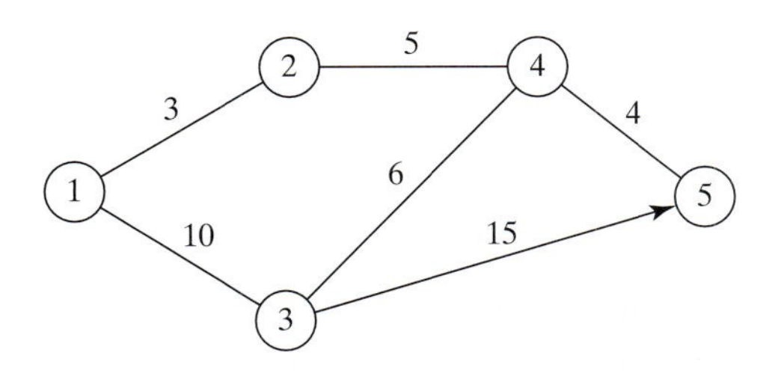

FIGURE 6.21
Network for Example 6.3-5

Iteration 0. The matrices D_0 and S_0 give the initial representation of the network. D_0 is symmetrical, except that $d_{53} = \infty$ because no traffic is allowed from node 5 to node 3.

D_0

	1	2	3	4	5
1	—	3	10	∞	∞
2	3	—	∞	5	∞
3	10	∞	—	6	15
4	∞	5	6	—	4
5	∞	∞	∞	4	—

S_0

	1	2	3	4	5
1	—	2	3	4	5
2	1	—	3	4	5
3	1	2	—	4	5
4	1	2	3	—	5
5	1	2	3	4	—

Iteration 1. Set $k = 1$. The pivot row and column are shown by the lightly shaded first row and first column in the D_0-matrix. The darker cells, d_{23} and d_{32}, are the only ones that can be improved by the *triple operation.* Thus, D_1 and S_1 are obtained from D_0 and S_0 in the following manner:

1. Replace d_{23} with $d_{21} + d_{13} = 3 + 10 = 13$ and set $s_{23} = 1$.
2. Replace d_{32} with $d_{31} + d_{12} = 10 + 3 = 13$ and set $s_{32} = 1$.

D_1

	1	2	3	4	5
1	—	3	10	∞	∞
2	3	—	**13**	5	∞
3	10	**13**	—	6	15
4	∞	5	6	—	4
5	∞	∞	∞	4	—

S_1

	1	2	3	4	5
1	—	2	3	4	5
2	1	—	**1**	4	5
3	1	**1**	—	4	5
4	1	2	3	—	5
5	1	2	3	4	—

Iteration 2. Set $k = 2$, as shown by the lightly shaded row and column in D_1. The *triple operation* is applied to the darker cells in D_1 and S_1. The resulting changes are shown in bold in D_2 and S_2.

D_2

	1	2	3	4	5
1	—	3	10	**8**	∞
2	3	—	13	5	∞
3	10	13	—	6	15
4	**8**	5	6	—	4
5	∞	∞	∞	4	—

S_2

	1	2	3	4	5
1	—	2	3	**2**	5
2	1	—	1	4	5
3	1	1	—	4	5
4	**2**	2	3	—	5
5	1	2	3	4	—

Iteration 3. Set $k = 3$, as shown by the shaded row and column in D_2. The new matrices are given by D_3 and S_3.

D_3

	1	2	3	4	5
1	—	3	10	8	**25**
2	3	—	13	5	**28**
3	10	13	—	6	15
4	8	5	6	—	4
5	∞	∞	∞	4	—

S_3

	1	2	3	4	5
1	—	2	3	2	**3**
2	1	—	1	4	**3**
3	1	1	—	4	5
4	2	2	3	—	5
5	1	2	3	4	—

Iteration 4. Set $k = 4$, as shown by the shaded row and column in D_3. The new matrices are given by D_4 and S_4.

D_4

	1	2	3	4	5
1	—	3	10	8	**12**
2	3	—	**11**	5	**9**
3	10	**11**	—	6	**10**
4	8	5	6	—	4
5	**12**	**9**	**10**	4	—

S_4

	1	2	3	4	5
1	—	2	3	2	**4**
2	1	—	**4**	4	**4**
3	1	**4**	—	4	**4**
4	2	2	3	—	5
5	**4**	**4**	**4**	4	—

Iteration 5. Set $k = 5$, as shown by the shaded row and column in D_4. No further improvements are possible in this iteration.

The final matrices D_4 and S_4 contain all the information needed to determine the shortest route between any two nodes in the network. For example, from D_4, the shortest distance from node 1 to node 5 is $d_{15} = 12$ miles. To determine the associated route, recall that a segment (i, j) represents a direct link only if $s_{ij} = j$. Otherwise, i and j are linked through at least one other intermediate node. Because $s_{15} = 4 \neq 5$, the route is initially given as $1 \rightarrow 4 \rightarrow 5$. Now, because $s_{14} = 2 \neq 4$, the segment $(1, 4)$ is not a *direct* link, and $1 \rightarrow 4$ is replaced with $1 \rightarrow 2 \rightarrow 4$, and the route $1 \rightarrow 4 \rightarrow 5$ now becomes $1 \rightarrow 2 \rightarrow 4 \rightarrow 5$. Next, because $s_{12} = 2$, $s_{24} = 4$, and $s_{45} = 5$, no further "dissecting" is needed, and $1 \rightarrow 2 \rightarrow 4 \rightarrow 5$ defines the shortest route.

TORA Moment

As in Dijkstra's algorithm, TORA can be used to generate Floyd's iterations. From SOLVE/MODIFY menu, select Solve problem ⇒ Iterations ⇒ Floyd's algorithm. File toraEx6.3-5.txt provides TORA's data for Example 6.3-5.

PROBLEM SET 6.3C

1. In Example 6.3-5, use Floyd's algorithm to determine the shortest routes between each of the following pairs of nodes:
   *(a) From node 5 to node 1.
   (b) From node 3 to node 5.

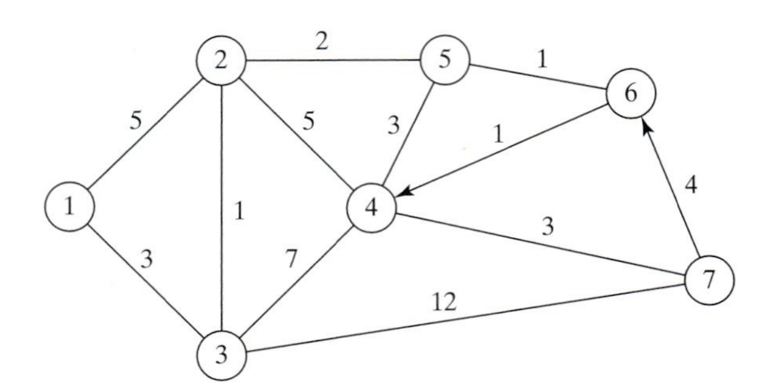

FIGURE 6.22
Network for Problem 2, Set 6.3c

(c) From node 5 to node 3.

(d) From node 5 to node 2.

2. Apply Floyd's algorithm to the network in Figure 6.22. Arcs (7, 6) and (6, 4) are unidirectional, and all the distances are in miles. Determine the shortest route between the following pairs of nodes:

(a) From node 1 to node 7.

(b) From node 7 to node 1.

(c) From node 6 to node 7.

3. The Tell-All mobile-phone company services six geographical areas. The satellite distances (in miles) among the six areas are given in Figure 6.23. Tell-All needs to determine the most efficient message routes that should be established between each two areas in the network.

***4.** Six kids, Joe, Kay, Jim, Bob, Rae, and Kim, play a variation of *hide and seek*. The hiding place of a child is known only to a select few of the other children. A child is then paired with another with the objective of finding the partner's hiding place. This may be achieved through a chain of other kids who eventually will lead to discovering where the designated child is hiding. For example, suppose that Joe needs to find Kim and that Joe knows where Jim is hiding, who in turn knows where Kim is. Thus, Joe can find Kim by first finding Jim, who in turn will lead Joe to Kim. The following list provides the whereabouts of the children:

Joe knows the hiding places of Bob and Kim.

Kay knows the hiding places of Bob, Jim, and Rae.

Jim and Bob each know the hiding place of Kay only.

Rae knows where Kim is hiding.

Kim knows where Joe and Bob are hiding.

FIGURE 6.23
Network for Problem 3, Set 6.3c

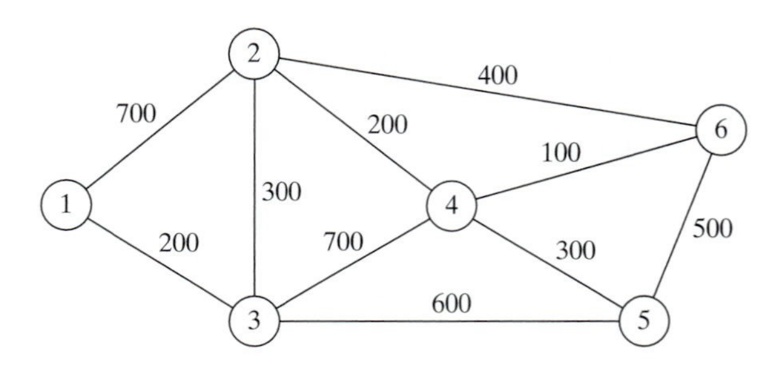

Devise a plan for each child to find every other child through the smallest number of contacts. What is the largest number of contacts?

6.3.3 Linear Programming Formulation of the Shortest-Route Problem

This section provides an LP model for the shortest-route problem. The model is general in the sense that it can be used to find the shortest route between any two nodes in the network. In this regard, it is equivalent to Floyd's algorithm.

Suppose that the shortest-route network includes n nodes and that we desire to determine the shortest route between any two nodes s and t in the network. The LP assumes that one unit of flow enters the network at node s and leaves at node t.

Define

$$x_{ij} = \text{amount of flow in arc } (i, j)$$
$$= \begin{cases} 1, \text{ if arc } (i, j) \text{ is on the shortest route} \\ 0, \text{ otherwise} \end{cases}$$
$$c_{ij} = \text{length of arc } (i, j)$$

Thus, the objective function of the linear program becomes

$$\text{Minimize } z = \sum_{\substack{\text{all defined} \\ \text{arcs } (i, j)}} c_{ij} x_{ij}$$

The constraints represent the *conservation-of-flow equation* at each node:

$$\text{Total input flow} = \text{Total output flow}$$

Mathematically, this translates for node j to

$$\begin{pmatrix} \text{External input} \\ \text{into node } j \end{pmatrix} + \sum_{\substack{i \\ \text{all defined} \\ \text{arcs } (i, j)}} x_{ij} = \begin{pmatrix} \text{External output} \\ \text{from node } j \end{pmatrix} + \sum_{\substack{k \\ \text{all defined} \\ \text{arcs } (j, k)}} x_{jk}$$

Example 6.3-6

Consider the shortest-route network of Example 6.3-4. Suppose that we want to determine the shortest route from node 1 to node 2—that is, $s = 1$ and $t = 2$. Figure 6.24 shows how the unit of flow enters at node 1 and leaves at node 2.

We can see from the network that the flow-conservation equation yields

$$\begin{aligned}
&\text{Node 1:} & 1 &= x_{12} + x_{13} \\
&\text{Node 2:} & x_{12} + x_{42} &= x_{23} + 1 \\
&\text{Node 3:} & x_{13} + x_{23} &= x_{34} + x_{35} \\
&\text{Node 4:} & x_{34} &= x_{42} + x_{45} \\
&\text{Node 5:} & x_{35} + x_{45} &= 0
\end{aligned}$$

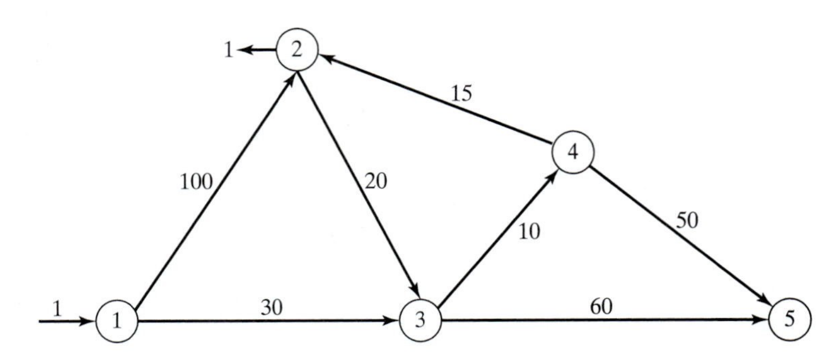

FIGURE 6.24
Insertion of unit flow to determine shortest route between node $s = 1$ and node $t = 2$

The complete LP can be expressed as

	x_{12}	x_{13}	x_{23}	x_{34}	x_{35}	x_{42}	x_{45}	
Minimize $z =$	100	30	20	10	60	15	50	
Node 1	1	1						$= 1$
Node 2	-1		1			-1		$= -1$
Node 3		-1	-1	1	1			$= 0$
Node 4				-1		1	1	$= 0$
Node 5					-1		-1	$= 0$

Notice that column x_{ij} has exactly one "1" entry in row i and one "-1" entry in row j, a typical property of a network LP.

The optimal solution (obtained by TORA, file toraEx6.3-6.txt) is

$$z = 55, x_{13} = 1, x_{34} = 1, x_{42} = 1$$

This solution gives the shortest route from node 1 to node 2 as $1 \rightarrow 3 \rightarrow 4 \rightarrow 2$, and the associated distance is $z = 55$ (miles).

PROBLEM 6.3D

1. In Example 6.3-6, use LP to determine the shortest routes between the following pairs of nodes:
   *(a) Node 1 to node 5.
   (b) Node 2 to node 5.

Solver Moment

Figure 6.25 provides the Excel Solver spreadsheet for finding the shortest route between *start* node N1 and *end* node N2 of Example 6.3-6 (file solverEx6.3-6.xls). The input data of the model is the distance matrix in cells B3:E6. Node N1 has no column because it has no incoming arcs, and node N5 has no row because it has no outgoing arcs. A blank entry means that the corresponding arc does not exist. Nodes N1 and N2 are designated as the *start* and *end* nodes by entering 1 in F3 and B7, respectively. These designations can be changed simply by moving the entry 1 to new cells. For example, to find the shortest route from node N2 to node N4, we enter 1 in each of F4 and D7.

	A	B	C	D	E	F	G	H
1	Solver Shortest-Route Model (Example 6.3-6)							
2	distance	N2	N3	N4	N5			
3	N1	100	30			1		
4	N2		20					
5	N3			10	60			
6	N4	15			50			
7		1						
8	solution	N2	N3	N4	N5	outFlow	inFlow	netFlow
9	N1	0	1	0	0	4.6E-12	0	5E-12
10	N2	0	0	0	0	0	5E-12	-5E-12
11	N3	0	0	1	0	1	1	0
12	N4	1	0	0	0	1	1	0
13	N5					0	0	0
14		0	1	1	0	totalDist=	55	

Cell	Formula	Copy to
B8	=B2	C8:E8
A9	=A3	A10:A12
F9	=SUMIF(B3:E3,">0",B9:E9)-F3	F10:F12
B14	=SUMIF(B3:B6,">0",B9:B12)-B7	C14:E14
G10	=OFFSET(A$14,0,ROW(A1))	G11:G13
H9	=F9-G9	H10:H13
G14	=SUMPRODUCT(distance,solution)	
F13	0	
G9	0	

Range	Cells
distance	B3:E6
solution	B9:E12
netFlow	H9:H13
totalDist	G14

Solver Parameters
Set Target Cell: totalDist
Equal To: Max Min Value of: 0
By Changing Cells:
solution
Subject to the Constraints:
netFlow = 0
solution >= 0
Solve
Close
Guess
Options
Add
Change
Delete
Reset All
Help

FIGURE 6.25

Excel Solver solution of the shortest route between nodes 1 and 2 in Example 6.3-6 (file solverEx6.3-6.xls)

The solution of the model is given cells B9:E12. A cell defines a leg connecting its designated nodes. For example, cell C10 defines the leg (N2, N3), and its associated variable is x_{23}. A cell variable $x_{ij} = 1$ if its leg (Ni, Nj) is on the route. Otherwise, its value is zero.

With the distance matrix given by the range B3:E6 (named *distance*) and the solution matrix given by the range B9:E12 (named *solution*), the objective function is computed in cell G14 as =SUMPRODUCT(B3:E6,B9:E12) or, equivalently, =SUMPRODUCT (distance,solution). You may wonder about the significance of the blank entries (which default to zero by Excel) in the distance matrix and their impact on the definition of the objective function. This point will be addressed shortly, after we have shown how the corresponding variables are totally excluded from the constraints of the problem.

As explained in the LP of Example 6.3-6, the constraints of the problem are of the general form:

$$\text{(Output flow)} - \text{(Input flow)} = 0$$

This definition is adapted to the spreadsheet layout by incorporating the external unit flow, if any, directly in either *Output flow* or *Input flow* of the equation. For example, in Example 6.3-6, an external flow unit enters at N1 and leaves at N2. Thus, the associated constraints are given as

$$\left.\begin{array}{ll}\text{(Output flow at N1)} & = x_{12} + x_{13} - 1 \\ \text{(Input flow at N1)} & = 0\end{array}\right\} \Rightarrow x_{12} + x_{13} - 1 = 0$$

$$\left.\begin{array}{ll}\text{(Output flow at N2)} & = x_{23} \\ \text{(Input flow at N2)} & = x_{12} + x_{42} - 1\end{array}\right\} \Rightarrow x_{23} - x_{12} - x_{42} - 1 = 0$$

Looking at the spreadsheet in Figure 6.25, the two constraints are expressed in terms of the cells as

(Output flow at N1)`=B9+C9-F3`

(Input flow at N1) `=0`

(Output flow at N2)`=C10`

(Input flow at N2) `=B9+B12-B7`

To identify the solution cells in the range B9:E12 that apply to each constraint, we note that a solution cell forms part of a constraint only if it has a positive entry in the distance matrix.[3] Thus, we use the following formulas to identify the output and input flows for each node:

1. Output flow: Enter`=SUMIF(B3:E3,">0",B9:E9)-F3` in cell F9 and copy it in cells F10:F12.
2. Input Flow: Enter`=SUMIF(B3:B6,">0",B9:B12)-B7` in cell B14 and copy it in cells C14:E14.
3. Enter `=OFFSET(A$14,0,ROW(A1))` in cell G10 and copy it in cells G11:G13 to transpose the input flow to column G.
4. Enter 0 in each of G9 and F13 to indicate that N1 has no input flow and N5 has no output flow (per spreadsheet definitions).
5. Enter `=F9-G9` in cell H9 and copy it in cells H10:H13 to compute the net flow.

The spreadsheet is now ready for the application of Solver as shown in Figure 6.25. There is one curious occurrence though: When you define the constraints within the **Solver Parameters** dialogue box as $outFlow = inFlow$, Solver does not locate a feasible solution, even after making adjustments in *precision* in the **Solver Option** box. (To reproduce this experience, *solution* cells B9:E12 must be reset to zero or blank.) More curious yet, if the constraints are replaced with $inFlow = outFlow$, the optimum solution is found. In file solverEx6.3-6.xls, we use the *netFlow* range in cells H9:H12 and express the constraint as $netFlow = 0$ with no problem. It is not clear why this peculiarity occurs, but the problem could be related to roundoff error.

The output in Figure 6.25 yields the solution (N1-N3 = 1, N3-N4 = 1, N4-N2 = 1) with a total distance of 55 miles. This means that the optimal route is $1 \rightarrow 3 \rightarrow 4 \rightarrow 2$.

Remarks. In most textbooks, the network is defined by its explicit arcs (node *i*, node *j*, distance), a lengthy and inconvenient process that may not be practical when the number of arcs is large. Our model is driven primarily by the compact distance matrix, which is all we need to develop the flow constraints. It may be argued that our model deals with $(n - 1 \times n - 1)x_{ij}$-variables, which could be much larger than the number of variables associated with the arcs of the model (for instance, Example 6.3-6 has 7 arcs

[3]If a problem happens to have a zero distance between two nodes, the zero distance can be replaced with a very small positive value.

and hence 7 x_{ij}-variables, as opposed to $4 \times 4 = 16$ in our formulation). Keep in mind that these additional variables appear only in the objective function and with zero coefficients (blank entries) and that the flow constraints are *exactly the same* as in other presentations (per the `SUMIF` function). As a result, *pre-solvers* in commercial software will spot this "oddity" and automatically exclude the additional variables prior to applying the simplex method, with no appreciable computational overhead.

AMPL Moment

Figure 6.26 provides the AMPL model for solving Example 6.3-6 (file amplEx6.3-6a.txt). The variable `x[i,j]` assumes the value 1 if arc `[i,j]` is on the shortest route and 0 otherwise. The model is general in the sense that it can be used to find the shortest route between any two nodes in a problem of any size.

As explained in Example 6.3-6, AMPL treats the problem as a network in which an external flow unit enters and exits at specified `start` and `end` nodes. The main input data of the model is an $n \times n$ matrix representing the distance `d[i,j]` of the arc joining nodes `i` and `j`. Per AMPL syntax, a dot entry in `d[i,j]` is a placeholder that signifies

FIGURE 6.26

AMPL shortest route model (file amplEx6.3-6a.txt)

```
#------ shortest route model (Example 6.3-6)-----
param n;
param start;
param end;
param M=999999; #infinity
param d{i in 1..n, j in 1..n} default M;
param rhs{i in 1..n}=if i=start then 1
                  else (if i=end then -1 else 0);
var x{i in 1..n,j in 1..n}>=0;
var outFlow{i in 1..n}=sum{j in 1..n}x[i,j];
var inFlow{j in 1..n}=sum{i in 1..n}x[i,j];

minimize z: sum{i in 1..n, j in 1..n}d[i,j]*x[i,j];
subject to limit{i in 1..n}:outFlow[i]-inFlow[i]=rhs[i];

data;
param n:=5;
param start:=1;
param end:=2;
param d:
  1  2   3  4  5:=
1 .  100 30 .  .
2 .  .   20 .  .
3 .  .   .  10 60
4 .  15  .  .  50
5 .  .   .  .  .;
solve;
print "Shortest length from",start,"to",end,"=",z;
printf "Associated route: %2i",start;
for {i in 1..n-1} for {j in 2..n}
    {if x[i,j]=1 then printf" - %2i",j;} print;
```

that no distance is specified for the corresponding arc. In the model, the dot entry is overridden by the infinite distance `M` (= 999999) in `param d{i in 1...n, j in 1...n} default M;` which will convert it into an infinite-distance route. The same result could be achieved by replacing the dot entry `(.)` with `999999` in the data section, which, in addition to "cluttering" the data, is inconvenient.

The constraints represent flow conservation through each node:

$$\text{(Input flow)} - \text{(Output flow)} = \text{(External flow)}$$

From `x[i, j]`, we can define the input and output flow for node i using the statements

```
var inFlow{j in 1..n}=sum{i in 1..n}x[i,j];
var outFlow{i in 1..n}=sum{j in 1..n}x[i,j];
```

The left-hand side of the constraint i is thus given as `outFlow[i]-inFlow[i]`.

The right-hand side of constraint i (external flow at node i) is defined as

```
param rhs{i in 1..n}=if i=start then 1 else(if i=end then -1 else 0);
```

(See Section A.3 for details of `if then else`.) With this statement, specifying `start` and `end` nodes automatically assigns `1, -1, or 0` to `rhs`, the right-hand side of the constraints.

The objective function seeks the minimization of the sum of `d[i,j]*x[i,j]` over all `i` and `j`.

In the present example, `start=1` and `end=2`, meaning that we want to determine the shortest route from node 1 to node 2. The associated output is

```
Shortest length from 1 to 2 = 55
Associated route:  1 -  3 -  4 -  2
```

Remarks. The AMPL model as given in Figure 6.26 has one flaw: The number of *active* variables x_{ij} is n^2, which could be significantly much larger than the actual number of (positive-distance) arcs in the network, thus resulting in a much larger problem. The reason is that the model accounts for the nonexisting arcs by assigning them an infinite distance `M` (= 999999) to guarantee that they will be zero in the optimum solution. This situation can be remedied by using a subset of `{i in 1..n,j in 1..n}` that excludes nonexisiting arcs, as the following statement shows:

```
var x{i in 1..n,j in 1..n:d[i,j]<M}>=0;
```

(See Section A.4 for the use of conditions to define subsets.) The same logic must be applied to the constraints as well by using the following statements:

```
var inFlow{j in 1..n}=sum{i in 1..n:d[i,j]<M}x[i,j];
var outFlow{i in 1..n}=sum{j in 1..n:d[i,j]<M}x[i,j];
```

File amplEx6.36b.txt gives the complete model.

PROBLEM 6.3E

1. Modify solverEx6.3-6.xls to find the shortest route between the following pairs of nodes:
   (a) Node 1 to node 5.
   (b) Node 4 to node 3.
2. Adapt amplEx6.3-6b.txt for Problem 2, Set 6.3a, to find the shortest route between node 1 and node 7. The input data must be the raw probabilities. Use AMPL programming facilities to print/display the optimum transmission route and its success probability.

6.4 MAXIMAL FLOW MODEL

Consider a network of pipelines that transports crude oil from oil wells to refineries. Intermediate booster and pumping stations are installed at appropriate design distances to move the crude in the network. Each pipe segment has a finite maximum discharge rate of crude flow (or capacity). A pipe segment may be uni- or bidirectional, depending on its design. Figure 6.27 demonstrates a typical pipeline network. How can we determine the maximum capacity of the network between the wells and the refineries?

The solution of the proposed problem requires equipping the network with a single source and a single sink by using unidirectional infinite capacity arcs as shown by dashed arcs in Figure 6.27.

Given arc (i, j) with $i < j$, we use the notation $(\overline{C}_{ij}, \overline{C}_{ji})$ to represent the flow capacities in the two directions $i \rightarrow j$ and $j \rightarrow i$, respectively. To eliminate ambiguity, we place $\overline{C}_{ij}$ on the arc next to node i with $\overline{C}_{ji}$ placed next to node j, as shown in Figure 6.28.

6.4.1 Enumeration of Cuts

A **cut** defines a set of arcs which when deleted from the network will cause a total disruption of flow between the source and sink nodes. The **cut capacity** equals the sum of the capacities of its arcs. Among *all* possible cuts in the network, the cut with the *smallest capacity* gives the maximum flow in the network.

Example 6.4-1

Consider the network in Figure 6.29. The bidirectional capacities are shown on the respective arcs using the convention in Figure 6.28. For example, for arc (3, 4), the flow limit is 10 units from 3 to 4 and 5 units from 4 to 3.

Figure 6.29 illustrates three cuts whose capacities are computed in the following table.

Cut	Associated arcs	Capacity
1	(1, 2), (1, 3), (1, 4)	20 + 30 + 10 = 60
2	(1, 3), (1, 4), (2, 3), (2, 5)	30 + 10 + 40 + 30 = 110
3	(2, 5), (3, 5), (4, 5)	30 + 20 + 20 = 70

FIGURE 6.27

Capacitated network connecting wells and refineries through booster stations

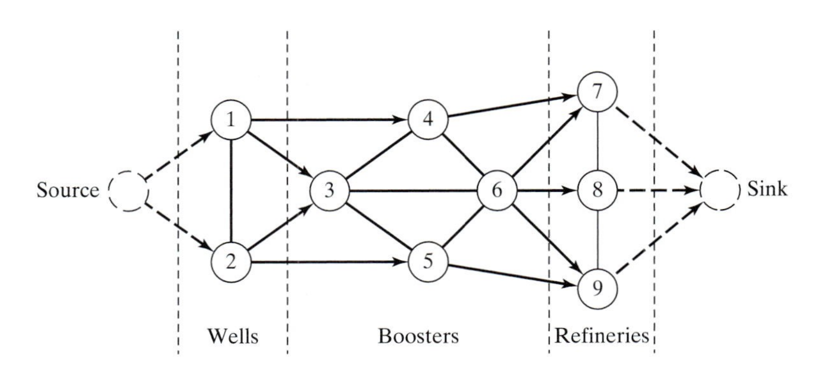

FIGURE 6.28
Arc flows $\overline{C}_{ij}$ from $i \rightarrow j$ and $\overline{C}_{ji}$ from $j \rightarrow i$

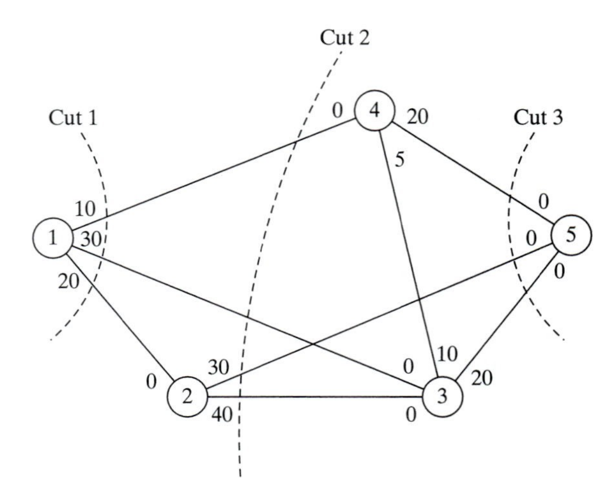

FIGURE 6.29
Examples of cuts in flow networks

The only information we can glean from the three cuts is that the maximum flow in the network cannot exceed 60 units. To determine the maximum flow, it is necessary to enumerate *all* the cuts, a difficult task for the general network. Thus, the need for an efficient algorithm is imperative.

PROBLEM SET 6.4A

***1.** For the network in Figure 6.29, determine two additional cuts, and find their capacities.

6.4.2 Maximal Flow Algorithm

The maximal flow algorithm is based on finding **breakthrough paths** with net *positive* flow between the source and sink nodes. Each path commits part or all of the capacities of its arcs to the total flow in the network.

Consider arc (i, j) with (initial) capacities $(\overline{C}_{ij}, \overline{C}_{ji})$. As portions of these capacities are committed to the flow in the arc, the **residuals** (or remaining capacities) of the arc are updated. We use the notation (c_{ij}, c_{ji}) to represent these residuals.

For a node j that receives flow from node i, we attach a label $[a_j, i]$, where a_j is the flow from node i to node j. The steps of the algorithm are thus summarized as follows.

Step 1. For all arcs (i, j), set the residual capacity equal to the initial capacity— that is $(c_{ij}, c_{ji}) = (\overline{C}_{ij}, \overline{C}_{ji})$. Let $a_1 = \infty$ and label source node 1 with $[\infty, -]$. Set $i = 1$, and go to step 2.

Step 2. Determine S_i, the set of unlabeled nodes j that can be reached directly from node i by arcs with *positive* residuals (that is, $c_{ij} > 0$ for all $j \in S_i$). If $S_i \neq \emptyset$, go to step 3. Otherwise, go to step 4.

Step 3. Determine $k \in S_i$ such that

$$c_{ik} = \max_{j \varepsilon S_i}\{c_{ij}\}$$

Set $a_k = c_{ik}$ and label node k with $[a_k, i]$. If $k = n$, the sink node has been labeled, and a *breakthrough path* is found, go to step 5. Otherwise, set $i = k$, and go to step 2.

Step 4. **(Backtracking).** If $i = 1$, no breakthrough is possible; go to step 6. Otherwise, let r be the node that has been labeled *immediately* before current node i and remove i from the set of nodes adjacent to r. Set $i = r$, and go to step 2.

Step 5. **(Determination of residuals).** Let $N_p = (1, k_1, k_2, \ldots, n)$ define the nodes of the pth breakthrough path from source node 1 to sink node n. Then the maximum flow along the path is computed as

$$f_p = \min\{a_1, a_{k_1}, a_{k_2}, \ldots, a_n\}$$

The residual capacity of each arc along the breakthrough path is *decreased* by f_p in the direction of the flow and *increased* by f_p in the reverse direction—that is, for nodes i and j on the path, the residual flow is changed from the current (c_{ij}, c_{ji}) to

(a) $(c_{ij} - f_p, c_{ji} + f_p)$ if the flow is from i to j

(b) $(c_{ij} + f_p, c_{ji} - f_p)$ if the flow is from j to i

Reinstate any nodes that were removed in step 4. Set $i = 1$, and return to step 2 to attempt a new breakthrough path.

Step 6. **(Solution).**

(a) Given that m breakthrough paths have been determined, the maximal flow in the network is

$$F = f_1 + f_2 + \cdots + f_m$$

(b) Using the *initial* and *final* residuals of arc (i, j), $(\overline{C}_{ij}, \overline{C}_{ji})$ and (c_{ij}, c_{ji}), respectively, the optimal flow in arc (i, j) is computed as follows: Let $(\alpha, \beta) = (\overline{C}_{ij} - c_{ij}, \overline{C}_{ji} - c_{ji})$. If $\alpha > 0$, the optimal flow from i to j is α. Otherwise, if $\beta > 0$, the optimal flow from j to i is β. (It is impossible to have both α and β positive.)

The backtracking process of step 4 is invoked when the algorithm becomes "dead-ended" at an intermediate node. The flow adjustment in step 5 can be explained via the simple flow network in Figure 6.30. Network (a) gives the first breakthrough path $N_1 = \{1, 2, 3, 4\}$ with its maximum flow $f_1 = 5$. Thus, the residuals of each of arcs (1, 2), (2, 3), and (3, 4) are changed from (5, 0) to (0, 5), per step 5. Network (b) now gives the second breakthrough path $N_2 = \{1, 3, 2, 4\}$ with $f_2 = 5$. After making the necessary flow adjustments, we get network (c), where no further breakthroughs are possible. What happened in the transition from (b) to (c) is nothing but a cancellation of a previously committed flow in the direction $2 \rightarrow 3$. The algorithm is able to "remember" that

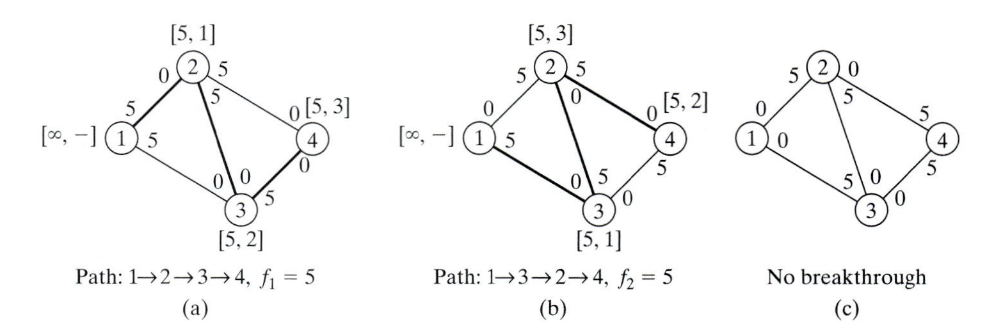

FIGURE 6.30
Use of residuals to calculate maximum flow

a flow from 2 to 3 has been committed previously only because we have increased the capacity in the reverse direction from 0 to 5 (per step 5).

Example 6.4-2

Determine the maximal flow in the network of Example 6.4-1 (Figure 6.29). Figure 6.31 provides a graphical summary of the iterations of the algorithm. You will find it helpful to compare the description of the iterations with the graphical summary.

Iteration 1. Set the initial residuals (c_{ij}, c_{ji}) equal to the initial capacities $(\overline{C}_{ij}, \overline{C}_{ji})$.

Step 1. Set $a_1 = \infty$ and label node 1 with $[\infty, —]$. Set $i = 1$.

Step 2. $S_1 = \{2, 3, 4\}$ $(\neq \emptyset)$.

Step 3. $k = 3$, because $c_{13} = \max\{c_{12}, c_{13}, c_{14}\} = \max\{20, 30, 10\} = 30$. Set $a_3 = c_{13} = 30$, and label node 3 with [30,1]. Set $i = 3$, and repeat step 2.

Step 2. $S_3 = (4, 5)$.

Step 3. $k = 5$ and $a_5 = c_{35} = \max\{10, 20\} = 20$. Label node 5 with [20, 3]. Breakthrough is achieved. Go to step 5.

Step 5. The breakthrough path is determined from the labels starting at node 5 and moving backward to node 1—that is, (**5**) → [20, **3**] → (**3**) → [30, **1**] → (**1**). Thus, $N_1 = \{1, 3, 5\}$ and $f_1 = \min\{a_1, a_3, a_5\} = \{\infty, 30, 20\} = 20$. The residual capacities along path N_1 are

$$(c_{13}, c_{31}) = (30 - 20, 0 + 20) = (10, 20)$$

$$(c_{35}, c_{53}) = (20 - 20, 0 + 20) = (0, 20)$$

Iteration 2

Step 1. Set $a_1 = \infty$, and label node 1 with $[\infty, —]$. Set $i = 1$.

Step 2. $S_1 = \{2, 3, 4\}$.

Step 3. $k = 2$ and $a_2 = c_{12} = \max\{20, 10, 10\} = 20$. Set $i = 2$, and repeat step 2.

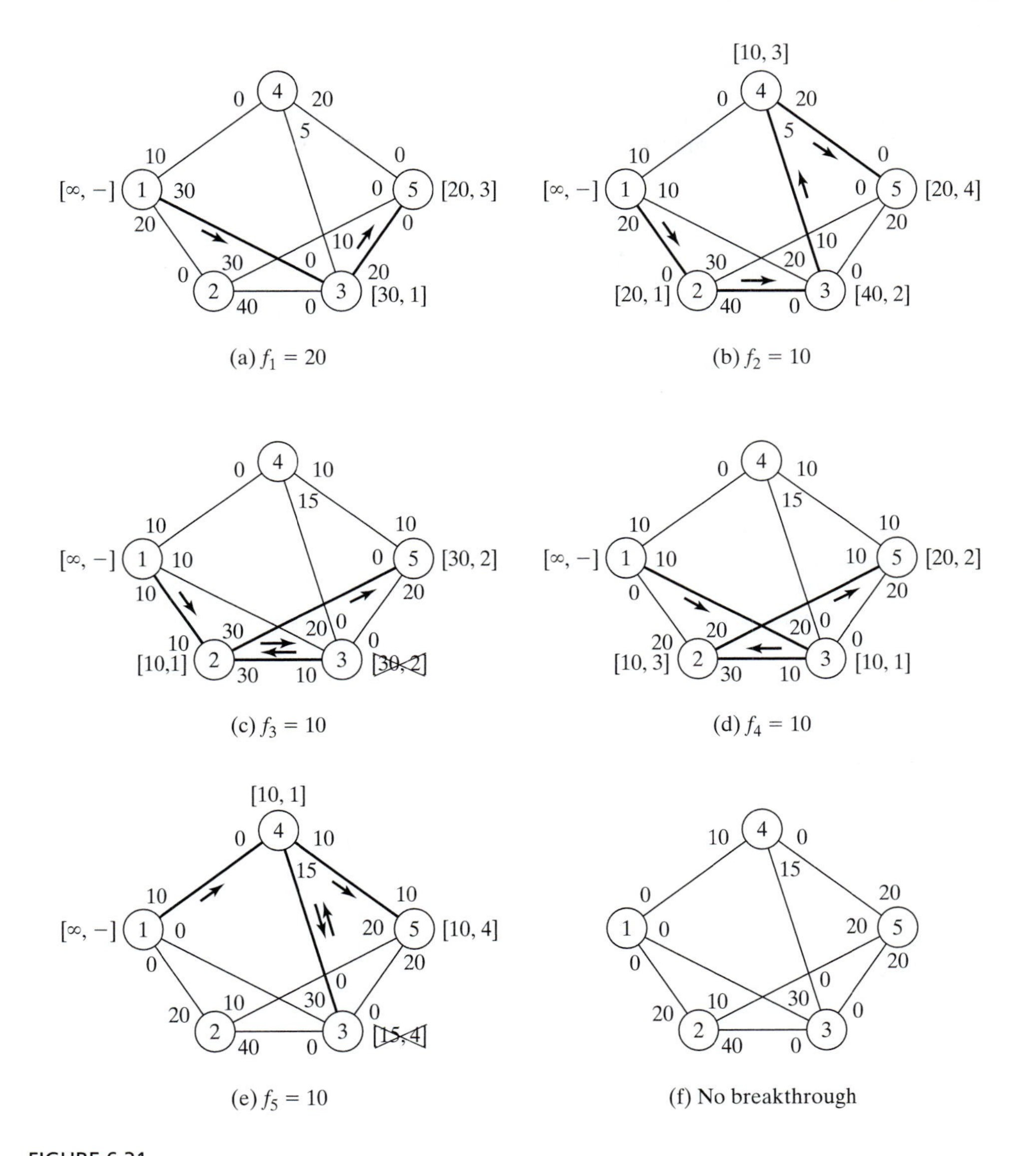

(a) $f_1 = 20$

(b) $f_2 = 10$

(c) $f_3 = 10$

(d) $f_4 = 10$

(e) $f_5 = 10$

(f) No breakthrough

FIGURE 6.31
Iterations of the maximum flow algorithm of Example 6.4-2

Step 2. $S_2 = \{3, 5\}$.

Step 3. $k = 3$ and $a_3 = c_{23} = 40$. Label node 3 with $[40, 2]$. Set $i = 3$, and repeat step 2.

Step 2. $S_3 = \{4\}$ (note that $c_{35} = 0$—hence, node 5 cannot be included in S_3).

Step 3. $k = 4$ and $a_4 = c_{34} = 10$. Label node 4 with $[10, 3]$. Set $i = 4$, and repeat step 2.

Step 2. $S_4 = \{5\}$ (note that nodes 1 and 3 are already labeled—hence, they cannot be included in S_4).

Step 3. $k = 5$ and $a_5 = c_{45} = 20$. Label node 5 with [20, 4]. Breakthrough has been achieved. Go to step 5.

Step 5. $N_2 = \{1, 2, 3, 4, 5\}$ and $f_2 = \min\{\infty, 20, 40, 10, 20\} = 10$. The residuals along the path of N_2 are

$$(c_{12}, c_{21}) = (20 - 10, 0 + 10) = (10, 10)$$

$$(c_{23}, c_{32}) = (40 - 10, 0 + 10) = (30, 10)$$

$$(c_{34}, c_{43}) = (10 - 10, 5 + 10) = (0, 15)$$

$$(c_{45}, c_{54}) = (20 - 10, 0 + 10) = (10, 10)$$

Iteration 3

Step 1. Set $a_1 = \infty$ and label node 1 with $[\infty, —]$. Set $i = 1$.

Step 2. $S_1 = \{2, 3, 4\}$.

Step 3. $k = 2$ and $a_2 = c_{12} = \max\{10, 10, 10\} = 10$. (Though ties are broken arbitrarily, TORA always selects the tied node with the smallest index. We will use this convention throughout the example.) Label node 2 with [10,1]. Set $i = 2$, and repeat step 2.

Step 2. $S_2 = \{3, 5\}$.

Step 3. $k = 3$ and $a_3 = c_{23} = 30$. Label node 3 with [30, 2]. Set $i = 3$, and repeat step 2.

Step 2. $S_3 = \emptyset$ (because $c_{34} = c_{35} = 0$). Go to step 4 to backtrack.

Step 3. *Backtracking.* The label [30, 2] at node 3 gives the immediately preceding node $r = 2$. Remove node 3 from further consideration *in this iteration* by crossing it out. Set $i = r = 2$, and repeat step 2.

Step 2. $S_2 = \{5\}$ (note that node 3 has been removed in the backtracking step).

Step 3. $k = 5$ and $a_5 = c_{25} = 30$. Label node 5 with [30, 2]. Breakthrough has been achieved; go to step 5.

Step 5. $N_3 = \{1, 2, 5\}$ and $c_5 = \min\{\infty, 10, 30\} = 10$. The residuals along the path of N_3 are

$$(c_{12}, c_{21}) = (10 - 10, 10 + 10) = (0, 20)$$

$$(c_{25}, c_{52}) = (30 - 10, 0 + 10) = (20, 10)$$

Iteration 4. This iteration yields $N_4 = \{1, 3, 2, 5\}$ with $f_4 = 10$ (verify!).

Iteration 5. This iteration yields $N_5 = \{1, 4, 5\}$ with $f_5 = 10$ (verify!).

Iteration 6. All the arcs out of node 1 have zero residuals. Hence, no further breakthroughs are possible. We turn to step 6 to determine the solution.

Step 6. Maximal flow in the network is $F = f_1 + f_2 + \cdots + f_5 = 20 + 10 + 10 + 10 + 10 = 60$ units. The flow in the different arcs is computed by subtracting the last residuals (c_{ij}, c_{ji}) in iterations 6 from the initial capacities $(\overline{C}_{ij}, \overline{C}_{ji})$, as the following table shows.

Arc	$(\overline{C}_{ij}, \overline{C}_{ji}) - (c_{ij}, c_{ji})_6$	Flow amount	Direction
(1, 2)	(20, 0) − (0, 20) = (20, −20)	20	$1 \rightarrow 2$
(1, 3)	(30, 0) − (0, 30) = (30, −30)	30	$1 \rightarrow 3$
(1, 4)	(10, 0) − (0, 10) = (10, −10)	10	$1 \rightarrow 4$
(2, 3)	(40, 0) − (40, 0) = (0, 0)	0	—
(2, 5)	(30, 0) − (10, 20) = (20, −20)	20	$2 \rightarrow 5$
(3, 4)	(10, 5) − (0, 15) = (10, −10)	10	$3 \rightarrow 4$
(3, 5)	(20, 0) − (0, 20) = (20, −20)	20	$3 \rightarrow 5$
(4, 5)	(20, 0) − (0, 20) = (20, −20)	20	$4 \rightarrow 5$

TORA Moment

You can use TORA to solve the maximal flow model in an automated mode or to produce the iterations outlined above. From the SOLVE/MODIFY menu, select Solve Problem. After specifying the output format, go to the output screen and select either Maximum Flows or Iterations. File toraEx6.4-2.txt provides TORA's data for Example 6.4-2.

PROBLEM SET 6.4B

*1. In Example 6.4-2,

(a) Determine the surplus capacities for all the arcs.

(b) Determine the amount of flow through nodes 2, 3, and 4.

(c) Can the network flow be increased by increasing the capacities in the directions $3 \rightarrow 5$ and $4 \rightarrow 5$?

2. Determine the maximal flow and the optimum flow in each arc for the network in Figure 6.32.

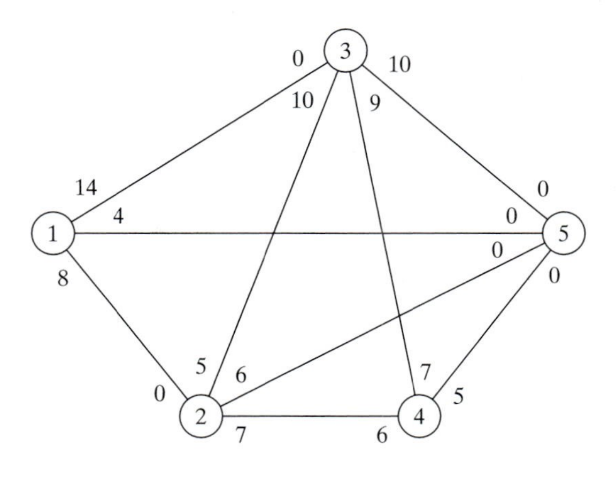

FIGURE 6.32

Network for Problem 2, Set 6.4b

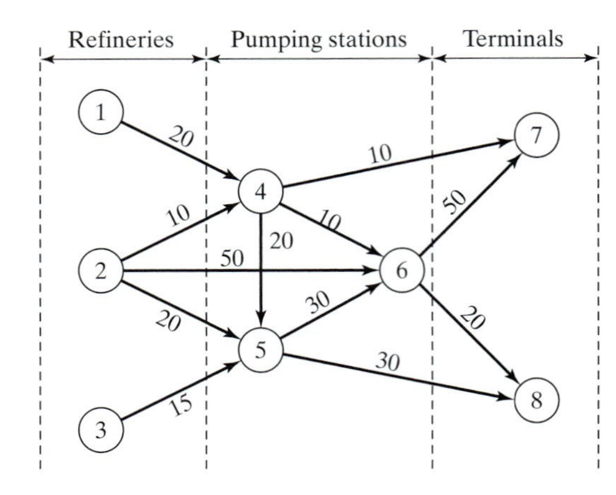

FIGURE 6.33
Network for Problem 3, Set 6.4b

3. Three refineries send a gasoline product to two distribution terminals through a pipeline network. Any demand that cannot be satisfied through the network is acquired from other sources. The pipeline network is served by three pumping stations, as shown in Figure 6.33. The product flows in the network in the direction shown by the arrows. The capacity of each pipe segment (shown directly on the arcs) is in million bbl per day. Determine the following:
   (a) The daily production at each refinery that matches the maximum capacity of the network.
   (b) The daily demand at each terminal that matches the maximum capacity of the network.
   (c) The daily capacity of each pump that matches the maximum capacity of the network.
4. Suppose that the maximum daily capacity of pump 6 in the network of Figure 6.33 is limited to 50 million bbl per day. Remodel the network to include this restriction. Then determine the maximum capacity of the network.
5. Chicken feed is transported by trucks from three silos to four farms. Some of the silos cannot ship directly to some of the farms. The capacities of the other routes are limited by the number of trucks available and the number of trips made daily. The following table shows the daily amounts of supply at the silos and demand at the farms (in thousands of pounds). The cell entries of the table specify the daily capacities of the associated routes.

		Farm				
		1	2	3	4	
Silo	1	30	5	0	40	**20**
	2	0	0	5	90	**20**
	3	100	40	30	40	**200**
		200	**10**	**60**	**20**	

   (a) Determine the schedule that satisfies the most demand.
   (b) Will the proposed schedule satisfy all the demand at the farms?

6. In Problem 5, suppose that transshipping is allowed between silos 1 and 2 and silos 2 and 3. Suppose also that transshipping is allowed between farms 1 and 2, 2 and 3, and 3 and 4. The maximum two-way daily capacity on the proposed transshipping routes is 50 (thousand) lb. What is the effect of transshipping on the unsatisfied demands at the farms?

***7.** A parent has five (teenage) children and five household chores to assign to them. Past experience has shown that forcing chores on a child is counterproductive. With this in mind, the children are asked to list their preferences among the five chores, as the following table shows:

Child	Preferred chore
Rif	3, 4, or 5
Mai	1
Ben	1 or 2
Kim	1, 2, or 5
Ken	2

The parent's modest goal now is to finish as many chores as possible while abiding by the children's preferences. Determine the maximum number of chores that can be completed and the assignment of chores to children.

8. Four factories are engaged in the production of four types of toys. The following table lists the toys that can be produced by each factory.

Factory	Toy productions mix
1	1, 2, 3
2	2, 3
3	1, 4
4	3, 4

All toys require approximately the same per-unit labor and material. The daily capacities of the four factories are 250, 180, 300, and 100 toys, respectively. The daily demands for the four toys are 200, 150, 350, and 100 units, respectively. Determine the factories' production schedules that will most satisfy the demands for the four toys.

9. The academic council at the U of A is seeking representation from among six students who are affiliated with four honor societies. The academic council representation includes three areas: mathematics, art, and engineering. At most two students in each area can be on the council. The following table shows the membership of the six students in the four honor societies:

Society	Affiliated students
1	1, 2, 3
2	1, 3, 5
3	3, 4, 5
4	1, 2, 4, 6

The students who are skilled in the areas of mathematics, art, and engineering are shown in the following table:

Area	Skilled students
Mathematics	1, 2, 4
Art	3, 4
Engineering	4, 5, 6

A student who is skilled in more than one area must be assigned exclusively to one area only. Can all four honor societies be represented on the council?

10. *Maximal/minimal flow in networks with lower bounds.* The maximal flow algorithm given in this section assumes that all the arcs have zero lower bounds. In some models, the lower bounds may be strictly positive, and we may be interested in finding the maximal or minimal flow in the network (see case 6-3 in Appendix E). The presence of the lower bound poses difficulty because the network may not have a feasible flow at all. The objective of this exercise is to show that any maximal and minimal flow model with positive lower bounds can be solved using two steps.

Step 1. Find an initial feasible solution for the network with positive lower bounds.

Step 2. Using the feasible solution in step 1, find the maximal or minimal flow in the original network.

(a) Show that an arc (i, j) with flow limited by $l_{ij} \leq x_{ij} \leq u_{ij}$ can be represented equivalently by a *sink* with demand l_{ij} at node i and a *source* with supply l_{ij} at node j with flow limited by $0 \leq x_{ij} \leq u_{ij} - l_{ij}$.

(b) Show that finding a feasible solution for the original network is equivalent to finding the maximal flow x'_{ij} in the network after (1) modifying the bounds on x_{ij} to $0 \leq x'_{ij} \leq u_{ij} - l_{ij}$, (2) "lumping" all the resulting sources into one supersource with outgoing arc capacities l_{ij}, (3) "lumping" all the resulting sinks into one supersink with incoming arc capacities l_{ij}, and (4) connecting the terminal node t to the source node s in the original network by a return infinite-capacity arc. A feasible solution exists if the maximal flow in the new network equals the sum of the lower bounds in the original network. Apply the procedure to the following network and find a feasible flow solution:

Arc (i, j)	(l_{ij}, u_{ij})
(1, 2)	(5, 20)
(1, 3)	(0, 15)
(2, 3)	(4, 10)
(2, 4)	(3, 15)
(3, 4)	(0, 20)

(c) Use the feasible solution for the network in (b) together with the maximal flow algorithm to determine the *minimal* flow in the original network. (*Hint:* First compute the residue network given the initial feasible solution. Next, determine the maximum flow *from the end node to the start node*. This is equivalent to finding the maximum flow that should be canceled from the start node to the end node. Now, combining the feasible and maximal flow solutions yields the minimal flow in the original network.)

(d) Use the feasible solution for the network in (b) together with the *maximal* flow model to determine the maximal flow in the original network. (*Hint*: As in part (c), start with the residue network. Next apply the breakthrough algorithm to the resulting residue network exactly as in the regular maximal flow model.)

6.4.3 Linear Programming Formulation of Maximal Flow Mode

Define x_{ij} as the amount of flow in arc (i, j) with capacity C_{ij}. The objective is to determine x_{ij} for all i and j that will maximize the flow between start node s and terminal node t subject to flow restrictions (input flow $=$ output flow) at all but nodes s and t.

Example 6.4-3

In the maximal flow model of Figure 6.29 (Example 6.4-2), $s = 1$ and $t = 5$. The following table summarizes the associated LP with two different, but equivalent, objective functions depending on whether we maximize the output from start node 1 ($= z_1$) or the input to terminal node 5 ($= z_2$).

	x_{12}	x_{13}	x_{14}	x_{23}	x_{25}	x_{34}	x_{35}	x_{43}	x_{45}	
Maximize $z_1 =$	1	1	1							
Maximize $z_2 =$					1		1		1	
Node 2	1			−1	−1					= 0
Node 3		1		1		−1	−1	1		= 0
Node 4			1			1		−1	−1	= 0
Capacity	20	30	10	40	30	10	20	5	20	

The optimal solution using either objective function is

$$x_{12} = 20, x_{13} = 30, x_{14} = 10, x_{25} = 20, x_{34} = 10, x_{35} = 20, x_{45} = 20$$

The associated maximum flow is $z_1 = z_2 = 60$.

Solver Moment

Figure 6.34 gives the Excel Solver model for the maximum flow model of Example 6.4-2 (file solverEx6.4-2.xls). The general idea of the model is similar to that used with the shortest-route model, which was detailed following Example 6.3-6. The main differences are: (1) there are no flow equations for the start node 1 and end node 5, and (2) the objective is to maximize the total outflow at start node 1 (F9) or, equivalently, the total inflow at terminal node 5 (G13). File solverEx6.4-2.xls uses G13 as the target cell. You are encouraged to execute the model with F9 as the target cell.

AMPL Moment

Figure 6.35 provides the AMPL model for the maximal flow problem. The data applies to Example 6.4-2 (file amplEx6.4-2.txt). The overall idea of determining the input and output flows at a node is similar to the one detailed following Example 6.3-6 of the shortest-route model (you will find it helpful to review files amplEx6.3-6a.txt and amplEx6.3-6b.txt first). However, because the model is designed to find the maximum flow between *any* two nodes, `start` and `end`, two additional constraints are needed to ensure that no flow *enters* `start` and no flow *leaves* `end`. Constraints `inStart` and

Solver Maximum Flow Model (Example 6.4-2)

capacity	N2	N3	N4	N5
N1	20	30	10	
N2		40		30
N3			10	20
N4		5		20

solution	N2	N3	N4	N5	outFlow	inFlow	netFlow
N1	20	30	10	0	60		
N2	0	0	0	20	20	20	0
N3	0	0	10	20	30	30	0
N4	0	0	0	20	20	20	0
	20	30	20	60	maxFlow=	60	

Cell	Formula	Copy to
F9	=SUMIF(B3:E3,">0",B9:E9)	F10:F12
B13	=SUMIF(B3:B6,">0",B9:B12)	C13:E13
G10	=OFFSET(A$13,0,ROW(A1))	G11:G13
H10	=F10-G10	H11:H12

Range	Cells
capacity	B3:E6
solution	B9:E12
net flow	H10:H12
maxFlow	G13

Solver Parameters

Set Target Cell: maxFlow

Equal To: Max Min Value of: 0

By Changing Cells: solution

Subject to the Constraints:

netFlow = 0
solution <= capacity
solution >= 0

Solve Close Guess Options Add Change Delete Reset All Help

FIGURE 6.34

Excel Solver solution of the maximal flow model of 6.4-2 (file solverEx6.4-2.xls)

`outEnd` in the model ensure this result. These two constraints are not needed when `start=1` and `end=5` because the nature of the data guarantees the desired result. However, for `start=3`, node 3 allows both input and output flow (arcs 4-3 and 3-4) and, hence, constraint `inStart` is needed (try the model without `inStart`!).

The objective function maximizes the sum of the output flow at node `start`. Equivalently, we can choose to maximize the sum of the input flow at node `end`. The model can find the maximum flow between any two designated `start` and `end` nodes in the network.

PROBLEM SET 6.4C

1. Model each of the following problems as a linear program, then solve using Solver and AMPL.
   (a) Problem 2, Set 6.4b.
   (b) Problem 5, Set 6.4b
   (c) Problem 9, Set 6.4b.
2. Jim lives in Denver, Colorado, and likes to spend his annual vacation in Yellowstone National Park in Wyoming. Being a nature lover, Jim tries to drive a different scenic route each year. After consulting the appropriate maps, Jim has represented his preferred routes between Denver (D) and Yellowstone (Y) by the network in Figure 6.36. Nodes 1 through 14 represent intermediate cities. Although driving distance is not an issue, Jim's stipulation is that selected routes between D and Y do not include any common cities.

```
#-------- Maximal Flow model (Example 6.4-2)---------
param n;
param start;
param end;
param c{i in 1..n, j in 1..n} default 0;

var x{i in 1..n,j in 1..n:c[i,j]>0}>=0,<=c[i,j];
var outFlow{i in 1..n}=sum{j in 1..n:c[i,j]>0}x[i,j];
var inFlow{i in 1..n}=sum{j in 1..n:c[j,i]>0}x[j,i];

maximize z: sum {j in 1..n:c[start,j]>0}x[start,j];
subject to
limit{i in 1..n:
     i<>start and i<>end}:outFlow[i]-inFlow[i]=0;
inStart:sum{i in 1..n:c[i,start]>0}x[i,start]=0;
outEnd:sum{j in 1..n:c[end,j]>0}x[end,j]=0;

data;
param n:=5;
param start:=1;
param end:=5;
param c:
  1  2  3  4  5 :=
1 .  20 30 10 0
2 .  .  40 0  30
3 .  .  0  10 20
4 .  .  5  .  20
5 .  .  .  .  .;

solve;
print "MaxFlow between nodes",start,"and",end, "=",z;
printf "Associated flows:\n";
for {i in 1..n-1} for {j in 2..n:c[i,j]>0}
    {if x[i,j]>0 then
     printf"(%2i-%2i)= %5.2f\n",i,j,x[i,j];} print;
```

FIGURE 6.35

AMPL model of the maximal flow problem of Example 6.4-2 (file amplEx6.4-2.txt)

Determine (using AMPL and Solver) all the distinct routes available to Jim. (*Hint*: Modify the maximal flow LP model to determine the maximum number of unique paths between *D* and *Y*.)

3. (Guéret and Associates, 2002, Section 12.1) A military telecommunication system connecting 9 sites is given in Figure 6.37. Sites 4 and 7 must continue to communicate even if as many as three other sites are destroyed by enemy actions. Does the present communication network meet this requirement? Use AMPL and Solver to work out the problem.

6.5 CPM AND PERT

CPM (Critical Path Method) and PERT (Program Evaluation and Review Technique) are network-based methods designed to assist in the planning, scheduling, and control of projects. A project is defined as a collection of interrelated activities with each activity

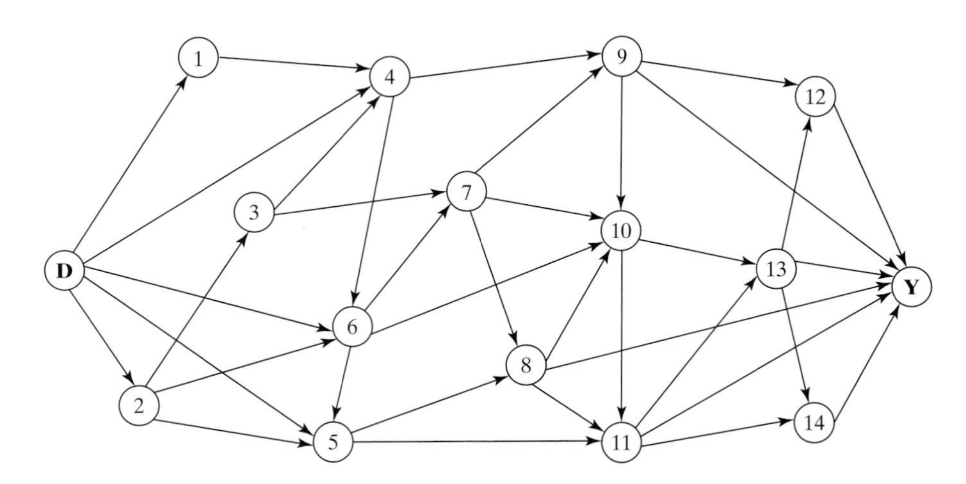

FIGURE 6.36
Network for Problem 2, Set 6.4c

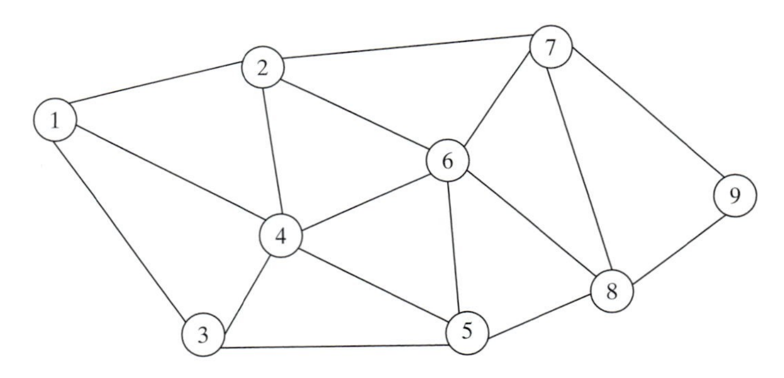

FIGURE 6.37
Network for Problem 3, Set 6.4c

consuming time and resources. The objective of CPM and PERT is to provide analytic means for scheduling the activities. Figure 6.38 summarizes the steps of the techniques. First, we define the activities of the project, their precedence relationships, and their time requirements. Next, the precedence relationships among the activities are represented by a network. The third step involves specific computations to develop the time schedule for the project. During the actual execution of the project things may not proceed as planned, as some of the activities may be expedited or delayed. When this happens, the schedule must be revised to reflect the realities on the ground. This is the reason for including a feedback loop between the time schedule phase and the network phase, as shown in Figure 6.38.

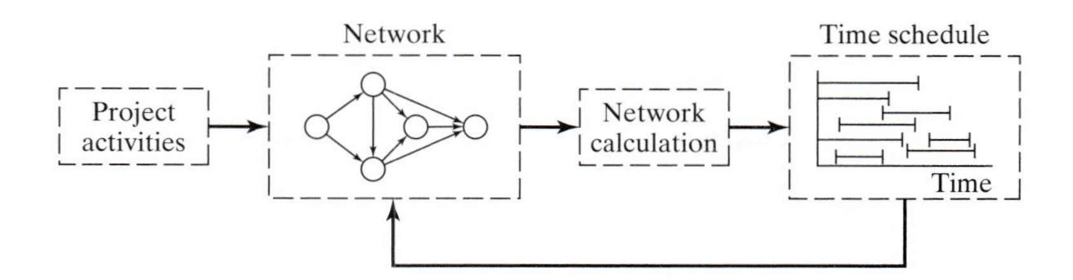

FIGURE 6.38

Phases for project planning with CPM-PERT

The two techniques, CPM and PERT, which were developed independently, differ in that CPM assumes deterministic activity durations and PERT assumes probabilistic durations. This presentation will start with CPM and then proceed with the details of PERT.

6.5.1 Network Representation

Each activity of the project is represented by an arc pointing in the direction of progress in the project. The nodes of the network establish the precedence relationships among the different activities.

Three rules are available for constructing the network.

Rule 1. *Each activity is represented by one, and only one, arc.*

Rule 2. *Each activity must be identified by two distinct end nodes.*

Figure 6.39 shows how a dummy activity can be used to represent two concurrent activities, A and B. By definition, a dummy activity, which normally is depicted by a

FIGURE 6.39

Use of dummy activity to produce unique representation of concurrent activities

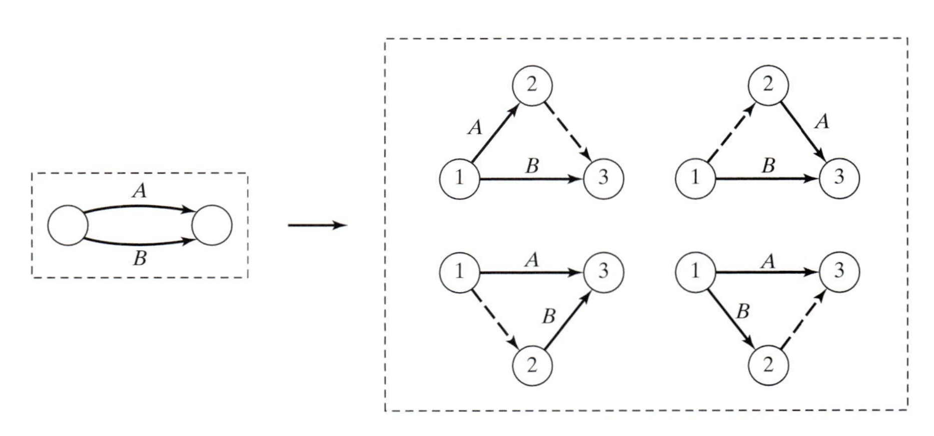

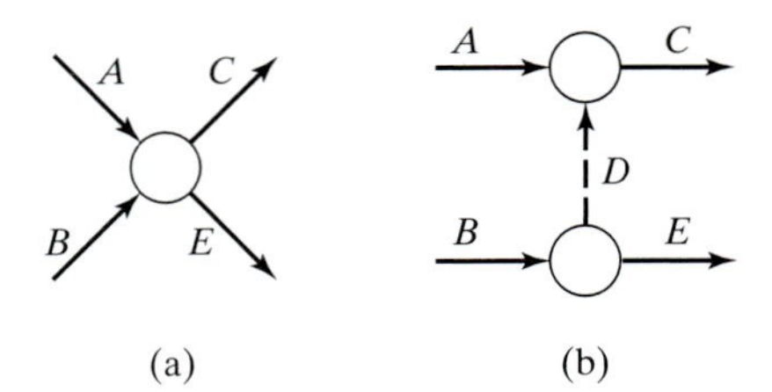

FIGURE 6.40
Use of dummy activity to ensure correct precedence relationship

dashed arc, consumes no time or resources. Inserting a dummy activity in one of the four ways shown in Figure 6.39, we maintain the concurrence of A and B, and provide unique end nodes for the two activities (to satisfy rule 2).

Rule 3. *To maintain the correct precedence relationships, the following questions must be answered as each activity is added to the network:*

(a) *What activities must immediately precede the current activity?*
(b) *What activities must follow the current activity?*
(c) *What activities must occur concurrently with the current activity?*

The answers to these questions may require the use of dummy activities to ensure correct precedences among the activities. For example, consider the following segment of a project:

1. Activity C starts immediately after A and B have been completed.
2. Activity E starts only after B has been completed.

Part (a) of Figure 6.40 shows the incorrect representation of the precedence relationship because it requires both A and B to be completed before E can start. In part (b), the use of a dummy activity rectifies the situation.

Example 6.5-1

A publisher has a contract with an author to publish a textbook. The (simplified) activities associated with the production of the textbook are given below. The author is required to submit to the publisher a hard copy and a computer file of the manuscript. Develop the associated network for the project.

Activity	Predecessor(s)	Duration (weeks)
A: Manuscript proofreading by editor	—	3
B: Sample pages preparation	—	2
C: Book cover design	—	4
D: Artwork preparation	—	3
E: Author's approval of edited manuscript and sample pages	*A*, *B*	2
F: Book formatting	*E*	4
G: Author's review of formatted pages	*F*	2
H: Author's review of artwork	*D*	1
I: Production of printing plates	*G*, *H*	2
J: Book production and binding	*C*, *I*	4

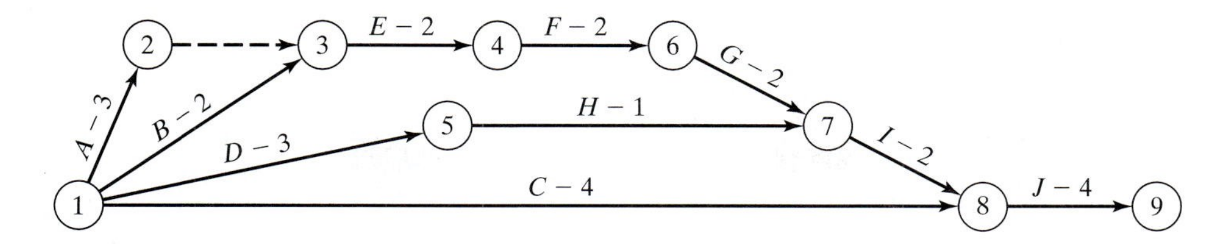

FIGURE 6.41

Project network for Example 6.5-1

Figure 6.41 provides the network describing the precedence relationships among the different activities. Dummy activity (2, 3) produces unique end nodes for concurrent activities A and B. It is convenient to number the nodes in ascending order in the direction of progress in the project.

PROBLEM SET 6.5A

1. Construct the project network comprised of activities A to L with the following precedence relationships:
 (a) A, B, and C, the first activities of the project, can be executed concurrently.
 (b) A and B precede D.
 (c) B precedes E, F, and H.
 (d) F and C precede G.
 (e) E and H precede I and J.
 (f) C, D, F, and J precede K.
 (g) K precedes L.
 (h) I, G, and L are the terminal activities of the project.

2. Construct the project network comprised of activities A to P that satisfies the following precedence relationships:
 (a) A, B, and C, the first activities of the project, can be executed concurrently.
 (b) D, E, and F follow A.
 (c) I and G follow both B and D.
 (d) H follows both C and G.
 (e) K and L follow I.
 (f) J succeeds both E and H.
 (g) M and N succeed F, but cannot start until both E and H are completed.
 (h) O succeeds M and I.
 (i) P succeeds J, L, and O.
 (j) K, N, and P are the terminal activities of the project.

***3.** The footings of a building can be completed in four consecutive sections. The activities for each section include (1) digging, (2) placing steel, and (3) pouring concrete. The digging of one section cannot start until that of the preceding section has been completed. The same restriction applies to pouring concrete. Develop the project network.

4. In Problem 3, suppose that 10% of the plumbing work can be started simultaneously with the digging of the first section but before any concrete is poured. After each section of the footings is completed, an additional 5% of the plumbing can be started provided that the preceding 5% portion is complete. The remaining plumbing can be completed at the end of the project. Construct the project network.

5. An opinion survey involves designing and printing questionnaires, hiring and training personnel, selecting participants, mailing questionnaires, and analyzing the data. Construct the project network, stating all assumptions.

6. The activities in the following table describe the construction of a new house. Construct the associated project network.

	Activity	Predecessor(s)	Duration (days)
A:	Clear site	—	1
B:	Bring utilities to site	—	2
C:	Excavate	*A*	1
D:	Pour foundation	*C*	2
E:	Outside plumbing	*B*, *C*	6
F:	Frame house	*D*	10
G:	Do electric wiring	*F*	3
H:	Lay floor	*G*	1
I:	Lay roof	*F*	1
J:	Inside plumbing	*E*, *H*	5
K:	Shingling	*I*	2
L:	Outside sheathing insulation	*F*, *J*	1
M:	Install windows and outside doors	*F*	2
N:	Do brick work	*L*, *M*	4
O:	Insulate walls and ceiling	*G*, *J*	2
P:	Cover walls and ceiling	*O*	2
Q:	Insulate roof	*I*, *P*	1
R:	Finish interior	*P*	7
S:	Finish exterior	*I*, *N*	7
T:	Landscape	*S*	3

7. A company is in the process of preparing a budget for launching a new product. The following table provides the associated activities and their durations. Construct the project network.

	Activity	Predecessor(s)	Duration (days)
A:	Forecast sales volume	—	10
B:	Study competitive market	—	7
C:	Design item and facilities	*A*	5
D:	Prepare production schedule	*C*	3
E:	Estimate cost of production	*D*	2
F:	Set sales price	*B*, *E*	1
G:	Prepare budget	*E*, *F*	14

8. The activities involved in a candlelight choir service are listed in the following table. Construct the project network.

	Activity	Predecessor(s)	Duration (days)
A:	Select music	—	2
B:	Learn music	*A*	14

C:	Make copies and buy books	*A*	14
D:	Tryouts	*B, C*	3
E:	Rehearsals	*D*	70
F:	Rent candelabra	*D*	14
G:	Decorate candelabra	*F*	1
H:	Set up decorations	*D*	1
I:	Order choir robe stoles	*D*	7
J:	Check out public address system	*D*	7
K:	Select music tracks	*J*	14
L:	Set up public address system	*K*	1
M:	Final rehearsal	*E, G, L*	1
N:	Choir party	*H, L, M*	1
O:	Final program	*I, N*	1

9. The widening of a road section requires relocating ("reconductoring") 1700 feet of 13.8-kV overhead primary line. The following table summarizes the activities of the project. Construct the associated project network.

	Activity	Predecessor(s)	Duration (days)
A:	Job review	—	1
B:	Advise customers of temporary outage	*A*	$\frac{1}{2}$
C:	Requisition stores	*A*	1
D:	Scout job	*A*	$\frac{1}{2}$
E:	Secure poles and material	*C, D*	3
F:	Distribute poles	*E*	$3\frac{1}{2}$
G:	Pole location coordination	*D*	$\frac{1}{2}$
H:	Re-stake	*G*	$\frac{1}{2}$
I:	Dig holes	*H*	3
J:	Frame and set poles	*F, I*	4
K:	Cover old conductors	*F, I*	1
L:	Pull new conductors	*J, K*	2
M:	Install remaining material	*L*	2
N:	Sag conductor	*L*	2
O:	Trim trees	*D*	2
P:	De-energize and switch lines	*B, M, N, O*	$\frac{1}{10}$
Q:	Energize and switch new line	*P*	$\frac{1}{2}$
R:	Clean up	*Q*	1
S:	Remove old conductor	*Q*	1
T:	Remove old poles	*S*	2
U:	Return material to stores	*R, T*	2

10. The following table gives the activities for buying a new car. Construct the project network.

	Activity	Predecessor(s)	Duration (days)
A:	Conduct feasibility study	—	3
B:	Find potential buyer for present car	*A*	14
C:	List possible models	*A*	1
D:	Research all possible models	*C*	3
E:	Conduct interview with mechanic	*C*	1
F:	Collect dealer propaganda	*C*	2
G:	Compile pertinent data	*D, E, F*	1
H:	Choose top three models	*G*	1

I:	Test-drive all three choices	*H*	3
J:	Gather warranty and financing data	*H*	2
K:	Choose one car	*I, J*	2
L:	Choose dealer	*K*	2
M:	Search for desired color and options	*L*	4
N:	Test-drive chosen model once again	*L*	1
O:	Purchase new car	*B, M, N*	3

6.5.2 Critical Path (CPM) Computations

The end result in CPM is the construction of the time schedule for the project (see Figure 6.38). To achieve this objective conveniently, we carry out special computations that produce the following information:

1. Total duration needed to complete the project.
2. Classification of the activities of the project as *critical* and *noncritical.*

An activity is said to be **critical** if there is no "leeway" in determining its start and finish times. A **noncritical** activity allows some scheduling slack, so that the start time of the activity can be advanced or delayed within limits without affecting the completion date of the entire project.

To carry out the necessary computations, we define an **event** as a point in time at which activities are terminated and others are started. In terms of the network, an event corresponds to a node. Define

$\square_j$ = Earliest occurrence time of event j
Δ_j = Latest occurrence time of event j
D_{ij} = Duration of activity (i, j)

The definitions of the *earliest* and *latest* occurrences of event j are specified relative to the start and completion dates of the entire project.

The critical path calculations involve two passes: The **forward pass** determines the *earliest* occurrence times of the events, and the **backward pass** calculates their *latest* occurrence times.

Forward Pass (Earliest Occurrence Times, $\square$). The computations start at node 1 and advance recursively to end node n.

Initial Step. Set $\square_1 = 0$ to indicate that the project starts at time 0.

General Step *j*. Given that nodes $p, q, \ldots,$ and v are linked *directly* to node j by incoming activities $(p, j), (q, j), \ldots,$ and (v, j) and that the earliest occurrence times of events (nodes) $p, q, \ldots,$ and v have already been computed, then the earliest occurrence time of event j is computed as

$$\square_j = \max\{\square_p + D_{pj}, \square_q + D_{qj}, \ldots, \square_v + D_{vj}\}$$

The forward pass is complete when $\square_n$ at node n has been computed. By definition $\square_j$ represents the longest path (duration) to node j.

Backward Pass (Latest Occurrence Times, Δ). Following the completion of the forward pass, the backward pass computations start at node n and end at node 1.

Initial Step. Set $\Delta_n = \square_n$ to indicate that the earliest and latest occurrences of the last node of the project are the same.

General Step *j*. Given that nodes $p, q, \ldots,$ and v are linked *directly* to node j by *outgoing* activities $(j, p), (j, q), \ldots,$ and (j, v) and that the latest occurrence times of nodes $p, q, \ldots,$ and v have already been computed, the latest occurrence time of node j is computed as

$$\Delta_j = \min\{\Delta_p - D_{jp}, \Delta_q - D_{jq}, \ldots, \Delta_v - D_{jv}\}$$

The backward pass is complete when Δ_1 at node 1 is computed. At this point, $\Delta_1 = \square_1 (= 0)$.

Based on the preceding calculations, an activity (i, j) will be *critical* if it satisfies three conditions.

1. $\Delta_i = \square_i$
2. $\Delta_j = \square_j$
3. $\Delta_j - \Delta_i = \square_j - \square_i = D_{ij}$

The three conditions state that the earliest and latest occurrence times of end nodes i and j are equal and the duration D_{ij} fits "tightly" in the specified time span. An activity that does not satisfy all three conditions is thus *noncritical.*

By definition, the critical activities of a network must constitute an uninterrupted path that spans the entire network from start to finish.

Example 6.5-2

Determine the critical path for the project network in Figure 6.42 . All the durations are in days.

Forward Pass

Node 1. Set $\square_1 = 0$
Node 2. $\square_2 = \square_1 + D_{12} = 0 + 5 = 5$
Node 3. $\square_3 = \max\{\square_1 + D_{13}, \square_2 + D_{23}\} = \max\{0 + 6, 5 + 3\} = 8$
Node 4. $\square_4 = \square_2 + D_{24} = 5 + 8 = 13$
Node 5. $\square_5 = \max\{\square_3 + D_{35}, \square_4 + D_{45}\} = \max\{8 + 2, 13 + 0\} = 13$

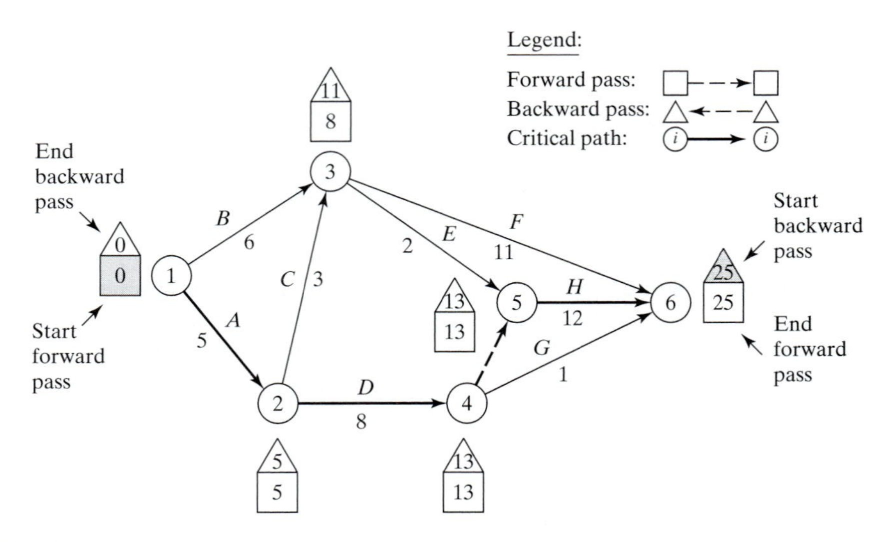

FIGURE 6.42
Forward and backward pass calculations for the project of Example 6.5-2

$$\textbf{Node 6.}\quad \square_6 = \max\{\square_3 + D_{36}, \square_4 + D_{46}, \square_5 + D_{56}\}$$
$$= \max\{8 + 11, 13 + 1, 13 + 12\} = 25$$

The computations show that the project can be completed in 25 days.

Backward Pass

Node 6. Set $\Delta_6 = \square_6 = 25$
Node 5. $\Delta_5 = \Delta_6 - D_{56} = 25 - 12 = 13$
Node 4. $\Delta_4 = \min\{\Delta_6 - D_{46}, \Delta_5 - D_{45}\} = \min\{25 - 1, 13 - 0\} = 13$
Node 3. $\Delta_3 = \min\{\Delta_6 - D_{36}, \Delta_5 - D_{35}\} = \min\{25 - 11, 13 - 2\} = 11$
Node 2. $\Delta_2 = \min\{\Delta_4 - D_{24}, \Delta_3 - D_{23}\} = \min\{13 - 8, 11 - 3\} = 5$
Node 1. $\Delta_1 = \min\{\Delta_3 - D_{13}, \Delta_2 - D_2\} = \min\{11 - 6, 5 - 5\} = 0$

Correct computations will always end with $\Delta_1 = 0$.

The forward and backward pass computations can be made directly on the network as shown in Figure 6.42. Applying the rules for determining the critical activities, the critical path is $1 \rightarrow 2 \rightarrow 4 \rightarrow 5 \rightarrow 6$, which, as should be expected, spans the network from start (node 1) to finish (node 6). The sum of the durations of the critical activities [(1, 2), (2, 4), (4, 5), and (5, 6)] equals the duration of the project (= 25 days). Observe that activity (4, 6) satisfies the first two conditions for a critical activity ($\Delta_4 = \square_4 = 13$ and $\Delta_5 = \square_5 = 25$) but not the third ($\square_6 - \square_4 \neq D_{46}$). Hence, the activity is noncritical.

PROBLEM SET 6.5B

*1. Determine the critical path for the project network in Figure 6.43.
2. Determine the critical path for the project networks in Figure 6.44.

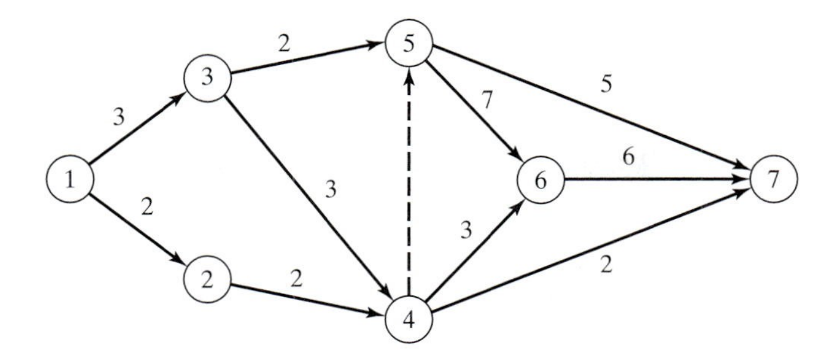

FIGURE 6.43
Project networks for Problem 1, Set 6.5b

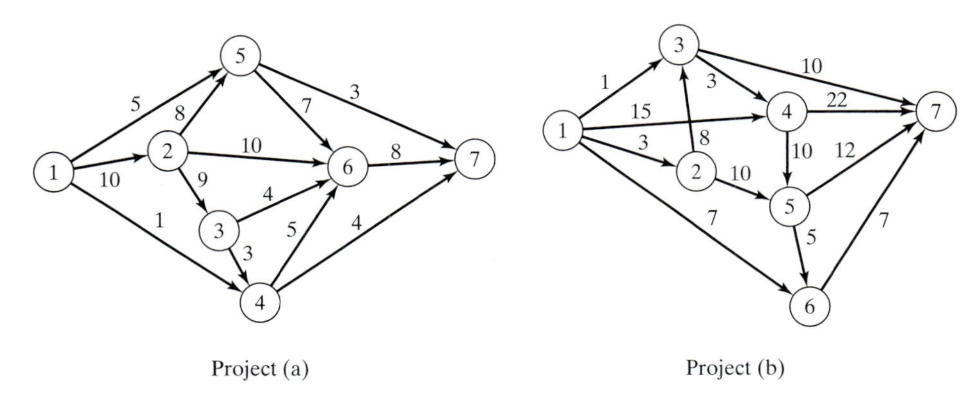

FIGURE 6.44
Project network for Problem 2, Set 6.5b

3. Determine the critical path for the project in Problem 6, Set6.5a.
4. Determine the critical path for the project in Problem 8, Set 6.5a.
5. Determine the critical path for the project in Problem 9, Set 6.5a.
6. Determine the critical path for the project in Problem 10, Set 6.5a.

6.5.3 Construction of the Time Schedule

This section shows how the information obtained from the calculations in Section 6.5.2 can be used to develop the time schedule. We recognize that for an activity (i, j), $\square_i$ represents the *earliest start time,* and Δ_j represents the *latest completion time.* This means that the interval $(\square_i, \Delta_j)$ delineates the (maximum) span during which activity (i, j) may be scheduled without delaying the entire project.

Construction of Preliminary Schedule. The method for constructing a preliminary schedule is illustrated by an example.

Example 6.5-3

Determine the time schedule for the project of Example 6.5-2 (Figure 6.42).

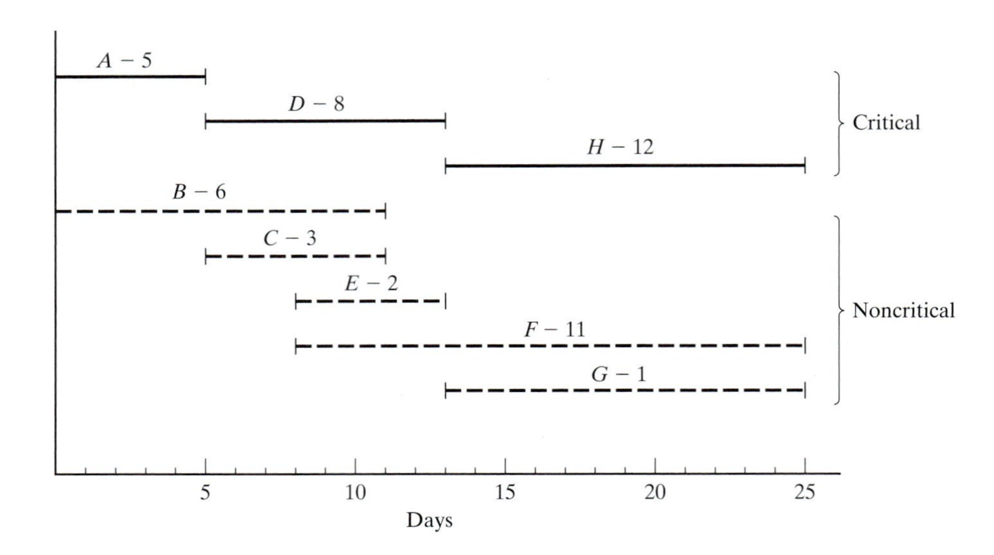

FIGURE 6.45
Preliminary schedule for the project of Example 6.5-2

We can get a preliminary time schedule for the different activities of the project by delineating their respective time spans as shown in Figure 6.45. Two observations are in order.

1. The critical activities (shown by solid lines) must be stacked one right after the other to ensure that the project is completed within its specified 25-day duration.
2. The noncritical activities (shown by dashed lines) have time spans that are larger than their respective durations, thus allowing slack (or "leeway") in scheduling them within their allotted time intervals.

How should we schedule the noncritical activities within their respective spans? Normally, it is preferable to start each noncritical activity as early as possible. In this manner, slack periods will remain opportunely available at the end of the allotted span where they can be used to absorb unexpected delays in the execution of the activity. It may be necessary, however, to delay the start of a noncritical activity past its earliest start time. For example, in Figure 6.45, suppose that each of the noncritical activities E and F requires the use of a bulldozer, and that only one is available. Scheduling both E and F as early as possible requires two bulldozers between times 8 and 10. We can remove the overlap by starting E at time 8 and pushing the start time of F to somewhere between times 10 and 14.

If all the noncritical activities can be scheduled as early as possible, the resulting schedule automatically is feasible. Otherwise, some precedence relationships may be violated if noncritical activities are delayed past their earliest time. Take for example activities C and E in Figure 6.45. In the project network (Figure 6.42), though C must be completed before E, the spans of C and E in Figure 6.45 allow us to schedule C between times 6 and 9, and E between times 8 and 10, which violates the requirement that C precede E. The need for a "red flag" that automatically reveals schedule conflict is thus evident. Such information is provided by computing the *floats* for the noncritical activities.

Determination of the Floats. Floats are the slack times available within the allotted span of the noncritical activity. The most common are the **total float** and the **free float.**

Figure 6.46 gives a convenient summary for computing the total float (TF_{ij}) and the free float (FF_{ij}) for an activity (i, j). The total float is the excess of the time span defined from the *earliest* occurrence of event i to the *latest* occurrence of event j over the duration of (i, j)—that is,

$$TF_{ij} = \Delta_j - \square_i - D_{ij}$$

The free float is the excess of the time span defined from the *earliest* occurrence of event i to the *earliest* occurrence of event j over the duration of (i, j)—that is,

$$FF_{ij} = \square_j - \square_i - D_{ij}$$

By definition, $FF_{ij} \leq TF_{ij}$.

Red-Flagging Rule. *For a noncritical activity (i, j)*

(a) *If $FF_{ij} = TF_{ij}$, then the activity can be scheduled anywhere within its $(\square_i, \Delta_j)$ span without causing schedule conflict.*

(b) *If $FF_{ij} < TF_{ij}$, then the start of the activity can be delayed by at most FF_{ij} relative to its earliest start time $(\square_i)$ without causing schedule conflict. Any delay larger than FF_{ij} (but not more than TF_{ij}) must be coupled with an equal delay relative to $\square_j$ in the start time of all the activities leaving node j.*

The implication of the rule is that a noncritical activity (i, j) will be red-flagged if its $FF_{ij} < TF_{ij}$. This red flag is important only if we decide to delay the start of the activity past its earliest start time, $\square_i$, in which case we must pay attention to the start times of the activities leaving node j to avoid schedule conflicts.

FIGURE 6.46

Computation of total and free floats

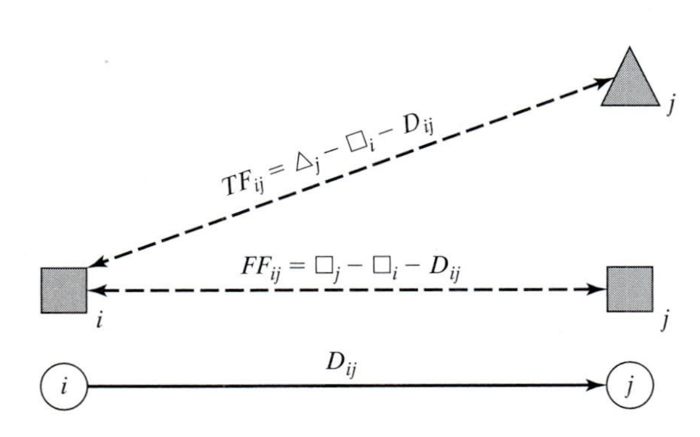

Example 6.5-4

Compute the floats for the noncritical activities of the network in Example 6.5-2, and discuss their use in finalizing a schedule for the project.

The following table summarizes the computations of the total and free floats. It is more convenient to do the calculations directly on the network using the procedure in Figure 6.42.

Noncritical activity	Duration	Total float (TF)	Free float (FF)
$B(1,3)$	6	$11 - 0 - 6 = 5$	$8 - 0 - 6 = 2$
$C(2,3)$	3	$11 - 5 - 3 = 3$	$8 - 5 - 3 = 0$
$E(3,5)$	2	$13 - 8 - 2 = 3$	$13 - 8 - 2 = 3$
$F(3,6)$	11	$25 - 8 - 11 = 6$	$25 - 8 - 11 = 6$
$G(4,6)$	1	$25 - 13 - 1 = 11$	$25 - 13 - 1 = 11$

The computations red-flag activities B and C because their $FF < TF$. The remaining activities (E, F, and G) have $FF = TF$, and hence may be scheduled anywhere between their earliest start and latest completion times.

To investigate the significance of the red-flagged activities, consider activity B. Because its $TF = 5$ days, this activity can start as early as time 0 or as late as time 5 (see Figure 6.45). However, because its $FF = 2$ days, starting B anywhere between time 0 and time 2 will have no effect on the succeeding activities E and F. If, however, activity B must start at time $2 + d\ (\leq 5)$, then the start times of the immediately succeeding activities E and F must be pushed forward past their earliest start time ($= 8$) by at least d. In this manner, the precedence relationship between B and its successors E and F is preserved.

Turning to red-flagged activity C, we note that its $FF = 0$. This means that *any* delay in starting C past its earliest start time ($= 5$) must be coupled with at least an equal delay in the start of its successor activities E and F.

TORA Moment

TORA provides useful tutorial tools for CPM calculations and for constructing the time schedule. To use these tools, select Project Planning ⇒ CPM-Critical Path Method from Main Menu. In the output screen, you have the option to select CPM Calculations to produce step-by-step computations of the forward pass, backward pass, and the floats or CPM Bar Chart to construct and experiment with the time schedule.

File toraEx6.5-2.txt provides TORA's data Example 6.5-2. If you elect to generate the output using the Next Step option, TORA will guide you through the details of the forward and backward pass calculations.

Figure 6.47 provides the TORA schedule produced by CPM Bar Chart option for the project of Example 6.5-2. The default bar chart automatically schedules all noncritical activities as early as possible. You can study the impact of delaying the start time of a noncritical activity by using the self-explanatory drop-down lists inside the bottom left frame of the screen. The impact of a delay of a noncritical activity will be shown directly on the bar chart together with an explanation. For example, if you delay the start of activity B by more than 2 time units, the succeeding activities E and F will be

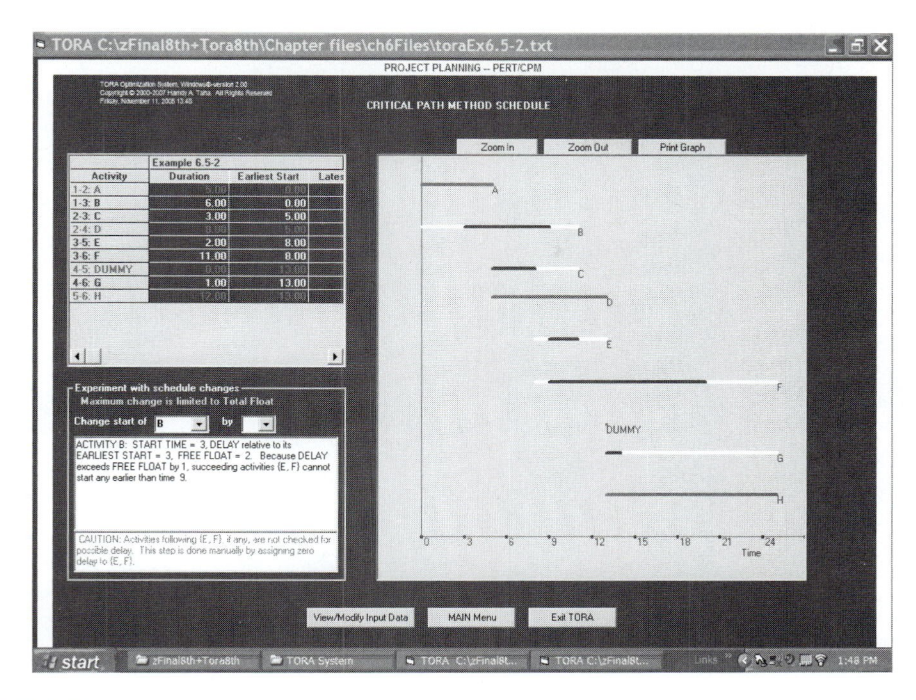

FIGURE 6.47

TORA bar chart output for Example 6.5-2 (file toraEx6.5-2.txt)

delayed by an amount equal to the difference between the delay over the free float of activity B. Specifically, given that the free float for B is 2 time units, if B is delayed by 3 time units, then E and F must be delayed by at least $3 - 2 = 1$ time unit. This situation is demonstrated in Figure 6.47.

AMPL Moment

Figure 6.48 provides the AMPL model for the CPM (file amplEx6.52.txt). The model is driven by the data of Example 6.5-2. This AMPL model is a unique application because it uses *indexed sets* (see Section A.4) and requires no optimization. In essence, no `solve` command is needed, and AMPL is implemented as a pure programming language similar to Basic or C.

The nature of the computations in CPM requires representing the network by associating two *indexed* sets with each node: `into` and `from`. For node `i`, the set `into[i]` defines all the nodes that feed into node `i`, and the set `from[i]` defines all the nodes that are reached from node `i`. For example, in Example 6.5-2, `from[1] = {2,3}` and `into[1]` is empty.

The determination of subsets `from` and `into` is achieved in the model as follows: Because `D[i,j]` can be zero when a CPM network uses dummy activities, the default value for `D[i,j]` is -1 for all nonexisting arcs. Thus, the set `from[i]` represents all the nodes `j` in the set `{1..n}` that can be reached from node `i`, which can happen only if `D[i,j]>=0`. This says that `from[i]` is defined by the subset `{j in 1..n:D[i,j]> =0}`.

```
#--------- CPM (Example 6.5-2)-----------
param n;
param D{1..n,1..n} default -1;

set into{1..n};
set from{1..n};

var x{i in 1..n,j in from[i]}>=0;
var ET{i in 1..n};
var LT{i in 1..n};
var TF{i in 1..n, j in from[i]};
var FF{i in 1..n, j in from[i]};

data;
param n:=6;
param D: 1 2 3 4 5 6:=
1 . 5 6 . . .
2 . . 3 8 . .
3 . . . . 2 11
4 . . . . 0 1
5 . . . . . 12
6 . . . . . .;

for {i in 1..n} {let from[i]:={j in 1..n:D[i,j]>=0}};
for {j in 1..n} {let into[j]:={i in 1..n:D[i,j]>=0}};

#-----nodes earliest and latest times and floats
let ET[1]:=0;          #earliest node time
for {i in 2..n}let ET[i]:=max{j in into[i]}(ET[j]+D[j,i]);

let LT[n]:=ET[n];      #latest node time
for{i in n-1..1 by -1}let LT[i]:=min{j in from[i]}(LT[j]-D[i,j]);

printf "%1s-%1s %5s %5s %5s %5s %5s %5s %5s \n\n",
     "i","j","D","ES","EC","LS","LC","TF","FF" >Ex6.6-2out.txt;
for {i in 1..n, j in from[i]}
{
let TF[i,j]:=LT[j]-ET[i]-D[i,j];
let FF[i,j]:=ET[j]-ET[i]-D[i,j];
printf "%1i-%1i %5i %5i %5i %5i %5i %5i %5i %3s\n",
  i,j,D[i,j],ET[i],ET[i]+D[i,j],LT[j]-D[i,j],LT[j],TF[i,j],FF[i,j],
  if TF[i,j]=0 then "c" else "" >Ex6.6-2out.txt;
}
```

FIGURE 6.48

AMPL model for Example 6.5-2 (file amplEx6.5-2.txt)

Similar reasoning applies to the determination of subsets `into[i]`. The following AMPL statements automate the determination of these sets and must follow the `D[i,j]` data, as shown in Figure 6.48:

```
for {i in 1..n} {let from[i]:={j in 1..n:D[i,j]>=0}};
for {j in 1..n} {let into[j]:={i in 1..n:D[i,j]>=0}};
```

Once the sets `from` and `into` have been determined, the model goes through the forward pass to compute the earliest time, `ET[i]`. With the completion of this pass, we can initiate the backward pass by using

```
let LT[n]:=ET[n];
```

The rest of the model is needed to obtain the output shown in Figure 6.49. This output determines all the data needed to construct the CPM chart. The logic of this segment is based on the computations given in Examples 6.5-2 and 6.5-4.

PROBLEM SET 6.5C

1. Given an activity (i, j) with duration D_{ij} and its earliest start time $\square_i$ and its latest completion time Δ_j, determine the earliest completion and the latest start times of (i, j).

2. What are the total and free floats of a critical activity? Explain.

***3.** For each of the following activities, determine the maximum delay in the starting time relative to its earliest start time that will allow all the immediately succeeding activities to be scheduled anywhere between their earliest and latest completion times.

(a) $TF = 10, FF = 10, D = 4$

(b) $TF = 10, FF = 5, D = 4$

(c) $TF = 10, FF = 0, D = 4$

4. In Example 6.5-4, use the floats to answer the following:

(a) If activity B is started at time 1, and activity C is started at time 5, determine the earliest start times for E and F.

(b) If activity B is started at time 3, and activity C is started at time 7, determine the earliest start times for E and F.

(c) How is the scheduling of other activities impacted if activity B starts at time 6?

***5.** In the project of Example 6.5-2 (Figure 6.42), assume that the durations of activities B and F are changed from 6 and 11 days to 20 and 25 days, respectively.

(a) Determine the critical path.

(b) Determine the total and free floats for the network, and identify the red-flagged activities.

FIGURE 6.49

Output of AMPL model for Example 6.5-2 (file amplEx6.5-2.txt)

i-j	D	ES	EC	LS	LC	TF	FF	
1-2	5	0	5	0	5	0	0	c
1-3	6	0	6	5	11	5	2	
2-3	3	5	8	8	11	3	0	
2-4	8	5	13	5	13	0	0	c
3-5	2	8	10	11	13	3	3	
3-6	11	8	19	14	25	6	6	
4-5	0	13	13	13	13	0	0	c
4-6	1	13	14	24	25	11	11	
5-6	12	13	25	13	25	0	0	c

(c) If activity A is started at time 5, determine the earliest possible start times for activities C, D, E, and G.

(d) If activities F, G, and H require the same equipment, determine the minimum number of units needed of this equipment.

6. Compute the floats and identify the red-flagged activities for the projects (a) and (b) in Figure 6.44, then develop the time schedules under the following conditions:

Project (a)

(i) Activity $(1, 5)$ cannot start any earlier than time 14.

(ii) Activities $(5, 6)$ and $(5, 7)$ use the same equipment, of which only one unit is available.

(iii) All other activities start as early as possible.

Project (b)

(i) Activity $(1, 3)$ must be scheduled at its earliest start time while accounting for the requirement that $(1, 2)$, $(1, 3)$, and $(1, 6)$ use a special piece of equipment, of which 1 unit only is available.

(ii) All other activities start as early as possible.

6.5.4 Linear Programming Formulation of CPM

A CPM problem can thought of as the opposite of the shortest-route problem (Section 6.3), in the sense that we are interested in finding the *longest* route of a unit flow entering at the start node and terminating at the finish node. We can thus apply the shortest route LP formulation in Section 6.3.3 to CPM in the following manner. Define

$$x_{ij} = \text{Amount of flow in activity } (i, j), \text{ for all defined } i \text{ and } j$$
$$D_{ij} = \text{Duration of activity } (i, j), \text{ for all defined } i \text{ and } j$$

Thus, the objective function of the linear program becomes

$$\text{Maximize } z = \sum_{\substack{\text{all defined} \\ \text{activities } (i, j)}} D_{ij}x_{ij}$$

(Compare with the shortest route LP formulation in Section 6.3.3 where the objective function is minimized.) For each node, there is one constraint that represents the conservation of flow:

$$\text{Total input flow} = \text{Total output flow}$$

All the variables, x_{ij}, are nonnegative.

Example 6.5-5

The LP formulation of the project of Example 6.5-2 (Figure 6.42) is given below. Note that nodes 1 and 6 are the start and finish nodes, respectively.

	A	B	C	D	E	F	Dummy	G	H	
	x_{12}	x_{13}	x_{23}	x_{24}	x_{35}	x_{36}	x_{45}	x_{46}	x_{56}	
Maximize z =	6	6	3	8	2	11	0	1	12	
Node 1	−1	−1								= −1
Node 2	1		−1	−1						= 0
Node 3		1	1		−1	−1				= 0
Node 4				1			−1	−1		= 0
Node 5					1		1		−1	= 0
Node 6						1		1	1	= 1

The optimum solution is

$$z = 25, x_{12}(A) = 1, x_{24}(D) = 1, x_{45}(\text{Dummy}) = 1, x_{56}(H) = 1, \text{all others} = 0$$

The solution defines the critical path as $A \rightarrow D \rightarrow$ Dummy $\rightarrow H$, and the duration of the project is 25 days. The LP solution is not complete, because it determines the critical path, but does not provide the data needed to construct the CPM chart. We have seen in Figure 6.48, however, that AMPL can be used to provide all the needed information without the LP optimization.

PROBLEM SET 6.5D

1. Use LP to determine the critical path for the project network in Figure 6.43.
2. Use LP to determine the critical path for the project networks in Figure 6.44.

6.5.5 PERT Networks

PERT differs from CPM in that it bases the duration of an activity on three estimates:

1. **Optimistic time,** a, which occurs when execution goes extremely well.
2. **Most likely time,** m, which occurs when execution is done under normal conditions.
3. **Pessimistic time,** b, which occurs when execution goes extremely poorly.

The range (a, b) encloses all possible estimates of the duration of an activity. The estimate m lies somewhere in the range (a, b). Based on the estimates, the average duration time, $\overline{D}$, and variance, v, are approximated as:

$$\overline{D} = \frac{a + 4m + b}{6}$$

$$v = \left(\frac{b - a}{6}\right)^2$$

CPM calculations given in Sections 6.5.2 and 6.5.3 may be applied directly, with $\overline{D}$ replacing the single estimate D.

It is possible now to estimate the probability that a node j in the network will occur by a prespecified scheduled time, S_j. Let e_j be the earliest occurrence time of node j. Because the durations of the activities leading from the start node to node j are random variables, e_j also must be a random variable. Assuming that all the activities in the network are statistically independent, we can determine the mean, $E\{e_j\}$, and variance, $\text{var}\{e_j\}$, in the following manner. If there is only one path from the start node to node j, then the mean is the sum of expected durations, $\overline{D}$, for all the activities along this path and the variance is the sum of the variances, v, of the same activities. On the other hand, if more than one path leads to node j, then it is necessary first to determine the statistical distribution of the duration of the longest path. This problem is rather difficult because it is equivalent to determining the distribution of the maximum of two or more random variables. A simplifying assumption thus calls for computing the mean and variance, $E\{e_j\}$ and $\text{var}\{e_j\}$, as those of the path to node j that has the largest sum of *expected* activity durations. If two or more paths have the same mean, the one with the largest variance is selected because it reflects the most uncertainty and, hence, leads to a more conservative estimate of probabilities.

Once the mean and variance of the path to node j, $E\{e_j\}$ and $\text{var}\{e_j\}$, have been computed, the probability that node j will be realized by a preset time S_j is calculated using the following formula:

$$P\{e_j \leq S_j\} = P\left\{\frac{e_j - E\{e_j\}}{\sqrt{\text{var}\{e_j\}}} \leq \frac{S_j - E\{e_j\}}{\sqrt{\text{var}\{e_j\}}}\right\} = P\{z \leq K_j\}$$

where

$$z = \text{Standard normal random variable}$$

$$K_j = \frac{S_j - E\{e_j\}}{\sqrt{\text{var}\{e_j\}}}$$

The standard normal variable z has mean 0 and standard deviation 1 (see Section 12.4.4). Justification for the use of the normal distribution is that e_j is the sum of independent random variables. According to the *central limit theorem* (see Section 12.4.4), e_j is approximately normally distributed.

Example 6.5-6

Consider the project of Example 6.5-2. To avoid repeating critical path calculations, the values of $a, m,$ and b in the table below are selected such that $\overline{D}_{ij} = D_{ij}$ for all i and j in Example 6.5-2.

Activity	i–j	(a, m, b)	Activity	i–j	(a, m, b)
A	1–2	$(3, 5, 7)$	E	3–5	$(1, 2, 3)$
B	1–3	$(4, 6, 8)$	F	3–6	$(9, 11, 13)$
C	2–3	$(1, 3, 5)$	G	4–6	$(1, 1, 1)$
D	2–4	$(5, 8, 11)$	H	5–6	$(10, 12, 14)$

The mean $\overline{D_{ij}}$ and variance v_{ij} for the different activities are given in the following table. Note that for a dummy activity $(a, m, b) = (0, 0, 0)$, hence its mean and variance also equal zero.

Activity	i–j	$\overline{D}_{ij}$	V_{ij}	Activity	i–j	$\overline{D}_{ij}$	V_{ij}
A	1–2	5	.444	E	3–5	2	.111
B	1–3	6	.444	F	3–6	11	.444
C	2–3	3	.444	G	4–6	1	.000
D	2–4	8	1.000	H	5–6	12	.444

The next table gives the longest path from node 1 to the different nodes, together with their associated mean and standard deviation.

Node	Longest path based on mean durations	Path mean	Path standard deviation
2	1–2	5.00	0.67
3	1–2–3	8.00	0.94
4	1–2–4	13.00	1.20
5	1–2–4–5	13.00	1.20
6	1–2–4–5–6	25.00	1.37

Finally, the following table computes the probability that each node is realized by time S_j specified by the analyst.

Node j	Longest path	Path mean	Path standard deviation	S_j	K_j	$P\{z \leq K_j\}$
2	1–2	5.00	0.67	5.00	0	.5000
3	1–2–3	8.00	0.94	11.00	3.19	.9993
4	1–2–4	13.00	1.20	12.00	−.83	.2033
5	1–2–4–5	13.00	1.20	14.00	.83	.7967
6	1–2–4–5–6	25.00	1.37	26.00	.73	.7673

TORA Moment.

TORA provides a module for carrying out PERT calculations. To use this module, select Project Planning ⇒ PERT-Program Evaluation and Review Technique from Main Menu. In the output screen, you have the option to select Activity Mean/Var to compute the mean and variance for each activity or PERT Calculations to compute the mean and variance of the longest path to each node in the network. File toraEx6.5-6.txt provides TORA's data for Example 6.5-6.

PROBLEM SET 6.5E

1. Consider Problem 2, Set 6.5b. The estimates (a, m, b) are listed below. Determine the probabilities that the different nodes of the project will be realized without delay.

Project (a)				Project (b)			
Activity	*(a, m, b)*	*Activity*	*(a, m, b)*	Activity	*(a, m, b)*	*Activity*	*(a, m, b)*
1-2	(5, 6, 8)	3-6	(3, 4, 5)	1-2	(1, 3, 4)	3-7	(12, 13, 14)
1-4	(1, 3, 4)	4-6	(4, 8, 10)	1-3	(5, 7, 8)	4-5	(10, 12, 15)
1-5	(2, 4, 5)	4-7	(5, 6, 8)	1-4	(6, 7, 9)	4-7	(8, 10, 12)
2-3	(4, 5, 6)	5-6	(9, 10, 15)	1-6	(1, 2, 3)	5-6	(7, 8, 11)
2-5	(7, 8, 10)	5-7	(4, 6, 8)	2-3	(3, 4, 5)	5-7	(2, 4, 8)
2-6	(8, 9, 13)	6-7	(3, 4, 5)	2-5	(7, 8, 9)	6-7	(5, 6, 7)
3-4	(5, 9, 19)			3-4	(10, 15, 20)		

REFERENCES

Ahuja, R., T. Magnati, and J. Orlin, *Network Flows: Theory, Algorithms, and Applications,* Prentice Hall, Upper Saddle River, NJ, 1993.

Bazaraa, M., J. Jarvis, and H. Sherali, *Linear Programming and Network Flow,* 2nd ed., Wiley, New York, 1990.

Bersetkas, D., *Network Optimization: Continuous and Discrete Models*, Athena Scientific, Nashua, NH, 1998.

Charnes, A. and W. Cooper, "Some Network Characterization for Mathematical Programming and Accounting Applications to Planning and Control," *The Accounting Review*, Vol. 42, No. 3, pp. 24–52, 1967.

Evans, J.R., and E. Minieka, *Optimization Algorithms for Networks and Graphs,* 2nd ed., Marcel Dekker, New York, 1992.

Guéret, C., C. Prins, and M. Sevaux, *Applications of Optimization with Xpress-MP*, translated and revised by Susanne Heipke, Dash Optimization Ltd., London, 2002.

Glover, F., D. Klingman, and N. Phillips, *Network Models and Their Applications in Practice*, Wiley, New York, 1992.

Glover, F., and M. Laguna, *Tabu Search*, Kulwer Academic Publishers, Boston, 1997.

Murty, K., *Network Programming,* Prentice Hall, Upper Saddle River, NJ, 1992.

Robinson, E.W, L. Gao, and S. Muggenborg, "Designing an Integrated Distribution System at DowBrands, Inc,"*Interfaces*, Vol. 23, No. 3, pp. 107–117, 1993.

CHAPTER 7

Advanced Linear Programming

Chapter Guide. This chapter presents the mathematical foundation of linear programming and duality theory. The presentation allows the development of a number of computationally efficient algorithms, including the revised simplex method, bounded variables, and parametric programming. Chapter 20 on the CD presents two additional algorithms that deal with large-scale LPs: decomposition and the Karmarkar interior-point algorithm.

The material in this chapter relies heavily on the use of matrix algebra. Appendix D on the CD provides a review of matrices.

The three topics that should receive special attention in this chapter are the revised simplex method, the bounded-variables algorithm, and parametric programming. The use of matrix manipulations in the revised simplex method allows a better control over machine roundoff error, an ever-present problem in the row operations method of Chapter 3. The bounded variables algorithm is used prominently with the integer programming branch-and-bound algorithm (Chapter 9). Parametric programming adds a dynamic dimension to the LP model that allows the determination of the changes in the optimum solution resulting from making continuous changes in the parameters of the model.

The task of understanding the details of the revised simplex method, bounded variables, decomposition, and parametric programming is improved by summarizing the results of matrix manipulations in the easy-to-read simplex tableau format of Chapter 3. Although matrix manipulations make the algorithms appear different, the theory is exactly the same as in Chapter 3.

This chapter includes 1 real-life application, 8 solved examples, 58 end-of-section problems, and 4 end-of-chapter comprehensive problems. The comprehensive problems are in Appendix E on the CD. The AMPL/Excel/Solver/TORA programs are in folder ch7Files.

Real-Life Application—Optimal Ship Routing and Personnel Assignment for Naval Recruitment in Thailand

Thailand Navy recruits are drafted four times a year. A draftee reports to one of 34 local centers and is then transported by bus to one of four navy branch bases. From there, recruits are transported to the main naval base by ship. The docking facilities at the branch bases may restrict the type of ship that can visit each base. Branch bases have limited capacities but, as a whole, the four bases have sufficient capacity to accommodate all the draftees. During the summer of 1983, a total of 2929 draftees were transported from the drafting centers to the four branch bases and eventually to the main base. The problem deals with determining the optimal schedule for transporting the draftees, first from the drafting centers to the branch bases and then from the branch bases to the main base. The study uses a combination of linear and integer programming. The details are given in Case 5, Chapter 24 on the CD.

7.1 SIMPLEX METHOD FUNDAMENTALS

In linear programming, the feasible solution space is said to form a **convex set** if the line segment joining any two *distinct* feasible points also falls in the set. An **extreme point** of the convex set is a feasible point that cannot lie on a line segment joining any two *distinct* feasible points in the set. Actually, extreme points are the same as corner point, the more apt name used in Chapters 2, 3, and 4.

Figure 7.1 illustrates two sets. Set (a), which is typical of the solution space of a linear program, is convex (with six extreme points), whereas set (b) is nonconvex.

In the graphical LP solution given in Section 2.3, we demonstrated that the optimum solution can always be associated with a feasible extreme (corner) point of the solution space. This result makes sense intuitively, because in the LP solution space every feasible point can be determined as a function of its feasible extreme points. For example, in convex set (a) of Figure 7.1, a feasible point $\mathbf{X}$ can be expressed as a **convex combination** of its extreme points $\mathbf{X}_1$, $\mathbf{X}_2$, $\mathbf{X}_3$, $\mathbf{X}_4$, $\mathbf{X}_5$, and $\mathbf{X}_6$ using

$$\mathbf{X} = \alpha_1\mathbf{X}_1 + \alpha_2\mathbf{X}_2 + \alpha_3\mathbf{X}_3 + \alpha_4\mathbf{X}_4 + \alpha_5\mathbf{X}_5 + \alpha_6\mathbf{X}_6$$

where

$$\alpha_1 + \alpha_2 + \alpha_3 + \alpha_4 + \alpha_5 + \alpha_6 = 1$$
$$\alpha_i \geq 0, i = 1, 2, \ldots, 6$$

This observation shows that extreme points provide all that is needed to define the solution space completely.

FIGURE 7.1

Examples of a convex and a nonconvex set

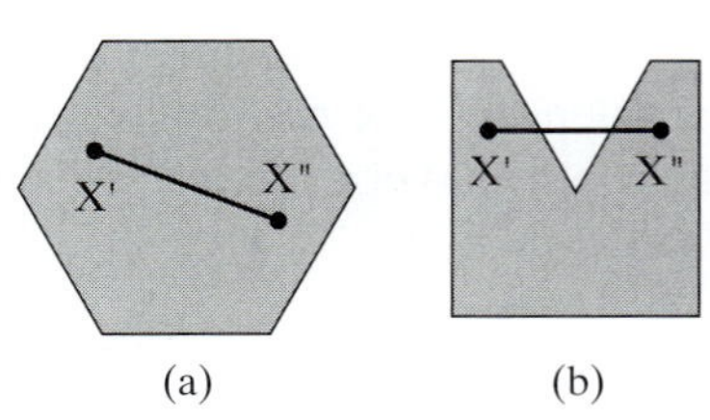

Example 7.1-1

Show that the following set is convex:

$$C = \{(x_1, x_2)|x_1 \leq 2, x_2 \leq 3, x_1 \geq 0, x_2 \geq 0\}$$

Let $\mathbf{X}_1 = \{x_1', x_2'\}$ and $\mathbf{X}_2 = \{x_1'', x_2''\}$ be any two distinct points in C. If C is convex, then $\mathbf{X} = (x_1, x_2) = \alpha_1\mathbf{X}_1 + \alpha_2\mathbf{X}_2$, $\alpha_1 + \alpha_2 = 1, \alpha_1, \alpha_2 \geq 0$, must also be in C. To show that this is true, we need to show that all the constraints of C are satisfied by the line segment $\mathbf{X}$; that is,

$$x_1 = \alpha_1 x_1' + \alpha_2 x_1'' \leq \alpha_1(2) + \alpha_2(2) = 2$$

$$x_2 = \alpha_1 x_2' + \alpha_2 x_2'' \leq \alpha_1(3) + \alpha_2(3) = 3$$

Thus, $x_1 \leq 2$ and $x_2 \leq 3$. Additionally, the nonnegativity conditions are satisfied because α_1 and α_2 are nonnegative.

PROBLEM SET 7.1A

1. Show that the set $Q = \{x_1, x_2|x_1 + x_2 \leq 1, x_1 \geq 0, x_2 \geq 0\}$ is convex. Is the nonnegativity condition essential for the proof?

*2. Show that the set $Q = \{x_1, x_2|x_1 \geq 1 \text{ or } x_2 \geq 2\}$ is not convex.

3. Determine graphically the extreme points of the following convex set:

$$Q = \{x_1, x_2|x_1 + x_2 \leq 2, x_1 \geq 0, x_2 \geq 0\}$$

Show that the entire feasible solution space can be determined as a convex combination of its extreme points. Hence conclude that any convex (bounded) solution space is totally defined once its extreme points are known.

4. In the solution space in Figure 7.2 (drawn to scale), express the interior point (3, 1) as a convex combination of the extreme points A, B, C, and D with each extreme point carrying a strictly positive weight.

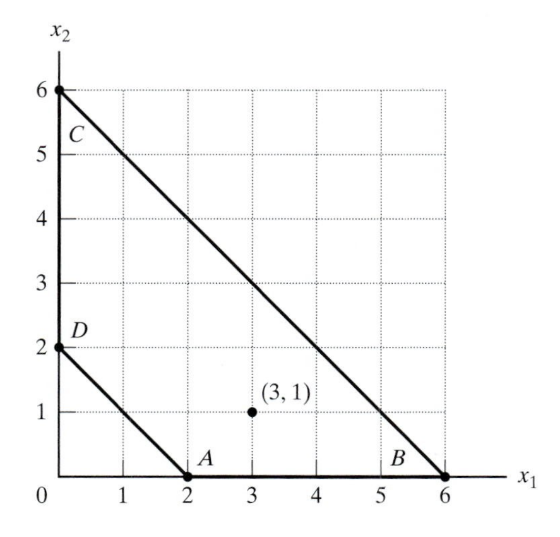

FIGURE 7.2

Solution space for Problem 4, Set 7.1a

7.1.1 From Extreme Points to Basic Solutions

It is convenient to express the general LP problem in equation form (see Section 3.1) using matrix notation. Define $\mathbf{X}$ as an n-vector representing the variables, $\mathbf{A}$ as an $(m \times n)$-matrix representing the constraint coefficients, $\mathbf{b}$ as a column vector representing the right-hand side, and $\mathbf{C}$ as an n-vector representing the objective-function coefficients. The LP is then written as

$$\text{Maximize or minimize } z = \mathbf{CX}$$

subject to

$$\mathbf{AX} = \mathbf{b}$$
$$\mathbf{X} \geq \mathbf{0}$$

Using the format of Chapter 3 (see also Figure 4.1), the rightmost m columns of $\mathbf{A}$ always can be made to represent the identity matrix $\mathbf{I}$ through proper arrangements of the slack/artificial variables associated with the starting basic solution.

A **basic solution** of $\mathbf{AX} = \mathbf{b}$ is determined by setting $n - m$ variables equal to zero, and then solving the resulting m equations in the remaining m unknowns, *provided that the resulting solution is unique*. Given this definition, the theory of linear programming establishes the following result between the geometric definition of extreme points and the algebraic definition of basic solutions:

$$\text{Extreme points of } \{\mathbf{X} | \mathbf{AX} = \mathbf{b}\} \Leftrightarrow \text{Basic solutions of } \mathbf{AX} = \mathbf{b}$$

The relationship means that the extreme points of the LP solution space are totally defined by the basic solutions of the system $\mathbf{AX} = \mathbf{b}$, and vice versa. Thus, we conclude that the basic solutions of $\mathbf{AX} = \mathbf{b}$ contain all the information we need to determine the optimum solution of the LP problem. Furthermore, if we impose the nonnegativity restriction, $\mathbf{X} \geq \mathbf{0}$, the search for the optimum solution is confined to the *feasible* basic solutions only.

To formalize the definition of a basic solution, the system $\mathbf{AX} = \mathbf{b}$ can be expressed in vector form as follows:

$$\sum_{j=1}^{n} \mathbf{P}_j x_j = \mathbf{b}$$

The vector $\mathbf{P}_j$ is the jth column of $\mathbf{A}$. A subset of m vectors is said to form a **basis, B,** if, and only if, the selected m vectors are **linearly independent**. In this case, the matrix $\mathbf{B}$ is **nonsingular**. If $\mathbf{X}_B$ is the set of m variables associated with the vectors of nonsingular $\mathbf{B}$, then $\mathbf{X}_B$ must be a basic solution. In this case, we have

$$\mathbf{BX}_B = \mathbf{b}$$

Given the inverse $\mathbf{B}^{-1}$ of $\mathbf{B}$, we then get the corresponding basic solution as

$$\mathbf{X}_B = \mathbf{B}^{-1}\mathbf{b}$$

If $\mathbf{B}^{-1}\mathbf{b} \geq \mathbf{0}$, then $\mathbf{X}_B$ is feasible. The definition assumes that the remaining $n - m$ variables are **nonbasic** at zero level.

The previous result shows that in a system of m equations and n unknowns, the maximum number of (feasible and infeasible) basic solutions is given by

$$\binom{n}{m} = \frac{n!}{m!(n-m)!}$$

Example 7.1-2

Determine and classify (as feasible and infeasible) all the basic solutions of the following system of equations.

$$\begin{pmatrix} 1 & 3 & -1 \\ 2 & -2 & -2 \end{pmatrix}\begin{pmatrix} x_1 \\ x_2 \\ x_3 \end{pmatrix} = \begin{pmatrix} 4 \\ 2 \end{pmatrix}$$

The following table summarizes the results. The inverse of **B** is determined by using one of the methods in Section D.2.7 on the CD.

B	$\mathbf{BX}_B = \mathbf{b}$	Solution	Status
$(\mathbf{P}_1, \mathbf{P}_2)$	$\begin{pmatrix} 1 & 3 \\ 2 & -2 \end{pmatrix}\begin{pmatrix} x_1 \\ x_2 \end{pmatrix} = \begin{pmatrix} 4 \\ 2 \end{pmatrix}$	$\begin{pmatrix} x_1 \\ x_2 \end{pmatrix} = \begin{pmatrix} \frac{1}{4} & \frac{3}{8} \\ \frac{1}{4} & -\frac{1}{8} \end{pmatrix}\begin{pmatrix} 4 \\ 2 \end{pmatrix} = \begin{pmatrix} \frac{7}{4} \\ \frac{3}{4} \end{pmatrix}$	Feasible
$(\mathbf{P}_1, \mathbf{P}_3)$	(Not a basis)	—	—
$(\mathbf{P}_2, \mathbf{P}_3)$	$\begin{pmatrix} 3 & -1 \\ -2 & -2 \end{pmatrix}\begin{pmatrix} x_2 \\ x_3 \end{pmatrix} = \begin{pmatrix} 4 \\ 2 \end{pmatrix}$	$\begin{pmatrix} x_2 \\ x_3 \end{pmatrix} = \begin{pmatrix} \frac{1}{4} & -\frac{1}{8} \\ -\frac{1}{4} & -\frac{3}{8} \end{pmatrix}\begin{pmatrix} 4 \\ 2 \end{pmatrix} = \begin{pmatrix} \frac{3}{4} \\ -\frac{7}{4} \end{pmatrix}$	Infeasible

We can also investigate the problem by expressing it in vector form as follows:

$$\begin{pmatrix} 1 \\ 2 \end{pmatrix}x_1 + \begin{pmatrix} 3 \\ -2 \end{pmatrix}x_2 + \begin{pmatrix} -1 \\ -2 \end{pmatrix}x_3 = \begin{pmatrix} 4 \\ 2 \end{pmatrix}$$

Each of $\mathbf{P}_1$, $\mathbf{P}_2$, $\mathbf{P}_3$, and $\mathbf{b}$ is a two-dimensional vector, which can be represented generically as $(a_1, a_2)^T$. Figure 7.3 graphs these vectors on the (a_1, a_2)-plane. For example, for $\mathbf{b} = (4, 2)^T$, $a_1 = 4$ and $a_2 = 2$.

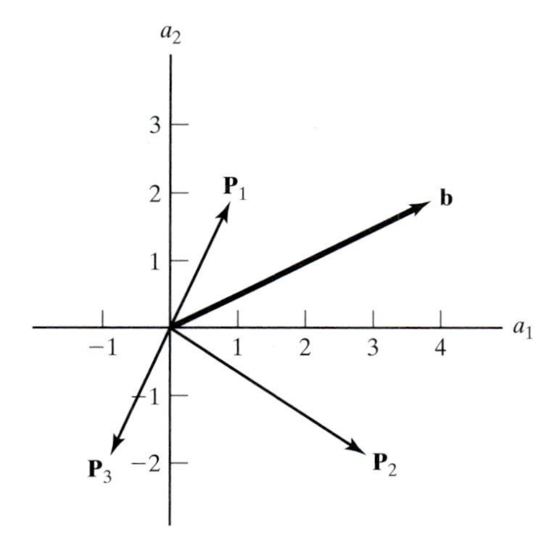

FIGURE 7.3

Vector representation of LP solution space

Because we are dealing with two equations ($m = 2$), a basis must include exactly two vectors, selected from among $\mathbf{P}_1$, $\mathbf{P}_2$, and $\mathbf{P}_3$. From Figure 7.3, the matrices $(\mathbf{P}_1, \mathbf{P}_2)$ and $(\mathbf{P}_2, \mathbf{P}_3)$ form bases because their associated vectors are independent. In the matrix $(\mathbf{P}_1, \mathbf{P}_3)$ the two vectors are dependent, and hence do not constitute a basis.

Algebraically, a (square) matrix forms a basis if its determinant is not zero (see Section D.2.5). The following computations show that the combinations $(\mathbf{P}_1, \mathbf{P}_2)$ and $(\mathbf{P}_2, \mathbf{P}_3)$ are bases, and the combination $(\mathbf{P}_1, \mathbf{P}_3)$ is not.

$$\det(\mathbf{P}_1, \mathbf{P}_2) = \det\begin{pmatrix} 1 & 3 \\ 2 & -2 \end{pmatrix} = (1 \times -2) - (2 \times 3) = -8 \neq 0$$

$$\det(\mathbf{P}_2, \mathbf{P}_3) = \det\begin{pmatrix} 3 & -1 \\ -2 & -2 \end{pmatrix} = (3 \times -2) - (-2 \times -1) = -8 \neq 0$$

$$\det(\mathbf{P}_1, \mathbf{P}_3) = \det\begin{pmatrix} 1 & -1 \\ 2 & -2 \end{pmatrix} = (1 \times -2) - (2 \times -1) = 0$$

We can take advantage of the vector representation of the problem to discuss how the starting solution of the simplex method is determined. From the vector representation in Figure 7.3, the basis $\mathbf{B} = (\mathbf{P}_1, \mathbf{P}_2)$ can be used to start the simplex iterations, because it produces the basic feasible solution $\mathbf{X}_B = (x_1, x_2)^T$. However, in the absence of the vector representation, the only available course of action is to try all possible bases (3 in this example, as shown above). The difficulty with using trial and error is that it is not suitable for automatic computations. In a typical LP with thousands of variables and constraints where the use of the computer is a must, trial and error is not a practical option because of its tremendous computational overhead. To alleviate this problem, the simplex method always uses an identity matrix, $\mathbf{B} = \mathbf{I}$, to start the iterations. Why does a starting $\mathbf{B} = \mathbf{I}$ offer an advantage? The answer is that it will always provide a *feasible* starting basic solution (provided that the right-hand side vector of the equations is nonnegative). You can see this result in Figure 7.3 by graphing the vectors of $\mathbf{B} = \mathbf{I}$ and noting that they coincide with the horizontal and vertical axes, thus always guaranteeing a starting basic feasible solution.

The basis $\mathbf{B} = \mathbf{I}$ automatically forms part of the LP equations if all the original constraints are $\leq$. In other cases, we simply add the unit vectors where needed. This is what the artificial variables accomplish (Section 3.4). We then penalize these variables in the objective function to force them to zero level in the final solution.

PROBLEM SET 7.1B

1. In the following sets of equations, (a) and (b) have unique (basic) solutions, (c) has an infinity of solutions, and (d) has no solution. Show how these results can be verified using graphical *vector* representation. From this exercise, state the general conditions for vector dependence-independence that lead to unique solution, infinity of solutions, and no solution.

(a) $x_1 + 3x_2 = 2$
$3x_1 + x_2 = 3$

(b) $2x_1 + 3x_2 = 1$
$2x_1 - x_2 = 2$

(c) $2x_1 + 6x_2 = 4$
$x_1 + 3x_2 = 2$

(d) $2x_1 - 4x_2 = 2$
$-x_1 + 2x_2 = 1$

2. Use vectors to determine graphically the type of solution for each of the sets of equations below: unique solution, an infinity of solutions, or no solution. For the cases of unique solutions, indicate from the vector representation (and without solving the equations algebraically) whether the values of the x_1 and x_2 are positive, zero, or negative.

(a) $\begin{pmatrix} 5 & 4 \\ 1 & -3 \end{pmatrix}\begin{pmatrix} x_1 \\ x_2 \end{pmatrix} = \begin{pmatrix} 1 \\ 1 \end{pmatrix}$ *__(b)__ $\begin{pmatrix} 2 & -2 \\ 1 & 3 \end{pmatrix}\begin{pmatrix} x_1 \\ x_2 \end{pmatrix} = \begin{pmatrix} 1 \\ 3 \end{pmatrix}$

(c) $\begin{pmatrix} 2 & 4 \\ 1 & 3 \end{pmatrix}\begin{pmatrix} x_1 \\ x_2 \end{pmatrix} = \begin{pmatrix} -2 \\ -1 \end{pmatrix}$ *__(d)__ $\begin{pmatrix} 2 & 4 \\ 1 & 2 \end{pmatrix}\begin{pmatrix} x_1 \\ x_2 \end{pmatrix} = \begin{pmatrix} 6 \\ 3 \end{pmatrix}$

(e) $\begin{pmatrix} -2 & 4 \\ 1 & -2 \end{pmatrix}\begin{pmatrix} x_1 \\ x_2 \end{pmatrix} = \begin{pmatrix} 2 \\ 1 \end{pmatrix}$ *__(f)__ $\begin{pmatrix} 1 & -2 \\ 0 & 0 \end{pmatrix}\begin{pmatrix} x_1 \\ x_2 \end{pmatrix} = \begin{pmatrix} 1 \\ 1 \end{pmatrix}$

3. Consider the following system of equations:

$$\begin{pmatrix} 1 \\ 2 \\ 3 \end{pmatrix} x_1 + \begin{pmatrix} 0 \\ 2 \\ 1 \end{pmatrix} x_2 + \begin{pmatrix} 1 \\ 4 \\ 2 \end{pmatrix} x_3 + \begin{pmatrix} 2 \\ 0 \\ 0 \end{pmatrix} x_4 = \begin{pmatrix} 3 \\ 4 \\ 2 \end{pmatrix}$$

Determine if any of the following combinations forms a basis.

*__(a)__ $(\mathbf{P}_1, \mathbf{P}_2, \mathbf{P}_3)$
(b) $(\mathbf{P}_1, \mathbf{P}_2, \mathbf{P}_4)$
(c) $(\mathbf{P}_2, \mathbf{P}_3, \mathbf{P}_4)$
*__(d)__ $(\mathbf{P}_1, \mathbf{P}_2, \mathbf{P}_3, \mathbf{P}_4)$

4. True or False?

(a) The system $\mathbf{BX} = \mathbf{b}$ has a unique solution if $\mathbf{B}$ is nonsingular.
(b) The system $\mathbf{BX} = \mathbf{b}$ has no solution if $\mathbf{B}$ is singular and $\mathbf{b}$ is independent of $\mathbf{B}$.
(c) The system $\mathbf{BX} = \mathbf{b}$ has an infinity of solutions if $\mathbf{B}$ is singular and $\mathbf{b}$ is dependent.

7.1.2 Generalized Simplex Tableau in Matrix Form

In this section, we use matrices to develop the general simplex tableau. This representation will be the basis for subsequent developments in the chapter.

Consider the LP in equation form:

$$\text{Maximize } z = \mathbf{CX}, \text{ subject to } \mathbf{AX} = \mathbf{b}, \mathbf{X} \geq \mathbf{0}$$

The problem can be written equivalently as

$$\begin{pmatrix} 1 & -\mathbf{C} \\ \mathbf{0} & \mathbf{A} \end{pmatrix}\begin{pmatrix} z \\ \mathbf{X} \end{pmatrix} = \begin{pmatrix} 0 \\ \mathbf{b} \end{pmatrix}$$

Suppose that $\mathbf{B}$ is a feasible basis of the system $\mathbf{AX} = \mathbf{b}, \mathbf{X} \geq \mathbf{0}$, and let $\mathbf{X}$ be the corresponding vector of basic variables with $\mathbf{C}_B$ as its associated objective vector. The corresponding solution may then be computed as follows (the method for inverting partitioned matrices is given in Section D.2.7):

$$\begin{pmatrix} z \\ \mathbf{X}_B \end{pmatrix} = \begin{pmatrix} 1 & -\mathbf{C}_B \\ \mathbf{0} & \mathbf{B} \end{pmatrix}^{-1}\begin{pmatrix} 0 \\ \mathbf{b} \end{pmatrix} = \begin{pmatrix} 1 & \mathbf{C}_B\mathbf{B}^{-1} \\ \mathbf{0} & \mathbf{B}^{-1} \end{pmatrix}\begin{pmatrix} 0 \\ \mathbf{b} \end{pmatrix} = \begin{pmatrix} \mathbf{C}_B\mathbf{B}^{-1}\mathbf{b} \\ \mathbf{B}^{-1}\mathbf{b} \end{pmatrix}$$

The general simplex tableau in matrix form can be derived from the original standard equations as follows:

$$\begin{pmatrix} 1 & \mathbf{C}_B\mathbf{B}^{-1} \\ \mathbf{0} & \mathbf{B}^{-1} \end{pmatrix}\begin{pmatrix} 1 & -\mathbf{C} \\ \mathbf{0} & \mathbf{A} \end{pmatrix}\begin{pmatrix} z \\ \mathbf{X} \end{pmatrix} = \begin{pmatrix} 1 & \mathbf{C}_B\mathbf{B}^{-1} \\ \mathbf{0} & \mathbf{B}^{-1} \end{pmatrix}\begin{pmatrix} 0 \\ \mathbf{b} \end{pmatrix}$$

Matrix manipulations yield the following equations:

$$\begin{pmatrix} 1 & \mathbf{C}_B\mathbf{B}^{-1}\mathbf{A} - \mathbf{C} \\ \mathbf{0} & \mathbf{B}^{-1}\mathbf{A} \end{pmatrix}\begin{pmatrix} z \\ \mathbf{X} \end{pmatrix} = \begin{pmatrix} \mathbf{C}_B\mathbf{B}^{-1}\mathbf{b} \\ \mathbf{B}^{-1}\mathbf{b} \end{pmatrix}$$

Given $\mathbf{P}_j$ is the jth vector of $\mathbf{A}$, the simplex tableau column associated with variable x_j can be represented as:

Basic	x_j	Solution
z	$\mathbf{C}_B\mathbf{B}^{-1}\mathbf{P}_j - \mathbf{c}_j$	$\mathbf{C}_B\mathbf{B}^{-1}\mathbf{b}$
$\mathbf{X}_B$	$\mathbf{B}^{-1}\mathbf{P}_j$	$\mathbf{B}^{-1}\mathbf{b}$

In fact, the tableau above is the same as the one we presented in Chapter 3 (see Problem 5 of Set 7.1c) and that of the primal-dual computations in Section 4.2.4. An important property of this table is that the inverse, $\mathbf{B}^{-1}$, is the only element that changes from one tableau to the next, and that the *entire* tableau can be generated once $\mathbf{B}^{-1}$ is known. This point is important, because the computational roundoff error in any tableau can controlled by controlling the accuracy of $\mathbf{B}^{-1}$. This result is the basis for the development of the revised simplex method in Section 7.2.

Example 7.1-3

Consider the following LP:

$$\text{Maximize } z = x_1 + 4x_2 + 7x_3 + 5x_4$$

subject to

$$2x_1 + x_2 + 2x_3 + 4x_4 = 10$$

$$3x_1 - x_2 - 2x_3 + 6x_4 = 5$$

$$x_1, x_2, x_3, x_4 \geq 0$$

Generate the simplex tableau associated with the basis $\mathbf{B} = (\mathbf{P}_1, \mathbf{P}_2)$.

Given $\mathbf{B} = (\mathbf{P}_1, \mathbf{P}_2)$, then $\mathbf{X}_B = (x_1, x_2)^T$ and $\mathbf{C}_B = (1, 4)$. Thus,

$$\mathbf{B}^{-1} = \begin{pmatrix} 2 & 1 \\ 3 & -1 \end{pmatrix}^{-1} = \begin{pmatrix} \frac{1}{5} & \frac{1}{5} \\ \frac{3}{5} & -\frac{2}{5} \end{pmatrix}$$

We then get

$$\mathbf{X}_B = \begin{pmatrix} x_1 \\ x_2 \end{pmatrix} = \mathbf{B}^{-1}\mathbf{b} = \begin{pmatrix} \frac{1}{5} & \frac{1}{5} \\ \frac{3}{5} & -\frac{2}{5} \end{pmatrix}\begin{pmatrix} 10 \\ 5 \end{pmatrix} = \begin{pmatrix} 3 \\ 4 \end{pmatrix}$$

To compute the constraint columns in the body of the tableau, we have

$$\mathbf{B}^{-1}(\mathbf{P}_1, \mathbf{P}_2, \mathbf{P}_3, \mathbf{P}_4) = \begin{pmatrix} \frac{1}{5} & \frac{1}{5} \\ \frac{3}{5} & -\frac{2}{5} \end{pmatrix}\begin{pmatrix} 2 & 1 & 2 & 4 \\ 3 & -1 & -2 & 6 \end{pmatrix} = \begin{pmatrix} 1 & 0 & 0 & 2 \\ 0 & 1 & 2 & 0 \end{pmatrix}$$

Next, we compute the objective row as follows:

$$\mathbf{C}_B(\mathbf{B}^{-1}(\mathbf{P}_1, \mathbf{P}_2, \mathbf{P}_2, \mathbf{P}_4)) - \mathbf{C} = (1, 4)\begin{pmatrix} 1 & 0 & 0 & 2 \\ 0 & 1 & 2 & 0 \end{pmatrix} - (1, 4, 7, 5) = (0, 0, 1, -3)$$

Finally, we compute the value of the objective function as follows:

$$z = \mathbf{C}_B\mathbf{B}^{-1}\mathbf{b} = \mathbf{C}_B\mathbf{X}_B = (1, 4)\begin{pmatrix} 3 \\ 4 \end{pmatrix} = 19$$

Thus, the entire tableau can be summarized as shown below.

Basic	x_1	x_2	x_3	x_4	Solution
z	0	0	1	-3	19
x_1	1	0	0	2	3
x_2	0	1	2	0	4

The main conclusion from this example is that once the inverse, $\mathbf{B}^{-1}$, is known, the entire simplex tableau can be generated from $\mathbf{B}^{-1}$ and the *original* data of the problem.

PROBLEM SET 7.1C

***1.** In Example 7.1-3, consider $\mathbf{B} = (\mathbf{P}_3, \mathbf{P}_4)$. Show that the corresponding basic solution is feasible, then generate the corresponding simplex tableau.

2. Consider the following LP:

$$\text{Maximize } z = 5x_1 + 12x_2 + 4x_3$$

subject to

$$\begin{aligned} x_1 + 2x_2 + x_3 + x_4 &= 10 \\ 2x_1 - 2x_2 - x_3 \quad\quad &= 2 \\ x_1, x_2, x_3, x_4 &\geq 0 \end{aligned}$$

Check if each of the following matrices forms a (feasible or infeasible) basis: $(\mathbf{P}_1, \mathbf{P}_2)$, $(\mathbf{P}_2, \mathbf{P}_3)$, $(\mathbf{P}_3, \mathbf{P}_4)$.

3. In the following LP, compute the entire simplex tableau associated with $\mathbf{X}_B = (x_1, x_2, x_5)^T$.

$$\text{Minimize } z = 2x_1 + x_2$$

subject to

$$\begin{aligned} 3x_1 + x_2 - x_3 \qquad\qquad &= 3 \\ 4x_1 + 3x_2 \qquad - x_4 \qquad &= 6 \\ x_1 + 2x_2 \qquad\qquad + x_5 &= 3 \\ x_1, x_2, x_3, x_4, x_5 \geq 0 \end{aligned}$$

*4. The following is an optimal LP tableau:

Basic	x_1	x_2	x_3	x_4	x_5	Solution
z	0	0	0	3	2	?
x_3	0	0	1	1	−1	2
x_2	0	1	0	1	0	6
x_1	1	0	0	−1	1	2

The variables x_3, x_4, and x_5 are slacks in the original problem. Use matrix manipulations to reconstruct the original LP, and then compute the optimum value.

5. In the generalized simplex tableau, suppose that the $\mathbf{X} = (\mathbf{X}_{\text{I}}, \mathbf{X}_{\text{II}})^T$, where $\mathbf{X}_{\text{II}}$ corresponds to a typical *starting* basic solution (consisting of slack and/or artificial variables) with $\mathbf{B} = \mathbf{I}$, and let $\mathbf{C} = (\mathbf{C}_{\text{I}}, \mathbf{C}_{\text{II}})$ and $\mathbf{A} = (\mathbf{D}, \mathbf{I})$ be the corresponding partitions of $\mathbf{C}$ and $\mathbf{A}$, respectively. Show that the matrix form of the simplex tableau reduces to the following form, which is exactly the form used in Chapter 3.

Basic	$\mathbf{X}_{\text{I}}$	$\mathbf{X}_{\text{II}}$	Solution
z	$\mathbf{C}_B\mathbf{B}^{-1}\mathbf{D} - \mathbf{C}_{\text{I}}$	$\mathbf{C}_B\mathbf{B}^{-1} - \mathbf{C}_{\text{II}}$	$\mathbf{C}_B\mathbf{B}^{-1}\mathbf{b}$
$\mathbf{X}_B$	$\mathbf{B}^{-1}\mathbf{D}$	$\mathbf{B}^{-1}$	$\mathbf{B}^{-1}\mathbf{b}$

7.2 REVISED SIMPLEX METHOD

Section 7.1.1 shows that the optimum solution of a linear program is always associated with a basic (feasible) solution. The simplex method search for the optimum starts by selecting a feasible basis, $\mathbf{B}$, and then moving to another basis, $\mathbf{B}_{\text{next}}$, that yields a better (or, at least, no worse) value of the objective function. Continuing in this manner, the optimum basis is eventually reached.

The iterative steps of the *revised* simplex method are exactly the same as in the *tableau* simplex method presented in Chapter 3. The main difference is that the computations in the revised method are based on matrix manipulations rather than on row operations. The use of matrix algebra reduces the adverse effect of machine roundoff error by controlling the accuracy of computing $\mathbf{B}^{-1}$. This result follows because, as Section 7.1.2 shows, the entire simplex tableau can be computed from the *original* data and the current $\mathbf{B}^{-1}$. In the tableau simplex method of Chapter 3, each tableau is generated from the immediately preceding one, which tends to worsen the problem of rounoff error.

7.2.1 Development of the Optimality and Feasibility Conditions

The general LP problem can be written as follows:

$$\text{Maximize or minimize } z = \sum_{j=1}^{n} c_j x_j \text{ subject to } \sum_{j=1}^{n} \mathbf{P}_j x_j = \mathbf{b},\ x_j \geq 0,\ j = 1, 2, \ldots, n$$

For a given basic vector $\mathbf{X}_B$ and its corresponding basis $\mathbf{B}$ and objective vector $\mathbf{C}_\mathrm{B}$, the general simplex tableau developed in Section 7.1.2 shows that any simplex iteration can be represented by the following equations:

$$z + \sum_{j=1}^{n} (z_j - c_j) x_j = \mathbf{C}_B \mathbf{B}^{-1} \mathbf{b}$$

$$(\mathbf{X}_B)_i + \sum_{j=1}^{n} (\mathbf{B}^{-1}\mathbf{P}_j)_i \, x_j = (\mathbf{B}^{-1}\mathbf{b})_i$$

$z_j - c_j$, the reduced cost of x_j (see Section 4.3.2), is defined as

$$z_j - c_j = \mathbf{C}_B \mathbf{B}^{-1} \mathbf{P}_j - c_j$$

The notation $(\mathbf{V})_i$ is used to represent the ith element of the vector $\mathbf{V}$.

Optimality Condition. From the z-equation given above, an increase in nonbasic x_j above its current zero value will improve the value of z relative to its current value $(= \mathbf{C}_B \mathbf{B}^{-1} \mathbf{b})$ only if its $z_j - c_j$ is strictly negative in the case of maximization and strictly positive in the case of minimization. Otherwise, x_j cannot improve the solution and must remain nonbasic at zero level. Though any nonbasic variable satisfying the given condition can be chosen to improve the solution, the simplex method uses a rule of thumb that calls for selecting the **entering variable** as the one with the *most* negative (*most* positive) $z_j - c_j$ in case of maximization (minimization).

Feasibility Condition. The determination of the **leaving vector** is based on examining the constraint equation associated with the ith *basic* variable. Specifically, we have

$$(\mathbf{X}_B)_i + \sum_{j=1}^{n} (\mathbf{B}^{-1}\mathbf{P}_j)_i \, x_j = (\mathbf{B}^{-1}\mathbf{b})_i$$

When the vector $\mathbf{P}_j$ is selected by the optimality condition to enter the basis, its associated variable x_j will increase above zero level. At the same time, all the remaining nonbasic variables remain at zero level. Thus, the ith constraint equation reduces to

$$(\mathbf{X}_B)_i = (\mathbf{B}^{-1}\mathbf{b})_i - (\mathbf{B}^{-1}\mathbf{P}_j)_i \, x_j$$

The equation shows that if $(\mathbf{B}^{-1}\mathbf{P}_j)_i > 0$, an increase in x_j can cause $(\mathbf{X}_B)_i$ to become negative, which violates the nonnegativity condition, $(\mathbf{X}_B)_i \geq \mathbf{0}$ for all i. Thus, we have

$$(\mathbf{B}^{-1}\mathbf{b})_i - (\mathbf{B}^{-1}\mathbf{P}_j)_i \, x_j \geq 0, \quad \text{for all } i$$

This condition yields the maximum value of the entering variable x_j as

$$x_j = \min_i \left\{ \frac{(\mathbf{B}^{-1}\mathbf{b})_i}{(\mathbf{B}^{-1}\mathbf{P}_j)_i} \middle| (\mathbf{B}^{-1}\mathbf{P}_j)_i > 0 \right\}$$

The basic variable responsible for producing the minimum ratio leaves the basic solution to become nonbasic at zero level.

PROBLEM SET 7.2A

*1. Consider the following LP:

$$\text{Maximize } z = c_1x_1 + c_2x_2 + c_3x_3 + c_4x_4$$

subject to

$$\mathbf{P}_1x_1 + \mathbf{P}_2x_2 + \mathbf{P}_3x_3 + \mathbf{P}_4x_4 = \mathbf{b}$$

$$x_1, x_2, x_3, x_4 \geq 0$$

The vectors $\mathbf{P}_1$, $\mathbf{P}_2$, $\mathbf{P}_3$, and $\mathbf{P}_4$ are shown in Figure 7.4. Assume that the basis $\mathbf{B}$ of the current iteration is comprised of $\mathbf{P}_1$ and $\mathbf{P}_2$.

(a) If the vector $\mathbf{P}_3$ enters the basis, which of the current two basic vectors must leave in order for the resulting basic solution to be feasible?

(b) Can the vector $\mathbf{P}_4$ be part of a feasible basis?

*2. Prove that, in any simplex iteration, $z_j - c_j = 0$ for all the associated *basic* variables.

3. Prove that if $z_j - c_j > 0\ (< 0)$ for all the nonbasic variables x_j of a maximization (minimization) LP problem, then the optimum is unique. Else, if $z_j - c_j$ equals zero for a nonbasic x_j, then the problem has an alternative optimum solution.

4. In an all-slack starting basic solution, show using the matrix form of the tableau that the mechanical procedure used in Section 3.3 in which the objective equation is set as

$$z - \sum_{j=1}^{n} c_jx_j = 0$$

automatically computes the proper $z_j - c_j$ for all the variables in the starting tableau.

5. Using the matrix form of the simplex tableau, show that in an all-artificial starting basic solution, the procedure employed in Section 3.4.1 that calls for substituting out

FIGURE 7.4

Vector representation of Problem 1, Set 7.2a

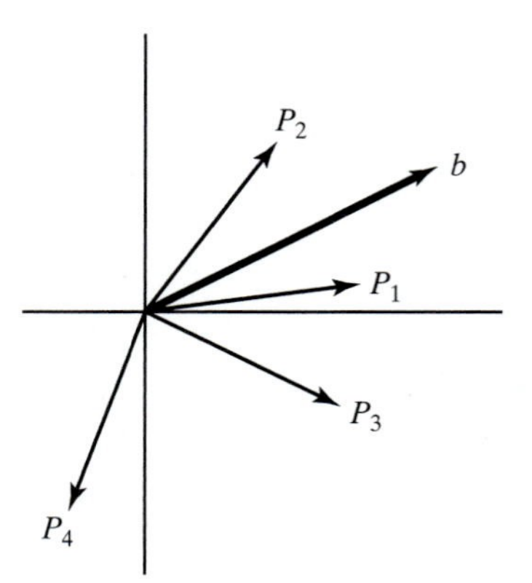

the artificial variables in the objective function (using the constraint equations) actually computes the proper $z_j - c_j$ for all the variables in the starting tableau.

6. Consider an LP in which the variable x_k is unrestricted in sign. Prove that by substituting $x_k = x_k^- - x_k^+$, where x_k^- and x_k^+ are nonnegative, it is impossible that the two variables will replace one another in an alternative optimum solution.

***7.** Given the general LP in equation form with m equations and n unknowns, determine the maximum number of *adjacent* extreme points that can be reached from a nondegenerate extreme point (all basic variable are >0) of the solution space.

8. In applying the feasibility condition of the simplex method, suppose that $x_r = 0$ is a basic variable and that x_j is the entering variable with $(\mathbf{B}^{-1}\mathbf{P}_j)_r \neq 0$. Prove that the resulting basic solution remains feasible even if $(\mathbf{B}^{-1}\mathbf{P}_j)_r$ is negative.

9. In the implementation of the feasibility condition of the simplex method, what are the conditions for encountering a degenerate solution (at least one basic variable $= 0$) for the first time? For continuing to obtain a degenerate solution in the next iteration? For removing degeneracy in the next iteration? Explain the answers mathematically.

***10.** What are the relationships between extreme points and basic solutions under degeneracy and nondegeneracy? What is the maximum number of iterations that can be performed at a given extreme point assuming no cycling?

***11.** Consider the LP, maximize $z = \mathbf{CX}$ subject to $\mathbf{AX} \leq \mathbf{b}, \mathbf{X} \geq \mathbf{0}$, where $\mathbf{b} \geq \mathbf{0}$. Suppose that the entering vector $\mathbf{P}_j$ is such that at least one element of $\mathbf{B}^{-1}\mathbf{P}_j$ is positive.

(a) If $\mathbf{P}_j$ is replaced with $\alpha\mathbf{P}_j$, where α is a positive scalar, and provided x_j remains the entering variable, find the relationship between the values of x_j corresponding to $\mathbf{P}_j$ and $\alpha\mathbf{P}_j$.

(b) Answer Part (a) if, additionally, $\mathbf{b}$ is replaced with $\beta\mathbf{b}$, where β is a positive scalar.

12. Consider the LP

$$\text{Maximize } z = \mathbf{CX} \text{ subject to } \mathbf{AX} \leq \mathbf{b}, \mathbf{X} \geq \mathbf{0}, \text{ where } \mathbf{b} \geq \mathbf{0}$$

After obtaining the optimum solution, it is suggested that a nonbasic variable x_j can be made basic (profitable) by reducing the (resource) requirements per unit of x_j for the different resources to $\frac{1}{\alpha}$ of their original values, $\alpha > 1$. Since the requirements per unit are reduced, it is expected that the profit per unit of x_j will also be reduced to $\frac{1}{\alpha}$ of its original value. Will these changes make x_j a profitable variable? Explain mathematically.

13. Consider the LP

$$\text{Maximize } z = \mathbf{CX} \text{ subject to } (\mathbf{A}, \mathbf{I})\mathbf{X} = \mathbf{b}, \mathbf{X} \geq \mathbf{0}$$

Define $\mathbf{X}_B$ as the current basic vector with $\mathbf{B}$ as its associated basis and $\mathbf{C}_B$ as its vector of objective coefficients. Show that if $\mathbf{C}_B$ is replaced with the new coefficients $\mathbf{D}_B$, the values of $z_j - c_j$ for the basic vector $\mathbf{X}_B$ will remain equal to zero. What is the significance of this result?

7.2.2 Revised Simplex Algorithm

Having developed the optimality and feasibility conditions in Section 7.2.1, we now present the computational steps of the revised simplex method.

Step 0. Construct a starting basic feasible solution and let $\mathbf{B}$ and $\mathbf{C}_B$ be its associated basis and objective coefficients vector, respectively.

Step 1. Compute the inverse $\mathbf{B}^{-1}$ by using an appropriate inversion method.[1]

Step 2. For each *nonbasic* variable x_j, compute

$$z_j - c_j = \mathbf{C}_B\mathbf{B}^{-1}\mathbf{P}_j - c_j$$

If $z_j - c_j \geq 0$ in maximization (≤ 0 in minimization) for all nonbasic x_j, stop; the optimal solution is given by

$$\mathbf{X}_B = \mathbf{B}^{-1}\mathbf{b},\ z = \mathbf{C}_B\mathbf{X}_B$$

Else, apply the optimality condition and determine the *entering* variable x_j as the nonbasic variable with the most negative (positive) $z_j - c_j$ in case of mazimization (minimization).

Step 3. Compute $\mathbf{B}^{-1}\mathbf{P}_j$. If all the elements of $\mathbf{B}^{-1}\mathbf{P}_j$ are negative or zero, stop; the problem has no bounded solution. Else, compute $\mathbf{B}^{-1}\mathbf{b}$. Then for all the *strictly positive* elements of $\mathbf{B}^{-1}\mathbf{P}_j$, determine the ratios defined by the feasibility condition. The basic variable x_i associated with the smallest ratio is the *leaving* variable.

Step 4. From the current basis $\mathbf{B}$, form a new basis by replacing the leaving vector $\mathbf{P}_i$ with the entering vector $\mathbf{P}_j$, Go to step 1 to start a new iteration.

Example 7.2-1

The Reddy Mikks model (Section 2.1) is solved by the revised simplex algorithm. The same model was solved by the tableau method in Section 3.3.2. A comparison between the two methods will show that they are one and the same.

The equation form of the Reddy Mikks model can be expressed in matrix form as

$$\text{maximize } z = (5, 4, 0, 0, 0, 0)(x_1, x_2, x_3, x_4, x_5, x_6)^T$$

subject to

$$\begin{pmatrix} 6 & 4 & 1 & 0 & 0 & 0 \\ 1 & 2 & 0 & 1 & 0 & 0 \\ -1 & 1 & 0 & 0 & 1 & 0 \\ 0 & 1 & 0 & 0 & 0 & 1 \end{pmatrix} \begin{pmatrix} x_1 \\ x_2 \\ x_3 \\ x_4 \\ x_5 \\ x_6 \end{pmatrix} = \begin{pmatrix} 24 \\ 6 \\ 1 \\ 2 \end{pmatrix}$$

$$x_1, x_2, \ldots, x_6 \geq 0$$

We use the notation $\mathbf{C} = (c_1, c_2, \ldots, c_6)$ to represent the objective-function coefficients and $(\mathbf{P}_1, \mathbf{P}_2, \ldots, \mathbf{P}_6)$ to represent the columns vectors of the constraint equations. The right-hand side of the constraints gives the vector $\mathbf{b}$.

---

[1] In most LP presentations, including the first six editions of this book, the *product form* method for inverting a basis (see Section D.2.7) is integrated into the revised simplex algorithm because the *product form* lends itself readily to the revised simplex computations, where successive bases differ in exactly one column. This detail is removed from this presentation because it makes the algorithm appear more complex than it really is. Moreover, the *product form* is rarely used in the development of LP codes because it is not designed for automatic computations, where machine round-off error can be a serious issue. Normally, some advanced numeric analysis method, such as the *LU decomposition* method, is used to obtain the inverse. (Incidentally, TORA matrix inversion is based on LU decomposition.)

In the computations below, we will give the algebraic formula for each step and its final numeric answer without detailing the arithmetic operations. You will find it instructive to fill in the gaps in each step.

Iteration 0

$$\mathbf{X}_{B_0} = (x_3, x_4, x_5, x_6), \mathbf{C}_{B_0} = (0, 0, 0, 0)$$
$$\mathbf{B}_0 = (\mathbf{P}_3, \mathbf{P}_4, \mathbf{P}_5, \mathbf{P}_6) = \mathbf{I}, \mathbf{B}_0^{-1} = \mathbf{I}$$

Thus,

$$\mathbf{X}_{B_0} = \mathbf{B}_0^{-1}\mathbf{b} = (24, 6, 1, 2)^T, z = \mathbf{C}_{B_0}\mathbf{X}_{B_0} = 0$$

Optimality computations:

$$\mathbf{C}_{B_0}\mathbf{B}_0^{-1} = (0, 0, 0, 0)$$
$$\{z_j - c_j\}_{j=1,2} = \mathbf{C}_{B_0}\mathbf{B}_0^{-1}(\mathbf{P}_1, \mathbf{P}_2) - (c_1, c_2) = (-5, -4)$$

Thus, $\mathbf{P}_1$ is the entering vector.

Feasibility computations:

$$\mathbf{X}_{B_0} = (x_3, x_4, x_5, x_6)^T = (24, 6, 1, 2)^T$$
$$\mathbf{B}_0^{-1}\mathbf{P}_1 = (6, 1, -1, 0)^T$$

Hence,

$$x_1 = \min\left\{\frac{24}{6}, \frac{6}{1}, -, -\right\} = \min\{\mathbf{4}, 6, -, -\} = 4$$

and $\mathbf{P}_3$ becomes the leaving vector.

The results above can be summarized in the familiar simplex tableau format. The presentation should help convince you that the two methods are essentially the same. You will find it instructive to develop similar tableaus in the succeeding iterations.

Basic	x_1	x_2	x_3	x_4	x_5	x_6	Solution
z	-5	-4	0	0	0	0	0
x_3	6						24
x_4	1						6
x_5	-1						1
x_6	0						2

Iteration 1

$$\mathbf{X}_{B_1} = (x_1, x_4, x_5, x_6), \mathbf{C}_{B_1} = (5, 0, 0, 0)$$
$$\mathbf{B}_1 = (\mathbf{P}_1, \mathbf{P}_4, \mathbf{P}_5, \mathbf{P}_6)$$
$$= \begin{pmatrix} 6 & 0 & 0 & 0 \\ 1 & 1 & 0 & 0 \\ -1 & 0 & 1 & 0 \\ 0 & 0 & 0 & 1 \end{pmatrix}$$

By using an appropriate inversion method (see Section D.2.7, in particular the *product form* method), the inverse is given as

$$\mathbf{B}_1^{-1} = \begin{pmatrix} \frac{1}{6} & 0 & 0 & 0 \\ -\frac{1}{6} & 1 & 0 & 0 \\ \frac{1}{6} & 0 & 1 & 0 \\ 0 & 0 & 0 & 1 \end{pmatrix}$$

Thus,

$$\mathbf{X}_{B_1} = \mathbf{B}_1^{-1}\mathbf{b} = (4, 2, 5, 2)^T, z = \mathbf{C}_{B_1}\mathbf{X}_{B_1} = 20$$

Optimality computations:

$$\mathbf{C}_{B_1}\mathbf{B}_1^{-1} = \left(\tfrac{5}{6}, 0, 0, 0\right)$$

$$\{z_j - c_j\}_{j=2,3} = \mathbf{C}_{B_1}\mathbf{B}_1^{-1}(\mathbf{P}_2, \mathbf{P}_3) - (c_2, c_3) = \left(-\tfrac{2}{3}, \tfrac{5}{6}\right)$$

Thus, $\mathbf{P}_2$ is the entering vector.

Feasibility computations:

$$\mathbf{X}_{B_1} = (x_1, x_4, x_5, x_6)^T = (4, 2, 5, 2)^T$$

$$\mathbf{B}_1^{-1}\mathbf{P}_2 = \left(\tfrac{2}{3}, \tfrac{4}{3}, \tfrac{5}{3}, 1\right)^T$$

Hence,

$$x_2 = \min\left\{\frac{4}{\frac{2}{3}}, \frac{2}{\frac{4}{3}}, \frac{5}{\frac{5}{3}}, \frac{2}{1}\right\} = \min\left\{6, \tfrac{3}{2}, 3, 2\right\} = \tfrac{3}{2}$$

and $\mathbf{P}_4$ becomes the leaving vector. (You will find it helpful to summarize the results above in the simplex tableau format as we did in iteration 0.)

Iteration 2

$$\mathbf{X}_{B_2} = (x_1, x_2, x_5, x_6)^T, \mathbf{C}_{B_2} = (5, 4, 0, 0)$$

$$\mathbf{B}_2 = (\mathbf{P}_1, \mathbf{P}_2, \mathbf{P}_5, \mathbf{P}_6)$$

$$= \begin{pmatrix} 6 & 4 & 0 & 0 \\ 1 & 2 & 0 & 0 \\ -1 & 1 & 1 & 0 \\ 0 & 1 & 0 & 1 \end{pmatrix}$$

Hence,

$$\mathbf{B}_2^{-1} = \begin{pmatrix} \frac{1}{4} & -\frac{1}{2} & 0 & 0 \\ -\frac{1}{8} & \frac{3}{4} & 0 & 0 \\ \frac{3}{8} & -\frac{5}{4} & 1 & 0 \\ \frac{1}{8} & -\frac{3}{4} & 0 & 1 \end{pmatrix}$$

Thus,

$$\mathbf{X}_{B_2} = \mathbf{B}_2^{-1}\mathbf{b} = \left(3, \tfrac{3}{2}, \tfrac{5}{2}, \tfrac{1}{2}\right)^T, z = \mathbf{C}_{B_2}\mathbf{X}_{B_2} = 21$$

Optimality computations:

$$\mathbf{C}_{B_2}\mathbf{B}_2^{-1} = \left(\tfrac{3}{4}, \tfrac{1}{2}, 0, 0\right)$$

$$\{z_j - c_j\}_{j=3,4} = \mathbf{C}_{B_2}\mathbf{B}_2^{-1}(\mathbf{P}_3, \mathbf{P}_4) - (c_3, c_4) = \left(\tfrac{3}{4}, \tfrac{1}{2}\right)$$

Thus, $\mathbf{X}_{B_2}$ is optimal and the computations end.

Summary of optimal solution:

$$x_1 = 3, x_2 = 1.5, z = 21$$

PROBLEM SET 7.2B

1. In Example 7.2-1, summarize the data of iteration 1 in the tableau format of Section 3.3.
2. Solve the following LPs by the revised simplex method:

(a) Maximize $z = 6x_1 - 2x_2 + 3x_3$
subject to

$$\begin{aligned} 2x_1 - x_2 + 2x_3 &\leq 2 \\ x_1 \quad\quad + 4x_3 &\leq 4 \\ x_1, x_2, x_3 &\geq 0 \end{aligned}$$

*__(b)__ Maximize $z = 2x_1 + x_2 + 2x_3$
subject to

$$\begin{aligned} 4x_1 + 3x_2 + 8x_3 &\leq 12 \\ 4x_1 + x_2 + 12x_3 &\leq 8 \\ 4x_1 - x_2 + 3x_3 &\leq 8 \\ x_1, x_2, x_3 &\geq 0 \end{aligned}$$

(c) Minimize $z = 2x_1 + x_2$
subject to

$$\begin{aligned} 3x_1 + x_2 &= 3 \\ 4x_1 + 3x_2 &\geq 6 \\ x_1 + 2x_2 &\leq 3 \\ x_1, x_2 &\geq 0 \end{aligned}$$

(d) Minimize $z = 5x_1 - 4x_2 + 6x_3 + 8x_4$

subject to

$$\begin{aligned} x_1 + 7x_2 + 3x_3 + 7x_4 &\leq 46 \\ 3x_1 - x_2 + x_3 + 2x_4 &\leq 20 \\ 2x_1 + 3x_2 - x_3 + x_4 &\geq 18 \\ x_1, x_2, x_3, x_4 &\geq 0 \end{aligned}$$

3. Solve the following LP by the revised simplex method given the starting basic feasible vector $\mathbf{X}_{B_0} = (x_2, x_4, x_5)^T$.

$$\text{Minimize } z = 7x_2 + 11x_3 - 10x_4 + 26x_6$$

subject to

$$\begin{aligned} x_2 - x_3 + x_5 + x_6 &= 6 \\ x_2 - x_3 + x_4 + 3x_6 &= 8 \\ x_1 + x_2 - 3x_3 + x_4 + x_5 &= 12 \\ x_1, x_2, x_3, x_4, x_5, x_6 &\geq 0 \end{aligned}$$

4. Solve the following using the two-phase revised simplex method:

(a) Problem 2-c.

(b) Problem 2-d.

(c) Problem 3 (ignore the given starting $\mathbf{X}_{B_0}$).

5. *Revised Dual Simplex Method.* The steps of the revised dual simplex method (using matrix manipulations) can be summarized as follows:

Step 0. Let $\mathbf{B}_0 = \mathbf{I}$ be the starting basis for which at least one of the elements of $\mathbf{X}_{B_0}$ is negative (infeasible).

Step 1. Compute $\mathbf{X}_B = \mathbf{B}^{-1}\mathbf{b}$, the current values of the basic variables. Select the leaving variable x_r as the one having the most negative value. If all the elements of $\mathbf{X}_B$ are nonnegative, stop; the current solution is feasible (and optimal).

Step 2. **(a)** Compute $z_j - c_j = \mathbf{C}_B\mathbf{B}^{-1}\mathbf{P}_j - c_j$ for all the nonbasic variables x_j.

(b) For all the nonbasic variables x_j, compute the constraint coefficients $(\mathbf{B}^{-1}\mathbf{P}_j)_r$ associated with the row of the leaving variable x_r.

(c) The entering variable is associated with

$$\theta = \min_i \left\{ \left| \frac{z_j - c_j}{(\mathbf{B}^{-1}\mathbf{P}_j)_r} \right|, (\mathbf{B}^{-1}\mathbf{P}_j)_r < 0 \right\}$$

If all $(\mathbf{B}^{-1}\mathbf{P}_j)_r \geq 0$, no feasible solution exists.

Step 3. Obtain the new basis by interchanging the entering and leaving vectors ($\mathbf{P}_j$ and $\mathbf{P}_r$). Compute the new inverse and go to step 1.

Apply the method to the following problem:

$$\text{Minimize } z = 3x_1 + 2x_2$$

subject to

$$\begin{aligned} 3x_1 + x_2 &\geq 3 \\ 4x_1 + 3x_2 &\geq 6 \\ x_1 + 2x_2 &\leq 3 \\ x_1, x_2 &\geq 0 \end{aligned}$$

7.3 BOUNDED-VARIABLES ALGORITHM

In LP models, variables may have explicit positive upper and lower bounds. For example, in production facilities, lower and upper bounds can represent the minimum and maximum demands for certain products. Bounded variables also arise prominently in the course of solving integer programming problems by the branch-and-bound algorithm (see Section 9.3.1).

The bounded algorithm is efficient computationally because it accounts for the bounds *implicitly*. We consider the lower bounds first because it is simpler. Given $\mathbf{X} \geq \mathbf{L}$, we can use the substitution

$$\mathbf{X} = \mathbf{L} + \mathbf{X}', \qquad \mathbf{X}' \geq \mathbf{0}$$

throughout and solve the problem in terms of $\mathbf{X}'$ (whose lower bound now equals zero). The original $\mathbf{X}$ is determined by back-substitution, which is legitimate because it guarantees that $\mathbf{X} = \mathbf{X}' + \mathbf{L}$ will remain nonnegative for all $\mathbf{X}' \geq \mathbf{0}$.

Next, consider the upper bounding constraints, $\mathbf{X} \leq \mathbf{U}$. The idea of direct substitution (i.e., $\mathbf{X} = \mathbf{U} - \mathbf{X}''$, $\mathbf{X}'' \geq \mathbf{0}$) is not correct because back-substitution, $\mathbf{X} = \mathbf{U} - \mathbf{X}''$, does not ensure that $\mathbf{X}$ will remain nonnegative. A different procedure is thus needed.

Define the upper bounded LP model as

$$\text{Maximize } z = \{\mathbf{CX} | (\mathbf{A}, \mathbf{I})\mathbf{X} = \mathbf{b}, \mathbf{0} \leq \mathbf{X} \leq \mathbf{U}\}$$

The bounded algorithm uses only the constraints $(\mathbf{A}, \mathbf{I})\mathbf{X} = \mathbf{b}$, $\mathbf{X} \geq \mathbf{0}$, while accounting for $\mathbf{X} \leq \mathbf{U}$ implicitly by modifying the simplex feasibility condition.

Let $\mathbf{X}_B = \mathbf{B}^{-1}\mathbf{b}$ be a current basic feasible solution of $(\mathbf{A}, \mathbf{I})\mathbf{X} = \mathbf{b}$, $\mathbf{X} \geq \mathbf{0}$, and suppose that, according to the (regular) optimality condition, $\mathbf{P}_j$ is the entering vector. Then, *given that all the nonbasic variables are zero,* the constraint equation of the ith basic variable can be written as

$$(\mathbf{X}_B)_i = (\mathbf{B}^{-1}\mathbf{b})_i - (\mathbf{B}^{-1}\mathbf{P}_j)_i \, x_j$$

When the entering variable x_j increases above zero level, $(\mathbf{X}_B)_i$ will *increase or decrease* depending on whether $(\mathbf{B}^{-1}\mathbf{P}_j)_i$ is negative or positive, respectively. Thus, in determining the value of the entering variable x_j, three conditions must be satisfied.

1. The basic variable $(\mathbf{X}_B)_i$ remains nonnegative—that is, $(\mathbf{X}_B)_i \geq 0$.
2. The basic variable $(\mathbf{X}_B)_i$ does not exceed its upper bound—that is, $(\mathbf{X}_B)_i \leq (\mathbf{U}_B)_i$, where $\mathbf{U}_B$ comprises the ordered elements of $\mathbf{U}$ corresponding to $\mathbf{X}_B$.
3. The entering variable x_j cannot assume a value larger than its upper bound—that is, $x_j \leq u_j$, where u_j is the jth element of $\mathbf{U}$.

The first condition $(\mathbf{X}_B)_i \geq 0$ requires that

$$(\mathbf{B}^{-1}\mathbf{b})_i - (\mathbf{B}^{-1}\mathbf{P}_j)_i \, x_j \geq 0$$

It is satisfied if

$$x_j \leq \theta_1 = \min_i \left\{ \frac{(\mathbf{B}^{-1}\mathbf{b})_i}{(\mathbf{B}^{-1}\mathbf{P}_j)_i} \middle| (\mathbf{B}^{-1}\mathbf{P}_j)_i > 0 \right\}$$

This condition is the same as the feasibility condition of the regular simplex method.

Next, the condition $(\mathbf{X}_B)_i \leq (\mathbf{U}_B)_i$ specifies that

$$(\mathbf{B}^{-1}\mathbf{b})_i - (\mathbf{B}^{-1}\mathbf{P}_j)_i\, x_j \leq (\mathbf{U}_B)_i$$

It is satisfied if

$$x_j \leq \theta_2 = \min_i \left\{ \frac{(\mathbf{B}^{-1}\mathbf{b})_i - (\mathbf{U}_B)_i}{(\mathbf{B}^{-1}\mathbf{P}_j)_i} \middle| (\mathbf{B}^{-1}\mathbf{P}_j)_i < 0 \right\}$$

Combining the three restrictions, x_j enters the solution at the level that satisfies all three conditions—that is,

$$x_j = \min\{\theta_1, \theta_2, u_j\}$$

The change of basis for the next iteration depends on whether x_j enters the solution at level θ_1, θ_2, or u_j. Assuming that $(\mathbf{X}_B)_r$ is the leaving variable, then we have the following rules:

1. $x_j = \theta_1$: $(\mathbf{X}_B)_r$ leaves the basic solution (becomes nonbasic) at level zero. The new iteration is generated using the normal simplex method with x_j and $(\mathbf{X}_B)_r$ as the entering and the leaving variables, respectively.
2. $x_j = \theta_2$: $(\mathbf{X}_B)_r$ becomes nonbasic *at its upper bound.* The new iteration is generated as in the case of $x_j = \theta_1$, with one modification that accounts for the fact that $(\mathbf{X}_B)_r$ will be nonbasic at *upper bound.* Because the values of θ_1 and θ_2 require *all nonbasic variables to be at zero level* (convince yourself that this is the case!), we must convert the new nonbasic $(\mathbf{X}_B)_r$ at upper bound to a nonbasic variable at zero level. This is achieved by using the substitution $(\mathbf{X}_B)_r = (\mathbf{U}_B)_r - (\mathbf{X}'_B)_r$, where $(\mathbf{X}'_B)_r \geq 0$. It is immaterial whether the substitution is made before or after the new basis is computed.
3. $x_j = u_j$: The basic vector $\mathbf{X}_B$ remains unchanged because $x_j = u_j$ stops short of forcing any of the current basic variables to reach its lower $(= 0)$ or upper bound. This means that x_j will remain nonbasic *but at upper bound.* Following the argument just presented, the new iteration is generated by using the substitution $x_j = u_j - x'_j$.

A tie among θ_1, θ_2, and u_j may be broken arbitrarily. However, it is preferable, where possible, to implement the rule for $x_j = u_j$ because it entails less computation.

The substitution $x_j = u_j - x'_j$ will change the original c_j, $\mathbf{P}_j$, and $\mathbf{b}$ to $c'_j = -c_j$, $\mathbf{P}'_j = -\mathbf{P}_j$, and $\mathbf{b}$ to $\mathbf{b}' = \mathbf{b} - u_j\mathbf{P}_j$. This means that if the revised simplex method is used, all the computations (e.g., $\mathbf{B}^{-1}$, $\mathbf{X}_B$, and $z_j - c_j$), should be based on the updated values of **C, A,** and **b** at each iteration (see Problem 5, Set 7.3a, for further details).

Example 7.3-1

Solve the following LP model by the upper-bounding algorithm.[2]

$$\text{Maximize } z = 3x_1 + 5y + 2x_3$$

---

[2]You can use TORA's Linear Programming ⇒ Solve problem ⇒ Algebraic ⇒ Iterations ⇒ Bounded simplex to produce the associated simplex iterations (file toraEx7.3-1.txt).

subject to

$$x_1 + \ \ y + 2x_3 \leq 14$$
$$2x_1 + 4y + 3x_3 \leq 43$$
$$0 \leq x_1 \leq 4, 7 \leq y \leq 10, 0 \leq x_3 \leq 3$$

The lower bound on y is accounted for using the substitution $y = x_2 + 7$, where $0 \leq x_2 \leq 10 - 7 = 3$.

To avoid being "sidetracked" by the computational details, we will not use the revised simplex method to carry out the computations. Instead, we will use the compact tableau form. Problems 5, 6, and 7, Set 7.3a address the revised version of the algorithm.

Iteration 0

Basic	x_1	x_2	x_3	x_4	x_5	Solution
z	-3	-5	-2	0	0	35
x_4	1	1	2	1	0	7
x_5	2	4	3	0	1	15

We have $\mathbf{B} = \mathbf{B}^{-1} = \mathbf{I}$ and $\mathbf{X}_B = (x_4, x_5)^T = \mathbf{B}^{-1}\mathbf{b} = (7, 15)^T$. Given that x_2 is the entering variable $(z_2 - c_2 = -5)$, we get

$$\mathbf{B}^{-1}\mathbf{P}_2 = (1, 4)^T$$

which yields

$$\theta_1 = \min\left\{\frac{7}{1}, \frac{15}{4}\right\} = 3.75, \text{ corresponding to } x_5$$

$$\theta_2 = \infty \text{(because all the elements of } \mathbf{B}^{-1}\mathbf{P}_2 > \mathbf{0})$$

Next, given the upper bound on the entering variable, $x_2 \leq 3$, it follows that

$$x_2 = \min\{3.75, \infty, 3\}$$
$$= 3 \ (= u_2)$$

Because x_2 enters at its upper bound $(= u_2 = 3)$, $\mathbf{X}_B$ remains unchanged, and x_2 becomes nonbasic *at its upper bound*. We use the substitution $x_2 = 3 - x_2'$ to obtain the new tableau as

Basic	x_1	x_2'	x_3	x_4	x_5	Solution
z	-3	5	-2	0	0	50
x_4	1	-1	2	1	0	4
x_5	2	-4	3	0	1	3

The substitution in effect changes the original right-hand side vector from $\mathbf{b} = (7, 15)^T$ to $\mathbf{b}' = (4, 3)^T$. This change should be considered in future computations.

Iteration 1. The entering variable is x_1. The basic vector $\mathbf{X}_B$ and $\mathbf{B}^{-1}$ $(= \mathbf{I})$ are the same as in iteration 0. Next,

$$\mathbf{B}^{-1}\mathbf{P}_1 = (1, 2)^T$$

$$\theta_1 = \min\left\{\frac{4}{1}, \frac{3}{2}\right\} = 1.5, \text{ corresponding to basic } x_5$$

$$\theta_2 = \infty (\text{because } \mathbf{B}^{-1}\mathbf{P}_1 > \mathbf{0})$$

Thus,

$$\begin{aligned} x_1 &= \min\{1.5, \infty, 4\} \\ &= 1.5\ (= \theta_1) \end{aligned}$$

Thus, the entering variable x_1 becomes basic, and the leaving variable x_5 becomes nonbasic at zero level, which yields

Basic	x_1	x_2'	x_3	x_4	x_5	Solution
z	0	-1	$\frac{5}{2}$	0	$\frac{3}{2}$	$\frac{109}{2}$
x_4	0	1	$\frac{1}{2}$	1	$-\frac{1}{2}$	$\frac{5}{2}$
x_1	1	-2	$\frac{3}{2}$	0	$\frac{1}{2}$	$\frac{3}{2}$

Iteration 2 The new inverse is

$$\mathbf{B}^{-1} = \begin{pmatrix} 1 & -\frac{1}{2} \\ 0 & \frac{1}{2} \end{pmatrix}$$

Now

$$\mathbf{X}_B = (x_4, x_1)^T = \mathbf{B}^{-1}\mathbf{b}' = \left(\tfrac{5}{2}, \tfrac{3}{2}\right)^T$$

where $\mathbf{b}' = (4, 3)^T$ as computed at the end of iteration 0. We select x_2' as the entering variable, and, noting that $\mathbf{P}_2' = -\mathbf{P}_2$, we get

$$\mathbf{B}^{-1}\mathbf{P}_2' = (1, -2)^T$$

Thus,

$$\theta_1 = \min\left\{\frac{\frac{5}{2}}{1}, -\right\} = 2.5, \text{ corresponding to basic } x_4$$

$$\theta_2 = \min\left\{-, \frac{\frac{3}{2} - 4}{-2}\right\} = 1.25, \text{ corresponding to basic } x_1$$

We then have

$$\begin{aligned} x_2' &= \min\{2.5, 1.25, 3\} \\ &= 1.25\ (= \theta_2) \end{aligned}$$

Because x_1 becomes nonbasic at its upper bound, we apply the substitution $x_1 = 4 - x_1'$ to obtain

Basic	x_1'	x_2'	x_3	x_4	x_5	Solution
z	0	-1	$\frac{5}{2}$	0	$\frac{3}{2}$	$\frac{109}{2}$
x_4	0	1	$\frac{1}{2}$	1	$-\frac{1}{2}$	$\frac{5}{2}$
x_1'	-1	-2	$\frac{3}{2}$	0	$\frac{1}{2}$	$-\frac{5}{2}$

Next, the entering variable x_2' becomes basic and the leaving variable x_1 becomes nonbasic at upper bound, which yields

Basic	x_1'	x_2'	x_3	x_4	x_5	Solution
z	$\frac{1}{2}$	0	$\frac{7}{4}$	0	$\frac{5}{4}$	$\frac{223}{4}$
x_4	$-\frac{1}{2}$	0	$\frac{5}{4}$	1	$-\frac{1}{4}$	$\frac{5}{4}$
x_2'	$\frac{1}{2}$	1	$-\frac{3}{4}$	0	$-\frac{1}{4}$	$\frac{5}{4}$

The last tableau is feasible and optimal. Note that the last two steps could have been reversed—meaning that we could first make x_2' basic and then apply the substitution $x_1 = 4 - x_1'$ (try it!). The sequence presented here involves less computation, however.

The optimal values of x_1, x_2, and x_3 are obtained by back-substitution as $x_1 = u_1 - x_1' = 4 - 0 = 4$, $x_2 = u_2 - x_2' = 3 - \frac{5}{4} = \frac{7}{4}$, and $x_3 = 0$. Finally, we get $y = l_2 + x_2 = 7 + \frac{7}{4} = \frac{35}{4}$. The associated optimal value of the objective function z is $\frac{223}{4}$.

PROBLEM SET 7.3A

1. Consider the following linear program:

Maximize $z = 2x_1 + x_2$

subject to

$$x_1 + x_2 \leq 3$$

$$0 \leq x_1 \leq 2, 0 \leq x_2 \leq 2$$

(a) Solve the problem graphically, and trace the sequence of extreme points leading to the optimal solution. (You may use TORA.)

(b) Solve the problem by the upper-bounding algorithm and show that the method produces the same sequence of extreme points as in the graphical optimal solution (you may use TORA to generate the iterations).

(c) How does the upper-bounding algorithm recognize the extreme points?

***2.** Solve the following problem by the bounded algorithm:

$$\text{Maximize } z = 6x_1 + 2x_2 + 8x_3 + 4x_4 + 2x_5 + 10x_6$$

subject to

$$8x_1 + x_2 + 8x_3 + 2x_4 + 2x_5 + 4x_6 \leq 13$$

$$0 \leq x_j \leq 1, j = 1, 2, \ldots, 6$$

3. Solve the following problems by the bounded algorithm:

(a) Minimize $z = 6x_1 - 2x_2 - 3x_3$
subject to

$$2x_1 + 4x_2 + 2x_3 \leq 8$$
$$x_1 - 2x_2 + 3x_3 \leq 7$$
$$0 \leq x_1 \leq 2, 0 \leq x_2 \leq 2, 0 \leq x_3 \leq 1$$

(b) Maximize $z = 3x_1 + 5x_2 + 2x_3$
subject to

$$x_1 + 2x_2 + 2x_3 \leq 10$$
$$2x_1 + 4x_2 + 3x_3 \leq 15$$
$$0 \leq x_1 \leq 4, 0 \leq x_2 \leq 3, 0 \leq x_3 \leq 3$$

4. In the following problems, some of the variables have positive lower bounds. Use the bounded algorithm to solve these problems.

(a) Maximize $z = 3x_1 + 2x_2 - 2x_3$
subject to

$$2x_1 + x_2 + x_3 \leq 8$$
$$x_1 + 2x_2 - x_3 \geq 3$$
$$1 \leq x_1 \leq 3, 0 \leq x_2 \leq 3, 2 \leq x_3$$

(b) Maximize $z = x_1 + 2x_2$
subject to

$$-x_1 + 2x_2 \geq 0$$
$$3x_1 + 2x_2 \leq 10$$
$$-x_1 + x_2 \leq 1$$
$$1 \leq x_1 \leq 3, 0 \leq x_2 \leq 1$$

(c) Maximize $z = 4x_1 + 2x_2 + 6x_3$
subject to

$$4x_1 - x_2 \leq 9$$
$$-x_1 + x_2 + 2x_3 \leq 8$$
$$-3x_1 + x_2 + 4x_3 \leq 12$$
$$1 \leq x_1 \leq 3, 0 \leq x_2 \leq 5, 0 \leq x_3 \leq 2$$

5. Consider the matrix definition of the bounded-variables problem. Suppose that the vector $\mathbf{X}$ is partitioned into $(\mathbf{X}_z, \mathbf{X}_u)$, where $\mathbf{X}_u$ represents the basic *and* nonbasic variables that will be substituted at upper bound during the course of the algorithm. The problem may thus be written as

$$\begin{pmatrix} 1 & -\mathbf{C}_z & -\mathbf{C}_u \\ 0 & \mathbf{D}_z & \mathbf{D}_u \end{pmatrix} \begin{pmatrix} z \\ \mathbf{X}_z \\ \mathbf{X}_u \end{pmatrix} = \begin{pmatrix} 0 \\ \mathbf{b} \end{pmatrix}$$

Using $\mathbf{X}_u = \mathbf{U}_u - \mathbf{X}'_u$ where $\mathbf{U}_u$ is a subset of $\mathbf{U}$ representing the upper bounds for $\mathbf{X}_u$, let $\mathbf{B}$ (and $\mathbf{X}_B$) be the basis of the current simplex iteration after $\mathbf{X}_u$ has been substituted out. Show that the associated general simplex tableau is given as

Basic	$\mathbf{X}_z^T$	$\mathbf{X}'^T_u$	Solution
z	$\mathbf{C}_B\mathbf{B}^{-1}\mathbf{D}_z - \mathbf{C}_z$	$-\mathbf{C}_B\mathbf{B}^{-1}\mathbf{D}_u + \mathbf{C}_u$	$\mathbf{C}_u\mathbf{B}^{-1}\mathbf{b}' + \mathbf{C}_u\mathbf{U}_u$
$\mathbf{X}_B$	$\mathbf{B}^{-1}\mathbf{D}_z$	$-\mathbf{B}^{-1}\mathbf{D}_u$	$\mathbf{B}^{-1}\mathbf{b}'$

where $\mathbf{b}' = \mathbf{b} - \mathbf{D}_u\mathbf{U}_u$.

6. In Example 7.3-1, do the following:

(a) In Iteration 1, verify that $\mathbf{X}_B = (x_4, x_1)^T = (\frac{5}{2}, \frac{3}{2})^T$ by using matrix manipulation.

(b) In Iteration 2, show how $\mathbf{B}^{-1}$ can be computed from the original data of the problem. Then verify the given values of basic x_4 and x'_2 using matrix manipulation.

7. Solve part (a) of Problem 3 using the revised simplex (matrix) version for upper-bounded variables.

8. *Bounded Dual Simplex Algorithm.* The dual simplex algorithm (Section 4.4.1) can be modified to accommodate the bounded variables as follows. Given the upper bound constraint $x_j \leq u_j$ for all j (if u_j is infinite, replace it with a sufficiently large upper bound M), the LP problem is converted to a dual feasible (i.e., primal optimal) form by using the substitution $x_j = u_j - x'_j$, where necessary.

Step 1. If any of the current basic variables $(\mathbf{X}_B)_i$ exceeds its upper bound, use the substitution $(\mathbf{X}_B)_i = (\mathbf{U}_B)_i - (\mathbf{X}_B)'_i$. Go to step 2.

Step 2. If all the basic variables are feasible, stop. Otherwise, select the leaving variable x_r as the basic variable having the most negative value. Go to step 3.

Step 3. Select the entering variable using the optimality condition of the regular dual simplex method (Section 4.4.1). Go to step 4.

Step 4. Perform a change of basis. Go to step 1.

Apply the given algorithm to the following problems:

(a) Minimize $z = -3x_1 - 2x_2 + 2x_3$

subject to

$$2x_1 + x_2 + x_3 \leq 8$$

$$-x_1 + 2x_2 + x_3 \geq 13$$

$$0 \leq x_1 \leq 2, 0 \leq x_2 \leq 3, 0 \leq x_3 \leq 1$$

(b) Maximize $z = x_1 + 5x_2 - 2x_3$

subject to

$$4x_1 + 2x_2 + 2x_3 \leq 26$$

$$x_1 + 3x_2 + 4x_3 \geq 17$$

$$0 \leq x_1 \leq 2, 0 \leq x_2 \leq 3, x_3 \geq 0$$

7.4 DUALITY

We have dealt with the dual problem in Chapter 4. This section presents a more rigorous treatment of duality and allows us to verify the primal-dual relationships that formed the basis for post-optimal analysis in Chapter 4. The presentation also lays the foundation for the development of parametric programming.

7.4.1 Matrix Definition of the Dual Problem

Suppose that the primal problem in equation form with m constraints and n variables is defined as

$$\text{Maximize } z = \mathbf{CX}$$

subject to

$$\mathbf{AX} = \mathbf{b}$$
$$\mathbf{X} \geq \mathbf{0}$$

Letting the vector $\mathbf{Y} = (y_1, y_2, \ldots, y_m)$ represent the dual variables, the rules in Table 4.2 produce the following dual problem:

$$\text{Minimize } w = \mathbf{Yb}$$

subject to

$$\mathbf{YA} \geq \mathbf{C}$$
$$\mathbf{Y} \text{ unrestricted}$$

Some of the constraints $\mathbf{YA} \geq \mathbf{C}$ may override unrestricted $\mathbf{Y}$.

PROBLEM SET 7.4A

1. Prove that the dual of the dual is the primal.

***2.** If the primal is given as min $z = \{\mathbf{CX} | \mathbf{AX} \geq \mathbf{b}, \mathbf{X} \geq \mathbf{0}\}$, define the corresponding dual problem.

7.4.2 Optimal Dual Solution

This section establishes relationships between the primal and dual problems and shows how the optimal dual solution can be determined from the optimal primal solution. Let $\mathbf{B}$ be the current *optimal* primal basis, and define $\mathbf{C}_B$ as the objective-function coefficients associated with the optimal vector $\mathbf{X}_B$.

Theorem 7.4-1. (Weak Duality Theory). *For any pair of* feasible *primal and dual solutions* **(X, Y)**, *the value of the objective function in the* minimization *problem sets an upper bound on the value of the objective function in the* maximization *problem. For the* optimal *pair* ($\mathbf{X}^*$, $\mathbf{Y}^*$), *the values of the objective functions are equal.*

Proof. The feasible pair **(X, Y)** satisfies all the restrictions of the two problems. Premultiplying both sides of the constraints of the maximization problem with (unrestricted) **Y**, we get

$$\mathbf{YAX} = \mathbf{YB} = w \tag{1}$$

Also, for the minimization problem, postmultiplying both sides of each of the first two sets of constraints by $\mathbf{X}(\geq \mathbf{0})$, we get

$$\mathbf{YAX} \geq \mathbf{CX}$$

or

$$\mathbf{YAX} \geq \mathbf{CX} = z \tag{2}$$

(The nonnegativity of the vector $\mathbf{X}$ is essential for preserving the direction of the inequality.) Combining (1) and (2), we get $z \leq w$ for any *feasible* pair $(\mathbf{X}, \mathbf{Y})$.

Note that the theorem does *not* depend on labeling the problems as primal or dual. What is important is the sense of optimization in each problem. Specifically, for any pair of feasible solutions, the objective value in the maximization problem does not exceed the objective value in the minimization problem.

The implication of the theorem is that, given $z \leq w$ for any feasible solutions, the maximum of z and the minimum of w are achieved when the two objective values are equal. A consequence of this result is that the "goodness" of any feasible primal and dual solutions relative to the optimum may be checked by comparing the difference $(w - z)$ to $\frac{z + w}{2}$. The smaller the ratio $\frac{2(w - z)}{z + w}$, the closer the two solutions are to being optimal. The suggested *rule of thumb* does *not* imply that the optimal objective value is $\frac{z + w}{2}$.

What happens if one of the two problems has an unbounded objective value? The answer is that the other problem must be infeasible. For if it is not, then both problems have feasible solutions, and the relationship $z \leq w$ must hold—an impossible result, because either $z = +\infty$ or $w = -\infty$ by assumption.

The next question is: If one problem is infeasible, is the other problem unbounded? Not necessarily. The following counterexample shows that both the primal and the dual can be infeasible (verify graphically!):

Primal. Maximize $z = \{x_1 + x_2 | x_1 - x_2 \leq -1, -x_1 + x_2 \leq -1, x_1, x_2 \geq 0\}$

Dual. Minimize $w = \{-y_1 - y_2 | y_1 - y_2 \geq 1, -y_1 + y_2 \geq 1, y_1, y_2 \geq 0\}$

Theorem 7.4-2. *Given the* optimal *primal basis* $\mathbf{B}$ *and its associated objective coefficient vector* $\mathbf{C}_B$, *the optimal solution of the dual problem is*

$$\mathbf{Y} = \mathbf{C}_B\mathbf{B}^{-1}$$

Proof. The proof rests on verifying two points: $\mathbf{Y} = \mathbf{C}_B\mathbf{B}^{-1}$ is a feasible dual solution and $z = w$, per Theorem 7.4-1.

The feasibility of $\mathbf{Y} = \mathbf{C}_B\mathbf{B}^{-1}$ is guaranteed by the optimality of the primal, $z_j - c_j \geq 0$ for all j—that is,

$$\mathbf{C}_B\mathbf{B}^{-1}\mathbf{A} - \mathbf{C} \geq \mathbf{0}$$

(See Section 7.2.1.) Thus, $\mathbf{YA} - \mathbf{C} \geq \mathbf{0}$ or $\mathbf{YA} \geq \mathbf{C}$, which shows that $\mathbf{Y} = \mathbf{C}_B\mathbf{B}^{-1}$ is a feasible dual solution.

Next, we show that the associated $w = z$ by noting that

$$w = \mathbf{Yb} = \mathbf{C}_B\mathbf{B}^{-1}\mathbf{b} \tag{1}$$

Similarly, given the primal solution $\mathbf{X}_B = \mathbf{B}^{-1}\mathbf{b}$, we get

$$z = \mathbf{C}_B\mathbf{X}_B = \mathbf{C}_B\mathbf{B}^{-1}\mathbf{b} \tag{2}$$

From relations (1) and (2), we conclude that $z = w$.

The dual variables $\mathbf{Y} = \mathbf{C}_B\mathbf{B}^{-1}$ are sometimes referred to as the **dual** or **shadow prices**, names that evolved from the economic interpretation of the dual variables in Section 4.3.1.

Given that $\mathbf{P}_j$ is the jth column of $\mathbf{A}$, we note from Theorem 7.4-2 that

$$z_j - c_j = \mathbf{C}_B\mathbf{B}^{-1}\mathbf{P}_j - c_j = \mathbf{Y}\mathbf{P}_j - c_j$$

represents the difference between the left- and right-hand sides of the dual constraints. The maximization primal starts with $z_j - c_j < 0$ for at least one j, which means that the corresponding dual constraint, $\mathbf{Y}\mathbf{P}_j \geq c_j$, is not satisfied. When the primal optimal is reached we get $z_j - c_j \geq 0$, for all j, which means that the corresponding dual solution $\mathbf{Y} = \mathbf{C}_B\mathbf{B}^{-1}$ becomes feasible. Thus, while the primal is seeking optimality, the dual is automatically seeking feasibility. This point is the basis for the development of the *dual simplex method* (Section 4.4.1) in which the iterations start better than optimal and infeasible and remain so until feasibility is acquired at the last iteration. This is in contrast with the (primal) simplex method (Chapter 3), which remains worse than optimal but feasible until the optimal iteration is reached.

Example 7.4-1

The *optimal* basis for the following LP is $\mathbf{B} = (\mathbf{P}_1, \mathbf{P}_4)$. Write the dual and find its optimum solution using the optimal primal basis.

$$\text{Maximize } z = 3x_1 + 5x_2$$

subject to

$$\begin{aligned} x_1 + 2x_2 + x_3 \quad\quad &= 5 \\ -x_1 + 3x_2 \quad\quad + x_4 &= 2 \\ x_1, x_2, x_3, x_4 &\geq 0 \end{aligned}$$

The dual problem is

$$\text{Minimize } w = 5y_1 + 2y_2$$

subject to

$$\begin{aligned} y_1 - y_2 &\geq 3 \\ 2y_1 + 3y_2 &\geq 5 \\ y_1, y_2 &\geq 0 \end{aligned}$$

We have $\mathbf{X}_B = (x_1, x_4)^T$ and $\mathbf{C}_B = (3, 0)$. The optimal basis and its inverse are

$$\mathbf{B} = \begin{pmatrix} 1 & 0 \\ -1 & 1 \end{pmatrix}, \mathbf{B}^{-1} = \begin{pmatrix} 1 & 0 \\ 1 & 1 \end{pmatrix}$$

The associated primal and dual values are

$$(x_1, x_4)^T = \mathbf{B}^{-1}\mathbf{b} = (5, 7)^T$$
$$(y_1, y_2) = \mathbf{C}_B\mathbf{B}^{-1} = (3, 0)$$

Both solutions are feasible and $z = w = 15$ (verify!). Thus, the two solutions are optimal.

PROBLEM SET 7.4B

1. Verify that the dual problem of the numeric example given at the end of Theorem 7.4-1 is correct. Then verify graphically that both the primal and dual problems have no feasible solution.
2. Consider the following LP:

$$\text{Maximize } z = 50x_1 + 30x_2 + 10x_3$$

subject to

$$\begin{aligned} 2x_1 + x_2 \qquad &= 1 \\ 2x_2 \qquad &= -5 \\ 4x_1 \qquad + x_3 &= 6 \\ x_1, x_2, x_3 &\geq 0 \end{aligned}$$

   (a) Write the dual.
   (b) Show by inspection that the primal is infeasible.
   (c) Show that the dual in (a) is unbounded.
   (d) From Problems 1 and 2, develop a general conclusion regarding the relationship between infeasibility and unboundedness in the primal and dual problems.
3. Consider the following LP:

$$\text{Maximize } z = 5x_1 + 12x_2 + 4x_3$$

subject to

$$\begin{aligned} 2x_1 - x_2 + 3x_3 \qquad &= 2 \\ x_1 + 2x_2 + x_3 + x_4 &= 5 \\ x_1, x_2, x_3, x_4 &\geq 0 \end{aligned}$$

   (a) Write the dual.
   (b) In each of the following cases, first verify that the given basis **B** is feasible for the primal. Next, using $\mathbf{Y} = \mathbf{C}_B\mathbf{B}^{-1}$, compute the associated dual values and verify whether or not the primal solution is optimal.
      (i) $\mathbf{B} = (\mathbf{P}_4, \mathbf{P}_3)$ **(iii)** $\mathbf{B} = (\mathbf{P}_1, \mathbf{P}_2)$
      (ii) $\mathbf{B} = (\mathbf{P}_2, \mathbf{P}_3)$ **(iv)** $\mathbf{B} = (\mathbf{P}_1, \mathbf{P}_4)$
4. Consider the following LP:

$$\text{Maximize } z = 2x_1 + 4x_2 + 4x_3 - 3x_4$$

subject to

$$x_1 + x_2 + x_3 = 4$$
$$x_1 + 4x_2 + \quad + x_4 = 8$$
$$x_1, x_2, x_3, x_4 \geq 0$$

(a) Write the dual problem.

(b) Verify that $\mathbf{B} = (\mathbf{P}_2, \mathbf{P}_3)$ is optimal by computing $z_j - c_j$ for all nonbasic $\mathbf{P}_j$.

(c) Find the associated optimal dual solution.

***5.** An LP model includes two variables x_1 and x_2 and three constraints of the type $\leq$. The associated slacks are x_3, x_4, and x_5. Suppose that the optimal basis is $\mathbf{B} = (\mathbf{P}_1, \mathbf{P}_2, \mathbf{P}_3)$, and its inverse is

$$\mathbf{B}^{-1} = \begin{pmatrix} 0 & -1 & 1 \\ 0 & 1 & 0 \\ 1 & 1 & -1 \end{pmatrix}$$

The optimal primal and dual solutions are

$$\mathbf{X}_B = (x_1, x_2, x_3)^T = (2, 6, 2)^T$$
$$\mathbf{Y} = (y_1, y_2, y_3) = (0, 3, 2)$$

Determine the optimal value of the objective function in two ways using the primal and dual problems.

6. Prove the following relationship for the optimal primal and dual solutions:

$$\sum_{i=1}^{m} c_i(\mathbf{B}^{-1}\mathbf{P}_k)_i = \sum_{i=1}^{m} y_i a_{ik}$$

where $\mathbf{C}_B = (c_1, c_2, \ldots, c_m)$ and $\mathbf{P}_k = (a_{1k}, a_{2k}, \ldots, a_{mk})^T$, for $k = 1, 2, \ldots, n$, and $(\mathbf{B}^{-1}\mathbf{P}_k)_i$ is the ith element of $\mathbf{B}^{-1}\mathbf{P}_k$.

***7.** Write the dual of

$$\text{Maximize } z = \{\mathbf{CX} | \mathbf{AX} = \mathbf{b}, \quad \mathbf{X} \text{ unrestricted}\}$$

8. Show that the dual of

$$\text{Maximize } z = \{\mathbf{CX} | \mathbf{AX} \leq \mathbf{b}, \quad \mathbf{0} < \mathbf{L} \leq \mathbf{X} \leq \mathbf{U}\}$$

always possesses a feasible solution.

7.5 PARAMETRIC LINEAR PROGRAMMING

Parametric linear programming is an extension of the post-optimal analysis presented in Section 4.5. It investigates the effect of *predetermined* continuous variations in the objective function coefficients and the right-hand side of the constraints on the optimum solution.

Let $\mathbf{X} = (x_1, x_2, \ldots, x_n)$ and define the LP as

$$\text{Maximize } z = \left\{\mathbf{CX} | \sum_{j=1}^{n} \mathbf{P}_j x_j = \mathbf{b}, \mathbf{X} \geq \mathbf{0}\right\}$$

In parametric analysis, the objective function and right-hand side vectors, **C** and **b**, are replaced with the parameterized functions $\mathbf{C}(t)$ and $\mathbf{b}(t)$, where t is the parameter of variation. Mathematically, t can assume any positive or negative value. In practice, however, t usually represents time, and hence it is nonnegative. In this presentation we will assume $t \geq 0$.

The general idea of parametric analysis is to start with the optimal solution at $t = 0$. Then, using the optimality and feasibility conditions of the simplex method, we determine the range $0 \leq t \leq t_1$ for which the solution at $t = 0$ remains optimal and feasible. In this case, t_1 is referred to as a **critical value**. The process continues by determining successive critical values and their corresponding optimal feasible solutions, and will terminate at $t = t_r$ when there is indication that either the last solution remains unchanged for $t > t_r$ or that no feasible solution exists beyond that critical value.

7.5.1 Parametric Changes in C

Let $\mathbf{X}_{B_i}$, $\mathbf{B}_i$, $\mathbf{C}_{B_i}(t)$ be the elements that define the optimal solution associated with critical t_i (the computations start at $t_0 = 0$ with $\mathbf{B}_0$ as its optimal basis). Next, the critical value t_{i+1} and its optimal basis, if one exists, is determined. Because changes in **C** can affect only the optimality of the problem, the current solution $\mathbf{X}_{B_i} = \mathbf{B}_i^{-1}\mathbf{b}$ will remain optimal for some $t \geq t_i$ so long as the reduced cost, $z_j(t) - c_j(t)$, satisfies the following optimality condition:

$$z_j(t) - c_j(t) = \mathbf{C}_{B_i}(t)\mathbf{B}_i^{-1}\mathbf{P}_j - c_j(t) \geq 0, \text{ for all } j$$

The value of t_{i+1} equals the largest $t > t_i$ that satisfies all the optimality conditions.

Note that *nothing* in the inequalities requires $\mathbf{C}(t)$ to be linear in t. Any function $\mathbf{C}(t)$, linear or nonlinear, is acceptable. However, with nonlinearity the numerical manipulation of the resulting inequalities may be cumbersome. (See Problem 5, Set 7.5a for an illustration of the nonlinear case.)

Example 7.5-1

$$\text{Maximize } z = (3 - 6t)x_1 + (2 - 2t)x_2 + (5 + 5t)x_3$$

subject to

$$\begin{aligned} x_1 + 2x_2 + x_3 &\leq 40 \\ 3x_1 \qquad + 2x_3 &\leq 60 \\ x_1 + 4x_2 \qquad &\leq 30 \\ x_1, x_2, x_3 &\geq 0 \end{aligned}$$

We have

$$\mathbf{C}(t) = (3 - 6t, 2 - 2t, 5 + 5t), t \geq 0$$

The variables x_4, x_5, and x_6 will be used as the slack variables associated with the three constraints.

Optimal Solution at $t = t_0 = 0$

Basic	x_1	x_2	x_3	x_4	x_5	x_6	Solution
z	4	0	0	1	2	0	160
x_2	$-\frac{1}{4}$	1	0	$\frac{1}{2}$	$-\frac{1}{4}$	0	5
x_3	$\frac{3}{2}$	0	1	0	$\frac{1}{2}$	0	30
x_6	2	0	0	-2	1	1	10

$$\mathbf{X}_{B_0} = (x_2, x_3, x_6)^T = (5, 30, 10)^T$$

$$\mathbf{C}_{B_0}(t) = (2 - 2t, 5 + 5t, 0)$$

$$\mathbf{B}_0^{-1} = \begin{pmatrix} \frac{1}{2} & -\frac{1}{4} & 0 \\ 0 & \frac{1}{2} & 0 \\ -2 & 1 & 1 \end{pmatrix}$$

The optimality conditions for the current nonbasic vectors, $\mathbf{P}_1$, $\mathbf{P}_4$, and $\mathbf{P}_5$, are

$$\{\mathbf{C}_{B_0}(t)\mathbf{B}_0^{-1}\mathbf{P}_j - c_j(t)\}_{j=1,4,5} = (4 + 14t, 1 - t, 2 + 3t) \geq \mathbf{0}$$

Thus, $\mathbf{X}_{B_0}$ remains optimal so long as the following conditions are satisfied:

$$4 + 14t \geq 0$$
$$1 - t \geq 0$$
$$2 + 3t \geq 0$$

Because $t \geq 0$, the second inequality gives $t \leq 1$ and the remaining two inequalities are satisfied for all $t \geq 0$. We thus have $t_1 = 1$, which means that $\mathbf{X}_{B_0}$ remains optimal (and feasible) for $0 \leq t \leq 1$.

The reduced cost $z_4(t) - c_4(t) = 1 - t$ equals zero at $t = 1$ and becomes negative for $t > 1$. Thus, $\mathbf{P}_4$ must enter the basis for $t > 1$. In this case, $\mathbf{P}_2$ must leave the basis (see the optimal tableau at $t = 0$). The new basic solution $\mathbf{X}_{B_1}$ is the alternative solution obtained at $t = 1$ by letting $\mathbf{P}_4$ enter the basis—that is, $\mathbf{X}_{B_1} = (x_4, x_3, x_6)^T$ and $\mathbf{B}_1 = (\mathbf{P}_4, \mathbf{P}_3, \mathbf{P}_6)$.

Alternative Optimal Basis at $t = t_1 = 1$

$$\mathbf{B}_1 = \begin{pmatrix} 1 & 1 & 0 \\ 0 & 2 & 0 \\ 0 & 0 & 1 \end{pmatrix}, \quad \mathbf{B}_1^{-1} = \begin{pmatrix} 1 & -\frac{1}{2} & 0 \\ 0 & \frac{1}{2} & 0 \\ 0 & 0 & 1 \end{pmatrix}$$

Thus,

$$\mathbf{X}_{B_1} = (x_4, x_3, x_6)^T = \mathbf{B}_1^{-1}\mathbf{b} = (10, 30, 30)^T$$

$$\mathbf{C}_{B_1}(t) = (0, 5 + 5t, 0)$$

The associated nonbasic vectors are $\mathbf{P}_1$, $\mathbf{P}_2$, and $\mathbf{P}_5$, and we have

$$\{\mathbf{C}_{B_1}(t)\mathbf{B}_1^{-1}\mathbf{P}_j - c_j(t)\}_{j=1,2,5} = \left(\tfrac{9 + 27t}{2}, -2 + 2t, \tfrac{5 + 5t}{2}\right) \geq \mathbf{0}$$

According to these conditions, the basic solution $\mathbf{X}_{B_1}$ remains optimal for all $t \geq 1$. Observe that the optimality condition, $-2 + 2t \geq 0$, automatically "remembers" that $\mathbf{X}_{B_1}$ is optimal for a range of t that starts from the last critical value $t_1 = 1$. This will always be the case in parametric programming computations.

The optimal solution for the entire range of t is summarized below. The value of z is computed by direct substitution.

t	x_1	x_2	x_3	z
$0 \leq t \leq 1$	0	5	30	$160 + 140t$
$t \geq 1$	0	0	30	$150 + 150t$

PROBLEM SET 7.5A

*1. In example 7.5-1, suppose that t is unrestricted in sign. Determine the range of t for which $\mathbf{X}_{B_0}$ remains optimal.

2. Solve Example 7.5-1, assuming that the objective function is given as
   - *(a) Maximize $z = (3 + 3t)x_1 + 2x_2 + (5 - 6t)x_3$
   - (b) Maximize $z = (3 - 2t)x_1 + (2 + t)x_2 + (5 + 2t)x_3$
   - (c) Maximize $z = (3 + t)x_1 + (2 + 2t)x_2 + (5 - t)x_3$

3. Study the variation in the optimal solution of the following parameterized LP given $t \geq 0$.

$$\text{Minimize } z = (4 - t)x_1 + (1 - 3t)x_2 + (2 - 2t)x_3$$

subject to

$$\begin{aligned} 3x_1 + x_2 + 2x_3 &= 3 \\ 4x_1 + 3x_2 + 2x_3 &\geq 6 \\ x_1 + 2x_2 + 5x_3 &\leq 4 \\ x_1, x_2, x_3 &\geq 0 \end{aligned}$$

4. The analysis in this section assumes that the optimal solution of the LP at $t = 0$ is obtained by the (primal) simplex method. In some problems, it may be more convenient to obtain the optimal solution by the dual simplex method (Section 4.4.1). Show how the parametric analysis can be carried out in this case, then analyze the LP of Example 4.4-1, assuming that the objective function is given as

$$\text{Minimize } z = (3 + t)x_1 + (2 + 4t)x_2 + x_3, t \geq 0$$

*5. In Example 7.5-1, suppose that the objective function is nonlinear in $t (t \geq 0)$ and is defined as

$$\text{Maximize } z = (3 + 2t^2)x_1 + (2 - 2t^2)x_2 + (5 - t)x_3$$

Determine the first critical value t_1.

7.5.2 Parametric Changes in b

The parameterized right-hand side $\mathbf{b}(t)$ can affect only the feasibility of the problem. The critical values of t are thus determined from the following condition:

$$\mathbf{X}_B(t) = \mathbf{B}^{-1}\mathbf{b}(t) \geq \mathbf{0}$$

Example 7.5-2

$$\text{Maximize } z = 3x_1 + 2x_2 + 5x_3$$

subject to

$$\begin{aligned} x_1 + 2x_2 + x_3 &\leq 40 - t \\ 3x_1 + 2x_3 &\leq 60 + 2t \\ x_1 + 4x_2 &\leq 30 - 7t \\ x_1, x_2, x_3 &\geq 0 \end{aligned}$$

Assume that $t \geq 0$.

At $t = t_0 = 0$, the problem is identical to that of Example 7.5-1. We thus have

$$\mathbf{X}_{B_0} = (x_2, x_3, x_6)^T = (5, 30, 10)^T$$

$$\mathbf{B}_0^{-1} = \begin{pmatrix} \frac{1}{2} & -\frac{1}{4} & 0 \\ 0 & \frac{1}{2} & 0 \\ -2 & 1 & 1 \end{pmatrix}$$

To determine the first critical value t_1, we apply the feasibility conditions $\mathbf{X}_{B_0}(t) = \mathbf{B}_0^{-1}\mathbf{b}(t) \geq 0$, which yields

$$\begin{pmatrix} x_2 \\ x_3 \\ x_6 \end{pmatrix} = \begin{pmatrix} 5 - t \\ 30 + t \\ 10 - 3t \end{pmatrix} \geq \begin{pmatrix} 0 \\ 0 \\ 0 \end{pmatrix}$$

These inequalities are satisfied for $t \leq \frac{10}{3}$, meaning that $t_1 = \frac{10}{3}$ and that the basis $\mathbf{B}_0$ remains feasible for the range $0 \leq t \leq \frac{10}{3}$. However, the values of the basic variables x_2, x_3, and x_6 will change with t as given above.

The value of the basic variable x_6 $(= 10 - 3t)$ will equal zero at $t = t_1 = \frac{10}{3}$, and will become negative for $t > \frac{10}{3}$. Thus, at $t = \frac{10}{3}$, we can determine the alternative basis $\mathbf{B}_1$ by applying the revised dual simplex method (see Problem 5, Set 7.2b, for details). The leaving variable is x_6.

Alternative Basis at $t = t_1 = \frac{10}{3}$

Given that x_6 is the leaving variable, we determine the entering variable as follows:

$$\mathbf{X}_{B_0} = (x_2, x_3, x_6)^T, \mathbf{C}_{B_0} = (2, 5, 0)$$

Thus,

$$\{z_j - c_j\}_{j=1,4,5} = \{\mathbf{C}_{B_0}\mathbf{B}_0^{-1}\mathbf{P}_j - c_j\}_{j=1,4,5} = (4, 1, 2)$$

Next, for nonbasic x_j, $j = 1, 4, 5$, we compute

$$\begin{aligned}(\text{Row of } \mathbf{B}_0^{-1} \text{ associated with } x_6)(\mathbf{P}_1, \mathbf{P}_4, \mathbf{P}_5) &= (\text{Third row of } \mathbf{B}_0^{-1})(\mathbf{P}_1, \mathbf{P}_4, \mathbf{P}_5) \\ &= (-2, 1, 1)(\mathbf{P}_1, \mathbf{P}_4, \mathbf{P}_5) \\ &= (2, -2, 1)\end{aligned}$$

The entering variable is thus associated with

$$\theta = \min\left\{-, \left|\frac{1}{-2}\right|, -\right\} = \tfrac{1}{2}$$

Thus, $\mathbf{P}_4$ is the entering vector. The alternative basic solution and its $\mathbf{B}_1$ and $\mathbf{B}_1^{-1}$ are

$$\mathbf{X}_{B_1} = (x_2, x_3, x_4)^T$$

$$\mathbf{B}_1 = (\mathbf{P}_2, \mathbf{P}_3, \mathbf{P}_4) = \begin{pmatrix} 2 & 1 & 1 \\ 0 & 2 & 0 \\ 4 & 0 & 0 \end{pmatrix}, \mathbf{B}_1^{-1} = \begin{pmatrix} 0 & 0 & \frac{1}{4} \\ 0 & \frac{1}{2} & 0 \\ 1 & -\frac{1}{2} & -\frac{1}{2} \end{pmatrix}$$

The next critical value t_2 is determined from the feasibility conditions, $\mathbf{X}_{B_1}(t) = \mathbf{B}_1^{-1}\mathbf{b}(t) \geq \mathbf{0}$, which yields

$$\begin{pmatrix} x_2 \\ x_3 \\ x_4 \end{pmatrix} = \begin{pmatrix} \frac{30 - 7t}{4} \\ 30 + t \\ \frac{-10 + 3t}{2} \end{pmatrix} \begin{pmatrix} 0 \\ 0 \\ 0 \end{pmatrix}$$

These conditions show that $\mathbf{B}_1$ remains feasible for $\frac{10}{3} \leq t \leq \frac{30}{7}$.

At $t = t_2 = \frac{30}{7}$, an alternative basis can be obtained by the revised dual simplex method. The leaving variable is x_2, because it corresponds to the condition yielding the critical value t_2.

Alternative Basis at $t = t_2 = \frac{30}{7}$

Given that x_2 is the leaving variable, we determine the entering variable as follows:

$$\mathbf{X}_{B_1} = (x_2, x_3, x_4)^T, \mathbf{C}_{B_1} = (2, 5, 0)$$

Thus,

$$\{z_j - c_j\}_{j=1,5,6} = \{\mathbf{C}_{B_1}\mathbf{B}_1^{-1}\mathbf{P}_j - c_j\}_{j=1,5,6} = \left(5, \tfrac{5}{2}, \tfrac{1}{2}\right)$$

Next, for nonbasic x_j, $j = 1, 5, 6$, we compute

$$\begin{aligned}(\text{Row of } \mathbf{B}_1^{-1} \text{ associated with } x_2)(\mathbf{P}_1, \mathbf{P}_5, \mathbf{P}_6) &= (\text{First row of } \mathbf{B}_1^{-1})(\mathbf{P}_1, \mathbf{P}_5, \mathbf{P}_6) \\ &= \left(0, 0, \tfrac{1}{4}\right)(\mathbf{P}_1, \mathbf{P}_5, \mathbf{P}_6) \\ &= \left(\tfrac{1}{4}, 0, \tfrac{1}{4}\right)\end{aligned}$$

Because all the denominator elements, $\left(\frac{1}{4}, 0, \frac{1}{4}\right)$, are ≥ 0, the problem has no feasible solution for $t > \frac{30}{7}$ and the parametric analysis ends at $t = t_2 = \frac{30}{7}$.

The optimal solution is summarized as

t	x_1	x_2	x_3	z
$0 \leq t \leq \frac{10}{3}$	0	$5 - t$	$30 + t$	$160 + 3t$
$\frac{10}{3} \leq t \leq \frac{30}{7}$	0	$\frac{30 - 7t}{4}$	$30 + t$	$165 + \frac{3}{2}t$
$t > \frac{30}{7}$		(No feasible solution exists)		

PROBLEM SET 7.5B

*1. In Example 7.5-2, find the first critical value, t_1, and define the vectors of $\mathbf{B}_1$ in each of the following cases:

*(a) $\mathbf{b}(t) = (40 + 2t, 60 - 3t, 30 + 6t)^{\mathrm{T}}$

(b) $\mathbf{b}(t) = (40 - t, 60 + 2t, 30 - 5t)^{\mathrm{T}}$

*2. Study the variation in the optimal solution of the following parameterized LP, given $t \geq 0$.

$$\text{Minimize } z = 4x_1 + x_2 + 2x_3$$

subject to

$$3x_1 + x_2 + 2x_3 = 3 + 3t$$
$$4x_1 + 3x_2 + 2x_3 \geq 6 + 2t$$
$$x_1 + 2x_2 + 5x_3 \leq 4 - t$$
$$x_1, x_2, x_3 \geq 0$$

3. The analysis in this section assumes that the optimal LP solution at $t = 0$ is obtained by the (primal) simplex method. In some problems, it may be more convenient to obtain the optimal solution by the dual simplex method (Section 4.4.1). Show how the parametric analysis can be carried out in this case, and then analyze the LP of Example 4.4-1, assuming that $t \geq 0$ and the right-hand side vector is

$$\mathbf{b}(t) = (3 + 2t, 6 - t, 3 - 4t)^T$$

4. Solve Problem 2 assuming that the right-hand side is changed to

$$\mathbf{b}(t) = (3 + 3t^2, 6 + 2t^2, 4 - t^2)^T$$

Further assume that t can be positive, zero, or negative.

REFERENCES

Bazaraa, M., J. Jarvis, and H. Sherali, *Linear Programming and Network Flows,* 2nd ed., Wiley, New York, 1990.

Chvàtal, V., *Linear Programming*, Freeman, San Francisco, 1983.

Nering, E., and A. Tucker, *Linear Programming and Related Problems,* Academic Press, Boston, 1992.

Saigal, R. *Linear Programming: A Modern Integrated Analysis,* Kluwer Academic Publishers, Boston, 1995.

Vanderbei, R., *Linear Programming: Foundation and Extensions,* 2nd ed, Kluwer Academic Publishers, Boston, 2001.

CHAPTER 8

Goal Programming

Chapter Guide. The LP models presented in the preceding chapters are based on the optimization of a *single* objective function. There are situations where multiple (conflicting) objectives may be more appropriate. For example, politicians promise to reduce the national debt and, simultaneously, offer income tax relief. In such situations, it is impossible to find a single solution that optimizes these two conflicting goals. What goal programming does is seek a *compromise* solution based on the relative importance of each objective.

The main prerequisite for this chapter is a basic understanding of the simplex method. There are two methods for solving goal programs: The *weights method* forms a single objective function consisting of the weighted sum of the goals and the *preemptive method* optimizes the goals one at a time starting with the highest-priority goal and terminating with the lowest, never degrading the quality of a higher-priority goal. The solution using the *weights method* is just like any ordinary linear program. The *preemptive method* entails "additional" algorithmic considerations that are very much within the realm of the simplex method of Chapter 3. The chapter includes an AMPL model that applies the preemptive method interactively to any goal program, simply by changing the input data. You are encouraged to study this model because it will assist you in understanding the details of the preemptive method.

This chapter includes a summary of 1 real-life application, 4 solved examples, 1 AMPL model, 25 end-of-section problems, and 2 cases. The cases are in Appendix E on the CD. The AMPL/Excel/Solver/TORA programs are in folder ch8Files.

Real-Life Application—Allocation of Operating Room Time in Mount Sinai Hospital

The situation takes place in Canada where health care insurance is mandatory and universal. Funding, which is based on a combination of premiums and taxes, is controlled by the individual provinces. Under this system, hospitals are advanced a fixed annual budget and each province pays physicians retrospectively using a fee-for-service funding mechanism. This funding arrangement limits the availability of hospital facilities (e.g., operating rooms), which in turn would curb physicians' tendency to boost personal gain through overservice to patients. The objective of the study is to determine an equitable daily schedule for the use of available operating rooms. The problem is modeled using a combination of goal and integer programming. Case 6 in Chapter 24 on the CD provides the details of the study.

8.1 A GOAL PROGRAMMING FORMULATION

The idea of goal programming is illustrated by an example.

Example 8.1-1 (Tax Planning)[1]

Fairville is a small city with a population of about 20,000 residents. The city council is in the process of developing an equitable city tax rate table. The annual taxation base for real estate property is \$550 million. The annual taxation bases for food and drugs and for general sales are \$35 million and \$55 million, respectively. Annual local gasoline consumption is estimated at 7.5 million gallons. The city council wants to develop the tax rates based on four main goals.

1. Tax revenues must be at least \$16 million to meet the city's financial commitments.
2. Food and drug taxes cannot exceed 10% of all taxes collected.
3. General sales taxes cannot exceed 20% of all taxes collected.
4. Gasoline tax cannot exceed 2 cents per gallon.

Let the variables x_p, x_f, and x_s represent the tax rates (expressed as proportions of taxation bases) for property, food and drug, and general sales, and define the variable x_g as the gasoline tax in cents per gallon. The goals of the city council are then expressed as

$$550x_p + 35x_f + 55x_s + .075x_g \geq 16 \quad \text{(Tax revenue)}$$

$$35x_f \leq .1(550x_p + 35x_f + 55x_3 + .075x_g) \quad \text{(Food/drug tax)}$$

$$55x_s \leq .2(550x_p + 35x_f + 55x_s + .075x_g) \quad \text{(General tax)}$$

$$x_g \leq 2 \quad \text{(Gasoline tax)}$$

$$x_p, x_f, x_s, x_g \geq 0$$

These constraints are then simplified as

$$550x_p + 35x_f + 55x_s + .075x_g \geq 16$$

$$55x_p - 31.5x_f + 5.5x_s + .0075x_g \geq 0$$

$$110x_p + 7x_f - 44x_s + .015x_g \geq 0$$

$$x_g \leq 2$$

$$x_p, x_f, x_s, x_g \geq 0$$

[1]This example is based on Chissman and Associates, 1989.

Each of the inequalities of the model represents a goal that the city council aspires to satisfy. Most likely, however, the best that can be done is a compromise solution among these conflicting goals.

The manner in which goal programming finds a compromise solution is to convert each inequality into a flexible goal in which the corresponding constraint may be violated, if necessary. In terms of the Fairville model, the flexible goals are expressed as follows:

$$550x_p + 35x_f + 55x_s + .075x_g + s_1^- - s_1^+ = 16$$

$$55x_p - 31.5x_f + 5.5x_s + .0075x_g + s_2^- - s_2^+ = 0$$

$$110x_p + 7x_f - 44x_s + .015x_g + s_3^- - s_3^+ = 0$$

$$x_g + s_4^- - s_4^+ = 2$$

$$x_p, x_f, x_s, x_g \geq 0$$

$$s_i^-, s_i^+ \geq 0, i = 1, 2, 3, 4$$

The nonnegative variables s_i^- and s_i^+, $i = 1, 2, 3, 4$, are called **deviational variables** because they represent the deviations *below* and *above* the right-hand side of constraint i.

The deviational variables s_i^- and s_i^+ are by definition dependent, and hence cannot be basic variables simultaneously. This means that in any simplex iteration, at most *one* of the two deviational variables can assume a positive value. If the original ith inequality is of the type $\leq$ and its $s_i^- > 0$, then the ith goal is satisfied; otherwise, if $s_i^+ > 0$, goal i is not satisfied. In essence, the definition of s_i^- and s_i^+ allows meeting or violating the ith goal at will. This is the type of flexibility that characterizes goal programming when it seeks a compromise solution. Naturally, a good compromise solution aims at minimizing, as much as possible, the amount by which each goal is violated.

In the Fairville model, given that the first three constraints are of the type $\geq$ and the fourth constraint is of the type $\leq$, the deviational variables s_1^-, s_2^-, s_3^- and s_4^+ (shown in the model in bold) represent the amounts by which the respective goals are violated. Thus, the compromise solution tries to satisfy the following four objectives as much as possible:

$$\text{Minimize } G_1 = s_1^-$$

$$\text{Minimize } G_2 = s_2^-$$

$$\text{Minimize } G_3 = s_3^-$$

$$\text{Minimize } G_4 = s_4^+$$

These functions are minimized subject to the constraint equations of the model.

How can we optimize a multiobjective model with possibly conflicting goals? Two methods have been developed for this purpose: (1) the weights method and (2) the preemptive method. Both methods, which are based on converting the multiple objectives into a single function, are detailed in Section 8.2.

PROBLEM SET 8.1A

*1. Formulate the Fairville tax problem, assuming that the town council is specifying an additional goal, G_5, that requires gasoline tax to equal at least 10% of the total tax bill.

2. The NW Shopping Mall conducts special events to attract potential patrons. Among the events that seem to attract teenagers, the young/middle-aged group, and senior citizens, the two most popular are band concerts and art shows. Their costs per presentation are \$1500 and \$3000, respectively. The total (strict) annual budget allocated to the two events is \$15,000. The mall manager estimates the attendance as follows:

	Number attending per presentation		
Event	*Teenagers*	*Young/middle age*	*Seniors*
Band concert	200	100	0
Art show	0	400	250

The manager has set minimum goals of 1000, 1200, and 800 for the attendance of teenagers, the young/middle-aged group, and seniors, respectively. Formulate the problem as a goal programming model.

*3. Ozark University admission office is processing freshman applications for the upcoming academic year. The applications fall into three categories: in-state, out-of-state, and international. The male-female ratios for in-state and out-of-state applicants are 1:1 and 3:2, respectively. For international students, the corresponding ratio is 8:1. The American College Test (ACT) score is an important factor in accepting new students. The statistics gathered by the university indicate that the average ACT scores for in-state, out-of-state, and international students are 27, 26, and 23, respectively. The committee on admissions has established the following desirable goals for the new freshman class:

(a) The incoming class is at least 1200 freshmen.

(b) The average ACT score for all incoming students is at least 25.

(c) International students constitute at least 10% of the incoming class.

(d) The female-male ratio is at least 3:4.

(e) Out-of-state students constitute at least 20% of the incoming class.

Formulate the problem as a goal programming model.

4. Circle K farms consumes 3 tons of special feed daily. The feed—a mixture of limestone, corn, and soybean meal—must satisfy the following nutritional requirements:

Calcium. At least 0.8% but not more than 1.2%.

Protein. At least 22%.

Fiber. At most 5%.

The following table gives the nutritional content of the feed ingredients.

	lb per lb of ingredient		
Ingredient	*Calcium*	*Protein*	*Fiber*
Limestone	.380	.00	.00
Corn	.001	.09	.02
Soybean meal	.002	.50	.08

Formulate the problem as a goal programming model, and state your opinion regarding the applicability of goal programming to this situation.

***5.** Mantel produces a toy carriage, whose final assembly must include four wheels and two seats. The factory producing the parts operates three shifts a day. The following table provides the amounts produced of each part in the three shifts.

	Units produced per run	
Shift	*Wheels*	*Seats*
1	500	300
2	600	280
3	640	360

Ideally, the number of wheels produced is exactly twice that of the number of seats. However, because production rates vary from shift to shift, exact balance in production may not be possible. Mantel is interested in determining the number of production runs in each shift that minimizes the imbalance in the production of the parts. The capacity limitations restrict the number of runs to between 4 and 5 for shift 1, 10 and 20 for shift 2, and 3 and 5 for shift 3. Formulate the problem as a goal programming model.

6. Camyo Manufacturing produces four parts that require the use of a lathe and a drill press. The two machines operate 10 hours a day. The following table provides the time in minutes required by each part:

	Production time in min	
Part	*Lathe*	*Drill press*
1	5	3
2	6	2
3	4	6
4	7	4

It is desired to balance the two machines by limiting the difference between their total operation times to at most 30 minutes. The market demand for each part is at least 10 units. Additionally, the number of units of part 1 may not exceed that of part 2. Formulate the problem as a goal programming model.

7. Two products are manufactured on two sequential machines. The following table gives the machining times in minutes per unit for the two products.

	Machining time in min	
Machine	*Product 1*	*Product 2*
1	5	3
2	6	2

The daily production quotas for the two products are 80 and 60 units, respectively. Each machine runs 8 hours a day. Overtime, though not desirable, may be used if necessary to meet the production quota. Formulate the problem as a goal programming model.

8. Vista City hospital plans the short-stay assignment of surplus beds (those that are not already occupied) 4 days in advance. During the 4-day planning period about 30, 25, and 20 patients will require 1-, 2-, or 3-day stays, respectively. Surplus beds during the same period are estimated at 20, 30, 30, and 30. Use goal programming to resolve the problem of overadmission and underadmission in the hospital.

9. The Von Trapp family is in the process of moving to a new city where both parents have accepted new jobs. In trying to find an ideal location for their new home, the Von Trapps list the following goals:

(a) It should be as close as possible to Mrs. Von Trapp's place of work (within $\frac{1}{4}$ mile).

(b) It should be as far as possible from the noise of the airport (at least 10 miles).

(c) It should be reasonably close to a shopping mall (within 1 mile).

Mr. and Mrs. Von Trapp use a landmark in the city as a reference point and locate the (x,y)-coordinates of work, airport, and shopping mall at $(1, 1)$, $(20, 15)$, and $(4, 7)$, respectively (all distances are in miles). Formulate the problem as a goal programming model. [*Note*: The resulting constraints are not linear.]

10. *Regression analysis*. In a laboratory experiment, suppose that y_i is the ith observed (independent) yield associated with the dependent observational measurements $x_{ij}, i = 1, 2, \ldots, m; j = 1, 2, \ldots, n$. It is desired to determine a linear regression fit into these data points. Let $b_j, j = 0, 1, \ldots, n$, be the regression coefficients. It is desired to determine all b_j such that the sum of the absolute deviations between the observed and the estimated yields is minimized. Formulate the problem as a goal programming model.

11. *Chebyshev Problem*. An alternative goal for the regression model in Problem 10 is to minimize over b_j the maximum of the absolute deviations. Formulate the problem as a goal programming model.

8.2 GOAL PROGRAMMING ALGORITHMS

This section presents two algorithms for solving goal programming. Both methods are based on representing the multiple goals by a single objective function. In the **weights method,** a single objective function is formed as the weighted sum of the functions representing the goals of the problem. The **preemptive method** starts by prioritizing the goals in order of importance. The model is then optimized, using one goal at a time such that the optimum value of a higher-priority goal is never degraded by a lower-priority goal.

The proposed two methods are different because they do not generally produce the same solution. Neither method, however, is superior to the other, because each technique is designed to satisfy certain decision-making preferences.

8.2.1 The Weights Method

Suppose that the goal programming model has n goals and that the ith goal is given as

$$\text{Minimize } G_i, i = 1, 2, \ldots, n$$

The combined objective function used in the weights method is then defined as

$$\text{Minimize } z = w_1G_1 + w_2G_2 + \cdots + w_nG_n$$

The parameters $w_i, i = 1, 2, \ldots, n$, are positive weights that reflect the decision maker's preferences regarding the relative importance of each goal. For example, $w_i = 1$, for all i, signifies that all goals carry equal weights. The determination of the specific values of these weights is subjective. Indeed, the apparently sophisticated analytic procedures developed in the literature (see, e.g., Cohon, 1978) are still rooted in subjective assessments.

Example 8.2-1

TopAd, a new advertising agency with 10 employees, has received a contract to promote a new product. The agency can advertise by radio and television. The following table gives the number of people reached by each type of advertisement and the cost and labor requirements.

	Data/min advertisement	
	Radio	*Television*
Exposure (in millions of persons)	4	8
Cost (in thousands of dollars)	8	24
Assigned employees	1	2

The contract prohibits TopAd from using more than 6 minutes of radio advertisement. Additionally, radio and television advertisements need to reach at least 45 million people. TopAd has a budget goal of \$100,000 for the project. How many minutes of radio and television advertisement should TopAd use?

Let x_1 and x_2 be the minutes allocated to radio and television advertisements. The goal programming formulation for the problem is given as

$$\text{Minimize } G_1 = s_1^- \text{ (Satisfy exposure goal)}$$

$$\text{Minimize } G_2 = s_2^+ \text{ (Satisfy budget goal)}$$

subject to

$$\begin{aligned}
4x_1 + 8x_2 + s_1^- - s_1^+ \qquad\qquad &= 45 \text{ (Exposure goal)} \\
8x_1 + 24x_2 \qquad\qquad + s_2^- - s_2^+ &= 100 \text{ (Budget goal)} \\
x_1 + 2x_2 \qquad\qquad\qquad\qquad &\leq 10 \text{ (Personnel limit)} \\
x_1 \qquad\qquad\qquad\qquad &\leq 6 \text{ (Radio limit)} \\
x_1, x_2, s_1^-, s_1^+, s_2^-, s_2^+ &\geq 0
\end{aligned}$$

TopAd's management assumes that the exposure goal is twice as important as the budget goal. The combined objective function thus becomes

$$\text{Minimize } z = 2G_1 + G_2 = 2s_1^- + s_2^+$$

The optimum solution is $z = 10$, $x_1 = 5$ minutes, $x_2 = 2.5$ minutes, $s_1^- = 5$ million persons. All the remaining variables are zero.

The fact that the optimum value of z is not zero indicates that at least one of the goals is not met. Specifically, $s_1^- = 5$ means that the exposure goal (of at least 45 million persons) is missed by 5 million individuals. Conversely, the budget goal (of not exceeding \$100,000) is not violated, because $s_2^+ = 0$.

Goal programming yields only an *efficient*, rather than optimum, solution to the problem. For example, the solution $x_1 = 6$ and $x_2 = 2$ yields the same exposure ($4 \times 6 + 8 \times 2 = 40$ million persons), but costs less ($8 \times 6 + 24 \times 2 = \$96{,}000$). In essence, what goal programming does is to find a solution that simply *satisfies* the goals of the model with no regard to optimization. Such "deficiency" in finding an optimum solution may raise doubts about the viability of goal programming as an optimizing technique (see Example 8.2-3 for further discussion).

PROBLEM SET 8.2A

*1. Consider Problem 1, Set 8.1a dealing with the Fairville tax situation. Solve the problem, assuming that all five goals have the same weight. Does the solution satisfy all the goals?

2. In Problem 2, Set 8.1a, suppose that the goal of attracting young/middle-aged people is twice as important as for either of the other two categories (teens and seniors). Find the associated solution, and check if all the goals have been met.

3. In the Ozark University admission situation described in Problem 3, Set 8.1a, suppose that the limit on the size of the incoming freshmen class must be met, but the remaining requirements can be treated as flexible goals. Further, assume that the ACT score goal is twice as important as any of the remaining goals.
   (a) Solve the problem, and specify whether or not all the goals are satisfied.
   (b) If, in addition, the size of the incoming class can be treated as a flexible goal that is twice as important as the ACT goal, how would this change affect the solution?

*4. In the Circle K model of Problem 4, Set 8.1a, is it possible to satisfy all the nutritional requirements?

5. In Problem 5, Set 8.1a, determine the solution, and specify whether or not the daily production of wheels and seats can be balanced.

6. In Problem 6, Set 8.1a, suppose that the market demand goal is twice as important as that of balancing the two machines, and that no overtime is allowed. Solve the problem, and determine if the goals are met.

*7. In Problem 7, Set 8.1a, suppose that production strives to meet the quota for the two products, using overtime if necessary. Find a solution to the problem, and specify the amount of overtime, if any, needed to meet the production quota.

8. In the Vista City Hospital of Problem 8, Set 8.1a, suppose that only the bed limits represent flexible goals and that all the goals have equal weights. Can all the goals be met?

9. The Malco Company has compiled the following table from the files of five of its employees to study the impact on income of three factors: age, education (expressed in number of college years completed), and experience (expressed in number of years in the business).

Age (yr)	Education (yr)	Experience (yr)	Annual income ($)
30	4	5	40,000
39	5	10	48,000
44	2	14	38,000
48	0	18	36,000
37	3	9	41,000

Use the goal programming formulation in Problem 10, Set 8.1a, to fit the data into the linear equation $y = b_0 + b_1x_1 + b_2x_2 + b_3x_3$.

10. Solve Problem 9 using the Chebyshev method proposed in Problem11, Set 8.1a.

8.2.2 The Preemptive Method

In the preemptive method, the decision maker must rank the goals of the problem in order of importance. Given an n-goal situation, the objectives of the problem are written as

$$\text{Minimize } G_1 = \rho_1 \text{ (Highest priority)}$$
$$\vdots$$
$$\text{Minimize } G_n = \rho_n \text{ (Lowest priority)}$$

The variable ρ_i is the component of the deviational variables, s_i^- or s_i^+, that represents goal i. For example, in the TopAd model (Example 8.2-1), $\rho_1 = s_1^-$ and $\rho_2 = s_2^+$.

The solution procedure considers one goal at a time, starting with the highest priority, G_1, and terminating with the lowest, G_n. *The process is carried out such that the solution obtained from a lower-priority goal never degrades any higher-priority solutions.*

The literature on goal programming presents a "special" simplex method that guarantees the nondegradation of higher-priority solutions. The method uses the **column-dropping rule** that calls for eliminating a *nonbasic* variable x_j with nonzero reduced cost $(z_j - c_j \neq 0)$ from the optimal tableau of goal G_k before solving the problem of goal G_{k+1}. The rule recognizes that such nonbasic variables, if elevated above zero level in the optimization of succeeding goals, can degrade (but never improve) the quality of a higher-priority goal. The procedure requires modifying the simplex tableau to include the objective functions of all the goals of the model.

The proposed *column-dropping* modification needlessly complicates goal programming. In this presentation, we show that the same results can be achieved in a more straightforward manner using the following steps:

Step 0. Identify the goals of the model and rank them in order of priority:

$$G_1 = \rho_1 > G_2 = \rho_2 > \cdots > G_n = \rho_n$$

Set $i = 1$.

Step 1. Solve LP_i that minimizes G_i, and let $\rho_i = \rho_i^*$ define the corresponding optimum value of the deviational variable ρ_i. If $i = n$, stop; LP_n solves the n-goal program. Otherwise, add the constraint $\rho_i = \rho_i^*$ to the constraints of the G_i-problem to ensure that the value of ρ_i will not be degraded in future problems. Set $i = i + 1$, and repeat step i.

The successive addition of the special constraints $\rho_i = \rho_i^*$ may not be as "elegant" theoretically as the *column-dropping rule.* Nevertheless, it achieves the exact same result. More important, it is easier to understand.

Some may argue that the *column-dropping rule* offers computational advantages. Essentially, the rule makes the problem smaller successively by removing variables, whereas our procedure makes the problem larger by adding new constraints. However, considering the nature of the additional constraints ($\rho_i = \rho_i^*$), we should be able to modify the simplex algorithm to implement the additional constraint implicitly through direct substitution of the variable ρ_i. This substitution affects only the constraint in which ρ_i appears and, in effect, reduces the number of variables as we move from one goal to the next. Alternatively, we can use the bounded simplex method of Section 7.4.2 by replacing $\rho_i = \rho_i^*$ with $\rho_i \leq \rho_i^*$, in which case the additional constraints are accounted for implicitly. In this regard, the *column-dropping rule*, theoretical appeal aside, does not appear to offer a particular computational advantage. For the sake of completeness, however, we will demonstrate in Example 8.2-3 how the *column-dropping rule* works.

Example 8.2-2

The problem of Example 8.2-1 is solved by the preemptive method. Assume that the exposure goal has a higher priority.

Step 0. $G_1 > G_2$

$$G_1\text{: Minimize } s_1^- \text{ (Satisfy exposure goal)}$$
$$G_2\text{: Minimize } s_2^+ \text{ (Satisfy budget goal)}$$

Step 1. Solve LP_1.

$$\text{Minimize } G_1 = s_1^-$$

subject to

$$\begin{aligned}
4x_1 + 8x_2 + s_1^- - s_1^+ \qquad\qquad &= 45 \text{ (Exposure goal)}\\
8x_1 + 24x_2 \qquad\qquad + s_2^- - s_2^+ &= 100 \text{ (Budget goal)}\\
x_1 + 2x_2 \qquad\qquad &\leq 10 \text{ (Personnel limit)}\\
x_1 \qquad\qquad &\leq 6 \text{ (Radio limit)}
\end{aligned}$$

$$x_1, x_2, s_1^-, s_1^+, s_2^-, s_2^+ \geq 0$$

The optimum solution (determined by TORA) is $x_1 = 5$ minutes, $x_2 = 2.5$ minutes, $s_1^- = 5$ million people, with the remaining variables equal to zero. The solution shows that the exposure goal, G_1, is violated by 5 million persons.

In LP_1, we have $\rho_1 = s_1^-$. Thus, the additional constraint we use with the G_2-problem is $s_1^- = 5$ (or, equivalently, $s_1^- \leq 5$).

Step 2. We need to solve LP_2, whose objective function is

$$\text{Minimize } G_2 = s_2^+$$

subject to the same set of constraint as in step 1 *plus* the additional constraint $s_1^- = 5$. (We can replace the new constraint conveniently in TORA's MODIFY option by assigning 5 to the lower and upper bounds of s_1^-.)

The optimization of LP_2 is not necessary, because the optimum solution to problem G_1 already yields $s_2^+ = 0$. Hence, the solution of LP_1 is automatically optimum for LP_2 as well. The solution $s_2^+ = 0$ shows that G_2 is satisfied.

The additional constraint $s_1^- = 5$ can also be accounted for by substituting out s_1^- in the first constraint. The result is that the right-hand side of the exposure goal constraint will be changed from 45 to 40, thus reducing LP_2 to

$$\text{Minimize } G_2 = s_2^+$$

subject to

$$\begin{aligned}
4x_1 + 8x_2 - s_1^+ \qquad\qquad &= 40 \text{ (Exposure goal)}\\
8x_1 + 24x_2 \qquad + s_2^- - s_2^+ &= 100 \text{ (Budget goal)}\\
x_1 + 2x_2 \qquad\qquad &\leq 10 \text{ (Personnel limit)}\\
x_1 \qquad\qquad &\leq 6 \text{ (Radio limit)}\\
x_1, x_2, s_1^+, s_2^-, s_2^+ &\geq 0
\end{aligned}$$

The new formulation is one variable less than the one in LP1, which is the general idea advanced by the *column-dropping rule*.

Example 8.2-3 (Column-Dropping Rule)

In this example, we show that a better solution for the problem of Examples 8.2-1 and 8.2-2 can be obtained if the preemptive method is used to *optimize* objectives rather than to *satisfy* goals. The example also serves to demonstrate the *column-dropping rule* for solving goal programs.

The goals of Example 8.2-1 can be restated as

Priority 1: Maximize exposure (P_1)

Priority 2: Minimize cost (P_2)

Mathematically, the two objectives are given as

$$\begin{aligned}
\text{Maximize } P_1 &= 4x_1 + 8x_2 \qquad \text{(Exposure)}\\
\text{Minimize } P_2 &= 8x_1 + 24x_2 \qquad \text{(Cost)}
\end{aligned}$$

The specific goal limits for exposure and cost (=45 and 100) in Examples 8.2-1 and 8.2-2 are removed, because we will allow the simplex method to determine these limits optimally.

The new problem can thus be stated as

$$\begin{aligned}
\text{Maximize } P_1 &= 4x_1 + 8x_2\\
\text{Minimize } P_2 &= 8x_1 + 24x_2
\end{aligned}$$

subject to

$$\begin{aligned}
x_1 + 2x_2 &\leq 10\\
x_1 \qquad &\leq 6\\
x_1, x_2 &\geq 0
\end{aligned}$$

We first solve the problem using the procedure introduced in Example 8.2-2.

Step 1. Solve LP_1.

$$\text{Maximize } P_1 = 4x_1 + 8x_2$$

subject to

$$x_1 + 2x_2 \le 10$$
$$x_1 \le 6$$
$$x_1, x_2 \ge 0$$

The optimum solution (obtained by TORA) is $x_1 = 0, x_2 = 5$ with $P_1 = 40$, which shows that the most exposure we can get is 40 million persons.

Step 2. Add the constraint $4x_1 + 8x_2 \ge 40$ to ensure that goal G_1 is not degraded. Thus, we solve LP_2 as

$$\text{Minimize } P_2 = 8x_1 + 24x_2$$

subject to

$$x_1 + 2x_2 \le 10$$
$$x_1 \le 6$$
$$4x_1 + 8x_2 \ge 40 \quad \text{(additional constraint)}$$
$$x_1, x_2 \ge 0$$

The optimum solution of LP_2 is $P_2 = \$96{,}000$, $x_1 = 6$ minutes, and $x_2 = 2$ minutes. It yields the same exposure ($P_1 = 40$ million people) but at a smaller cost than the one in Example 8.2-2, where the main objective is to satisfy rather than optimize the goals.

The same problem is solved now by using the *column-dropping rule*. The rule calls for carrying the objective rows associated with all the goals in the simplex tableau, as we will show below.

LP_1 (Exposure Maximization): The LP_1 simplex tableau carries both objective rows, P_1 and P_2. The optimality condition applies to the P_1-objective row only. The P_2-row plays a passive role in LP_1, but must be updated with the rest of the simplex tableau in preparation for the optimization of LP_2.

LP_1 is solved in two iterations as follows:

Iteration	Basic	x_1	x_2	s_1	s_2	Solution
1	P_1	−4	−8	0	0	0
	P_2	−8	−24	0	0	0
	s_1	1	2	1	0	10
	s_2	1	0	0	1	6
2	P_1	0	0	4	0	40
	P_2	4	0	12	0	120
	x_2	$\frac{1}{2}$	1	$\frac{1}{2}$	0	5
	s_2	1	0	0	1	6

The last tableau yields the optimal solution $x_1 = 0, x_2 = 5$, and $P_1 = 40$.

The *column-dropping rule* calls for eliminating any *nonbasic* variable x_j with $z_j - c_j \neq 0$ from the optimum tableau of LP_1 before LP_2 is optimized. The reason for doing so is that these variables, if left unchecked, could become positive in lower-priority optimization problems, which would degrade the quality of higher-priority solutions.

LP_2 (Cost Minimization): The column-dropping rule eliminates s_1 (with $z_j - c_j = 4$). We can see from the P_2-row that if s_1 is not eliminated, it will be the entering variable at the start of the P_2-iterations and will yield the optimum solution $x_1 = x_2 = 0$, which will degrade the optimum objective value of the P_1-problem from $P_1 = 40$ to $P_1 = 0$. (Try it!)

The P_2-problem is of the minimization type. Following the elimination of s_1, the variable x_1 with $z_j - c_j = 4\ (> 0)$ can improve the value of P_2. The following table shows the LP_2 iterations. The P_1-row has been deleted because it no longer serves a purpose in the optimization of LP_2.

Iteration	Basic	x_1	x_2	s_1	s_2	Solution
1	P_1					40
	P_2	4	0		0	120
	x_2	$\frac{1}{2}$	1		0	5
	s_2	1	0		1	6
2	P_1					40
	P_2	0	0		-4	96
	x_2	0	1		$-\frac{1}{2}$	2
	x_1	1	0		1	6

The optimum solution $(x_1 = 6, x_2 = 2)$ with a total exposure of $P_1 = 40$ and a total cost of $P_2 = 96$ is the same as obtained earlier.

AMPL Moment

AMPL lends itself readily to applying the idea presented in Example 8.2-2, where simple constraints are added to ensure that higher-priority solutions are not degraded. Figure 8.1 provides a generic AMPL code that allows the application of the preemptive method interactively (file amplEx8.1-1.txt).

The design of the model is standard except for the provisions that allow applying the preemptive method interactively. Specifically, the model assumes that the first r constraints are goal constraints and the remaining $m - r - 1$ are strict constraints. The model has r distinct goal objective functions, which can be included in the same model by using the following indexed AMPL statement (only an *indexed* name is allowed for multiple objective functions):

```
minimize z{i in 1..r}:p*sminus[i]+q*splus[i];
```

The given definition of the objective function accounts for minimizing $z_i = s_i^-$ and $z_i = s_i^+$ by setting $(p = 1, q = 0)$ and $(p = 0, q = 1)$, respectively.

Instead of adding a new constraint each time we move from one priority level to the next, we use a programming trick that allows modifying the upper bounds on the deviational variables. The parameters `um[i]` and `up[i]` represent the upper bounds on `sminus[i]` (s_i^-) and `splus[i]` (s_i^+), respectively. These parameters are modified to impose implicit constraints of the type `sminus[i]<=um[i]` and `splus[i]<=up[i]`, respectively. The values of `um` and `up` in priority goal `i` are determined from the solutions of the problems of priority goals 1, 2, and $i - 1$. The initial (default) value for `um` and `up` is infinity.

```
param n;
param m;
param r;
param p;
param q;
param a{1..m,1..n};
param b{1..m};
param um{1..m} default 100000;
param up{1..m} default 100000;
var x{1..n} >=0;
var sminus{i in 1..r}>=0,<=um[i];
var splus{i in 1..r}>=0,<=up[i];

minimize z{i in 1..r}: p*sminus[i]+q*splus[i];
subject to
c1{i in 1..r}:sum{j in 1..n}a[i,j]*x[j]+sminus[i]-splus[i]=b[i];
c2{i in r+1..m}: sum{j in 1..n} a[i,j]*x[j]<=b[i];
data;
param m:=4;
param n:=4;
param r:=4;
param a: 1      2       3       4:=
      1  550    35      55      .075
      2  55     -31.5   5.5     .0075
      3  110    7       -44     .015
      4  0      0       0       1;
param b:=1 16 2 0  3 0  4 2;
```

FIGURE 8.1

AMPL model for interactive application of the preemptive method (file amplEx8.1-1.txt)

We will show shortly how AMPL activates any of the r objective functions, specifies the values of p and q, and sets the upper limits on s_i^- and s_i^+, all interactively, which makes AMPL ideal for carrying out goal programming computations.

Using the data of Example 8.1-1, the goals of the model are

$$\text{Minimize } G_1 = s_1^-$$

$$\text{Minimize } G_2 = s_2^-$$

$$\text{Minimize } G_3 = s_3^-$$

$$\text{Minimize } G_4 = s_4^+$$

Suppose that the goals are prioritized as

$$G_2 > G_1 > G_3 > G_4$$

The implementation of AMPL model thus proceeds in the following manner: For G_2, set $p = 1$ and $q = 0$ because we are minimizing z[2]=sminus[2]. The following commands are used to carry out the calculations:

```
ampl: model amplEx8.1-1.txt;
ampl: let p:=1;let q:=0;
ampl: objective z[2];
ampl: solve; display z[2], x, sminus, splus;
```

These commands produce the following results:

```
z[2] = 0
:          x        sminus splus    :=
1   2.62361e-28     16      0
2   0                0      0
3   8.2564e-28       0      0
4   0                2      0
```

The solution ($x_1 = 0, x_2 = 0, x_3 = 0$, and $x_4 = 0$) shows that goal G_2 is satisfied because `z[2] = 0` (that is, $s_2^- = 0$). However, the next-priority goal, G_1, is not satisfied because $s_1^- = 16$. Hence, we need to optimize goal G_1 without degrading the solution of G_2. This requires changing the upper bounds on s_2^- to the value specified by the solution of G_2—namely, zero. For goal G_1, current $p = 1$ and $q = 0$ from G_2 remain unchanged because we are minimizing s_1^-. The following interactive AMPL commands achieve this result:

```
ampl: let um[2]:=0;
ampl: objective z[1]; solve; display z[1],x, sminus, splus;
```

The ouput is

```
z[1] = 0
:          x      sminus splus    :=
1   0.0203636     0      0
2   0.0457143     0      0
3   0.0581818     0      0
4   0             2      0
```

The solution shows that all the remaining goals are satisfied. Hence, no further optimization is needed. The goal programming solution is $x_p = .0203636$, $x_f = .0457143$, $x_s = .0581818$, and $x_g = 0$.

Remarks.

1. We can replace `let um[2]:=0;` with either `fix sminus[2]:=0;` or `let sminus[2]:=0;` with equal end result.
2. The interactive session can be totally automated using a `commands` file that automatically selects the current goal to be optimized and imposes the proper restrictions before solving the next priority goal. The use of this file (which we name amplCmds.txt) requires making some modifications in the original model as shown in file amplEx8.1-1A.txt. To be completely versatile, the data of the model are stored in a separate file named amplData.txt. In this case, the execution of the model requires issuing three command lines:

```
ampl: model amplEx8.1-1A.txt;
ampl: data amplData.txt;
ampl: commands amplCmds.txt;
```

See Section A.7 for more information about the use of `commands`.

PROBLEM SET 8.2B[2]

1. In Example 8.2-2, suppose that the budget goal is increased to $110,000. The exposure goal remains unchanged at 45 million persons. Show how the preemptive method will reach a solution.
2. Solve Problem 1, Set 8.1a, using the following priority ordering for the goals:

$$G_1 > G_2 > G_3 > G_4 > G_5.$$

3. Consider Problem 2, Set 8.1a, which deals with the presentation of band concerts and art shows at the NW Mall. Suppose that the goals set for teens, the young/middle-aged group, and seniors are referred to as G_1, G_2, and G_3, respectively. Solve the problem for each of the following priority orders:
   (a) $G_1 > G_2 > G_3$
   (b) $G_3 > G_2 > G_1$
   Show that the satisfaction of the goals (or lack of it) can be a function of the priority order.
4. Solve the Ozark University model (Problem 3, Set 8.1a) using the preemptive method, assuming that the goals are prioritized in the same order given in the problem.

REFERENCES

Chissman, J., T. Fey, G. Reeves, H. Lewis, and R. Weinstein, "A Multiobjective Linear Programming Methodology for Public Sector Tax Planning," *Interfaces*, Vol. 19, No. 5, pp. 13–22, 1989.

Cohon, T. L., *Multiobjective Programming and Planning,* Academic Press, New York, 1978.

Ignizio, J. P., and T. M. Cavalier, *Linear Programming,* Prentice Hall, Upper Saddle River, NJ 1994.

Steuer, R. E., *Multiple Criteria Optimization: Theory, Computations, and Application,* Wiley, New York, 1986.

[2]You may find it computationally convenient to use interactive AMPL to solve the problems of this set.

C H A P T E R 9

Integer Linear Programming

Chapter Guide. Integer linear programs (ILPs) are linear programs with some or all the variables restricted to integer (or discrete) values. When you study ILP, you need to concentrate on three areas: application, theory, and computation. The chapter starts with a number of applications that demonstrate the rich use of ILP in practice. Then it presents the two prominent algorithms of ILP: branch and bound (B&B) and cutting plane. Of the two algorithms, B&B is decidedly more efficient computationally. Indeed, practically all commercial codes are rooted in B&B. The chapter closes with a presentation of the traveling salesperson problem (TSP), a problem that has important practical applications.

A drawback of ILP algorithms is their lack of consistency in solving integer problems. Although these algorithms are proven theoretically to converge in a finite number of iterations, their implementation on the computer (with its inherent machine roundoff error) is a different experience. You should keep this point in mind as you study the ILP algorithms.

The chapter shows how AMPL and Solver are used with ILP. You will find TORA's user-guided option useful in detailing the B&B computations.

This chapter includes a summary of 1 real-life application, 12 solved examples, 5 AMPL models, 1 Excel spreadsheet, 65 end-of-section problems, and 10 cases. The cases are in Appendix E on the CD. The AMPL/Excel/Solver/TORA programs are in folder ch9Files.

Real-Life Application—Optimizing Trailer Payloads at PFG Building Glass

PFG uses specially equipped (fifth-wheel) trailers to deliver packs of sheets of flat glass to customers. The packs vary in both size and weight, and a single trailer load may include different packs, depending on received orders. Government regulations set maximum limits on axle weights, and the actual positioning of the packs on the trailer is crucial in determining these weights. The problem deals with determining the optimal loading of the packs on the trailer bed to satisfy axle-weight limits. The problem is solved as an integer program. Case 7 in Chapter 24 on the CD provides the details of the study.

9.1 ILLUSTRATIVE APPLICATIONS

This section presents a number of ILP applications. The applications generally fall into two categories: *direct* and *transformed*. In the *direct* category, the variables are naturally integer and may assume binary (0 or 1) or general discrete values. For example, the problem may involve determining whether or not a project is selected for execution (binary) or finding the optimal number of machines needed to perform a task (general discrete value). In the *transformed* category, the original problem, which may not involve any integer variables, is analytically intractable. Auxiliary integer variables (usually binary) are used to make it tractable. For example, in sequencing two jobs, A and B, on a single machine, job A may precede job B *or* job B may precede job A. The "or" nature of the constraints is what makes the problem analytically intractable, because all mathematical programming algorithms deal with "and" constraints only. The situation is remedied by using auxiliary binary variables to transform the "or" constraints into *equivalent* "and" constraints.

For convenience, a **pure** integer problem is defined to have *all* integer variables. Otherwise, a problem is a **mixed** integer program if it deals with both continuous and integer variables.

9.1.1 Capital Budgeting

This section deals with decisions regarding whether or not investments should be made in individual projects. The decision is made under limited-budget considerations as well as priorities in the execution of the projects.

Example 9.1-1 (Project Selection)

Five projects are being evaluated over a 3-year planning horizon. The following table gives the expected returns for each project and the associated yearly expenditures.

	Expenditures (million $)/yr			
Project	*1*	*2*	*3*	Returns (million $)
1	5	1	8	20
2	4	7	10	40
3	3	9	2	20
4	7	4	1	15
5	8	6	10	30
Available funds (million $)	25	25	25	

Which projects should be selected over the 3-year horizon?

The problem reduces to a "yes-no" decision for each project. Define the binary variable x_j as

$$x_j = \begin{cases} 1, \text{ if project } j \text{ is selected} \\ 0, \text{ if project } j \text{ is not selected} \end{cases}$$

The ILP model is

$$\text{Maximize } z = 20x_1 + 40x_2 + 20x_3 + 15x_4 + 30x_5$$

subject to

$$\begin{aligned} 5x_1 + \ 4x_2 + 3x_3 + 7x_4 + \ 8x_5 &\leq 25 \\ x_1 + \ 7x_2 + 9x_3 + 4x_4 + \ 6x_5 &\leq 25 \\ 8x_1 + 10x_2 + 2x_3 + \ x_4 + 10x_5 &\leq 25 \\ x_1, x_2, x_3, x_4, x_5 &= (0, 1) \end{aligned}$$

The optimum integer solution (obtained by AMPL, Solver, or TORA)[1] is $x_1 = x_2 = x_3 = x_4 = 1, x_5 = 0$, with $z = 95$ (million \$). The solution shows that all but project 5 must be selected.

Remarks. It is interesting to compare the continuous LP solution with the ILP solution. The LP optimum, obtained by replacing $x_j = (0, 1)$ with $0 \leq x_j \leq 1$ for all j, yields $x_1 = .5789$, $x_2 = x_3 = x_4 = 1, x_5 = .7368$, and $z = 108.68$ (million \$). The solution is meaningless because two of the variables assume fractional values. We may *round* the solution to the closest integer values, which yields $x_1 = x_5 = 1$. However, the resulting solution is infeasible because the constraints are violated. More important, the concept of *rounding* is meaningless here because x_j represents a "yes-no" decision.

PROBLEM SET 9.1A[2]

1. Modify and solve the capital budgeting model of Example 9.1-1 to account for the following additional restrictions:
   (a) Project 5 must be selected if either project 1 or project 3 is selected.
   (b) Projects 2 and 3 are mutually exclusive.
2. Five items are to be loaded in a vessel. The weight w_i, volume v_i, and value r_i for item i are tabulated below.

Item i	Unit weight, w_i (tons)	Unit volume, v_i (yd^3)	Unit worth, r_i (100 \$)
1	5	1	4
2	8	8	7
3	3	6	6
4	2	5	5
5	7	4	4

[1]To use TORA, select Integer Programming from Main Menu. After entering the problem data, go to output screen and select Automated B&B to obtain the optimum solution. Solver use is the same as in LP except that the targeted variables must be declared integer. The integer option (*int* or *bin*) is available in the **Solver Parameters** dialogue box when you add a new constraint. AMPL implementation for integer programming is the same as in linear programming, except that some or all the variables are declared integer by adding the key word `integer` (or `binary`) in the definition statement of the targeted variables. For example, the statement `var x{J}>=0,integer;` declares x_j as nonnegative integer for all $j \in J$. If x_j is binary, the statement is changed to `var x{J} binary;`. For execution, the statement `option solver cplex;` must precede `solve;`.
[2]Problems 3 to 6 are adapted from Malba Tahan, *El Hombre que Calculaba,* Editorial Limusa, Mexico City, pp. 39–182, 1994.

The maximum allowable cargo weight and volume are 112 tons and 109 yd^3, respectively. Formulate the ILP model, and find the most valuable cargo.

***3.** Suppose that you have 7 full wine bottles, 7 half-full, and 7 empty. You would like to divide the 21 bottles among three individuals so that each will receive exactly 7. Additionally, each individual must receive the same quantity of wine. Express the problem as ILP constraints, and find a solution. (*Hint*: Use a dummy objective function in which all the objective coefficients are zeros.)

4. An eccentric sheikh left a will to distribute a herd of camels among his three children: Tarek receives at least one-half of the herd, Sharif gets at least one third, and Maisa gets at least one-ninth. The remainder goes to charity. The will does not specify the size of the herd except to say that it is an odd number of camels and that the named charity receives exactly one camel. Use ILP to determine how many camels the sheikh left in the estate and how many each child got.

5. A farm couple are sending their three children to the market to sell 90 apples with the objective of educating them about money and numbers. Karen, the oldest, carries 50 apples; Bill, the middle one, carries 30; and John, the youngest, carries only 10. The parents have stipulated five rules: (a) The selling price is either \$1 for 7 apples or \$3 for 1 apple, or a combination of the two prices. (b) Each child may exercise one or both options of the selling price. (c) Each of the three children must return with exactly the same amount of money. (d) Each child's income must be in whole dollars (no cents allowed). (e) The amount received by each child must be the largest possible under the stipulated conditions. Given that the three kids are able to sell all they have, use ILP to show how they can satisfy the parents' conditions.

***6.** Once upon a time, there was a captain of a merchant ship who wanted to reward three crew members for their valiant effort in saving the ship's cargo during an unexpected storm in the high seas. The captain put aside a certain sum of money in the purser's office and instructed the first officer to distribute it equally among the three mariners after the ship had reached shore. One night, one of the sailors, unbeknown to the others, went to the purser's office and decided to claim (an equitable) one-third of the money in advance. After he had divided the money into three equal shares, an extra coin remained, which the mariner decided to keep (in addition to one-third of the money). The next night, the second mariner got the same idea and, repeating the same three-way division with what was left, ended up keeping an extra coin as well. The third night, the third mariner also took a third of what was left, plus an extra coin that could not be divided. When the ship reached shore, the first officer divided what was left of the money equally among the three mariners, again to be left with an extra coin. To simplify things, the first officer put the extra coin aside and gave the three mariners their allotted equal shares. How much money was in the safe to start with? Formulate the problem as an ILP, and find the solution. (*Hint*: The problem has a countably infinite number of integer solutions. For convenience, assume that we are interested in determining the smallest sum of money that satisfies the problem conditions. Then, boosting the resulting sum by 1, add it as a lower bound and obtain the next smallest sum. Continuing in this manner, a general solution pattern will evolve.)

7. (Weber, 1990) You have the following three-letter words: AFT, FAR, TVA, ADV, JOE, FIN, OSF, and KEN. Suppose that we assign numeric values to the alphabet starting with

$A = 1$ and ending with $Z = 26$. Each word is scored by adding numeric codes of its three letters. For example, AFT has a score of $1 + 6 + 20 = 27$. You are to select five of the given eight words that yield the maximum total score. Simultaneously, the selected five words must satisfy the following conditions:

$$\begin{pmatrix}\text{sum of letter 1}\\ \text{scores}\end{pmatrix} < \begin{pmatrix}\text{sum of letter 2}\\ \text{scores}\end{pmatrix} < \begin{pmatrix}\text{sum of letter 3}\\ \text{scores}\end{pmatrix}$$

Formulate the problem as an ILP, and find the optimum solution.

8. Solve Problem 7 given that, in addition to the total sum being the largest, the sum of column 1 and the sum of column 2 will be the largest as well. Find the optimum solution.

9. (Weber, 1990) Consider the following two groups of words:

Group 1	Group 2
AREA	ERST
FORT	FOOT
HOPE	HEAT
SPAR	PAST
THAT	PROF
TREE	STOP

All the words in groups 1 and 2 can be formed from the nine letters A, E, F, H, O, P, R, S, and T. Develop a model to assign a unique numeric value from 1 through 9 to these letters such that the difference between the total scores of the two groups will be as small as possible. [*Note*: The score for a word is the sum of the numeric values assigned to its individual letters.]

***10.** The Record-a-Song Company has contracted with a rising star to record eight songs. The durations of the different songs are 8, 3, 5, 5, 9, 6, 7, and 12 minutes, respectively. Record-a-Song uses a two-sided cassette tape for the recording. Each side has a capacity of 30 minutes. The company would like to distribute the songs between the two sides such that the length of the songs on each side is about the same. Formulate the problem as an ILP, and find the optimum solution.

11. In Problem 10, suppose that the nature of the melodies dictates that songs 3 and 4 cannot be recorded on the same side. Formulate the problem as an ILP. Would it be possible to use a 25-minute tape (each side) to record the eight songs? If not, use ILP to determine the minimum tape capacity needed to make the recording.

***12.** (Graves and Associates, 1993) Ulern University uses a mathematical model that optimizes student preferences taking into account the limitation of classroom and faculty resources. To demonstrate the application of the model, consider the simplified case of 10 students who are required to select two courses out of six offered electives. The table below gives scores that represent each student's preference for individual courses, with a score of 100 being the highest. For simplicity, it is assumed that the preference score for a two-course selection is the sum of the individual score. Course capacity is the maximum number of students allowed to take the class.

	Preference score for course					
Student	*1*	*2*	*3*	*4*	*5*	*6*
1	20	40	50	30	90	100
2	90	100	80	70	10	40
3	25	40	30	80	95	90
4	80	50	60	80	30	40
5	75	60	90	100	50	40
6	60	40	90	10	80	80
7	45	40	70	60	55	60
8	30	100	40	70	90	55
9	80	60	100	70	65	80
10	40	60	80	100	90	10
Course capacity	6	8	5	5	6	5

Formulate the problem as an ILP and find the optimum solution.

9.1.2 Set-Covering Problem

In this class of problems, overlapping services are offered by a number of installations to a number of facilities. The objective is to determine the minimum number of installations that will *cover* (i.e., satisfy the service needs) of each facility. For example, water treatment plants can be constructed at various locations, with each plant serving different sets of cities. The overlapping arises when a given city can receive service from more than one plant.

Example 9.1-2 (Installing Security Telephones)

To promote on-campus safety, the U of A Security Department is in the process of installing emergency telephones at selected locations. The department wants to install the minimum number of telephones, provided that each of the campus main streets is served by at least one telephone. Figure 9.1 maps the principal streets (A to K) on campus.

It is logical to place the telephones at street intersections so that each telephone will serve at least two streets. Figure 9.1 shows that the layout of the streets requires a maximum of eight telephone locations.

Define

$$x_j = \begin{cases} 1, \text{ a telephone is installed in location } j \\ 0, \text{ otherwise} \end{cases}$$

The constraints of the problem require installing at least one telephone on each of the 11 streets (A to K). Thus, the model becomes

$$\text{Minimize } z = x_1 + x_2 + x_3 + x_4 + x_5 + x_6 + x_7 + x_8$$

subject to

$$x_1 + x_2 \geq 1 \quad (\text{Street } A)$$

$$x_2 + x_3 \geq 1 \quad (\text{Street } B)$$

$$x_4 + x_5 \geq 1 \quad (\text{Street } C)$$

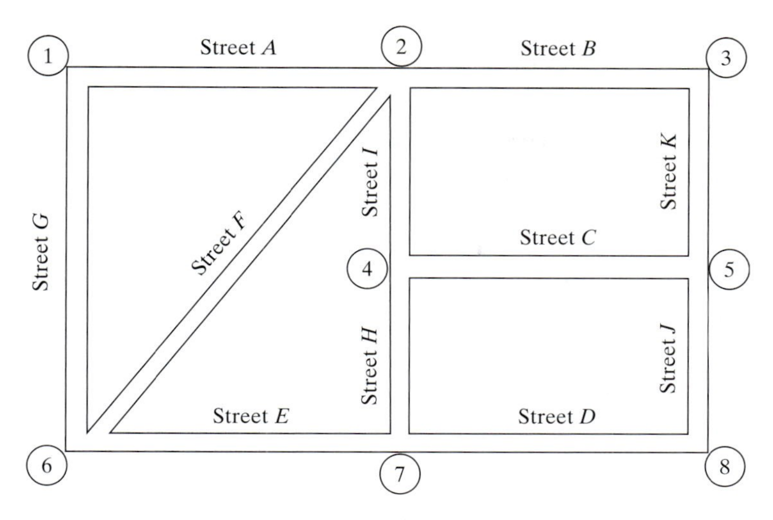

FIGURE 9.1
Street Map of the U of A Campus

$$
\begin{array}{lr}
x_7 + x_8 \geq 1 & \text{(Street } D\text{)} \\
x_6 + x_7 \geq 1 & \text{(Street } E\text{)} \\
x_2 + x_6 \geq 1 & \text{(Street } F\text{)} \\
x_1 + x_6 \geq 1 & \text{(Street } G\text{)} \\
x_4 + x_7 \geq 1 & \text{(Street } H\text{)} \\
x_2 + x_4 \geq 1 & \text{(Street } I\text{)} \\
x_5 + x_8 \geq 1 & \text{(Street } J\text{)} \\
x_3 + x_5 \geq 1 & \text{(Street } K\text{)} \\
x_j = (0, 1), j = 1, 2, \ldots, 8 &
\end{array}
$$

The optimum solution of the problem requires installing four telephones at intersections 1, 2, 5, and 7.

Remarks. In the strict sense, set-covering problems are characterized by (1) the variables $x_j, j = 1, 2, \ldots, n$, are binary, (2) the left-hand-side coefficients of the constraints are 0 or 1, (3) the right-hand side of each constraint is of the form (≥ 1), and (4) the objective function minimizes $c_1x_1 + c_2x_2 + \cdots + c_nx_n$, where $c_j > 0$ for all $j = 1, 2, \ldots, n$. In the present example, $c_j = 1$ for all j. If c_j represents the installation cost in location j, then these coefficients may assume values other than 1. Variations of the set-covering problem include additional side conditions, as some of the situations in Problem Set 9.1b show.

AMPL Moment

Figure 9.2 presents a general AMPL model for any set-covering problem (file amplEx9.1-2.txt). The formulation is straightforward, once the use of *indexed set* is understood (see Section A.4). The model defines `street` as a (regular) set whose elements

```
#--------------Example 9.1-2--------------------
param n;   #maximum number of corners
set street;
set corner{street};
var x{1..n}binary;
minimize z: sum {j in 1..n} x[j];
subject to limit {i in street}:
        sum {j in corner[i]} x[j]>=1;
data;
param n:=8;
set street:=A B C D E F G H I J K;
set corner[A]:=1 2;
set corner[B]:=2 3;
set corner[C]:=4 5;
set corner[D]:=7 8;
set corner[E]:=6 7;
set corner[F]:=2 6;
set corner[G]:=1 6;
set corner[H]:=4 7;
set corner[I]:=2 4;
set corner[J]:=5 8;
set corner[K]:=3 5;

option solver cplex;
solve;
display z,x;
```

FIGURE 9.2

General AMPL model for the set-covering problem (file ampl Ex 9.1-2.txt)

are `A` through `K`. Next, the *indexed* set `corner{street}` defines the corners as a function of street. With these two sets, the constraints of the model can be formulated directly. The data of the model give the elements of the indexed sets that are specific to the situation in Example 9.1-2. Any other situation is handled by changing the data of the model.

PROBLEM SET 9.1B

***1.** ABC is an LTL (less-than-truckload) trucking company that delivers loads on a daily basis to five customers. The following list provides the customers associated with each route:

Route	Customers served on the route
1	1, 2, 3, 4
2	4, 3, 5
3	1, 2, 5
4	2, 3, 5
5	1, 4, 2
6	1, 3, 5

The segments of each route are dictated by the capacity of the truck delivering the loads. For example, on route 1, the capacity of the truck is sufficient to deliver the loads to customers 1, 2, 3, and 4 only. The following table lists distances (in miles) among the truck terminal (ABC) and the customers.

	Miles from i to j					
i \ j	ABC	*1*	*2*	*3*	*4*	*5*
ABC	0	10	12	16	9	8
1	10	0	32	8	17	10
2	12	32	0	14	21	20
3	16	8	14	0	15	18
4	9	17	21	15	0	11
5	8	10	20	18	11	0

The objective is to determine the least distance needed to make the daily deliveries to all five customers. Though the solution may result in a customer being served by more than one route, the implementation phase will use only one such route. Formulate the problem as an ILP and find the optimum solution.

*2. The U of A is in the process of forming a committee to handle students' grievances. The administration wants the committee to include at least one female, one male, one student, one administrator, and one faculty member. Ten individuals (identified, for simplicity, by the letters a to j) have been nominated. The mix of these individuals in the different categories is given as follows:

Category	Individuals
Females	a, b, c, d, e
Males	f, g, h, i, j
Students	a, b, c, j
Administrators	e, f
Faculty	d, g, h, i

The U of A wants to form the smallest committee with representation from each of the five categories. Formulate the problem as an ILP and find the optimum solution.

3. Washington County includes six towns that need emergency ambulance service. Because of the proximity of some of the towns, a single station may serve more than one community. The stipulation is that the station must be within 15 minutes of driving time from the towns it serves. The table below gives the driving times in minutes among the six towns.

	Time in minutes from i to j					
i \ j	*1*	*2*	*3*	*4*	*5*	*6*
1	0	23	14	18	10	32
2	23	0	24	13	22	11
3	14	24	0	60	19	20
4	18	13	60	0	55	17
5	10	22	19	55	0	12
6	32	11	20	17	12	0

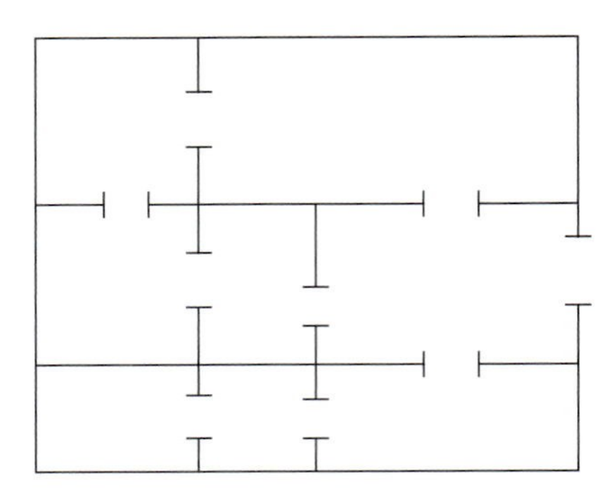

FIGURE 9.3
Museum Layout for Problem 4, Set 9.1c

Formulate an ILP whose solution will produce the smallest number of stations and their locations. Find the optimum solution.

4. The treasures of King Tut are on display in a museum in New Orleans. The layout of the museum is shown in Figure 9.3, with the different rooms joined by open doors. A guard standing at a door can watch two adjoining rooms. The museum wants to ensure guard presence in every room, using the minimum number possible. Formulate the problem as an ILP and find the optimum solution.

5. Bill has just completed his exams for the academic year and wants to celebrate by seeing every movie showing in theaters in his town and in six other neighboring cities. If he travels to another town, he will stay there until he has seen all the movies he wants. The following table provides the information about the movie offerings and the round-trip distance to the neighboring town.

Theater location	Movie offerings	Round-trip miles	Cost per show ($)
In-town	1, 3	0	7.95
City A	1, 6, 8	25	5.50
City B	2, 5, 7	30	5.00
City C	1, 8, 9	28	7.00
City D	2, 4, 7	40	4.95
City E	1, 3, 5, 10	35	5.25
City F	4, 5, 6, 9	32	6.75

The cost of driving is 75 cents per mile. Bill wishes to determine the towns he needs to visit to see all the movies while minimizing his total cost.

6. Walmark Stores is in the process of expansion in the western United States. During the next year, Walmark is planning to construct new stores that will serve 10 geographically dispersed communities. Past experience indicates that a community must be within 25 miles of a store to attract customers. In addition, the population of a community plays an important role in where a store is located, in the sense that bigger communities generate more participating customers. The following tables provide the populations as well as the distances (in miles) between the communities:

	Miles from community i to community j										
i \ j	*1*	*2*	*3*	*4*	*5*	*6*	*7*	*8*	*9*	*10*	*Population*
1		20	40	35	17	24	50	58	33	12	10,000
2	20		23	68	40	30	20	19	70	40	15,000
3	40	23		36	70	22	45	30	21	80	28,000
4	35	68	36		70	80	24	20	40	10	30,000
5	17	40	70	70		23	70	40	13	40	40,000
6	24	30	22	80	23		12	14	50	50	30,000
7	50	20	45	24	70	12		26	40	30	20,000
8	58	19	30	20	40	14	26		20	50	15,000
9	33	70	21	40	13	50	40	20		22	60,000
10	12	40	80	10	40	50	30	50	22		12,000

The idea is to construct the least number of stores, taking into account the distance restriction and the concentration of populations.

Specify the communities where the stores should be located.

*7. (Guéret and Associates, 2002, Section 12.6) MobileCo is budgeting 15 million dollars to construct as many as 7 transmitters to cover as much population as possible in 15 contiguous geographical communities. The communities covered by each transmitter and the budgeted construction costs are given below.

Transmitter	Covered communities	Cost (million $)
1	1, 2	3.60
2	2, 3, 5	2.30
3	1, 7, 9, 10	4.10
4	4, 6, 8, 9	3.15
5	6, 7, 9, 11	2.80
6	5, 7, 10, 12, 14	2.65
7	12, 13, 14, 15	3.10

The following table provides the populations of the different communities:

Community	1	2	3	4	5	6	7	8	9	10	11	12	13	14	15
Population (in 1000s)	4	3	10	14	6	7	9	10	13	11	6	12	7	5	16

Which of the proposed transmitters should be constructed?

8. (Gavernini and Associates, 2004) In modern electric networks, automated electric utility meter reading replaces the costly labor-intensive system of manual meter reading. In the automated system, meters from several customers are linked wirelessly to a single receiver. The meter sends monthly signals to a designated receiver to report the customer's consumption of electricity. The receiver then sends the data to a central computer to generate the electricity bills. The problem reduces to determining the least number of receivers needed to serve a number of customers. In real life, the problem encompasses thousands of meters and receivers. However, for the purpose of this problem, consider the case of 10 meters and 8 receivers, using the following configurations:

Receiver	1	2	3	4	5	6	7	8
Meters	1, 2, 3	2, 3, 9	5, 6, 7	7, 9, 10	3, 6, 8	1, 4, 7, 9	4, 5, 9	1, 4, 8

Determine the minimum number of receivers.

9. Solve Problem 8 if, additionally, each receiver can handle at most 3 meters.

9.1.3 Fixed-Charge Problem

The fixed-charge problem deals with situations in which the economic activity incurs two types of costs: an initial "flat" fee that must be incurred to start the activity and a variable cost that is directly proportional to the level of the activity. For example, the initial tooling of a machine prior to starting production incurs a fixed setup cost regardless of how many units are manufactured. Once the setup is done, the cost of labor and material is proportional to the amount produced. Given that F is the fixed charge, c is the variable unit cost, and x is the level of production, the cost function is expressed as

$$C(x) = \begin{cases} F + cx, \text{ if } x > 0 \\ 0, \text{ otherwise} \end{cases}$$

The function $C(x)$ is intractable analytically because it involves a discontinuity at $x = 0$. The next example shows how binary variables are used to remove this intractability.

Example 9.1-3 (Choosing a Telephone Company)

I have been approached by three telephone companies to subscribe to their long distance service in the United States. MaBell will charge a flat \$16 per month plus \$.25 a minute. PaBell will charge \$25 a month but will reduce the per-minute cost to \$.21. As for BabyBell, the flat monthly charge is \$18, and the cost per minute is \$.22. I usually make an average of 200 minutes of long-distance calls a month. Assuming that I do not pay the flat monthly fee unless I make calls and that I can apportion my calls among all three companies as I please, how should I use the three companies to minimize my monthly telephone bill?

This problem can be solved readily without ILP. Nevertheless, it is instructive to formulate it as an integer program.

Define

$$x_1 = \text{MaBell long-distance minutes per month}$$

$$x_2 = \text{PaBell long-distance minutes per month}$$

$$x_3 = \text{BabyBell long-distance minutes per month}$$

$$y_1 = 1 \text{ if } x_1 > 0 \text{ and } 0 \text{ if } x_1 = 0$$

$$y_2 = 1 \text{ if } x_2 > 0 \text{ and } 0 \text{ if } x_2 = 0$$

$$y_3 = 1 \text{ if } x_3 > 0 \text{ and } 0 \text{ if } x_3 = 0$$

We can ensure that y_j will equal 1 if x_j is positive by using the constraint

$$x_j \leq My_j, j = 1, 2, 3$$

The value of M should be selected sufficiently large so as not restrict to the variable x_j artificially. Because I make about 200 minutes of calls a month, then $x_j \leq 200$ for all j, and it is safe to select $M = 200$.

The complete model is

$$\text{Minimize } z = .25x_1 + .21x_2 + .22x_3 + 16y_1 + 25y_2 + 18y_3$$

subject to

$$\begin{aligned} x_1 + x_2 + x_3 &= 200 \\ x_1 &\leq 200y_1 \\ x_2 &\leq 200y_2 \\ x_3 &\leq 200y_3 \\ x_1, x_2, x_3 &\geq 0 \\ y_1, y_2, y_3 &= (0, 1) \end{aligned}$$

The formulation shows that the jth monthly flat fee will be part of the objective function z only if $y_j = 1$, which can happen only if $x_j > 0$ (per the last three constraints of the model). If $x_j = 0$ at the optimum, then the minimization of z, together with the fact that the objective coefficient of y_j is strictly positive, will force y_j to equal zero, as desired.

The optimum solution yields $x_3 = 200$, $y_3 = 1$, and all the remaining variables equal to zero, which shows that BabyBell should be selected as my long-distance carrier. Remember that the information conveyed by $y_3 = 1$ is redundant because the same result is implied by $x_3 > 0$ $(= 200)$. Actually, the main reason for using y_1, y_2, and y_3 is to account for the monthly flat fee. In effect, the three binary variables convert an ill-behaved (nonlinear) model into an analytically tractable formulation. This conversion has resulted in introducing the integer (binary) variables in an otherwise continuous problem.

PROBLEM SET 9.1C

1. Leatherco is contracted to manufacture batches of pants, vests, and jackets. Each product requires a special setup of the machines needed in the manufacturing processes. The following table provides the pertinent data regarding the use of raw material (leather) and labor time together with cost and revenue estimates. Current supply of leather is estimated at 3000 ft^2 and available labor time is limited to 2500 hours.

	Pants	Vests	Jackets
Leather material per unit (ft^2)	5	3	8
Labor time per unit (hrs)	4	3	5
Production cost per unit ($)	30	20	80
Equipment setup cost per batch ($)	100	80	150
Price per unit ($)	60	40	120
Minimum number of units needed	100	150	200

Determine the optimum number of units that Leatherco must manufacture of each product.

*2. Jobco is planning to produce at least 2000 widgets on three machines. The minimum lot size on any machine is 500 widgets. The following table gives the pertinent data of the situation.

Machine	Cost	Production cost/unit	Capacity (units)
1	300	2	600
2	100	10	800
3	200	5	1200

Formulate the problem as an ILP, and find the optimum solution.

*3. Oilco is considering two potential drilling sites for reaching four targets (possible oil wells). The following table provides the preparation costs at each of the two sites and the cost of drilling from site i to target j ($i = 1, 2; j = 1, 2, 3, 4$).

	Drilling cost (million \$) to target				
Site	*1*	*2*	*3*	*4*	Preparation cost (million \$)
1	2	1	8	5	5
2	4	6	3	1	6

Formulate the problem as an ILP, and find the optimum solution.

4. Three industrial sites are considered for locating manufacturing plants. The plants send their supplies to three customers. The supply at the plants, the demand at the customers, and the unit transportation cost from the plants to the customers are given in the following table.

	Unit transportations cost (\$)			
Plant \ *Customer*	*1*	*2*	*3*	*Supply*
1	10	15	12	1800
2	17	14	20	1400
3	15	10	11	1300
Demand	1200	1700	1600	

In addition to the transportation costs, fixed costs are incurred at the rate of \$12,000, \$11,000, and \$12,000 for plants 1, 2, and 3, respectively. Formulate the problem as an ILP and find the optimum solution.

5. Repeat Problem 4 assuming that the demands at each of customers 2 and 3 are changed to 800.

6. (Liberatore and Miller, 1985) A manufacturing facility uses two production lines to produce three products over the next 6 months. Backlogged demand is not allowed. However, a product may be overstocked to meet demand in later months. The following table provides the data associated with the demand, production, and storage of the three products.

Product	Demand in period						Unit holding cost ($)/month	Initial inventory
	1	*2*	*3*	*4*	*5*	*6*		
1	50	30	40	60	20	45	.50	55
2	40	60	50	30	30	55	.35	75
3	30	40	20	70	40	30	.45	60

There is a fixed cost for switching a line from one product to another. The following tables give the switching cost, the production rates, and the unit production cost for each line:

	Line switching cost ($)		
	Product 1	*Product 2*	*Product 3*
Line 1	200	180	300
Line 2	250	200	174

	Production rate (units/month)			Unit production cost ($)		
	Product 1	*Product 2*	*Product 3*	*Product 1*	*Product 2*	*Product 3*
Line 1	40	60	80	10	8	15
Line 2	90	70	60	12	6	10

Develop a model for determining the optimal production schedule.

7. (Jarvis and Associates, 1978) Seven cities are being considered as potential locations for the construction of at most four wastewater treatment plants. The table below provides the data for the situation. Missing links indicate that a pipeline cannot be constructed.

	Cost ($) of pipeline construction between cities per 1000 gal/hr capacity						
From \ *To*	*1*	*2*	*3*	*4*	*5*	*6*	*7*
1		100		200		50	
2				120		150	
3	400				120		90
4			120		120		
5		200				100	200
6			110	180			70
7	200			150			
Cost ($million) of plant construction	1.00	1.20	2.00	1.60	1.80	.90	1.40
Population (1000s)	50	100	45	90	75	60	30

The capacity of a pipeline (in gallons per hour) is a direct function of the amount of wastewater generated, which is a function of the populations. Approximately 500 gallons per 1000 residents are discharged in the sewer system per hour. The maximum plant capacity is 100,000 gal/hr. Determine the optimal location and capacity of the plants.

8. (Brown and Associates, 1987) A company uses four special tank trucks to deliver four different gasoline products to customers. Each tank has five compartments with different capacities: 500, 750, 1200, 1500, and 1750 gallons. The daily demands for the four products are estimated at 10, 15, 12, and 8 thousand gallons. Any quantities that cannot be delivered by the company's four trucks must be subcontracted at the additional costs of 5, 12, 8, and 10 cents per gallon for products 1, 2, 3, and 4, respectively. Develop the optimal daily loading schedule for the four trucks that will minimize the additional cost of subcontracting.

9. A household uses at least 3000 minutes of long-distance telephone calls monthly and can choose to use the services of any of three companies: A, B, and C. Company A charges a fixed monthly fee of $10 and 5 cents per minute for the first 1000 minutes and 4 cents per minute for all additional minutes. Company B's monthly fee is $20 with a flat 4 cents per minute. Company C's monthly charge is $25 with 5 cents per minute for the first 1000 minutes and 3.5 cents per minute beyond that limit. Which company should be selected to minimize the total monthly charge?

***10.** (Barnett, 1987) Professor Yataha needs to schedule six round-trips between Boston and Washington, D.C. The route is served by three airlines: Eastern, US Air, and Continental and there is no penalty for the purchase of one-way tickets. Each airline offers bonus miles for frequent fliers. Eastern gives 1000 miles per (one-way) ticket plus 5000 extra miles if the number of tickets in a month reaches 2 and another 5000 miles if the number exceeds 5. US Air gives 1500 miles per trip plus 10,000 extra for each 6 tickets. Continental gives 1800 miles plus 7000 extra for each 5 tickets. Professor Yataha wishes to allocate the 12 one-way tickets among the three airlines to maximize the total number of bonus miles earned.

9.1.4 Either-Or and If-Then Constraints

In the fixed-charge problem (Section 9.1.3), we used binary variables to handle the discontinuity in the objective cost function. In this section, we deal with models in which constraints are not satisfied simultaneously (either-or) or are dependent (if-then), again using binary variables. The transformation does not change the "or" or "dependence" nature of the constraints. It simply uses a mathematical trick to present them in the desired format of "and" constraints.

Example 9.1-4 (Job-Sequencing Model)

Jobco uses a single machine to process three jobs. Both the processing time and the due date (in days) for each job are given in the following table. The due dates are measured from zero, the assumed start time of the first job.

Job	Processing time (days)	Due date (days)	Late penalty $/day
1	5	25	19
2	20	22	12
3	15	35	34

The objective of the problem is to determine the minimum late-penalty sequence for processing the three jobs.

Define

$$x_j = \text{Start date in days for job } j \text{ (measured from zero)}$$

The problem has two types of constraints: the noninterference constraints (guaranteeing that no two jobs are processed concurrently) and the due-date constraints. Consider the noninterference constraints first.

Two jobs i and j with processing time p_i and p_j will not be processed concurrently if either $x_i \geq x_j + p_j$ or $x_j \geq x_i + p_i$, depending on whether job j precedes job i, or vice versa. Because all mathematical programs deal with *simultaneous* constraints only, we transform the either-or constraints by introducing the following auxiliary binary variable:

$$y_{ij} = \begin{cases} 1, \text{ if } i \text{ precedes } j \\ 0, \text{ if } j \text{ precedes } i \end{cases}$$

For M sufficiently large, the **either-or constraint** is converted to the following two *simultaneous* constraints

$$My_{ij} + (x_i - x_j) \geq p_j \text{ and } M(1 - y_{ij}) + (x_j - x_i) \geq p_i$$

The conversion guarantees that only one of the two constraints can be active at any one time. If $y_{ij} = 0$, the first constraint is active, and the second is redundant (because its left-hand side will include M, which is much larger than p_i). If $y_{ij} = 1$, the first constraint is redundant, and the second is active.

Next, the due-date constraint is considered. Given that d_j is the due date for job j, let s_j be an unrestricted variable. Then, the associated constraint is

$$x_j + p_j + s_j = d_j$$

If $s_j \geq 0$, the due date is met, and if $s_j < 0$, a late penalty applies. Using the substitution

$$s_j = s_j^- - s_j^+, s_j^-, s_j^+ \geq 0$$

the constraint becomes

$$x_j + s_j^- - s_j^+ = d_j - p_j$$

The late-penalty cost is proportional to s_j^+.

The model for the given problem is

$$\text{Minimize } z = 19s_1^+ + 12s_2^+ + 34s_3^+$$

subject to

$$\begin{array}{lllllllll}
x_1 - x_2 & & + My_{12} & & & & & & \geq 20 \\
-x_1 + x_2 & & - My_{12} & & & & & & \geq 5 - M \\
x_1 & - x_3 & & + My_{13} & & & & & \geq 15 \\
-x_1 & + x_3 & & - My_{13} & & & & & \geq 5 - M \\
 & x_2 - x_3 & & & + My_{23} & & & & \geq 15 \\
 & - x_2 + x_3 & & & - My_{23} & & & & \geq 20 - M \\
x_1 & & & & & + s_1^- - s_1^+ & & & = 25 - 5 \\
 & x_2 & & & & & + s_2^- - s_2^+ & & = 22 - 20 \\
 & & x_3 & & & & & + s_3^- - s_3^+ & = 35 - 15
\end{array}$$

$$x_1, x_2, x_3, s_1^-, s_1^+, s_2^-, s_2^+, s_3^-, s_3^+ \geq 0$$

$$y_{12}, y_{13}, y_{23} = (0, 1)$$

The integer variables, y_{12}, y_{13}, and y_{23}, are introduced to convert the either-or constraints into simultaneous constraints. The resulting model is a *mixed* ILP.

To solve the model, we choose $M = 100$, a value that is larger than the sum of the processing times for all three activities.

The optimal solution is $x_1 = 20$, $x_2, = 0$, and $x_3 = 25$, This means that job 2 starts at time 0, job 1 starts at time 20, and job 3 starts at time 25, thus yielding the optimal processing sequence $2 \rightarrow 1 \rightarrow 3$. The solution calls for completing job 2 at time $0 + 20 = 20$, job 1 at time $= 20 + 5 = 25$, and job 3 at $25 + 15 = 40$ days. Job 3 is delayed by $40 - 35 = 5$ days past its due date at a cost of $5 \times \$34 = \$170$.

AMPL Moment

File amplEx9.1-4.txt provides the AMPL model for the problem of Example 9.1-4. The model is self-explanatory because it is a direct translation of the general mathematical model given above. It can handle any number of jobs by changing the input data. Note that the model is a direct function of the raw data: processing time `p`, due date `d`, and delay penalty `perDayPenalty`.

FIGURE 9.4

AMPL model of the job sequencing problem (file amplEx9.1-4.txt)

```
#------------------Example 9.1-4-------------------
param n;
set I={1..n};
set J={1..n};  #I is the same as J
param p{I};
param d{I};
param perDayPenalty{I};
param M=1000;
var x{J}>=0;              #continuous
var y{I,J} binary;        #0-1
var sMinus{J}>=0;         # s=sMinus-sPlus
var sPlus{J}>=0;
minimize penalty: sum {j in J}
           perDayPenalty[j]*sPlus[j];
subject to
eitherOr1{i in I,j in J:i<>j}:
          M*y[i,j]+x[i]-x[j]>=p[j];
eitherOr2{i in I,j in J:i<>j}:
          M*(1-y[i,j])+x[j]-x[i]>=p[i];
dueDate{j in J}:x[j]+sMinus[j]-sPlus[j]=d[j]-p[j];
data;
param n:=3;
param p:= 1 5  2 20  3 15;
param d:= 1 25  2 22  3 35;
param perDayPenalty := 1 19  2 12 3 34;
option solver cplex; solve;
display penalty,x;
```

Example 9.1-5 (Job Sequencing Model Revisited)

In Example 9.1-4, suppose that we have the following additional condition: If job i precedes job j then job k must precede job m. Mathematically, this **if-then condition** is translated as

$$\text{if } x_i + p_i \leq x_j \text{ then } x_k + p_k \leq x_m$$

Given $\varepsilon > 0$ and infinitesimally small and M sufficiently large, this condition is equivalent to the following two simultaneous constraints:

$$x_j - (x_i + p_i) \leq M(1 - w) - \varepsilon$$

$$(x_k + p_k) - x_m \leq Mw$$

$$w = (0, 1)$$

If $x_i + p_i \leq x_j$, then $x_j - (x_i + p_i) \geq 0$, which requires $w = 0$, and the second constraint becomes $x_k + p_k \leq x_m$, as desired. Else, w may assume the value 0 or 1, in which case the second constraint may or may not be satisfied, depending on other conditions in the model.

PROBLEM SET 9.1D

*1. A game board consists of nine equal squares. You are required to fill each square with a number between 1 and 9 such that the sum of the numbers in each row, each column, and each diagonal equals 15. Additionally, the numbers in all the squares must be distinct. Use ILP to determine the assignment of numbers to squares.

2. A machine is used to produce two interchangeable products. The daily capacity of the machine can produce at most 20 units of product 1 and 10 units of product 2. Alternatively, the machine can be adjusted to produce at most 12 units of product 1 and 25 units of product 2 daily. Market analysis shows that the maximum daily demand for the two products combined is 35 units. Given that the unit profits for the two respective products are \$10 and \$12, which of the two machine settings should be selected? Formulate the problem as an ILP and find the optimum. [*Note:* This two-dimensional problem can be solved by inspecting the graphical solution space. This is not the case for the n-dimensional problem.]

*3. Gapco manufactures three products, whose daily labor and raw material requirements are given in the following table.

Product	Required daily labor (hr/unit)	Required daily raw material (lb/unit)
1	3	4
2	4	3
3	5	6

The profits per unit of the three products are \$25, \$30, and \$22, respectively. Gapco has two options for locating its plant. The two locations differ primarily in the availability of labor and raw material, as shown in the following table:

Location	Available daily labor (hr)	Available daily raw material (lb)
1	100	100
2	90	120

Formulate the problem as an ILP, and determine the optimum location of the plant.

4. Jobco Shop has 10 outstanding jobs to be processed on a single machine. The following table provides processing times and due dates. All times are in days and due time is measured from time 0:

Job	Processing time	Due time
1	10	20
2	3	98
3	13	100
4	15	34
5	9	50
6	22	44
7	17	32
8	30	60
9	12	80
10	16	150

If job 4 precedes job 3, then job 9 must precede job 7. The objective is to process all 10 jobs in the shortest possible time. Formulate the model as an ILP and determine the optimum solution by modifying AMPL file amplEx9.1-4.txt.

5. In Problem 4, suppose that job 4 cannot be processed until job 3 has been completed. Also, machine settings for jobs 7 and 8 necessitate processing them one right after the other (i.e., job 7 immediately succeeds or immediately precedes 8). Jobco's objective is to process all ten jobs with the smallest sum of due-time violations. Formulate the model mathematically and determine the optimum solution.

6. Jaco owns a plant in which three products are manufactured. The labor and raw material requirements for the three products are given in the following table.

Product	Required daily labor (hr/unit)	Required daily raw material (lb/unit)
1	3	4
2	4	3
3	5	6
Daily availability	100	100

The profits per unit for the three products are \$25, \$30, and \$45, respectively. If product 3 is to be manufactured at all, then its production level must be at least 5 units daily. Formulate the problem as a mixed ILP, and find the optimal mix.

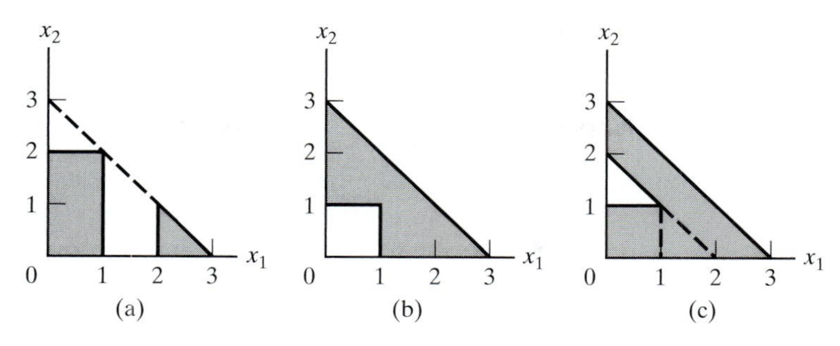

FIGURE 9.5

Solution spaces for Problem 9, Set 9.1d

7. UPak is a subsidiary of an LTL (less-than-truck-load) transportation company. Customers bring their shipments to the UPak terminal to be loaded on the trailer and can rent space up to 36 ft. The customer pays for the exact linear space (in foot increments) the shipment occupies. No partial shipment is allowed, in the sense that the entire shipment per customer must be on the same trailer. A movable barrier, called bulkhead, is installed to separate different shipments. The per-foot fee UPak collects depends on the destination of the shipment: The longer the trip, the higher the fee. The following table provides the outstanding orders UPak needs to process.

Order	1	2	3	4	5	6	7	8	9	10
Size (ft)	5	11	22	15	7	9	18	14	10	12
Rate ($)	120	93	70	85	125	104	98	130	140	65

The terminal currently has two trailers ready to be loaded. Determine the priority orders that will maximize the total income from the two trailers. (*Hint*: A formulation using binary x_{ij} to represent load i on trailer j is straightforward. However, you are challenged to define x_{ij} as *feet* assigned to load i in trailer j. The use *if-then* constraint to prevent partial load shipping.)

8. Show how the nonconvex shaded solution spaces in Figure 9.5 can be represented by a set of simultaneous constraints. Find the optimum solution that maximizes $z = 2x_1 + 3x_2$ subject to the solution space given in (a).

9. Suppose that it is required that *any k* out of the following *m* constraints must be active:

$$g_i(x_1, x_2, \ldots, x_n) \leq b_i, i = 1, 2, \ldots, m$$

Show how this condition may be represented.

10. In the following constraint, the right-hand side may assume one of values, $b_1, b_2, \ldots,$ and b_m.

$$g(x_1, x_2, \ldots, x_n) \leq (b_1, b_2, \ldots, \text{ or } b_m)$$

Show how this condition is represented.

9.2 INTEGER PROGRAMMING ALGORITHMS

The ILP algorithms are based on exploiting the tremendous computational success of LP. The strategy of these algorithms involves three steps.

Step 1. Relax the solution space of the ILP by deleting the integer restriction on all integer variables and replacing any binary variable y with the continuous range $0 \leq y \leq 1$. The result of the relaxation is a regular LP.

Step 2. Solve the LP, and identify its continuous optimum.

Step 3. Starting from the continuous optimum point, add special constraints that iteratively modify the LP solution space in a manner that will eventually render an optimum extreme point satisfying the integer requirements.

Two general methods have been developed for generating the special constraints in step 3.

1. Branch-and-bound (B&B) method
2. Cutting-plane method

Although neither method is consistently effective computationally, experience shows that the B&B method is far more successful than the cutting-plane method. This point is discussed further in this chapter.

9.2.1 Branch-and-Bound (B&B) Algorithm[3]

The first B&B algorithm was developed in 1960 by A. Land and G. Doig for the general mixed and pure ILP problem. Later, in 1965, E. Balas developed the **additive algorithm** for solving ILP problems with pure binary (zero or one) variables[4]. The additive algorithm's computations were so simple (mainly addition and subtraction) that it was hailed as a possible breakthrough in the solution of general ILP. Unfortunately, it failed to produce the desired computational advantages. Moreover, the algorithm, which initially appeared unrelated to the B&B technique, was shown to be but a special case of the general Land and Doig algorithm.

This section will present the general Land-Doig B&B algorithm only. A numeric example is used to explain the details.

Example 9.2-1

$$\text{Maxmize } z = 5x_1 + 4x_2$$

subject to

$$x_1 + x_2 \leq 5$$

$$10x_1 + 6x_2 \leq 45$$

$$x_1, x_2 \text{ nonnegative integer}$$

---

[3]TORA integer programming module is equipped with a facility for generating the B&B tree interactively. To use this facility, select User-guided B&B in the output screen of the integer programming module. The resulting screen provides all the information needed to create the B&B tree.

[4]A general ILP can be expressed in terms of binary (0–1) variables as follows. Given an integer variable x with a finite upper bound u (i.e., $0 \leq x \leq u$), then

$$x = 2^0 y_0 + 2^1 y_1 + 2^2 y_2 + \cdots + 2^k y_k$$

The variables $y_0, y_1, \ldots,$ and y_k are binary and the index k is the smallest integer satisfying $2^{k+1} - 1 \geq u$.

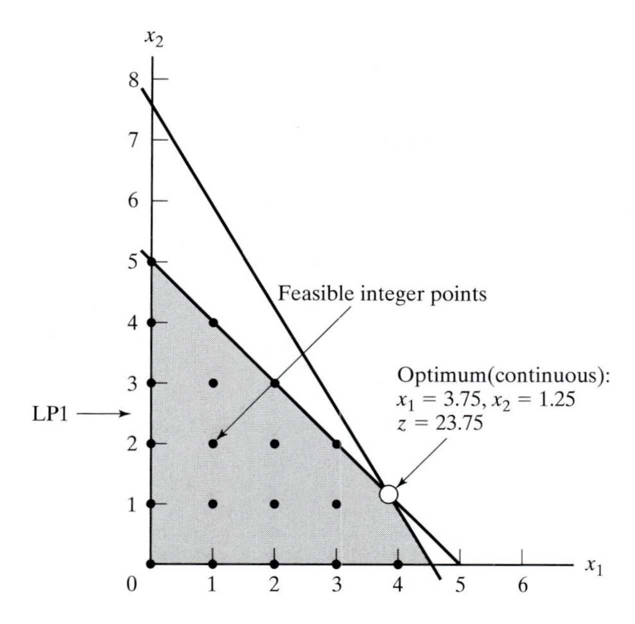

FIGURE 9.6

Solution spaces for ILP (lattice points) and LP1 (shaded area) of Example 9.2-1

The lattice points (dots) in Figure 9.6 define the ILP solution space. The associated continuous LP1 problem at node 1 (shaded area) is defined from ILP by removing the integer restrictions. The optimum solution of LP1 is $x_1 = 3.75$, $x_2 = 1.25$, and $z = 23.75$.

Because the optimum LP1 solution does not satisfy the integer requirements, the B&B algorithm modifies the solution space in a manner that eventually identifies the ILP optimum. First, we select one of the integer variables whose optimum value at LP1 is not integer. Selecting x_1 $(= 3.75)$ arbitrarily, the region $3 < x_1 < 4$ of the LP1 solution space contains no integer values of x_1, and thus can be eliminated as nonpromising. This is equivalent to replacing the original LP1 with two new LPs:

$$\text{LP2 space} = \text{LP1 space} + (x_1 \leq 3)$$

$$\text{LP3 space} = \text{LP1 space} + (x_1 \geq 4)$$

Figure 9.7 depicts the LP2 and LP3 spaces. The two spaces combined contain the same feasible integer points as the original ILP, which means that, from the standpoint of the integer solution, dealing with LP2 and LP3 is the same as dealing with the original LP1; no information is lost.

If we *intelligently* continue to remove the regions that do not include integer solutions (e.g., $3 < x_1 < 4$ at LP1) by imposing the appropriate constraints, we will eventually produce LPs whose optimum extreme points satisfy the integer restrictions. In effect, we will be solving the ILP by dealing with a sequence of (continuous) LPs.

The new restrictions, $x_1 \leq 3$ and $x_1 \geq 4$, are mutually exclusive, so that LP2 and LP3 at nodes 2 and 3 must be dealt with as separate LPs, as Figure 9.8 shows. This dichotomization gives rise to the concept of **branching** in the B&B algorithm. In this case, x_1 is called the **branching variable**.

The optimum ILP lies in *either* LP2 *or* LP3. Hence, both subproblems must be examined. We arbitrarily examine LP2 (associated with $x_1 \leq 3$) first:

$$\text{Maxmize } z = 5x_1 + 4x_2$$

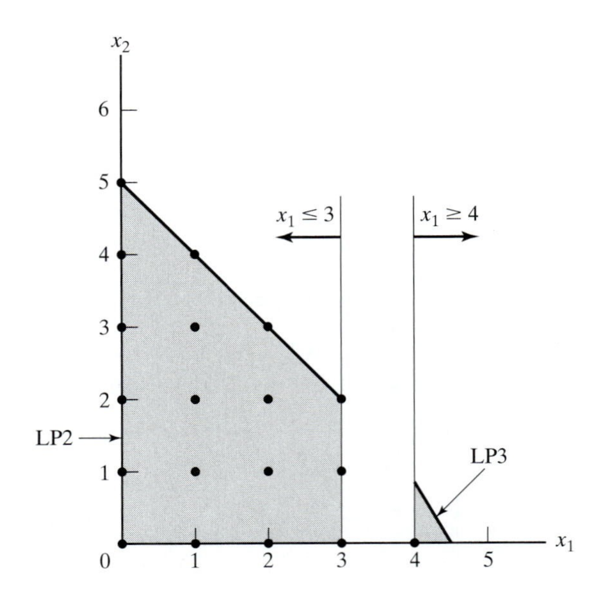

FIGURE 9.7
Solution Spaces of LP2 and LP3 for Example 9.2-1

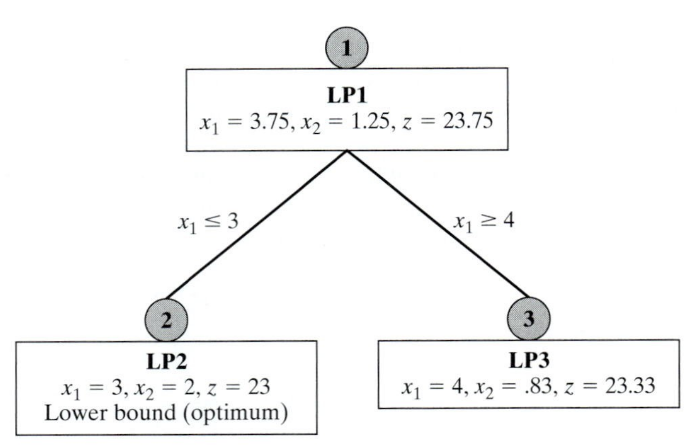

FIGURE 9.8
Using branching variable x_1 to create LP2 and LP3 for Example 9.2-1

subject to

$$
\begin{aligned}
x_1 + x_2 &\le 5 \\
10x_1 + 6x_2 &\le 45 \\
x_1 &\le 3 \\
x_1, x_2 &\ge 0
\end{aligned}
$$

The solution of LP2 (which can be solved efficiently by the upper-bounded algorithm of Section 7.3) yields the solution

$$x_1 = 3, x_2 = 2, \text{ and } z = 23$$

The LP2 solution satisfies the integer requirements for x_1 and x_2. Hence, LP2 is said to be **fathomed**, meaning that it need not be investigated any further because it cannot yield any *better* ILP solution.

We cannot at this point say that the integer solution obtained from LP2 is optimum for the original problem, because LP3 may yield a better integer solution with a higher value of z. All we can say is that $z = 23$ is **a lower bound** on the optimum (maximum) objective value of the original ILP. This means that any unexamined subproblem that cannot yield a better objective value than the lower bound must be discarded as nonpromising. If an unexamined subproblem produces a better integer solution, then the lower bound must be updated accordingly.

Given the lower bound $z = 23$, we examine LP3 (the only remaining unexamined subproblem at this point). Because optimum $z = 23.75$ at LP1 *and all the coefficients of the objective function happen to be integers,* it is impossible that LP3 (which is more restrictive than LP1) will produce a better integer solution with $z > 23$. As a result, we discard LP3 and conclude that it has been *fathomed.*

The B&B algorithm is now complete because both LP2 and LP3 have been examined and fathomed (the first for producing an integer solution and the second for failing to produce a *better* integer solution). We thus conclude that the optimum ILP solution is the one associated with the lower bound—namely, $x_1 = 3$, x_2, and $z = 23$.

Two questions remain unanswered regarding the procedure.

1. At LP1, could we have selected x_2 as the *branching variable* in place of x_1?
2. When selecting the next subproblem to be examined, could we have solved LP3 first instead of LP2?

The answer to both questions is "yes," but ensuing computations could differ dramatically. Figure 9.9 demonstrates this point. Suppose that we examine LP3 first (instead of LP2 as we did in Figure 9.8). The solution is $x_1 = 4$, $x_2 = .83$, and $z = 23.33$ (verify!). Because x_2 $(= .83)$ is noninteger, LP3 is examined further by creating subproblems LP4 and LP5 using the branches $x_2 \leq 0$ and $x_2 \geq 1$, respectively. This means that

$$\begin{aligned} \text{LP4 space} &= \text{LP3 space} + (x_2 \leq 0) \\ &= \text{LP1 space} + (x_1 \geq 4) + (x_2 \leq 0) \\ \text{LP5 space} &= \text{LP3 space} + (x_2 \geq 1) \\ &= \text{LP1 space} + (x_1 \geq 4) + (x_2 \geq 1) \end{aligned}$$

We now have three "dangling" subproblems to be examined: LP2, LP4, and LP5. Suppose that we arbitrarily examine LP5 first. LP5 has no solution, and hence it is fathomed. Next, let us examine LP4. The optimum solution is $x_1 = 4.5$, $x_2 = 0$, and $z = 22.5$. The noninteger value of x_1 leads to the two branches $x_1 \leq 4$ and $x_1 \geq 5$, and the creation of subproblems LP6 and LP7 from LP4.

$$\begin{aligned} \text{LP6 space} &= \text{LP1 space} + (x_1 \geq 4) + (x_2 \leq 0) + (x_1 \leq 4) \\ \text{LP7 space} &= \text{LP1 space} + (x_1 \geq 4) + (x_2 \leq 0) + (x_1 \geq 5) \end{aligned}$$

Now, subproblems LP2, LP6, and LP7 remain unexamined. Selecting LP7 for examination, the problem has no feasible solution, and thus is fathomed. Next, we select LP6. The problem yields the first integer solution $(x_1 = 4, x_2 = 0, z = 20)$, and thus provides the first lower bound

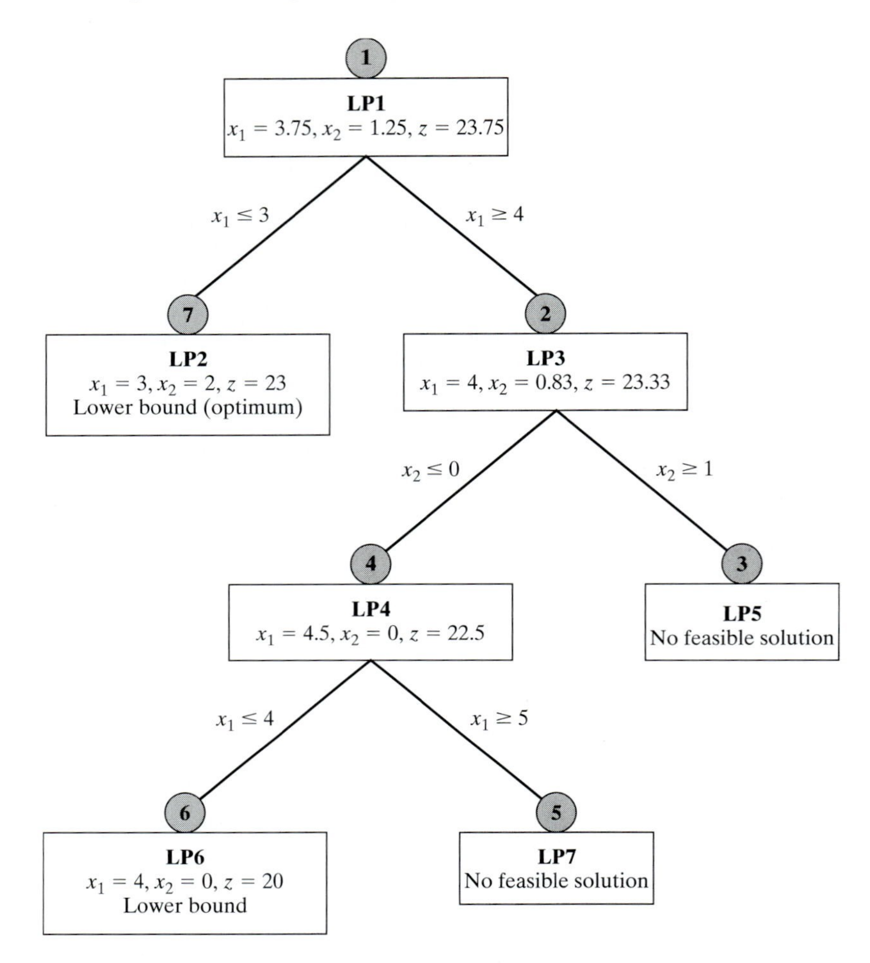

FIGURE 9.9
Alternative B&B tree for Example 9.2-1

$(= 20)$ on the optimum ILP objective value. We are now left with subproblem LP2, which yields a better integer solution $(x_1 = 3, x_2 = 2, z = 23)$. Thus, the lower bound is updated from $z = 20$ to $z = 23$. At this point, *all* the subproblems have been fathomed (examined) and the optimum solution is the one associated with the most up-to-date lower bound—namely, $x_1 = 3$, $x_2 = 2$, and $z = 23$.

The solution sequence in Figure 9.9 (LP1 → LP3 → LP5 → LP4 → LP7 → LP6 → LP2) is a worst-case scenario that, nevertheless, may well occur in practice. In Figure 9.8, we were lucky to "stumble" upon a good lower bound at the very first subproblem we examined (LP2), thus allowing us to fathom LP3 without investigating it. In essence, we completed the procedure by solving a total of two LPs. In Figure 9.9, the story is different: We needed to solve seven LPs before the B&B algorithm could be terminated.

Remarks. The example points to a principal weakness in the B&B algorithm: Given multiple choices, how do we select the next subproblem and its branching variable? Although there are heuristics for enhancing the ability of B&B to "foresee" which branch can lead to an improved ILP solution (see Taha, 1975, pp. 154–171), solid theory with consistent results does not exist, and herein lies the difficulty that plagues computations in ILP. Indeed, Problem 7, Set 9.2a, demonstrates the bizarre behavior of the B&B algorithm in investigating over 25,000 LPs before optimality is verified, even though the problem is quite small (16 binary variables and 1 constraint). Unfortunately, to date, and after more than four decades of research coupled with tremendous advances in computing power, available ILP codes (commercial and academic alike) are not totally reliable, in the sense that they may not find the optimum ILP solution regardless of how long they execute on the computer. What is even more frustrating is that this behavior can apply just the same to some relatively small problems.

AMPL Moment

AMPL can be used interactively to generate the B&B search tree. The following table shows the sequence of commands needed to generate the tree of Example 9.2-1 (Figure 9.9) starting with the continuous LP0. AMPL model (file amplEx9.2-1.txt) has two variables `x1` and `x2` and two constraints `c0` and `c1`. You will find it helpful to synchronize the AMPL commands with the branches in Figure 9.9.

AMPL command	Result
ampl: `model amplEx9.2-1.txt;solve;display x1,x2;`	LP1 ($x_1 = 3.75, x_2 = 1.25$)
ampl: `c2:x1>=4;solve;display x1,x2;`	LP3 ($x_1 = 4, x_2 = .83$)
ampl: `c3:x2>=1;solve;display x1,x2;`	LP5 (no solution)
ampl: `drop c3;c4:x2<=0;solve;display x1,x2;`	LP4 ($x_1 = 4.5, x_2 = 0$)
ampl: `c5:x1>=5;solve;display x1,x2;`	LP7 (no solution)
ampl: `drop c5;c6:x1<=4;solve;display x1,x2;`	LP6 ($x_1 = 4, x_2 = 0$)
ampl: `drop c2;drop c4;drop c6;c7:x1<=3;` `solve;display x1,x2;`	LP2 ($x_1 = 3, x_2 = 2$)

Solver Moment

Solver can be used to obtain the solution of the different subproblems by using the add/change/delete options in the **Solver Parameters** dialogue box.

Summary of the B&B Algorithm. We now summarize the B&B algorithm. Assuming a maximization problem, set an initial lower bound $\underline{z} = -\infty$ on the optimum objective value of ILP. Set $i = 0$.

Step 1. (*Fathoming/bounding*). Select LP*i*, the next subproblem to be examined. Solve LP*i*, and attempt to fathom it using one of three conditions:

(a) The optimal z-value of LP*i* cannot yield a better objective value than the current lower bound.

(b) LP*i* yields a better feasible integer solution than the current lower bound.

(c) LP*i* has no feasible solution.

Two cases will arise.

(a) If LPi is fathomed and a better solution is found, update the lower bound. If all subproblems have been fathomed, stop; the optimum ILP is associated with the current finite lower bound. If no finite lower bound exists, the problem has no feasible solution. Else, set $i = i + 1$, and repeat step 1.

(b) If LPi is not fathomed, go to step 2 for branching.

Step 2. (*Branching*). Select one of the integer variables x_j, whose optimum value x_j^* in the LPi solution is not integer. Eliminate the region

$$[x_j^*] < x_j < [x_j^*] + 1$$

(where $[v]$ defines the largest integer $\leq v$) by creating two LP subproblems that correspond to

$$x_j \leq [x_j^*] \text{ and } x_j \geq [x_j^*] + 1$$

Set $i = i + 1$, and go to step 1.

The given steps apply to maximization problems. For minimization, we replace the lower bound with an upper bound (whose initial value is $z = +\infty$).

The B&B algorithm can be extended directly to mixed problems (in which only some of the variables are integer). If a variable is continuous, we simply never select it as a branching variable. A feasible subproblem provides a new bound on the objective value if the values of the discrete variables are integer and the objective value is improved relative to the current bound.

PROBLEM SET 9.2A[5]

1. Solve the ILP of Example 9.2-1 by the B&B algorithm starting with x_2 as the branching variable. Start the procedure by solving the subproblem associated with $x_2 \leq [x_2^*]$.

2. Develop the B&B tree for each of the following problems. For convenience, always select x_1 as the branching variable at node 0.

***(a)** Maximize $z = 3x_1 + 2x_2$

subject to

$$2x_1 + 5x_2 \leq 9$$
$$4x_1 + 2x_2 \leq 9$$
$$x_1, x_2 \geq 0 \text{ and integer}$$

[5]In this set, you may solve the subproblems interactively with AMPL or Solver or using TORA's MODIFY option for the upper and lower bounds.

(b) Maximize $z = 2x_1 + 3x_2$
subject to

$$5x_1 + 7x_2 \leq 35$$
$$4x_1 + 9x_2 \leq 36$$
$$x_1, x_2 \geq 0 \text{ and integer}$$

(c) Maximize $z = x_1 + x_2$
subject to

$$2x_1 + 5x_2 \leq 16$$
$$6x_1 + 5x_2 \leq 27$$
$$x_1, x_2 \geq 0 \text{ and integer}$$

***(d)** Minimize $z = 5x_1 + 4x_2$
subject to

$$3x_1 + 2x_2 \geq 5$$
$$2x_1 + 3x_2 \geq 7$$
$$x_1, x_2 \geq 0 \text{ and integer}$$

(e) Maximize $z = 5x_1 + 7x_2$
subject to

$$2x_1 + x_2 \leq 13$$
$$5x_1 + 9x_2 \leq 41$$
$$x_1, x_2 \geq 0 \text{ and integer}$$

***3.** Repeat Problem 2, assuming that x_1 is continuous.

4. Show graphically that the following ILP has no feasible solution, and then verify the result using B&B.

$$\text{Maximize } z = 2x_1 + x_2$$

subject to

$$10x_1 + 10x_2 \leq 9$$
$$10x_1 + 5x_2 \geq 1$$
$$x_1, x_2 \geq 0 \text{ and integer}$$

5. Solve the following problems by B&B.

$$\text{Maximize } z = 18x_1 + 14x_2 + 8x_3 + 4x_4$$

subject to

$$15x_1 + 12x_2 + 7x_3 + 4x_4 + x_5 \leq 37$$
$$x_1, x_2, x_3, x_4, x_5 = (0, 1)$$

6. Convert the following problem into a mixed ILP and find the optimum solution.

$$\text{Maximize } z = x_1 + 2x_2 + 5x_3$$

subject to

$$|-x_1 + 10x_2 - 3x_3| \geq 15$$
$$2x_1 + x_2 + x_3 \leq 10$$
$$x_1, x_2, x_3 \geq 0$$

7. *TORA/Solver/AMPL Experiment.* The following problem is designed to demonstrate the bizarre behavior of the B&B algorithm even for small problems. In particular, note how many subproblems are examined before the optimum is found and how many are needed to verify optimality.

$$\text{Minimize y}$$

subject to

$$2(x_1 + x_2 + \cdots + x_{15}) + y = 15$$
$$\text{All variables are } (0, 1)$$

(a) Use TORA's automated option to show that although the optimum is found after only 9 subproblems, over 25,000 subproblems are examined before optimality is confirmed.

(b) Show that Solver exhibits an experience similar to TORA's. [*Note*: In Solver, you can watch the change in the number of generated branches (subproblems) at the bottom of the spreadsheet.]

(c) Solve the problem with AMPL and show that the solution is obtained instantly with 0 MIP simplex iterations and 0 B&B nodes. The reason for this superior performance can only be attributed to preparatory steps performed by AMPL and/or the CPLEX solver prior to solving the problem.

8. *TORA Experiment.* Consider the following ILP:

$$\text{Maximize } z = 18x_1 + 14x_2 + 8x_3$$

subject to

$$15x_1 + 12x_2 + 7x_3 \leq 43$$
$$x_1, x_2, x_3 \text{ nonnegative integers}$$

Use TORA's B&B user-guided option to generate the search tree with and without activating the objective-value bound. What is the impact of activating the objective-value bound on the number of generated subproblems? For consistency, always select the branching variable as the one with the lowest index and investigate all the subproblems in a current row from left to right before moving to the next row.

***9.** *TORA Experiment.* Reconsider Problem 8 above. Convert the problem into an equivalent 0-1 ILP, then solve it with TORA's automated option. Compare the size of the search trees in the two problems.

10. *AMPL Experiment.* In the following 0-1 ILP use interactive AMPL to generate the associated search tree. In each case, show how the z-bound is used to fathom subproblems.

$$\text{Maximize } z = 3x_1 + 2x_2 - 5x_3 - 2x_4 + 3x_5$$

subject to

$$\begin{aligned} x_1 + x_2 + x_3 + 2x_4 + x_5 &\leq 4 \\ 7x_1 + 3x_3 - 4x_4 + 3x_5 &\leq 8 \\ 11x_1 - 6x_2 + 3x_4 - 3x_5 &\geq 3 \\ x_1, x_2, x_3, x_4, x_5 &= (0, 1) \end{aligned}$$

9.2.2 Cutting-Plane Algorithm

As in the B&B algorithm, the cutting-plane algorithm also starts at the continuous optimum LP solution. Special constraints (called **cuts**) are added to the solution space in a manner that renders an integer optimum extreme point. In Example 9.2-2, we first demonstrate graphically how cuts are used to produce an integer solution and then implement the idea algebraically.

Example 9.2-2

Consider the following ILP.

$$\text{Maximize } z = 7x_1 + 10x_2$$

subject to

$$\begin{aligned} -x_1 + 3x_2 &\leq 6 \\ 7x_1 + x_2 &\leq 35 \\ x_1, x_2 &\geq 0 \text{ and integer} \end{aligned}$$

The cutting-plane algorithm modifies the solution space by adding *cuts* that produce an optimum integer extreme point. Figure 9.10 gives an example of two such cuts. Initially, we start with the continuous LP optimum $z = 66\frac{1}{2}$, $x_1 = 4\frac{1}{2}$, $x_2 = 3\frac{1}{2}$. Next, we add cut I, which produces the (continuous) LP optimum solution $z = 62$, $x_1 = 4\frac{4}{7}$, $x_2 = 3$. Then, we add cut II, which, together with cut I and the original constraints, produces the LP optimum $z = 58$, $x_1 = 4$, $x_2 = 3$. The last solution is all integer, as desired.

The added cuts do not eliminate any of the original feasible integer points, but must pass through at least one feasible or infeasible integer point. These are basic requirements of any cut.

FIGURE 9.10

Illustration of the use of cuts in ILP

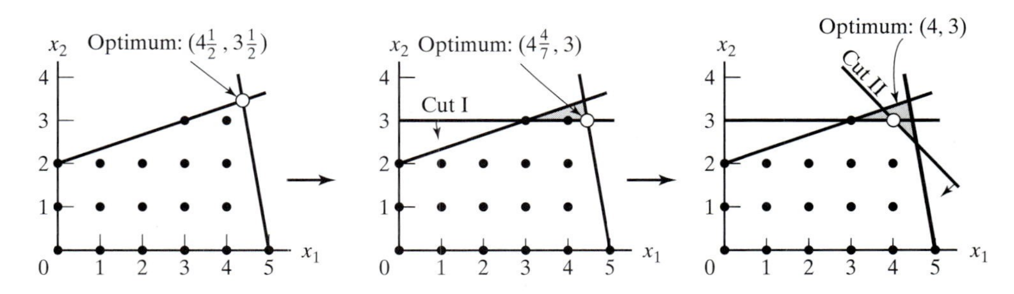

It is purely accidental that a 2-variable problem used exactly 2 cuts to reach the optimum integer solution. In general, the number of cuts, though finite, is independent of the size of the problem, in the sense that a problem with a small number of variables and constraints may require more cuts than a larger problem.

Next, we use the same example to show how the cuts are constructed and implemented algebraically.

Given the slacks x_3 and x_4 for constraints 1 and 2, the optimum LP tableau is given as

Basic	x_1	x_2	x_3	x_4	Solution
z	0	0	$\frac{63}{22}$	$\frac{31}{22}$	$66\frac{1}{2}$
x_2	0	1	$\frac{7}{22}$	$\frac{1}{22}$	$3\frac{1}{2}$
x_1	1	0	$-\frac{1}{22}$	$\frac{3}{22}$	$4\frac{1}{2}$

The optimum continuous solution is $z = 66\frac{1}{2}$, $x_1 = 4\frac{1}{2}$, $x_2 = 3\frac{1}{2}$, $x_3 = 0$, $x_4 = 0$. The cut is developed under the assumption that *all* the variables (including the slacks x_3 and x_4) are integer. Note also that because all the original objective coefficients are integer in this example, the value of z is integer as well.

The information in the optimum tableau can be written explicitly as

$$z + \tfrac{63}{22}x_3 + \tfrac{31}{22}x_4 = 66\tfrac{1}{2} \qquad (z\text{-equation})$$

$$x_2 + \tfrac{7}{22}x_3 + \tfrac{1}{22}x_4 = 3\tfrac{1}{2} \qquad (x_2\text{-equation})$$

$$x_1 - \tfrac{1}{22}x_3 + \tfrac{3}{22}x_4 = 4\tfrac{1}{2} \qquad (x_1\text{-equation})$$

A constraint equation can be used as a **source row** for generating a cut, provided its right-hand side is fractional. We also note that the z-equation can be used as a source row because z happens to be integer in this example. We will demonstrate how a cut is generated from each of these source rows, starting with the z-equation.

First, we factor out all the noninteger coefficients of the equation into an integer value and a fractional component, *provided that the resulting fractional component is strictly positive*. For example,

$$\tfrac{5}{2} = \left(2 + \tfrac{1}{2}\right)$$

$$-\tfrac{7}{3} = \left(-3 + \tfrac{2}{3}\right)$$

The factoring of the z-equation yields

$$z + \left(2 + \tfrac{19}{22}\right)x_3 + \left(1 + \tfrac{9}{22}\right)x_4 = \left(66 + \tfrac{1}{2}\right)$$

Moving all the integer components to the left-hand side and all the fractional components to the right-hand side, we get

$$z + 2x_3 + 1x_4 - 66 = -\tfrac{19}{22}x_3 - \tfrac{9}{22}x_4 + \tfrac{1}{2} \qquad (1)$$

Because x_3 and x_4 are nonnegative and all fractions are originally strictly positive, the right-hand side must satisfy the following inequality:

$$-\frac{19}{22}x_3 - \frac{9}{22}x_4 + \frac{1}{2} \leq \frac{1}{2} \tag{2}$$

Next, because the left-hand side in Equation (1), $z + 2x_3 + 1x_4 - 66$, is an integer value by construction, the right-hand side, $-\frac{19}{22}x_3 - \frac{19}{22}x_4 + \frac{1}{2}$, must also be integer. It then follows that (2) can be replaced with the inequality:

$$-\frac{19}{22}x_3 - \frac{9}{22}x_4 + \frac{1}{2} \leq 0$$

This result is justified because an integer value $\leq\frac{1}{2}$ must necessarily be ≤ 0.

The last inequality is the desired cut and it represents a *necessary* (but not sufficient) condition for obtaining an integer solution. It is also referred to as the **fractional cut** because all its coefficients are fractions.

Because $x_3 = x_4 = 0$ in the optimum continuous LP tableau given above, the current continuous solution violates the cut (because it yields $\frac{1}{2} \leq 0$). Thus, if we add this cut to the optimum tableau, the resulting optimum extreme point moves the solution toward satisfying the integer requirements.

Before showing how a cut is implemented in the optimal tableau, we will demonstrate how cuts can also be constructed from the constraint equations. Consider the x_1-row:

$$x_1 - \frac{1}{22}x_3 + \frac{3}{22}x_4 = 4\frac{1}{2}$$

Factoring the equation yields

$$x_1 + \left(-1 + \frac{21}{22}\right)x_3 + \left(0 + \frac{3}{22}\right)x_4 = \left(4 + \frac{1}{2}\right)$$

The associated cut is

$$-\frac{21}{22}x_2 - \frac{3}{22}x_4 + \frac{1}{2} \leq 0$$

Similarly, the x_2-equation

$$x_2 + \frac{7}{22}x_3 + \frac{1}{22}x_4 = 3\frac{1}{2}$$

is factored as

$$x_2 + \left(0 + \frac{7}{22}\right)x_3 + \left(0 + \frac{1}{22}\right)x_4 = 3 + \frac{1}{2}$$

Hence, the associated cut is given as

$$-\frac{7}{22}x_3 - \frac{1}{22}x_4 + \frac{1}{2} \leq 0$$

Any one of three cuts given above can be used in the first iteration of the cutting-plane algorithm. It is not necessary to generate all three cuts before selecting one.

Arbitrarily selecting the cut generated from the x_2-row, we can write it in equation form as

$$-\frac{7}{22}x_3 - \frac{1}{22}x_4 + s_1 = -\frac{1}{2}, \quad s_1 \geq 0 \qquad \text{(Cut I)}$$

This constraint is added to the LP optimum tableau as follows:

Basic	x_1	x_2	x_3	x_4	s_1	Solution
z	0	0	$\frac{63}{22}$	$\frac{31}{22}$	0	$66\frac{1}{2}$
x_2	0	1	$\frac{7}{22}$	$\frac{1}{22}$	0	$3\frac{1}{2}$
x_1	1	0	$-\frac{1}{22}$	$\frac{3}{22}$	0	$4\frac{1}{2}$
s_1	0	0	$-\frac{7}{22}$	$-\frac{1}{22}$	1	$-\frac{1}{2}$

The tableau is optimal but infeasible. We apply the dual simplex method (Section 4.4.1) to recover feasibility, which yields

Basic	x_1	x_2	x_3	x_4	s_1	Solution
z	0	0	0	1	9	62
x_2	0	1	0	0	1	3
x_1	1	0	0	$\frac{1}{7}$	$-\frac{1}{7}$	$4\frac{4}{7}$
x_3	0	0	1	$\frac{1}{7}$	$-\frac{22}{7}$	$1\frac{4}{7}$

The last solution is still noninteger in x_1 and x_3. Let us arbitrarily select x_1 as the next source row—that is,

$$x_1 + \left(0 + \tfrac{1}{7}\right)x_4 + \left(-1 + \tfrac{6}{7}\right)s_1 = 4 + \tfrac{4}{7}$$

The associated cut is

$$-\tfrac{1}{7}x_4 - \tfrac{6}{7}s_1 + s_2 = -\tfrac{4}{7}, \quad s_2 \geq 0 \qquad \text{(Cut II)}$$

Basic	x_1	x_2	x_3	x_4	s_1	s_2	Solution
z	0	0	0	1	9	0	62
x_2	0	1	0	0	1	0	3
x_1	1	0	0	$\frac{1}{7}$	$-\frac{1}{7}$	0	$4\frac{4}{7}$
x_3	0	0	1	$\frac{1}{7}$	$-\frac{22}{7}$	0	$1\frac{4}{7}$
s_2	0	0	0	$-\frac{1}{7}$	$-\frac{6}{7}$	1	$-\frac{4}{7}$

The dual simplex method yields the following tableau:

Basic	x_1	x_2	x_3	x_4	s_1	s_2	Solution
z	0	0	0	0	3	7	58
x_2	0	1	0	0	1	0	3
x_1	1	0	0	0	−1	1	4
x_3	0	0	1	0	−4	1	1
x_4	0	0	0	1	6	−7	4

The optimum solution ($x_1 = 4, x_2 = 3, z = 58$) is all integer. It is not accidental that all the coefficients of the last tableau are integers, a property of the implementation of the fractional cut.

Remarks. It is important to point out that the fractional cut assumes that *all* the variables, *including slack and surplus,* are integer. This means that the cut deals with pure integer problems only. The importance of this assumption is illustrated by an example.

Consider the constraint

$$x_1 + \tfrac{1}{3}x_2 \leq \tfrac{13}{2}$$

$$x_1, x_2 \geq 0 \text{ and integer}$$

From the standpoint of solving the associated ILP, the constraint is treated as an equation by using the nonnegative slack s_1—that is,

$$x_1 + \tfrac{1}{3}x_2 + s_1 = \tfrac{13}{2}$$

The application of the fractional cut assumes that the constraint has a feasible integer solution in all x_1, x_2, and s_1. However, the equation above will have a feasible integer solution in x_1 and x_2 *only if* s_1 *is noninteger.* This means that the cutting-plane algorithm will show that the problem has no feasible integer solution, even though the variables of concern, x_1 and x_2, can assume feasible integer values.

There are two ways to remedy this situation.

1. Multiply the entire constraint by a proper constant to remove all the fractions. For example, multiplying the constraint above by 6, we get

$$6x_1 + 2x_2 \leq 39$$

Any integer solution of x_1 and x_2 automatically yields integer slack. However, this type of conversion is appropriate for only simple constraints, because the magnitudes of the integer coefficients may become excessively large in some cases.

2. Use a special cut, called the **mixed cut**, which allows only a subset of variables to assume integer values, with all the other variables (including slack and surplus) remaining continuous. The details of this cut will not be presented in this chapter (see Taha, 1975, pp. 198–202).

PROBLEM SET 9.2B

1. In Example 9.2-2, show graphically whether or not each of the following constraints can form a legitimate cut:

*__(a)__ $x_1 + 2x_2 \leq 10$

(b) $2x_1 + x_2 \leq 10$

(c) $3x_2 \leq 10$

(d) $3x_1 + x_2 \leq 15$

2. In Example 9.2-2, show graphically how the following two (legitimate) cuts can lead to the optimum integer solution:

$$x_1 + 2x_2 \leq 10 \quad \text{(cut I)}$$
$$3x_1 + x_2 \leq 15 \quad \text{(cut II)}$$

3. Express cuts I and II of Example 9.2-2 in terms of x_1 and x_2 and show that they are the same ones used graphically in Figure 9.10.

4. In Example 9.2-2, derive cut II from the x_3-row. Use the new cut to complete the solution of the example.

5. Show that, even though the following problem has a feasible integer solution in x_1 and x_2, the fractional cut would not yield a feasible solution unless all the fractions in the constraint were eliminated.

$$\text{Maximize } z = x_1 + 2x_2$$

subject to

$$x_1 + \tfrac{1}{2}x_2 \leq \tfrac{13}{4}$$
$$x_1, x_2 \geq 0 \text{ and integer}$$

6. Solve the following problems by the fractional cut, and compare the true optimum integer solution with the solution obtained by rounding the continuous optimum.

*(a) Maximize $z = 4x_1 + 6x_2 + 2x_3$
subject to

$$4x_1 - 4x_2 \leq 5$$
$$-x_1 + 6x_2 \leq 5$$
$$-x_1 + x_2 + x_3 \leq 5$$
$$x_1, x_2, x_3 \geq 0 \text{ and integer}$$

(b) Maximize $z = 3x_1 + x_2 + 3x_3$
subject to

$$-x_1 + 2x_2 + x_3 \leq 4$$
$$4x_2 - 3x_3 \leq 2$$
$$x_1 - 3x_2 + 2x_3 \leq 3$$
$$x_1, x_2, x_3 \geq 0 \text{ and integer}$$

9.2.3 Computational Considerations in ILP

To date, and despite over 40 years of research, there does not exist a computer code that can solve ILP consistently. Nevertheless, of the two solution algorithms presented in this chapter, B&B is more reliable. Indeed, practically all commercial ILP codes are B&B based. Cutting-plane methods are generally difficult and uncertain, and the roundoff error presents a serious problem. This is true because the "accuracy" of the cut depends on the accuracy of a true representation of its fractions on the computer. For instance, in Example 9.2-2, the fraction $\frac{1}{7}$ cannot be represented exactly as a floating point regardless of the level of precision that may be used. Though attempts have been made to improve the cutting-plane computational efficacy, the end results are not encouraging. In most cases, the cutting-plane method is used in a secondary capacity to improve B&B performance at each subproblem by eliminating a portion of the solution space associated with a subproblem.

The most important factor affecting computations in integer programming is the number of integer variables and the feasible range in which they apply. Because available algorithms are not consistent in producing a numeric ILP solution, it may be advantageous computationally to reduce the number of integer variables in the ILP model as much as possible. The following suggestions may prove helpful:

1. Approximate integer variables by continuous ones wherever possible.
2. For the integer variables, restrict their feasible ranges as much as possible.
3. Avoid the use of nonlinearity in the model.

The importance of the integer problem in practice is not yet matched by the development of reliable solution algorithms. The nature of discrete mathematics and the fact that the integer solution space is a nonconvex set make it unlikely that new theoretical breakthroughs will be achieved in the area of integer programming. Instead, new technological advances in computers (software and hardware) remain the best hope for improving the efficiency of ILP codes.

9.3 TRAVELING SALESPERSON (TSP) PROBLEM

Historically, the TSP problem deals with finding the shortest (closed) tour in an n-city situation where each city is visited exactly once. The problem, in essence, is an assignment model that excludes subtours. Specifically, in an n-city situation, define

$$x_{ij} = \begin{cases} 1, \text{ if city } j \text{ is reached from city } i \\ 0, \text{ otherwise} \end{cases}$$

Given that d_{ij} is the distance from city i to city j, the TSP model is given as

$$\text{Minimize } z = \sum_{i=1}^{n} \sum_{j=1}^{n} d_{ij} x_{ij},\ d_{ij} = \infty \text{ for all } i = j$$

subject to

$$\sum_{j=1}^{n} x_{ij} = 1, i = 1, 2, \ldots, n \quad (1)$$

$$\sum_{i=1}^{n} x_{ij} = 1, j = 1, 2, \ldots, n \quad (2)$$

$$x_{ij} = (0, 1) \quad (3)$$

$$\text{Solution forms an } n\text{-city tour} \quad (4)$$

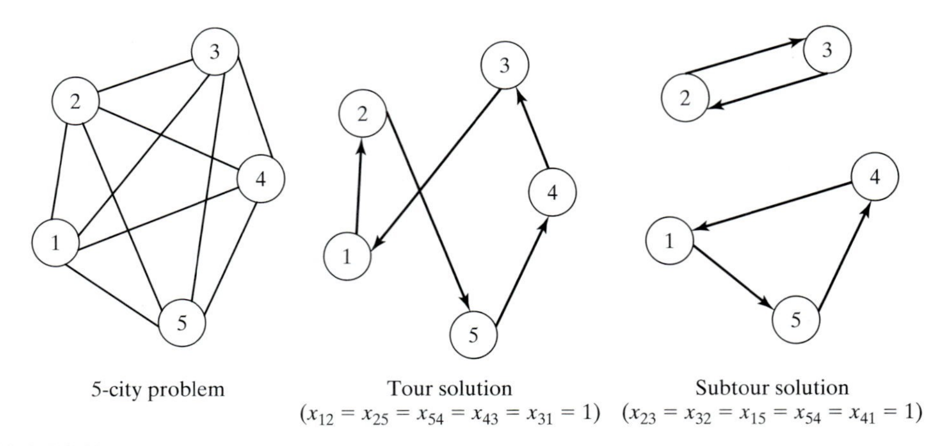

FIGURE 9.11

A 5-city TSP example with a tour and subtour solutions of the associated assignment model

Constraints (1), (2), and (3) define a regular assignment model (Section 5.4). Figure 9.11 demonstrates a 5-city problem. The arcs represent two-way routes. The figure also illustrates a tour and a subtour solution of the associated assignment model. If the optimum solution of the assignment model (i.e., excluding constraint 4) happens to produce a tour, then it is also optimum for the TSP. Otherwise, restriction (4) must be accounted for to ensure a tour solution.

Exact solutions of the TSP problem include branch-and-bound and cutting-plane algorithms. Both are rooted in the ideas of the general B&B and cutting plane algorithms presented in Section 9.2. Nevertheless, the problem is typically difficult computationally, in the sense that either the size or the computational time needed to obtain a solution may become inordinately large. For this reason, heuristics are sometimes used to provide a "good" solution for the problem.

Before presenting the heuristic and exact solution algorithms, we present an example that demonstrates the versatility of the TSP model in representing other practical situations (see also Problem Set 9.3a).

Example 9.3-1

The daily production schedule at the Rainbow Company includes batches of white (W), yellow (Y), red (R), and black (B) paints. Because Rainbow uses the same facilities for all four types of paint, proper cleaning between batches is necessary. The table below summarizes the clean-up time in minutes. Because each color is produced in a single batch, diagonal entries in the table are assigned infinite setup time. The objective is to determine the optimal sequencing for the daily production of the four colors that will minimize the associated total clean-up time.

	Cleanup min given next paint is			
Current paint	*White*	*Yellow*	*Black*	*Red*
White	∞	10	17	15
Yellow	20	∞	19	18
Black	50	44	∞	25
Red	45	40	20	∞

Each paint is thought of as a "city" and the "distances" are represented by the clean-up time needed to switch from one paint batch to the next. The situation reduces to determining the *shortest loop* that starts with one paint batch and passes through each of the remaining three paint batches exactly once before returning back to the starting paint.

We can solve this problem by exhaustively enumerating the six $[(4 - 1)! = 3! = 6]$ possible loops of the network. The following table shows that $W \rightarrow Y \rightarrow R \rightarrow B \rightarrow W$ is the optimum loop.

Production loop	Total clean-up time
$W \rightarrow Y \rightarrow B \rightarrow R \rightarrow W$	$10 + 19 + 25 + 45 = 99$
$W \rightarrow Y \rightarrow R \rightarrow B \rightarrow W$	$10 + 18 + 20 + 50 = 98$
$W \rightarrow B \rightarrow Y \rightarrow R \rightarrow W$	$17 + 44 + 18 + 45 = 124$
$W \rightarrow B \rightarrow R \rightarrow Y \rightarrow W$	$17 + 25 + 40 + 20 = 102$
$W \rightarrow R \rightarrow B \rightarrow Y \rightarrow W$	$15 + 20 + 44 + 20 = 99$
$W \rightarrow R \rightarrow Y \rightarrow B \rightarrow W$	$15 + 40 + 19 + 50 = 124$

Exhaustive enumeration of the loops is not practical in general. Even a modest size 11-city problem will require enumerating $10! = 3{,}628{,}800$ tours, a daunting task indeed. For this reason, the problem must be formulated and solved in a different manner, as we will show later in this section.

To develop the assignment-based formulation for the paint problem, define

$$x_{ij} = 1 \text{ if paint } j \text{ follows paint } i \text{ and zero otherwise}$$

Letting M be a sufficiently large positive value, we can formulate the Rainbow problem as

$$\text{Minimize } z = Mx_{WW} + 10x_{WY} + 17x_{WB} + 15x_{WR} + 20x_{YW} + Mx_{YY} + 19x_{YB} + 18x_{YR} + 50x_{BW} + 44x_{BY} + Mx_{BB} + 25x_{BR} + 45x_{RW} + 40x_{RY} + 20x_{RB} + Mx_{RR}$$

subject to

$$x_{WW} + x_{WY} + x_{WB} + x_{WR} = 1$$
$$x_{YW} + x_{YY} + x_{YB} + x_{YR} = 1$$
$$x_{BW} + x_{BY} + x_{BB} + x_{BR} = 1$$
$$x_{RW} + x_{RY} + x_{RB} + x_{RR} = 1$$
$$x_{WW} + x_{YW} + x_{BW} + x_{RW} = 1$$
$$x_{WY} + x_{YY} + x_{BY} + x_{RY} = 1$$
$$x_{WB} + x_{YB} + x_{BB} + x_{RB} = 1$$
$$x_{WR} + x_{YR} + x_{BR} + x_{RR} = 1$$
$$x_{ij} = (0, 1) \quad \text{for all } i \text{ and } j$$

Solution is a tour (loop)

The use of M in the objective function guarantees that a paint job cannot follow itself. The same result can be realized by deleting x_{WW}, x_{YY}, x_{BB}, and x_{RR} from the entire model.

PROBLEM SET 9.3A

***1.** A manager has a total of 10 employees working on six projects. There are overlaps among the assignments as the following table shows:

	Project 1	2	3	4	5	6
Employee 1		x		x	x	
2	x		x		x	
3		x	x	x		x
4			x	x	x	
5	x	x	x			
6	x	x	x	x		x
7	x	x			x	x
8	x		x	x		
9					x	x
10	x	x		x	x	x

The manager meets with each employee individually once a week for a progress report. Each meeting lasts about 20 minutes for a total of 3 hours and 20 minutes for all 10 employees. To reduce the total time, the manager wants to hold group meetings depending on shared projects. The objective is to schedule the meetings in a way that will reduce the traffic (number of employees) in and out of the meeting room. Formulate the problem as a mathematical model.

2. A book salesperson who lives in Basin must call once a month on four customers located in Wald, Bon, Mena, and Kiln. The following table gives the distances in miles among the different cities.

	Miles between cities				
	Basin	*Wald*	*Bon*	*Mena*	*Kiln*
Basin	0	120	220	150	210
Wald	120	0	80	110	130
Bon	220	80	0	160	185
Mena	150	110	160	0	190
Kiln	210	130	185	190	0

The objective is to minimize the total distance traveled by the salesperson. Formulate the problem as an assignment-based ILP.

3. Circuit boards (such as those used with PCs) are fitted with holes for mounting different electronic components. The holes are drilled with a movable drill. The following table provides the distances (in centimeters) between pairs of 6 holes of a specific circuit board.

$$\|d_{ij}\| = \begin{pmatrix} - & 1.2 & .5 & 2.6 & 4.1 & 3.2 \\ 1.2 & - & 3.4 & 4.6 & 2.9 & 5.2 \\ .5 & 3.4 & - & 3.5 & 4.6 & 6.2 \\ 2.6 & 4.6 & 3.5 & - & 3.8 & .9 \\ 4.1 & 2.9 & 4.6 & 3.8 & - & 1.9 \\ 3.2 & 5.2 & 6.2 & .9 & 1.9 & - \end{pmatrix}$$

Formulate the assignment portion of an ILP representing this problem.

9.3.1 Heuristic Algorithms

This section presents two heuristics: the *nearest-neighbor* and the *subtour-reversal* algorithms. The first is easy to implement and the second requires more computations. The tradeoff is that the second algorithm generally yields better results. Ultimately, the two heuristics are combined into one heuristic, in which the output of the nearest-neighbor algorithm is used as input to the reversal algorithm.

The Nearest-Neighbor Heuristic. As the name of the heuristic suggests, a "good" solution of the TSP problem can be found by starting with any city (node) and then connecting it with the closest one. The just-added city is then linked to its nearest unlinked city (with ties broken arbitrarily). The process continues until a tour is formed.

Example 9.3-2

The matrix below summarizes the distances in miles in a 5-city TSP problem.

$$\|d_{ij}\| = \begin{pmatrix} \infty & 120 & 220 & 150 & 210 \\ 120 & \infty & 100 & 110 & 130 \\ 220 & 80 & \infty & 160 & 185 \\ 150 & \infty & 160 & \infty & 190 \\ 210 & 130 & 185 & \infty & \infty \end{pmatrix}$$

The heuristic can start from any of the five cities. Each starting city may lead to a different tour. The following table provides the steps of the heuristic starting at city 3.

Step	Action	(Partial) tour
1	Start with city 3	3
2	Link to city 2 because it is closest to city 3 ($d_{32} = \min\{220, \mathbf{80}, \infty, 160, 185\}$)	3-2
3	Link to node 4 because it is closest to node 2 ($d_{24} = \min\{120, \infty, —, \mathbf{110}, 130\}$)	3-2-4
4	Link to node 1 because it is closest to node 4 ($d_{41} = \min\{\mathbf{150}, \infty, —, —, 190\}$)	3-2-4-1
5	Link to node 5 by default and connect back to node 3 to complete the tour	3-2-4-1-5-3

Notice the progression of the steps: Comparisons exclude distances to nodes that are part of a constructed partial tour. These are indicated by (—) in the *Action* column of the table.

The resulting tour 3-2-4-1-5-3 has a total length of 80 + 110 + 150 + 210 + 185 = 735 miles. Observe that the quality of the heuristic solution is starting-node dependent. For example, starting from node 1, the constructed tour is 1-2-3-4-5-1 with a total length of 780 miles (try it!).

Subtour Reversal Heuristic. In an n-city situation, the subtour reversal heuristic starts with a feasible tour and then tries to improve on it by reversing 2-city subtours, followed by 3-city subtours, and continuing until reaching subtours of size $n - 1$.

Example 9.3-3

Consider the problem of Example 9.3-2. The reversal steps are carried out in the following table using the feasible tour 1-4-3-5-2-1 of length 745 miles:

Type	Reversal	Tour	Length
Start	—	(1-4-3-5-2-1)	**745**
Two-at-a-time reversal	4-3	1-3-4-5-2-1	820
	3-5	(1-4-5-3-2-1)	**725**
	5-2	1-4-3-2-5-1	730
Three-at-a-time reversal	4-5-3	1-3-5-4-2-1	∞
	5-3-2	1-4-2-3-5-1	∞
Four-at-a-time reversal	4-5-3-2	1-2-3-5-4-1	∞

The two-at-a-time reversals of the initial tour 1-4-3-5-2-1 are 4-3, 3-5, and 5-2, which leads to the given tours with their associated lengths of 820, 725, and 730. Since 1-4-5-3-2-1 yields a smaller length (= 725), it is used as the starting tour for making the three-at-a-time reversals. As shown in the table, these reversals produce no better results. The same result applies to the four-at-a-time reversal. Thus, 1-4-5-3-2-1 (with length 725 miles) provides the best solution of heuristic.

Notice that the three-at-a-time reversals did not produce a better tour, and, for this reason, we continued to use the best two-at-a-time tour with the four-at-a-time reversal. Notice also that the reversals do not include the starting city of the tour (= 1 in this example) because the process does not yield a tour. For example, the reversal 1-4 leads to 4-1-3-5-2-1, which is not a tour.

The solution determined by the reversal heuristic is a function of the initial feasible tour used to start the algorithm. For example, if we start with 2-3-4-1-5-2 with length 750 miles, the heuristic produces the tour 2-1-4-3-5-2 with length 745 miles (verify!), which is inferior to the solution we have in the table above. For this reason, it may be advantageous to first utilize the nearest-neighbor heuristic to determine *all* the tours that result from using each city as a starting node and then select the best as the starting tour for the reversal heuristic. This combined heuristic should, in general, lead to superior solutions than if either heuristic is applied separately. The following table shows the application of the composite heuristic to the present example.

Heuristic	Starting city	Tour	Length
Nearest neighbor	1	1-2-3-4-5-1	780
	2	2-3-4-1-5-2	750
	3	**(3-2-4-1-5-3)**	**735**
	4	4-1-2-3-5-4	∞
	5	5-2-3-4-1-5	750
Reversals	2-4	3-4-2-1-5-3	∞
	4-1	(3-2-1-4-5-3)	**725**
	1-5	3-2-4-5-1-3	810
	2-1-4	3-4-1-2-5-3	745
	1-4-5	3-2-5-4-1-3	∞
	2-1-4-5	3-5-4-1-2-3	∞

Excel Moment.

Figure 9.12 provides a general Excel template (file excelTSP.xls) for the heuristics. It uses three execution options depending on the entry in cell H3:

1. If you enter a city number, the nearest-neighbor heuristic is used to find a tour starting with the designated city.
2. If you enter the word "tour" (without the quotes), you must simultaneously provide an initial feasible tour in the designated space. In this case, only the reversal heuristic is applied to the tour you provided.
3. If you enter the word "all," the nearest-neighbor heuristic is used first, and its best tour is then used to execute the reversal heuristic.

File excelTSP.v2.xls automates the operations of Step 3.

FIGURE 9.12

Execution of the TSP heuristic using Excel spreadsheet (file excelTSP.xls)

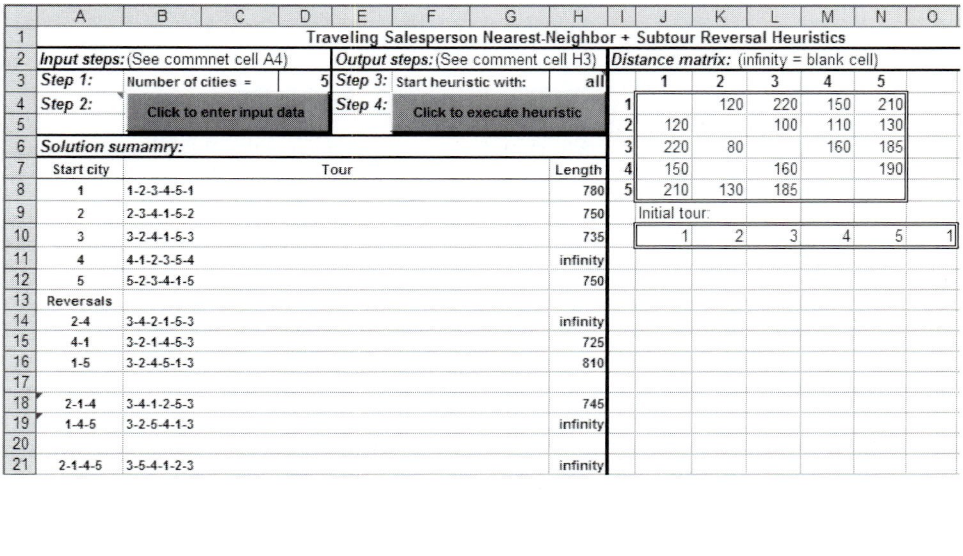

	A	B	C	D	E	F	G	H	I	J	K	L	M	N	O
1	Traveling Salesperson Nearest-Neighbor + Subtour Reversal Heuristics														
2	*Input steps:* (See commnet cell A4)				*Output steps:* (See comment cell H3)					*Distance matrix:* (infinity = blank cell)					
3	*Step 1:*	Number of cities =		5	*Step 3:*	Start heuristic with:		all		1	2	3	4	5	
4	*Step 2:*	Click to enter input data			*Step 4:*	Click to execute heuristic			1		120	220	150	210	
5									2	120		100	110	130	
6	*Solution sumamry:*								3	220	80		160	185	
7	Start city					Tour		Length	4	150		160		190	
8	1	1-2-3-4-5-1						780	5	210	130	185			
9	2	2-3-4-1-5-2						750		Initial tour:					
10	3	3-2-4-1-5-3						735		1	2	3	4	5	1
11	4	4-1-2-3-5-4						infinity							
12	5	5-2-3-4-1-5						750							
13	Reversals														
14	2-4	3-4-2-1-5-3						infinity							
15	4-1	3-2-1-4-5-3						725							
16	1-5	3-2-4-5-1-3						810							
17															
18	2-1-4	3-4-1-2-5-3						745							
19	1-4-5	3-2-5-4-1-3						infinity							
20															
21	2-1-4-5	3-5-4-1-2-3						infinity							

PROBLEM SET 9.3B

1. Apply the heuristic to the following problems:
   (a) The paint sequencing problem of Example 9.3-1.
   (b) Problem 1 of Set 9.3a.
   (c) Problem 2 of Set 9.3a.
   (d) Problem 3 of Set 9.3a.

9.3.2 B&B Solution Algorithm

The idea of the B&B algorithm is to start with the optimum solution of the associated assignment problem. If the solution is a tour, the process ends. Otherwise, restrictions are imposed to remove the subtours. This can be achieved by creating as many branches as the number of x_{ij}-variables associated with one of the subtours. Each branch will correspond to setting one of the variables of the subtour equal to zero (recall that all the variables associated with a subtour equal 1). The solution of the resulting assignment problem may or may not produce a tour. If it does, we use its objective value as an upper bound on the true minimum tour length. If it does not, further branching is necessary, again creating as many branches as the number of variables in one of the subtours. The process continues until all unexplored subproblems have been fathomed, either by producing a better (smaller) *upper bound* or because there is evidence that the subproblem cannot produce a better solution. The optimum tour is the one associated with the best upper bound.

The following example provides the details of the TSP B&B algorithm.

Example 9.3-4

Consider the following 5-city TSP problem:

$$\|d_{ij}\| = \begin{pmatrix} \infty & 10 & 3 & 6 & 9 \\ 5 & \infty & 5 & 4 & 2 \\ 4 & 9 & \infty & 7 & 8 \\ 7 & 1 & 3 & \infty & 4 \\ 3 & 2 & 6 & 5 & \infty \end{pmatrix}$$

We start by solving the associated assignment, which yields the following solution:

$$z = 15, (x_{13} = x_{31} = 1), (x_{25} = x_{54} = x_{42} = 1), \text{all others} = 0$$

This solution yields two subtours: (1-3-1) and (2-5-4-2), as shown at node 1 in Figure 9.13. The associated total distance is $z = 15$, which provides a lower bound on the optimal length of the 5-city tour.

A straightforward way to determine an upper bound is to select any tour and use its length as an upper bound estimate. For example, the tour 1-2-3-4-5-1 (selected totally arbitrarily) has a total length of $10 + 5 + 7 + 4 + 3 = 29$. Alternatively, a better upper bound can be found by applying the heuristic of Section 9.3.1. For the moment, we will use the upper bound of length 29 to apply the B&B algorithm. Later, we use the "improved" upper bound obtained by the heuristic to demonstrate its impact on the search tree.

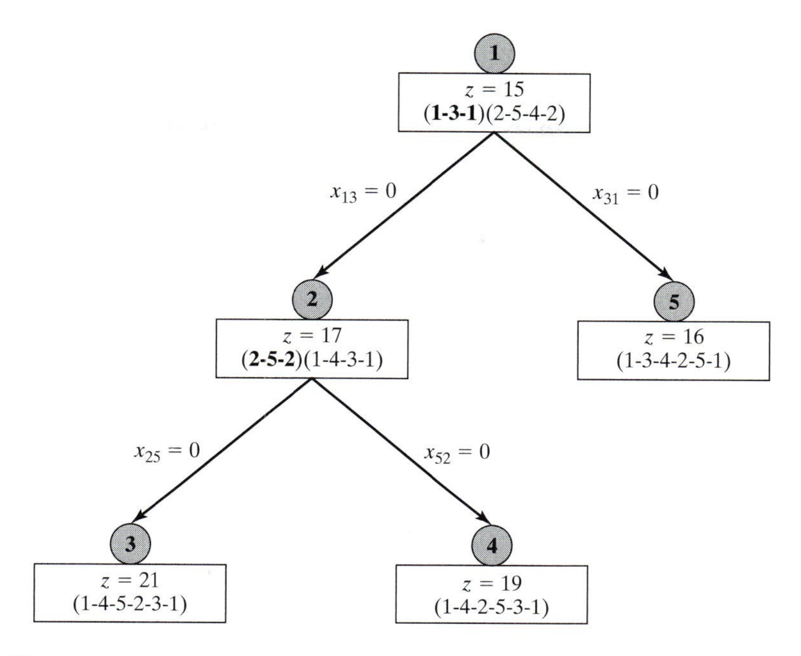

FIGURE 9.13
B&B solution of the TSP problem of Example 9.3-4

The computed lower and upper bounds indicate that the optimum tour length lies in range (15, 29). A solution that yields a tour length larger than (or equal to) 29 is discarded as nonpromising.

To eliminate the subtours at node 1, we need to "disrupt" its loop by forcing its member variables, x_{ij}, to be zero. Subtour 1-3-1 is disrupted if we impose the restriction $x_{13} = 0$ or $x_{31} = 0$ (i.e., one at a time) on the assignment problem at node 1. Similarly, subtour 2-5-4-2 is eliminated by imposing one of the restrictions $x_{25} = 0$, $x_{54} = 0$, or $x_{42} = 0$. In terms of the B&B tree, each of these restrictions gives rise to a branch and hence a new subproblem. It is important to notice that branching *both* subtours at node 1 is *not* necessary. Instead, only *one* subtour needs to be disrupted at any one node. The idea is that a breakup of one subtour automatically alters the member variables of the other subtour and hence produces conditions that are favorable to creating a tour. Under this argument, it is more efficient to select the subtour with the smallest number of cities because it creates the smallest number of branches.

Targeting subtour (1-3-1), two branches $x_{13} = 0$ and $x_{31} = 0$ are created at node 1. The associated assignment problems are constructed by removing the row and column associated with the zero variable, which makes the assignment problem smaller. Another way to achieve the same result is to leave the size of the assignment problem unchanged and simply assign an infinite distance to the branching variable. For example, the assignment problem associated with $x_{13} = 0$ requires substituting $d_{13} = \infty$ in the assignment model at node 1. Similarly, for $x_{31} = 0$, we substitute $d_{31} = \infty$.

In Figure 9.13, we arbitrarily start by solving the subproblem associated with $x_{13} = 0$ by setting $d_{13} = \infty$. Node 2 gives the solution $z = 17$ but continues to produce the subtours (2-5-2) and (1-4-3-1). Repeating the procedure we applied at node 1 gives rise to two branches: $x_{25} = 0$ and $x_{52} = 0$.

We now have three unexplored subproblems, one from node 1 and two from node 2, and we are free to investigate any of them at this point. Arbitrarily exploring the subproblem associated with $x_{25} = 0$ from node 2, we set $d_{13} = \infty$ and $d_{25} = \infty$ in the *original* assignment problem, which yields the solution $z = 21$ and the tour solution 1-4-5-2-3-1 at node 3. The tour solution at node 3 lowers the upper bound from $z = 29$ to $z = 21$. This means that any unexplored subproblem that can be shown to yield a tour length larger than 21 is discarded as nonpromising.

We now have two unexplored subproblems. Selecting the subproblem 4 for exploration, we set $d_{13} = \infty$ and $d_{52} = \infty$ in the *original* assignment, which yields the tour solution 1-4-2-5-3-1 with $z = 19$. The new solution provides a better tour than the one associated with the current upper bound of 21. Thus, the new upper bound is updated to $z = 19$ and its associated tour, 1-4-2-5-3-1, is the best available so far.

Only subproblem 5 remains unexplored. Substituting $d_{31} = \infty$ in the *original* assignment problem at node 1, we get the tour solution 1-3-4-2-5-1 with $z = 16$, at node 5. Once again, this is a better solution than the one associated with node 4 and thus requires updating the upper bound to $z = 16$.

There are no remaining unfathomed nodes, which completes the search tree. The optimal tour is the one associated with the current upper bound: 1-3-4-2-5-1 with length 16 miles.

Remarks. The solution of the example reveals two points:

1. Although the search sequence $1 \rightarrow 2 \rightarrow 3 \rightarrow 4 \rightarrow 5$ was selected deliberately to demonstrate the mechanics of the B&B algorithm and the updating of its upper bound, we generally have no way of predicting which sequence should be adopted to improve the efficiency of the search. Some rules of thumb can be of help. For example, at a given node we can start with the branch associated with the *largest* d_{ij} among all the created branches. By canceling the tour leg with the largest d_{ij}, the hope is that a "good" tour with a smaller total length will be found. In the present example, this rule calls for exploring branch $x_{31} = 0$ to node 5 before branch x_{13} to node 2 because $(d_{31} = 4) > (d_{13} = 3)$, and this would have produced the upper bound $z = 16$, which automatically fathoms node 2 and, hence, eliminates the need to create nodes 3 and 4. Another rule calls for sequencing the exploration of the nodes in a horizontal tier (rather than vertically). The idea is that nodes closer to the starting node are *more likely* to produce a tighter upper bound because the number of additional constraints (of the type $x_{ij} = 0$) is smaller. This rule would have also discovered the solution at node 5 sooner.

2. The B&B should be applied in conjunction with the heuristic in Section 9.3.1. The heuristic provides a "good" upper bound which can be used to fathom nodes in the search tree. In the present example, the heuristic yields the tour 1-3-4-2-5-1 with a length of 16 distance units.

AMPL Moment

Interactive AMPL commands are ideal for the implementation of the TSP B&B algorithm using the general assignment model (file amplAssignment.txt). The following table summarizes the AMPL commands needed to create the B&B tree in Figure 9.13 (Example 9.3-4):

AMPL command	Result
ampl: `model amplAssignment.txt;display x;`	Node 1 solution
ampl: `fix x[1,3]:=0;solve;display x;`	Node 2 solution
ampl: `fix x[2,5]:=0;solve;display x;`	Node 3 solution
ampl: `unfix x[2,5];fix x[5,2]:=0;solve;display x;`	Node 4 solution
ampl: `unfix x[5,2];unfix x[1,3];fix x[3,1]:=0;` `solve;display x;`	Node 5 solution

PROBLEM SET 9.3C

1. Solve Example 9.3-3 using subtour 2-5-4-2 to start the branching process at node 1, using the following sequences for exploring the nodes.
   (a) Explore all the subproblems horizontally from left to right in each tier before preceeding to the next tier.
   (b) Follow each path vertically from node 1 until it ends with a fathomed node.

*2. Solve Problem 1, Set 9.3a using B&B.

3. Solve Problem 2, Set 9.3a using B&B.

4. Solve Problem 3, Set 9.3a using B&B.

9.3.3 Cutting-Plane Algorithm

The idea of the cutting plane algorithm is to add a set of constraints to the assignment problem that prevent the formation of a subtour. The additional constraints are defined as follows. In an n-city situation, associate a continuous variable $u_j\ (\geq 0)$ with cities $2, 3, \ldots,$ and n. Next, define the required set of additional constraints as

$$u_i - u_j + nx_{ij} \leq n - 1, i = 2, 3, \ldots, n; j = 2, 3, \ldots, n; i \neq j$$

These constraints, when added to the assignment model, will automatically remove all subtour solutions.

Example 9.3-5

Consider the following distance matrix of a 4-city TSP problem.

$$\|d_{ij}\| = \begin{pmatrix} - & 13 & 21 & 26 \\ 10 & - & 29 & 20 \\ 30 & 20 & - & 5 \\ 12 & 30 & 7 & - \end{pmatrix}$$

The associated LP consists of the assignment model constraints plus the additional constraints in the table below. All $x_{ij} = (0, 1)$ and all $u_j \geq 0$.

No.	x_{11}	x_{12}	x_{13}	x_{14}	x_{21}	x_{22}	x_{23}	x_{24}	x_{31}	x_{32}	x_{33}	x_{34}	x_{41}	x_{42}	x_{43}	x_{44}	u_2	u_3	u_4	
1							4										1	−1		≤3
2								4									1		−1	≤3
3										4							−1	1		≤3
4												4						1	−1	≤3
5														4			−1		1	≤3
6															4			−1	1	≤3

The optimum solution is

$$u_2 = 0, u_3 = 2, u_4 = 3, x_{12} = x_{23} = x_{34} = x_{41} = 1, \text{tour length} = 59.$$

This corresponds to the tour solution 1-2-3-4-1. The solution satisfies all the additional constraints in u_j (verify!).

To demonstrate that subtour solutions do not satisfy the additional constraints, consider (1-2-1, 3-4-3), which corresponds to $x_{12} = x_{21} = 1, x_{34} = x_{43} = 1$. Now, consider constraint 6 in the tableau above:

$$4x_{43} + u_4 - u_3 \leq 3$$

Substituting $x_{43} = 1, u_3 = 2, u_4 = 3$ yields $5 \leq 3$, which is impossible, thus disallowing $x_{43} = 1$ and subtour 3-4-3.

The disadvantage of the cutting-plane model is that the number of variables grows exponentially with the number of cities, making it difficult to obtain a numeric solution for practical situations. For this reason, the B&B algorithm (coupled with the heuristic) may be a more feasible alternative for solving the problem.

AMPL Moment

Figure 9.14 provides the AMPL model of the cutting-plane algorithm (file amplEx9.3-5.txt). The data of the 4-city TSP of Example 9.3-5 are used to drive the model. The formulation is straightforward: The first two sets of constraints define the assignment model associated with the problem, and the third set represents the cuts needed to remove subtour solutions. Notice that the assignment-model variables must be binary and that `option solver cplex;` must precede `solve;` to ensure that the obtained solution is integer.

The `for` and `if-then` statements at the bottom of the model are used to present the output in the following readable format:

```
Optimal tour length = 59.00
Optimal tour:  1- 2- 3- 4- 1
```

PROBLEM SET 9.3D

1. An automatic guided vehicle (AGV) is used to deliver mail to 5 departments located on a factory floor. The trip starts at the mail sorting room and makes the delivery round to the different departments before returning to the mailroom. Using the mailroom as the origin $(0, 0)$, the (x, y) locations of the delivery spots are $(10, 30)$, $(10, 50)$, $(30, 10)$, $(40, 40)$, and $(50, 60)$ for departments 1 through 5, respectively. All distances are in meters. The AGV can move along horizontal and vertical aisles only. The objective is to minimize the length of the round trip.

   Formulate the problem as a TSP, including the cuts.
2. Write down the cuts associated with the following TSP:

$$\|d_{ij}\| = \begin{pmatrix} \infty & 43 & 21 & 20 & 10 \\ 12 & \infty & 9 & 22 & 30 \\ 20 & 10 & \infty & 5 & 13 \\ 14 & 30 & 42 & \infty & 20 \\ 44 & 7 & 9 & 10 & \infty \end{pmatrix}$$

```
param k;
param n;
param c{1..n,1..n} default 10000;
var x{i in 1..n,j in 1..n} binary;
var u{i in 1..n:i>1}>=0;

minimize tourLength:sum{i in 1..n,j in 1..n}c[i,j]*x[i,j];
subject to
 fromCity {i in 1..n}:sum {j in 1..n} x[i,j] = 1;
 toCity {j in 1..n}:sum {i in 1..n} x[i,j] = 1;
 cut{i in 1..n,j in 1..n:i>1 and j>1 and i<>j}:
                    u[i]-u[j]+n*x[i,j] <= n-1;
data;
param n:=4;
param c:
       1    2    3    4:=
  1    .    13   21   26
  2    10   .    29   20
  3    30   20   .    5
  4    12   30   7    .;

option solver cplex; solve;
display u;
#-------------------------------print formatted output
printf "\n\nOptimal tour length = %7.2f\n",tourLength;
printf "Optimal tour:";
let k:=1;                       #tour starts at city k=1
for {i in 1..n}
    {
    printf "%3i", k;
    for {j in 1..n}  #search for next city following k
        {
        if x[k,j]=1 then
           {
           let k:=j;  #next city found, set k=j
           break;
           }
        }
    printf "-";             #insert last hyphen
    }
printf "  1\n\n";
```

FIGURE 9.14

AMPL cutting-plane model of the TSP problem (file amplEx9.3-5.txt)

3. *AMPL experiment.* Use AMPL to solve the following TSP problem by the cutting plane algorithm.

(a) Problem 2, Set 9.3a.

(b) Problem 3, Set 9.3a.

REFERENCES

Barnett, A., "Misapplication Review: High Road to Glory," *Interfaces*, Vol. 17, No. 5, pp. 51–54, 1987.

Graves, R., L. Schrage, and J. Sankaran, "An Auction Method for Course Registration," *Interfaces*, Vol. 23, No. 5, pp. 81–97, 1993.

Guéret, C., C. Prins, and M. Sevaux, *Applications of Optimization with Xpress-MP*, Dash Optimization, London, 2002.

Jarvis, J., R. Rardin, V. Unger, R. Moore, and C. Schimpeler, "Optimal Design of Regional Wastewater System: A Fixed Charge Network Flow Model," *Operations Research*, Vol. 26, No. 4, pp. 538–550, 1978.

Lee, J., *A First Course in Combinatorial Optimization*, Cambridge University Press, 2004.

Liberatore, M., and T. Miller, "A Hierarchial Production Planning System," *Interfaces*, Vol. 15, No. 4, pp. 1–11, 1985.

Nemhauser, G., and L. Wolsey, *Integer and Combinatorial Optimization*, Wiley, New York, 1988.

Salkin, H., and K. Mathur, *Foundations of Integer Programming*, North-Holland, New York, 1989.

Schrijver, A. *Theory of Linear and Integer Programming*, Wiley, New York, 1998.

Taha, H., *Integer Programming: Theory, Applications, and Computations*, Academic Press, Orlando, FL, 1975.

Weber, G., "Puzzle Contests in MS/OR Education," *Interfaces*, Vol. 20, No. 2, pp. 72–76, 1990.

Wolsey, L., *Integer Programming*, Wiley, New York, 1998.

CHAPTER 10

Deterministic Dynamic Programming

Chapter Guide. Dynamic programming (DP) determines the optimum solution of a multivariable problem by decomposing it into *stages,* each stage comprising a single-variable subproblem. The advantage of the decomposition is that the optimization process at each stage involves one variable only, a simpler task computationally than dealing with all the variables simultaneously. A DP model is basically a recursive equation linking the different stages of the problem in a manner that guarantees that each stage's optimal feasible solution is also optimal and feasible for the entire problem. The notation and the conceptual framework of the recursive equation are unlike any you have studied so far. Experience has shown that the structure of the recursive equation may not appear "logical" to a beginner. Should you have a similar experience, the best course of action is to try to implement what may appear logical to you, and then carry out the computations accordingly. You will soon discover that the definitions in the book are the correct ones and, in the process, will learn how DP works. We have also included two partially automated Excel spreadsheets for some of the examples in which the user must provide key information to drive the DP computations. The exercise should help you understand some of the subtleties of DP.

Although the recursive equation is a common framework for formulating DP models, the solution details differ. Only through exposure to different formulations will you be able to gain experience in DP modeling and DP solution. A number of *deterministic* DP applications are given in this chapter. Chapter 22 on the CD presents *probabilistic* DP applications. Other applications in the important area of inventory modeling are presented in Chapters 11 and 14.

This chapter includes a summary of 1 real-life application, 7 solved examples, 2 Excel spreadsheet models, 32 end-of-section problems, and 1 case. The case is in Appendix E on the CD. The AMPL/Excel/Solver/TORA programs are in folder ch10Files.

Real-Life Application—Optimization of Crosscutting and Log Allocation at Weyerhaeuser.

Mature trees are harvested and crosscut into logs to manufacture different end products (such as construction lumber, plywood, wafer boards, or paper). Log specifications (e.g., length and end diameters) differ depending on the mill where the logs are used. With harvested trees measuring up to 100 feet in length, the number of crosscut combinations meeting mill requirements can be large, and the manner in which a tree is disassembled into logs can affect revenues. The objective is to determine the crosscut combinations that maximize the total revenue. The study uses dynamic programming to optimize the process. The proposed system was first implemented in 1978 with an annual increase in profit of at least $7 million. Case 8 in Chapter 24 on the CD provides the details of the study.

10.1 RECURSIVE NATURE OF COMPUTATIONS IN DP

Computations in DP are done recursively, so that the optimum solution of one subproblem is used as an input to the next subproblem. By the time the last subproblem is solved, the optimum solution for the entire problem is at hand. The manner in which the recursive computations are carried out depends on how we decompose the original problem. In particular, the subproblems are normally linked by common constraints. As we move from one subproblem to the next, the feasibility of these common constraints must be maintained.

Example 10.1-1 (Shortest-Route Problem)

Suppose that you want to select the shortest highway route between two cities. The network in Figure 10.1 provides the possible routes between the starting city at node 1 and the destination city at node 7. The routes pass through intermediate cities designated by nodes 2 to 6.

We can solve this problem by exhaustively enumerating all the routes between nodes 1 and 7 (there are five such routes). However, in a large network, exhaustive enumeration may be intractable computationally.

To solve the problem by DP, we first decompose it into **stages** as delineated by the vertical dashed lines in Figure 10.2. Next, we carry out the computations for each stage separately.

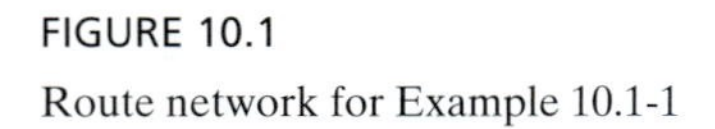

FIGURE 10.1

Route network for Example 10.1-1

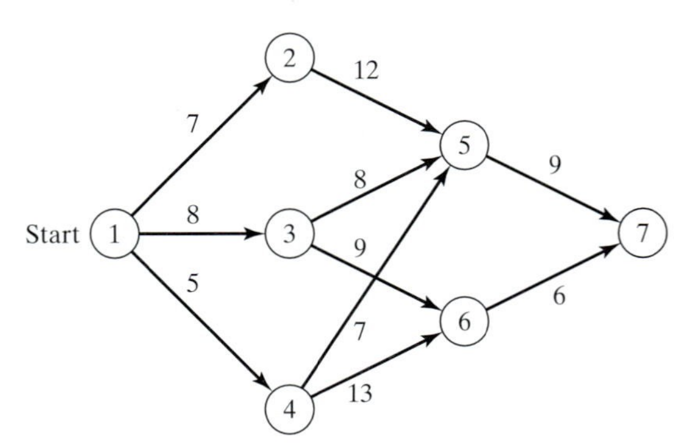

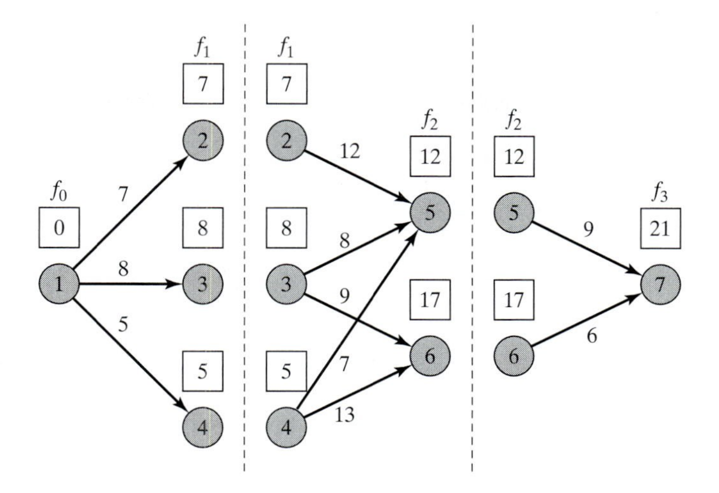

FIGURE 10.2

Decomposition of the shortest-route problem into stages

The general idea for determining the shortest route is to compute the shortest (cumulative) distances to all the terminal nodes of a stage and then use these distances as input data to the immediately succeeding stage. Starting from node 1, stage 1 includes three end nodes (2, 3, and 4) and its computations are simple.

Stage 1 Summary.

$$\text{Shortest distance from node 1 to node 2} = 7 \text{ miles } (\textit{from node } 1)$$

$$\text{Shortest distance from node 1 to node 3} = 8 \text{ miles } (\textit{from node } 1)$$

$$\text{Shortest distance from node 1 to node 4} = 5 \text{ miles } (\textit{from node } 1)$$

Next, stage 2 has two end nodes, 5 and 6. Considering node 5 first, we see from Figure 10.2 that node 5 can be reached from three nodes, 2, 3, and 4, by three different routes: (2, 5), (3, 5), and (4, 5). This information, together with the shortest distances to nodes 2, 3, and 4, determines the shortest (cumulative) distance to node 5 as

$$\begin{pmatrix}\text{Shortest distance}\\ \text{to node 5}\end{pmatrix} = \min_{i=2,3,4}\left\{\begin{pmatrix}\text{Shortest distance}\\ \text{to node } i\end{pmatrix} + \begin{pmatrix}\text{Distance from}\\ \text{node } i \text{ to node 5}\end{pmatrix}\right\}$$

$$= \min\begin{Bmatrix} 7 + 12 = 19 \\ 8 + \ \ 8 = 16 \\ 5 + \ \ 7 = 12 \end{Bmatrix} = 12\ (\textit{from node } 4)$$

Node 6 can be reached from nodes 3 and 4 only. Thus

$$\begin{pmatrix}\text{Shortest distance}\\ \text{to node 6}\end{pmatrix} = \min_{i=3,4}\left\{\begin{pmatrix}\text{Shortest distance}\\ \text{to node } i\end{pmatrix} + \begin{pmatrix}\text{Distance from}\\ \text{node } i \text{ to node 6}\end{pmatrix}\right\}$$

$$= \min\begin{Bmatrix} 8 + 9 = 17 \\ 5 + 13 = 18 \end{Bmatrix} = 17\ (\textit{from node } 3)$$

Stage 2 Summary.

$$\text{Shortest distance from node 1 to node 5} = 12 \text{ miles } (\textit{from node } 4)$$

$$\text{Shortest distance from node 1 to node 6} = 17 \text{ miles } (\textit{from node } 3)$$

The last step is to consider stage 3. The destination node 7 can be reached from either nodes 5 or 6. Using the summary results from stage 2 and the distances from nodes 5 and 6 to node 7, we get

$$\begin{aligned}\begin{pmatrix}\text{Shortest distance}\\ \text{to node 7}\end{pmatrix} &= \min_{i=5,6}\left\{\begin{pmatrix}\text{Shortest distance}\\ \text{to node } i\end{pmatrix} + \begin{pmatrix}\text{Distance from}\\ \text{node } i \text{ to node 7}\end{pmatrix}\right\} \\ &= \min\begin{Bmatrix}12 + 9 = 21\\ 17 + 6 = 23\end{Bmatrix} = 21 \ (\textit{from node } 5)\end{aligned}$$

Stage 3 Summary.

$$\text{Shortest distance from node 1 to node 7} = 21 \text{ miles } (\textit{from node } 5)$$

Stage 3 summary shows that the shortest distance between nodes 1 and 7 is 21 miles. To determine the optimal route, stage 3 summary links node 7 to node 5, stage 2 summary links node 4 to node 5, and stage 1 summary links node 4 to node 1. Thus, the shortest route is $1 \rightarrow 4 \rightarrow 5 \rightarrow 7$.

The example reveals the basic properties of computations in DP:

1. The computations at each stage are a function of the feasible routes of that stage, and that stage alone.
2. A current stage is linked to the *immediately preceding* stage only without regard to earlier stages. The linkage is in the form of the shortest-distance summary that represents the output of the immediately preceding stage.

Recursive Equation. We now show how the recursive computations in Example 10.1-1 can be expressed mathematically. Let $f_i(x_i)$ be the shortest distance to node x_i at stage i, and define $d(x_{i-1}, x_i)$ as the distance from node x_{i-1} to node x_i; then f_i is computed from f_{i-1} using the following recursive equation:

$$f_i(x_i) = \min_{\substack{\text{all feasible}\\ (x_{i-1},\, x_i) \text{ routes}}} \{d(x_{i-1}, x_i) + f_{i-1}(x_{i-1})\}, i = 1, 2, 3$$

Starting at $i = 1$, the recursion sets $f_0(x_0) = 0$. The equation shows that the shortest distances $f_i(x_i)$ at stage i must be expressed in terms of the next node, x_i. In the DP terminology, x_i is referred to as the **state** of the system at stage i. In effect, the *state* of the system at stage i is the information that links the stages together, so that optimal decisions for the remaining stages can be made without reexamining how the decisions for the previous stages are reached. The proper definition of the *state* allows us to consider each stage separately and guarantee that the solution is feasible for all the stages.

The definition of the *state* leads to the following unifying framework for DP.

Principle of Optimality

Future decisions for the remaining stages will constitute an optimal policy regardless of the policy adopted in previous stages.

The implementation of the principle is evident in the computations in Example 10.1-1. For example, in stage 3, we only use the shortest distances to nodes 5 and 6, and do not concern ourselves with how these nodes are reached from node 1. Although the principle of optimality is "vague" about the details of how each stage is optimized, its application greatly facilitates the solution of many complex problems.

PROBLEM SET 10.1A

***1.** Solve Example 10.1-1, assuming the following routes are used:

$$d(1,2) = 5, d(1,3) = 9, d(1,4) = 8$$
$$d(2,5) = 10, d(2,6) = 17$$
$$d(3,5) = 4, d(3,6) = 10$$
$$d(4,5) = 9, d(4,6) = 9$$
$$d(5,7) = 8$$
$$d(6,7) = 9$$

2. I am an avid hiker. Last summer, I went with my friend G. Don on a 5-day hike-and-camp trip in the beautiful White Mountains in New Hampshire. We decided to limit our hiking to an area comprising three well-known peaks: Mounts Washington, Jefferson, and Adams. Mount Washington has a 6-mile base-to-peak trail. The corresponding base-to-peak trails for Mounts Jefferson and Adams are 4 and 5 miles, respectively. The trails joining the bases of the three mountains are 3 miles between Mounts Washington and Jefferson, 2 miles between Mounts Jefferson and Adams, and 5 miles between Mounts Adams and Washington. We started on the first day at the base of Mount Washington and returned to the same spot at the end of 5 days. Our goal was to hike as many miles as we could. We also decided to climb exactly one mountain each day and to camp at the base of the mountain we would be climbing the next day. Additionally, we decided that the same mountain could not be visited in any two consecutive days. How did we schedule our hike?

10.2 FORWARD AND BACKWARD RECURSION

Example 10.1-1 uses **forward recursion** in which the computations proceed from stage 1 to stage 3. The same example can be solved by **backward recursion**, starting at stage 3 and ending at stage 1.

Both the forward and backward recursions yield the same solution. Although the forward procedure appears more logical, DP literature invariably uses backward recursion. The reason for this preference is that, in general, backward recursion may be more efficient computationally. We will demonstrate the use of backward recursion by applying it to Example 10.1-1. The demonstration will also provide the opportunity to present the DP computations in a compact tabular form.

Example 10.2-1

The backward recursive equation for Example 10.2-1 is

$$f_i(x_i) = \min_{\substack{\text{all feasible} \\ \text{routes } (x_i, x_{i+1})}} \{d(x_i, x_{i+1}) + f_{i+1}(x_{i+1})\}, i = 1, 2, 3$$

where $f_4(x_4) = 0$ for $x_4 = 7$. The associated order of computations is $f_3 \rightarrow f_2 \rightarrow f_1$.

Stage 3. Because node 7 ($x_4 = 7$) is connected to nodes 5 and 6 ($x_3 = 5$ and 6) with exactly one route each, there are no alternatives to choose from, and stage 3 results can be summarized as

	$d(x_3, x_4)$	Optimum solution	
x_3	$x_4 = 7$	$f_3(x_3)$	x_4^*
5	9	9	7
6	6	6	7

Stage 2. Route (2, 6) is blocked because it does not exist. Given $f_3(x_3)$ from stage 3, we can compare the feasible alternatives as shown in the following tableau:

	$d(x_2, x_3) + f_3(x_3)$		Optimum solution	
x_2	$x_3 = 5$	$x_3 = 6$	$f_2(x_2)$	x_3^*
2	$12 + 9 = 21$	—	21	5
3	$8 + 9 = 17$	$9 + 6 = 15$	15	6
4	$7 + 9 = 16$	$13 + 6 = 19$	16	5

The optimum solution of stage 2 reads as follows: If you are in cities 2 or 4, the shortest route passes through city 5, and if you are in city 3, the shortest route passes through city 6.

Stage 1. From node 1, we have three alternative routes: (1, 2), (1, 3), and (1, 4). Using $f_2(x_2)$ from stage 2, we can compute the following tableau.

	$d(x_1, x_2) + f_2(x_2)$			Optimum solution	
x_1	$x_2 = 2$	$x_2 = 3$	$x_2 = 4$	$f_1(x_1)$	x_2^*
1	$7 + 21 = 28$	$8 + 15 = 23$	$5 + 16 = 21$	21	4

The optimum solution at stage 1 shows that city 1 is linked to city 4. Next, the optimum solution at stage 2 links city 4 to city 5. Finally, the optimum solution at stage 3 connects city 5 to city 7. Thus, the complete route is given as $1 \rightarrow 4 \rightarrow 5 \rightarrow 7$, and the associated distance is 21 miles.

PROBLEM SET 10.2A

1. For Problem 1, Set 10.1a, develop the backward recursive equation, and use it to find the optimum solution.

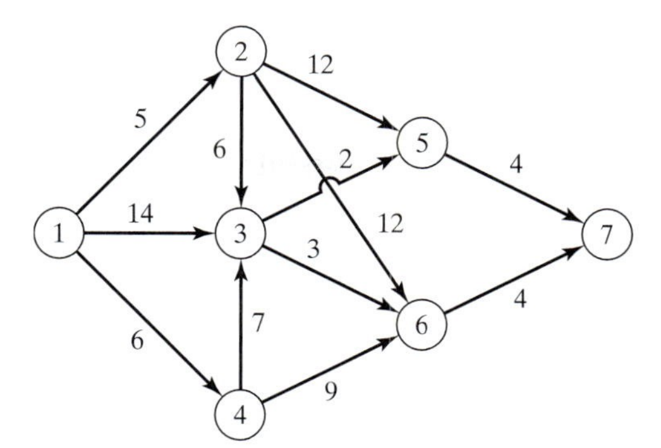

FIGURE 10.3
Network for Problem 3, Set 10.2a

2. For Problem 2, Set 10.1a, develop the backward recursive equation, and use it to find the optimum solution.
*3. For the network in Figure 10.3, it is desired to determine the shortest route between cities 1 to 7. Define the stages and the states using backward recursion, and then solve the problem.

10.3 SELECTED DP APPLICATIONS

This section presents four applications, each with a new idea in the implementation of dynamic programming. As you study each application, pay special attention to the three basic elements of the DP model:

1. Definition of the *stages*
2. Definition of the *alternatives* at each stage
3. Definition of the *states* for each stage

Of the three elements, the definition of the *state* is usually the most subtle. The applications presented here show that the definition of the state varies depending on the situation being modeled. Nevertheless, as you investigate each application, you will find it helpful to consider the following questions:

1. What relationships bind the stages together?
2. What information is needed to make feasible decisions at the current stage without reexamining the decisions made at previous stages?

My teaching experience indicates that understanding the concept of the *state* can be enhanced by questioning the validity of the way it is defined in the book. Try a different definition that may appear "more logical" to you, and use it in the recursive computations. You will eventually discover that the definitions presented here provide the correct way for solving the problem. Meanwhile, the proposed mental process should enhance your understanding of the concept of the state.

10.3.1 Knapsack/Fly-Away/Cargo-Loading Model

The knapsack model classically deals with the situation in which a soldier (or a hiker) must decide on the most valuable items to carry in a backpack. The problem paraphrases a general resource allocation model in which a single limited resource is assigned to a number of alternatives (e.g., limited funds assigned to projects) with the objective of maximizing the total return.

Before presenting the DP model, we remark that the *knapsack* problem is also known in the literature as the *fly-away kit* problem, in which a jet pilot must determine the most valuable (emergency) items to take aboard a jet; and the *cargo-loading* problem, in which a vessel with limited volume or weight capacity is loaded with the most valuable cargo items. It appears that the three names were coined to ensure equal representation of three branches of the armed forces: Air Force, Army, and Navy!

The (backward) recursive equation is developed for the general problem of an n-item W-lb knapsack. Let m_i be the number of units of item i in the knapsack and define r_i and w_i as the revenue and weight per unit of item i. The general problem is represented by the following ILP:

$$\text{Maximize } z = r_1m_1 + r_2m_2 + \cdots + r_nm_n$$

subject to

$$w_1m_1 + w_2m_2 + \cdots + w_nm_n \leq W$$
$$m_1, m_2, \ldots, m_n \geq 0 \text{ and integer}$$

The three elements of the model are

1. *Stage* i is represented by item i, $i = 1, 2, \ldots, n$.

2. The *alternatives* at stage i are represented by m_i, the number of units of item i included in the knapsack. The associated return is r_im_i. Defining $\left[\frac{W}{w_i}\right]$ as the largest integer less than or equal to $\frac{W}{w_i}$, it follows that $m_i = 0, 1, \ldots, \left[\frac{W}{w_i}\right]$.

3. The *state* at stage i is represented by x_i, the total weight assigned to stages (items) $i, i + 1, \ldots,$ and n. This definition reflects the fact that the weight constraint is the only restriction that links all n stages together.

Define

$$f_i(x_i) = \text{maximum return for stages } i, i + 1, \text{ and } n, \text{ given state } x_i$$

The simplest way to determine a recursive equation is a two-step procedure:

Step 1. Express $f_i(x_i)$ as a function of $f_i(x_{i+1})$ as follows:

$$f_i(x_i) = \min_{\substack{m_i=0,1,\ldots,\left[\frac{W}{w_i}\right] \\ x_i \leq W}} \{r_im_i + f_{i+1}(x_{i+1})\}, i = 1, 2, \ldots, n$$

$$f_{n+1}(x_{n+1}) \equiv 0$$

Step 2. Express x_{i+1} as a function of x_i to ensure that the left-hand side, $f_i(x_i)$, is a function of x_i only. By definition, $x_i - x_{i+1} = w_im_i$ represents the weight used at stage i. Thus, $x_{i+1} = x_i - w_im_i$, and the proper recursive equation is given as

$$f_i(x_i) = \max_{\substack{m_i=0,1,\ldots,\left[\frac{W}{w_i}\right] \\ x_i \leq W}} \{r_im_i + f_{i+1}(x_i - w_im_i)\}, i = 1, 2, \ldots, n$$

Example 10.3-1

A 4-ton vessel can be loaded with one or more of three items. The following table gives the unit weight, w_i, in tons and the unit revenue in thousands of dollars, r_i, for item i. How should the vessel be loaded to maximize the total return?

Item i	w_i	r_i
1	2	31
2	3	47
3	1	14

Because the unit weights w_i and the maximum weight W are integer, the state x_i assumes integer values only.

Stage 3. The exact weight to be allocated to stage 3 (item 3) is not known in advance, but can assume one of the values 0, 1, ..., and 4 (because $W = 4$ tons). The states $x_3 = 0$ and $x_3 = 4$, respectively, represent the extreme cases of not shipping item 3 at all and of allocating the entire vessel to it. The remaining values of x_3 ($= 1, 2,$ and 3) imply a partial allocation of the vessel capacity to item 3. In effect, the given range of values for x_3 covers all possible allocations of the vessel capacity to item 3.

Given $w_3 = 1$ ton per unit, the maximum number of units of item 3 that can be loaded is $\frac{4}{1} = 4$, which means that the possible values of m_3 are 0, 1, 2, 3, and 4. An alternative m_3 is feasible only if $w_3 m_3 \leq x_3$. Thus, all the infeasible alternatives (those for which $w_3 m_3 > x_3$) are excluded. The following equation is the basis for comparing the alternatives of stage 3.

$$f_3(x_3) = \max_{m_3=0,1,\ldots,4} \{14m_3\}$$

The following tableau compares the feasible alternatives for each value of x_3.

	$14m_3$					Optimum solution	
x_3	$m_3 = 0$	$m_3 = 1$	$m_3 = 2$	$m_3 = 3$	$m_3 = 4$	$f_3(x_3)$	m_3^*
0	0	—	—	—	—	0	0
1	0	14	—	—	—	14	1
2	0	14	28	—	—	28	2
3	0	14	28	42	—	42	3
4	0	14	28	42	56	56	4

Stage 2. $\max\{m_2\} = \left[\frac{4}{3}\right] = 1$, or $m_3 = 0, 1$

$$f_2(x_2) = \max_{m_2=0,1} \{47m_2 + f_3(x_2 - 3m_2)\}$$

	$47m_2 + f_3(x_2 - 3m_2)$		Optimum solution	
x_2	$m_2 = 0$	$m_2 = 1$	$f_2(x_2)$	m_2^*
0	$0 + 0 = 0$	—	0	0
1	$0 + 14 = 14$	—	14	0
2	$0 + 28 = 28$	—	28	0
3	$0 + 42 = 42$	$47 + 0 = 47$	47	1
4	$0 + 56 = 56$	$47 + 14 = 61$	61	1

Stage 1. $\max\{m_1\} = \left[\frac{4}{2}\right] = 2$ or $m_1 = 0, 1, 2$

$$f_1(x_1) = \max_{m_1=0,1,2}\{31m_1 + f_2(x_1 - 2m_1)\}, \max\{m_1\} = \left[\tfrac{4}{2}\right] = 2$$

	$31m_1 + f_2(x_1 - 2m_1)$			Optimum solution	
x_1	$m_1 = 0$	$m_1 = 1$	$m_1 = 2$	$f_1(x_1)$	m_1^*
0	0 + 0 = 0	—	—	0	0
1	0 + 14 = 14	—	—	14	0
2	0 + 28 = 28	31 + 0 = 31	—	31	1
3	0 + 47 = 47	31 + 14 = 45	—	47	0
4	0 + 61 = 61	31 + 28 = 59	62 + 0 = 62	62	2

The optimum solution is determined in the following manner: Given $W = 4$ tons, from stage 1, $x_1 = 4$ gives the optimum alternative $m_1^* = 2$, which means that 2 units of item 1 will be loaded on the vessel. This allocation leaves $x_2 = x_1 - 2m_2^* = 4 - 2 \times 2 = 0$. From stage 2, $x_2 = 0$ yields $m_2^* = 0$, which, in turn, gives $x_3 = x_2 - 3m_2 = 0 - 3 \times 0 = 0$. Next, from stage 3, $x_3 = 0$ gives $m_3^* = 0$. Thus, the complete optimal solution is $m_1^* = 2$, $m_2^* = 0$, and $m_3^* = 0$. The associated return is $f_1(4) = \$62{,}000$.

In the table for stage 1, we actually need to obtain the optimum for $x_1 = 4$ only because this is the last stage to be considered. However, the computations for $x_1 = 0, 1, 2$, and 3 are included to allow carrying out *sensitivity analysis*. For example, what happens if the vessel capacity is 3 tons in place of 4 tons? The new optimum solution can be determined as

$$(x_1 = 3) \rightarrow (m_1^* = 0) \rightarrow (x_2 = 3) \rightarrow (m_2^* = 1) \rightarrow (x_3 = 0) \rightarrow (m_3^* = 0)$$

Thus the optimum is $(m_1^*, m_2^*, m_3^*) = (0, 1, 0)$ and the optimum revenue is $f_1(3) = \$47{,}000$.

Remarks. The cargo-loading example represents a typical *resource allocation* model in which a limited resource is apportioned among a finite number of (economic) activities. The objective maximizes an associated return function. In such models, the definition of the state at each stage will be similar to the definition given for the cargo-loading model. Namely, the state at stage i is the total resource amount allocated to stages $i, i + 1, \ldots,$ and n.

Excel moment

The nature of dynamic programming computations makes it impossible to develop a general computer code that can handle all DP problems. Perhaps this explains the persistent absence of commercial DP software.

In this section, we present a Excel-based algorithm for handling a subclass of DP problems: the single-constraint knapsack problem (file Knapsack.xls). The algorithm is not data specific and can handle problems in this category with 10 alternatives or less.

Figure 10.4 shows the starting screen of the knapsack (backward) DP model. The screen is divided into two sections: The right section (columns Q:V) is used to summarize

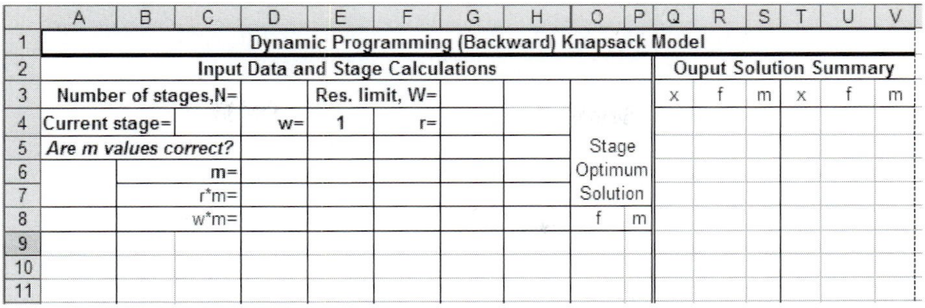

	A	B	C	D	E	F	G	H	O	P	Q	R	S	T	U	V
1	Dynamic Programming (Backward) Knapsack Model															
2	Input Data and Stage Calculations										Ouput Solution Summary					
3	Number of stages,N=				Res. limit, W=						x	f	m	x	f	m
4	Current stage=			w=	1		r=									
5	*Are m values correct?*								Stage							
6			m=						Optimum							
7			r*m=						Solution							
8			w*m=						f	m						
9																
10																
11																

FIGURE 10.4

Excel starting screen of the general DP knapsack model (file excelKnapsack.xls)

the output solution. In the left section (columns A:P), rows 3, 4, and 6 provide the input data for the current stage, and rows 7 and down are reserved for stage computations. The input data symbols correspond to the mathematical notation in the DP model, and are self-explanatory. To fit the spreadsheet conveniently on one screen, the maximum feasible value for alternative m_i at stage i is 10 (cells D6:N6).

Figure 10.5 shows the stage computations generated by the algorithm for Example 10.3-1. The computations are carried out one stage at a time, and the user provides the basic data that drive each stage. Engaging you in this manner will enhance your understanding of the computational details in DP.

Starting with stage 3, and using the notation and data in Example 10.3-1, the input cells are updated as the following list shows:

Cell(s)	Entry
D3	Number of stages, $N = 3$
G3	Resource limit, $W = 4$
C4	Current stage $= 3$
E4	$w_3 = 1$
G4	$r_3 = 14$
D6:H6	$m_3 = (0, 1, 2, 3, 4)$

Note that the feasible values of m_3 are $0, 1, \ldots,$ and $\left[\frac{W}{w_3}\right] = \left[\frac{4}{1}\right] = 4$, as in Example 10.3-1. The spreadsheet automatically tells you how many m_3-values are needed and checks the validity of the values you enter by issuing self-explanatory messages in row 5: "yes," "no," and "delete."

As stage 3 data are entered and verified, the spreadsheet will "come alive" and will generate all the necessary computations of the stage (columns B through P) automatically. The value -1111111 is used to indicate that the corresponding entry is not feasible. The optimum solution (f_3, m_3) for the stage is given in columns O and P. Column A provides the values of f_4. Because the computations start at stage 3, $f_4 = 0$ for all values of x_3. You can leave A9:A13 blank or enter all zero values.

Stage 3:

	A	B	C	D	E	F	G	H	O	P	Q	R	S	T	U	V
1	Dynamic Programming (Backward) Knapsack Model															
2	Input Data and Stage Calculations										Ouput Solution Summary					
3	Number of stages,N=			3	Res. limit, W=		4				x	f	m	x	f	m
4	Current stage=		3	w3=	1	r3=	14				stage 3					
5	*Are m3 values correct?*			yes	yes	yes	yes	yes	Stage		0	0	0			
6		m3=		0	1	2	3	4	Optimum		1	14	1			
7	Stage4	r3*m3=		0	14	28	42	56	Solution		2	28	2			
8	f4	w3*m3=		0	1	2	3	4	f3	m3	3	42	3			
9		x3=	0	0	111111	111111	111111	111111	0	0	4	56	4			
10		x3=	1	0	14	111111	111111	111111	14	1						
11		x3=	2	0	14	28	111111	111111	28	2						
12		x3=	3	0	14	28	42	111111	42	3						
13		x3=	4	0	14	28	42	56	56	4						

Stage 2:

	A	B	C	D	E	F	G	H	O	P	Q	R	S	T	U	V
1	Dynamic Programming (Backward) Knapsack Model															
2	Input Data and Stage Calculations										Ouput Solution Summary					
3	Number of stages,N=			3	Res. limit, W=		4				x	f	m	x	f	m
4	Current stage=		2	w2=	3	r2=	47				stage 3			stage 2		
5	*Are m2 values correct?*			yes	yes	delete	delete	delete	Stage		0	0	0	0	0	0
6		m2=		0	1	2	3	4	Optimum		1	14	1	1	14	0
7	Stage3	r2*m2=		0	47				Solution		2	28	2	2	28	0
8	f3	w2*m2=		0	3				f2	m2	3	42	3	3	47	1
9	0	x2=	0	0	111111				0	0	4	56	4	4	61	1
10	14	x2=	1	14	111111				14	0						
11	28	x2=	2	28	111111				28	0						
12	42	x2=	3	42	47				47	1						
13	56	x2=	4	56	61				61	1						

Stage 1:

	A	B	C	D	E	F	G	H	O	P	Q	R	S	T	U	V
1	Dynamic Programming (Backward) Knapsack Model															
2	Input Data and Stage Calculations										Ouput Solution Summary					
3	Number of stages,N=			3	Res. limit, W=		4				x	f	m	x	f	m
4	Current stage=		1	w1=	2	r1=	31				stage 3			stage 2		
5	*Are m1 values correct?*			yes	yes	yes	delete	delete	Stage		0	0	0	0	0	0
6		m1=		0	1	2	3	4	Optimum		1	14	1	1	14	0
7	Stage2	r1*m1=		0	31	62			Solution		2	28	2	2	28	0
8	f2	w1*m1=		0	2	4			f1	m1	3	42	3	3	47	1
9	0	x1=	0	0	111111	111111			0	0	4	56	4	4	61	1
10	14	x1=	1	14	111111	111111			14	0				stage 1		
11	28	x1=	2	28	31	111111			31	1				0	0	0
12	47	x1=	3	47	45	111111			47	0				1	14	0
13	61	x1=	4	61	59	62			62	2				2	31	1
14														3	47	0
15														4	62	2

FIGURE 10.5

Excel DP model for the knapsack problem of Example 10.3-1 (file excelKnapsack.xls)

Now that stage 3 calculations are at hand, take the following steps to create a *permanent* record of the optimal solution of the current stage and to prepare the spreadsheet for next stage calculations:

Step 1. Copy the x_3-values, C9:C13, and paste them in Q5:Q9 in the optimum solution summary section. Next, copy the (f_3, m_3)-values, O9:P13, and paste them in R5:S9. Remember that you need to paste values only, which requires selecting *Paste Special* from Edit menu and *Values* from the dialogue box.

Step 2. Copy the f_3-values in R5:R9 and paste them in A9:A13 (you do *not* need *Paste Special* in this step).

Step 3. Change cell C4 to 2 and enter the new values of w_2, r_2, and m_2 to record the data of stage 2.

Step 2 places $f_{i+1}(x_i - w_i m_i)$ in column A in preparation for calculating $f_i(x_i)$ at stage i (see the recursive formula for the knapsack problem in Example 10.3-1). This explains the reason for entering zero values, representing f_4, in column A of stage 3 tableau.

Once stage 2 computations are available, you can prepare the screen for stage 1 in a similar manner. When stage 1 is complete, the optimum solution summary can be used to read the solution, as was explained in Example 10.3-1. Note that the organization of the output solution summary area (right section of the screen, columns Q:V) is free-formatted and you can organize its contents in any convenient manner you desire.

PROBLEM SET 10.3A[1]

1. In Example 10.3-1, determine the optimum solution, assuming that the maximum weight capacity of the vessel is 2 tons then 5 tons.
2. Solve the cargo-loading problem of Example 10.3-1 for each of the following sets of data:
   *(a) $w_1 = 4, r_1 = 70, w_2 = 1, r_2 = 20, w_3 = 2, r_3 = 40, W = 6$
   (b) $w_1 = 1, r_1 = 30, w_2 = 2, r_2 = 60, w_3 = 3, r_3 = 80, W = 4$
3. In the cargo-loading model of Example 10.3-1, suppose that the revenue per item includes a constant amount that is realized only if the item is chosen, as the following table shows:

Item	Revenue
1	$\begin{cases} -5 + 31m_1, & \text{if } m_1 > 0 \\ 0, & \text{otherwise} \end{cases}$
2	$\begin{cases} -15 + 47m_2, & \text{if } m_2 > 0 \\ 0, & \text{otherwise} \end{cases}$
3	$\begin{cases} -4 + 14m_3, & \text{if } m_3 > 0 \\ 0, & \text{otherwise} \end{cases}$

   Find the optimal solution using DP. (*Hint*: You can use the Excel file excelSetupKnapsack.xls to check your calculations.)
4. A wilderness hiker must pack three items: food, first-aid kits, and clothes. The backpack has a capacity of 3 ft^3. Each unit of food takes 1 ft^3. A first-aid kit occupies $\frac{1}{4}$ft^3 and each piece of cloth takes about $\frac{1}{2}$ft^3. The hiker assigns the priority weights 3, 4, and 5 to food, first aid, and clothes, which means that clothes are the most valuable of the three items. From experience, the hiker must take at least one unit of each item and no more than two first-aid kits. How many of each item should the hiker take?

*5. A student must select 10 electives from four different departments, with at least one course from each department. The 10 courses are allocated to the four departments in a manner that maximizes "knowledge." The student measures knowledge on a 100-point scale and comes up with the following chart:

	No. of courses						
Department	*1*	*2*	*3*	*4*	*5*	*6*	*≥7*
I	25	50	60	80	100	100	100
II	20	70	90	100	100	100	100
III	40	60	80	100	100	100	100
IV	10	20	30	40	50	60	70

   How should the student select the courses?

[1]In this Problem Set, you are encouraged where applicable to work out the computations by hand and then verify the results using the template excelKnapsack.xls.

6. I have a small backyard garden that measures 10×20 feet. This spring I plan to plant three types of vegetables: tomatoes, green beans, and corn. The garden is organized in 10-foot rows. The corn and tomatoes rows are 2 feet wide, and the beans rows are 3 feet wide. I like tomatoes the most and beans the least, and on a scale of 1 to 10, I would assign 10 to tomatoes, 7 to corn, and 3 to beans. Regardless of my preferences, my wife insists that I plant at least one row of green beans and no more than two rows of tomatoes. How many rows of each vegetable should I plant?

***7.** Habitat for Humanity is a wonderful charity organization that builds homes for needy families using volunteer labor. An eligible family can chose from three home sizes: 1000, 1100, and 1200 ft^2. Each size house requires a certain number of labor volunteers. The Fayetteville chapter has received five applications for the upcoming 6 months. The committee in charge assigns a score to each application based on several factors. A higher score signifies more need. For the next 6 months, the Fayetteville chapter can count on a maximum of 23 volunteers. The following data summarize the scores for the applications and the required number of volunteers. Which applications should the committee approve?

Application	House size (ft^2)	Score	Required no. of volunteers
1	1200	78	7
2	1000	64	4
3	1100	68	6
4	1000	62	5
5	1200	85	8

8. Sheriff Bassam is up for reelection in Washington county. The funds available for the campaign are about $10,000. Although the reelection committee would like to launch the campaign in all five precincts of the county, limited funds dictate otherwise. The following table lists the voting population and the amount of funds needed to launch an effective campaign in each precinct. The choice for each precinct is to receive either all allotted funds or none. How should the funds be allocated?

Precinct	Population	Required funds ($)
1	3100	3500
2	2600	2500
3	3500	4000
4	2800	3000
5	2400	2000

9. An electronic device consists of three components. The three components are in series so that the failure of one component causes the failure of the device. The reliability (probability of no failure) of the device can be improved by installing one or two standby units in each component. The following table charts the reliability, r, and the cost, c. The total capital available for the construction of the device is $10,000. How should the device be constructed? (*Hint*: The objective is to maximize the reliability, $r_1r_2r_3$, of the device. This means that the decomposition of the objective function is multiplicative rather than additive.)

No. of parallel units	Component 1		Component 2		Component 3	
	r_1	c_1($)	r_2	c_2($)	r_3	c_3($)
1	.6	1000	.7	3000	.5	2000
2	.8	2000	.8	5000	.7	4000
3	.9	3000	.9	6000	.9	5000

10. Solve the following model by DP:

$$\text{Maximize } z = \prod_{i=1}^{n} y_i$$

subject to

$$y_1 + y_2 + \cdots + y_n = c$$

$$y_j \geq 0, j = 1, 2, \ldots, n$$

(*Hint*: This problem is similar to Problem 9, except that the variables, y_j, are continuous.)

11. Solve the following problem by DP:

$$\text{Minimize } z = y_1^2 + y_2^2 + \cdots + y_n^2$$

subject to

$$\prod_{i=1}^{n} y_i = c$$

$$y_i > 0, i = 1, 2, \ldots, n$$

12. Solve the following problem by DP:

$$\text{Maximize } z = (y_1 + 2)^2 + y_2 y_3 + (y_4 - 5)^2$$

subject to

$$y_1 + y_2 + y_3 + y_4 \leq 5$$

$$y_i \geq 0 \text{ and integer}, i = 1, 2, 3, 4$$

13. Solve the following problem by DP:

$$\text{Minimize } z = \max\{f(y_1), f(y_2), \ldots, f(y_n)\}$$

subject to

$$y_1 + y_2 + \cdots + y_n = c$$

$$y_i \geq 0, i = 1, 2, \ldots, n$$

Provide the solution for the special case of $n = 3, c = 10$, and $f(y_1) = y_1 + 5$, $f(y_2) = 5y_2 + 3$, and $f(y_3) = y_3 - 2$.

10.3.2 Work-Force Size Model

In some construction projects, hiring and firing are exercised to maintain a labor force that meets the needs of the project. Given that the activities of hiring and firing both incur additional costs, how should the labor force be maintained throughout the life of the project?

Let us assume that the project will be executed over the span of n weeks and that the minimum labor force required in week i is b_i laborers. Theoretically, we can use hiring and firing to keep the work-force in week i exactly equal to b_i. Alternatively, it may be more economical to maintain a labor force larger than the minimum requirements through new hiring. This is the case we will consider here.

Given that x_i is the actual number of laborers employed in week i, two costs can be incurred in week i: $C_1(x_i - b_i)$, the cost of maintaining an excess labor force $x_i - b_i$, and $C_2(x_i - x_{i-1})$, the cost of hiring additional laborers, $x_i - x_{i-1}$. It is assumed that no additional cost is incurred when employment is discontinued.

The elements of the DP model are defined as follows:

1. *Stage* i is represented by week i, $i = 1, 2, \ldots, n$.
2. The *alternatives* at stage i are x_i, the number of laborers in week i.
3. The *state* at stage i is represented by the number of laborers available at stage (week) $i - 1$, x_{i-1}.

The DP recursive equation is given as

$$f_i(x_{i-1}) = \min_{x_i \geq b_i}\{C_1(x_i - b_i) + C_2(x_i - x_{i-1}) + f_{i+1}(x_i)\}, i = 1, 2, \ldots, n$$
$$f_{n+1}(x_n) \equiv 0$$

The computations start at stage n with $x_n = b_n$ and terminate at stage 1.

Example 10.3-2

A construction contractor estimates that the size of the work force needed over the next 5 weeks to be 5, 7, 8, 4, and 6 workers, respectively. Excess labor kept on the force will cost \$300 per worker per week, and new hiring in any week will incur a fixed cost of \$400 plus \$200 per worker per week.

The data of the problem are summarized as

$$b_1 = 5, b_2 = 7, b_3 = 8, b_4 = 4, b_5 = 6$$
$$C_1(x_i - b_i) = 3(x_i - b_i), x_i > b_i, i = 1, 2, \ldots, 5$$
$$C_2(x_i - x_{i-1}) = 4 + 2(x_i - x_{i-1}), x_i > x_{i-1}, i = 1, 2, \ldots, 5$$

Cost functions C_1 and C_2 are in hundreds of dollars.

Stage 5 $(b_5 = 6)$

	$C_1(x_5 - 6) + C_2(x_5 - x_4)$	Optimum solution	
x_4	$x_5 = 6$	$f_5(x_4)$	x_5^*
4	$3(0) + 4 + 2(2) = 8$	8	6
5	$3(0) + 4 + 2(1) = 6$	6	6
6	$3(0) + 0 = 0$	0	6

Stage 4 $(b_4 = 4)$

	$C_1(x_4 - 4) + C_2(x_4 - x_3) + f_5(x_4)$			Optimum solution	
x_3	$x_4 = 4$	$x_4 = 5$	$x_4 = 6$	$f_4(x_3)$	x_4^*
8	$3(0) + 0 + 8 = 8$	$3(1) + 0 + 6 = 9$	$3(2) + 0 + 0 = 6$	6	6

Stage 3 $(b_3 = 8)$

	$C_1(x_3 - 8) + C_2(x_3 - x_2) + f_4(x_3)$	Optimum solution	
x_2	$x_3 = 8$	$f_3(x_2)$	x_3^*
7	$3(0) + 4 + 2(1) + 6 = 12$	12	8
8	$3(0) + 0 + 6 = 6$	6	8

Stage 2 $(b_2 = 7)$

	$C_1(x_2 - 7) + C_2(x_3 - x_2) + f_3(x_2)$		Optimum solution	
x_1	$x_2 = 7$	$x_2 = 8$	$f_2(x_1)$	x_2^*
5	$3(0) + 4 + 2(2) + 12 = 20$	$3(1) + 4 + 2(3) + 6 = 19$	19	8
6	$3(0) + 4 + 2(1) + 12 = 18$	$3(1) + 4 + 2(2) + 6 = 17$	17	8
7	$3(0) + 0 + 12 = 12$	$3(1) + 4 + 2(1) + 6 = 15$	12	7
8	$3(0) + 0 + 12 = 12$	$3(1) + 0 + 6 = 9$	9	8

Stage 1 $(b_1 = 5)$

	$C_1(x_1 - 5) + C_2(x_1 - x_0) + f_2(x_1)$				Optimum solution	
x_0	$x_1 = 5$	$x_1 = 6$	$x_1 = 7$	$x_1 = 8$	$f_1(x_0)$	x_1^*
0	$3(0) + 4 + 2(5) + 19 = 33$	$3(1) + 4 + 2(6) + 17 = 36$	$3(2) + 4 + 2(7) + 12 = 36$	$3(2) + 4 + 2(8) + 9 = 35$	33	5

The optimum solution is determined as

$$x_0 = 0 \rightarrow x_1^* = 5 \rightarrow x_2^* = 8 \rightarrow x_3^* = 8 \rightarrow x_4^* = 6 \rightarrow x_5^* = 6$$

The solution can be translated to the following plan:

Week i	Minimum labor force (b_i)	Actual labor force (x_i)	Decision	Cost
1	5	5	Hire 5 workers	$4 + 2 \times 5 = 14$
2	7	8	Hire 3 workers	$4 + 2 \times 3 + 1 \times 3 = 13$
3	8	8	No change	0
4	4	6	Fire 2 workers	$3 \times 2 = 6$
5	6	6	No change	0

The total cost is $f_1(0) = \$3300$.

PROBLEM SET 10.3B

1. Solve Example 10.3.2 for each of the following minimum labor requirements:
   *(a) $b_1 = 6, b_2 = 5, b_3 = 3, b_4 = 6, b_5 = 8$
   (b) $b_1 = 8, b_2 = 4, b_3 = 7, b_4 = 8, b_5 = 2$
2. In Example 10.3–2, if a severance pay of $100 is incurred for each fired worker, determine the optimum solution.
*3. Luxor Travel arranges 1-week tours to southern Egypt. The agency is contracted to provide tourist groups with 7, 4, 7, and 8 rental cars over the next 4 weeks, respectively. Luxor Travel subcontracts with a local car dealer to supply rental needs. The dealer charges a rental fee of $220 per car per week, plus a flat fee of $500 for any rental transaction. Luxor, however, may elect not to return the rental cars at the end of the week, in which case the agency will be responsible only for the weekly rental ($220). What is the best way for Luxor Travel to handle the rental situation?
4. GECO is contracted for the next 4 years to supply aircraft engines at the rate of four engines a year. Available production capacity and production costs vary from year to year. GECO can produce five engines in year 1, six in year 2, three in year 3, and five in year 4. The corresponding production costs per engine over the next 4 years are $300,000, $330,000, $350,000, and $420,000, respectively. GECO can elect to produce more than it needs in a certain year, in which case the engines must be properly stored until shipment date. The storage cost per engine also varies from year to year, and is estimated to be $20,000 for year 1, $30,000 for year 2, $40,000 for year 3, and $50,000 for year 4. Currently, at the start of year 1, GECO has one engine ready for shipping. Develop an optimal production plan for GECO.

10.3.3 Equipment Replacement Model

The longer a machine stays in service, the higher is its maintenance cost, and the lower its productivity. When a machine reaches a certain age, it may be more economical to replace it. The problem thus reduces to determining the most economical age of a machine.

Suppose that we are studying the machine replacement problem over a span of n years. At the *start* of each year, we decide whether to keep the machine in service an extra year or to replace it with a new one. Let $r(t)$, $c(t)$, and $s(t)$ represent the yearly revenue, operating cost, and salvage value of a t-year-old machine. The cost of acquiring a new machine in any year is I.

The elements of the DP model are

1. *Stage i* is represented by year i, $i = 1, 2, \ldots, n$.
2. The *alternatives* at stage (year) i call for either *keeping* or *replacing* the machine at the *start* of year i.
3. The *state* at stage i is the age of the machine at the start of year i.

Given that the machine is t years old at the *start* of year i, define

$$f_i(t) = \text{maximum net income for years } i, i + 1, \ldots, \text{ and } n$$

The recursive equation is derived as

$$f_i(t) = \max\begin{cases} r(t) - c(t) + f_{i+1}(t + 1), & \text{if KEEP} \\ r(0) + s(t) - I - c(0) + f_{i+1}(1), & \text{if REPLACE} \end{cases}$$

$$f_{n+1}(.) \equiv 0$$

Example 10.3-3

A company needs to determine the optimal replacement policy for a current 3-year-old machine over the next 4 years ($n = 4$). The company requires that a 6-year-old machine be replaced. The cost of a new machine is \$100,000. The following table gives the data of the problem.

Age, t (yr)	Revenue, $r(t)$ (\$)	Operating cost, $c(t)$ (\$)	Salvage value, $s(t)$ (\$)
0	20,000	200	—
1	19,000	600	80,000
2	18,500	1200	60,000
3	17,200	1500	50,000
4	15,500	1700	30,000
5	14,000	1800	10,000
6	12,200	2200	5000

The determination of the feasible values for the age of the machine at each stage is somewhat tricky. Figure 10.6 summarizes the network representing the problem. At the *start* of year 1, we have a 3-year-old machine. We can either replace it (R) or keep it (K) for another year. At the start of year 2, if replacement occurs, the new machine will be 1 year old; otherwise, the old machine will be 4 years old. The same logic applies at the start of years 2 to 4. If a 1-year-old machine is replaced at the start of year 2, 3, or 4, its replacement will be 1 year old at the start of the following year. Also, at the start of year 4, a 6-year-old machine must be replaced, and at the end of year 4 (end of the planning horizon), we salvage (S) the machines.

FIGURE 10.6

Representation of machine age as a function of decision year in Example 10.3-3

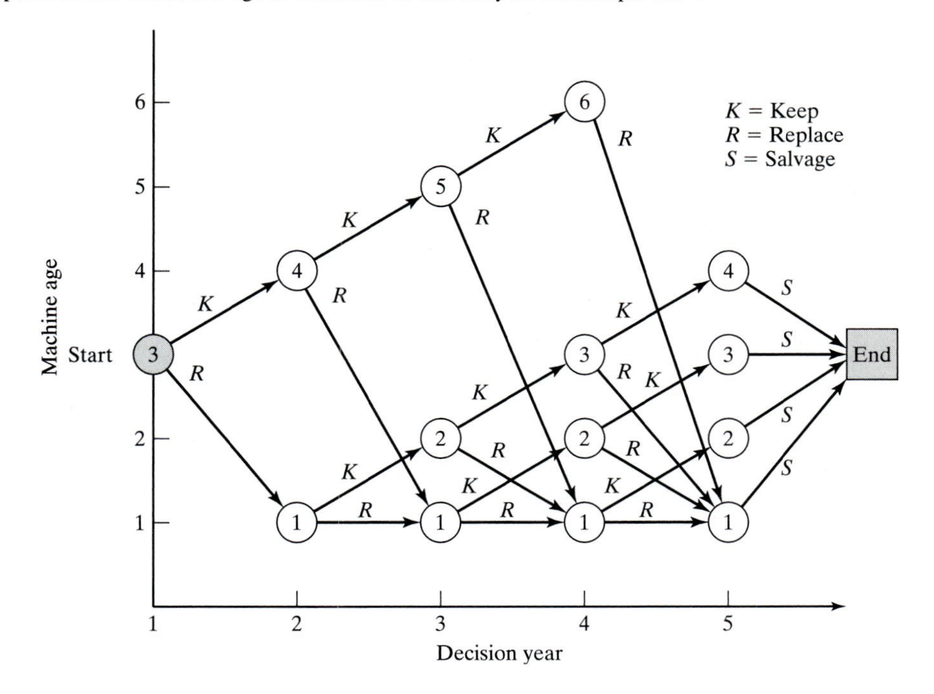

The network shows that at the start of year 2, the possible ages of the machine are 1 and 4 years. For the start of year 3, the possible ages are 1, 2, and 5 years, and for the start of year 4, the possible ages are 1, 2, 3, and 6 years.

The solution of the network in Figure 10.6 is equivalent to finding the longest route (i.e., maximum revenue) from the start of year 1 to the end of year 4. We will use the tabular form to solve the problem. All values are in thousands of dollars. Note that if a machine is replaced in year 4 (i.e., end of the planning horizon), its revenue will include the salvage value, $s(t)$, of the *replaced* machine and the salvage value, $s(1)$, of the *replacement* machine.

Stage 4

	K	*R*	Optimum solution	
t	$r(t) + s(t + 1) - c(t)$	$r(0) + s(t) + s(1) - c(0) - I$	$f_4(t)$	*Decision*
1	19.0 + 60 − .6 = 78.4	20 + 80 + 80 − .2 − 100 = 79.8	79.8	*R*
2	18.5 + 50 − 1.2 = 67.3	20 + 60 + 80 − .2 − 100 = 59.8	67.3	*K*
3	17.2 + 30 − 1.5 = 45.7	20 + 50 + 80 − .2 − 100 = 49.8	49.8	*R*
6	(Must replace)	20 + 5 + 80 − .2 − 100 = 4.8	4.8	*R*

Stage 3

	K	*R*	Optimum solution	
t	$r(t) - c(t) + f_4(t + 1)$	$r(0) + s(t) - c(0) - I + f_4(1)$	$f_3(t)$	*Decision*
1	19.0 − .6 + 67.3 = 85.7	20 + 80 − .2 − 100 + 79.8 = 79.6	85.7	*K*
2	18.5 − 1.2 + 49.8 = 67.1	20 + 60 − .2 − 100 + 79.8 = 59.6	67.1	*K*
5	14.0 − 1.8 + 4.8 = 17.0	20 + 10 − .2 − 100 + 79.8 = 19.6	19.6	*R*

Stage 2

	K	*R*	Optimum solution	
t	$r(t) - c(t) + f_3(t + 1)$	$r(0) + s(t) - c(0) - I + f_3(1)$	$f_2(t)$	*Decision*
1	19.0 − .6 + 67.1 = 85.5	20 + 80 − .2 − 100 + 85.7 = 85.5	85.5	*K* or *R*
4	15.5 − 1.7 + 19.6 = 33.4	20 + 30 − .2 − 100 + 85.7 = 35.5	35.5	*R*

Stage 1

	K	*R*	Optimum solution	
t	$r(t) - c(t) + f_2(t + 1)$	$r(0) + s(t) - c(0) - I + f_2(1)$	$f_1(t)$	*Decision*
3	17.2 − 1.5 + 35.5 = 51.2	20 + 50 − .2 − 100 + 85.5 = 55.3	55.3	*R*

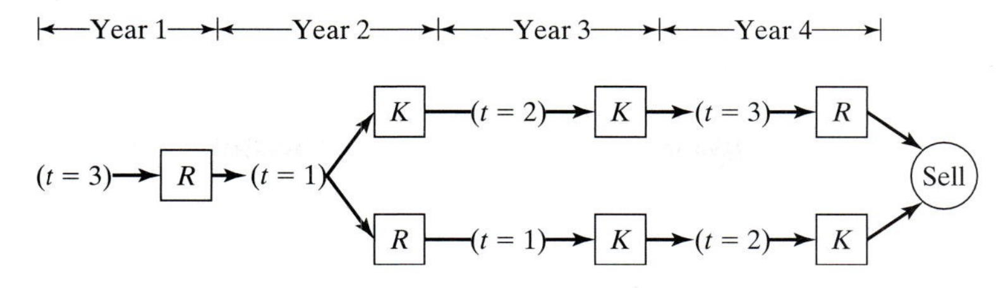

FIGURE 10.7
Solution of Example 10.3-3

Figure 10.7 summarizes the optimal solution. At the start of year 1, given $t = 3$, the optimal decision is to replace the machine. Thus, the new machine will be 1 year old at the start of year 2, and $t = 1$ at the start of year 2 calls for either keeping or replacing the machine. If it is replaced, the new machine will be 1 year old at the start of year 3; otherwise, the kept machine will be 2 years old. The process is continued in this manner until year 4 is reached.

The alternative optimal policies starting in year 1 are (R, K, K, R) and (R, R, K, K). The total cost is $55,300.

PROBLEM SET 10.3C

1. In each of the following cases, develop the network, and find the optimal solution for the model in Example 10.3-3:

(a) The machine is 2 years old at the start of year 1.

(b) The machine is 1 year old at the start of year 1.

(c) The machine is bought new at the start of year 1.

***2.** My son, age 13, has a lawn-mowing business with 10 customers. For each customer, he cuts the grass 3 times a year, which earns him $50 for each mowing. He has just paid $200 for a new mower. The maintenance and operating cost of the mower is $120 for the first year in service, and increases by 20% a year thereafter. A 1-year-old mower has a resale value of $150, which decreases by 10% a year thereafter. My son, who plans to keep his business until he is 16, thinks that it is more economical to buy a new mower every 2 years. He bases his decision on the fact that the price of a new mower will increase only by 10% a year. Is his decision justified?

3. Circle Farms wants to develop a replacement policy for its 2-year-old tractor over the next 5 years. A tractor must be kept in service for at least 3 years, but must be disposed of after 5 years. The current purchase price of a tractor is $40,000 and increases by 10% a year. The salvage value of a 1-year-old tractor is $30,000 and decreases by 10% a year. The current annual operating cost of the tractor is $1300 but is expected to increase by 10% a year.

(a) Formulate the problem as a shortest-route problem.

(b) Develop the associated recursive equation.

(c) Determine the optimal replacement policy of the tractor over the next 5 years.

4. Consider the equipment replacement problem over a period of n years. A new piece of equipment costs c dollars, and its resale value after t years in operation is $s(t) = n - t$

for $n > t$ and zero otherwise. The annual revenue is a function of the age t and is given by $r(t) = n^2 - t^2$ for $n > t$ and zero otherwise.

(a) Formulate the problem as a DP model.

(b) Find the optimal replacement policy given that $c = \$10{,}000$, $n = 5$, and the equipment is 2 years old.

5. Solve Problem 4, assuming that the equipment is 1 year old and that $n = 4$, $c = \$6000$, $r(t) = \frac{n}{1 + t}$.

10.3.4 Investment Model

Suppose that you want to invest the amounts $P_1, P_2, \ldots, P_n$ at the start of each of the next n years. You have two investment opportunities in two banks: First Bank pays an interest rate r_1 and Second Bank pays r_2, both compounded annually. To encourage deposits, both banks pay bonuses on new investments in the form of a percentage of the amount invested. The respective bonus percentages for First Bank and Second Bank are q_{i1} and q_{i2} for year i. Bonuses are paid at the end of the year in which the investment is made and may be reinvested in either bank in the immediately succeeding year. This means that only bonuses and fresh new money may be invested in either bank. However, once an investment is deposited, it must remain in the bank until the end of the n-year horizon. Devise the investment schedule over the next n years.

The elements of the DP model are

1. *Stage* i is represented by year i, $i = 1, 2, \ldots, n$.

2. The *alternatives* at stage i are I_i and $\overline{I}_i$, the amounts invested in First Bank and Second Bank, respectively.

3. The *state*, x_i, at stage i is the amount of capital available for investment at the start of year i.

We note that $\overline{I}_i = x_i - I_i$ by definition. Thus

$$
\begin{aligned}
x_1 &= P_1 \\
x_i &= P_i + q_{i-1,1}I_{i-1} + q_{i-1,2}(x_{i-1} - I_{i-1}) \\
&= P_i + (q_{i-1,1} - q_{i-1,2})I_{i-1} + q_{i-1,2}x_{i-1}, i = 2, 3, \ldots, n
\end{aligned}
$$

The reinvestment amount x_i includes only new money plus any bonus from investments made in year $i - 1$.

Define

$f_i(x_i)$ = optimal value of the investments for years $i, i + 1, \ldots$, and n, given x_i

Next, define s_i as the accumulated sum at the end of year n, given that I_i and $(x_i - I_i)$ are the investments made in year i in First Bank and Second Bank, respectively. Letting $\alpha_k = (1 + r_k)$, $k = 1, 2$, the problem can be stated as

$$\text{Maximize } z = s_1 + s_2 + \cdots + s_n$$

where

$$
\begin{aligned}
s_i &= I_i\alpha_1^{n+1-i} + (x_i - I_i)\alpha_2^{n+1-i} \\
&= (\alpha_1^{n+1-i} - \alpha_2^{n+1-i})I_i + \alpha_2^{n+1-i}x_i, i = 1, 2, \ldots, n - 1 \\
s_n &= (\alpha_1 + q_{n1} - \alpha_2 - q_{n2})I_n + (\alpha_2 + q_{n2})x_n
\end{aligned}
$$

The terms q_{n1} and q_{n2} in s_n are added because the bonuses for year n are part of the final accumulated sum of money from the investment.

The backward DP recursive equation is thus given as

$$f_i(x_i) = \max_{0 \le I_i \le x_i} \{s_i + f_{i+1}(x_{i+1})\}, i = 1, 2, \ldots, n - 1$$
$$f_{n+1}(x_{n+1}) \equiv 0$$

As given previously, x_{i+1} is defined in terms of x_i.

Example 10.3-4

Suppose that you want to invest \$4000 now and \$2000 at the start of years 2 to 4. The interest rate offered by First Bank is 8% compounded annually, and the bonuses over the next 4 years are 1.8%, 1.7%, 2.1%, and 2.5%, respectively. The annual interest rate offered by Second Bank is .2% lower than that of First Bank, but its bonus is .5% higher. The objective is to maximize the accumulated capital at the end of 4 years.

Using the notation introduced previously, we have

$$P_1 = \$4{,}000, P_2 = P_3 = P_4 = \$2000$$
$$\alpha_1 = (1 + .08) = 1.08$$
$$\alpha_2 = (1 + .078) = 1.078$$
$$q_{11} = .018, q_{21} = .017, q_{31} = .021, q_{41} = .025$$
$$q_{12} = .023, q_{22} = .022, q_{32} = .026, q_{42} = .030$$

Stage 4

$$f_4(x_4) = \max_{0 \le I_4 \le x_4} \{s_4\}$$

where

$$s_4 = (\alpha_1 + q_{41} - \alpha_2 - q_{42})I_4 + (\alpha_2 + q_{42})x_4 = -.003I_4 + 1.108x_4$$

The function s_4 is linear in I_4 in the range $0 \le I_4 \le x_4$ and its maximum occurs at $I_4 = 0$ because of the negative coefficient of I_4. Thus, the optimum solution for stage 5 can be summarized as

	Optimum solution	
State	$f_4(x_4)$	I_4^*
x_4	$1.108x_4$	0

Stage 3

$$f_3(x_3) = \max_{0 \le I_3 \le x_3} \{s_3 + f_4(x_4)\}$$

where

$$s_3 = (1.08^2 - 1.078^2)I_3 + 1.078^2 x_3 = .00432I_3 + 1.1621x_3$$
$$x_4 = 2000 - .005I_3 + .026x_3$$

Thus,

$$f_3(x_3) = \max_{0 \le I_3 \le x_3} \{.00432I_3 + 1.1621x_3 + 1.108(2000 - .005I_3 + 0.026x_3\}$$
$$= \max_{0 \le I_3 \le x_3} \{2216 - .00122I_3 + 1.1909x_3\}$$

	Optimum solution	
State	$f_3(x_3)$	I_3^*
x_3	$2216 + 1.1909x_3$	0

Stage 2

$$f_2(x_2) = \max_{0 \le I_2 \le x_2} \{s_2 + f_3(x_3)\}$$

where

$$s_2 = (1.08^3 - 1.078^3)I_2 + 1.078^3x_2 = .006985I_2 + 1.25273x_2$$
$$x_3 = 2000 - .005I_2 + .022x_2$$

Thus,

$$f_2(x_2) = \max_{0 \le I_2 \le x_2} \{.006985I_2 + 1.25273x_2 + 2216 + 1.1909(2000 - .005I_2 + .022x_2)\}$$
$$= \max_{0 \le I_2 \le x_2} \{4597.8 + .0010305I_2 + 1.27893x_2\}$$

	Optimum solution	
State	$f_2(x_2)$	I_2^*
x_2	$4597.8 + 1.27996x_2$	x_2

Stage 1

$$f_1(x_1) = \max_{0 \le I_1 \le x_1} \{s_1 + f_2(x_2)\}$$

where

$$s_1 = (1.08^4 - 1.078^4)I_1 + 1.078^4x_1 = .01005I_2 + 1.3504x_1$$
$$x_2 = 2000 - .005I_1 + .023x_1$$

Thus,

$$f_1(x_1) = \max_{0 \le I_1 \le x_1} \{.01005I_1 + 1.3504x_1 + 4597.8 + 1.27996(2000 - .005I_1 + .023x_1)\}$$
$$= \max_{0 \le I_1 \le x_1} \{7157.7 + .00365I_1 + 1.37984x_1\}$$

	Optimum solution	
State	$f_1(x_1)$	I_1^*
$x_1 = \$4000$	$7157.7 + 1.38349x_1$	\$4000

Working backward and noting that $I_1^* = 4000$, $I_2^* = x_2$, $I_3^* = I_4^* = 0$, we get

$$x_1 = 4000$$
$$x_2 = 2000 - .005 \times 4000 + .023 \times 4000 = \$2072$$
$$x_3 = 2000 - .005 \times 2072 + .022 \times 2072 = \$2035.22$$
$$x_4 = 2000 - .005 \times 0 + .026 \times \$2035.22 = \$2052.92$$

The optimum solution is thus summarized as

Year	Optimum solution	Decision	Accumulation
1	$I_1^* = x_1$	Invest $x_1 = \$4000$ in First Bank	$s_1 = \$5441.80$
2	$I_2^* = x_2$	Invest $x_2 = \$2072$ in First Bank	$s_2 = \$2610.13$
3	$I_3^* = 0$	Invest $x_3 = \$2035.22$ in Second Bank	$s_3 = \$2365.13$
4	$I_4^* = 0$	Invest $x_4 = \$2052.92$ in Second Bank	$s_4 = \$2274.64$

Total accumulation $= f_1(x_1) = 7157.7 + 1.38349(4000) = \$12,691.66$ $(= s_1 + s_2 + s_3 + s_4)$

PROBLEM SET 10.3D

1. Solve Example 10.3-4, assuming that $r_1 = .085$ and $r_2 = .08$. Additionally, assume that $P_1 = \$5000$, $P_2 = \$4000$, $P_3 = \$3000$, and $P_4 = \$2000$.
2. An investor with an initial capital of \$10,000 must decide at the end of each year how much to spend and how much to invest in a savings account. Each dollar invested returns $\alpha = \$1.09$ at the end of the year. The satisfaction derived from spending $\$y$ in any one year is quantified by the equivalence of owning $\$\sqrt{y}$. Solve the problem by DP for a span of 5 years.
3. A farmer owns k sheep. At the end of each year, a decision is made as to how many to sell or keep. The profit from selling a sheep in year i is P_i. The sheep kept in year i will double in number in year $i + 1$. The farmer plans to sell out completely at the end of n years.
   ***(a)** Derive the general recursive equation for the problem.
   (b) Solve the problem for $n = 3$ years, $k = 2$ sheep, $p_1 = \$100$, $p_2 = \$130$. and $p_3 = \$120$.

10.3.5 Inventory Models

DP has important applications in the area of inventory control. Chapters 11 and 14 present some of these applications. The models in Chapter 11 are deterministic, and those in Chapter 14 are probabilistic.

10.4 PROBLEM OF DIMENSIONALITY

In all the DP models we presented, the *state* at any stage is represented by a single element. For example, in the knapsack model (Section 10.3.1), the only restriction is the weight of the item. More realistically, the volume of the knapsack may also be another viable restriction. In such a case, the *state* at any stage is said to be two-dimensional because it consists of two elements: weight and volume.

The increase in the number of state variables increases the computations at each stage. This is particularly clear in DP tabular computations because the number of rows in each tableau corresponds to all possible combinations of state variables. This computational difficulty is sometimes referred to in the literature as the **curse of dimensionality**.

The following example is chosen to demonstrate the *problem of dimensionality*. It also serves to show the relationship between linear and dynamic programming.

Example 10.4-1

Acme Manufacturing produces two products. The daily capacity of the manufacturing process is 430 minutes. Product 1 requires 2 minutes per unit, and product 2 requires 1 minute per unit. There is no limit on the amount produced of product 1, but the maximum daily demand for product 2 is 230 units. The unit profit of product 1 is \$2 and that of product 2 is \$5. Find the optimal solution by DP.

The problem is represented by the following linear program:

$$\text{Maximize } zx = 2x_1 + 5x_2$$

subject to

$$\begin{aligned} 2x_1 + x_2 &\leq 430 \\ x_2 &\leq 230 \\ x_1, x_2 &\geq 0 \end{aligned}$$

The elements of the DP model are

1. *Stage* i corresponds to product $i, i = 1, 2$.
2. *Alternative* x_i is the amount of product $i, i = 1, 2$.
3. *State* (v_2, w_2) represents the amounts of resources 1 and 2 (production time and demand limits) used in stage 2.
4. *State* (v_1, w_1) represents the amounts of resources 1 and 2 (production time and demand limits) used in stages 1 *and* 2.

Stage 2. Define $f_2(v_2, w_2)$ as the maximum profit for stage 2 (product 2), given the state (v_2, w_2). Then

$$f_2(v_2, w_2) = \max_{\substack{0 \leq x_2 \leq v_2 \\ 0 \leq x_2 \leq w_2}} \{5x_2\}$$

Thus, $\max\{5x_2\}$ occurs at $x_2 = \min\{v_2, w_2\}$, and the solution for stage 2 is

	Optimum solution	
State	$f_2(v_2, w_2)$	x_2
(v_2, w_2)	$5\min\{v_2, w_2\}$	$\min\{v_2, w_2\}$

Stage 1

$$\begin{aligned} f_1(v_1, w_1) &= \max_{0 \le 2x_1 \le v_1} \{2x_1 + f_2(v_1 - 2x_1, w_1)\} \\ &= \max_{0 \le 2x_1 \le v_1} \{2x_1 + 5\min(v_1 - 2x_1, w_1)\} \end{aligned}$$

The optimization of stage 1 requires the solution of a (generally difficult) minimax problem. For the present problem, we set $v_1 = 430$ and $w_1 = 230$, which gives $0 \le 2x_1 \le 430$. Because $\min(430 - 2x_1, 230)$ is the lower envelope of two intersecting lines (verify!), it follows that

$$\min(430 - 2x_1, 230) = \begin{cases} 230, & 0 \le x_1 \le 100 \\ 430 - 2x_1, & 100 \le x_1 \le 215 \end{cases}$$

and

$$\begin{aligned} f_1(430, 230) &= \max_{0 \le x_1 \le 215} \{2x_1 + 5\min(430 - 2x_1, 230)\} \\ &= \max_{x_1} \begin{cases} 2x_1 + 1150, & 0 \le x_1 \le 100 \\ -8x_1 + 2150, & 100 \le x_1 \le 215 \end{cases} \end{aligned}$$

You can verify graphically that the optimum value of $f_1(430, 230)$ occurs at $x_1 = 100$. Thus, we get

	Optimum solution	
State	$f_1(v_1, w_1)$	x_1
(430, 230)	1350	100

To determine the optimum value of x_2, we note that

$$v_2 = v_1 - 2x_1 = 430 - 200 = 230$$

$$w_2 = w_1 - 0 = 230$$

Consequently,

$$x_2 = \min(v_2, w_2) = 230$$

The complete optimum solution is thus summarized as

$$x_1 = 100 \text{ units}, x_2 = 230 \text{ units}, z = \$1350$$

PROBLEM SET 10.4A

1. Solve the following problems by DP.

(a) Maximize $z = 4x_1 + 14x_2$

subject to

$$2x_1 + 7x_2 \leq 21$$
$$7x_1 + 2x_2 \leq 21$$
$$x_1, x_2 \geq 0$$

(b) Maximize $z = 8x_1 + 7x_2$

subject to

$$2x_1 + x_2 \leq 8$$
$$5x_1 + 2x_2 \leq 15$$
$$x_1, x_2 \geq 0 \text{ and integer}$$

(c) Maximize $z = 7x_1^2 + 6x_1 + 5x_2^2$

subject to

$$x_1 + 2x_2 \leq 10$$
$$x_1 - 3x_2 \leq 9$$
$$x_1, x_2 \geq 0$$

2. In the n-item knapsack problem of Example 10.3-1, suppose that the weight and volume limitations are W and V, respectively. Given that w_i, v_i, and r_i are the weight, value, and revenue per unit of item i, write the DP backward recursive equation for the problem.

REFERENCES

Bertsekas, D., *Dynamic Programming: Deterministic and Stochastic Models*, Prentice Hall, Upper Saddle River, NJ, 1987.

Denardo, E., *Dynamic Programming Theory and Applications*, Prentice Hall, Upper Saddle River, NJ 1982.

Dreyfus, S., and A. Law, *The Art and Theory of Dynamic Programming*, Academic Press, New York, 1977.

Sntedovich, M., *Dynamic Programming*, Marcel Dekker, New York, 1991.

CHAPTER 11

Deterministic Inventory Models

Chapter Guide. Inventory modeling deals with determining the level of a commodity that a business must maintain to ensure smooth operation. The basis for the decision is a model that balances the cost of capital resulting from holding too much inventory against the penalty cost resulting from inventory shortage. The principal factor affecting the solution is the nature of the demand: deterministic or probabilistic. In real life, demand is usually probabilistic, but in some cases the simpler deterministic approximation may be acceptable. This chapter deals with deterministic models. Probabilistic models are covered in Chapter 14.

The complexity of the inventory problem does not allow the development of a general model that covers all possible situations. This chapter includes representative models of different situations. When you study the different models, you will notice that the solution uses different algorithms, including calculus, linear, nonlinear, and dynamic programming. Regardless of the tool used to solve the model, you should always keep in mind that any inventory model seeks two basic results: *how much* and *when* to order.

The computations associated with some of the models may be tedious. To alleviate this difficulty, a number of Excel spreadsheets, Solver, and AMPL models are included in the chapter. They can be used either for experimentation (e.g., carrying out sensitivity analysis by making changes in the model parameters) or to check your calculations when you work problems.

This chapter includes 8 solved examples, 1 Solver model, 1 AMPL model, 4 Excel spreadsheets, 33 end-of-section problems, and 3 cases. The cases are in Appendix E on the CD. The AMPL/Excel/Solver/TORA programs are in folder ch11Files.

11.1 GENERAL INVENTORY MODEL

The inventory problem involves placing and receiving orders of given sizes periodically. From this standpoint, an **inventory policy** answers two questions:

1. *How much* to order?
2. *When* to order?

The basis for answering these questions is the minimization of the following inventory cost function:

$$\begin{pmatrix}\text{Total}\\ \text{inventory}\\ \text{cost}\end{pmatrix} = \begin{pmatrix}\text{Purchasing}\\ \text{cost}\end{pmatrix} + \begin{pmatrix}\text{Setup}\\ \text{cost}\end{pmatrix} + \begin{pmatrix}\text{Holding}\\ \text{cost}\end{pmatrix} + \begin{pmatrix}\text{Shortage}\\ \text{cost}\end{pmatrix}$$

1. *Purchasing cost* is the price per unit of an inventory item. At times the item is offered at a discount if the order size exceeds a certain amount, which is a factor in deciding *how much to order*.
2. *Setup cost* represents the fixed charge incurred when an order is placed regardless of its size. Increasing the order quantity reduces the setup cost associated with a given demand, but will increase the average inventory level and hence the cost of tied capital. On the other hand, reducing the order size increases the frequency of ordering and the associated setup cost. An inventory cost model balances the two costs.
3. *Holding cost* represents the cost of maintaining inventory in stock. It includes the interest on capital and the cost of storage, maintenance, and handling.
4. *Shortage cost* is the penalty incurred when we run out of stock. It includes potential loss of income and the more subjective cost of loss in customer's goodwill.

An inventory system may be based on **periodic review** (e.g., ordering every week or every month), in which new orders are placed at the start of each period. Alternatively, the system may be based on **continuous review**, where a new order is placed when the inventory level drops to a certain level, called the **reorder point**. An example of periodic review can occur in a gas station where new deliveries arrive at the start of each week. Continuous review occurs in retail stores where items (such as cosmetics) are replenished only when their level on the shelf drops to a certain level.

11.2 ROLE OF DEMAND IN THE DEVELOPMENT OF INVENTORY MODELS

In general, the analytic complexity of inventory models depends on whether the demand for an item is deterministic or probabilistic. Within either category, the demand may or may not vary with time. For example, the consumption of natural gas used in heating homes is a function of the time of the year, reaching its maximum in midwinter and tapering off during spring and summer months. Though this seasonal pattern repeats itself annually, the same-month consumption may vary from year to year, depending, for example, on the severity of weather.

In practical situations the demand pattern in an inventory model may assume one of four types:

1. Deterministic and constant (static) with time.
2. Deterministic and variable (dynamic) with time.
3. Probabilistic and stationary over time.
4. Probabilistic and nonstationary over time.

This categorization assumes the availability of data that are representative of future demand.

In terms of the development of inventory models, the first category is the simplest analytically and the fourth is the most complex. On the other hand, the first category is the least likely to occur in practice and the fourth is the most prevalent. In practice, we seek a balance between model simplicity and model accuracy, in the sense that we do not want to use a simplified model that does not reflect reality, or a complex that it is analytically intractable.

How can we determine if a certain approximation of demand is acceptable? We can start by computing the mean and standard deviation of consumption for a specific period, say monthly. The coefficient of variation $V = \frac{\text{Standard deviation}}{\text{Mean}} \times 100$ can then be used to determine the nature of demand using the following general guideline:[1]

1. If the *average* monthly demand is "approximately" constant for all months and V is reasonably small ($<20\%$), then the demand may be considered deterministic and constant, with its value equal to the average of all monthly demands.
2. If the *average* monthly demand varies appreciably among the different months but V remains reasonably small, then the demand is considered deterministic but variable.
3. If, in Case 1, V is high ($>20\%$) but approximately constant, then the demand is probabilistic and stationary.
4. The only remaining case is the probabilistic nonstationary demand which occurs when the means and coefficients of variation vary appreciably over time.

In cases 3 and 4 additional data usually are needed to determine the associated probability distributions.

Example 11.2-1

The data in Table 11.1 provide the monthly (January through December) consumption of natural gas in a rural residential home over a span of 10 years (1990–1999). Whenever requested by a home owner, the natural-gas supplier sends a truck to the site to fill a tank. The owner of the home decides the time and size of a delivery.

From the standpoint of inventory modeling, it is reasonable to assume that each month represents a decision period in which the owner places an order. Our main concern here, however, is to analyze the nature of the demand.

[1]The coefficient of variation, V, measures the relative variation or spread of the data around the mean. In general, higher values of V indicate higher uncertainty in the use of the mean as an approximation of the monthly consumption. For deterministic demand, $V = 0$, because the associated standard deviation is zero.

TABLE 11.1

Year	Natural-Gas consumption in cubic feet											
	Jan	*Feb*	*Mar*	*Apr*	*May*	*Jun*	*Jul*	*Aug*	*Sep*	*Oct*	*Nov*	*Dec*
1990	100	110	90	70	65	50	40	42	56	68	88	95
1991	110	125	98	80	60	53	44	45	63	77	92	99
1992	90	100	88	79	56	57	38	39	60	70	82	90
1993	121	130	95	90	70	58	41	44	70	80	95	100
1994	109	119	99	75	68	55	43	41	65	79	88	94
1995	130	122	100	85	73	58	42	43	64	75	80	101
1996	115	100	103	90	76	55	45	40	67	78	98	97
1997	130	115	100	95	80	60	49	48	64	85	96	105
1998	125	100	94	86	79	59	46	39	69	90	100	110
1999	87	80	78	75	69	48	39	41	50	70	88	93
Mean	111.7	110	95	82.5	69.6	55.3	42.7	42.2	62.8	77.2	90.7	98
Std Dev	15.54	15.2	7.5	7.99	7.82	3.95	3.4	2.86	6.09	6.91	6.67	6
$V(\%)$	13.91	13.8	7.9	9.68	11.24	7.13	7.96	6.78	9.69	8.95	7.35	6.1

An examination of the mean and the coefficient of variation, V, in Table 11.1 reveals two results:

1. Average consumption is dynamic (not constant) because it shows high average consumption during winter months relative to summer months.
2. The coefficient of variation, V, is reasonably small ($<15\%$) so that the monthly demand can be considered approximately deterministic.

These two results thus lead to the development of an inventory model in which the monthly demand is (approximately) deterministic but variable.

11.3 STATIC ECONOMIC-ORDER-QUANTITY (EOQ) MODELS

This section presents three variations of the economic-order-quantity model with static (constant) demand. These models are characteristically simple from the analytic standpoint.

11.3.1 Classic EOQ Model

The simplest of the inventory models involves constant-rate demand with instantaneous order replenishment and no shortage. Define

$$y = \text{Order quantity (number of units)}$$

$$D = \text{Demand rate (units per unit time)}$$

$$t_0 = \text{Ordering cycle length (time units)}$$

The inventory level follows the pattern depicted in Figure 11.1. An order of size y units is placed and received instantaneously when the inventory reaches zero level. The

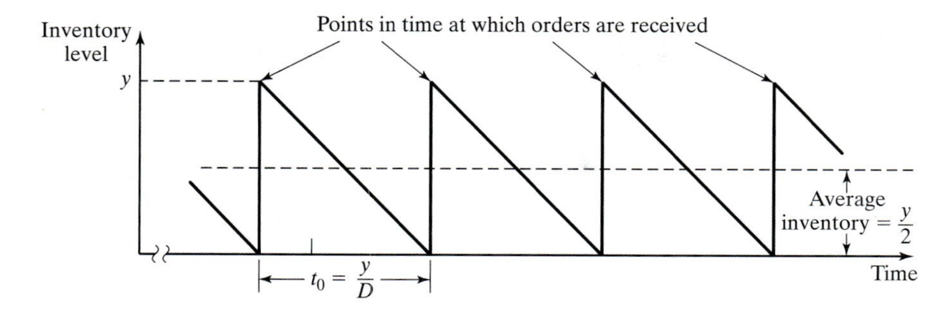

FIGURE 11.1

Inventory pattern in the classic EOQ model

stock is then depleted uniformly at the constant demand rate D. The ordering cycle for this pattern is

$$t_0 = \frac{y}{D}\text{time units}$$

The cost model requires two cost parameters.

K = Setup cost associated with the placement of an order (dollars per order)

h = Holding cost (dollars per inventory unit per unit time)

Given that the average inventory level is $\frac{y}{2}$, the total cost *per unit time* (TCU) is thus computed as

$$\begin{aligned}
\text{TCU}(y) &= \text{Setup cost per unit time} + \text{Holding cost per unit time} \\
&= \frac{\text{Setup cost} + \text{Holding cost per cycle } t_0}{t_0} \\
&= \frac{K + h\left(\frac{y}{2}\right)t_0}{t_0} \\
&= \frac{K}{\left(\frac{y}{D}\right)} + h\left(\frac{y}{2}\right)
\end{aligned}$$

The optimum value of the order quantity y is determined by minimizing TCU(y) with respect to y. Assuming y is continuous, a necessary condition for finding the optimal value of y is

$$\frac{d\,\text{TCU}(y)}{dy} = -\frac{KD}{y^2} + \frac{h}{2} = 0$$

The condition is also sufficient because TCU(y) is convex.

The solution of the equation yields the EOQ y^* as

$$y^* = \sqrt{\frac{2KD}{h}}$$

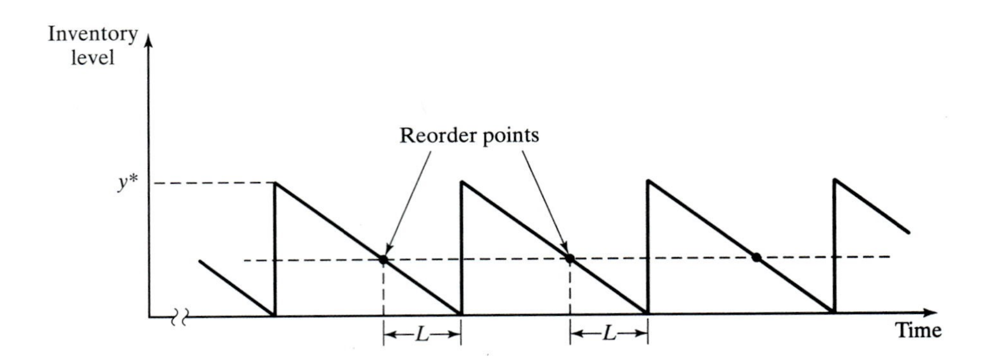

FIGURE 11.2
Reorder point in the classic EOQ model

Thus, the optimum inventory policy for the proposed model is

$$\text{Order } y^* = \sqrt{\frac{2KD}{h}} \text{ units every } t_0^* = \frac{y^*}{D} \text{ time units}$$

Actually, a new order need not be received at the instant it is ordered. Instead, a positive **lead time**, L, may occur between the placement and the receipt of an order as Figure 11.2, demonstrates. In this case, the **reorder point** occurs when the inventory level drops to LD units.

Figure 11.2 assumes that the lead time L is less than the cycle length t_0^*, which may not be the case in general. To account for this situation, we define the **effective lead time** as

$$L_e = L - nt_0^*$$

where n is the largest integer not exceeding $\frac{L}{t_0^*}$. This result is justified because after n cycles of t_0^* each, the inventory situation acts as if the interval between placing an order and receiving another is L_e. Thus, the reorder point occurs at L_eD units, and the inventory policy can be restated as

Order the quantity y^* whenever the inventory level drops to L_eD units

Example 11.3-1

Neon lights on the U of A campus are replaced at the rate of 100 units per day. The physical plant orders the neon lights periodically. It costs \$100 to initiate a purchase order. A neon light kept in storage is estimated to cost about \$.02 per day. The lead time between placing and receiving an order is 12 days. Determine the optimal inventory policy for ordering the neon lights.

From the data of the problem, we have

$$D = 100 \text{ units per day}$$

$$K = \$100 \text{ per order}$$

$$h = \$.02 \text{ per unit per day}$$
$$L = 12 \text{ days}$$

Thus,

$$y^* = \sqrt{\frac{2KD}{h}} = \sqrt{\frac{2 \times \$100 \times 100}{.02}} = 1000 \text{ neon lights}$$

The associated cycle length is

$$t_0^* = \tfrac{y^*}{D} = \tfrac{1000}{100} = 10 \text{ days}$$

Because the lead time $L = 12$ days exceeds the cycle length t_0^* (= 10 days), we must compute L_e. The number of integer cycles included in L is

$$\begin{aligned} n &= \left(\text{Largest integer} \leq \tfrac{L}{t_0^*}\right) \\ &= \left(\text{Largest integer} \leq \tfrac{12}{10}\right) \\ &= 1 \end{aligned}$$

Thus,

$$L_e = L - nt_0^* = 12 - 1 \times 10 = 2 \text{ days}$$

The reorder point thus occurs when the inventory level drops to

$$L_eD = 2 \times 100 = 200 \text{ neon lights}$$

The inventory policy for ordering the neon lights is

Order 1000 units whenever the inventory level drops to 200 units.

The daily inventory cost associated with the proposed inventory policy is

$$\begin{aligned} \text{TCU}(y) &= \frac{K}{\left(\frac{y}{D}\right)} + h\left(\tfrac{y}{2}\right) \\ &= \frac{\$100}{\left(\frac{1000}{100}\right)} + \$.02\left(\tfrac{1000}{2}\right) = \$20 \text{ per day} \end{aligned}$$

Excel Moment

Template exelEOQ.xls is designed to carry out the EOQ computations. The model solves the general EOQ described in Problem 10, Set 11.3a, with shortage and simultaneous production-consumption operation, of which the present model is a special case. It also solves the price-breaks situation presented in Section 11.3.2. To use the template with Example 11.3-1, enter −1 in cells C3:C5, C8, and C10 to indicate that the corresponding data are not applicable, as shown in Figure 11.3.

	B	C	D
1	**General Economic Order Quantity (EOQ)**		
2	**Input data:** Enter -1 in column C if data element does not apply		
3	**Item cost, c1 =**	-1	
4	**Qty discount limit, q =**	-1	
5	**Item cost, c2 =**	-1	
6	**Setup cost, K =**	100	
7	**Demand rate, D =**	100	
8	**Production rate, a =**	-1	
9	**Unit holding cost, h =**	0.02	
10	**Unit penalty cost, p =**	-1	
11	**Lead time, L =**	12	
12	**Model output results:**		
13	Order qty, y* =	1000.00	
14	Shortage qty, w* =	0.00	
15	Reorder point, R =	200.00	
16	TCU(y*) =	20.00	
17	Purchase/prod. Cost =	0.00	
18	Setup cost/unit time =	10.00	
19	Holding cost /unit time =	10.00	
20	shortage cost/unit time =	0.00	
21	*Optimal inventory policy: Order 1000.00 units when level drops to 200.00 units*		
22	**Model intermediate calculations:**		
23	ym =	1000.00	
24	TCU1(ym)=	Not applicable	
25	Q-equation:	Not applicable	
26	Q =	Not applicable	
27	cycle length, t0 =	10.00	
28	Optimization zone =	Not applicable	
29	Effectice lead time, Le =	2.00	

FIGURE 11.3

Excel solution of Example 11.3-1 (file excelEOQ.xls)

PROBLEM SET 11.3A

1. In each of the following cases, no shortage is allowed, and the lead time between placing and receiving an order is 30 days. Determine the optimal inventory policy and the associated cost per day.

   (a) $K = \$100, h = \$.05, D = 30$ units per day

   (b) $K = \$50, h = \$.05, D = 30$ units per day

   (c) $K = \$100, h = \$.01, D = 40$ units per day

   (d) $K = \$100, h = \$.04, D = 20$ units per day

***2.** McBurger orders ground meat at the start of each week to cover the week's demand of 300 lb. The fixed cost per order is \$20. It costs about \$.03 per lb per day to refrigerate and store the meat.

   (a) Determine the inventory cost per week of the present ordering policy.

   (b) Determine the optimal inventory policy that McBurger should use, assuming zero lead time between the placement and receipt of an order.

3. A company stocks an item that is consumed at the rate of 50 units per day. It costs the company \$20 each time an order is placed. An inventory unit held in stock for a week will cost \$.35.

   (a) Determine the optimum inventory policy, assuming a lead time of 1 week.

   (b) Determine the optimum number of orders per year (based on 365 days per year).

*4. Two inventory policies have been suggested by the purchasing department of a company:

Policy 1. Order 150 units. The reorder point is 50 units and the time between placing and receiving an order is 10 days.

Policy 2. Order 200 units. The reorder point is 75 units and the time between placing and receiving an order is 15 days.

The setup cost per order is $20, and the holding cost per unit inventory per day is $.02.

(a) Which of the two policies should the company adopt?

(b) If you were in charge of devising an inventory policy for the company, what would you recommend assuming that the supplier requires a lead time of 22 days?

5. Walmark Store compresses and palletizes empty merchandise cartons for recycling. The store generates five pallets a day. The cost of storing a pallet in the store's back lot is $.10 per day. The company that moves the pallets to the recycling center charges a flat fee of $100 for the rental of its loading equipment plus a variable transportation cost of $3 per pallet. Graph the change in number of pallets with time, and devise an optimal policy for hauling the pallets to the recycling center.

6. A hotel uses an external laundry service to provide clean towels. The hotel generates 600 soiled towels a day. The laundry service picks up the soiled towels and replaces them with clean ones at regular intervals. There is a fixed charge of $81 per pickup and delivery service, in addition to the variable cost of $.60 per towel. It costs the hotel $.02 a day to store a soiled towel and $.01 per day to store a clean one. How often should the hotel use the pickup and delivery service? (*Hint*: There are two types of inventory items in this situation. As the level of the soiled towels increases, that of clean towels decreases at an equal rate.)

7. (Lewis, 1996) An employee of a multinational company is on loan from the United States to the company's subsidiary in Europe. During that year, the employee's financial obligations in the United States (e.g., mortgage and insurance premium payments) amount to $12,000, distributed evenly over the months of the year. The employee can meet these obligations by depositing the entire sum in a U.S. bank prior to departure for Europe. However, at present the interest rate in the United States is quite low (about 1.5% per year) in comparison with the interest rate in Europe (6.5% per year). The cost of sending funds from overseas is $50 per transaction. Determine an optimal policy for transferring funds from Europe to the United States and discuss the practical implementation of the solution. State all the assumptions.

8. Consider the inventory situation in which the stock is replenished uniformly (rather than instantaneously) at the rate a. Consumption occurs at the constant rate D. Because consumption also occurs during the replenishment period, it is necessary that $a > D$. The setup cost is K per order, and the holding cost is h per unit per unit time. If y is the order size and no shortage is allowed, show that

(a) The maximum inventory level is $y\left(1 - \frac{D}{a}\right)$.

(b) The total cost per unit time given y is

$$\text{TCU}(y) = \frac{KD}{y} + \frac{h}{2}\left(1 - \frac{D}{a}\right)y$$

(c) The economic order quantity is

$$y^* = \sqrt{\frac{2KD}{h\left(1 - \frac{D}{a}\right)}},\ D < a$$

(d) Show that the EOQ under instantaneous replenishment can be derived from the formula in (c).

9. A company can produce an item or buy it from a contractor. If it is produced, it will cost \$20 each time the machines are set up. The production rate is 100 units per day. If it is bought from a contractor, it will cost \$15 each time an order is placed. The cost of maintaining the item in stock, whether bought or produced, is \$.02 per unit per day. The company's usage of the item is estimated at 26,000 units annually. Assuming that no shortage is allowed, should the company buy or produce?

10. In Problem 8, suppose that shortage is allowed at a penalty cost of p per unit per unit time.

    (a) If w is the maximum shortage during the inventory cycle, show that

$$\text{TCU}(y, w) = \frac{KD}{y} + \frac{h\{y(1 - \frac{D}{a}) - w\}^2 + pw^2}{2(1 - \frac{D}{a})y}$$

$$y^* = \sqrt{\frac{2KD(p + h)}{ph(1 - \frac{D}{a})}}$$

$$w^* = \sqrt{\frac{2KDh(1 - \frac{D}{a})}{p(p + h)}}$$

    (b) Show that the EOQ results in Section 11.3.1 can be derived from the general formulas in (a).

11.3.2 EOQ with Price Breaks

This model is the same as in Section 11.3.1, except that the inventory item may be purchased at a discount if the size of the order, y, exceeds a given limit, q. Mathematically, the unit purchasing price, c, is given as

$$c = \begin{Bmatrix} c_1, \text{ if } y \leq q \\ c_2, \text{ if } y > q \end{Bmatrix}, c_1 > c_2$$

Hence,

$$\text{Purchasing cost per unit time} = \begin{cases} \dfrac{c_1 y}{t_0} = \dfrac{c_1 y}{\left(\frac{y}{D}\right)} = Dc_1, \; y \leq q \\ \dfrac{c_2 y}{t_0} = \dfrac{c_2 y}{\left(\frac{y}{D}\right)} = Dc_2, \; y > q \end{cases}$$

Using the notation in Section 11.3.1, the total cost per unit time is

$$\text{TCU}(y) = \begin{cases} \text{TCU}_1(y) = Dc_1 + \dfrac{KD}{y} + \dfrac{h}{2}y, \; y \leq q \\ \text{TCU}_2(y) = Dc_2 + \dfrac{KD}{y} + \dfrac{h}{2}y, \; y > q \end{cases}$$

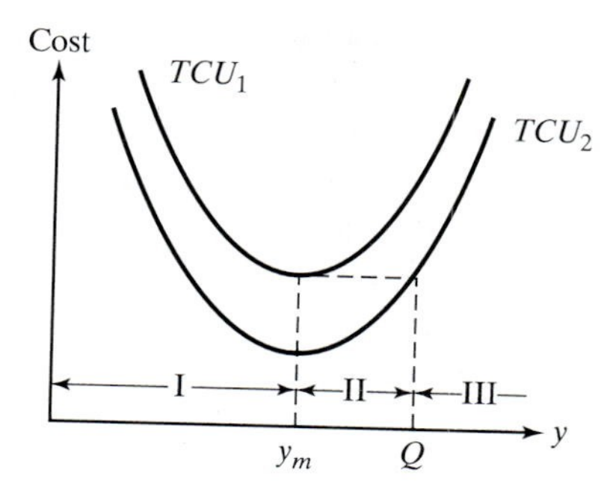

FIGURE 11.4
Inventory cost function with price breaks

The functions TCU_1 and TCU_2 are graphed in Figure 11.4. Because the two functions differ only by a constant amount, their minima must coincide at

$$y_m = \sqrt{\frac{2KD}{h}}$$

The cost function $\text{TCU}(y)$ starts on the left with $\text{TCU}_1(y)$ and drops to $\text{TCU}_2(y)$ at the price breakpoint q. The determination of the optimum order quantity y^* depends on where the price breakpoint, q, lies with respect to zones I, II, and III delineated in Figure 11.4 by $(0, y_m)$, (y_m, Q), and (Q, ∞), respectively. The value of Q ($> y_m$) is determined from the equation

$$\text{TCU}_2(Q) = \text{TCU}_1(y_m)$$

or

$$c_2D + \frac{KD}{Q} + \frac{hQ}{2} = \text{TCU}_1(y_m)$$

which simplifies to

$$Q^2 + \left(\frac{2(c_2D - \text{TCU}_1(y_m))}{h}\right)Q + \frac{2KD}{h} = 0$$

Figure 11.5 shows that the desired optimum quantity y^* is

$$y^* = \begin{cases} y_m, \text{ if } q \text{ is in zones I or III} \\ q, \text{ if } q \text{ is in zone II} \end{cases}$$

The steps for determining y^* are

Step 1. Determine $y_m = \sqrt{\frac{2KD}{h}}$. If q is in zone I, then $y^* = y_m$. Otherwise, go to step 2.

Step 2. Determine Q ($> y_m$) from the Q-equation

$$Q^2 + \left(\frac{2(c_2D - \text{TCU}_1(y_m))}{h}\right)Q + \frac{2KD}{h} = 0$$

Define zones II and III. If q is in zone II, $y^* = q$. Otherwise, q is in zone III, and $y^* = y_m$.

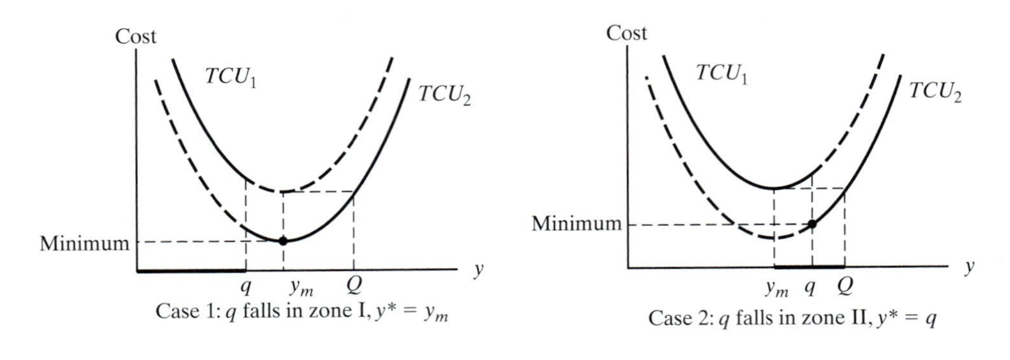

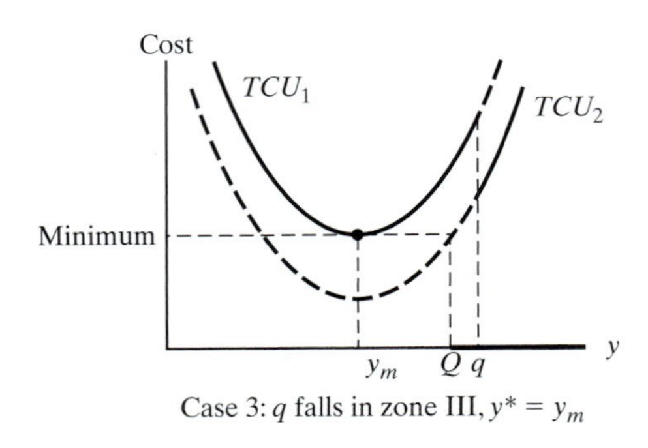

FIGURE 11.5
Optimum solution for inventory problems with price breaks

Example 11.3-2

LubeCar specializes in fast automobile oil change. The garage buys car oil in bulk at \$3 per gallon. A discount price of \$2.50 per gallon is available if LubeCar purchases more than 1000 gallons. The garage services approximately 150 cars per day, and each oil change takes 1.25 gallons. LubeCar stores bulk oil at the cost of \$.02 per gallon per day. Also, the cost of placing an order for bulk oil is \$20. There is a 2-day lead time for delivery. Determine the optimal inventory policy.

The consumption of oil per day is

$$D = 150 \text{ cars per day} \times 1.25 \text{ gallons per car} = 187.5 \text{ gallons per day}$$

We also have

$$h = \$.02 \text{ per gallon per day}$$
$$K = \$20 \text{ per order}$$
$$L = 2 \text{ days}$$
$$c_1 = \$3 \text{ per gallon}$$
$$c_2 = \$2.50 \text{ per gallon}$$
$$q = 1000 \text{ gallons}$$

Step 1. Compute

$$y_m = \sqrt{\frac{2KD}{h}} = \sqrt{\frac{2 \times 20 \times 187.5}{.02}} = 612.37 \text{ gallons}$$

Because $q = 1000$ is larger than $y_m = 612.37$, we move to step 2.

Step 2. Determine Q.

$$\begin{aligned} TCU_1(y_m) &= c_1 D + \frac{KD}{y_m} + \frac{h y_m}{2} \\ &= 3 \times 187.5 + \frac{20 \times 187.5}{612.37} + \frac{.02 \times 612.37}{2} \\ &= 574.75 \end{aligned}$$

Hence, the Q-equation is calculated as

$$Q^2 + \left(\frac{2 \times (2.5 \times 187.5 - 574.75)}{.02}\right)Q + \frac{2 \times 20 \times 187.5}{.02} = 0$$

or

$$Q^2 - 10{,}599.74Q + 375{,}000 = 0$$

This yields $Q = 10{,}564.25$ $(> y_m)$. Thus,

$$\begin{aligned} \text{Zone II} &= (612.37, 10{,}564.25) \\ \text{Zone III} &= (10{,}564.25, \infty) \end{aligned}$$

Because q $(= 1000)$ falls in zone II, the optimal order quantity is $y^* = q = 1000$ gallons.

Given a 2-day lead time, the reorder point is $2D = 2 \times 187.5 = 375$ gallons. Thus, the optimal inventory policy is

Order 1000 gallons when the inventory level drops to 375 gallons.

Excel Moment

Excel template excelEOQ.xls solves the discount price situation given above. The use of the model is straightforward. Enter the data of the model in the input data section of the spreadsheet (C3:C11). Appropriate error messages will be displayed to resolve input data conflicts. The output of the model gives the optimal inventory policy as well as all the intermediate calculations of the problem.

PROBLEM SET 11.3B

1. Consider the hotel laundry service situation in Problem 6, Set 11.3a. The normal charge for washing a soiled towel is $.60, but the laundry service will charge only $.50 if the hotel supplies them in lots of at least 2500 towels. Should the hotel take advantage of the discount?

***2.** An item is consumed at the rate of 30 items per day. The holding cost per unit per day is $.05, and the setup cost is $100. Suppose that no shortage is allowed and that the purchasing cost per unit is \$10 for any quantity not exceeding 500 units and \$8 otherwise. The lead time is 21 days. Determine the optimal inventory policy.

3. An item sells for \$25 a unit, but a 10% discount is offered for lots of 150 units or more. A company uses this item at the rate of 20 units per day. The setup cost for ordering a lot is \$50, and the holding cost per unit per day is \$.30. The lead time is 12 days. Should the company take advantage of the discount?

*4. In Problem 3, determine the range on the price discount percentage that, when offered for lots of size 150 units or more, will not result in any financial advantage to the company.

5. In the inventory model discussed in this section, suppose that the holding cost per unit per unit time is h_1 for quantities below q and h_2 otherwise, $h_1 > h_2$. Show how the economic lot size is determined.

11.3.3 Multi-Item EOQ with Storage Limitation

This model deals with n (> 1) items, whose individual inventory fluctuations follow the same pattern as in Figure 11.1 (no shortage allowed). The difference is that the items are competing for a limited storage space.

Define for item i, $i = 1, 2, \ldots, n$,

D_i = Demand rate

K_i = Setup cost

h_i = Unit holding cost per unit time

y_i = Order quantity

a_i = Storage area requirement per inventory unit

A = Maximum available storage area for all n items

Under the assumption of no shortage, the mathematical model representing the inventory situation is given as

$$\text{Minimize TCU}(y_1, y_2, \ldots, y_n) = \sum_{i=1}^{n}\left(\frac{K_i D_i}{y_i} + \frac{h_i y_i}{2}\right)$$

subject to

$$\sum_{i=1}^{n} a_i y_i \leq A$$

$$y_i > 0, i = 1, 2, \ldots, n$$

To solve the problem, we try the unconstrained solution first:

$$y_i^* = \sqrt{\frac{2K_i D_i}{h_i}}, i = 1, 2, \ldots, n$$

If this solution satisfies the constraint, then we are done. Otherwise, the constraint must be activated.

In previous editions of this book, we used the (rather involved) Lagrangian algorithm and trial-and-error calculations to find the constrained optimum solution. With the availability of powerful packages (such as AMPL and Solver), the problem can be solved directly as a nonlinear program, as will be demonstrated in the following example.

Example 11.3-3

The following data describe three inventory items.

Item i	K_i (\$)	D_i (units per day)	h_i (\$)	a_i (ft^2)
1	10	2	.30	1
2	5	4	.10	1
3	15	4	.20	1
	Total available storage area $= 25\ ft^2$			

The unconstrained optimum values, $y_i^* = \sqrt{\frac{2K_iD_i}{h_i}}, i = 1, 2, 3$, are 11.55, 20.00, and 24.49 units, respectively. These values violate the storage constraint

$$y_1 + y_2 + y_3 \leq 25$$

Thus, the problem is solved as a nonlinear program using Solver or AMPL, as explained below. The Solver model must be adjusted to fit the size of the problem. The AMPL model can be applied to any number of items simply by changing the input data.

The optimum solution is $y_1^* = 6.34$ units, $y_2^* = 7.09$ units, $y_3^* = 11.57$ units, cost = \$13.62/day.

Solver Moment

Figure 11.6 shows how Solver can be used to solve Example 11.3-3 as a nonlinear program (file solverConstrEOQ.xls). Details of the formulas used in the template and of the Solver parameters are shown in the figure. As with most nonlinear programs, initial solution values must be given (in this template, $y_1 = y_2 = y_3 = 1$ in row 9). A *nonzero* initial value is mandatory because the objective function includes division by y_i. Indeed, it may be a good idea to replace K_iD_i/y_i with $K_iD_i/(y_i + \Delta)$, where $\Delta > 0$ and is very small to suppress division by zero during the iterations. In general, different initial values may be needed before a (local optimum) solution is found. The optimum solution at the bottom of the figure is global because the objective function and the constraints are well behaved (convex objective function and convex solution space).

AMPL Moment

The AMPL nonlinear model for the general multi-item EOQ with storage limitation is given in Figure 11.7 (file amplConstrEOQ.txt). The model follows the same rules used in solving linear programs. However, as with Solver, AMPL nonlinear models exhibit peculiarities that may impede reaching a solution. In particular, "judicious" initial values must be specified for the variables. In Figure 11.7, the definition statement

```
var y{1..n}>=0, :=10;          #initial trial value = 10;
```

includes the code `:=10` that assigns the initial value 10 to all the variables. If you use an initial value of 1 in the present example, division by zero will result during the iterations.

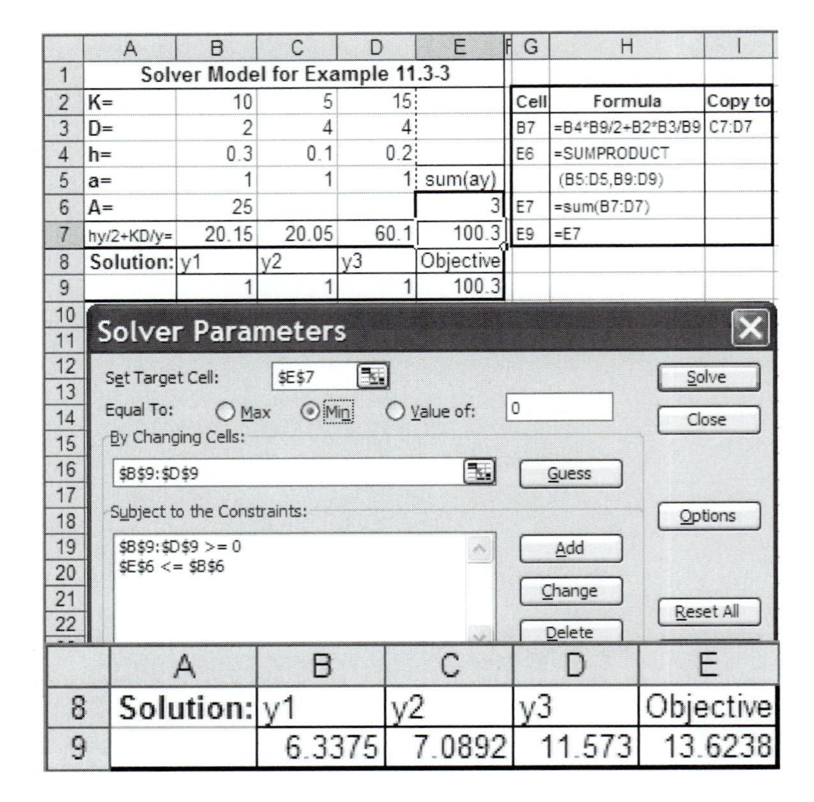

FIGURE 11.6
Solver template for Example 11.3-3 (file solverConstrEOQ.xls)

Thus, as in Solver, you may need to replace K_iD_i/y_i with $K_iD_i/(y_i + \Delta)$, where $\Delta > 0$ and is very small to prevent division by zero during the iterative process. Indeed, Problems 1 and 4, Set 11.3c, could not be solved with AMPL without invoking this trick.

PROBLEM SET 11.3C[2]

*1. The following data describe five inventory items.

Item i	K_i (\$)	D_i (units per day)	h_i (\$)	a_i (ft^2)
1	20	22	0.35	1.0
2	25	34	0.15	0.8
3	30	14	0.28	1.1
4	28	21	0.30	0.5
5	35	26	0.42	1.2

Total available storage area = 25 ft^2

Determine the optimal order quantities.

[2]You will find files solverConstrEOQ.xls and amplConstrEOQ.txt useful in solving the problems of this set.

```
param n;
param K{1..n};
param D{1..n};
param h{1..n};
param a{1..n};
param A;

var y{1..n}>=0, :=10;  #initial trial value = 10

minimize z: sum{j in 1..n}(K[j]*D[j]/y[j]+h[j]*y[j]/2);
subject to storage:sum{j in 1..n}a[j]*y[j]<=A;
data;
param n:=3;
param K:= 1 10   2 5   3 15;
param D:=1 2   2 4   3 4;
param h:=1 .3   2 .1   3 .2;
param a:=1 1   2 1   3 1;
param A:=25;

solve;display z,y;

printf"SOLUTION:\n">a.out;
printf" Total cost = %4.2f\n",z>a.out;
for {i in 1..n}
        printf" y%1i = %4.2f\n",i,y[i]>a.out;
```

FIGURE 11.7

AMPL model for Example 11.3-3 (file amplConstrEOQ.txt)

2. Solve the model of Example 11.3-3, assuming that we require the sum of the average inventories for all the items to be less than 25 units.

3. In Problem 2, assume that the only restriction is a limit of $1000 on the amount of capital that can be invested in inventory. The purchase costs per unit of items 1, 2, and 3 are $100, $55, and $100, respectively. Determine the optimum solution.

*__4.__ The following data describe four inventory items.

Item i	K_i ($)	D_i (units per day)	h_i ($)
1	100	10	.1
2	50	20	.2
3	90	5	.2
4	20	10	.1

The company wishes to determine the economic order quantity for each of the four items such that the total number of orders per 365-day year is at most 150. Formulate the problem as a nonlinear program and find the optimum solution.

11.4 DYNAMIC EOQ MODELS

The models presented here differ from those in Section 11.3 in two respects: (1) the inventory level is reviewed periodically over a finite number of equal periods; and (2) the demand per period, though deterministic, is dynamic, in the sense that it varies from one period to the next.

A situation in which dynamic deterministic demand occurs is **materials requirement planning** (MRP). The idea of MRP is described by an example. Suppose that the quarterly demands over the next year for two final models, $M1$ and $M2$, of a given product are 100 and 150 units, respectively. Deliveries of the quarterly lots are made at the end of each quarter. The production lead time is 2 months for $M1$ and 1 month for $M2$. Each unit of $M1$ and $M2$ uses 2 units of a subassembly S. The lead time for the production of S is 1 month.

Figure 11.8 depicts the production schedules for $M1$ and $M2$. The schedules start with the quarterly demand for the two models (shown by solid arrows) occurring at the end of months 3, 6, 9, and 12. Given the lead times of 2 and 1 months for $M1$ and $M2$, the dashed arrows then show the planned starts of each production lot.

To start the production of the two models on time, the delivery of subassembly S must coincide with the occurrence of the dashed $M1$ and $M2$ arrows. This information is shown by the solid arrows in the S-chart, where the resulting S-demand is 2 units per unit of $M1$ or $M2$. Using a lead time of 1 month, the dashed arrows on the S-chart give the production schedules for S. From these two schedules, the combined demand for S corresponding to $M1$ and $M2$ can then be determined as shown at the bottom of Figure 11.8. The resulting *variable* (but known) demand for S is typical of the situation where dynamic EOQ occurs. In essence, given the indicated variable demand for S, how much should be produced at the start of each month to reduce the total production-inventory cost?

FIGURE 11.8

Example of dynamic demand generated by MRP

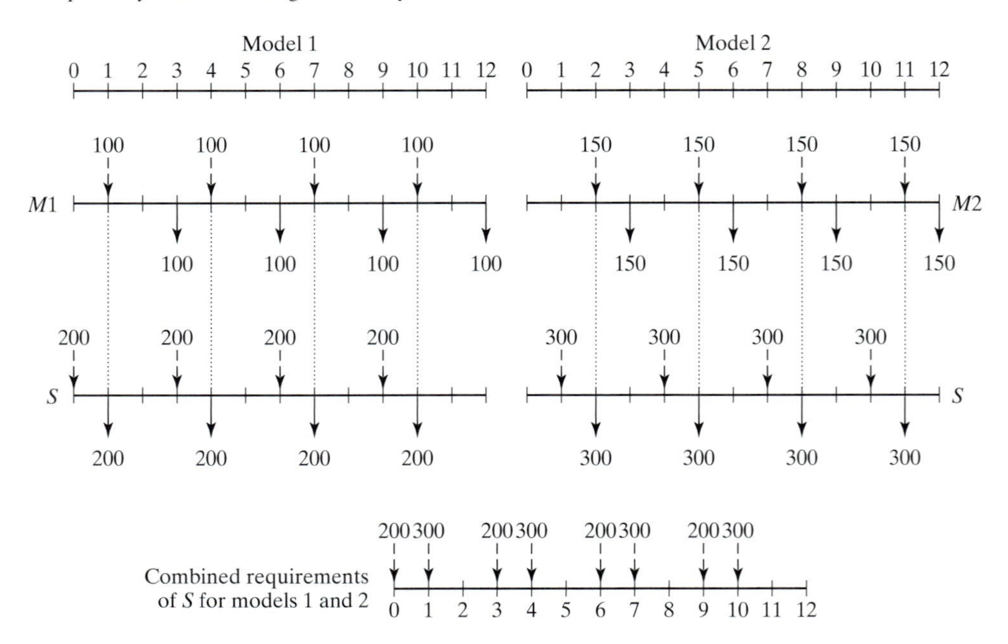

Two models are presented in this section. The first model does not assume a setup (ordering) cost, and the second one does. This seemingly "small" detail makes a difference in the complexity of the model.

PROBLEM SET 11.4A

1. In Figure 11.8 determine the combined requirements for subassembly S in each of the following cases:
   *(a) Lead time for $M1$ is only one period.
   (b) Lead time for $M1$ is three periods.

11.4.1 No-Setup Model

This model involves a planning horizon with n equal periods. Each period has a limited production capacity that can include several production levels (e.g., regular time and overtime represent two production levels). A current period may produce more than its immediate demand to satisfy demand for later periods, in which case an inventory holding cost must be charged.

The general assumptions of the model are

1. No setup cost is incurred in any period.
2. No shortage is allowed.
3. The unit production cost function in any period either is constant or has increasing (convex) marginal costs.
4. The unit holding cost in any period is constant.

The absence of shortage signifies that production in future periods cannot fill the demand in a current period. This assumption requires the cumulative production capacity for periods 1, 2, ..., and i to equal at least the cumulative demand for the same inclusive periods.

Figure 11.9 illustrates the unit production cost function with increasing margins. For example, regular time and overtime production correspond to two levels in which unit production cost during overtime is higher than during regular time.

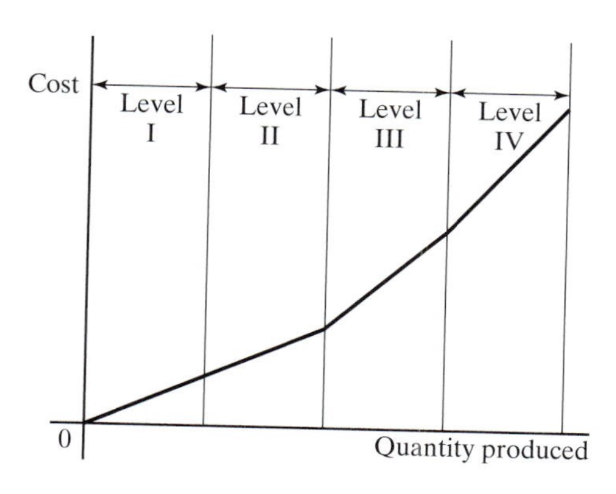

FIGURE 11.9

Convex unit production cost function

The n-period problem can be formulated as a transportation model (see Chapter 5) with kn sources and n destinations, where k is the number of production levels per period (e.g., if each period uses regular time and overtime, then $k = 2$). The production capacity of each of the kn production-level sources provides the supply amounts. The demand amounts are specified by each period's demand. The unit "transportation" cost from a source to a destination is the sum of the applicable production and holding costs per unit. The solution of the problem as a transportation model determines the minimum-cost production amounts in each production level.

The resulting transportation model can be solved without using the familiar transportation technique presented in Chapter 5. The validity of the new solution algorithm rests on the special assumptions of no shortage and a convex production-cost function.

Example 11.4-1

Metalco produces draft deflectors for use in home fireplaces during the months of December to March. The demand starts slow, peaks in the middle of the season, and tapers off toward the end. Because of the popularity of the product, Metalco may use overtime to satisfy the demand. The following table provides the production capacities and the demands for the four winter months.

	Capacity		
Month	*Regular (units)*	*Overtime (units)*	Demand (units)
1	90	50	100
2	100	60	190
3	120	80	210
4	110	70	160

Unit production cost in any period is \$6 during regular time and \$9 during overtime. Holding cost per unit per month is \$.10.

To ensure that the model has a feasible solution when no shortage is allowed, the cumulative supply (production capacity) up to any month must equal at least the associated cumulative demand, as the following table shows.

Month	Cumulative supply	Cumulative demand
1	90 + 50 = 140	100
2	140 + 100 + 60 = 300	100 + 190 = 290
3	300 + 120 + 80 = 500	290 + 210 = 500
4	500 + 110 + 70 = 680	500 + 160 = 660

Table 11.2 summarizes the model and its solution. The symbols R_i and O_i represent regular and overtime production levels in period i, $i = 1, 2, 3, 4$. Because cumulative supply at period 4 exceeds cumulative demand, a dummy surplus destination is added to balance the model as

TABLE 11.2

	1	2	3	4	Surplus	
R_1	6 **90**	6.1	6.2	6.3	0	**90**
O_1	9 **10**	9.1 **30**	9.2 **10**	9.3	0	**50 → 40 → 10**
R_2		6 **100**	6.1	6.2	0	**100**
O_2		9 **60**	9.1	9.2	0	**60**
R_3			6 **120**	6.1	0	**120**
O_3			9 **80**	9.1	0	**80**
R_4				6 **110**	0	**110**
O_4				9 **50**	0 **20**	**70 → 20**
	100 ↓ **10**	**190** ↓ **90** ↓ **30**	**210** ↓ **90** ↓ **10**	**160** ↓ **50**	**20**	

shown in Table 11.1. All the "transportation" routes from a previous to a current period are blocked because no shortage is allowed.

The unit "transportation" costs are the sum of applicable production and holding costs. For example, unit cost from R_1 to period 1 equals unit production cost only (= \$6) Unit cost from O_1 to period 4 equals unit production cost plus unit holding cost from periods 1 to 4—that is, $\$9 + (\$.1 + \$.1 + \$.1) = \$9.30$. Finally, unit costs to *surplus* destination are zero.

The optimal solution is obtained in one pass by starting from column 1 and moving, one column at a time, toward the *surplus* column. For each column, the demand is satisfied using the cheapest routes in that column.[3]

Starting with column 1, route $(R_1, 1)$ has the cheapest unit cost, and we assign the most we can to it—namely, $\min\{90, 100\} = 90$ units, which leaves 10 unsatisfied units in column 1. The next-cheapest route in column 1 is $(O_1, 1)$, to which we assign $\min\{50, 10\} = 10$. The demand for period 1 is now satisfied.

Next, we move to column 2. The assignments in this column occur in the following order: 100 units to $(R_2, 2)$, 60 units to $(O_2, 2)$, and 30 units to $(O_1, 2)$. The respective unit "transportation" costs of these assignments are \$6, \$9, and \$9.10. We did not use the route $(R_1, 2)$, whose unit cost is \$6.10, because all the supply of R_1 has been assigned to period 1.

Continuing in the same manner, we satisfy the demands of column 3 and then column 4. The optimum solution, shown in boldface in Table 11.1, is summarized as follows:

[3]For a proof of the optimality of this procedure, see S.M. Johnson, "Sequential Production Planning over Time at Minimum Cost," *Management Science*, Vol. 3, pp. 435–437, 1957.

Period	Production Schedule
Regular 1	Produce 90 units for period 1.
Overtime 1	Produce 50 units: 10 units for period 1, 30 for 2, and 10 for 3.
Regular 2	Produce 100 units for period 2.
Overtime 2	Produce 60 units for period 2.
Regular 3	Produce 120 units for period 3.
Overtime 3	Produce 80 units for period 3.
Regular 4	Produce 110 units for period 4.
Overtime 4	Produce 50 units for period 4, with 20 units idle capacity.

The associated total cost is $90 \times \$6 + 10 \times \$9 + 30 \times \$9.10 + 100 \times \$6 + 60 \times \$9 + 10 \times \$9.20 + 120 \times \$6 + 80 \times \$9 + 110 \times \$6 + 50 \times \$9 = \$4685$.

PROBLEM SET 11.4B

1. Solve Example 11.4-1, assuming that the unit production and holding costs are as given in the following table.

Period i	Regular time unit cost ($)	Overtime unit cost ($)	Unit holding cost ($) to period $i + 1$
1	5.00	7.50	.10
2	3.00	4.50	.15
3	4.00	6.00	.12
4	1.00	1.50	.20

2. An item is manufactured to meet known demand for four periods according to the following data:

	Unit production cost ($) for period			
Production range (units)	*1*	*2*	*3*	*4*
1–3	1	2	2	3
4–11	1	4	5	4
12–15	2	4	7	5
16–25	5	6	10	7
Unit holding cost to next period ($)	.30	.35	.20	.25
Total demand (units)	11	4	17	29

   (a) Find the optimal solution, indicating the number of units to be produced in each period.

   (b) Suppose that 10 additional units are needed in period 4. Where should they be produced?

*3. The demand for a product over the next five periods may be filled from regular production, overtime production, or subcontracting. Subcontracting may be used only if the overtime capacity has been used. The following table gives the supply, demand, and cost data of the situation.

	Production capacity (units)			
Period	*Regular time*	*Overtime*	*Subcontracting*	Demand
1	100	50	30	153
2	40	60	80	200
3	90	80	70	150
4	60	50	20	200
5	70	50	100	203

The unit production costs for the three levels in each period are \$4, \$6, and \$7, respectively. The unit holding cost per period is \$.50. Determine the optimal solution.

11.4.2 Setup Model

In this situation, no shortage is allowed and a setup cost is incurred each time a new production lot is started. Two solution methods will be presented: an exact dynamic programming algorithm and a heuristic.

Figure 11.10 summarizes the inventory situation schematically. The symbols shown in the figure are defined for period i, $i = 1, 2, \ldots, n$, as

$$z_i = \text{Amount ordered}$$

$$D_i = \text{Demand for period } i$$

$$x_i = \text{Inventory at the start of period } i$$

The cost elements of the situation are defined as

$$K_i = \text{Setup cost in period } i$$

$$h_i = \text{Unit inventory holding cost from period } i \text{ to } i + 1$$

The associated production cost function for period i is

$$C_i(z_i) = \begin{cases} 0, & z_i = 0 \\ K_i + c_i(z_i), & z_i > 0 \end{cases}$$

The function $c_i(z_i)$ is the marginal production cost function, given z_i

FIGURE 11.10

Elements of the dynamic inventory model with setup cost

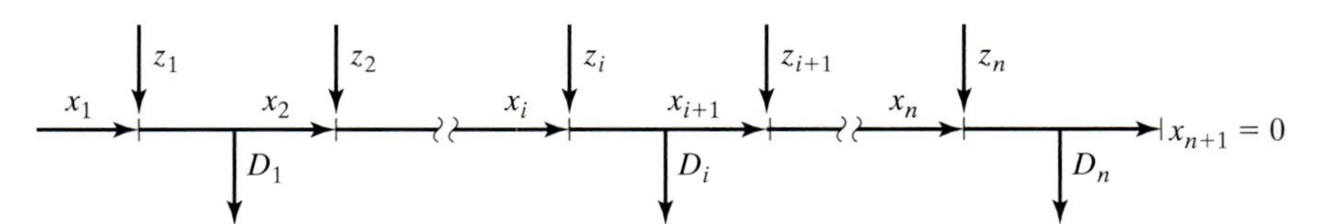

General Dynamic Programming Algorithm. In the absence of shortage, the inventory model is based on minimizing the sum of production and holding costs for all n periods. For simplicity, we will assume that the holding cost for period i is based on end-of-period inventory, defined as

$$x_{i+1} = x_i + z_i - D_i$$

For the forward recursive equation, the *state* at *stage* (period) i is defined as x_{i+1}, the end-of-period inventory level, where, as Figure 11.10 shows,

$$0 \leq x_{i+1} \leq D_{i+1} + \cdots + D_n$$

This inequality recognizes that, in the extreme case, the remaining inventory, x_{i+1}, can satisfy the demand for all the remaining periods.

Let $f_i(x_{i+1})$ be the minimum inventory cost for periods $1, 2, \ldots,$ and i given the end-of-period inventory x_{i+1}. The forward recursive equation is thus given as

$$f_1(x_2) = \min_{z_1 = D_1 + x_2 - x_1} \{C_1(z_1) + h_1 x_2\}$$

$$f_i(x_{i+1}) = \min_{0 \leq z_i \leq D_i + x_{i+1}} \{C_i(z_i) + h_i x_{i+1} + f_{i-1}(x_{i+1} + D_i - z_i)\}, i = 2, 3, \ldots, n$$

Note that for period 1, z_1 must exactly equal $D_1 + x_2 - x_1$. For $i > 1$, z_i can be as low as zero because D_i can be satisfied from the production in preceding periods, $1, 2, \ldots,$ and $i - 1$.

Example 11.4-2

The following table provides the data for a 3-period inventory situation.

Period i	Demand D_i (units)	Setup cost K_i (\$)	Holding cost h_i (\$)
1	3	3	1
2	2	7	3
3	4	6	2

The demand occurs in discrete units, and the starting inventory is $x_1 = 1$ unit. The unit production cost is \$10 for the first 3 units and \$20 for each additional unit, which is translated mathematically as

$$c_i(z_i) = \begin{cases} 10z_i, & 0 \leq z_i \leq 3 \\ 30 + 20(z_i - 3), & z_i \geq 4 \end{cases}$$

Determine the optimal inventory policy.

Period 1: $D_1 = 3, 0 \le x_2 \le 2 + 4 = 6, z_1 = x_2 + D_1 - x_1 = x_2 + 2$

		$C_1(z_1) + h_1x_2$							Optimal solution	
		$z_1 = 2$	3	4	5	6	7	8		
x_2	h_1x_2	$C_1(z_1) = 23$	33	53	73	93	113	133	$f_1(x_2)$	z_1^*
0	0	23							23	2
1	1		34						34	3
2	2			55					55	4
3	3				76				76	5
4	4					97			97	6
5	5						118		118	7
6	6							139	139	8

Note that because $x_1 = 1$, the smallest value of z_1 is $D_1 - x_1 = 3 - 1 = 2$.

Period 2: $D_2 = 2, 0 \le x_3 \le 4, 0 \le z_2 \le D_2 + x_3 = x_3 + 2$

		$C_2(z_2) + h_2x_3 + f_1(x_3 + D_2 - z_2)$							Optimal solution	
		$z_2 = 0$	1	2	3	4	5	6		
x_3	h_2x_3	$C_2(z_2) = 0$	17	27	37	57	77	97	$f_2(x_3)$	z_2^*
0	0	0 + 55 = 55	17 + 34 = 51	27 + 23 = 50					50	2
1	3	3 + 76 = 79	20 + 55 = 75	30 + 34 = 64	40 + 23 = 63				63	3
2	6	6 + 97 = 103	23 + 76 = 99	33 + 55 = 88	43 + 34 = 77	63 + 23 = 86			77	3
3	9	9 + 118 = 127	26 + 97 = 123	36 + 76 = 112	46 + 55 = 101	66 + 34 = 100	86 + 23 = 109		100	4
4	12	12 + 139 = 151	29 + 118 = 147	39 + 97 = 136	49 + 76 = 125	69 + 55 = 124	89 + 34 = 123	109 + 23 = 132	123	5

Period 3: $D_3 = 4, x_4 = 0, 0 \le z_3 \le D_3 + x_4 = 4$

		$C_3(z_3) + h_3x_4 + f_2(x_4 + D_3 - z_3)$					Optimal solution	
		$z_3 = 0$	1	2	3	4		
x_4	h_3x_4	$C_3(z_3) = 0$	16	26	36	56	$f_3(x_4)$	z_3^*
0	0	0 + 123 = 123	16 + 100 = 116	26 + 77 = 103	36 + 63 = 99	56 + 50 = 106	99	3

The optimum solution is read in the following manner:

$$(x_4 = 0) \rightarrow \boxed{z_3 = 3} \rightarrow (x_3 = 0 + 4 - 3 = 1) \rightarrow \boxed{z_2 = 3}$$

$$\rightarrow (x_2 = 1 + 2 - 3 = 0) \rightarrow \boxed{z_1 = 2}.$$

Thus, the solution is summarized as $z_1^* = 2$, $z_2^* = 3$, and $z_3^* = 3$, with a total cost of \$99.

Excel Moment

Template excelDPInv.xls is designed to solve the general DP inventory problem with up to 10 periods. The design of the spreadsheet is similar to that of excelKanpsack.xls given in Section 10.3.1, where the model carries out the computations one stage a time and user input is needed to link successive stages.

Figure 11.11 shows the application of excelDPInv.xls to Example 11.4-2. The input data are entered for each stage. The computations start with period 1. Note how the cost function $c_i(z_i)$ is entered in row 3: $(G3 = 10, H3 = 20, I3 = 3)$ means that the unit cost is \$10 for the first three items and \$20 for additional items. Note also that the

FIGURE 11.11

Excel DP solution of Example 11.4-2 (file excelDPInv.xls)

Period 1:

	A	B	C	D	E	F	G	H	I	J	K	S	T	U	V	W	X	Y	Z
1					General (Forward) Dynamic Programming Inventory Model														
2	I	Number of periods, N=			3	Current period=		1						Optimum solution					
3	N	K1=	3	h1=	1	c1(z1)=	10	20	3					Summary					
4	P	Period	1	2	3									x	f	z	x	f	z
5	U	D(1 to 3)=	2	2	4									Period 1					
6	T	*Are z1 values correct?*			yes	yes	yes	yes	yes	yes	yes	Optimum		0	23	2			
7		Period 0		z1=	2	3	4	5	6	7	8	Period1		1	34	3			
8		f 0	C1(z1)=		23	33	53	73	93	113	133	f1	z1	2	55	4			
9	S		x2=	0	23	1111111	1111111	1111111	1111111	1111111	1111111	23	2	3	76	5			
10	T		x2=	1	1111111	34	1111111	1111111	1111111	1111111	1111111	34	3	4	97	6			
11	A		x2=	2	1111111	1111111	55	1111111	1111111	1111111	1111111	55	4	5	118	7			
12	G		x2=	3	1111111	1111111	1111111	76	1111111	1111111	1111111	76	5	6	139	8			
13	E		x2=	4	1111111	1111111	1111111	1111111	97	1111111	1111111	97	6						
14			x2=	5	1111111	1111111	1111111	1111111	1111111	118	1111111	118	7						
15	C		x2=	6	1111111	1111111	1111111	1111111	1111111	1111111	139	139	8						

Period 2:

	A	B	C	D	E	F	G	H	I	J	K	S	T	U	V	W	X	Y	Z
1					General (Forward) Dynamic Programming Inventory Model														
2	I	Number of periods, N=			3	Current period=		2						Optimum solution					
3	N	K2=	7	h2=	3	c2(z2)=	10	20	3					Summary					
4	P	Period	1	2	3									x	f	z	x	f	z
5	U	D(1 to 3)=	2	2	4									Period 1			Period 2		
6	T	*Are z2 values correct?*			yes	yes	yes	yes	yes	yes	yes	Optimum		0	23	2	0	50	2
7		Period 1		z2=	0	1	2	3	4	5	6	Period2		1	34	3	1	63	3
8		f 1	C2(z2)=		0	17	27	37	57	77	97	f2	z2	2	55	4	2	77	3
9	S	23	x3=	0	55	51	50	1111111	1111111	1111111	1111111	50	2	3	76	5	3	100	4
10	T	34	x3=	1	79	75	64	63	1111111	1111111	1111111	63	3	4	97	6	4	123	5
11	A	55	x3=	2	103	99	88	77	86	1111111	1111111	77	3	5	118	7			
12	G	76	x3=	3	127	123	112	101	100	109	1111111	100	4	6	139	8			
13	E	97	x3=	4	151	147	136	125	124	123	132	123	5						
14		118																	
15	C	139																	

Period 3:

	A	B	C	D	E	F	G	H	I	J	K	S	T	U	V	W	X	Y	Z
1					General (Forward) Dynamic Programming Inventory Model														
2	I	Number of periods, N=			3	Current period=		3						Optimum solution					
3	N	K3=	6	h3=	2	c3(z3)=	10	20	3					Summary					
4	P	Period	1	2	3									x	f	z	x	f	z
5	U	D(1 to 3)=	2	2	4									Period 1			Period 2		
6	T	*Are z3 values correct?*			yes	yes	yes	yes	yes			Optimum		0	23	2	0	50	2
7		Period 2		z3=	0	1	2	3	4			Period3		1	34	3	1	63	3
8		f 2	C3(z3)=		0	16	26	36	56			f3	z3	2	55	4	2	77	3
9	S	50	x4=	0	123	116	103	99	106			99	3	3	76	5	3	100	4
10	T	63												4	97	6	4	123	5
11	A	77												5	118	7	Period 3		
12	G	100												6	139	8	0	99	3
13	E	123																	

amount entered for D_1 must be the net after the initial inventory has been written off $(= 3 - x_1 = 3 - 1 = 2)$. Additionally, you need to create the feasible values of the variable z_1. The spreadsheet automatically checks if the values you enter are correct, and issues self-explanatory messages in row 6: yes, no, or delete.

Once all input data have been entered, optimum values of f_i and z_i for the stage are given in column S and T. Next, a permanent record for period 1 solution, (x_1, f_1, z_1), is created in the optimum solution summary section of the spreadsheet, as Figure 11.11 shows. This requires copying D9:D15 and S9:T15 and then pasting them using *paste special + values* (you may need to review the proper procedure for creating the permanent record given in conjunction with excelKnapsack.xls in Section 10.3.1).

Next, to prepare for stage 2, copy f_1 from the permanent record and paste it in column B as shown in Figure 11.11. All that is needed now is to update the input data for period 2. The process is repeated for period 3.

PROBLEM SET 11.4C

***1.** Consider Example 11.4-2.

(a) Does it make sense to have $x_4 > 0$?

(b) For each of the following two cases, determine the feasible ranges for z_1, z_2, z_3, x_1, x_2, and x_3. (You will find it helpful to represent each situation as in Figure 11.10.)

(i) $x_1 = 4$ and all the remaining data are the same.

(ii) $x_1 = 0, D_1 = 5, D_2 = 3$, and $D_3 = 4$.

2. *(a) Find the optimal solution for the following four-period inventory model.

Period i	Demand D_i (units)	Setup cost K_i (\$)	Holding cost h_i (\$)
1	5	5	1
2	2	7	1
3	3	9	1
4	3	7	1

The unit production cost is \$1 each for the first 6 units and \$2 each for additional units.

(b) Verify the computations using excelDPInv.xls.

3. Suppose that the inventory-holding cost is based on the *average* inventory during the period. Develop the corresponding forward recursive equation.

4. Develop the backward recursive equation for the model, and then use it to solve Example 11.4-2.

5. Develop the backward recursive equation for the model, assuming that the inventory-holding cost is based on the *average* inventory in the period.

Dynamic Programming Algorithm with Constant or Decreasing Marginal Costs. The general DP given above can be used with any cost function. However, the nature of the algorithm dictates that the state x_i and the alternatives z_i at stage i assume values in increments of 1. This means that for large demand amounts, the tableau at each stage could be extremely large, and hence computationally unwieldy.

A special case of the general DP model holds promise in reducing the volume of computations. In this special situation, both the unit production and unit holding costs are *nonincreasing* (concave) functions of the production quantity and the inventory level, respectively. This situation typically occurs when the unit cost function is constant or when quantity discount is allowed.

Under the given conditions, it can be proved that[4]

1. Given zero initial inventory ($x_1 = 0$), it is optimal to satisfy the demand in any period i either from new production or from entering inventory, but never from both—that is, $z_i x_i = 0$. (For the case with positive initial inventory, $x_1 > 0$, the amount can be written off from the demands of the successive periods until it is exhausted.)

2. The optimal production quantity, z_i, for period i must either be zero or satisfy the exact demand for one or more contiguous succeeding periods.

Example 11.4-3

A four-period inventory model operates with the following data:

Period i	Demand D_i (units)	Setup cost K_i ($)
1	76	98
2	26	114
3	90	185
4	67	70

The initial inventory $x_1 = 15$ units. The unit production cost is $2, and the unit holding cost per period is $1 for all the periods. (For simplicity, the unit production and holding costs are assumed unchanged for all the periods.)

The solution is determined by the forward algorithm given previously, except that the values of x_{i+1} and z_i assume "lump" sums rather than increments of 1. Because $x_1 = 15$, the demand for the first period is adjusted to $76 - 15 = 61$ units.

Period 1: $D_1 = 61$

		$C_1(z_1) + h_1 x_2$				Optimal solution	
		$z_1 = 61$	87	177	244		
x_2	$h_1 x_2$	$C_1(z_1) = 220$	272	452	586	$f_1(x_2)$	z_1^*
0	0	220				220	61
26	26		298			298	87
116	116			568		568	177
183	183				769	769	244
Order in 1 for		1	1, 2	1, 2, 3	1, 2, 3, 4		

[4]See H. Wagner and T. Whitin, "Dynamic Version of the Economic Lot Size Model," *Management Science*, Vol. 5, pp. 89–96, 1958. The optimality proof imposes the restrictive assumption of constant and identical cost functions for all the periods. The assumption was later relaxed by A. Veinott Jr. to include distinct concave cost functions.

Period 2: $D_2 = 26$

		$C_2(z_2) + h_2x_3 + f_1(x_3 + D_2 - z_2)$				Optimal solution	
		$z_2 = 0$	26	116	183		
x_3	h_2x_3	$C_2(z_2) = 0$	166	346	480	$f_2(x_3)$	z_2^*
0	0	0 + 298 = 298	166 + 220 = 386			298	0
90	90	90 + 568 = 658		436 + 220 = 656		656	116
157	157	157 + 769 = 926			637 + 220 = 857	857	183
Order in 2 for		—	2	2, 3	2, 3, 4		

Period 3: $D_3 = 90$

		$C_3(z_3) + h_3x_4 + f_2(x_4 + D_3 - z_3)$			Optimal solution	
		$z_3 = 0$	90	157		
x_4	h_3x_4	$C_3(z_3) = 0$	365	499	$f_3(x_4)$	z_3^*
0	0	0 + 656 = 656	365 + 298 = 663		656	0
67	67	67 + 857 = 924		566 + 298 = 864	864	157
Order in 3 for		—	3	3, 4		

Period 4: $D_4 = 67$

		$C_4(z_4) + h_4x_5 + f_3(x_5 + D_4 - z_4)$		Optimal solution	
		$z_4 = 0$	67		
x_5	h_4x_5	$C_4(z_4) = 0$	204	$f_4(x_5)$	z_4^*
0	0	0 + 864 = 864	204 + 656 = 860	860	67
Order in 4 for		—	4		

The optimal policy is determined from the tableaus as follows:

$$(x_5 = 0) \rightarrow \boxed{z_4 = 67} \rightarrow (x_4 = 0) \rightarrow \boxed{z_3 = 0}$$

$$\rightarrow (x_3 = 90) \rightarrow \boxed{z_2 = 116} \rightarrow (x_2 = 0) \rightarrow \boxed{z_1 = 61}$$

This gives $z_1^* = 61$, $z_2^* = 116$, $z_3^* = 0$, and $z_4^* = 67$, at a total cost of \$860.

Period 1:

	A	B	C	D	E	F	G	H	I	O	P	Q	R	S
1		**Wagner-Whitin (Forward) Dynamic Programming Inventory Model**												
2		**Number of periods, N=**		**4**				**Current period=**	**1**					
3	I	Period	1	2	3	4								
4	N	**c(1 to 4) =**	**2**	**2**	**2**	**2**						Optimum Solution		
5	P	**K(1 to 4) =**	**98**	**114**	**185**	**70**							Summary	
6	U	**h(1 to 4) =**	**1**	**1**	**1**	**3**						x	f	z
7	T	**D(1 to 4) =**	**61**	**26**	**90**	**67**				Current			period 1	
8		***Are z1 values correct?***			yes	yes	yes	yes		optimum		0	220	61
9		**Period 0**		**z1=**	**61**	**87**	**177**	**244**		Period 1		26	298	87
10	4	**f 0**		C1(z1)=	220	272	452	586		f1	z1	116	568	177
11	S		x2=	0	220	111111	1111111	1111111		220	61	183	769	244
12	T		x2=	26	1111111	298	1111111	1111111		298	87			
13	A		x2=	116	1111111	111111	568	1111111		568	177			
14	G		x2=	183	1111111	111111	1111111	769		769	244			

FIGURE 11.12

Wagner-Whitin Excel DP model applied to Period 1 of Example 11.4-3 (file excelWagnerWhitin.xls)

Excel Moment

Template excelWagnerWhitin.xls is similar to that of the general model excelDPInv.xls. The only difference is that lump sums for the state x and alternative z are used. Also, for simplicity, the new spreadsheet does not allow for quantity discount. Figure 11.12 produces period 1 calculations for Example 11.4-3. The template is limited to a maximum of 10 periods. Remember to use *paste special + values* when creating the output solution summary (columns Q:V).

PROBLEM SET 11.4D

*__1.__ Solve Example 11.4-3, assuming that the initial inventory is 80 units. You may use excelWagnerWhitin.xls to check your calculations.

2. Solve the following 10-period deterministic inventory model. Assume an initial inventory of 50 units.

Period i	Demand D_i (units)	Unit production cost ($)	Unit holding cost ($)	Setup cost ($)
1	150	6	1	100
2	100	6	1	100
3	20	4	2	100
4	40	4	1	200
5	70	6	2	200
6	90	8	3	200
7	130	4	1	300
8	180	4	4	300
9	140	2	2	300
10	50	6	1	300

3. Find the optimal inventory policy for the following five-period model. The unit production cost is \$10 for all periods. The unit holding cost is \$1 per period.

Period i	Demand D_i (units)	Setup cost K_i (\$)
1	50	80
2	70	70
3	100	60
4	30	80
5	60	60

4. Find the optimal inventory policy for the following six-period inventory situation: The unit production cost is \$2 for all the periods.

Period I	D_i (units)	K_i (\$)	h_i (\$)
1	10	20	1
2	15	17	1
3	7	10	1
4	20	18	3
5	13	5	1
6	25	50	1

Silver-Meal Heuristic. This heuristic is valid only for the inventory situations in which the unit production cost is constant and identical for all the periods. For this reason, it balances only the setup and holding costs.

The heuristic identifies the successive future periods whose demand can be filled from the demand of current period. The objective is to minimize the associated setup and holding costs per period.

Suppose that we produce in period i for periods $i, i + 1, \ldots,$ and $t, i \leq t$, and define $\text{TC}(i, t)$ as the associated setup and holding costs for the same periods. Mathematically, using the same notation of the DP models, we have

$$\text{TC}(i, t) = \begin{cases} K_i, & t = i \\ K_i + h_i D_{i+1} + (h_i + h_{i+1}) D_{i+2} + \cdots + \left(\sum_{k=i}^{t-1} h_k \right) D_t, & t > i \end{cases}$$

Next, define $\text{TCU}(i, t)$ as the associated cost per period—that is,

$$\text{TCU}(i, t) = \frac{\text{TC}(i, t)}{t - i + 1}$$

Thus, given a current period i, the heuristic determines t^* that minimizes $\text{TCU}(i, t)$.

The function $\text{TC}(i, t)$ can be computed recursively as follows:

$$\text{TC}(i, i) = K_i$$

$$\text{TC}(i, t) = \text{TC}(i, t - 1) + \left(\sum_{k=i}^{t-1} h_k \right) D_t, \; t = i + 1, i + 2, \ldots, n$$

Step 0. Set $i = 1$.

Step 1. Determine the local minimum t^* that satisfies the following two conditions:

$$\text{TCU}(i, t^* - 1) \geq \text{TCU}(i, t^*)$$

$$\text{TCU}(i, t^* + 1) \geq \text{TCU}(i, t^*)$$

If the conditions are satisfied, then the heuristic calls for ordering the amount $(D_i + D_{i+1} + \cdots + D_{t^*})$ in period i for periods $i, i + 1, \ldots,$ and t^*.

Step 3. Set $i = t^* + 1$. If $i > n$, stop; the entire planning horizon has been covered. Otherwise, go to step 1.

Example 11.4-4

Find the optimal inventory policy for the following six-period inventory situation:

Period i	D_i (units)	K_i (\$)	h_i (\$)
1	10	20	1
2	15	17	1
3	7	10	1
4	20	18	3
5	13	5	1
6	25	50	1

The unit production cost is $2 for all the periods.

Iteration 1 ($i = 1, K_1 = \$20$). The function TC(1, t) is computed recursively in t. For example, given TC(1, 1) = \$20, TC(1, 2) = TC(1, 1) + h_1D_2 = 20 + 1 × 15 = \$35.

Period t	D_i	TC(1, t)	TCU(1, t)
1	10	\$20	$\frac{20}{1}$ = \$20.00
2	15	20 + 1 × 15 = \$35	$\frac{35}{2}$ = \$17.50
3	7	35 + (1 + 1) × 7 = 94	$\frac{49}{3}$ = **\$16.33**
4	20	49 + (1 + 1 + 1) × 20 = \$109	$\frac{109}{4}$ = \$27.25

The local minimum occurs at $t^* = 3$, which calls for ordering 10 + 15 + 7 = 32 units in period 1 for periods 1 to 3. Set $i = t^* + 1 = 3 + 1 = 4$.

Iteration 2 ($i = 4, K_4 = \$18$).

Period t	D_i	TC(4, t)	TCU(4, t)
4	20	\$18	$\frac{18}{1}$ = **\$18.00**
5	13	18 + 3 × 13 = \$57	$\frac{57}{2}$ = \$28.50

The calculations show that $t^* = 4$, which calls for ordering 20 units in period 4 for period 4. Set $i = 4 + 1 = 5$.

Iteration 3 ($i = 5, K_5 = \$5$).

Period t	D_i	TC(5, t)	TCU(5, t)
5	13	\$5	$\frac{5}{1} = \boxed{\$5}$
6	25	$5 + 1 \times 25 = \$30$	$\frac{30}{2} = \$15$

The minimum occurs at $t^* = 5$, which requires ordering 13 units in period 5 for period 5. Next, we set $i = 5 + 1 = 6$. However, because $i = 6$ is the last period of the planning horizon, we must order 25 units in period 6 for period 6.

Remarks. The following table compares the heuristic and the exact DP solution. We have deleted the unit production cost in the dynamic programming model because it is not included in the heuristic computations.

	Heuristic		Dynamic programming	
Period	*Units produced*	*Cost (\$)*	*Units produced*	*Cost (\$)*
1	32	49	10	20
2	0	0	22	24
3	0	0	0	0
4	20	18	20	18
5	13	5	38	30
6	25	50	0	0
Total	90	122	90	92

The production schedule given by the heuristic costs about 32% more than that of the DP solution (\$122 versus \$92). The "inadequate" performance of the heuristic may be the result of the data used in the problem. Specifically, the problem may lie in the extreme variations in the setup costs for periods 5 and 6. Nevertheless, the example shows that the heuristic does not have the capability to "look ahead" for better scheduling opportunities. For example, ordering in period 5 for periods 5 and 6 (instead of ordering for each period separately) can save \$25, which will bring the total heuristic cost down to \$97.

Excel Moment

Excel template excelSilverMeal.xls is designed to carry out all the iterative computations as well as provide the final solution. The procedure starts with entering the data needed to drive the calculations, including N, K, h, and D for all the periods (these entries are highlighted in turquoise in the spreadsheet). Then, the user must initiate the start of each iteration manually until all the periods have been covered.

Figure 11.13 shows the application of the Excel heuristic to Example 11.4-4. The first iteration is initiated by entering the value 1 in cell J11, signaling that iteration 1 starts at period 1. The spreadsheet will then generate as many rows as the number of periods, N (= 6 in this example). The period number will be listed in ascending order in

	J	K	L	M	N	O	P	Q–V
1	Silver-Meal Heuristic Inventory Model							
2	Input data:							
3	Number of periods, N =	6	<<Maximum 14 periods					
4	Period t=	1	2	3	4	5	6	
5	Setup cost, Kt =	20	17	10	18	5	50	
6	Holding cost, ht =	1	1	1	3	1	1	
7	Demand, Dt =	10	15	7	20	13	25	
8								
9	Solution complete	Model calculations:						Optimum solution (Total cost = $122.00):
10	Start Iteration at Period	Period	D_t	$\sum D_t$	$\sum h_t$	TC	TCU	
11	1	1	10	10	0.00	20.00	20.00	
12		2	15	25	1.00	35.00	17.50	
13		3	7	32	2.00	49.00	16.33	
14		4	20	52	3.00	109.00	27.25	
15		5	13	65	6.00	187.00	37.40	
16		6	25	90	7.00	362.00	60.33	Order 32 in period 1 for periods 1 to 3, cost = $49.00
17								
18	4	4	20	20	0.00	18.00	18.00	
19		5	13	33	3.00	57.00	28.50	
20		6	25	58	4.00	157.00	52.33	Order 20 in period 4 for periods 4 to 4, cost = $18.00
21								
22	5	5	13	13	0.00	5.00	5.00	
23		6	25	38	1.00	30.00	15.00	Order 13 in period 5 for periods 5 to 5, cost = $5.00
24								
25	6	6	25	25	0.00	50.00	50.00	Order 25 in period 6 for periods 6 to 6, cost = $50.00
26								

FIGURE 11.13

Excel solution of Example 11.4-4 using Silver-Meal heuristic (file ExcelSilverMeal.xls)

cells K11:K16. Now, examine TCU in column P (highlighted in turquoise) and locate the period that corresponds to the local minimum at $t = 3$ with TCU $=$ \$16.33. This means that the next iteration will start at period 4. Now, skip a blank row and enter the value 4 in J18. This action will produce the calculations for iteration 2, will show that its local minimum will be at period 4 (TCU $=$ \$18.00), and will signal the start of iteration 3 at period 5. Again, entering 5 in J22, the local minimum for iteration 3 will occur at node 5. Next, entering the value 6 in J25 will produce the terminating iteration of the problem. As you go through each iteration, the spreadsheet will automatically display the associated optimal policy and its total cost, as shown in Figure 11.13.

PROBLEM SET 11.4E

*1. The demand for fishing poles is at its minimum during the month of December and reaches its maximum during the month of April. Fishing Hole, Inc., estimates the December demand at 50 poles. It increases by 10 poles a month until it reaches 90 in April. Thereafter, the demand decreases by 5 poles a month. The setup cost for a production lot is \$250, except during the peak demand months of February to April, where it increases to \$300. The production cost per pole is approximately constant at \$15 throughout the year, and the holding cost per pole per month is \$1. Fishing Hole is developing next year's (January through December) production plan. How should it schedule its production facilities?

2. A small publisher reprints a novel to satisfy the demand over the next 12 months. The demand estimates for the successive months are 100, 120, 50, 70, 90, 105, 115, 95, 80, 85, 100, and 110. The setup cost for reprinting the book is \$200.00 and the holding cost per book per month is \$1.20. Determine the optimal reprint schedule.

REFERENCES

Bishop, J. "Experience with a Successful System for Forecasting and Inventory Control," *Operations Research*, Vol. 22, No. 6, pp. 1224–1231, 1974.

Edwards, J, H. Wagner, and W. Wood, "Blue Bell trims its inventory," *Interfaces*, Vol. 15, No. 1, pp. 34–52, 1985.

Lewis, T., "Personal Operations Research: Practicing OR on Ourselves," *Interfaces*, Vol. 26, No. 5, pp. 34–41, 1996.

Nahmias, S., *Production and Operations Analysis*, 5th ed., Irwin, Homewood, IL, 2005.

Silver, E., D. Pyke, and R. Peterson, *Decision Systems for Inventory Management and Production Control,* 3rd ed., Wiley, New York, 1998.

Tersine, R., *Principles of Inventory and Materials Management,* 3rd ed., North Holland, New York, 1988.

Waters, C., *Inventory Control and Management,* Wiley, New York, 1992.

CHAPTER 12

Review of Basic Probability

Chapter Guide. This chapter provides a review of probability laws, random variables, and probability distributions. If you already have had a course in basic probability and statistics, you may skip this chapter. Nevertheless, the chapter provides a useful summary of five common distributions that are used frequently in the book: binomial, Poisson, uniform, exponential, and normal. We have also developed a spreadsheet-based statistical table (file StatTables.xls) that automates the computations of the mean, standard deviation, probabilities, and percentiles of 16 different distributions. Another spreadsheet is provided for histogramming empirical data (file excelMeanVar.xls).

This chapter includes 12 solved examples, 2 spreadsheets, and 44 end-of-section problems. The AMPL/Excel/Solver/TORA programs are in folder ch12Files.

12.1 LAWS OF PROBABILITY

Probability deals with random outcomes of an **experiment**. The conjunction of all the outcomes is referred to as the **sample space**, and a subset of the sample space is known as an **event**. As an illustration, the outcomes of rolling a (six-faced) die are 1, 2, 3, 4, 5, and 6. The set $\{1, 2, 3, 4, 5, 6\}$ defines the associated sample space. An example of an event is that a roll turns up an even value (2, 4, or 6).

An experiment may deal with a continuous sample space as well. For example, the time between failures of an electronic component may assume any nonnegative value.

If an event E occurs m times in an n-trial experiment, then the probability, $P\{E\}$, of realizing the event E is defined as

$$P\{E\} = \lim_{n \to \infty} \frac{m}{n}$$

The definition implies that if the experiment is repeated *indefinitely* ($n \to \infty$), then the desired probability is represented by $\frac{m}{n}$. You can verify this definition by flipping a coin and observing its outcome: head (H) or tail (T). The longer you repeat the experiment, the closer will be the estimate of $P\{H\}$ (or P$\{T\}$) to the theoretical value of 0.5.

By definition,

$$0 \leq P\{E\} \leq 1$$

An event E is impossible if $P\{E\} = 0$, and certain if $P\{E\} = 1$. For example, in a six-faced die experiment, rolling a 7 is impossible, whereas rolling an integer value from 1 to 6, inclusive, is certain.

PROBLEM SET 12.1A

*1. In a survey conducted in the State of Arkansas high schools to study the correlation between senior year scores in mathematics and enrollment in engineering colleges, 400 out of 1000 surveyed seniors have studied mathematics. Engineering enrollment shows that, of the 1000 seniors, 150 students have studied mathematics and 29 have not. Determine the probabilities of the following events:
   (a) A student who studied mathematics is enrolled in engineering. Is not enrolled in engineering.
   (b) A student neither studied mathematics nor enrolled in engineering.
   (c) A student is not studying engineering.

*2. Consider a random gathering of n persons. Determine the smallest n such that it is more likely than not that two persons or more have the same birthday. (*Hint*: Assume no leap years and that all days of the year are equally likely to be a person's birthday.)

*3. Answer Problem 2 assuming that two or more persons share your birthday.

12.1.1 Addition Law of Probability

For two events, E and F, $E + F$ (or $E \cup F$) represents the **union** of E and F, and EF (or $E \cap F$) represents their **intersection**. The events E and F are **mutually exclusive** if they do not intersect—that is, if the occurrence of one event precludes the occurrence of the other. Based on these definitions, the addition law of probability can be stated as

$$P\{E + F\} = \begin{cases} P\{E\} + P\{F\}, & E \text{ and } F \text{ mutually exclusive} \\ P\{E\} + P\{F\} - P\{EF\}, & \text{otherwise} \end{cases}$$

$P\{EF\}$ is the probability that events E and F occur simultaneously.

Example 12.1-1

Consider the experiment of rolling a die. The sample space of the experiment is $\{1, 2, 3, 4, 5, 6\}$. For a fair die, we have

$$P\{1\} = P\{2\} = P\{3\} = P\{4\} = P\{5\} = P\{6\} = \tfrac{1}{6}$$

Define

$$E = \{1, 2, 3, \text{ or } 4\}$$

$$F = \{3, 4, \text{ or } 5\}$$

The outcomes 3 and 4 are common between E and F—hence, $EF = \{3 \text{ or } 4\}$. Thus,

$$P\{E\} = P\{1\} + P\{2\} + P\{3\} + P\{4\} = \tfrac{1}{6} + \tfrac{1}{6} + \tfrac{1}{6} + \tfrac{1}{6} = \tfrac{2}{3}$$

$$P\{F\} = P\{3\} + P\{4\} + P\{5\} = \tfrac{1}{2}$$

$$P\{EF\} = P\{3\} + P\{4\} = \tfrac{1}{3}$$

It then follows that

$$P\{E + F\} = P\{E\} + P\{F\} - P\{EF\} = \tfrac{2}{3} + \tfrac{1}{2} - \tfrac{1}{3} = \tfrac{5}{6}$$

Intuitively, the result makes sense because $(E + F) = \{1, 2, 3, 4, 5\}$, whose probability of occurrence is $\frac{5}{6}$.

PROBLEM SET 12.1B

1. A fair 6-faced die is tossed twice. Letting E and F represent the outcomes of the two tosses, compute the following probabilities:
   - **(a)** The sum of E and F is 11.
   - **(b)** The sum of E and F is even.
   - **(c)** The sum of E and F is odd and greater than 3.
   - **(d)** E is even less than 6 and F is odd greater than 1.
   - **(e)** E is greater than 2 and F is less than 4.
   - **(f)** E is 4 and the sum of E and F is odd.

2. Suppose that you roll two dice independently and record the number that turns up for each die. Determine the following:
   - **(a)** The probability that both numbers are even.
   - **(b)** The probability that the sum of the two numbers is 10.
   - **(c)** The probability that the two numbers differ by at least 3.

***3.** You can toss a fair coin up to 7 times. You will win $100 if three tails appear before a head is encountered. What are your chances of winning?

***4.** Ann, Jim, John, and Liz are scheduled to compete in a racquetball tournament. Ann is twice as likely to beat Jim, and Jim is at the same level as John. Liz's past winning record against John is one out of three. Determine the following:
   - **(a)** The probability that Jim will win the tournament.
   - **(b)** The probability that a woman will win the tournament.
   - **(c)** The probability that no woman will win.

12.1.2 Conditional Law of Probability

Given the two events E and F with $P\{F\} > 0$, the conditional probability of E given F, $P\{E|F\}$, is defined as

$$P\{E|F\} = \frac{P\{EF\}}{P\{F\}}, \quad P\{F\} > 0$$

If E is a subset of (i.e., contained in) F, then $P\{EF\} = P\{E\}$.

The two events, E and F, are *independent* if, and only if,

$$P\{E|F\} = P\{E\}$$

In this case, the conditional probability law reduces to

$$P\{EF\} = P\{E\}P\{F\}$$

Example 12.1-2

You are playing a game in which another person is rolling a die. You cannot see the die, but you are given information about the outcomes. Your job is to predict the outcome of each roll. Determine the probability that the outcome is a 6, given that you are told that the roll has turned up an even number.

Let $E = \{6\}$, and define $F = \{2, 4, \text{or } 6\}$. Thus,

$$P\{E|F\} = \frac{P\{EF\}}{P\{F\}} = \frac{P\{E\}}{P\{F\}} = \left(\frac{1/6}{1/2}\right) = \frac{1}{3}$$

Note that $P\{EF\} = P\{E\}$ because E is a subset of F.

PROBLEM SET 12.1C

1. In Example 12.1-2, suppose that you are told that the outcome is less than 6.
   (a) Determine the probability of getting an even number.
   (b) Determine the probability of getting an odd number larger than one.
2. The stock of WalMark Stores, Inc., trades on the New York Stock Exchange under the symbol WMS. Historically, the price of WMS goes up with the increase in the Dow average 60% of the time and goes down with the Dow 25% of the time. There is also a 5% chance that WMS will go up when the Dow goes down and 10% that it will go down when the Dow goes up.
   (a) Determine the probability that WMS will go up regardless of the Dow.
   (b) Find the probability that WMS goes up given that the Dow is up.
   (c) What is the probability WMS goes down given that Dow is down?

*3. Graduating high school seniors with an ACT score of at least 26 can apply to two universities, A and B, for admission. The probability of being accepted in A is .4 and in B .25. The chance of being accepted in both universities is only 15%.
   (a) Determine the probability that the student is accepted in B given that A has granted admission as well.
   (b) What is the probability that admission will be granted in A given that the student was accepted in B?

4. Prove that if the probability $P\{A|B\} = P\{A\}$, then A and B must be independent.

5. *Bayes' theorem.*[1] Given the two events A and B, show that

$$P\{A|B\} = \frac{P\{B|A\}P\{A\}}{P\{B\}}, P\{B\} > 0$$

6. A retailer receives 75% of its batteries from Factory A and 25% from Factory B. The percentages of defectives produced by A and B are known to be 1% and 2%, respectively. A customer has just bought a battery randomly from the retailer.

(a) What is the probability that the battery is defective?

(b) If the battery you bought is defective, what is the probability that it came from Factory A? (*Hint*: Use Bayes' theorem in Problem 5.)

***7.** Statistics show that 70% of all men have some form of prostate cancer. The PSA test will show positive 90% of the time for afflicted men and 10% of the time for healthy men. What is the probability that a man who tested positive does have prostate cancer?

12.2 RANDOM VARIABLES AND PROBABILITY DISTRIBUTIONS

The outcomes of an experiment either are naturally numeric or can be coded numerically. For example, the outcomes of rolling a die are naturally numeric—namely, 1, 2, 3, 4, 5, or 6. Conversely, the testing of an item produces two outcomes: bad and good. In such a case, we can use the numeric code (0, 1) to represent (bad, good). The numeric representation of the outcomes produces what is known as a **random variable**.

A random variable, x, may be **discrete** or **continuous**. For example, the random variable associated with the die-rolling experiment is discrete with $x = 1, 2, 3, 4, 5,$ or 6, whereas the interarrival time at a service facility is continuous with $x \geq 0$.

Each continuous or discrete random variable x is quantified by a **probability density function** (pdf), $f(x)$ or $p(x)$. These functions must satisfy the conditions in the following table:

	Random variable, x	
Characteristic	*Discrete*	*Continuous*
Applicability range	$x = a, a + 1, \ldots, b$	$a \leq x \leq b$
Conditions for the pdf	$p(x) \geq 0, \sum_{x=a}^{b} p(x) = 1$	$f(x) \geq 0, \int_a^b f(x)dx = 1$

A pdf, $p(x)$ or $f(x)$, must be nonnegative (otherwise, the probability of some event may be negative!). Also, the probability of the entire sample space must equal 1.

An important probability measure is the **cumulative distribution function** (CDF), defined as

$$P\{x \leq X\} = \begin{cases} P(X) = \sum_{x=a}^{X} p(x), & x \text{ discrete} \\ F(X) = \int_a^X f(x)\,dx, & x \text{ continuuous} \end{cases}$$

[1]Section 13.2.2 provides a more detailed presentation of Bayes' theorem.

Example 12.2-1

Consider the case of rolling a fair die. The random variable $x = \{1, 2, 3, 4, 5, 6\}$ represents the face of the die that turns up. The associated pdf and CDF are

$$p(x) = \tfrac{1}{6}, x = 1, 2, \ldots, 6$$
$$P(X) = \tfrac{X}{6}, X = 1, 2, \ldots, 6$$

Figure 12.1 graphs the two functions. The pdf $p(x)$ is a **uniform discrete function** because all the values of the random variables occur with equal probabilities.

The continuous counterpart of uniform $p(x)$ is illustrated by the following experiment. A needle of length l is pivoted in the center of a circle whose diameter also equals l. After marking an arbitrary reference point on the circumference, we spin the needle clockwise and measure the circumference distance x from where the pointer stops to the marked point. Thus, the random variable x is continuous in the range $0 \leq x \leq \pi l$. There is no reason to believe that the needle will tend to stop more often in a specific region of the circumference. Hence, all the values of x in the specified range are equally likely to occur, and the distribution of x must be uniform.

The pdf of x, $f(x)$, is defined as

$$f(x) = \frac{1}{\pi l}, 0 \leq x \leq \pi l$$

The associated CDF, $F(X)$, is computed as

$$F(X) = P\{x \leq X\} = \int_0^X f(x)\,dx = \int_0^X \frac{1}{\pi l}dx = \frac{X}{\pi l}, 0 \leq X \leq \pi l$$

Figure 12.2 graphs the two functions.

PROBLEM SET 12.2A

1. The number of units, x, needed of an item is discrete from 1 to 5. The probability, $p(x)$, is directly proportional to the number of units needed. The constant of proportionality is K.
   (a) Determine the pdf and CDF of x, and graph the resulting functions.
   (b) Find the probability that x is an even value.

FIGURE 12.1
CDF and pdf for rolling a fair die

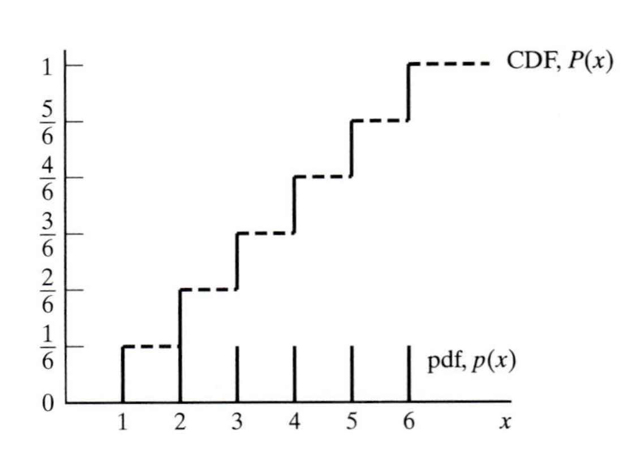

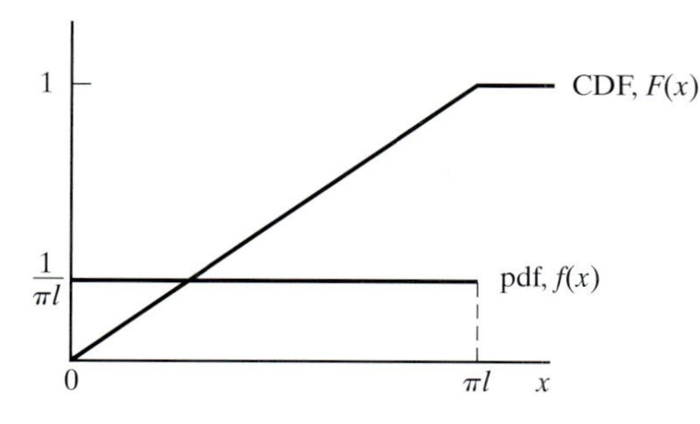

FIGURE 12.2
CDF and pdf for spinning a needle

2. Consider the following function:

$$f(x) = \frac{k}{x^2},\ 10 \le x \le 20$$

*(a) Find the value of the constant k that will make $f(x)$ a pdf.

(b) Determine the CDF, and find the probability that x is (i) larger than 12, and (ii) between 13 and 15.

***3.** The daily demand for unleaded gasoline is uniformly distributed between 750 and 1250 gallons. The gasoline tank, with a capacity of 1100 gallons, is refilled daily at midnight. What is the probability that the tank will be empty just before a refill?

12.3 EXPECTATION OF A RANDOM VARIABLE

Given that $h(x)$ is a real function of a random variable x, we define the **expected value** of $h(x)$, $E\{h(x)\}$, as the (long-run) weighted average with respect to the pdf of x. Mathematically, given that $p(x)$ and $f(x)$ are, respectively, the discrete and continuous pdfs of x, $E\{h(x)\}$ is computed as

$$E\{h(x)\} = \begin{cases} \sum_{x=a}^{b} h(x)p(x), & x \text{ discrete} \\ \int_a^b h(x)f(x)\,dx, & x \text{ continuous} \end{cases}$$

Example 12.3-1

During the first week of each month, I (like many people) pay all my bills and answer a few letters. I usually buy 20 first-class mail stamps each month for this purpose. The number of stamps I will be using varies randomly between 10 and 24, with equal probabilities. What is the average number of stamps left?

The pdf of the number of stamps used is

$$p(x) = \tfrac{1}{15},\ x = 10, 11, \ldots, 24.$$

The number of stamps left is given as

$$h(x) = \begin{cases} 20 - x, x = 10, 11, \ldots, 19 \\ 0, \quad \text{otherwise} \end{cases}$$

Thus,

$$E\{h(x)\} = \tfrac{1}{15}[(20 - 10) + (20 - 11) + (20 - 12) + \cdots + (20 - 19)] + \tfrac{5}{15}(0)$$
$$= 3\tfrac{2}{3}$$

The product $\frac{5}{15}(0)$ is needed to complete the expected value of $h(x)$. Specifically, the probability of being left with *zero* extra stamps equals the probability of needing 20 stamps or more—that is,

$$P\{x \geq 20\} = p(20) + p(21) + p(22) + p(23) + p(24) = 5\left(\tfrac{1}{15}\right) = \tfrac{5}{15}$$

PROBLEM SET 12.3A

1. In Example 12.3-1, compute the average number of extra stamps needed to meet your maximum possible demand.

2. The results of Example 12.3-1 and of Problem 1 show *positive* averages for *both* the surplus and shortage of stamps. Are these results inconsistent? Explain.

***3.** The owner of a newspaper stand receives 50 copies of *Al Ahram* newspaper every morning. The number of copies sold daily, x, varies randomly according to the following probability distribution:

$$p(x) = \begin{cases} \frac{1}{45}, & x = 35, 36, \ldots, 49 \\ \frac{1}{30}, & x = 50, 51, \ldots, 59 \\ \frac{1}{33}, & x = 60, 61, \ldots, 70 \end{cases}$$

(a) Determine the probability that the owner will sell out completely.

(b) Determine the expected number of unsold copies per day.

(c) If the owner pays 50 cents a copy and sells it for \$1.00. Determine the owner's expected net income per day.

12.3.1 Mean and Variance (Standard Deviation) of a Random Variable

The **mean** of x, $E\{x\}$, is a numeric measure of the central tendency (or weighted sum) of the random variable. The **variance**, $\text{var}\{x\}$, is a measure of the dispersion or deviation of x around the mean $E\{x\}$. Its square root is known as the **standard deviation** of x, stdDev $\{x\}$. A larger standard deviation means a higher degree of uncertainty regarding the random variable. Specifically, when the value of a variable is known with certainty, its standard deviation is zero.

The formulas for the mean and variance can be derived from the general definition of $E\{h(x)\}$ as follows: For $E\{x\}$, use $h(x) = x$, and for $\text{var}\{x\}$ use $h(x) = (x - E\{x\})^2$. Thus,

$$E\{x\} = \begin{cases} \sum_{x=a}^{b} x p(x), & x \text{ discrete} \\ \int_a^b x f(x)\, dx, & x \text{ continuous} \end{cases}$$

$$\text{var}\{x\} = \begin{cases} \sum_{x=a}^{b} (x - E\{x\})^2 p(x), & x \text{ discrete} \\ \int_a^b (x - E\{x\})^2 f(x)\, dx, & x \text{ continuous} \end{cases}$$

$$\text{stdDev}\{x\} = \sqrt{\text{var}\{x\}}$$

We can see the basis for the development of the formulas more readily by examining the discrete case. Here, $E\{x\}$ is the *weighted sum* of the discrete values of x. Also, $\text{var}\{x\}$ is the *weighted sum* of the square of the deviation around $E\{x\}$. The continuous case can be interpreted similarly, with integration replacing summation.

Example 12.3-2

We compute the mean and variance for each of the two experiments in Example 12.2-1.

Case 1 (Die Rolling). The pdf is $p(x) = \frac{1}{6}, x = 1, 2, \ldots, 6$. Thus,

$$E\{x\} = 1\left(\tfrac{1}{6}\right) + 2\left(\tfrac{1}{6}\right) + 3\left(\tfrac{1}{6}\right) + 4\left(\tfrac{1}{6}\right) + 5\left(\tfrac{1}{6}\right) + 6\left(\tfrac{1}{6}\right) = 3.5$$

$$\text{var}\{x\} = \left(\tfrac{1}{6}\right)\{(1 - 3.5)^2 + (2 - 3.5)^2 + (3 - 3.5)^2 + (4 - 3.5)^2 + (5 - 3.5)^2 + (6 - 3.5)^2\} = 2.917$$

$$\text{stdDev}(x) = \sqrt{2.917} = 1.708$$

Case 2 (Needle Spinning). Suppose that the length of the needle is 1 inch. Then,

$$f(x) = \frac{1}{3.14}, \qquad 0 \le x \le 3.14$$

The mean and variance are computed as

$$E(x) = \int_0^{3.14} x\left(\tfrac{1}{3.14}\right) dx = 1.57 \text{ inch}$$

$$\text{var}(x) = \int_0^{3.14} (x - 1.57)^2\left(\tfrac{1}{3.14}\right) dx = .822 \text{ inch}^2$$

$$\text{stdDev}(x) = \sqrt{.822} = .906 \text{ inch}$$

Excel Moment

Template exelStatTables.xls is designed to compute the mean, standard deviation, probabilities, and percentiles for 16 common pdfs, including the discrete and continuous uniform distributions of Example 12.3-2. The use of the template is self-explanatory.

PROBLEM SET 12.3B

*1. Compute the mean and variance of the random variable defined in Problem 1, Set 12.2a.

2. Compute the mean and variance of the random variable in Problem 2, Set 12.2a.

3. Show that the mean and variance of a uniform random variable x, $a \leq x \leq b$, are

$$E\{x\} = \frac{b + a}{2}$$

$$\text{var}\{x\} = \frac{(b - a)^2}{12}$$

4. Given the pdf $f(x)$, $a \leq x \leq b$, prove that

$$\text{var}\{x\} = E\{x^2\} - (E\{x\})^2$$

5. Given the pdf $f(x)$, $a \leq x \leq b$, and $y = cx + d$, where c and d are constants. Prove that

$$E\{y\} = cE\{x\} + d$$

$$\text{var}\{y\} = c^2 \text{ var}\{x\}$$

12.3.2 Mean and Variance of Joint Random Variables

Consider the two continuous random variables x_1, $a_1 \leq x_1 \leq b_1$, and x_2, $a_2 \leq x_2 \leq b_2$. Define $f(x_1, x_2)$ as the **joint pdf** of x_1 and x_2 and $f_1(x_1)$ and $f_2(x_2)$ as the **marginal pdfs** of x_1 and x_2, respectively. Then

$$f(x_1, x_2) \geq 0, a_1 \leq x_1 \leq b_1, a_2 \leq x_2 \leq b_2$$

$$\int_{a_1}^{b_1} dx_1 \int_{a_2}^{b_2} dx_2 f(x_1, x_2) = 1$$

$$f_1(x_1) = \int_{a_2}^{b_2} f(x_1, x_2)\, dx_2$$

$$f_2(x_2) = \int_{a_1}^{b_1} f(x_1, x_2)\, dx_1$$

$$f(x_1, x_2) = f_1(x_1)f_2(x_2), \text{ if } x_1 \text{ and } x_2 \text{ are independent}$$

The same formulas apply to discrete pdfs, replacing integration with summation.

For the special case $y = c_1x_1 + c_2x_2$, where the random variables x_1 and x_2 are jointly distributed according to the pdf $f(x_1, x_2)$, we can prove that

$$E\{c_1x_1 + c_2x_2\} = c_1E\{x_1\} + c_2E\{x_2\}$$

$$\text{var}\{c_1x_1 + c_2x_2\} = c_1^2 \text{ var}\{x_1\} + c_2^2 \text{ var}\{x_2\} + 2c_1c_2 \text{ cov}\{x_1, x_2\}$$

where

$$\begin{aligned}
\text{cov}\{x_1, x_2\} &= E\{(x_1 - E\{x_1\})(x_2 - E\{x_2\}) \\
&= E(x_1x_2 - x_1E\{x_2\} - x_2E\{x_1\} + E\{x_1\}E\{x_2\}) \\
&= E\{x_1x_2\} - E\{x_1\}E\{x_2\}
\end{aligned}$$

If x_1 and x_2 are *independent*, then $E\{x_1 x_2\} = E\{x_1\}E\{x_2\}$ and $\text{cov}\{x_1, x_2\} = 0$. The converse is not true, in the sense that two *dependent* variables may have zero covariance.

Example 12.3-3

A lot includes four defective (D) items and six good (G) ones. You select one item randomly and test it. Then, without replacement, you test a second item. Let the random variables x_1 and x_2 represent the outcomes for the first and second items, respectively.

a. Determine the joint and marginal pdfs of x_1 and x_2.

b. Suppose that you get \$5 for each good item you select but must pay \$6 if it is defective. Determine the mean and variance of your revenue after two items have been selected.

Let $p(x_1, x_2)$ be the joint pdf of x_1 and x_2, and define $p_1(x_1)$ and $p_2(x_2)$ as the respective marginal pdfs. First, we determine $p_1(x_1)$ as

$$p_1(G) = \tfrac{6}{10} = .6,\ p_1(D) = \tfrac{4}{10} = .4$$

Next, we know that x_2, the second outcome, depends on x_1. Hence, to determine $p_2(x_2)$, we first determine the joint pdf $p(x_1, x_2)$, from which we can determine the marginal distribution $p_2(x_2)$.

$$\begin{aligned}
P\{x_2 = G|x_1 = G\} &= \tfrac{5}{9} \\
P\{x_2 = G|x_1 = B\} &= \tfrac{6}{9} \\
P\{x_2 = B|x_1 = G\} &= \tfrac{4}{9} \\
P\{x_2 = B|x_1 = B\} &= \tfrac{3}{9}
\end{aligned}$$

To determine $p(x_1, x_2)$, we use the formula $P\{AB\} = P\{A|B\}P\{B\}$ (see Section 12.1.2).

$$\begin{aligned}
p\{x_2 = G, x_1 = G\} &= \tfrac{5}{9} \times \tfrac{6}{10} = \tfrac{5}{15} \\
p\{x_2 = G, x_1 = B\} &= \tfrac{6}{9} \times \tfrac{4}{10} = \tfrac{4}{15} \\
p\{x_2 = B, x_1 = G\} &= \tfrac{4}{9} \times \tfrac{6}{10} = \tfrac{4}{15} \\
p\{x_2 = B, x_1 = B\} &= \tfrac{3}{9} \times \tfrac{4}{10} = \tfrac{2}{15}
\end{aligned}$$

The marginal distributions, $p_1(x_1)$ and $p_2(x_2)$, can be determined by first summarizing the joint distribution, $p(x_1, x_2)$, in a table format and then adding the respective rows and columns, as the following table shows.

	$x_2 = G$	$x_2 = B$	$p_1(x_1)$
$x_1 = G$	$\frac{5}{15}$	$\frac{4}{15}$	$\frac{9}{15} = .6$
$x_1 = B$	$\frac{4}{15}$	$\frac{2}{15}$	$\frac{6}{15} = .4$
$p_2(x_2)$	$\frac{9}{15} = .6$	$\frac{6}{15} = .4$	

It is interesting that, contrary to intuition, $p_1(x_1) = p_2(x_2)$.

The expected revenue can be determined from the joint distribution by recognizing that G produces \$5 and B yields $-\$6$. Thus,

$$\text{Expected revenue} = (5+5)\tfrac{5}{15} + (5-6)\tfrac{4}{15} + (-6+5)\tfrac{4}{15} + (-6-6)\tfrac{2}{15} = \$1.20$$

The same result can be determined by recognizing that the expected revenue for both selections equals the sum of the expected revenue for each individual selection (even though the two variables are *not* independent). This means that

$$\begin{aligned}\text{Expected revenue} &= \text{Selection 1 expected revenue} + \text{Selection 2 expected revenue}\\ &= (5 \times .6 - 6 \times .4) + (5 \times .6 - 6 \times .4) = \$1.20\end{aligned}$$

To compute the variance of the total revenue, we note that

$$\text{var}\{\text{revenue}\} = \text{var}\{\text{revenue 1}\} + \text{var}\{\text{revenue 2}\} + 2\,\text{cov}\{\text{revenue 1, revenue 2}\}$$

Because $p_1(x_1) = p_2(x_2)$, then $\text{var}\{\text{revenue 1}\} = \text{var}\{\text{revenue 2}\}$. To compute the variance, we use the formula

$$\text{var}\{x\} = E\{x^2\} - (E\{x\})^2$$

(See Problem 4, Set 12.3b.) Thus,

$$\text{var}\{\text{revenue 1}\} = [5^2 \times .6 + (-6)^2 \times .4] - .6^2 = 29.04$$

Next, to compute the covariance, we use the formula

$$\text{cov}\{x_1, x_2) = E\{x_1x_2\} - E\{x_1\}E\{x_2\}$$

The term $E\{x_1x_2\}$ can be computed from the joint pdf of x_1 and x_2. Thus, we have

$$\begin{aligned}\text{Convariance} &= [(5 \times 5)(\tfrac{5}{15}) + (5 \times -6)(\tfrac{4}{15}) + (-6 \times 5)(\tfrac{4}{15})\\ &\quad + (-6 \times -6)(\tfrac{2}{15})] - .6 \times .6 = -3.23\end{aligned}$$

Thus,

$$\text{Variance} = 29.04 + 29.04 + 2(-3.23) = 51.62$$

PROBLEM SET 12.3C

1. The joint pdf of x_1 and x_2, $p(x_1, x_2)$, is

	$x_2 = 3$	$x_2 = 5$	$x_2 = 7$
$x_1 = 1$	.2	0	.2
$x_1 = 2$	0	.2	0
$x_1 = 3$	.2	0	.2

***(a)** Find the marginal pdfs $p_1(x_1)$ and $p_2(x_2)$.
***(b)** Are x_1 and x_2 independent?
(c) Compute $E\{x_1 + x_2\}$.
(d) Compute $\text{cov}\{x_1, x_2\}$.
(e) Compute $\text{var}\{5x_1 - 6x_2\}$.

12.4 FOUR COMMON PROBABILITY DISTRIBUTIONS

In Sections 12.2 and 12.3 we discussed the (discrete and continuous) uniform distribution. This section presents four additional pdfs that are encountered often in operations research studies: the discrete binomial and Poisson, and the continuous exponential and normal.

12.4.1 Binomial Distribution

Suppose that a manufacturer produces a certain product in lots of n items each. The fraction of defective items in each lot, p, is estimated from historical data. We are interested in determining the pdf of the number of defectives in a lot.

There are $C_x^n = \frac{n!}{x!(n - x)!}$ distinct combinations of x defectives in a lot of n items, and the probability of getting each combination is $p^x(1 - p)^{n-x}$. It follows (from the addition law of probability) that the probability of k defectives in a lot of n items is

$$P\{x = k\} = C_k^n p^k(1 - p)^{n-k}, k = 0, 1, 2, \ldots, n$$

This is the binomial distribution with parameters n and p. Its mean and variance are given by

$$E\{x\} = np$$
$$\text{var}\{x\} = np(1 - p)$$

Example 12.4-1

John Doe's daily chores require making 10 round trips by car between two towns. Once through with all 10 trips, Mr. Doe can take the rest of the day off, a good enough motivation to drive above the speed limit. Experience shows that there is a 40% chance of getting a speeding ticket on any round trip.

a. What is the probability that the day will end without a speeding ticket?
b. If each speeding ticket costs $80, what is the average daily fine?

The probability of getting a ticket on any one trip is $p = .4$. Thus, the probability of not getting a ticket in any one day is

$$P\{x = 0\} = C_0^{10}(.4)^0(.6)^{10} = .006$$

This means that there is less than 1% chance of finishing the day without a fine. In fact, the average fine per day can be computed as

$$\text{Average fine} = \$80E\{x\} = \$80(np) = 80 \times 10 \times .4 = \$320$$

Remark. $P\{x = 0\}$ can be computed using excelStatTables.xls. Enter 10 in F7, .4 in G7, and 0 in J7. The answer, $P\{x = 0\} = .006047$, is given in M7.

PROBLEM SET 12.4A

*1. A fair die is rolled 10 times. What is the probability that the rolled die will not show an even number?

2. Suppose that five fair coins are tossed independently. What is the probability that exactly one of the coins will be different from the remaining four?

*3. A fortune teller claims to predict whether or not people will amass financial wealth in their lifetime by examining their handwriting. To verify this claim, 10 millionaires and 10 university professors were asked to provide samples of their handwriting. The samples are then paired, one millionaire and one professor, and presented to the fortune teller. We say that the claim is true if the fortune teller makes at least eight correct predictions. What is the probability that the claim is proved true by a "fluke"?

4. In a gambling casino game you are required to select a number from 1 to 6 before the operator rolls three fair dice simultaneously. The casino pays you as many dollars as the number of dice that match your selection. If there is no match, you pay the casino only $1. What is your long-run expected payoff from this game?

5. Suppose that you play the following game: You throw 2 fair dice. If there is no match, you pay 10 cents. If there is a match, you get 50 cents. What is the expected payoff for the game?

6. Prove the formulas for the mean and variance of the binomial distribution.

12.4.2 Poisson Distribution

Customers arrive at a bank or a grocery store in a "totally random" fashion, meaning that we cannot predict when someone will arrive. The pdf describing the *number* of such arrivals during a specified period is the Poisson distribution.

Let x be the number of events (e.g., arrivals) that take place during a specified time unit (e.g., a minute or an hour). Given that λ is a known constant, the Poisson pdf is defined as

$$P\{x = k\} = \frac{\lambda^k e^{-\lambda}}{k!}, k = 0, 1, 2, \ldots$$

The mean and variance of the Poisson are

$$E\{x\} = \lambda$$

$$\text{var}\{x\} = \lambda$$

The formula for the mean reveals that λ must represent the rate at which events occur.

The Poisson distribution figures prominently in the study of queues (see Chapter 15).

Example 12.4-2

Repair jobs arrive at a small-engine repair shop in a totally random fashion at the rate of 10 per day.

a. What is the average number of jobs that are received daily at the shop?

b. What is the probability that no jobs will arrive during any 1 hour, assuming that the shop is open 8 hours a day?

The average number of jobs received per day equals $\lambda = 10$ jobs per day. To compute the probability of no arrivals per *hour*, we need to compute the arrival rate per hour—namely, $\lambda_{\text{hour}} = \frac{10}{8} = 1.25$ jobs per hour. Thus

$$P\{\text{no arrivals per hour}\} = \frac{(\lambda_{\text{hour}})^0 e^{-\lambda_{\text{hour}}}}{0!}$$

$$= \frac{1.25^0 e^{-1.25}}{0!} = .2865$$

Remark. The probability above can be computed with excelStatTables.xls. Enter 1.25 in F16 and 0 in J16. The answer, .286505, appears in M16.

PROBLEM SET 12.4B

*1. Customers arrive at a service facility according to a Poisson distribution at the rate of four per minute. What is the probability that at least one customer will arrive in any given 30-second interval?

2. The Poisson distribution with parameter λ approximates the binomial distribution with parameters (n, p) when $n \to \infty$, $p \to 0$, and $np \to \lambda$. Demonstrate this result for the situation where a manufactured lot is known to contain 1% defective items. If a sample of 10 items is taken from the lot, compute the probability of at most one defective item in a sample, first by using the (exact) binomial distribution and then by using the (approximate) Poisson distribution. Show that the approximation will not be acceptable if the value of p is increased to, say, 0.5.

*3. Customers arrive randomly at a checkout counter at the average rate of 20 per hour.
   (a) Determine the probability that the counter is idle.
   (b) What is the probability that at least two people are in line awaiting service?

4. Prove the formulas for the mean and variance of the Poisson distribution.

12.4.3 Negative Exponential Distribution

If the *number* of arrivals at a service facility during a specified period follows the Poisson distribution (Section 12.4.2), then, automatically, the distribution of the *time interval* between successive arrivals must follow the negative exponential (or, simply, the exponential) distribution. Specifically, if λ is the rate at which Poisson events occur, then the distribution of time between successive arrivals, x, is

$$f(x) = \lambda e^{-\lambda x}, x > 0$$

Figure 12.3 graphs $f(x)$.

The mean and variance of the exponential distribution are

$$E\{x\} = \frac{1}{\lambda}$$

$$\text{var}\{x\} = \frac{1}{\lambda}$$

The mean $E\{x\}$ is consistent with the definition of λ. If λ is the *rate* at which events occur, then $\frac{1}{\lambda}$ is the average time interval between successive events.

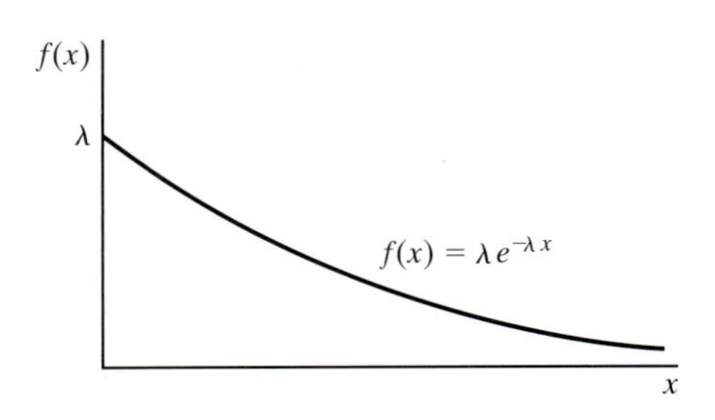

FIGURE 12.3
Probability density function of the exponential distribution

Example 12.4-3

Cars arrive at a gas station randomly every 2 minutes, on the average. Determine the probability that the interarrival time of cars does not exceed 1 minute.

The desired probability is of the form $P\{x \leq A\}$, where $A = 1$ minute in the present example. The determination of this probability is the same as computing the CDF of x—namely,

$$\begin{aligned} P\{x \leq A\} &= \int_0^A \lambda e^{-\lambda x}\, dx \\ &= -e^{-\lambda x}\big|_o^A \\ &= 1 - e^{-\lambda A} \end{aligned}$$

The arrival rate for the example is computed as

$$\lambda = \tfrac{1}{2} \text{ arrival per minute}$$

Thus,

$$P\{x \leq 1\} = 1 - e^{-\left(\frac{1}{2}\right)(1)} = .3934$$

Remark. You can use excelStatTables.xls to compute the probability above. Enter .5 in F9, 1 in J9. The answer, .393468, appears in O9.

PROBLEM SET 12.4C

*1. Walmark Store gets its customers from within town and the surrounding rural areas. Town customers arrive at the rate of 5 per minute, and rural customers arrive at the rate of 7 per minute. Arrivals are totally random. Determine the probability that the interarrival time for all customers is less than 5 seconds.

2. Prove the formulas for the mean and variance of the exponential distribution.

12.4.4 Normal Distribution

The normal distribution describes many random phenomena that occur in everyday life, including test scores, weights, heights, and many others. The pdf of the normal distribution is defined as

$$f(x) = \frac{1}{\sqrt{2\pi\sigma^2}} e^{-\frac{1}{2}\left(\frac{x-\mu}{\sigma}\right)^2}, -\infty < x < \infty$$

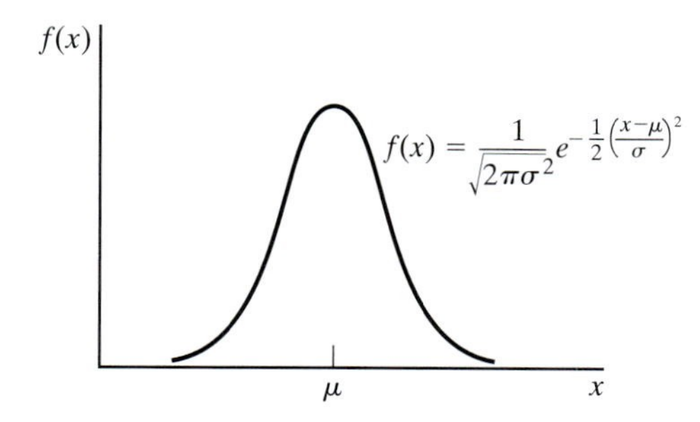

FIGURE 12.4
Probability density function of the normal random variable

The mean and variance are

$$E\{x\} = \mu$$
$$\text{var}\{x\} = \sigma^2$$

The notation $N(\mu, \sigma)$ is usually used to represent a normal distribution with mean μ and standard deviation σ.

Figure 12.4 graphs the normal pdf, $f(x)$. The function is always symmetrical around the mean μ.

An important property of the normal random variable is that it approximates the distribution of the average of a sample taken from *any* distribution. This remarkable result is based on the following theorem:

Central Limit Theorem. *Let $x_1, x_2, \ldots,$ and x_n be independent and identically distributed random variables, each with mean μ and standard deviation σ, and define*

$$s_n = x_1 + x_2 + \cdots + x_n$$

As n becomes large ($n \to \infty$), the distribution of s_n becomes asymptotically normal with mean $n\mu$ and variance $n\sigma^2$, regardless of the original distribution of $x_1, x_2, \ldots,$ and x_n.

The central limit theorem particularly tells us that the distribution of the *average* of a sample of size n drawn from any distribution is asymptotically normal with mean μ and variance $\frac{\sigma^2}{n}$. This result has important applications in statistical quality control.

The CDF of the normal random variable cannot be determined in a closed form. As a result, normal tables (Table 1 in Appendix B or excelStatTables.xls) have been prepared for this purpose. These tables apply to the **standard normal** with mean zero and standard deviation 1—that is, $N(0, 1)$. Any normal random variable, x (with mean μ and standard deviation σ), can be converted to a standard normal, z, by using the transformation

$$z = \frac{x - \mu}{\sigma}$$

Over 99% of the area under any normal distribution is enclosed in the range $\mu - 3\sigma \le x \le \mu + 3\sigma$, known as the **6-sigma limits**.

Example 12.4-4

The inside diameter of a cylinder has the specifications 1 ± .03 in. The machining process output follows a normal distribution with mean 1 cm and standard deviation .1 cm. Determine the percentage of production that will meet the specifications.

Let x represent the output of the process. The probability that a cylinder will meet specifications is

$$P\{1 - .03 \leq x \leq 1 + .03\} = P\{.97 \leq x \leq 1.03\}$$

Given $\mu = 1$ and $\sigma = .1$, the equivalent standard normal probability statement is

$$\begin{aligned} P\{.97 \leq x \leq 1.03\} &= P\left\{\tfrac{.97-1}{.1} \leq z \leq \tfrac{1.03-1}{.1}\right\} \\ &= P\{-.3 \leq z \leq .3\} \\ &= P\{z \leq .3\} - P\{z \leq -.3\} \\ &= P\{z \leq .3\} - P\{z \geq .3\} \\ &= P\{z \leq .3\} - [1 - P\{z \leq .3\}] \\ &= 2P\{z \leq .3\} - 1 \\ &= 2 \times .6179 - 1 \\ &= .2358 \end{aligned}$$

The given probability statements can be justified by picturing the shaded area in Figure 12.5. Notice that $P\{z \leq -.3\} = 1 - P\{z \leq .3\}$ because of the symmetry of the pdf. The value .6179 ($= P\{z \leq .3\}$ is obtained from the standard normal table (Table 1 in Appendix B).

Remark. $P\{a \leq x \leq b\}$ can be computed directly from excelStatTables.xls. Enter 1 in F15, .1 in G15, .97 in J15, and 1.03 in K15. The answer, .235823, appears in Q15.

PROBLEM SET 12.4D

1. The college of engineering at U of A requires a minimum ACT score of 26. The test score among high school seniors in a given school district is normally distributed with mean 22 and standard deviation 2.
   (a) Determine the percentage of high school seniors who are potential engineering recruits.
   (b) If U of A does not accept any student with an ACT score less than 17, determine the percentage of students that will not be eligible for admission at U of A.

FIGURE 12.5

Calculation of $P\{-.3 \leq z \leq .3\}$ in a standard normal distribution

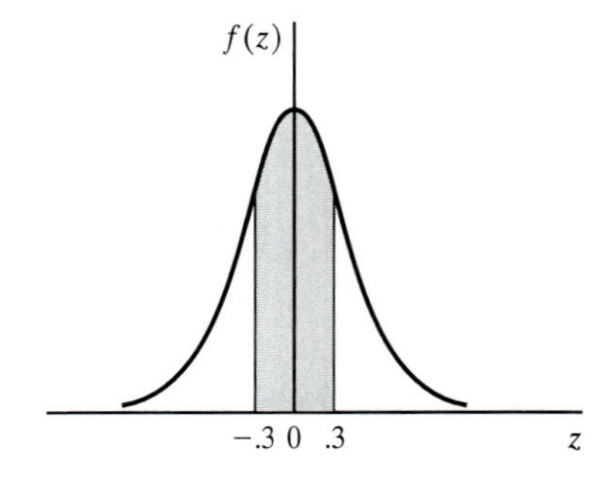

*2. The weights of individuals who seek a helicopter ride in an amusement park have a mean of 180 lb and a standard deviation of 15 lb. The helicopter can carry five persons but has a maximum weight capacity of 1000 1b. What is the probability that the helicopter will not take off with five persons aboard? (*Hint*: Apply the central limit theorem.)

3. The inside diameter of a cylinder is normally distributed with a mean of 1 cm and a standard deviation of .01 cm. A solid rod is assembled inside each cylinder. The diameter of the rod is also normally distributed with a mean of .99 cm and a standard deviation of .01 cm. Determine the percentage of rod-cylinder pairs that will not fit in an assembly. (*Hint*: The difference between two normal random variables is also normal.)

12.5 EMPIRICAL DISTRIBUTIONS

In the preceding sections, we presented the properties of the pdfs and CDFs of random variables and gave examples of five common distributions (uniform, binomial, Poisson, exponential, and normal). How do we determine such distributions in practice?

The determination, actually estimation, of any pdf is rooted in the raw data we collect about the situation under study. For example, to estimate the pdf of the interarrival time of customers at a grocery store, we first record the clock time of arriving customers. The desired interarrival data are the differences between successive arrival times.

This section shows how sampled data can be converted into a pdf:

Step 1. Summarize the raw data in the form of an appropriate frequency histogram, and determine the associated empirical pdf.

Step 2. Use the *goodness-of-fit test* to test if the resulting empirical pdf is sampled from a known theoretical pdf.

Frequency Histogram. A frequency histogram is constructed from raw data by dividing the range of the data (minimum value to maximum value) into nonoverlapping bins. Given the boundaries (I_{i-1}, I_i) for bin i, the corresponding frequency is determined as the count (or tally) of all the raw data, x, that satisfy $I_{i-1} < x \leq I_i$.

Example 12.5-1

The data in the following table represent the service time (in minutes) in a service facility for a sample of 60 customers.

.7	.4	3.4	4.8	2.0	1.0	5.5	6.2	1.2	4.4
1.5	2.4	3.4	6.4	3.7	4.8	2.5	5.5	.3	8.7
2.7	.4	2.2	2.4	.5	1.7	9.3	8.0	4.7	5.9
.7	1.6	5.2	.6	.9	3.9	3.3	.2	.2	4.9
9.6	1.9	9.1	1.3	10.6	3.0	.3	2.9	2.9	4.8
8.7	2.4	7.2	1.5	7.9	11.7	6.3	3.8	6.9	5.3

The minimum and maximum values of the data are .2 and 11.7, respectively. This means that all data can be covered by the range (0, 12). We arbitrarily divide the range (0, 12) into 12 bins, each of width 1 minute. The proper selection of the bin width is crucial in capturing the shape of the empirical distribution. Although there are no hard rules for determining the optimal bin width, a general rule of thumb is to use from 10 to 20 bins. In practice, it may be necessary to try different bin widths before encountering an acceptable histogram.

The following table summarizes the histogram information for the given raw data. The relative-frequency column, f_i, is computed by dividing the entries of the observed-frequency column, o_i, into the total number of observations ($n = 60$). For example, $f_1 = \frac{11}{60} = .1833$. The cumulative-frequency column, F_i, is generated by summing the values of f_i recursively. Thus, $F_1 = f_1 = .1833$ and $F_2 = F_1 + f_2 = .1833 + .1333 = .3166$.

i	Bin interval	Observations tally	Observed frequency, O_i	Relative frequency, f_i	Cumulative relative frequency, F_i
1	(0, 1)	卌 卌 |	11	.1833	.1833
2	(1, 2)	卌 |||	8	.1333	.3166
3	(2, 3)	卌 ||||	9	.1500	.4666
4	(3, 4)	卌 ||	7	.1167	.5833
5	(4, 5)	卌 |	6	.1000	.6833
6	(5, 6)	卌	5	.0833	.7666
7	(6, 7)	||||	4	.0667	.8333
8	(7, 8)	||	2	.0333	.8666
9	(8, 9)	|||	3	.0500	.9166
10	(9, 10)	|||	3	.0500	.9666
11	(10, 11)	|	1	.0167	.9833
12	(11, 12)	|	1	.0167	1.0000
Totals			60	1.0000	

The values of f_i and F_i provide the equivalences of the pdf and the CDF for the service time, t. Because the frequency histogram provides a "discretized" version of the continuous service time, we can convert the resulting CDF into a piecewise-continuous function by joining the resulting points with linear segments. Figure 12.6 provides the empirical pdf and CDF for the example. The CDF, as given by the histogram, is defined at midpoints of the bins.

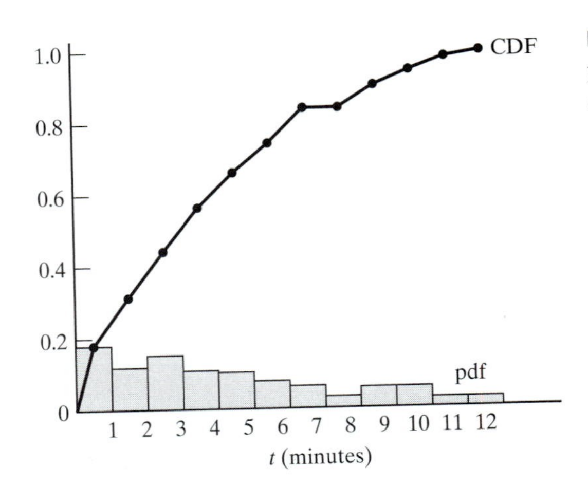

FIGURE 12.6
Piecewise-linear CDF of an empirical distribution

We can now estimate the mean, $\bar{t}$, and variance, s_t^2, of the empirical distribution as follows. Let N be the number of bins in the histogram, and define $\bar{t}_i$ as the midpoint of bin i, then

$$\bar{t} = \sum_{i=1}^{N} f_i \bar{t}_i$$

$$s_t^2 = \sum_{i=1}^{N} f_i (\bar{t}_i - \bar{t})^2$$

Applying these formulas to the present example, we get

$$\bar{t} = .1833 \times .5 + .133 \times 1.5 + \cdots + 11.5 \times .0167 = 3.934 \text{ minutes}$$

$$s_t^2 = .1883 \times (.5 - 3.934)^2 + .1333 \times (1.5 - 3.934)^2 + \cdots$$
$$+ .0167 \times (11.5 - 3.934)^2 = 8.646 \text{ minutes}^2$$

Excel Moment

Histograms can be constructed conveniently using Excel spreadsheet. From the menu bar, select Tools ⇒ Data Analysis ⇒ Histogram, then enter the pertinent data in the dialogue box. However, the *Histogram* tool in Excel does not produce the mean and standard deviation of the frequency histogram directly as part of the output.[2] For this reason, Excel template excelMeanVar.xls is designed to calculate the sample mean, variance, maximum, and minimum, as well as allow the use of Excel *Histogram* tool.

Figure 12.7 stores the input data for Example 12.5-1 in cells A8:E19. The template automatically updates the sample statistics (mean, standard deviation, minimum, and maximum) as the data are entered into the spreadsheet.

To construct the histogram, first create the upper bin limits and enter them in column F starting at row 8. In the present example, cells F8:F19 are used to specify bin limits. The location of the sample data and bin limits must then be entered in the Histogram dialogue box (as shown in the bottom section of Figure 12.7):

Input Range: A8:E19

Bin Range: F8:F19

Output Options: Check *Cumulative Percentage* and *Chart Output*.

Now, click OK. The output is as shown in Figure 12.8.

Goodness-of-fit Test. The goodness-of-fit test evaluates whether the sample used in determining the empirical distribution is drawn from a specific theoretical distribution. An initial evaluation of the data can be made by comparing the empirical CDF with the CDF of the assumed theoretical distribution. If the two CDFs do not deviate excessively, then it is likely that the sample is drawn from the proposed theoretical distribution. This initial "hunch" can be supported further by applying the goodness-of-fit test. The following example provides the details of the proposed procedure.

---

[2] *Data Analysis* in Excel does provide the separate tool, *Descriptive Statistics,* which can be used to compute the mean and variance (as well as volumes of other statistics which you may never use!).

	A	B	C	D	E	F
1	**Sample Mean and Variance +Histogram**					
2	**Output:**					
3	Sample size	60		Mean	3.9367	
4	Minimum	0.2000		Variance	8.9105	
5	Maximum	11.7000		Std Dev.	2.9850	
6	**Input:**					
7	**Enter data in A8:E100**					**Bin**
8	0.7	3.4	2	5.5	1.2	0.9999
9	1.5	3.4	3.7	2.5	0.3	1.9999
10	2.7	2.2	0.5	9.3	4.7	2.9999
11	0.7	5.2	0.9	3.3	0.2	3.9999
12	9.6	9.1	10.6	0.3	2.9	4.9999
13	8.7	7.2	7.9	6.3	6.9	5.9999
14	0.4	4.8	1	6.2	4.4	6.9999
15	2.4	6.4	4.8	5.5	8.7	7.9999
16	0.4	2.4	1.7	8	5.9	8.9999
17	1.6	0.6	3.9	0.2	4.9	9.9999
18	1.9	1.3	3	2.9	4.8	10.9999
19	2.4	1.5	11.7	3.8	5.3	11.9999

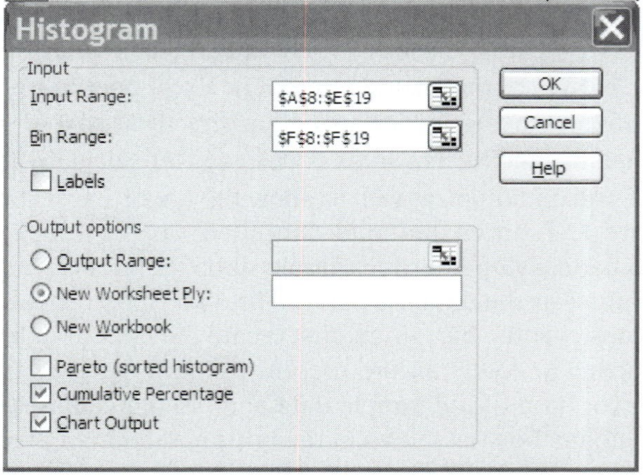

Histogram

Input
Input Range: $A$8:$E$19
Bin Range: $F$8:$F$19
Labels

OK
Cancel
Help

Output options
Output Range:
New Worksheet Ply:
New Workbook

Pareto (sorted histogram)
Cumulative Percentage
Chart Output

FIGURE 12.7

Excel histogram input data and dialogue box for Example 12.5-1

Example 12.5-2

Test the data in Example 12.5-1 for a hypothesized exponential distribution.

Our first task is to specify the function that defines the theoretical distribution. From Example 12.5-1, $\bar{t} = 3.934$ minutes. Hence, $\lambda = \frac{1}{3.934} = .2542$ service per minute for the hypothesized exponential distribution (see Section 12.4.3), and the associated pdf and CDF are given as

$$f(t) = .2542e^{-.2542t}, t > 0$$

$$F(T) = \int_0^T f(t)dt = 1 - e^{-.2542T}, T > 0$$

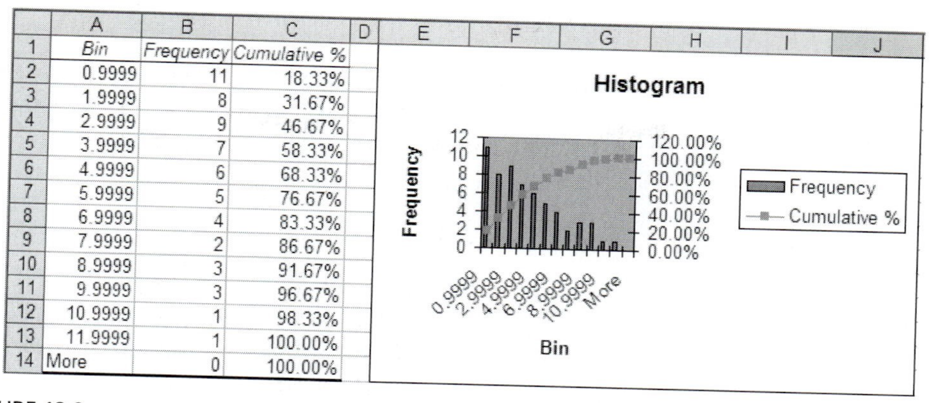

	A	B	C
1	Bin	Frequency	Cumulative %
2	0.9999	11	18.33%
3	1.9999	8	31.67%
4	2.9999	9	46.67%
5	3.9999	7	58.33%
6	4.9999	6	68.33%
7	5.9999	5	76.67%
8	6.9999	4	83.33%
9	7.9999	2	86.67%
10	8.9999	3	91.67%
11	9.9999	3	96.67%
12	10.9999	1	98.33%
13	11.9999	1	100.00%
14	More	0	100.00%

FIGURE 12.8

Excel histogram output of Example 12.5-1

We can use the CDF, $F(T)$, to compute the theoretical CDF for $T = .5, 1.5, \ldots,$ and 11.5, and then compare them graphically with empirical value F_i, $i = 1, 2, \ldots, 12$, as computed in Example 12.5-1. For example,

$$F(.5) = 1 - e^{-(.2542 \times .5)} \approx .12$$

Figure 12.9 provides the resulting comparison. A cursory examination of the two graphs suggests that the exponential distribution may indeed provide a reasonable fit for the observed data.

The next step is to implement a goodness-of-fit test. Two such tests exist: (1) the **Kolmogrov-Smirnov** test, and (2) the **chi-square** test. We will limit our presentation to the chi-square test.

The chi-square test is based on a measurement of the deviation between the empirical and theoretical frequencies corresponding to the different bins of the developed histogram.

FIGURE 12.9

Comparison of the empirical CDF and theoretical exponential CDF

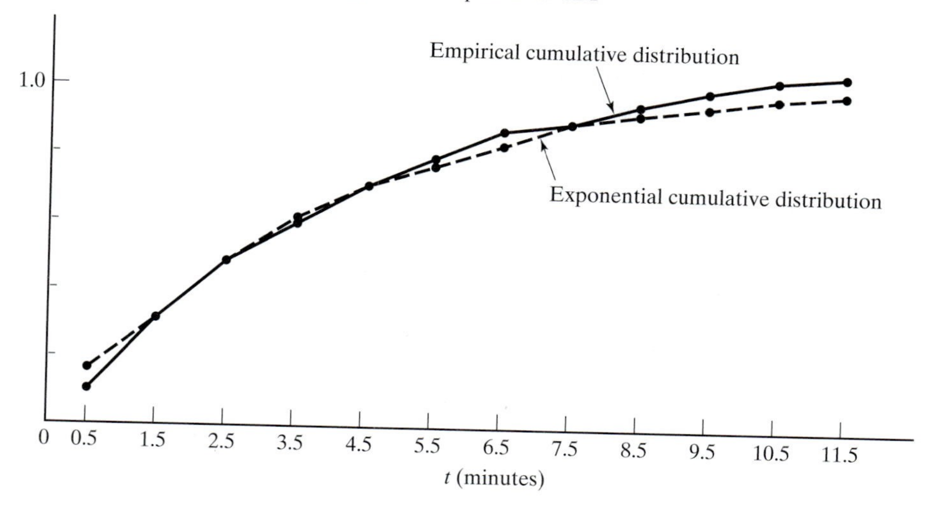

Specifically, the theoretical frequency, n_i, corresponding to the observed frequency, o_i, of bin i, is computed as

$$\begin{aligned} n_i &= n \int_{I_{i-1}}^{I_i} f(t)\, dt \\ &= n(F(I_i) - F(I_{i-1})) \\ &= 60(e^{-.2542 I_{i-1}} - e^{-.2542 I_i}) \end{aligned}$$

Given o_i and n_i for bin i of the histogram, a measure of the deviation between the empirical and observed frequencies is computed as

$$\chi^2 = \sum_{i=1}^{N} \frac{(o_i - n_i)^2}{n_i}$$

As $N \rightarrow \infty$, χ^2 is asymptotically a chi-square pdf with $N - k - 1$ degrees of freedom, where k is the number of parameters estimated from the raw (or histogrammed) data and used for defining the theoretical distribution.

The null hypothesis stating that the observed sample is drawn from the theoretical distribution $f(t)$ is accepted if

$$\chi^2 < \chi^2_{N-k-1,1-\alpha}$$

where $\chi^2_{N-k-1,1-\alpha}$ is the chi-square value for $N - k - 1$ degrees of freedom and α significance level.

The computations of the test are shown in the following table:

i	Bin	Observed frequency, o_i		Theoretical frequency, n_i		$\frac{(o_i-n_i)^2}{n_i}$
1	(0, 1)	11		13.448		.453
2	(1, 2)	8		10.435		.570
3	(2, 3)	9		8.095		.100
4	(3, 4)	7		6.281		.083
5	(4, 5)	6	11	4.873	8.654	.636
6	(5, 6)	5		3.781		
7	(6, 7)	4		2.933		
8	(7, 8)	2	9	2.276	6.975	.588
9	(8, 9)	3		1.766		
10	(9, 10)	3		1.370		
11	(10, 11)	1	5	1.063	6.111	.202
12	(10, ∞)	1		3.678		
Totals		$n = 60$		$n = 60$		χ^2-value $= 2.623$

As a rule of thumb, the expected theoretical frequency count in any bin must be at least 5. This requirement is usually resolved by combining successive bins until the rule is satisfied, as shown in the table. The resulting number of bins becomes $N = 7$. Because we are estimating one parameter from the observed data (namely, λ), the degrees of freedom for the chi-square must equal 7 − 1 − 1 = 5. If we assume a significance level $\alpha = .05$, we get the critical value $\chi^2_{5,.05} = 11.07$ (use Table 3 in Appendix B or, in excelStatTables.xls, enter 5 in F8 and .05 in L8 and receive the answer in R8). Because the χ^2-value (= 2.623) is less than the critical value, we accept the hypothesis that the sample is drawn from the hypothesized exponential pdf.

PROBLEM SET 12.5A

1. The following data represent the interarrival time (in minutes) at a service facility:

4.3	3.4	.9	.7	5.8	3.4	2.7	7.8
4.4	.8	4.4	1.9	3.4	3.1	5.1	1.4
.1	4.1	4.9	4.8	15.9	6.7	2.1	2.3
2.5	3.3	3.8	6.1	2.8	5.9	2.1	2.8
3.4	3.1	.4	2.7	.9	2.9	4.5	3.8
6.1	3.4	1.1	4.2	2.9	4.6	7.2	5.1
2.6	.9	4.9	2.4	4.1	5.1	11.5	2.6
.1	10.3	4.3	5.1	4.3	1.1	4.1	6.7
2.2	2.9	5.2	8.2	1.1	3.3	2.1	7.3
3.5	3.1	7.9	.9	5.1	6.2	5.8	1.4
.5	4.5	6.4	1.2	2.1	10.7	3.2	2.3
3.3	3.3	7.1	6.9	3.1	1.6	2.1	1.9

(a) Use Excel to develop three histograms for the data based on bin widths of .5, 1, and 1.5 minutes, respectively.

(b) Compare graphically the cumulative distribution of the empirical CDF and that of a corresponding exponential distribution.

(c) Test the hypothesis that the given sample is drawn from an exponential distribution. Use a 95% confidence level.

(d) Which of the three histograms is "best" for the purpose of testing the null hypothesis?

2. The following data represent the period (in seconds) needed to transmit a message.

25.8	67.3	35.2	36.4	58.7
47.9	94.8	61.3	59.3	93.4
17.8	34.7	56.4	22.1	48.1
48.2	35.8	65.3	30.1	72.5
5.8	70.9	88.9	76.4	17.3
77.4	66.1	23.9	23.8	36.8
5.6	36.4	93.5	36.4	76.7
89.3	39.2	78.7	51.9	63.6
89.5	58.6	12.8	28.6	82.7
38.7	71.3	21.1	35.9	29.2

Use Excel to construct a suitable histogram. Test the hypothesis that these data are drawn from a uniform distribution at a 95% confidence level, given the following additional information about the theoretical uniform distribution:

(a) The range of the distribution is between 0 and 100.

(b) The range of the distribution is estimated from the sample data.

(c) The maximum limit on the range of the distribution is 100, but the minimum limit must be estimated from the sample data.

3. An automatic device is used to count the volume of traffic at a busy intersection. The device records the time a car arrives at the intersection on a continuous time scale, starting from zero. The following table provides the (cumulative) arrival time (in minutes) for the first 60 cars. Use Excel to construct a suitable histogram, then test the hypothesis that the interarrival time is drawn from an exponential distribution. Use a 95% confidence level.

Arrival	Arrival time (min)	Arrival	Arrival time (min)	Arrival	Arrival time (min)	Arrival	Arrival time (min)
1	5.2	16	67.6	31	132.7	46	227.8
2	6.7	17	69.3	32	142.3	47	233.5
3	9.1	18	78.6	33	145.2	48	239.8
4	12.5	19	86.6	34	154.3	49	243.6
5	18.9	20	91.3	35	155.6	50	250.5
6	22.6	21	97.2	36	166.2	51	255.8
7	27.4	22	97.9	37	169.2	52	256.5
8	29.9	23	111.5	38	169.5	53	256.9
9	35.4	24	116.7	39	172.4	54	270.3
10	35.7	25	117.3	40	175.3	55	275.1
11	44.4	26	118.2	41	180.1	56	277.1
12	47.1	27	124.1	42	188.8	57	278.1
13	47.5	28	1127.4	43	201.2	58	283.6
14	49.7	29	127.6	44	218.4	59	299.8
15	67.1	30	127.8	45	219.9	60	300.0

REFERENCES

Feller, W., *An Introduction to Probability Theory and Its Applications*, 2nd ed., Vols. 1 and 2, Wiley, New York, 1967.

Paulos, J. A., *Innumeracy: Mathematical Illiteracy and Its Consequences*, Hill and Wang, New York, 1988.

Papoulis, A., *Probability and Statistics*, Prentice Hall, Upper Saddle River, NJ, 1990.

Parzen, E., *Modern Probability Theory and Its Applications*, Wiley, New York, 1960.

Ross, S., *Introduction to Probability Models*, 5th ed., Academic Press, New York, 1993.

CHAPTER 13

Decision Analysis and Games

Chapter Guide. Decision problems involving a finite number of alternatives arise frequently in practice. The tools used to solve these problems depend largely on the type of data available (deterministic, probabilistic, or uncertain). The analytic hierarchy process (AHP) is a prominent tool for dealing with decisions under certainty, where subjective judgment is quantified in a logical manner and then used as a basis for reaching a decision. For probabilistic data, decision trees comparing the expected cost (or profit) for the different alternatives are the basis for reaching a decision. Decisions under uncertainty use criteria reflecting the decision maker's attitude toward risk, ranging from optimism to pessimism. Another tool of decision under uncertainty is game theory, where two opponents with conflicting goals aim to achieve the best out of the worst conditions available to each. To demonstrate the importance of these tools in practice, four case analyses in Chapter 24 on the CD deal with using AHP to determine the layout of a CIM laboratory, using decision-tree analysis to determine booking limits in hotel reservations, applying Bayes probabilities to evaluate the results of a medical test, and using game theory to rank golfers in Ryder Cup matches. To assist you in understanding the details of the different tools, the chapter provides 4 spreadsheets. You will also find TORA useful in carrying out the graphical and algebraic solution of games. A basic knowledge of probability and statistics is needed for this chapter.

This chapter includes summaries of 4 real-life applications, 10 solved examples, 4 spreadsheets, 63 end-of-section problems, and 5 cases. The cases are in Appendix E on the CD. The AMPL/Excel/Solver/TORA programs are in folder ch13Files.

Real-Life Application—Layout Planning of a Computer Integrated Manufacturing (CIM) Facility

The engineering college in an academic institution wants to establish a CIM laboratory in a vacated building. The new lab will serve as a teaching and research facility and a center of technical excellence for industry. Recommendations are solicited from the faculty regarding a layout plan for the new laboratory, from which the ideal and absolute minimum square footage for each unit are compiled. The study uses both AHP (analytic hierarchy process) and goal programming to reach a satisfactory compromise solution that meets the needs for teaching, research, and service to industry. The details of the study are given in Case 9, Chapter 24 on the CD.

13.1 DECISION MAKING UNDER CERTAINTY—ANALYTIC HIERARCHY PROCESS (AHP)

The LP models presented in Chapters 2 through 9 are examples of decision making under certainty in which all the functions are well defined. AHP is designed for situations in which ideas, feelings, and emotions affecting the decision process are quantified to provide a numeric scale for prioritizing the alternatives.

Example 13.1-1 (Overall Idea of AHP)

Martin Hans, a bright high school senior, has received full academic scholarships from three institutions: U of A, U of B, and U of C. To select a university, Martin specifies two main criteria: location and academic reputation. Being the excellent student he is, he judges academic reputation to be five times as important as location, giving a weight of approximately 17% to location and 83% to reputation. He then uses a systematic analysis (which will be detailed later) to rank the three universities from the standpoint of location and reputation. The following table ranks the two criteria for the three universities:

	Percent weight estimates for		
Criterion	*U of A*	*U of B*	*U of C*
Location	12.9	27.7	59.4
Reputation	54.5	27.3	18.2

The structure of the decision problem is summarized in Figure 13.1. The problem involves a single hierarchy (level) with two criteria (location and reputation) and three decision alternatives (U of A, U of B, and U of C).

The ranking of each university is based on computing the following *composite* weights:

$$\text{U of A} = .17 \times .129 + .83 \times .545 = \mathbf{.4743}$$

$$\text{U of B} = .17 \times .277 + .83 \times .273 = .2737$$

$$\text{U of C} = .17 \times .594 + .83 \times .182 = .2520$$

Based on these calculations, U of A has the highest composite weight, and hence represents the best choice for Martin.

Remarks. The general structure of AHP may include several hierarchies of criteria. Suppose in Example 13.1-1 that Martin's twin sister, Jane, was also accepted with full scholarship to the three universities. Their parents stipulate that they both must attend the same university so they can share one car. Figure 13.2 summarizes the decision

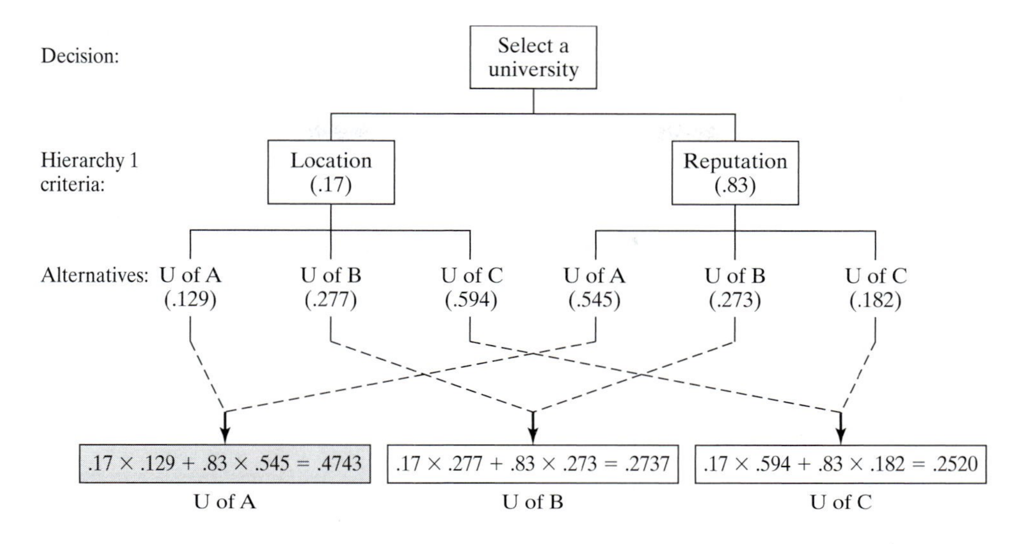

FIGURE 13.1

Summary of AHP calculations for Example 13.1-1

FIGURE 13.2

Embellishment of the decision problem of Example 13.1-1

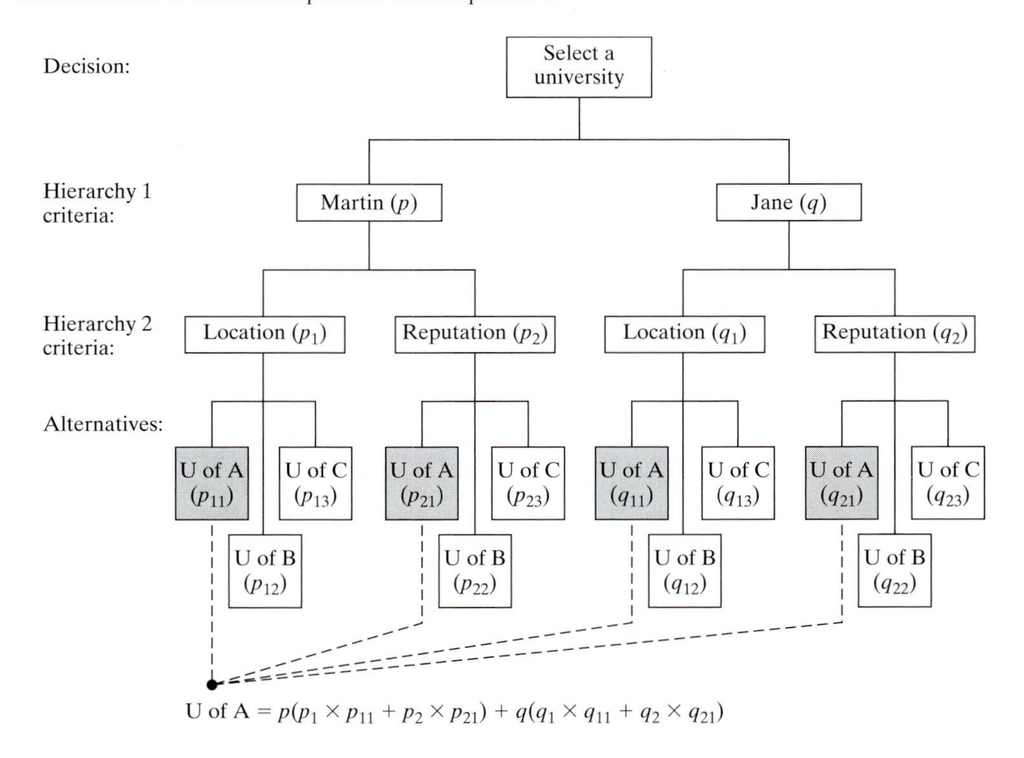

problem, which now involves two hierarchies. The values p and q (presumably equal) at the first hierarchy represent the relative weights given to Martin's and Jane's opinions about the selection process. The second hierarchy uses the weights (p_1, p_2) and (q_1, q_2) to reflect Martin's and Jane's preferences regarding location and reputation of each university. The remainder of the decision-making chart can be interpreted similarly. Note that $p + q = 1, p_1 + p_2 = 1, q_1 + q_2 = 1, p_{11} + p_{12} + p_{13} = 1, p_{21} + p_{22} + p_{23} = 1, q_{11} + q_{12} + q_{13} = 1, q_{21} + q_{22} + q_{23} = 1$. The determination of the U of A composite weight, shown in Figure 13.2, demonstrates the manner in which the computations are carried out.

PROBLEM SET 13.1A

*1. Suppose that the following weights are specified for the situation of Martin and Jane:

$$p = .5, q = .5$$

$$p_1 = .17, p_2 = .83$$

$$p_{11} = .129, p_{12} = .277, p_{13} = .594$$

$$p_{21} = .545, p_{22} = .273, p_{23} = .182$$

$$q_1 = .3, q_2 = .7$$

$$q_{11} = .2, q_{12} = .3, q_{13} = .5$$

$$q_{21} = .5, q_{22} = .2, q_{23} = .3$$

Based on this information, rank the three universities.

Determination of the Weights. The crux of AHP is the determination of the relative weights (such as those used in Example 13.1-1) to rank the decision alternatives. Assuming that we are dealing with n criteria at a given hierarchy, the procedure establishes an $n \times n$ *pairwise* **comparison matrix, A**, that quantifies the decision maker's judgment regarding the relative importance of the different criteria. The pairwise comparison is made such that the criterion in row i $(i = 1, 2, \ldots, n)$ is ranked relative to every other criterion. Letting a_{ij} define the element (i, j) of **A**, AHP uses a discrete scale from 1 to 9 in which $a_{ij} = 1$ signifies that i and j are of *equal importance*, $a_{ij} = 5$ indicates that i is *strongly more important* than j, and $a_{ij} = 9$ indicates that i is *extremely more important* than j. Other intermediate values between 1 and 9 are interpreted correspondingly. Consistency in judgement requires that $a_{ij} = k$ automatically implies that $a_{ji} = \frac{1}{k}$. Also, all the diagonal elements a_{ii} of **A** must equal 1, because they rank a criterion against itself.

Example 13.1-2

To show how the comparison matrix **A** is determined for Martin's decision problem of Example 13.1-1, we start with the main hierarchy dealing with the criteria of reputation and location of a university. In Martin's judgment, the reputation is *strongly more important* than the location, and hence $a_{12} = 5$. This assignment automatically implies that $a_{21} = \frac{1}{5}$. Using the symbols R and L to represent reputation and location, the associated comparison matrix is given as

$$A = \begin{matrix} & L & R \\ L & 1 & \frac{1}{5} \\ R & 5 & 1 \end{matrix}$$

The relative weights of R and L can be determined from **A** by normalizing it into a new matrix **N**. The process requires dividing the elements of each column by the sum of the elements of the same column. Thus, to compute **N**, we divide the elements of columns 1 by $(5 + 1 = 6)$ and those of column 2 by $\left(1 + \frac{1}{5} = 1.2\right)$. The desired relative weights, w_R and w_L, are then computed as the row average:

$$N = \begin{matrix} & L & R \\ L & .17 & .17 \\ R & .83 & .83 \end{matrix} \qquad \begin{matrix} \text{Row average} \\ w_L = \frac{.17 + .17}{2} = .17 \\ w_R = \frac{.83 + .83}{2} = .83 \end{matrix}$$

The computations yield $w_L = .17$ and $w_R = .83$, the weight used in Figure 13.1. The columns of **N** are identical, a characteristic that occurs only when the decision maker exhibits perfect *consistency* in specifying the entries of the comparison matrix **A**. This point is discussed further later in this section.

The relative weights of the alternatives U of A, U of B, and U of C are determined within each of the L and R criteria using the following two comparison matrices, whose elements are based on Martin's judgment regarding the relative importance of the three universities.

$$\mathbf{A}_L = \begin{matrix} & A & B & C \\ A & 1 & \frac{1}{2} & \frac{1}{5} \\ B & 2 & 1 & \frac{1}{2} \\ C & 5 & 2 & 1 \end{matrix}, \quad \mathbf{A}_R = \begin{matrix} & A & B & C \\ A & 1 & 2 & 3 \\ B & \frac{1}{2} & 1 & \frac{3}{2} \\ C & \frac{1}{3} & \frac{2}{3} & 1 \end{matrix}$$

Summing the columns, we get

$$\mathbf{A}_L\text{-column sum} = (8, 3.5, 1.7)$$

$$\mathbf{A}_R\text{-column sum} = (1.83, 3.67, 5.5)$$

The following normalized matrices are determined by dividing all the entries by the respective column-sums:

$$\mathbf{N}_L = \begin{matrix} & A & B & C \\ A & .125 & .143 & .118 \\ B & .250 & .286 & .294 \\ C & .625 & .571 & .588 \end{matrix} \qquad \begin{matrix} \text{Row averages} \\ w_{LA} = \frac{.125 + .143 + .118}{3} = .129 \\ w_{LB} = \frac{.250 + .286 + .294}{3} = .277 \\ w_{LC} = \frac{.625 + .571 + .588}{3} = .594 \end{matrix}$$

$$\mathbf{N}_R = \begin{matrix} & A & B & C \\ A & .545 & .545 & .545 \\ B & .273 & .273 & .273 \\ C & .182 & .182 & .182 \end{matrix} \qquad \begin{matrix} \text{Row averages} \\ w_{RA} = \frac{.545 + .545 + .545}{3} = .545 \\ w_{RB} = \frac{.273 + .273 + .273}{3} = .273 \\ w_{RC} = \frac{.182 + .182 + .182}{3} = .182 \end{matrix}$$

The values $(w_{LA}, w_{LB}, w_{LC}) = (.129, .277, .594)$ provide the respective location weights for U of A, U of B, and U of C. Similarly, $(w_{RA}, w_{RB}, w_{RC}) = (.545, .273, .182)$ give the relative weights regarding academic reputation.

Consistency of the Comparison Matrix. In Example 13.1-2, all the columns of the normalized matrices $\mathbf{N}$ and $\mathbf{N}_R$ are identical, and those of $\mathbf{N}_L$ are not. As such, the original comparison matrices $\mathbf{A}$ and $\mathbf{A}_R$ are said to be *consistent*, whereas $\mathbf{A}_L$ is not.

Consistency implies coherent judgment on the part of the decision maker regarding the pairwise comparisons. Mathematically, we say that a comparison matrix $\mathbf{A}$ is consistent if

$$a_{ij}a_{jk} = a_{ik}, \text{ for all } i, j, \text{ and } k$$

For example, in matrix $\mathbf{A}_R$ of Example 13.1-2, $a_{13} = 3$ and $a_{12}a_{23} = 2 \times \frac{3}{2} = 3$. This property requires all the columns (and rows) of $\mathbf{A}$ to be linearly dependent. In particular, the columns of any 2×2 comparison matrix are by definition dependent, and hence a 2×2 matrix is always consistent.

It is unusual for all comparison matrices to be consistent. Indeed, given that human judgment is the basis for the construction of these matrices, some "reasonable" degree of inconsistency is expected and tolerated.

To determine whether or not a level of consistency is "reasonable," we need to develop a quantifiable measure for the comparison matrix $\mathbf{A}$. We have seen in Example 13.1-2 that a perfectly consistent $\mathbf{A}$ produces a normalized matrix $\mathbf{N}$ in which all the columns are identical—that is,

$$\mathbf{N} = \begin{pmatrix} w_1 & w_1 & \cdots & w_1 \\ w_2 & w_2 & \cdots & w_2 \\ \vdots & \vdots & \vdots & \vdots \\ w_n & w_n & \cdots & w_n \end{pmatrix}$$

It then follows that the original comparison matrix $\mathbf{A}$ can be determined from $\mathbf{N}$ by dividing the elements of column i by w_i (which is the reverse of the process of determining $\mathbf{N}$ from $\mathbf{A}$). We thus have

$$\mathbf{A} = \begin{pmatrix} 1 & \frac{w_1}{w_2} & \cdots & \frac{w_1}{w_n} \\ \frac{w_2}{w_1} & 1 & \cdots & \frac{w_2}{w_n} \\ \vdots & \vdots & \vdots & \vdots \\ \frac{w_n}{w_1} & \frac{w_n}{w_2} & \cdots & 1 \end{pmatrix}$$

From the given definition of $\mathbf{A}$, we have

$$\begin{pmatrix} 1 & \frac{w_1}{w_2} & \cdots & \frac{w_1}{w_n} \\ \frac{w_2}{w_1} & 1 & \cdots & \frac{w_2}{w_n} \\ \vdots & \vdots & \vdots & \vdots \\ \frac{w_n}{w_1} & \frac{w_n}{w_2} & \cdots & 1 \end{pmatrix} \begin{pmatrix} w_1 \\ w_2 \\ \vdots \\ w_n \end{pmatrix} = \begin{pmatrix} nw_1 \\ nw_2 \\ \vdots \\ nw_n \end{pmatrix} = n \begin{pmatrix} w_1 \\ w_2 \\ \vdots \\ w_n \end{pmatrix}$$

More compactly, given that $\mathbf{w}$ is the column vector of the relative weights w_i, $i = 1, 2, \ldots, n$, $\mathbf{A}$ is consistent if,

$$\mathbf{Aw} = n\mathbf{w}$$

For the case where $\mathbf{A}$ is not consistent, the relative weight, w_i, is approximated by the average of the n elements of row i in the normalized matrix $\mathbf{N}$ (see Example 13.1-2). Letting $\overline{\mathbf{w}}$ be the computed average vector, it can be shown that

$$\mathbf{A}\overline{\mathbf{w}} = n_{\max}\overline{\mathbf{w}},\ n_{\max} \geq n$$

In this case, the closer $n_{\max}$ is to n, the more consistent is the comparison matrix $\mathbf{A}$. Based on this observation, AHP computes the **consistency ratio** as

$$CR = \frac{CI}{RI}$$

where

$$\begin{aligned} CI &= \text{Consistency index of } \mathbf{A} \\ &= \frac{n_{\max} - n}{n - 1} \\ RI &= \text{Random consistency of } \mathbf{A} \\ &= \frac{1.98(n - 2)}{n} \end{aligned}$$

The random consistency index, RI, was determined empirically as the average CI of a large sample of randomly generated comparison matrices, $\mathbf{A}$.

If $CR \leq .1$, the level of inconsistency is acceptable. Otherwise, the inconsistency is high and the decision maker may need to reestimate the elements a_{ij} of $\mathbf{A}$ to realize better consistency.

We compute the value of $n_{\max}$ from $\mathbf{A}\overline{\mathbf{w}} = n_{\max}\overline{\mathbf{w}}$ by noting that the ith equation is

$$\sum_{j=1}^{n} a_{ij}\overline{w}_j = n_{\max}\overline{w}_i,\ i = 1, 2, \ldots, n$$

Given $\sum_{i=1}^{n}\overline{w}_i = 1$, we get

$$\sum_{i=1}^{n}\left(\sum_{j=1}^{n} a_{ij}\overline{w}_j\right) = n_{\max}\sum_{i=1}^{n}\overline{w}_i = n_{\max}$$

This means that the value of $n_{\max}$ can be determined by first computing the column vector $\mathbf{A}\overline{\mathbf{w}}$ and then summing its elements.

Example 13.1-3

In Example 13.1-2, the matrix $\mathbf{A}_L$ is inconsistent because the columns of its $\mathbf{N}_L$ are not identical. Test the degree of consistency of $\mathbf{N}_L$.

We start by computing n_{max}. From Example 13.1-2, we have

$$\overline{w}_1 = .129, \overline{w}_2 = .277, \overline{w}_3 = .594$$

Thus,

$$\mathbf{A}_L\overline{\mathbf{w}} = \begin{pmatrix} 1 & \frac{1}{2} & \frac{1}{5} \\ 2 & 1 & \frac{1}{2} \\ 5 & 2 & 1 \end{pmatrix}\begin{pmatrix} .129 \\ .277 \\ .594 \end{pmatrix} = \begin{pmatrix} 0.3863 \\ 0.8320 \\ 1.7930 \end{pmatrix}$$

This yields

$$n_{max} = .3863 + .8320 + 1.7930 = 3.0113$$

Hence, for $n = 3$,

$$CI = \frac{n_{max} - n}{n - 1} = \frac{3.0113 - 3}{3 - 1} = .00565$$

$$RI = \frac{1.98(n - 2)}{n} = \frac{1.98 \times 1}{3} = .66$$

$$CR = \frac{CI}{RI} = \frac{.00565}{.66} = .00856$$

Because CR $< .1$, the level of inconsistency in $\mathbf{A}_L$ is acceptable.

Excel Moment

Template excelAHP.xls is designed to handle comparison matrices with sizes up to 8×8. As in the Excel models in Chapters 10 and 11, user input drives the model. Figure 13.3 demonstrates the application of the model to Example 13.1-2.[1] The comparison matrices of the problem are entered *one at a time* in the (top) input data section of the spreadsheet. The order in which the comparison matrices are entered is unimportant, though it makes more sense to consider them in their natural hierarchal order. Upon entering the data for a comparison matrix, the output (bottom) section of the spreadsheet will provide the associated normalized matrix together with its consistency ratio, *CR*. The user must copy the weights, *w*, from column J and paste them into the solution summary area (the right section of the spreadsheet). Remember to use Paste Special $\Rightarrow$ Values when performing this step to guarantee a permanent record. The process is repeated until all the comparison matrices have been stored in columns K:R.

In Figure 13.3, the final ranking is given in cells (K20:K22). The formula in cell K20 is

=$L$4*$L7+$L$5*$N7

This formula provides the final evaluation of alternative UA, and is copied in cells K21 and K22 to evaluate alternatives UB and UC. Note how the formula in K20 is constructed: Cell reference to the alternative UA must be column-fixed (namely, $L7 and $N7), whereas *all* other references must be row-and-column-fixed (namely, $L$4 and

---

[1]The more accurate results of the spreadsheet differ from those in Example 13.1-2 and 13.1-3 because of manual roundoff approximation. Columns F:I and rows 11:13 are suppressed to conserve space.

	A	B	C	D	E	J	K	L	M	N
1	AHP-Analytic Hierarchy Process									
2		Input: Comparison matrix					Solution summary			
3	Matrix name:	AL								
4	Matrix size=	3	<<Maximum 8					A		
5	Matrix data:	UA	UB	UC			R	0.83333		
6	UA	1	0.5	0.2			L	0.16667		
7	UB	2	1	0.5				AR		AL
8	UC	5	2	1			UA	0.54545	UA	0.1285
9							UB	0.27273	UB	0.27661
13							UC	0.18182	UC	0.59489
14	Col sum	8	3.5	1.7						
15		Output: Normalized martix								
16		nMax=	3.00746	CR=	0.0056					
17		UA	UB	UC		Weight				
18	UA	0.12500	0.14286	0.11765		0.12850				
19	UB	0.25000	0.28571	0.29412		0.27661				
20	UC	0.62500	0.57143	0.58824		0.59489	Final ranking			
21							UA= 0.47596			
22							UB= 0.27337			
26							UC= 0.25066			

FIGURE 13.3

Excel solution of Example 13.1-2 (file excelAHP.xls)

$L$5). The validity of the copied formulas requires that the (column-fixed) *alternative* weights of each matrix appear in the same column with no intervening empty cells. For example, in Figure 13.3, the AR-weights in column L cannot be broken between two columns. The same applies to the AL-weights in column N. There are no restrictions on the placement of the A-weights because they are row- and column-fixed in the formula.

You can embellish the formula to capture the names of the alternatives directly. Here is how the formula for alternative UA should be entered:

=$K7&"="&TEXT($L$4*$L7+$L$5*$N7,"####0.00000")

The procedure for evaluating alternatives can be extended readily to any number of hierarchy levels. Once you develop the formula correctly for the first alternative, the same formula applies to the remaining alternatives simply by copying it into (same column) succeeding rows. Remember that *all* cell references in the formula must be row-and-column-fixed, except for references to the alternatives, which must be column-fixed only. Problem 1, Set 13.1b, asks you to develop the formula for a 3-level problem.

PROBLEM SET 13.1B[2]

1. Consider the data of Problem 1, Set 13.1a. Copy the weights in a logical order into the solution summary section of the spreadsheet excelAHP.xls, then develop the formula for evaluating the first alternative, UA, and copy it to evaluate the remaining two alternatives.

[2]Spreadsheet excelAHP.xls should be helpful in verifying your calculations.

***2.** The personnel department at C&H has narrowed the search for a prospective employee to three candidates: Steve (S), Jane (J), and Maisa (M). The final selection is based on three criteria: personal interview (I), experience (E), and references (R). The department uses matrix $\mathbf{A}$ (given below) to establish the preferences among the three criteria. After interviewing the three candidates and compiling the data regarding their experiences and references, the matrices $\mathbf{A}_I$, $\mathbf{A}_E$, and $\mathbf{A}_R$ are constructed. Which of the three candidates should be hired? Assess the consistency of the data.

$$\mathbf{A} = \begin{array}{c} \\ I \\ E \\ R \end{array}\begin{array}{c} \begin{array}{ccc} I & E & R \end{array} \\ \begin{pmatrix} 1 & 2 & \frac{1}{4} \\ \frac{1}{2} & 1 & \frac{1}{5} \\ 4 & 5 & 1 \end{pmatrix} \end{array} \quad \mathbf{A}_I = \begin{array}{c} \\ S \\ J \\ M \end{array}\begin{array}{c} \begin{array}{ccc} S & J & M \end{array} \\ \begin{pmatrix} 1 & 3 & 4 \\ \frac{1}{3} & 1 & \frac{1}{5} \\ \frac{1}{4} & 5 & 1 \end{pmatrix} \end{array}$$

$$\mathbf{A}_E = \begin{array}{c} \\ S \\ J \\ M \end{array}\begin{array}{c} \begin{array}{ccc} S & J & M \end{array} \\ \begin{pmatrix} 1 & \frac{1}{3} & 2 \\ 3 & 1 & \frac{1}{2} \\ \frac{1}{2} & 2 & 1 \end{pmatrix} \end{array} \quad \mathbf{A}_R = \begin{array}{c} \\ S \\ J \\ M \end{array}\begin{array}{c} \begin{array}{ccc} S & J & M \end{array} \\ \begin{pmatrix} 1 & \frac{1}{2} & 1 \\ 2 & 1 & \frac{1}{2} \\ 1 & 2 & 1 \end{pmatrix} \end{array}$$

3. Kevin and June Park (K and J) are in the process of buying a new house. Three houses, A, B, and C, are available. The Parks have agreed on two criteria for the selection of the house: yard work (Y) and proximity to work (W), and have developed the following comparison matrices. Rank the three houses in order of priority, and compute the consistency ratio for each matrix.

$$\mathbf{A} = \begin{array}{c} \\ K \\ J \end{array}\begin{array}{c} \begin{array}{cc} K & J \end{array} \\ \begin{pmatrix} 1 & 2 \\ \frac{1}{2} & 1 \end{pmatrix} \end{array}$$

$$\mathbf{A}_K = \begin{array}{c} \\ Y \\ W \end{array}\begin{array}{c} \begin{array}{cc} Y & W \end{array} \\ \begin{pmatrix} 1 & \frac{1}{3} \\ 3 & 1 \end{pmatrix} \end{array} \quad \mathbf{A}_J = \begin{array}{c} \\ Y \\ W \end{array}\begin{array}{c} \begin{array}{cc} Y & W \end{array} \\ \begin{pmatrix} 1 & 4 \\ \frac{1}{4} & 1 \end{pmatrix} \end{array}$$

$$\mathbf{A}_{KY} = \begin{array}{c} \\ A \\ B \\ C \end{array}\begin{array}{c} \begin{array}{ccc} A & B & C \end{array} \\ \begin{pmatrix} 1 & 2 & 3 \\ \frac{1}{2} & 1 & 2 \\ \frac{1}{3} & \frac{1}{2} & 1 \end{pmatrix} \end{array} \quad \mathbf{A}_{KW} = \begin{array}{c} \\ A \\ B \\ C \end{array}\begin{array}{c} \begin{array}{ccc} A & B & C \end{array} \\ \begin{pmatrix} 1 & 2 & \frac{1}{2} \\ \frac{1}{2} & 1 & \frac{1}{3} \\ 2 & 3 & 1 \end{pmatrix} \end{array} \quad \mathbf{A}_{JY} = \begin{array}{c} \\ A \\ B \\ C \end{array}\begin{array}{c} \begin{array}{ccc} A & B & C \end{array} \\ \begin{pmatrix} 1 & 4 & 2 \\ \frac{1}{4} & 1 & 3 \\ \frac{1}{2} & \frac{1}{3} & 1 \end{pmatrix} \end{array} \quad \mathbf{A}_{JW} = \begin{array}{c} \\ A \\ B \\ C \end{array}\begin{array}{c} \begin{array}{ccc} A & B & C \end{array} \\ \begin{pmatrix} 1 & \frac{1}{2} & 4 \\ \frac{1}{2} & 1 & 3 \\ \frac{1}{4} & \frac{1}{3} & 1 \end{pmatrix} \end{array}$$

***4.** A new author sets three criteria for selecting a publisher for an OR textbook: royalty percentage (R), marketing (M), and advance payment (A). Two publishers, H and P, have expressed interest in the book. Using the following comparison matrices, rank the two publishers and assess the consistency of the decision.

$$\mathbf{A} = \begin{array}{c} \\ R \\ M \\ A \end{array}\begin{array}{c} \begin{array}{ccc} R & M & A \end{array} \\ \begin{pmatrix} 1 & 1 & \frac{1}{4} \\ 1 & 1 & \frac{1}{5} \\ 4 & 5 & 1 \end{pmatrix} \end{array}$$

$$\mathbf{A}_R = \begin{array}{c} \\ H \\ P \end{array}\begin{array}{c} \begin{array}{cc} H & P \end{array} \\ \begin{pmatrix} 1 & 2 \\ \frac{1}{2} & 1 \end{pmatrix} \end{array} \quad \mathbf{A}_M = \begin{array}{c} \\ H \\ P \end{array}\begin{array}{c} \begin{array}{cc} H & P \end{array} \\ \begin{pmatrix} 1 & \frac{1}{2} \\ 2 & 1 \end{pmatrix} \end{array} \quad \mathbf{A}_A = \begin{array}{c} \\ H \\ P \end{array}\begin{array}{c} \begin{array}{cc} H & P \end{array} \\ \begin{pmatrix} 1 & 1 \\ 1 & 1 \end{pmatrix} \end{array}$$

5. A professor of political science wants to predict the outcome of a school board election. Three candidates, Ivy (I), Bahrn (B), and Smith (S) are running for one position. The professor places the voters into three categories: left (L), center (C), and right (R). The candidates are judged based on three factors: educational experience (E), stand on issues (S), and personal character (P). The following are the comparison matrices for the first hierarchy of left, center, and right.

$$\mathbf{A} = \begin{matrix} & L & C & R \\ L & 1 & 2 & \frac{1}{2} \\ C & \frac{1}{2} & 1 & \frac{1}{5} \\ R & 2 & 5 & 1 \end{matrix} \qquad \mathbf{A}_L = \begin{matrix} & E & S & P \\ E & 1 & 3 & \frac{1}{2} \\ S & \frac{1}{3} & 1 & \frac{1}{3} \\ P & 2 & 3 & 1 \end{matrix}$$

$$\mathbf{A}_C = \begin{matrix} & E & S & P \\ E & 1 & 2 & 2 \\ S & \frac{1}{2} & 1 & 1 \\ P & \frac{1}{2} & 1 & 1 \end{matrix} \qquad \mathbf{A}_R = \begin{matrix} & E & S & P \\ E & 1 & 1 & 9 \\ S & 1 & 1 & 8 \\ P & \frac{1}{9} & \frac{1}{8} & 1 \end{matrix}$$

The professor was able to generate nine more comparison matrices to account for the three candidates at the second hierarchy representing experience, stand on issues, and personal character. The AHP process was then used to reduce these matrices to the following relative weights:

	Left			Center			Right		
Candidate	E	S	P	E	S	P	E	S	P
Ivy	.1	.2	.3	.3	.5	.2	.7	.1	.3
Bahrn	.5	.4	.2	.4	.2	.4	.1	.4	.2
Smith	.4	.4	.5	.3	.3	.4	.2	.5	.5

Determine the winning candidate and assess the consistency of the decision.

6. A school district is in dire need to reduce expenses to meet new budgetary restrictions at its elementary schools. Two options are available: Delete the physical education program (E), or delete the music program (M). The superintendent of the district has formed a committee with equal-vote representation from the School Board (S) and the Parent-Teacher Association (P) to study the situation and make a recommendation. The committee has decided to study the issue from the standpoint of budget restriction (B) and students needs (N). The analysis produced the following comparison matrices:

$$\mathbf{A}_S = \begin{matrix} & B & N \\ B & 1 & 1 \\ N & 1 & 1 \end{matrix} \qquad \mathbf{A}_P = \begin{matrix} & B & N \\ B & 1 & \frac{1}{2} \\ N & 2 & 1 \end{matrix}$$

$$\mathbf{A}_{SB} = \begin{matrix} & E & M \\ E & 1 & \frac{1}{2} \\ M & 2 & 1 \end{matrix} \qquad \mathbf{A}_{SN} = \begin{matrix} & E & M \\ E & 1 & \frac{1}{3} \\ M & 3 & 1 \end{matrix}$$

$$\mathbf{A}_{PB} = \begin{matrix} & E & M \\ E & 1 & \frac{1}{3} \\ M & 3 & 1 \end{matrix} \qquad \mathbf{A}_{PN} = \begin{matrix} & E & M \\ E & 1 & 2 \\ M & \frac{1}{2} & 1 \end{matrix}$$

Analyze the decision problem, and make a recommendation.

7. An individual is in the process of buying a car and has narrowed the choices to three models, *M*1, *M*2, and *M*3. The deciding factors include purchase price (PP), maintenance cost (MC), cost of city driving (CD), and cost of rural driving (RD). The following table provides the relevant data for 3 years of operation:

Car model	PP($)	MC($)	CD($)	RD($)
*M*1	6,000	1800	4500	1500
*M*2	8,000	1200	2250	750
*M*3	10,000	600	1125	600

Use the cost data to develop the comparison matrices. Assess the consistency of the matrices, and determine the choice of model.

13.2 DECISION MAKING UNDER RISK

Under conditions of risk, the payoffs associated with each decision alternative are described by probability distributions. For this reason, decision making under risk can be based on the *expected value criterion*, in which decision alternatives are compared based on the maximization of expected profit or the minimization of expected cost. However, because the approach has limitations, the expected value criterion can be modified to encompass other situations.

Real-Life Application—Booking Limits in Hotel Reservations

Hotel La Posada has a total of 300 guest rooms. Its clientele includes both business and leisure travelers. Rooms can be sold in advance (usually to leisure travelers) at a discount price. Business travelers, who usually are late in booking their rooms, pay full price. La Posada must thus set a *booking limit* on the number of discount rooms sold to leisure travelers in order to take advantage of the full-price business customers. Decision-tree analysis is used in Case 10, Chapter 24 on the CD to determine the booking limit.

13.2.1 Decision Tree-Based Expected Value Criterion

The expected value criterion seeks the maximization of expected (average) profit or the minimization of expected cost. The data of the problem assumes that the payoff (or cost) associated with each decision alternative is probabilistic.

Decision Tree Analysis. The following example considers simple decision situations with a finite number of decision alternatives and explicit payoff matrices.

Example 13.2-1

Suppose that you want to invest $10,000 in the stock market by buying shares in one of two companies: *A* and *B*. Shares in Company *A*, though risky, could yield a 50% return on investment during the next year. If the stock market conditions are not favorable (i.e., "bear" market), the stock may lose 20% of its value. Company *B* provides safe investments with 15% return in a "bull" market and only 5% in a "bear" market. All the publications you have consulted (and there is always a flood of them at the end of the year!) are predicting a 60% chance for a "bull" market and 40% for a "bear" market. Where should you invest your money?

The decision problem can be summarized as follows:

	1-year return on $10,000 investment	
Decision alternative	*"Bull" market ($)*	*"Bear" market ($)*
Company *A* stock	5000	−2000
Company *B* stock	1500	500
Probability of occurrence	.6	.4

The problem can also be represented as a **decision tree** as shown in Figure 13.4. Two types of nodes are used in the tree: a square (□) represents a *decision point* and a circle (○) represents a *chance event*. Thus, two branches emanate from decision point 1 to represent the two alternatives of investing in stock *A* or stock *B*. Next, the two branches emanating from chance events 2 and 3 represent the "bull" and the "bear" markets with their respective probabilities and payoffs.

From Figure 13.4, the expected 1-year returns for the two alternatives are

$$\text{For stock } A = \$5000 \times .6 + (-2000) \times .4 = \$2200$$

$$\text{For stock } B = \$1500 \times .6 + \$500 \times .4 = \$1100$$

Based on these computations, your decision is to invest in stock *A*.

Remarks. In the terminology of decision theory, the "bull" and the "bear" markets in the preceding example are referred to as **states of nature**, whose chances of occurrence are probabilistic (.6 versus .4). In general, a decision problem may include n states of nature and m alternatives. If $p_j\ (> 0)$ is the probability of occurrence for state of nature j and a_{ij} is the payoff of alternative i, given state of nature $j (i = 1, 2, \ldots, m;$ $j = 1, 2, \ldots, n)$, then the expected payoff for alternative i is computed as

$$EV_i = a_{i1}p_1 + a_{i2}p_2 + \cdots + a_{in}p_n, i = 1, 2, \ldots, n$$

By definition, $p_1 + p_2 + \cdots + p_n = 1$.

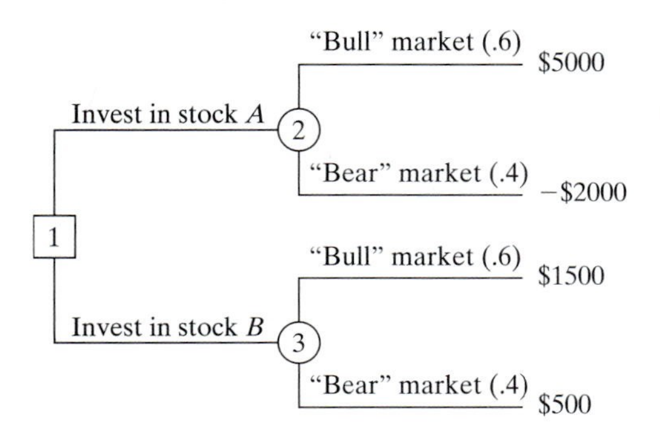

FIGURE 13.4

Decision-tree representation of the stock market problem

The best alternative is the one associated with $EV^* = \max_i\{EV_i\}$ or $EV^* = \min_i\{EV_i\}$, depending, respectively, on whether the payoff of the problem represents profit (income) or loss (expense).

PROBLEM SET 13.2A

1. You have been invited to play the Fortune Wheel game on television. The wheel operates electronically with two buttons that produce hard (H) or soft (S) spin of the wheel. The wheel itself is divided into white (W) and red (R) half-circle regions. You have been told that the wheel is designed to stop with a probability of .3 in the white region and .7 in the red region. The payoff you get for the game is

	W	R
H	\$800	\$200
S	−\$2500	\$1000

Draw the associated decision tree, and specify a course of action.

***2.** Farmer McCoy can plant either corn or soybeans. The probabilities that the next harvest prices of these commodities will go up, stay the same, or go down are .25, .30, and .45, respectively. If the prices go up, the corn crop will net \$30,000 and the soybeans will net \$10,000. If the prices remain unchanged, McCoy will (barely) break even. But if the prices go down, the corn and soybeans crops will sustain losses of \$35,000 and \$5000, respectively.

(a) Represent McCoy's problem as a decision tree.

(b) Which crop should McCoy plant?

3. You have the chance to invest in three mutual funds: utility, aggressive growth, and global. The value of your investment will change depending on the market conditions. There is a 10% chance the market will go down, 50% chance it will remain moderate, and 40% chance it will perform well. The following table provides the percentage change in the investment value under the three conditions:

	Percent return on investment		
Alternative	*Down market (%)*	*Moderate market (%)*	*Up market (%)*
Utility	+5	+7	+8
Aggressive growth	−10	+5	+30
Global	+2	+7	+20

(a) Represent the problem as a decision tree.

(b) Which mutual fund should you select?

4. You have the chance to invest your money in either a 7.5% bond that sells at face value or an aggressive growth stock that pays only 1% dividend. If inflation is feared, the interest rate will go up to 8%, in which case the principal value of the bond will go down by

10%, and the stock value will go down by 20%. If recession materializes, the interest rate will go down to 6%. Under this condition, the principal value of the bond is expected to go up by 5%, and the stock value will increase by 20%. If the economy remains unchanged, the stock value will go up by 8% and the bond principal value will remain the same. Economists estimate a 20% chance that inflation will rise and 15% that recession will set in. Assume that you are basing your investment decision on next year's economic conditions.

(a) Represent the problem as a decision tree.

(b) Would you invest in stocks or bonds?

5. AFC is about to launch its new Wings 'N Things fast food nationally. The research department is convinced that Wings 'N Things will be a great success and wants to introduce it immediately in all AFC outlets without advertising. The marketing department sees "things" differently and wants to unleash an intensive advertising campaign. The advertising campaign will cost \$100,000 and if successful will produce \$950,000 revenue. If the campaign is unsuccessful (there is a 30% chance it won't be), the revenue is estimated at only \$200,000. If no advertising is used, the revenue is estimated at \$400,000 with probability .8 if customers are receptive and \$200,000 with probability .2 if they are not.

(a) Draw the associated decision tree.

(b) What course of action should AFC follow in launching the new product?

***6.** A fair coin is flipped three successive times. You receive \$1.00 for each head (H) that turns up and an additional \$.25 for each two successive heads that appear (remember that HHH includes two sets of HH). However, you give back \$1.10 for each tail that shows up. You have the options to either play or not play the game.

(a) Draw the decision tree for the game.

(b) Would you favor playing this game?

7. You have the chance to play the following game in a gambling casino. A fair die is rolled twice, leading to four outcomes: (1) both rolls show an even match, (2) both rolls show an odd match, (3) the outcomes are either even-odd or odd-even, and (4) all other outcomes. You are allowed to bet your money on exactly two outcomes with equal dollar amounts. For example, you can bet equal dollars on even-match (outcome 1) and odd-match (outcome 2). The payoff for each dollar you bet is \$2.00 for the first outcome, \$1.95 for the second and the third outcomes, and \$1.50 for the fourth outcome.

(a) Draw the decision tree for the game.

(b) Which two choices would you make?

(c) Do you ever come out ahead in this game?

8. Acme Manufacturing produces widget batches with .8%, 1%, 1.2%, and 1.4% defectives according to the respective probabilities .4, .3, .25, and .05. Three customers, A, B, and C, are contracted to receive batches with no more than .8%, 1.2%, and 1.4% defectives, respectively. If the defectives are higher than contracted, Acme will be penalized \$100 for each .1% increase. Conversely, supplying higher-quality batches than required costs Acme \$50 for each .1%. Assume that the batches are not inspected before shipment.

(a) Draw the associated decision tree.

(b) Which of the three customers should have the highest priority to receive their order?

9. TriStar plans to open a new plant in Arkansas. The company can open a full-sized plant now or a small-sized plant that can be expanded 2 years later if high demand conditions prevail. The time horizon for the decision problem is 10 years. TriStar estimates that the probabilities for high and low demands over the next 10 years is .75 and .25, respectively.

The cost of immediate construction of a large plant is $5 million, and a small plant costs $1 million. Expansion of a small plant 2 years from now costs $4.2 million. The income from the operation over the next 10 years is given in the following table:

	Estimated annual income (in $1000)	
Alternative	*High demand*	*Low demand*
Full-sized plant now	1000	300
Small-sized plant now	250	200
Expanded plant in 2 years	900	200

(a) Develop the associated decision tree, given that after 2 years TriStar has the options to expand or not expand the small plant.

(b) Develop a construction strategy for TriStar over the next 10 years. (For simplicity, ignore the time value of money.)

10. Rework Problem 9, assuming that the annual interest rate is 10% and that the decisions are made considering the time value of money. [*Note*: You need compound interest tables to solve this problem. You can use Excel function NPV(i, R) to compute the present value of cash flows stored in range R for an interest rate i. NPV assumes that each cash flow occurs at the end of the year.]

11. Rework Problem 9, assuming that the demand can be high, medium, and low with probabilities .7, .2, and .1, respectively. Expansion of a small plant will occur only if demand in the first 2 years is high. The following table gives the annual income. Ignore the time value of money.

	Estimated annual income (in $1000)		
Alternative	*High demand*	*Medium demand*	*Low demand*
Full-sized plant now	1000	500	300
Small-sized plant now	400	280	150
Expanded plant in 2 years	900	600	200

***12.** Sunray Electric Coop uses a fleet of 20 trucks to service its electric network. The company wants to develop a schedule of periodic preventive maintenance for the fleet. The probability of a breakdown in year 1 is zero. For year 2, the breakdown probability is .03, and it increases annually by .01 for years 3 through 10. Beyond year 10, the breakdown probability is constant at .13. A random breakdown costs $200 per truck, and a scheduled maintenance costs only $75 per truck. Sunray wants to determine the optimal period (in months) between scheduled preventive maintenances.

(a) Develop the associated decision tree.

(b) Determine the optimal maintenance-cycle length.

13. Daily demands for loaves of bread at a grocery store are specified by the following probability distribution:

n	100	150	200	250	300
p_n	.20	.25	.30	.15	.10

The store buys a loaf for 55 cents and sells it for $1.20 each. Any unsold loaves at the end of the day are disposed of at 25 cents each. Assume that the stock level is restricted to one of the demand levels specified for p_n.

(a) Develop the associated decision tree.

(b) How many loaves should be stocked daily?

14. In Problem 13, suppose that the store wishes to extend the decision problem to a 2-day horizon. The alternatives for the second day depend on the demand in the first day. If demand on day 1 equals the amount stocked, the store will continue to order the same quantity for day 2; if it exceeds the amount stocked, the store can order any of the higher-level stocks; and if it is less than the amount stocked, the store can only order any of the lower-level stocks. Develop the associated decision tree, and determine the optimal ordering strategy.

***15.** An automatic machine produces α (thousands of) units of a product per day. As α increases, the proportion of defectives, p, goes up according to the following probability density function

$$f(p) = \begin{cases} \alpha p^{\alpha-1}, & 0 \leq p \leq 1 \\ 0, & \text{otherwise} \end{cases}$$

Each defective item incurs a loss of $50. A good item yields $5 profit.

(a) Develop a decision tree for this problem.

(b) Determine the value of α that maximizes the expected profit.

16. The outer diameter, d, of a cylinder is processed on an automatic machine with upper and lower tolerance limits of $d + t_U$ and $d - t_L$. The production process follows a normal distribution with mean μ and standard deviation σ. An oversized cylinder is reworked at the cost of c_1 dollars. An undersized cylinder must be salvaged at the cost of c_2 dollars. Develop the decision tree and determine the optimal setting d for the machine.

17. (Cohan and Associates, 1984) Modern forest management uses controlled fires to reduce fire hazards and to stimulate new forest growth. Management has the option to postpone or plan a burning. In a specific forest tract, if burning is postponed, a general administrative cost of $300 is incurred. If a controlled burning is planned, there is a 50% chance that good weather will prevail and burning will be carried out at a cost of $3200. The results of the burning may be either successful with probability .6 or marginal with probability .4. Successful execution will result in an estimated benefit of $6000 and marginal execution will provide only $3000 in benefits. If the weather is poor, burning will be cancelled, and the associated planning cost is $1200 with no benefit.

(a) Develop a decision tree to determine whether burning should be planned or postponed.

(b) Study the sensitivity of the solution to changes in the probability of good weather.

18. (Rappaport, 1967) A manufacturer has used linear programming to determine the optimum production mix of the various TV models it produces. Recent information received by the manufacturer indicates that there is a 40% chance that the supplier of a component used in one of the models may raise the price by $35. The manufacturer thus can take one of two actions: Continue to use the original (optimum) product mix (A1), or use a new (optimum) mix based on the higher component price (A2). Action A1 is ideal if the price is not raised and action A2 will also be ideal if the price is raised. The following table provides the resulting total profit per month as a function of the action taken and the random outcome regarding the component price.

	Price increase (O1)	No price increase (O2)
Original mix (A1)	$400,000	$295,500
New mix (A2)	$372,000	$350,000

(a) Develop the associated decision tree and determine which action should be adopted.

(b) The manufacturer can invest $1000 to obtain additional information about whether or not the price will increase. This information says that there is a 58% chance that the probability of price increase will be .9 and a 42% chance that the probability of price increase will be .3. Would you recommend the additional investment?

***19.** *Aspiration Level Criterion.* Acme Manufacturing uses an industrial chemical in one of its processes. The shelf life of the chemical is 1 month, following which any amount left is destroyed. The use of the chemical by Acme (in gallons) occurs randomly according to the following distribution:

$$f(x) = \begin{cases} \frac{200}{x^2}, & 100 \le x \le 200 \\ 0, & \text{otherwise} \end{cases}$$

The actual consumption of the chemical occurs instantaneously at the start of the month. Acme wants to determine the level of the chemical that satisfies two conflicting criteria (or aspiration levels): The average excess quantity for the month does not exceed 20 gallons and the average shortage quantity for the month does not exceed 40 gallons.

13.2.2 Variations of the Expected Value Criterion

This section addresses three issues relating to the expected value criterion. The first issue deals with the determination of *posterior probabilities* based on experimentation, and the second deals with the *utility* versus the actual value of money.

Posterior (Bayes') Probabilities. The probabilities used in the expected value criterion are usually determined from historical data (see Section 12.5). In some cases, these probabilities can be adjusted using additional information based on sampling or experimentation. The resulting probabilities are referred to as **posterior (or Bayes') probabilities**, as opposed to the **prior probabilities** determined from raw data.

Real-Life Application—Casey's Problem: Interpreting and Evaluating a New Test

A screening test of a newborn baby, named Casey, indicated a C14:1 enzyme deficiency. The enzyme is required to digest a particular form of long-chain fats, and its absence could lead to severe illness or mysterious death (broadly categorized under sudden infant death syndrome or SIDS). The test had been administered previously to approximately 13,000 newborns and Casey was the first to test positive. Though the screening test does not in itself constitute a definitive diagnosis, the extreme rarity of the condition led her doctors to conclude that that there was an 80–90% chance that she was suffering from this deficiency. Given that Casey tested positive, Bayes' posterior probability is used to assess whether or not the child has the C14:1 deficiency. The analysis of this situation is detailed in Case 11, Chapter 24 on the CD.

Example 13.2-2

This example demonstrates how the expected-value criterion is modified to take advantage of the posterior probabilities. In Example 13.2-1, the (prior) probabilities of .6 and .4 of a "bull" and a "bear" market are determined from available financial publications. Suppose that rather than relying solely on these publications, you have decided to conduct a more "personal" investigation by consulting a friend who has done well in the stock market. The friend offers the general opinion of "for" or "against" investment quantified in the following manner: If it is a "bull" market, there is a 90% chance the vote will be "for." If it is a "bear" market, the chance of a "for" vote is lowered to 50%. How do you make use of this additional information?

The statement made by the friend provides conditional probabilities of "for/against," given that the states of nature are "bull" and "bear" markets. To simplify the presentation, let us use the following symbols:

$$v_1 = \text{"For" vote}$$
$$v_2 = \text{"Against" vote}$$
$$m_1 = \text{"Bull" market}$$
$$m_2 = \text{"Bear" market}$$

The friend's statement may be written in the form of probability statements as

$$P\{v_1|m_1\} = .9,\ P\{v_2|m_1\} = .1$$
$$P\{v_1|m_2\} = .5,\ P\{v_2|m_2\} = .5$$

With this additional information, the decision problem can be summarized as follows:

1. If the friend's recommendation is "for," would you invest in stock A or in stock B?
2. If the friend's recommendation is "against," would you invest in stock A or in stock B?

The problem can be summarized in the form of a decision tree as shown in Figure 13.5. Node 1 is a chance event representing the "for" and "against" prossibilities. Nodes 2 and 3 are decision points for choosing between stocks A and B, given that the friend's votes are "for" and "against," respectively. Finally, nodes 4 to 7 are chance events representing the "bull" and "bear" markets.

To evaluate the different alternatives in Figure 13.5, it is necessary to compute the *posterior* probabilities $P\{m_i|v_j\}$ shown on the m_1- and m_2-branches of chance nodes 4, 5, 6, and 7. These posterior probabilities take into account the additional information provided by the friend's "for/against" recommendation and are computed according to the following general steps:

Step 1. The conditional probabilities $P\{v_j|m_i\}$ of the problem can be summarized as

	v_1	v_2
m_1	.9	.1
m_2	.5	.5

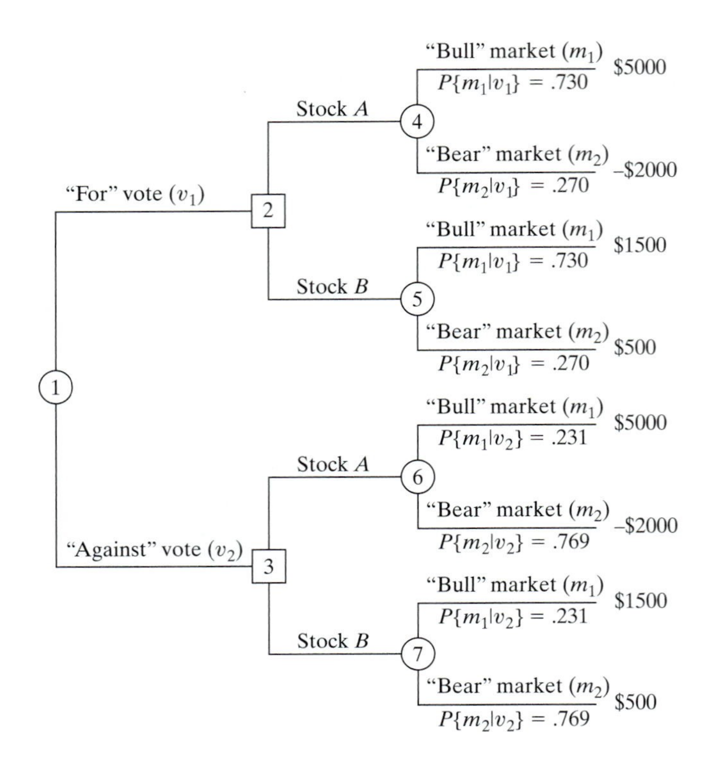

FIGURE 13.5
Decision tree for the stock market problem with posterior probabilities

Step 2. Compute the joint probabilities as

$$P\{m_i, v_j\} = P\{v_j|m_i\}P\{m_i\}, \text{ for all } i \text{ and } j$$

Given the *prior* probabilities $P\{m_1\} = .6$ and $P\{m_2\} = .4$, the joint probabilities are determined by multiplying the first and the second rows of the table in step 1 by .6 and .4, respectively. We thus get

	v_1	v_2
m_1	.54	.06
m_2	.20	.20

The sum of all the entries in the table equals 1.

Step 3. Compute the absolute probabilities as

$$P\{v_j\} = \sum_{\text{all } i} P\{m_i, v_j\}, \text{ for all } j$$

These probabilities are computed from the table in step 2 by summing the rows of each column, which yields

$P\{v_1\}$	$P\{v_2\}$
.74	.26

Step 4. Determine the desired posterior probabilities as

$$P\{m_i|v_j\} = \frac{P\{m_i, v_j\}}{P\{v_j\}}$$

These probabilities are computed by dividing the rows of each column in the table of step 2 by the element of the corresponding column in the table of step 3, which (rounded to three digits) yields

	v_1	v_2
m_1	.730	.231
m_2	.270	.769

These are the probabilities shown in Figure 13.5. They are different from the original prior probabilities $P\{m_1\} = .6$ and $P\{m_2\} = .4$.

We are now ready to evaluate the alternatives based on the expected payoffs for nodes 4, 5, 6, and 7—that is,

"For" Vote

$$\text{Stock } A \text{ at node } 4 = 5000 \times .730 + (-2000) \times .270 \times = \mathbf{\$3110}$$

$$\text{Stock } B \text{ at node } 5 = 1500 \times .730 + 500 \times .270 = 1230$$

Decision. Invest in stock A.

"Against" Vote

$$\text{Stock } A \text{ at node } 6 = 5000 \times .231 + (-2000) \times .769 = -\$383$$

$$\text{Stock } B \text{ at node } 7 = 1500 \times .231 + 500 \times .769 = \mathbf{\$731}$$

Decision. Invest in stock B.

The preceding decisions are equivalent to saying that the expected payoffs at decision nodes 2 and 3 are \$3110 and \$731, respectively (see Figure 13.5). Thus, given the probabilities $P\{v_1\} = .74$ and $P\{v_2\} = .26$ as computed in step 3, we can compute the expected payoff for the entire decision tree. (See Problem 3, Set 13.2b.)

	A	B	C	D	E	L	M	N
1	Bayes Posterior Probabilities							
2	Input Data						Output Results	
3			P{v|m} (10 x 10) maximum					P{v,m}
4	P{m}		v1	v2			v1	v2
5	m1	0.6	0.9	0.1			0.5400	0.0600
6	m2	0.4	0.5	0.5			0.2000	0.2000
15	Input Data Error Messages						P{v}	
16							0.7400	0.2600
17							P{m|v}	
18						m1	0.7297	0.2308
19						m2	0.2703	0.7692
20								
21								
22								
23								
24								
25								
26								
27								

FIGURE 13.6

Excel calculation of Bayes posterior probabilities for Example 13.2-2 (file excelBayes.xls)

Excel Moment

Excel file excelBayes.xls is designed to determine the Bayes posterior probabilities for prior probability matrices of sizes up to 10×10 (rows 7:14 and columns F:K and O:V are hidden to conserve space). The input data include $P\{m\}$ and $P\{v|m\}$. The spreadsheet checks input data errors and displays an appropriate error message. Figure 13.6 demonstrates the application of the model to the problem of Example 13.2-2.

PROBLEM SET 13.2B

1. Data in a community college show that 75% of new students who took calculus in high school do well, compared with 50% of those who did not take calculus. Admissions for the current academic year show that only 30% of the new students have completed a course in calculus. What is the probability that a new student will do well in college?

*2. Elektra receives 75% of its electronic components from vendor A and the remaining 25% from vendor B. The percentage of defectives from vendors A and B are 1% and 2%, respectively. When a random sample of size 5 from a received lot is inspected, only one defective unit is found. Determine the probability that the lot is received from vendor A. Repeat the same for vendor B. (*Hint:* The probability of a defective item in a sample is binomial.)

3. In Example 13.2-2, suppose that you have the additional option of investing the original $10,000 in a safe certificate of deposit that yields 8% interest. Your friend's advice applies to investing in the stock market only.
   (a) Develop the associated decision tree.
   (b) What is the optimal decision in this case? (*Hint:* Make use of $P\{v_1\}$ and $P\{v_2\}$ given in step 3 of Example 13.2-2 to determine the expected value of investing in the stock market.)

*4. You are the author of what promises to be a successful novel. You have the option to either publish the novel yourself or through a publisher. The publisher is offering you

$20,000 for signing the contract. If the novel is successful, it will sell 200,000 copies. If it isn't, it will sell only 10,000 copies. The publisher pays a $1 royalty per copy. A market survey by the publisher indicates that there is a 70% chance that the novel will be successful. If you publish the novel yourself, you will incur an initial cost of $90,000 for printing and marketing, but each copy sold will net you $2.

(a) Based on the given information, would you accept the publisher's offer or publish the book yourself?

(b) Suppose that you contract a literary agent to conduct a survey concerning the potential success of the novel. From past experience, the agent advises you that when a novel is successful, the survey will predict the wrong outcome 20% of the time. When the novel is not successful, the survey will give the correct prediction 85% of the time. How would this information affect your decision?

5. Consider Farmer McCoy's decision situation in Problem 2, Set 13.2a. The farmer has the additional option of using the land as a grazing range, in which case he is guaranteed a payoff of $7500. The farmer has also secured additional information from a broker regarding the degree of stability of future commodity prices. The broker's assessment of "favorable" and "unfavorable" is further quantified by the following conditional probabilities:

$P\{a_j|s_i\} =$

	a_1	a_2
s_1	.85	.15
s_2	.50	.50
s_3	.15	.85

The symbols a_1 and a_2 represent the "favorable" and "unfavorable" assessment by the broker, and s_1, s_2, and s_3 represent, respectively, the up, same, and down change in future prices.

(a) Draw the associated decision tree.

(b) Specify the optimal decision for the problem.

6. In Problem 5, Set 13.2a, suppose that AFC management has decided to test-market its Wings 'N Things in selective locations. The outcome of the test is either "good" (a_1) or "bad" (a_2). The test yields the following conditional probabilities with and without the advertising campaign:

$P\{a_j|v_i\}$-With campaign

	a_1	a_2
v_1	.95	.05
v_2	.3	.7

$P\{a_j|w_i\}$-No campaign

	a_1	a_2
w_1	.8	.2
w_2	.4	.6

The symbols v_1 and v_2 represent "success" and "no success," and w_1 and w_2 represent "receptive" and "not receptive."

(a) Develop the associated decision tree.

(b) Determine the best course of action for AFC.

7. Historical data at Acme Manufacturing estimate a 5% chance that a manufactured batch of widgets will be unacceptable (bad). A bad batch has 15% defective items, and a good batch includes only 4% defective items. Letting $a = \theta_1$ ($= \theta_2$) indicate that the batch is good (bad), the associated prior probabilities are given as

$$P\{a = \theta_1\} = .95 \text{ and } P\{a = \theta_2\} = .05$$

Instead of shipping batches based solely on prior probabilities, a test sample of two items is used, which gives rise to three possible outcomes: (1) both items are good (z_1), (2) one item is good (z_2), and (3) both items are defective (z_3).

(a) Determine the posterior probabilities $P\{\theta_i|z_j\}, i = 1, 2; j = 1, 2, 3$.

***(b)** Suppose that the manufacturer ships batches to two customers, A and B. The contracts specify that the defectives for A and B should not exceed 5% and 8%, respectively. A penalty of $100 is incurred per percentage point above the maximum limit. Supplying better-quality batches than specified by the contract costs the manufacturer $50 per percentage point. Develop the associated decision tree, and determine a priority strategy for shipping the batches.

Utility Functions. In the preceding presentation, the expected value criterion has been applied to situations where the payoff is *real* money. There are cases where the *utility* rather than the real value should be used in the analysis. To illustrate this point, suppose there is a 50-50 chance that a $20,000 investment will produce a profit of $40,000 or be lost completely. The associated expected profit is $40{,}000 \times .5 - 20{,}000 \times .5 = \$10{,}000$. Although there is a net expected profit, different individuals may vary in interpreting the result. An investor who is willing to accept risk may undertake the investment for a 50% chance to make a $40,000 profit. Conversely, a conservative investor may not be willing to risk losing $20,000. From this standpoint, we say that different individuals exhibit different attitudes toward risk, meaning that individuals exhibit different *utility* regarding risk.

The determination of the utility is subjective. It depends on our attitude toward accepting risk. In this section, we present a procedure for quantifying the degree of tolerance of the decision maker toward risk. The end result is a utility function that takes the place of real money.

In the preceding investment illustration, the best payoff is $40,000, and the worst is −$20,000. We thus establish an arbitrary, but logical, utility scale, U, from 0 to 100, in which $U(-\$20{,}000) = 0$ and $U(\$40{,}000) = 100$. The utilities for values between −$20,000 and $40,000 are determined as follows: If the decision maker's attitude is indifferent toward risk, then the resulting utility function will be a straight line joining (0, −$20,000) and (100, $40,000). In this case, both the real money and its utility will produce the same decisions. More realistically, the utility function takes other forms that reflect the attitude of the decision maker toward risk. Figure 13.7 illustrates the cases of individuals *X*, *Y*, and *Z*. Individual *X* is **risk averse** (or cautious) because of exhibiting higher sensitivity to loss than to profit. Individual *Z* is the opposite, and hence is a **risk seeker**. The figure demonstrates that for the risk-averse individual, X, the drop in utility *bc* corresponding to a loss of $10,000 is larger than the increase *ab* associated with a gain of $10,000. For the same ±$10,000 changes, the risk seeker, *Z*, exhibits an opposite behavior because $de > ef$. Further, individual *Y* is **risk neutral** because the

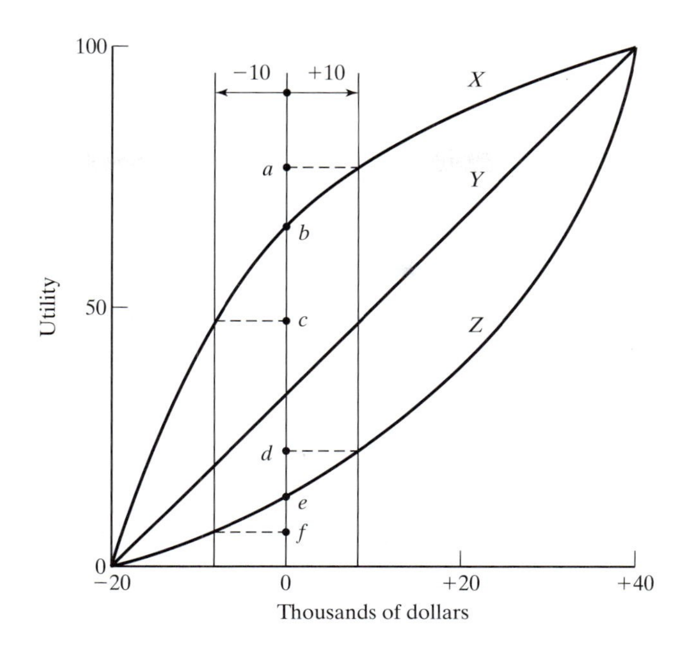

FIGURE 13.7

Utility functions for risk averse (X), indifferent (Y), and risk seeker (Z) decision makers

suggested changes yield equal changes in utility. In general, an individual may be both risk averse and risk seekeing, in which case the associated utility curve will follow an elongated S-shape.

Utility curves similar to the ones demonstrated in Figure 13.7 are determined by "quantifying" the decision maker's attitude toward risk for different levels of cash money. In our example, the desired range is (−\$20,000 to \$40,000), and the corresponding utility range is (0 to 100). What we would like to do is specify the utility associated with intermediate cash values, such as −\$10,000, \$0, \$10,000, \$20,000, and \$30,000. The procedure starts by establishing a lottery for a cash amount x whose expected utility is given as:

$$\begin{aligned} U(x) &= pU(-20{,}000) + (1 - p)U(\$40{,}000), 0 \leq p \leq 1 \\ &= 0p + 100(1 - p) \\ &= 100 - 100p \end{aligned}$$

To determine $U(x)$, we ask the decision maker to state a preference between a *guaranteed* cash amount x and the chance to play a lottery in which a loss of −\$20,000 occurs with probability p, and a profit of \$40,000 is realized with probability $1 - p$. The decision maker translates the preference by specifying the value of p that will render him indifferent between the two choices. For example, if $x = \$20{,}000$, the decision maker may say that a guaranteed \$20,000 cash and the lottery are equally attractive if $p = .8$. In this case, we can compute the utility of $x = \$20{,}000$ as

$$U(\$20{,}000) = 100 - 100 \times .8 = 20$$

We continue in this manner until we generate enough points [x versus $U(x)$] to identify the shape of the utility function. We may then determine the desired utility function by using regression analysis or simply by using a piecewise-linear function.

Although we are using a quantitative procedure to determine the utility function, the approach is far from being scientific. The fact that the procedure is totally driven by the contributed opinion of the decision maker casts doubt on the reliability of the process. In particular, the procedure implicitly assumes that the decision maker is rational, a requirement that cannot always be reconciled with the wide changes in behavior and mood that typify human beings. In this regard, decision makers should take the concept of utility in the broad sense that monetary values should not be the only critical factor in decision making.

PROBLEM SET 13.2C

*1. You are a student at the University of Arkansas and desperately want to attend the next Razorbacks basketball game. The problem is that the admission ticket costs $10, and you have only $5. You can bet your $5 in a poker game, with a 50-50 chance of either doubling your money or losing all of it.
   (a) Based on the real value of money, would you be tempted to participate in the poker game?
   (b) Based on your ardent desire to see the game, translate the actual money into a utility function.
   (c) Based on the utility function you developed in (b), would you be tempted to participate in the poker game?

*2. The Golden family have just moved to a location where earthquakes are known to occur. They must decide whether they should build their house according to the high-standard earthquake code. The construction cost using the earthquake code is $850,000; otherwise, a comparable house can be constructed for only $350,000. If an earthquake occurs (and there is a probability of .001 it might happen), a substandard home will cost $900,000 to repair. Develop the lottery associated with this situation, assuming a utility scale from 0 to 100.

3. An investment of $10,000 in a high-risk venture has a 50-50 chance over the next year of increasing to $14,000 or decreasing to $8000. Thus the net return can be either $4000 or −$2000.
   (a) Assuming a risk-neutral investor and a utility scale from 0 to 100, determine the utility of $0 *net* return on investment and the associated indifference probability.
   (b) Suppose that two investors A and B have exhibited the following indifference probabilities:

	Indifference probability	
Net return ($)	*Investor A*	*Investor B*
−2000	1.00	1.00
−1000	0.30	0.90
0	0.20	0.80
1000	0.15	0.70
2000	0.10	0.50
3000	0.05	0.40
4000	0.00	0.00

Graph the utility functions for investors A and B and categorize each investor as either a risk-averse person or a risk seeker.

(c) Suppose that investor A has the chance to invest in one of two ventures, I or II. Venture I can produce a net return of \$3000 with probability .4 or a net loss of \$1000 with probability .6. Venture II can produce a net return of \$2000 with probability .6 and no return with probability .4. Based on the utility function in (b), use the expected utility criterion to determine the venture investor A should select. What is the expected monetary value associated with the selected venture? (*Hint*: Use linear interpolation of the utility function.)

(d) Repeat part (c) for investor B.

13.3 DECISION UNDER UNCERTAINTY

Decision making under uncertainty, as under risk, involves alternative actions whose payoffs depend on the (random) *states of nature*. Specifically, the payoff matrix of a decision problem with m alternative actions and n states of nature can be represented as

	s_1	s_2	$\cdots$	s_n
a_1	$v(a_1, s_1)$	$v(a_1, s_2)$	$\cdots$	$v(a_1, s_n)$
a_2	$v(a_2, s_1)$	$v(a_2, s_2)$	$\cdots$	$v(a_2, s_n)$
$\vdots$	$\vdots$	$\vdots$	$\vdots$	$\vdots$
a_m	$v(a_m, s_1)$	$v(a_m, s_2)$	$\cdots$	$v(a_m, s_n)$

The element a_i represents action i and the element s_j represents state of nature j. The payoff or outcome associated with action a_i and state s_j is $v(a_i, s_j)$.

The difference between making a decision under risk and under uncertainty is that in the case of uncertainty, the probability distribution associated with the states $s_j, j = 1, 2, \ldots, n$, is either unknown or cannot be determined. This lack of information has led to the development of the following criteria for analyzing the decision problem:

1. Laplace
2. Minimax
3. Savage
4. Hurwicz

These criteria differ in how conservative the decision maker is in the face of uncertainty.

The **Laplace** criterion is based on the **principle of insufficient reason**. Because the probability distributions are not known, there is no reason to believe that the probabilities associated with the states of nature are different. The alternatives are thus evaluated using the *optimistic* assumption that all states are equally likely to occur—that is, $P\{s_1\} = P\{s_2\} = \cdots = P\{s_n\} = \frac{1}{n}$. Given that payoff $v(a_i, s_j)$ represents gain, the best alternative is the one that yields

$$\max_{a_i}\left\{\frac{1}{n}\sum_{j=1}^{n} v(a_i, s_j)\right\}$$

If $v(a_i, s_j)$ represents loss, then minimization replaces maximization.

The **maximin (minimax)** criterion is based on the conservative attitude of making the best of the worst possible conditions. If $v(a_i, s_j)$ is loss, then we select the action that corresponds to the *minimax* criterion

$$\min_{a_i}\left\{\max_{s_j} v(a_i, s_j)\right\}$$

If $v(a_i, s_j)$ is gain, we use the *maximin* criterion given by

$$\max_{a_i}\left\{\min_{s_j} v(a_i, s_j)\right\}$$

The **Savage regret** criterion aims at moderating conservatism in the *minimax (maximin)* criterion by replacing the (gain or loss) payoff matrix $v(a_i, s_j)$ with a *loss* (or regret) $r(a_i, s_j)$ matrix, using the following transformation:

$$r(a_i, s_j) = \begin{cases} v(a_i, s_j) - \min_{a_k}\{v(a_k, s_j)\}, \text{ if } v \text{ is loss} \\ \max_{a_k}\{v(a_k, s_j)\} - v(a_i, s_j), \text{ if } v \text{ is gain} \end{cases}$$

To show why the Savage criterion "moderates" the minimax (maximin) criterion, consider the following *loss*, $v(a_i, s_j)$, matrix

	s_1	s_2	Row max
a_1	\$11,000	\$90	\$11,000
a_2	\$10,000	\$10,000	**\$10,000** ← Minimax

The application of the minimax criterion shows that a_2, with a definite loss of \$10,000, is preferable. However, we may choose a_1, because there is a chance of limiting the loss to \$90 only if s_2 is realized.

Let us see what happens if we use the following regret, $r(a_i, v_j)$, matrix instead:

	s_1	s_2	Row max
a_1	\$1000	\$0	**\$1000** ← Minimax
a_2	\$0	\$9910	\$9910

The minimax criterion, when applied to the regret matrix, will select a_1, as desired.

The last test to be considered is **Hurwicz** criterion, which is designed to reflect decision-making attitudes, ranging from the most optimistic to the most pessimistic (or conservative). Define $0 \leq \alpha \leq 1$, and assume that $v(a_i, s_j)$ represents gain. Then the selected action must be associated with

$$\max_{a_i}\left\{\alpha \max_{s_j} v(a_i, s_j) + (1 - \alpha)\min_{s_j} v(a_i, s_j)\right\}$$

The parameter α is called the **index of optimism**. If $\alpha = 0$, the criterion is conservative because it applies the regular minimax criterion. If $\alpha = 1$, the criterion produces optimistic results because it seeks *the best of the best* conditions. We can adjust the degree of optimism (or pessimism) through a proper selection of the value of α in the specified $(0, 1)$ range. In the absence of strong feeling regarding optimism and pessimism, $\alpha = .5$ may be an appropriate choice.

If $v(a_i, s_j)$ represents loss, then the criterion is changed to

$$\min_{a_i}\left\{\alpha \min_{s_j} v(a_i, s_j) + (1 - \alpha)\max_{s_j} v(a_i, s_j)\right\}$$

Example 13.3-1

National Outdoors School (NOS) is preparing a summer campsite in the heart of Alaska to train individuals in wilderness survival. NOS estimates that attendance can fall into one of four categories: 200, 250, 300, and 350 persons. The cost of the campsite will be the smallest when its size meets the demand exactly. Deviations above or below the ideal demand levels incur additional costs resulting from building surplus (unused) capacity or losing income opportunities when the demand is not met. Letting a_1 to a_4 represent the sizes of the campsites (200, 250, 300, and 350 persons) and s_1 to s_4 the level of attendance, the following table summarizes the cost matrix (in thousands of dollars) for the situation.

	s_1	s_2	s_3	s_4
a_1	5	10	18	25
a_2	8	7	12	23
a_3	21	18	12	21
a_4	30	22	19	15

The problem is analyzed using all four criteria.

Laplace. Given $P\{s_j\} = \frac{1}{4}$, $j = 1$ to 4, the expected values for the different actions are computed as

$$E\{a_1\} = \tfrac{1}{4}(5 + 10 + 18 + 25) = \$14{,}500$$

$$E\{a_2\} = \tfrac{1}{4}(8 + 7 + 12 + 23) = \mathbf{\$12{,}500} \leftarrow \text{Optimum}$$

$$E\{a_3\} = \tfrac{1}{4}(21 + 18 + 12 + 21) = \$18{,}000$$

$$E\{a_4\} = \tfrac{1}{4}(30 + 22 + 19 + 15) = \$21{,}500$$

Minimax. The minimax criterion produces the following matrix:

	s_1	s_2	s_3	s_4	Row max
a_1	5	10	18	25	25
a_2	8	7	12	23	23
a_3	21	18	12	21	**21** ← Minimax
a_4	30	22	19	15	30

Savage. The regret matrix is determined by subtracting 5, 7, 12, and 15 from columns 1 to 4, respectively. Thus,

	s_1	s_2	s_3	s_4	Row max
a_1	0	3	6	10	10
a_2	3	0	0	8	**8** ← Minimax
a_3	16	11	0	6	16
a_4	25	15	7	0	25

Hurwicz. The following table summarizes the computations.

Alternative	Row min	Row max	α(Row min) + $(1 - \alpha)$(Row max)
a_1	5	25	$25 - 20\alpha$
a_2	7	23	$23 - 16\alpha$
a_3	12	21	$21 - 9\alpha$
a_4	15	30	$30 - 15\alpha$

Using an appropriate α, we can determine the optimum alternative. For example, at $\alpha = .5$, either a_1 or a_2 will yield the optimum, and at $\alpha = .25$, a_3 is the optimum.

Excel Moment

Template excelUncertainty.xls can be used to automate the computations of Laplace, maximin, Savage, and Hurwicz criteria. The spreadsheet assumes a *cost* matrix. To use a reward matrix, all entries must be multiplied by -1. Figure 13.8 demonstrates the application of the template to Example 13.3-1. The maximum matrix size is (10×10) (columns F:K are hidden to conserve space).

	A	B	C	D	E	L	M	N	O
1	Decision Under Uncertainty								
2	Enter x to select method:					Output Results			
3		Laplace	x						
4		Minimax	x						
5		Savage	x						
6		Hurwicz	x	Alpha=	0.5	Optimum strategies			
7	Input (cost) Matrix: Maximum size = (10x10)					a2	a3	a2	a2
8		s1	s2	s3	s4	Laplace	Minimax	Savage	Hurwicz
9	a1	5	10	18	25	14.5	25	10	15
10	a2	8	7	12	23	12.5	23	8	15
11	a3	21	18	12	21	18	21	16	16.5
12	a4	30	22	19	15	21.5	30	25	22.5
13									
14									
15									
16									
17									
18									
19									

FIGURE 13.8

Excel solution of Example 13.3-1 (file excelUncertainty.xls)

PROBLEM SET 13.3A

***1.** Hank is an intelligent student and usually makes good grades, provided that he can review the course material the night before the test. For tomorrow's test, Hank is faced with a small problem. His fraternity brothers are having an all-night party in which he would like to participate. Hank has three options:

$$a_1 = \text{Party all night}$$

$$a_2 = \text{Divide the night equally between studying and partying}$$

$$a_3 = \text{Study all night}$$

Tomorrow's exam can be easy (s_1), moderate (s_2), or tough (s_3), depending on the professor's unpredictable mood. Hank anticipates the following scores:

	s_1	s_2	s_3
a_1	85	60	40
a_2	92	85	81
a_3	100	88	82

(a) Recommend a course of action for Hank (based on each of the four criteria of decisions under uncertainty).

(b) Suppose that Hank is more interested in the letter grade he will get. The dividing scores for the passing letter grades A to D are 90, 80, 70, and 60, respectively. Would this attitude toward grades call for a change in Hank's course of action?

2. For the upcoming planting season, Farmer McCoy can plant corn (a_1), plant wheat (a_2), plant soybeans (a_3), or use the land for grazing (a_4). The payoffs associated with the different actions are influenced by the amount of rain: heavy rainfall (s_1), moderate rainfall (s_2), light rainfall (s_3), or drought season (s_4).

The payoff matrix (in thousands of dollars) is estimated as

	s_1	s_2	s_3	s_4
a_1	−20	60	30	−5
a_2	40	50	35	0
a_3	−50	100	45	−10
a_4	12	15	15	10

Develop a course of action for Farmer McCoy.

3. One of N machines must be selected for manufacturing Q units of a specific product. The minimum and maximum demands for the product are Q^* and Q^{**}, respectively. The total production cost for Q items on machine i involves a fixed cost K_i and a variable cost per unit c_i, and is given as

$$TC_i = K_i + c_i Q$$

(a) Devise a solution for the problem under each of the four criteria of decisions under uncertainty.

(b) For $1000 \leq Q \leq 4000$, solve the problem for the following set of data:

Machine i	K_i (\$)	C_i (\$)
1	100	5
2	40	12
3	150	3
4	90	8

13.4 GAME THEORY

Game theory deals with decision situations in which two *intelligent* opponents with conflicting objectives are trying to outdo one another. Typical examples include launching advertising campaigns for competing products and planning strategies for warring armies.

In a game conflict, two opponents, known as **players**, will each have a (finite or infinite) number of alternatives or **strategies**. Associated with each pair of strategies is a **payoff** that one player receives from the other. Such games are known as **two-person zero-sum games** because a gain by one player signifies an equal loss to the other. It suffices, then, to summarize the game in terms of the payoff to one player. Designating the two players as A and B with m and n strategies, respectively, the game is usually represented by the payoff matrix to player A as

	B_1	B_2	$\dots$	B_n
A_1	a_{11}	a_{12}	$\dots$	a_{1m}
A_2	a_{21}	a_{22}	$\dots$	a_{2m}
$\vdots$	$\vdots$	$\vdots$	$\vdots$	$\vdots$
A_m	a_{m1}	a_{m2}	$\dots$	a_{mn}

The representation indicates that if A uses strategy i and B uses strategy j, the payoff to A is a_{ij}, which means that the payoff to B is $-a_{ij}$.

Real-Life Application—Ordering Golfers on the Final Day of Ryder Cup Matches

In the final day of a golf tournament, two teams compete for the championship. Each team captain must submit an ordered list of golfers (a *slate*) that automatically determines the matches. It is plausible to assume that if two competing players occupy the same order in their respective slates then there is 50-50 chance that either golfer will win the match. This probability will increase when a higher-order golfer is matched with a lower-order one. The goal is to develop an analytical procedure that will support or refute the idea of using slates. Case 12, Chapter 24 on the CD provides details on the study.

13.4.1 Optimal Solution of Two-Person Zero-Sum Games

Because games are rooted in conflict of interest, the optimal solution selects one or more strategies for each player such that any change in the chosen strategies does not improve the payoff to either player. These solutions can be in the form of a single pure strategy or several strategies mixed according to specific probabilities. The following two examples demonstrate the two cases.

Example 13.4-1

Two companies, A and B, sell two brands of flu medicine. Company A advertises in radio (A_1), television (A_2), and newspapers (A_3). Company B, in addition to using radio (B_1), television (B_2), and newspapers (B_3), also mails brochures (B_4). Depending on the effectiveness of each advertising campaign, one company can capture a portion of the market from the other. The following matrix summarizes the percentage of the market captured or lost by company A.

	B_1	B_2	B_3	B_4	Row min
A_1	8	−2	9	−3	−3
A_2	6	**5**	6	8	**5** ← Maximin
A_3	−2	4	−9	5	−9
Column max	8	**5**	9	8	
		↑ Minimax			

The solution of the game is based on the principle of securing the *best of the worst* for each player. If Company A selects strategy A_1, then regardless of what B does, the worst that can happen is that A loses 3% of the market share to B. This is represented by the minimum value of the entries in row 1. Similarly, the strategy A_2 worst outcome is for A to capture 5% of the market from B, and the strategy A_3 worst outcome is for A to lose 9% to B. These result are listed in the "row min" column. To achieve the *best* of the *worst*, Company A chooses strategy A_2 because it corresponds to the maximin value, or the largest element in the "row min" column.

Next, consider Company B's strategy. Because the given payoff matrix is for A, *B's best of the worst* criterion requires determining the minimax value. The result is that Company B should select strategy B_2.

The optimal solution of the game calls for selecting strategies A_2 and B_2, which means that both companies should use television advertising. The payoff will be in favor of company A, because its market share will increase by 5%. In this case, we say that the **value of the game** is 5%, and that A and B are using a **saddle-point** solution.

The saddle-point solution precludes the selection of a better strategy by either company. If B moves to another strategy $(B_1, B_3, \text{or } B_4)$, Company A can stay with strategy A_2, which ensures that B will lose a worse share of the market (6% or 8%). By the same token, A does not want to use a different strategy because if A moves to strategy A_3, B can move to B_3 and realize a 9% increase in market share. A similar conclusion is realized if A moves to A_1, as B can move to B_4 and realize a 3% increase in market share.

The optimal saddle-point solution of a game need not be a pure strategy. Instead, the solution may require mixing two or more strategies randomly, as the following example illustrates.

Example 13.4-2

Two players, A and B, play the coin-tossing game. Each player, unbeknownst to the other, chooses a head (H) or a tail (T). Both players would reveal their choices simultaneously. If they match (HH or TT), player A receives \$1 from B. Otherwise, A pays B \$1.

The following payoff matrix for player A gives the row-min and the column-max values corresponding to A's and B's strategies, respectively.

	B_H	B_T	Row min
A_H	1	−1	−1
A_T	−1	1	−1
Column max	1	1	

The maximin and the minimax values of the games are $-\$1$ and \$1, respectively. Because the two values are not equal, the game does not have a pure strategy solution. In particular, if A_H is used by player A, player B will select B_T to receive \$1 from A. If this happens, A can move to strategy A_T to reverse the outcome of the game by receiving \$1 from B. The constant temptation to switch to another strategy shows that a pure strategy solution is not acceptable. Instead, both players can randomly mix their respective pure strategies. In this case, the optimal value of the game will occur somewhere between the maximin and the minimax values of the game—that is,

$$\text{maximin (lower) value} \leq \text{value of the game} \leq \text{minimax (upper) value}$$

(See Problem 5, Set 13.4a.) Thus, in the coin-tossing example, the value of the game must lie between $-\$1$ and $+\$1$.

PROBLEM SET 13.4A

1. Determine the saddle-point solution, the associated pure strategies, and the value of the game for each of the following games. The payoffs are for player A.

*(a)

	B_1	B_2	B_3	B_4
A_1	8	6	2	8
A_2	8	9	4	5
A_3	7	5	3	5

(b)

	B_1	B_2	B_3	B_4
A_1	4	−4	−5	6
A_2	−3	−4	−9	−2
A_3	6	7	−8	−9
A_4	7	3	−9	5

2. The following games give A's payoff. Determine the values of p and q that will make the entry $(2, 2)$ of each game a saddle point:

(a)

	B_1	B_2	B_3
A_1	1	q	6
A_2	p	5	10
A_3	6	2	3

(b)

	B_1	B_2	B_3
A_1	2	4	5
A_2	10	7	q
A_3	4	p	6

3. Specify the range for the value of the game in each of the following cases, assuming that the payoff is for player A:

*(a)

	B_1	B_2	B_3	B_4
A_1	1	9	6	0
A_2	2	3	8	4
A_3	−5	−2	10	−3
A_4	7	4	−2	−5

(b)

	B_1	B_2	B_3	B_4
A_1	−1	9	6	8
A_2	−2	10	4	6
A_3	5	3	0	7
A_4	7	−2	8	4

(c)

	B_1	B_2	B_3
A_1	3	6	1
A_2	5	2	3
A_3	4	2	−5

(d)

	B_1	B_2	B_3	B_4
A_1	3	7	1	3
A_2	4	8	0	−6
A_3	6	−9	−2	4

4. Two companies promote two competing products. Currently, each product controls 50% of the market. Because of recent improvements in the two products, each company is preparing to launch an advertising campaign. If neither company advertises, equal market

shares will continue. If either company launches a stronger campaign, the other is certain to lose a proportional percentage of its customers. A survey of the market shows that 50% of potential customers can be reached through television, 30% through newspapers, and 20% through radio.

(a) Formulate the problem as a two-person zero-sum game, and select the appropriate advertising media for each company.

(b) Determine a range for the value of the game. Can each company operate with a single pure strategy?

5. Let a_{ij} be the (i, j)th element of a payoff matrix with m strategies for player A and n strategies for player B. The payoff is for player A. Prove that

$$\max_i \min_j a_{ij} \le \min_j \max_i a_{ij}$$

13.4.2 Solution of Mixed Strategy Games

Games with mixed strategies can be solved either graphically or by linear programming. The graphical solution is suitable for games in which at least one player has exactly two pure strategies. The method is interesting because it explains the idea of a saddle point graphically. Linear programming can be used to solve any two-person zero-sum game.

Graphical Solution of Games. We start with the case of $(2 \times n)$ games in which player A has two strategies.

		y_1 B_1	y_2 B_2	$\cdots$	y_n B_n
x_1:	A_1	a_{11}	a_{12}	$\cdots$	a_{1m}
$1 - x_1$:	A_2	a_{21}	a_{22}	$\cdots$	a_{2m}

The game assumes that player A mixes strategies A_1 and A_2 with the respective probabilities x_1 and $1 - x_1, 0 \le x_1 \le 1$. Player B mixes strategies B_1 to B_n with the probabilities $y_1, y_2, \ldots,$ and y_n, where $y_j \ge 0$ for $j = 1, 2, \ldots, n$, and $y_1 + y_2 + \cdots + y_n = 1$. In this case, A's expected payoff corresponding to B's jth pure strategy is computed as

$$(a_{1j} - a_{2j})x_1 + a_{2j}, j = 1, 2, \ldots, n$$

Player A thus seeks to determine the value of x_1 that maximizes the minimum expected payoffs—that is,

$$\max_{x_1} \min_j \{(a_{1j} - a_{2j})x_1 + a_{2j}\}$$

Example 13.4-3

Consider the following 2 × 4 game. The payoff is for player A.

	B_1	B_2	B_3	B_4
A_1	2	2	3	−1
A_2	4	3	2	6

The game has no pure strategy solution. A's expected payoffs corresponding to B's pure strategies are given as

B's pure strategy	A's expected payoff
1	$-2x_1 + 4$
2	$-x_1 + 3$
3	$x_1 + 2$
4	$-7x_1 + 6$

Figure 13.9 provides TORA plot of the four straight lines associated with B's pure strategies (file toraEx13.4-3.txt).[3] To determine the *best of the worst*, the lower envelope of the four lines (delineated by vertical stripes) represents the minimum (worst) expected payoff for A regardless of what B does. The maximum (best) of the lower envelope corresponds to the maximin solution point at $x_1^* = .5$. This point is the intersection of lines associated with strategies B_3 and B_4. Player A's optimal solution thus calls for mixing A_1 and A_2 with probabilities .5, and .5, respectively.

FIGURE 13.9

TORA graphical solution of the two-person zero-sum game of Example 13.4-3 (file toraEx13.4-3.txt)

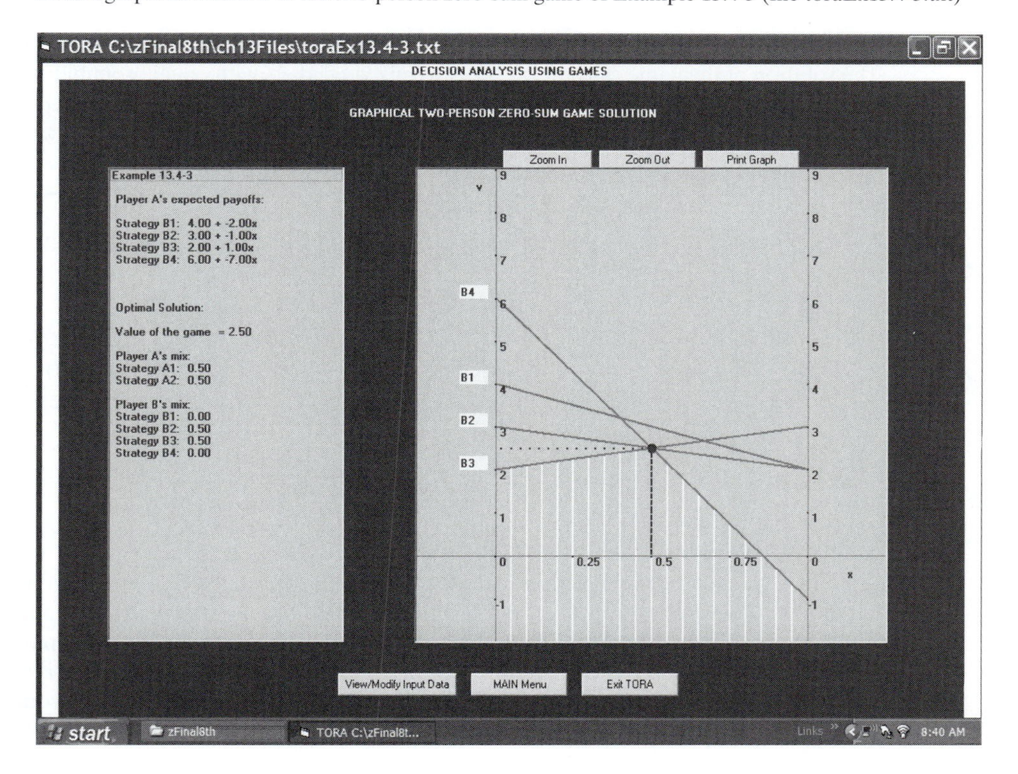

[3]From Main menu select Zero-sum Games and input the problem data, then select Graphical from the SOLVE/MODIFY menu.

The corresponding value of the game, v, is determined by substituting $x_1 = .5$ in either of the functions for lines 3 and 4, which gives

$$v = \begin{cases} \frac{1}{2} + 2 = \frac{5}{2}, & \text{from line 3} \\ -7\left(\frac{1}{2}\right) + 6 = \frac{5}{2}, & \text{from line 4} \end{cases}$$

Player B's optimal mix is determined by the two strategies that define the lower envelope of the graph. This means that B can mix strategies B_3 and B_4, in which case $y_1 = y_2 = 0$ and $y_4 = 1 - y_3$. As a result, B's expected payoffs corresponding to A's pure strategies are given as

A's pure strategy	B's expected payoff
1	$4y_3 - 1$
2	$-4y_3 + 6$

The *best of the worst* solution for B is the minimum point on the *upper* envelope of the given two lines (you will find it instructive to graph the two lines and identify the upper envelope). This process is equivalent to solving the equation

$$4y_3 - 1 = -4y_3 + 6$$

The solution gives $y_3 = \frac{7}{8}$, which yields the value of the game as $v = 4 \times \left(\frac{7}{8}\right) - 1 = \frac{5}{2}$.

The solution of the game calls for player A to mix A_1 and A_2 with equal probabilities and for player B to mix B_3 and B_4 with probabilities $\frac{7}{8}$ and $\frac{1}{8}$. (Actually, the game has alternative solutions for B, because the maximin point in Figure 13.9 is determined by more than two lines. Any nonnegative combination of these alternative solutions is also a legitimate solution.)

Remarks. Games in which player A has m strategies and player B has only two can be treated similarly. The main difference is that we will be plotting B's expected payoff corresponding to A's pure strategies. As a result, we will be seeking the minimax, rather than the maximin, point of the *upper envelope* of the plotted lines. However, to solve the problem with TORA, it is necessary to express the payoff in terms of the player that has two strategies by multiplying the payoff matrix by -1, if necessary.

PROBLEM SET 13.4B[4]

*1. Solve the coin-tossing game of Example 13.4-2 graphically.

*2. Robin, who travels frequently between two cities, has two route options: Route A is a fast four-lane highway, and route B is a long winding road. The highway patrol has a limited police force. If the full force is allocated to either route, Robin, with her passionate desire for driving "superfast," is certain to receive a $100 speeding ticket. If the force is split 50-50 between the two routes, there is a 50% chance she will get a $100 ticket on route A and only a 30% chance that she will get the same fine on route B. Develop a strategy for both Robin and the police.

[4]The TORA Zero-sum Games module can be used to verify your answer.

3. Solve the following games graphically. The payoff is for Player A.

(a)

	B_1	B_2	B_3
A_1	1	−3	7
A_2	2	4	−6

(b)

	B_1	B_2
A_1	5	8
A_2	6	5
A_3	5	7

4. Consider the following two-person, zero-sum game:

	B_1	B_2	B_3
A_1	5	50	50
A_2	1	1	.1
A_3	10	1	10

(a) Verify that the strategies $\left(\frac{1}{6}, 0, \frac{5}{6}\right)$ for A and $\left(\frac{49}{54}, \frac{5}{54}, 0\right)$ for B are optimal, and determine the value of the game.

(b) Show that the optimal value of the game equals

$$\sum_{i=1}^{3}\sum_{j=1}^{3} a_{ij}x_i y_j$$

Linear Programming Solution of Games. Game theory bears a strong relationship to linear programming, in the sense that a two-person zero-sum game can be expressed as a linear program, and vice versa. In fact, G. Dantzig (1963, p. 24) states that J. von Neumann, father of game theory, when first introduced to the simplex method in 1947, immediately recognized this relationship and further pinpointed and stressed the concept of *duality* in linear programming. This section illustrates the solution of games by linear programming.

Player A's optimal probabilities, $x_1, x_2, \ldots,$ and x_m, can be determined by solving the following maximin problem:

$$\max_{x_i}\left\{\min\left(\sum_{i=l}^{m} a_{i1}x_i, \sum_{i=1}^{m} a_{i2}x_i, \ldots, \sum_{i=1}^{m} a_{in}x_i\right)\right\}$$

$$x_1 + x_2 + \cdots + x_m = 1$$

$$x_i \geq 0, i = 1, 2, \ldots, m$$

Now, let

$$v = \min\left\{\sum_{i=1}^{m} a_{i1}x_i, \sum_{i=1}^{m} a_{i2}x_i, \ldots, \sum_{i=1}^{m} a_{in}x_i\right\}$$

The equation implies that

$$\sum_{i=1}^{m} a_{ij}x_i \geq v, j = 1, 2, \ldots, n$$

Player A's problem thus can be written as

$$\text{Maximize } z = v$$

subject to

$$v - \sum_{i=1}^{m} a_{ij}x_i \leq 0, j = 1, 2, \ldots, n$$

$$x_1 + x_2 + \cdots + x_m = 1$$

$$x_i \geq 0, i = 1, 2, \ldots, m$$

$$v \text{ unrestricted}$$

Note that the value of the game, v, is unrestricted in sign.

Player B's optimal strategies, $y_1, y_2, \ldots,$ and y_n, are determined by solving the problem

$$\min_{y_j}\left\{\max\left(\sum_{j=1}^{n} a_{1j}y_j, \sum_{j=1}^{n} a_{2j}y_j, \ldots, \sum_{j=1}^{n} a_{mj}y_j\right)\right\}$$

$$y_1 + y_2 + \cdots + y_n = 1$$

$$y_j \geq 0, j = 1, 2, \ldots, n$$

Using a procedure similar to that of player A, B's problem reduces to

$$\text{Minimize } w = v$$

subject to

$$v - \sum_{j=1}^{n} a_{ij}y_j \geq 0, i = 1, 2, \ldots, m$$

$$y_1 + y_2 + \cdots + y_n = 1$$

$$y_j \geq 0, j = 1, 2, \ldots, n$$

$$v \text{ unrestricted}$$

The two problems optimize the same (unrestricted) variable v, the value of the game. The reason is that B's problem is the dual of A's problem (verify this claim using the definition of duality in Chapter 4). This means that the optimal solution of one problem automatically yields the optimal solution of the other.

Example 13.4-4

Solve the following game by linear programming.

	B_1	B_2	B_3	Row min
A_1	3	−1	−3	−3
A_2	−2	4	−1	**−2**
A_3	−5	−6	2	−6
Column max	3	4	**2**	

The value of the game, v, lies between -2 and 2.

Player A's Linear Program

$$\text{Maximize } z = v$$

subject to

$$\begin{aligned}
v - 3x_1 + 2x_2 + 5x_3 &\leq 0 \\
v + x_1 - 4x_2 + 6x_3 &\leq 0 \\
v + 3x_1 + x_2 - 2x_2 &\leq 0 \\
x_1 + x_2 + x_3 &= 1 \\
x_1, x_2, x_3 &\geq 0 \\
v \text{ unrestricted}
\end{aligned}$$

The optimum solution[5] is $x_1 = .39$, $x_2 = .31$, $x_3 = .29$, and $v = -0.91$.

Player B's Linear Program

$$\text{Minimize } z = v$$

subject to

$$\begin{aligned}
v - 3y_1 + y_2 + 3y_3 &\geq 0 \\
v + 2y_1 - 4y_2 + y_3 &\geq 0 \\
v + 5y_1 + 6y_2 - 2y_3 &\geq 0 \\
y_1 + y_2 + y_3 &= 1 \\
v \text{ unrestricted}
\end{aligned}$$

The solution yields $y_1 = .32$, $y_2 = .08$, $y_3 = .60$, and $v = -0.91$.

PROBLEM SET 13.4C

1. On a picnic outing, 2 two-person teams are playing hide-and-seek. There are four hiding locations (A, B, C, and D), and the two members of the hiding team can hide separately in any two of the four locations. The other team will then have the chance to search any two locations. The searching team gets a bonus point if they find both members of the hiding team. If they miss both, they lose a point. Otherwise, the outcome is a draw.
   *(a) Set up the problem as a two-person zero-sum game.
   (b) Determine the optimal strategy and the value of the game.
2. UA and DU are setting up their strategies for the 1994 national championship college basketball game. Assessing the strengths of their respective "benches," each coach comes up with four strategies for rotating his players during the game. The ability of each team to score 2-pointers, 3-pointers, and free throws is a key factor in determining the final

[5]TORA Zero-sum Games ⇒ Solve ⇒ LP-based can be used to solve any two-person zero-sum game.

score of the game. The following table summarizes the net points UA will score per possession as a function of the different strategies available to each team:

	DU_1	DU_2	DU_3	DU_4
UA_1	3	−2	1	2
UA_2	2	3	−3	0
UA_3	−1	2	−2	2
UA_4	−1	−2	4	1

(a) Solve the game by linear programming and determine a strategy for the championship game.

(b) Based on the given information, which of the two teams is projected to win the championship?

(c) Suppose that the entire game will have a total of 60 possessions (30 for each team). Predict the expected number of points by which the championship will be won.

3. Colonel Blotto's army is fighting for the control of two strategic locations. Blotto has two regiments and the enemy has three. A location will fall to the army that attacks with more regiments. Otherwise, the result of the battle is a draw.

*__(a)__ Formulate the problem as a two-person zero-sum game, and solve by linear programming.

(b) Which army will win the battle?

4. In the two-player, two-finger Morra game, each player shows one or two fingers, and simultaneously guesses the number of fingers the opponent will show. The player making the correct guess wins an amount equal to the total number of fingers shown. Otherwise, the game is a draw. Set up the problem as a two-person zero-sum game, and solve by linear programming.

REFERENCES

Chen, S., and Hwang, C., *Fuzzy Multiple Attribute Decision Making*, Springer-Verlag, Berlin, 1992.

Clemen, R. J., and Reilly, T., *Making Hard Decisions: An Introduction to Decision Analysis*, 2nd ed., Duxbury, Pacific Grove, CA, 1996.

Cohan, D., S. Haas, D. Radloff, and R. Yancik, "Using Fire in Forest Management: Decision Making under Uncertainty," *Interfaces*, Vol. 14, No. 5, pp. 8–19, 1984.

Dantzig, G. B., *Linear Programming and Extensions*, Princeton University Press, Princeton, NJ, 1963.

Meyerson, R., *Game Theory: Analysis of Conflict*, Harvard University Press, Cambridge, MA, 1991.

Rapport, A., "Sensitivity Analysis in Decision Making," *The Accounting Review*, Vol. 42, No. 3, pp. 441–456, 1967.

Saaty, T. L., *Fundamentals of Decision Making*, RWS Publications, Pittsburgh, 1994.

Zahedi, F., "The Analytic Hierarchy Process–A Survey of the Method and its Applications," *Intefaces*, Vol. 16, No. 4, pp. 96–108, 1986.

Zeleny, M., *Multiple Criteria Decision Making*, McGraw-Hill, New York, 1982.

CHAPTER 14

Probabilistic Inventory Models

Chapter Guide. This chapter is a continuation of the material in Chapter 11 on deterministic inventory models. It deals with inventory situations in which the demand is probabilistic. The developed models are categorized broadly under *continuous* and *periodic* review situations. The periodic review models include both single-period and multiperiod cases. The proposed solutions range from the use of a probabilistic version of the deterministic EOQ to more complex situations solved by dynamic programming. It may appear that the probabilistic models presented here are "too theoretical" to be practical. But, in fact, a case analysis in Chapter 24 on the CD uses one of these models to help Dell, Inc. manage its inventory situation and realize sizable savings.

This chapter includes a summary of 1 real-life application, 4 solved examples, 1 Excel template, 22 end-of-section problems, and 2 cases. The cases are in Appendix E on the CD. The AMPL/Excel/Solver/TORA programs are in folder ch14Files.

Real-Life Application—Inventory Decisions in Dell's Supply Chain

Dell, Inc., implements a direct-sales business model in which personal computers are sold directly to customers in the United States. When an order arrives from a customer, the specifications are sent to a manufacturing plant in Austin, Texas, where the computer is built, tested, and packaged in about eight hours. Dell carries little inventory. Instead, its suppliers, normally located in Southeast Asia, are required to keep what is known as "revolving" inventory on hand in *revolvers* (warehouses) near the manufacturing plants. These revolvers are owned by Dell and leased to the suppliers. Dell then "pulls" parts as needed from the revolvers, and it is the suppliers' responsibility to replenish the inventory to meet Dell's forecasted demand. Although Dell does not own the inventory in the revolvers, its cost is indirectly passed on to customers through component pricing. Thus, any reduction in inventory directly benefits Dell's customers by reducing product prices. The proposed solution has resulted in an estimated $2.7 million in annual savings. Case 13 in Chapter 24 on the CD provides the details of the study.

14.1 CONTINUOUS REVIEW MODELS

This section presents two models: (1) a "probabilitized" version of the deterministic EOQ (Section 11.2.1) that uses a buffer stock to account for probabilistic demand, and (2) a more exact probabilistic EOQ model that includes the probabilistic demand directly in the formulation.

14.1.1 "Probabilitized" EOQ Model

Some practitioners have sought to adapt the deterministic EOQ model (Section 11.2.1) to reflect the probabilistic nature of demand by using an approximation that superimposes a constant buffer stock on the inventory level throughout the entire planning horizon. The size of the buffer is determined such that the probability of running out of stock *during lead time* (the period between placing and receiving an order) does not exceed a prespecified value.

Let

L = Lead time between placing and receiving an order

x_L = Random variable representing demand during lead time

μ_L = Average demand during lead time

σ_L = Standard deviation of demand during lead time

B = Buffer stock size

α = Maximum allowable probability of running out of stock during lead time

The main assumption of the model is that the demand, x_L, during lead time L is normally distributed with mean μ_L and standard deviation σ_L—that is, $N(\mu_L, \sigma_L)$.

Figure 14.1 depicts the relationship between the buffer stock, B, and the parameters of the deterministic EOQ model that include the lead time L, the average demand

FIGURE 14.1

Buffer stock imposed on the classical EOQ model

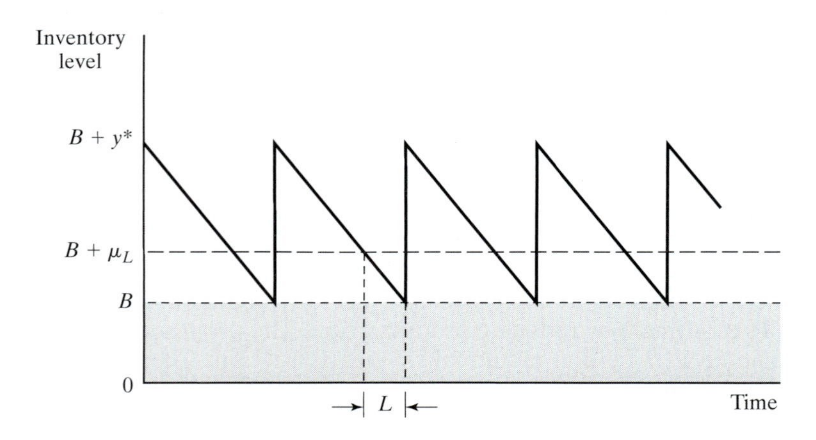

during lead time, μ_L, and the EOQ, y^*. Note that L must equal the *effective* lead time as defined in Section 11.2.1.

The probability statement used to determine B can be written as

$$P\{x_L \geq B + \mu_L\} \leq \alpha$$

We can convert x_L into a standard $N(0, 1)$ random variable by using the following substitution (see Section 12.5.4):

$$z = \frac{x_L - \mu_L}{\sigma_L}$$

Thus, we have

$$P\left\{z \geq \frac{B}{\sigma_L}\right\} \leq \alpha$$

Figure 14.2 defines K_α (which is determined from the standard normal tables in Appendix B or by using file excelStatTables.xls) such that

$$P\{z \geq K_\alpha\} = \alpha$$

Hence, the buffer size must satisfy

$$B \geq \sigma_L K_\alpha$$

The demand during the lead time L usually is described by a probability density function *per unit time* (e.g., per day or week), from which the distribution of the demand during L can be determined. Given that the demand per unit time is normal with mean D and standard deviation σ, the mean and standard deviation, μ_L and σ_L, of demand during lead time, L, are computed as

$$\mu_L = DL$$
$$\sigma_L = \sqrt{\sigma^2 L}$$

The formula for σ_L requires L to be (rounded to) an integer value.

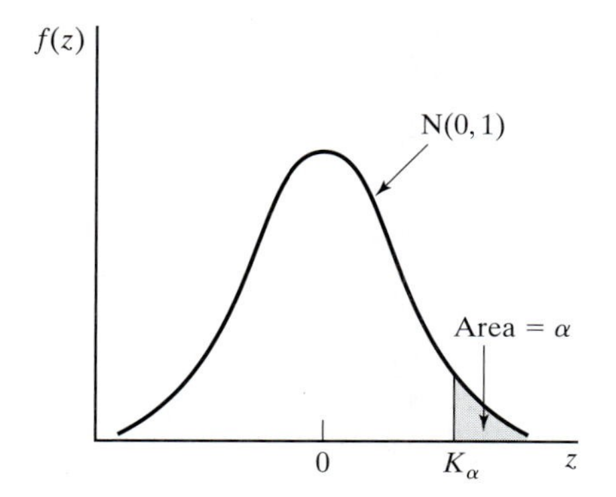

FIGURE 14.2
Probability of running out of stock, $P\{z \geq K_\alpha\} = \alpha$

Example 14.1-1

In Example 11.2-1 dealing with determining the inventory policy of neon lights, EOQ = 1000 *units*. If the *daily* demand is normal with mean $D = 100$ lights and standard deviation $\sigma = 10$ lights—that is, $N(100, 10)$—determine the buffer size so that the probability of running out of stock is below $\alpha = .05$.

From Example 11.2-1, the effective lead time is $L = 2$ days. Thus,

$$\mu_L = DL = 100 \times 2 = 200 \text{ units}$$

$$\sigma_L = \sqrt{\sigma^2 L} = \sqrt{10^2 \times 2} = 14.14 \text{ units}$$

Given $K_{.05} = 1.645$, the buffer size is computed as

$$B \geq 14.14 \times 1.645 \approx 23 \text{ neon lights}$$

Thus, the optimal inventory policy with buffer B calls for ordering 1000 units whenever the inventory level drops to 223 ($= B + \mu_L = 23 + 2 \times 100$) units.

PROBLEM SET 14.1A

1. In Example 14.1-1, determine the optimal inventory policy for each of the following cases:
   *(a) Lead time = 15 days.
   (b) Lead time = 23 days.
   (c) Lead time = 8 days.
   (d) Lead time = 10 days.
2. A music store sells a best-selling compact disc. The daily demand (in number of units) for the disc is approximately normally distributed with mean 200 discs and standard deviation 20 discs. The cost of keeping the discs in the store is $.04 per disc per day. It costs the store $100 to place a new order. There is a 7-day lead time for delivery. Assuming that the store wants to limit the probability of running out of discs during the lead time to no more than .02, determine the store's optimal inventory policy.
3. The daily demand for camera films at a gift shop in a resort area is normally distributed with mean 300 rolls and standard deviation 5 rolls. The cost of holding a roll in the shop is $.02. A fixed cost of $30 is incurred each time a new order of films is placed by the shop. The shop's inventory policy calls for ordering 150 rolls whenever the inventory level drops to 80 units while simultaneously maintaining a constant buffer of 20 rolls at all times.
   (a) For the stated inventory policy, determine the probability of running out of stock during lead time.
   (b) Given the data of the situation, recommend an inventory policy for the shop, assuming that the probability of running out of films during the lead time does not exceed .10.

14.1.2 Probabilistic EOQ Model

There is no reason to believe that the "probabilitized" EOQ model in Section 14.1.1 will produce an optimal inventory policy. The fact that pertinent information regarding the probabilistic nature of demand is initially ignored, only to be "revived" in a totally independent manner at a later stage of the calculations, is sufficient to refute optimality.

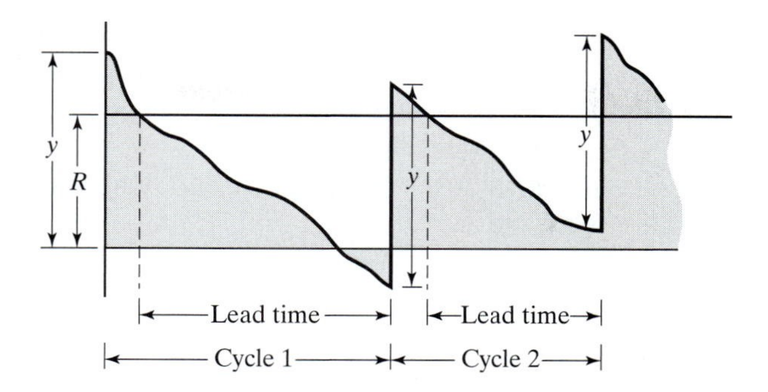

FIGURE 14.3

Probabilistic inventory model with shortage

To remedy the situation, a more accurate model is presented in which the probabilistic nature of the demand is included directly in the formulation of the model.

Unlike the case in Section 14.1.1, the new model allows shortage of demand, as Figure 14.3 demonstrates. The policy calls for ordering the quantity y whenever the inventory drops to level R. As in the deterministic case, the reorder level R is a function of the lead time between placing and receiving an order. The optimal values of y and R are determined by minimizing the expected cost per unit time that includes the sum of the setup, holding, and shortage costs.

The model has three assumptions.

1. Unfilled demand during lead time is backlogged.
2. No more than one outstanding order is allowed.
3. The distribution of demand during lead time remains stationary (unchanged) with time.

To develop the total cost function per unit time, let

$$\begin{aligned} f(x) &= \text{pdf of demand, } x\text{, during lead time} \\ D &= \text{Expected demand per unit time} \\ h &= \text{Holding cost per inventory unit per unit time} \\ p &= \text{Shortage cost per inventory unit} \\ K &= \text{Setup cost per order} \end{aligned}$$

Based on these definitions, the elements of the cost function are now determined.

1. *Setup cost.* The approximate number of orders per unit time is $\frac{D}{y}$, so that the setup cost per unit time is approximately $\frac{KD}{y}$.

2. *Expected holding cost.* The average inventory is

$$I = \frac{(y + E\{R - x\}) + E\{R - x\}}{2} = \frac{y}{2} + R - E\{x\}$$

The formula is based on the average of the beginning and ending expected inventories of a cycle, $y + E\{R - x\}$ and $E\{R - x\}$, respectively. As an approximation, the expression ignores the case where $R - E\{x\}$ may be negative. The expected holding cost per unit time thus equals hI.

3. *Expected shortage cost.* Shortage occurs when $x > R$. Thus, the expected shortage quantity per cycle is

$$S = \int_R^{\infty} (x - R)f(x)\,dx$$

Because p is assumed to be proportional to the shortage quantity only, the expected shortage cost per cycle is pS, and, based on $\frac{D}{y}$ cycles per unit time, the shortage cost per unit time is $\frac{pDS}{y}$.

The resulting total cost function per unit time is

$$\text{TCU}(y, R) = \frac{DK}{y} + h\left(\frac{y}{2} + R - E\{x\}\right) + \frac{pD}{y}\int_R^{\infty} (x - R)f(x)\,dx$$

The solutions for optimal y^* and R^* are determined from

$$\frac{\partial \text{TCU}}{\partial y} = -\left(\frac{DK}{y^2}\right) + \frac{h}{2} - \frac{pDS}{y^2} = 0$$

$$\frac{\partial \text{TCU}}{\partial R} = h - \left(\frac{pD}{y}\right)\int_R^{\infty} f(x)\,dx = 0$$

We thus get

$$y^* = \sqrt{\frac{2D(K + pS)}{h}} \qquad (1)$$

$$\int_{R^*}^{\infty} f(x)\,dx = \frac{hy^*}{pD} \qquad (2)$$

Because y^* and R^* cannot be determined in closed forms from (1) and (2), a numeric algorithm, developed by Hadley and Whitin (1963, pp. 169–174), is used to find the solutions. The algorithm converges in a finite number of iterations, provided a feasible solution exists.

For $R = 0$, (1) and (2) above yield

$$\hat{y} = \sqrt{\frac{2D(K + pE\{x\})}{h}}$$

$$\widetilde{y} = \frac{PD}{h}$$

If $\widetilde{y} \geq \hat{y}$, unique optimal values of y and R exist. The solution procedure recognizes that the smallest value of y^* is $\sqrt{\frac{2KD}{h}}$, which is achieved when $S = 0$.

The steps of the algorithm are

Step 0. Use the initial solution $y_1 = y^* = \sqrt{\frac{2KD}{h}}$, and let $R_0 = 0$. Set $i = 1$, and go to step i.

Step i. Use y_i to determine R_i from Equation (2). If $R_i \approx R_{i-1}$, stop; the optimal solution is $y^* = y_i$, and $R^* = R_i$. Otherwise, use R_i in Equation (1) to compute y_i. Set $i = i + 1$, and repeat step i.

Example 14.1-2

Electro uses resin in its manufacturing process at the rate of 1000 gallons per month. It costs Electro \$100 to place an order for a new shipment. The holding cost per gallon per month is \$2, and the shortage cost per gallon is \$10. Historical data show that the demand during lead time is uniform over the range (0, 100) gallons. Determine the optimal ordering policy for Electro.

Using the symbols of the model, we have

$$D = 1000 \text{ gallons per month}$$
$$K = \$100 \text{ per order}$$
$$h = \$2 \text{ per gallon per month}$$
$$p = \$10 \text{ per gallon}$$
$$f(x) = \tfrac{1}{100},\ 0 \le x \le 100$$
$$E\{x\} = 50 \text{ gallons}$$

First, we need to check whether the problem has a feasible solution. Using the equations for $\hat{y}$ and $\widetilde{y}$ we get

$$\hat{y} = \sqrt{\frac{2 \times 1000(100 + 10 \times 50)}{2}} = 774.6 \text{ gallons}$$
$$\widetilde{y} = \frac{10 \times 1000}{2} = 5000 \text{ gallons}$$

Because $\widetilde{y} \ge \hat{y}$, a unique solution exists for y^* and R^*.

The expression for S is computed as

$$S = \int_R^{100} (x - R)\frac{1}{100}dx = \frac{R^2}{200} - R + 50$$

Using S in Equations (1) and (2), we obtain

$$y_i = \sqrt{\frac{2 \times 1000(100 + 10S)}{2}} = \sqrt{100{,}000 + 10{,}000S} \text{ gallons} \tag{3}$$

$$\int_R^{100} \frac{1}{100}dx = \frac{2y_i}{10 \times 1000}$$

The last equation yields

$$R_i = 100 - \frac{y_i}{50} \quad (4)$$

We now use Equations (3) and (4) to determine the solution.

Iteration 1

$$y_1 = \sqrt{\frac{2KD}{h}} = \sqrt{\frac{2 \times 1000 \times 100}{2}} = 316.23 \text{ gallons}$$

$$R_1 = 100 - \frac{316.23}{50} = 93.68 \text{ gallons}$$

Iteration 2

$$S = \frac{R_1^2}{200} - R_1 + 50 = .19971 \text{ gallons}$$

$$y_2 = \sqrt{100{,}000 + 10{,}000 \times .19971} = 319.37 \text{ gallons}$$

Hence,

$$R_2 = 100 - \frac{319.39}{50} = 93.612$$

Iteration 3

$$S = \frac{R_2^2}{200} - R_2 + 50 = .20399 \text{ gallon}$$

$$y_3 = \sqrt{100{,}000 + 10{,}000 \times .20399} = 319.44 \text{ gallons}$$

Thus,

$$R_3 = 100 - \frac{319.44}{50} = 93.611 \text{ gallons}$$

Because $y_3 \approx y_2$ and $R_3 \approx R_2$, the optimum is $R^* \approx 93.611$ gallons, $y^* \approx 319.44$ gallons. File excelContRev.xls can be used to determine the solution to any desired degree of accuracy. The optimal inventory policy calls for ordering approximately 320 gallons whenever the inventory level drops to 94 gallons.

PROBLEM SET 14.1B

1. For the data given in Example 14.1-2, determine the following:
   (a) The approximate number of orders per month.
   (b) The expected monthly setup cost.
   (c) The expected holding cost per month.
   (d) The expected shortage cost per month.
   (e) The probability of running out of stock during lead time.

***2.** Solve Example 14.1-2, assuming that the demand during lead time is uniform between 0 and 50 gallons.

***3.** In Example 14.1-2, suppose that the demand during lead time is uniform between 40 and 60 gallons. Compare the solution with that obtained in Example 14.1-2, and interpret the results. (*Hint:* In both problems $E\{x\}$ is the same, but the variance in the present problem is smaller.)

4. Find the optimal solution for Example 14.1-2, assuming that the demand during lead time is $N(100, 2)$. Assume that $D = 10{,}000$ gallons per month, $h = \$2$ per gallon per month, $p = \$4$ per gallon, and $K = \$20$.

14.2 SINGLE-PERIOD MODELS

Single-item inventory models occur when an item is ordered only once to satisfy the demand for the period. For example, fashion items become obsolete at the end of the season. This section presents two models representing the no-setup and the setup cases.

The symbols used in the development of the models include

$$
\begin{aligned}
K &= \text{Setup cost per order} \\
h &= \text{Holding cost per held unit during the period} \\
p &= \text{Penalty cost per shortage unit during the period} \\
D &= \text{Random variable representing demand during the period} \\
f(D) &= \text{pdf of demand during the period} \\
y &= \text{Order quantity} \\
x &= \text{Inventory on hand before an order is placed.}
\end{aligned}
$$

The model determines the optimal value of y that minimizes the sum of the expected holding and shortage costs. Given optimal y $(= y^*)$, the inventory policy calls for ordering $y^* - x$ if $x < y$; otherwise, no order is placed.

14.2.1 No-Setup Model (Newsvendor Model)

This model has come to be known in the literature as the *newsvendor* model (the original classical name is the *newsboy* model) because it deals with items with short life such as newspapers.

The assumptions of this model are

1. Demand occurs instantaneously at the start of the period immediately after the order is received.
2. No setup cost is incurred.

Figure 14.4 demonstrates the inventory position after the demand, D, is satisfied. If $D < y$, the quantity $y - D$ is held during the period. Otherwise, a shortage amount $D - y$ will result if $D > y$.

The expected cost for the period, $E\{C(y)\}$, is expressed as

$$
E\{C(y)\} = h\int_0^y (y - D)f(D)\,dD + p\int_y^\infty (D - y)f(D)\,dD
$$

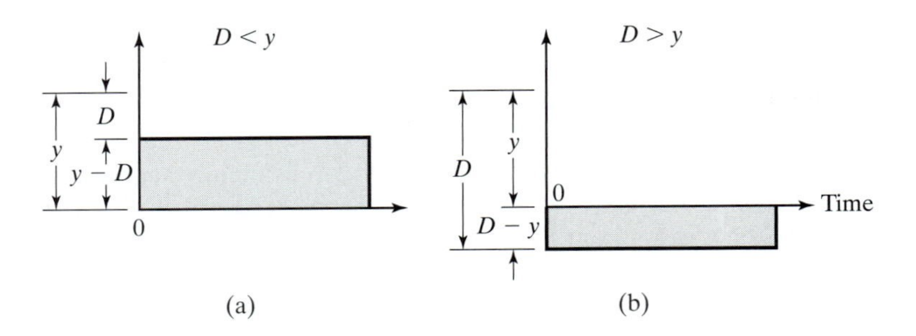

FIGURE 14.4
Holding and shortage inventory in a single-period model

The function $E\{C(y)\}$ can be shown to have a unique minimum because it is convex in y. Taking the first derivative of $E\{C(y)\}$ with respect to y and equating it to zero, we get

$$h \int_0^y f(D)\,dD - p \int_y^\infty f(D)\,dD = 0$$

or

$$hP\{D \le y\} - p(1 - P\{D \le y\}) = 0$$

or

$$P\{D \le y^*\} = \frac{p}{p + h}$$

The preceding development assumes that the demand D is continuous. If D is discrete, then $f(D)$ is defined only at discrete points and the associated cost function is

$$E\{C(y)\} = h \sum_{D=0}^{y} (y - D)f(D) + p \sum_{D=y+1}^{\infty} (D - y)f(D)$$

The necessary conditions for optimality are

$$E\{C(y - 1)\} \ge E\{C(y)\} \text{ and } E\{C(y + 1)\} \ge E\{C(y)\}$$

These conditions are sufficient because $E\{C(y)\}$ is a convex function. After some algebraic manipulations, the application of these conditions yields the following inequalities for determining y^*:

$$P\{D \le y^* - 1\} \le \frac{p}{p + h} \le P\{D \le y^*\}$$

Example 14.2-1

The owner of a newsstand wants to determine the number of *USA Now* newspapers that must be stocked at the start of each day. The owner pays 30 cents for a copy and sells it for 75 cents. The sale of the newspaper typically occurs between 7:00 and 8:00 A.M. Newspapers left at the end of the day are recycled for an income of 5 cents a copy. How many copies should the owner stock every morning, assuming that the demand for the day can be described as

(a) A normal distribution with mean 300 copies and standard deviation 20 copies.

(b) A discrete pdf, $f(D)$, defined as

D	200	220	300	320	340
$f(D)$	.1	.2	.4	.2	.1

The holding and penalty costs are not defined directly in this situation. The data of the problem indicate that each unsold copy will cost the owner $30 - 5 = 25$ cents and that the penalty for running out of stock is $75 - 30 = 45$ cents per copy. Thus, in terms of the parameters of the inventory problem, we have $h = 25$ cents per copy per day and $p = 45$ cents per copy per day.

First, we determine the critical ratio as

$$\frac{p}{p + h} = \frac{45}{45 + 25} = .643$$

Case (a). The demand D is $N(300, 20)$. We can use excelStatTables.xls to determine the optimum order quantity by entering 300 in F15, 20 in G15, and .643 in L15, which gives the desired answer of 307.33 newspapers in R15. Alternatively, we can use the standard normal tables in Appendix B. Define

$$z = \frac{D - 300}{20}$$

Then from the tables

$$P\{z \leq .366\} \approx .643$$

or

$$\frac{y^* - 300}{20} = .366$$

Thus, $y^* = 307.3$. The optimal order is approximately 308 copies.

Case (b). The demand D follows a discrete pdf, $f(D)$. First, we determine the CDF $P\{D \leq y\}$ as

y	200	220	300	320	340
$P\{D \leq y\}$	.1	.3	.7	.9	1.0

For the computed critical ratio of .643, we have

$$P(D \leq 220) \leq .643 \leq P(D \leq 300)$$

It only follows that $y^* = 300$ copies.

PROBLEM SET 14.2A

1. For the single-period model, show that for the discrete demand the optimal order quantity is determined from

$$P\{D \leq y^* - 1\} \leq \frac{p}{p + h} \leq P\{D \leq y^*\}$$

2. The demand for an item during a single period occurs instantaneously at the start of the period. The associated pdf is uniform between 10 and 15 units. Because of the difficulty in estimating the cost parameters, the order quantity is determined such that the probability of either surplus or shortage does not exceed .1. Is it possible to satisfy both conditions simultaneously?

***3.** The unit holding cost in a single-period inventory situation is $1. If the order quantity is 4 units, find the permissible range of the unit penalty cost implied by the optimal conditions. Assume that the demand occurs instantaneously at the start of the period and that demand pdf is given in the following table:

D	0	1	2	3	4	5	6	7	8
$f(D)$	.05	.1	.1	.2	.25	.15	.05	.05	.05

4. The U of A Bookstore offers a program of reproducing class notes for participating professors. Professor Yataha teaches a freshmen-level class, where an enrollment of between 200 and 250 students, uniformly distributed, is expected. It costs the bookstore $10 to produce each copy, which it then sells to the students for $25 a copy. The students purchase their books at the start of the semester. Any unsold copies of Professor Yataha's notes are shredded for recycling. In the meantime, once the bookstore runs out of copies, no additional copies are printed, and the students are responsible for securing the notes from other sources. If the bookstore wants to maximize its revenues, how many copies should it print?

5. QuickStop provides its customers with coffee and donuts at 6:00 A.M. each day. The convenience store buys the donuts for 7 cents apiece and sells them for 25 cents apiece until 8:00 A.M. After 8:00 A.M., the donuts sell for 5 cents apiece. The number of customers buying donuts between 6:00 and 8:00 is uniformly distributed between 30 and 50. Each customer usually orders 3 donuts with coffee. Approximately how many dozen donuts should QuickStop stock every morning to maximize revenues?

***6.** Colony Shop is stocking heavy coats for next winter. Colony pays $50 for a coat and sells it for $110. At the end of the winter season, Colony offers the coats at $55 each. The demand for coats during the winter season is more than 20 but less than or equal to 30, all with equal probabilities. Because the winter season is short, the unit holding cost is negligible. Also, Colony's manager does not believe that any penalty would result from coat shortages. Determine the optimal order quantity that will maximize the revenue for Colony Shop. You may use continuous approximation.

7. For the single-period model, suppose that the item is consumed uniformly during the period (rather than instantaneously at the start of the period). Develop the associated cost model, and find the optimal order quantity.

8. Solve Example 14.2-1, assuming that the demand is continuous and uniform during the period and that the pdf of demand is uniform between 0 and 100. (*Hint*: Use the results of Problem 7.)

14.2.2 Setup Model (*s*-*S* Policy)

The present model differs from the one in Section 14.2.1 in that a setup cost K is incurred. Using the same notation, the total expected cost per period is

$$\begin{aligned} E\{\overline{C}(y)\} &= K + E\{C(y)\} \\ &= K + h\int_0^y (y - D)f(D)\,dD + p\int_y^\infty (D - y)f(D)\,dD \end{aligned}$$

As shown in Section 14.2.1, the optimum value y^* must satisfy

$$P\{y \leq y^*\} = \frac{p}{p + h}$$

Because K is constant, the minimum value of $E\{\overline{C}(y)\}$ must also occur at y^*.

In Figure 14.5, $S = y^*$ and the value of s $(< S)$ is determined from the equation

$$E\{C(s)\} = E\{\overline{C}(S)\} = K + E\{C(S)\},\ s < S$$

The equation yields another value s_1 $(> S)$, which is discarded.

Given that the amount on hand before an order is placed is x units, how much should be ordered? This question is investigated under three conditions:

1. $x < s$.
2. $s \leq x \leq S$.
3. $x > S$.

Case 1 ($x < s$). Because x is already on hand, its equivalent cost is given by $E\{C(x)\}$. If any additional amount $y - x$ $(y > x)$ is ordered, the corresponding cost given y is $E\{\overline{C}(y)\}$, which includes the setup cost K. From Figure 14.5, we have

$$\min_{y>x} E\{\overline{C}(y)\} = E(\overline{C}(S)) < E\{C(x)\}$$

Thus, the optimal inventory policy in this case is to order $S - x$ units.

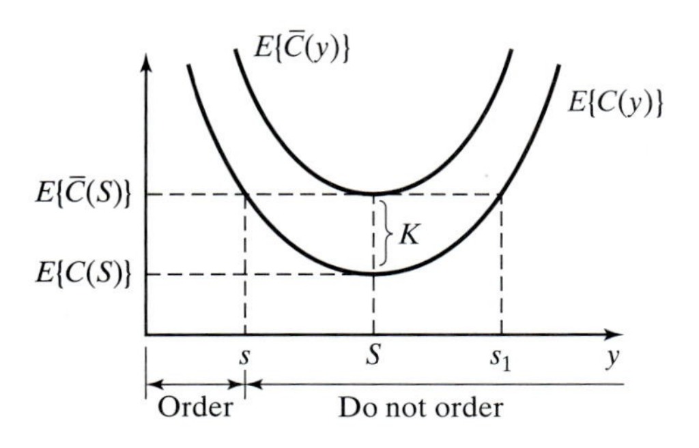

FIGURE 14.5
(s-S) optimal ordering policy in a single-period model with setup cost

Case 2 ($s \le x \le S$). From Figure 14.5, we have

$$E\{C(x)\} \le \min_{y>x} E\{\overline{C}(y)\} = E(\overline{C}(S))$$

Thus, it is *not* advantageous to order in this case. Hence, $y^* = x$.

Case 3 ($x > S$). From Figure 14.5, we have for $y > x$,

$$E\{C(x)\} < E\{\overline{C}(y)\}$$

This condition indicates that it is not advantageous to order in this case—that is, $y^* = x$.

The optimal inventory policy, frequently referred to as the *s-S* policy, is summarized as

$$\text{If } x < s, \text{ order } S - x$$

$$\text{If } x \ge s, \text{ do not order}$$

The optimality of the *s-S* policy is guaranteed because the associated cost function is convex.

Example 14.2-2

The daily demand for an item during a single period occurs instantaneously at the start of the period. The pdf of the demand is uniform between 0 and 10 units. The unit holding cost of the item during the period is \$.50, and the unit penalty cost for running out of stock is \$4.50. A fixed cost of \$25 is incurred each time an order is placed. Determine the optimal inventory policy for the item.

To determine y^*, consider

$$\frac{p}{p+h} = \frac{4.5}{4.5 + .5} = .9$$

Also,

$$P\{D \le y^*\} = \int_0^{y^*} \frac{1}{10} dD = \frac{y^*}{10}$$

Thus, $S = y^* = 9$.

The expected cost function is given as

$$\begin{aligned} E\{C(y)\} &= .5\int_0^{y} \frac{1}{10}(y - D)\,dD + 4.5\int_y^{10} \frac{1}{10}(D - y)\,dD \\ &= .25y^2 - 4.5y + 22.5 \end{aligned}$$

The value of s is determined by solving

$$E\{C(s)\} = K + E\{C(S)\}$$

This yields

$$.25s^2 - 4.5s + 22.5 = 25 + .25S^2 - 4.5S + 22.5$$

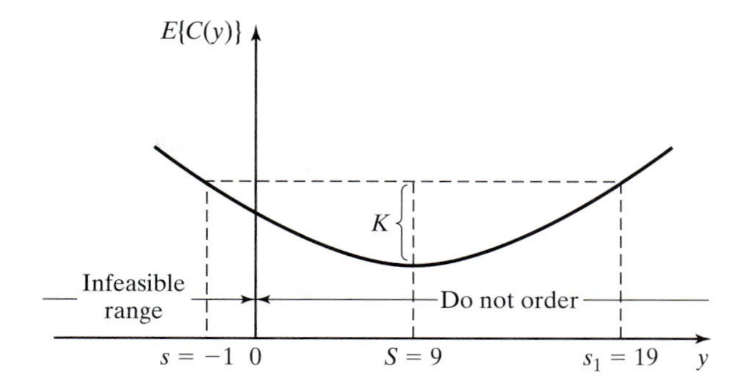

FIGURE 14.6
s-S policy applied to Example 14.2-2

Given $S = 9$, the preceding equation reduces to

$$s^2 - 18s - 19 = 0$$

The solution of this equation is $s = -1$ or $s = 19$. The value of $s > S$ is discarded. Because the remaining value is negative $(= -1)$, s has no feasible value (Figure 14.6). This conclusion usually happens when the cost function is "flat" or when the setup cost is high relative to the other costs of the model.

PROBLEM SET 14.2B

*1. Determine the optimal inventory policy for the situation in Example 14.2-2, assuming that the setup cost is \$5.

2. In the single-period model in Section 14.2.1, suppose instead that the model maximizes profit and that a setup cost K is incurred. Given that r is the unit selling price and using the information in Section 14.2.1, develop an expression for the expected profit and determine the optimal order quantity. Solve the problem numerically for $r = \$3$, $c = \$2$, $p = \$4$, $h = \$1$, and $K = \$10$. The demand pdf is uniform between 0 and 10.

3. Work Problem 5, Set 14.2a, assuming that there is a fixed cost of \$10 associated with the delivery of donuts.

14.3 MULTIPERIOD MODEL

This section presents a multiperiod model under the assumption of no setup cost. Additionally, the model allows backlog of demand and assumes a zero-delivery lag. It further assumes that the demand D in any period is described by a stationary pdf $f(D)$.

The multiperiod model considers the discounted value of money. If α (< 1) is the discount factor per period, then an amount \$$A$ available n periods from now has a present value of \$$\alpha^n A$.

Suppose that the inventory situation encompasses n periods and that unfilled demand can be backlogged exactly one period. Define

$F_i(x_i)$ = Maximum expected profit for periods $i, i + 1, \ldots,$ and n, given that x_i is the amount on hand before an order is placed in period i

Using the notation in Section 14.2 and assuming that c and r are the cost and revenue per unit, respectively, the inventory situation can be formulated using the following dynamic programming model (see Chapter 22 on the CD):

$$F_i(x_i) = \max_{y_i \geq x_i}\left\{-c(y_i - x_i) + \int_0^{y_i}[rD - h(y_i - D)]f(D)\,dD\right.$$

$$+ \int_{y_i}^{\infty}[ry_i + \alpha r(D - y_i) - p(D - y_i)]f(D)\,dD$$

$$\left.+ \alpha\int_0^{\infty}F_{i+1}(y_i - D)f(D)\,dD\right\}, i = 1, 2, \ldots, n$$

where $F_{n+1}(y_n - D) = 0$. The value of x_i may be negative because unfilled demand is backlogged. The quantity $\alpha r(D - y_i)$ in the second integral is included because $(D - y_i)$ is the unfilled demand in period i that must be filled in period $i + 1$.

The problem can be solved recursively. For the case where the number of periods is infinite, the recursive equation reduces to

$$F(x) = \max_{y \geq x}\left\{-c(y - x) + \int_0^{y}[rD - h(y - D)]f(D)\,dD\right.$$

$$+ \int_{y}^{\infty}[ry + \alpha r(D - y) - p(D - y)]f(D)\,dD$$

$$\left.+ \alpha\int_0^{\infty}F(y - D)f(D)\,dD\right\}$$

where x and y are the inventory levels for each period before and after an order is received, respectively.

The optimal value of y can be determined from the following necessary condition, which also happens to be sufficient because the expected revenue function $F(x)$ is concave.

$$\frac{\partial(.)}{\partial y} = -c - h\int_0^{y}f(D)\,dD + \int_{y}^{\infty}[(1 - \alpha)r + p]f(D)\,dD$$

$$+ \alpha\int_0^{\infty}\frac{\partial F(y - D)}{\partial y}f(D)\,dD = 0$$

The value of $\frac{\partial F(y - D)}{\partial y}$ is determined as follows. If there are β (> 0) more units on hand at the start of the next period, the profit for the next period will increase by $c\beta$, because this much less has to be ordered. This means that

$$\frac{\partial F(y - D)}{\partial y} = c$$

The necessary condition thus becomes

$$-c - h\int_0^y f(D)\,dD + [(1-\alpha)r + p]\left(1 - \int_0^y f(D)\,dD\right) + \alpha c\int_0^\infty f(D)\,dD = 0$$

The optimum inventory level y^* is thus determined from

$$\int_0^{y^*} f(D)\,dD = \frac{p + (1-\alpha)(r-c)}{p + h + (1-\alpha)r}$$

The optimal inventory policy for each period, given its entering inventory level x, is thus given as

$$\begin{aligned} &\text{If } x < y^*, \quad \text{order } y^* - x \\ &\text{If } x \geq y^*, \quad \text{do not order} \end{aligned}$$

PROBLEM SET 14.3A

1. Consider a two-period probabilistic inventory model in which the demand is backlogged, and orders are received with zero delivery lag. The demand pdf per period is uniform between 0 and 10, and the cost parameters are given as

Unit selling price = \$2

Unit purchase price = \$1

Unit holding cost per month = \$.10

Unit penalty cost per month = \$3

Discount factor = .8

Find the optimal inventory policy for the two periods, assuming that the initial inventory for period 1 is zero.

***2.** The pdf of the demand per period in an infinite-horizon inventory model is given as

$$f(D) = .08D, 0 \leq D \leq 5$$

The unit cost parameters are

Unit selling price = \$10

Unit purchase price = \$8

Unit holding cost per month = \$1

Unit penalty cost per month = \$10

Discount factor = .9

Determine the optimal inventory policy assuming zero delivery lag and that the unfilled demand is backlogged.

3. Consider the infinite-horizon inventory situation with zero delivery lag and backlogged demand. Develop the optimal inventory policy based on the minimization of cost given that

$$\text{Holding cost for } z \text{ units} = hz^2$$
$$\text{Penalty cost for } z \text{ units} = px^2$$

Show that for the special case where $h = p$, the optimal solution is independent of pdf of demand.

REFERENCES

Cohen, R. and F. Dunford, "Forecasting for Inventory Control: An Example of When 'Simple' Means 'Better,'" *Interfaces*, Vol. 16, No. 6, pp. 95–99, 1986.

Hadley, G., and T. Whitin, *Analysis of Inventory Systems*, Prentice Hall, Upper Saddle River, NJ, 1963.

Nahmias, S., *Production and Operations Analysis*, 5th ed., Irwin, Homewood, IL, 2005.

Silver, E., D. Pyke, and R. Peterson, *Decision Systems for Inventory Management and Production Control*, 3rd ed., Wiley, New York, 1998.

Tersine, R., *Principles of Inventory and Materials Management*, North Holland, New York, 1982.

Zipken, P., *Foundations of Inventory Management*, McGraw-Hill, Boston, 2000.

C H A P T E R 1 5

Queuing Systems

Chapter Guide. The objective of queuing analysis is to offer a reasonably satisfactory service to waiting customers. Unlike the other tools of OR presented in the preceding chapters, queuing theory is not an optimization technique. Rather, it determines the measures of performance of waiting lines, such as the average waiting time in queue and the productivity of the service facility, which can then be used to design the service installation. This chapter emphasizes the implementation of queuing results in practice. However, to fully appreciate the practical side of queuing, you will need a reasonable background in the underlying theory. For this reason, the chapter starts with a presentation of the "total randomness" property of two important distributions: the Poisson and the exponential. This point is important because it helps identify the situations where queuing results apply in practice.

Queuing results involve computationally difficult formulas, and it is recommended that you use exelPoissonQ.xls or TORA to carry out these calculations. You will find TORA helpful in comparing multiple scenarios. Throughout the chapter, TORA is used to carry out the computations. The bulk of the discussion concentrates on the practical interpretations of the results. We recommend that you follow the same procedure when you work out the problems in this chapter. In this manner, you are not "bogged down" in the tedious computational details and can readily test different scenarios conveniently.

This chapter includes a summary of 2 real-life applications, 17 solved examples, 2 Excel templates, 137 end-of-section problems, and 5 cases. The cases are in Appendix E on the CD. The AMPL/Excel/Solver/TORA programs are in folder ch15Files.

Real-Life Application—Analysis of an Internal Transport System in a Manufacturing Plant

Three trucks are used in a manufacturing plant to transport materials. The trucks wait in a central parking lot until requested. A truck answering a request will travel to the customer location, carry a load to its destination, and then return to the central parking lot. The principal user of the service is production, followed by the workshop and maintenance. Other departments occasionally may request the use of the trucks.

Complaints about the long wait for a free truck have prompted users, especially production, to request adding a fourth truck to the fleet. This is an unusual application, because queuing theory is used to show that the source of the long delays is mainly logistical and that with a simple change in the operating procedure of the truck pool, a fourth truck is not needed. Case 14 in Chapter 24 on the CD provides the details of the study.

15.1 WHY STUDY QUEUES?

Waiting for service is part of our daily life. We wait to eat in restaurants, we "queue up" at the check-out counters in grocery stores, and we "line up" for service in post offices. And the waiting phenomenon is not an experience limited to human beings only: Jobs wait to be processed on a machine, planes circle in a stack before given permission to land at an airport, and cars stop at traffic lights. Waiting cannot be eliminated completely without incurring inordinate expenses, and the goal is to reduce its adverse impact to "tolerable" levels.

The study of queues deals with quantifying the phenomenon of waiting in lines using representative measures of performance, such as average queue length, average waiting time in queue, and average facility utilization. The following example demonstrates how these measures can be used to design a service facility.

Example 15.1-1

McBurger is a fast-food restaurant with three service counters. The manager has commissioned a study to investigate complaints about slow service. The study reveals the following relationship between the number of service counters and the waiting time for service:

No. of cashiers	1	2	3	4	5	6	7
Average waiting time (min)	16.2	10.3	6.9	4.8	2.9	1.9	1.3

An examination of these data shows a 7-minute average waiting time for the present 3-counter situation. Five counters are needed to reduce the waiting time to about 3 minutes.

Remarks. The results of queuing analysis can be used in the context of a cost optimization model, where we seek the minimization of the sum of two costs: the cost of offering the service and the cost of waiting. Figure 15.1 depicts a typical cost model (in dollars per unit time) where the cost of service increases with the increase in the level of service (e.g., the number of service counters). At the same time, the cost of waiting decreases with the increase in level of service. The main obstacle in implementing cost models is the difficulty of obtaining reliable estimates of the cost of waiting, particularly when human behavior is an integral part of the operation. This point is discussed in Section 15.9.

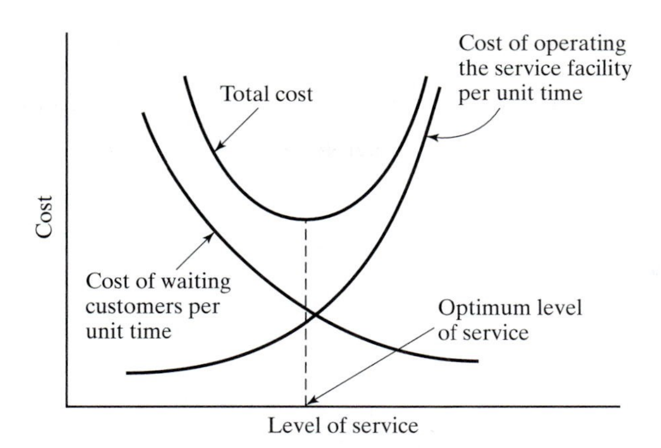

FIGURE 15.1
Cost-based queuing decision model

PROBLEM SET 15.1A

*1. Suppose that further analysis of the McBurger restaurant reveals the following additional results:

No. of cashiers	1	2	3	4	5	6	7
Idleness (%)	0	8	12	18	29	36	42

(a) What is the productivity of the operation (expressed as the percentage of time the employees are busy) when the number of cashiers is five?

(b) The manager wants to keep the average waiting time around 3 minutes and, simultaneously, maintain the efficiency of the facility at approximately 90%. Can the two goals be achieved? Explain.

2. Acme Metal Jobshop is in the process of purchasing a multipurpose drill press. Two models, A and B, are available with hourly operating costs of \$18 and \$25, respectively. Model A is slower than model B. Queuing analysis of similar machines shows that when A is used, the average number of jobs in the queue is 4, which is 30% higher than the queue size in B. A delayed job represents lost income, which is estimated by Acme at \$10 per waiting job per hour. Which model should Acme purchase?

15.2 ELEMENTS OF A QUEUING MODEL

The principal actors in a queuing situation are the **customer** and the **server**. Customers are generated from a **source**. On arrival at a service **facility**, they can start service immediately or wait in a **queue** if the facility is busy. When a facility completes a service, it automatically "pulls" a waiting customer, if any, from the queue. If the queue is empty, the facility becomes idle until a new customer arrives.

From the standpoint of analyzing queues, the arrival of customers is represented by the **interarrival time** between successive customers, and the service is described by the **service time** per customer. Generally, the interarrival and service times can be probabilistic, as in the operation of a post office, or deterministic, as in the arrival of applicants for job interviews.

Queue size plays a role in the analysis of queues, and it may have a finite size, as in the buffer area between two successive machines, or it may be infinite, as in mail order facilities.

Queue discipline, which represents the order in which customers are selected from a queue, is an important factor in the analysis of queuing models. The most common discipline is **first come, first served** (FCFS). Other disciplines include **last come, first served** (LCFS) and **service in random order** (SIRO). Customers may also be selected from the queue based on some order of **priority**. For example, rush jobs at a shop are processed ahead of regular jobs.

The queuing behavior of customers plays a role in waiting-line analysis. "Human" customers may **jockey** from one queue to another in the hope of reducing waiting time. They may also **balk** from joining a queue altogether because of anticipated long delay, or they may **renege** from a queue because they have been waiting too long.

The design of the service facility may include parallel servers (e.g., post office or bank operation). The servers may also be arranged in series (e.g., jobs processed on successive machines), or they may be networked (e.g., routers in a computer network).

The source from which customers are generated may be finite or infinite. A **finite source** limits the customers arriving for service (e.g., machines requesting the service of a repairperson). An **infinite source** is forever abundant (e.g., calls arriving at a telephone exchange).

Variations in the elements of a queuing situation give rise to a variety of queuing models. This chapter provides examples of these models.

PROBLEM SET 15.2A

1. In each of the following situations, identify the customer and the server:

*__(a)__ Planes arriving at an airport.

*__(b)__ Taxi stand serving waiting passengers.

(c) Tools checked out from a crib in a machining shop.

(d) Letters processed in a post office.

(e) Registration for classes in a university.

(f) Legal court cases.

(g) Check-out operation in a supermarket.

*__(h)__ Parking lot operation.

2. For each of the situations in Problem 1, identify the following: (a) nature of the calling source (finite or infinite), (b) nature of arriving customers (individually or in bulk), (c) type of the interarrival time (probabilistic or deterministic), (d) definition and type of service time, (f) queue capacity (finite or infinite), and (g) queue discipline.

3. Study the following system and identify the associated queuing situations. For each situation, define the customers, the server(s), the queue discipline, the service time, the maximum queue length, and the calling source.

Orders for jobs are received at a workshop for processing. On receipt, the supervisor decides whether it is a rush or a regular job. Some orders require the use of one of several identical machines. The remaining orders are processed in a two-stage production line, of which two are available. In each group, one facility is assigned to handle rush jobs.

Jobs arriving at any facility are processed in order of arrival. Completed orders are shipped on arrival from a shipping zone having a limited capacity.

Sharpened tools for the different machines are supplied from a central tool crib. When a machine breaks down, a repairperson is summoned from the service pool to make the repair. Machines working on rush orders always receive priorities both in acquiring new tools from the crib and in receiving repair service.

4. True or False?

(a) An impatient waiting customer may elect to renege.

(b) If a long waiting time is anticipated, an arriving customer may elect to balk.

(c) Jockeying from one queue to another is exercised to reduce waiting time.

5. In each of the situations in Problem 1, discuss the possibility of the customers jockeying, balking, and reneging.

15.3 ROLE OF EXPONENTIAL DISTRIBUTION

In most queuing situations, the arrival of customers occurs in a *totally random* fashion. Randomness here means that the occurrence of an event (e.g., arrival of a customer or completion of a service) is not influenced by the length of time that has elapsed since the occurrence of the last event.

Random interarrival and service times are described quantitatively in queuing models by the **exponential distribution**, which is defined as

$$f(t) = \lambda e^{-\lambda t}, t > 0$$

Section 12.4.3 shows that for the exponential distribution

$$E\{t\} = \frac{1}{\lambda}$$

$$P\{t \leq T\} = \int_0^T \lambda e^{-\lambda t}\, dt$$

$$= 1 - e^{-\lambda T}$$

The definition of $E\{t\}$ shows that λ is the rate per unit time at which events (arrivals or departures) are generated. The fact that the exponential distribution is **completely random** is illustrated by the following example: If the time now is 8:20 A.M. and the last arrival has occurred at 8:02 A.M., the probability that the next arrival will occur by 8:29 is a function of the interval from 8:20 to 8:29 only, and is totally independent of the length of time that has elapsed since the occurrence of the last event (8:02 to 8:20). This result is referred to as the **forgetfulness** or **lack of memory** of the exponential.

Let the exponential distribution, $f(t)$, represent the time, t, between successive events. If S is the interval since the occurrence of the last event, then the *forgetfulness property* implies that

$$P\{t > T + S | t > S\} = P\{t > T\}$$

To prove this result, we note that for the exponential with mean $\frac{1}{\lambda}$,

$$P\{t > Y\} = 1 - P\{t < Y\} = e^{-\lambda Y}$$

Thus,

$$P\{t > T + S | t > S\} = \frac{P\{t > T + S, t > S\}}{P\{t > S\}} = \frac{P\{t > T + S\}}{P\{t > S\}}$$

$$= \frac{e^{-\lambda(T+S)}}{e^{-\lambda S}} = e^{-\lambda T}$$

$$= P\{t > T\}$$

Example 15.3-1

A service machine always has a standby unit for immediate replacement upon failure. The time to failure of the machine (or its standby unit) is exponential and occurs every 5 hours, on the average. The machine operator claims that the machine "has the habit" of breaking down every night around 8:30 P.M. Analyze the operator's claim.

The average failure rate of the machine is $\lambda = \frac{1}{5} = .2$ failure per hour. Thus, the exponential distribution of the time to failure is

$$f(t) = .2e^{-.2t}, t > 0$$

Regarding the operator's claim, we know offhand that it cannot be correct because it conflicts with the fact that the time between breakdowns is exponential and, hence, totally random. The probability that a failure will occur by 8:30 P.M. cannot be used to support or refute the operator's claim, because the value of such probability depends on the time of the day (relative to 8:30 P.M.) at which it is computed. For example, if the time now is 8:20 P.M., the probability that the operator's claim will be right tonight is

$$p\left\{t < \tfrac{10}{60}\right\} = 1 - e^{-.2\left(\frac{10}{60}\right)} = .03278$$

which is low. If the time now is 1:00 P.M., the probability that a failure will occur by 8:30 P.M. increases to approximately .777 (verify!). These two extreme values show that the operator's claim cannot be supported.

PROBLEM SET 15.3A

1. **(a)** Explain your understanding of the relationship between the arrival rate λ and the average interarrival time. What are the units describing each variable?
   (b) In each of the following cases, determine the average arrival rate per hour, λ, and the average interarrival time in hours.
   - ***(i)** One arrival occurs every 10 minutes.
   - **(ii)** Two arrivals occur every 6 minutes.
   - **(iii)** Number of arrivals in a 30-minute period is 10.
   - **(iv)** The average interval between successive arrivals is .5 hour.

(c) In each of the following cases, determine the average service rate per hour, μ, and the average service time in hours.

*** (i)** One service is completed every 12 minutes.

(ii) Two departures occur every 15 minutes.

(iii) Number of customers served in a 30-minute period is 5.

(iv) The average service time is .3 hour.

2. In Example 15.3-1, determine the following:

(a) The average number of failures in 1 week, assuming the service is offered 24 hours a day, 7 days a week.

(b) The probability of at least one failure in a 2-hour period.

(c) The probability that the next failure will *not* occur within 3 hours.

(d) If no failure has occurred 3 hours after the last failure, what is the probability that interfailure time is at least 4 hours?

3. The time between arrivals at the State Revenue Office is exponential with mean value .05 hour. The office opens at 8:00 A.M.

*** (a)** Write the exponential distribution that describes the interarrival time.

*** (b)** Find the probability that no customers will arrive at the office by 8:15 A.M.

(c) It is now 8:35 A.M. The last customer entered the office at 8:26. What is the probability that the next customer will arrive before 8:38 A.M.? That the next customer will not arrive by 8:40 A.M.?

(d) What is the average number of arriving customers between 8:10 and 8:45 A.M.?

4. Suppose that the time between breakdowns for a machine is exponential with mean 6 hours. If the machine has worked without failure during the last 3 hours, what is the probability that it will continue without failure during the next hour? That it will break down during the next .5 hour?

5. The time between arrivals at the game room in the student union is exponential with mean 10 minutes.

(a) What is the arrival rate per hour?

(b) What is the probability that no students will arrive at the game room during the next 15 minutes?

(c) What is the probability that at least one student will visit the game room during the next 20 minutes?

6. The manager of a new fast-food restaurant wants to quantify the arrival process of customers by estimating the fraction of interarrival time intervals that will be (a) less than 2 minutes, (b) between 2 and 3 minutes, and (c) more than 3 minutes. Arrivals in similar restaurants occur at the rate of 35 customers per hour. The interarrival time is exponentially distributed.

*** 7.** Ann and Jim, two employees in a fast-food restaurant, play the following game while waiting for customers to arrive: Jim pays Ann 2 cents if the next customer does not arrive within 1 minute; otherwise, Ann pays Jim 2 cents. Determine Jim's average payoff in an 8-hour period. The interarrival time is exponential with mean 1.5 minute.

8. Suppose that in Problem 7 the rules of the game are such that Jim pays Ann 2 cents if the next customer arrives after 1.5 minutes, and Ann pays Jim an equal amount if the next arrival is within 1 minute. For arrivals within the range 1 to 1.5 minutes, the game is a draw. Determine Jim's expected payoff in an 8-hour period.

9. In Problem 7, suppose that Ann pays Jim 2 cents if the next arrival occurs within 1 minute and 3 cents if the interarrival time is between 1 and 1.5 minutes. Ann receives from Jim 5 cents if the interarrival time is between 1.5 and 2 minutes and 6 cents if it is larger than 2 minutes. Determine Ann's expected payoff in an 8-hour period.

***10.** A customer arriving at a McBurger fast-food restaurant within 4 minutes of the immediately preceding customer will receive a 10% discount. If the interarrival time is between 4 and 5 minutes, the discount is 6%. If the interarrival time is longer than 5 minutes, the customer gets 2% discount. The interarrival time is exponential with mean 6 minutes.

(a) Determine the probability that an arriving customer will receive the 10% discount.

(b) Determine the average discount per arriving customer.

11. The time between failures of a Kencore refrigerator is known to be exponential with mean value 9000 hours (about 1 year of operation) and the company issues a 1-year warranty on the refrigerator. What are the chances that a breakdown repair will be covered by the warranty?

12. The U of A runs two bus lines on campus: red and green. The red line serves north campus, and the green line serves south campus with a transfer station linking the two lines. Green buses arrive randomly (exponential interarrival time) at the transfer station every 10 minutes. Red buses also arrive randomly every 7 minutes.

(a) What is the probability distribution of the waiting time for a student arriving on the red line to get on the green line?

(b) What is the probability distribution of the waiting time for a student arriving on the green line to get on the red line?

13. Prove that the mean and standard deviation of the exponential distribution are equal.

15.4 PURE BIRTH AND DEATH MODELS (RELATIONSHIP BETWEEN THE EXPONENTIAL AND POISSON DISTRIBUTIONS)

This section presents two queuing situations: the **pure birth** model in which arrivals only are allowed, and the **pure death** model in which departures only can take place. An example of the pure birth model is the creation of birth certificates for newly born babies. The pure death model may be demonstrated by the random withdrawal of a stocked item in a store.

The exponential distribution is used to describe the interarrival time in the pure birth model and the interdeparture time in the pure death model. A by-product of the development of the two models is to show the close relationship between the exponential and the Poisson distributions, in the sense that one distribution automatically defines the other.

15.4.1 Pure Birth Model

Define

$$p_0(t) = \text{Probability of no arrivals during a period of time } t$$

Given that the interarrival time is exponential and that the arrival rate is λ customers per unit time, then

$$\begin{aligned} p_0(t) &= P\{\text{interarrival time} \geq t\} \\ &= 1 - P\{\text{interarrival time} \leq t\} \\ &= 1 - (1 - e^{-\lambda t}) \\ &= e^{-\lambda t} \end{aligned}$$

For a sufficiently small time interval $h > 0$, we have

$$p_0(h) = e^{-\lambda h} = 1 - \lambda h + \frac{(\lambda h)^2}{2!} - \ldots = 1 - \lambda h + 0(h^2)$$

The exponential distribution is based on the assumption that during $h > 0$, at most one event (arrival) can occur. Thus, as $h \rightarrow 0$,

$$p_1(h) = 1 - p_0(h) \approx \lambda h$$

This result shows that the probability of an arrival during h is directly proportional to h, with the arrival rate, λ, being the constant of proportionality.

To derive the distribution of the *number* of arrivals during a period t when the interarrival time is exponential with mean $\frac{1}{\lambda}$, define

$$p_n(t) = \text{Probability of } n \text{ arrivals during } t$$

For a sufficiently small $h > 0$,

$$\begin{aligned} p_n(t + h) &\approx p_n(t)(1 - \lambda h) + p_{n-1}(t)\lambda h, \ n > 0 \\ p_0(t + h) &\approx p_0(t)(1 - \lambda h), \qquad\qquad\qquad n = 0 \end{aligned}$$

In the first equation, n arrivals will be realized during $t + h$ if there are n arrivals during t and no arrivals during h, or $n - 1$ arrivals during t and one arrival during h. All other combinations are not allowed because, according to the exponential distribution, at most one arrival can occur during a very small period h. The product law of probability is applicable to the right-hand side of the equation because arrivals are independent. For the second equation, zero arrivals during $t + h$ can occur only if no arrivals occur during t and h.

Rearranging the terms and taking the limits as $h \rightarrow 0$, we get

$$p'_n(t) = \lim_{h \rightarrow 0} \frac{p_n(t + h) - p_n(t)}{h} = -\lambda p_n(t) + \lambda p_{n-1}(t), \ n > 0$$

$$p'_0(t) = \lim_{h \rightarrow 0} \frac{p_0(t - h) - p_0(t)}{h} = -\lambda p_0(t), \qquad n = 0$$

where $p'_n(t)$ is the first derivative of $p_n(t)$ with respect to t.

The solution of the preceding difference-differential equations yields

$$p_n(t) = \frac{(\lambda t)^n e^{-\lambda t}}{n!}, n = 0, 1, 2, \ldots$$

This is a **Poisson distribution** with mean $E\{n|t\} = \lambda t$ arrivals during t.

The preceding result shows that if the time between arrivals is exponential with mean $\frac{1}{\lambda}$ then the number of arrivals during a specific period t is Poisson with mean λt. The converse is true also.

The following table summarizes the strong relationships between the exponential and the Poisson given an arrival rate of λ arrivals per unit time:

	Exponential	Poisson
Random variable	*Time* between successive arrivals, t	*Number* of arrivals, n, during a specified period T
Range	$t \geq 0$	$n = 0, 1, 2, \ldots$
Density function	$f(t) = \lambda e^{-\lambda t}, t \geq 0$	$p_n(T) = \frac{(\lambda T)^n e^{-\lambda T}}{n!}, n = 0, 1, 2, \ldots$
Mean value	$\frac{1}{\lambda}$ time units	λT arrivals during T
Cumulative probability	$P\{t \leq A\} = 1 - e^{-\lambda A}$	$p_{n \leq N}(T) = p_0(T) + p_1(T) + \ldots + p_N(T)$
$P\{$no arrivals during period $A\}$	$P\{t > A\} = e^{-\lambda A}$	$p_0(A) = e^{-\lambda A}$

Example 15.4-1

Babies are born in a sparsely populated state at the rate of one birth every 12 minutes. The time between births follows an exponential distribution. Find the following:

(a) The average number of births per year.
(b) The probability that no births will occur in any one day.
(c) The probability of issuing 50 birth certificates in 3 hours given that 40 certificates were issued during the first 2 hours of the 3-hour period.

The birth rate per day is computed as

$$\lambda = \frac{24 \times 60}{12} = 120 \text{ births/day}$$

The number of births per year in the state is

$$\lambda t = 120 \times 365 = 43{,}800 \text{ births/year}$$

The probability of no births in any one day is computed from the Poisson distribution as

$$p_0(1) = \frac{(120 \times 1)^0 e^{-120 \times 1}}{0!} = e^{-120} = 0$$

Another way to compute the same probability is to note that no birth in any one day is equivalent to saying that the *time between successive births* exceeds one day. We can thus use the exponential distribution to compute the desired probability as

$$P\{t > 1\} = e^{-120} = 0$$

To compute the probability of issuing 50 certificates by the end of 3 hours given that 40 certificates were issued during the first 2 hours is equivalent to having 10 $(= 50 - 40)$ births in one $(= 3 - 2)$ hour because the distribution of the number of births is Poisson.
Given $\lambda = \frac{60}{12} = 5$ births per hour, we get

$$p_{10}(1) = \frac{(5 \times 1)^{10} e^{-5\times 1}}{10!} = .01813$$

Excel Moment

The calculations associated with the Poisson distribution and, indeed, all queuing formulas are tedious and require special programming skill to secure reasonable computational accuracy. You can use Excel POISSON, POISSONDIST, and EXPONDIST functions to compute the individual and cumulative probabilities Poisson and exponential probabilities. These functions are also automated in exceStatTables.xls. For example, for a birth rate of 5 babies per hour, the probability of *exactly* 10 births in .5 hour is computed by entering 2.5 in F16, 10 in J16 to obtain the answer .000216 in M16. The cumulative probability of *at most* 10 births is given in O16 $(= .999938)$. To determine the probability of the time between births being less than or equal to 18 minutes, use the exponential distribution by entering 2.5 in F9 and .3 in J9. The answer, .527633, is found in O9.

TORA/Excel Moment

You can also use TORA (file toraEx15.4-1.txt) or template excelPoissonQ.xls to determine all significant ($>10^{-5}$ in TORA and 10^{-7} in Excel) Poisson probabilities automatically. In both cases, the input data are the same. For the pure birth model of Example 15.4-1 the data are entered as follows:

Lambda	Mu	c	System limit	Source limit
5	0	(not applicable)	infinity	infinity

Note the entry under Lambda is $\lambda t = 5 \times 1 = 5$ births per day.

PROBLEM SET 15.4A

*1. In Example 15.4-1, suppose that the clerk who enters the information from birth certificates into the computer normally waits until at least 5 certificates have accumulated. Find the probability that the clerk will be entering a new batch every hour.

2. An art collector travels to art auctions once a month on the average. Each trip is guaranteed to produce one purchase. The time between trips is exponentially distributed. Determine the following:
   (a) The probability that no purchase is made in a 3-month period.
   (b) The probability that no more than 8 purchases are made per year.
   (c) The probability that the time between successive trips will exceed 1 month.
3. In a bank operation, the arrival rate is 2 customers per minute. Determine the following:
   (a) The average number of arrivals during 5 minutes.
   (b) The probability that no arrivals will occur during the next .5 minute.
   (c) The probability that at least one arrival will occur during the next .5 minute.
   (d) The probability that the time between two successive arrivals is at least 3 minutes.
4. The time between arrivals at L&J restaurant is exponential with mean 5 minutes. The restaurant opens for business at 11:00 A.M. Determine the following:
   ***(a)** The probability of having 10 arrivals in the restaurant by 11:12 A.M. given that 8 customers arrived by 11:05 A.M.
   (b) The probability that a new customer will arrive between 11:28 and 11:33 A.M. given that the last customer arrived at 11:25 A.M.
5. The Springdale Public Library receives new books according to a Poisson distribution with mean 25 books per day. Each shelf in the stacks holds 100 books. Determine the following:
   (a) The average number of shelves that will be stacked with new books each (30-day) month.
   (b) The probability that more than 10 bookcases will be needed each month, given that a bookcase has 5 shelves.
6. The U of A runs two bus lines on campus: red and green. The red line serves north campus and the green line serves south campus with a transfer station linking the two lines. Green buses arrive randomly (according to a Poisson distribution) at the transfer station every 10 minutes. Red buses also arrive randomly every 7 minutes.
   ***(a)** What is the probability that two buses will stop at the station during a 5-minute interval?
   (b) A student whose dormitory is located next to the station has a class in 10 minutes. Either bus will take the student to the classroom building. The ride takes 5 minutes, after which the student will walk for about 3 minutes to reach the classroom. What is the probability that the student will make it to class on time?
7. Prove that the mean and variance of the Poisson distribution during an interval t equal λt, where λ is the arrival rate.
8. Derive the Poisson distribution from the difference-differential equations of the pure birth model. *Hint*: The solution of the general differential equation

$$y' + a(t)y = b(t)$$

is

$$y = e^{-\int a(t)\,dt}\left\{\int b(t)e^{\int a(t)}\,dt + \text{constant}\right.$$

15.4.2 Pure Death Model

In the pure death model, the system starts with N customers at time 0 and no new arrivals are allowed. Departures occur at the rate μ customers per unit time. To develop the difference-differential equations for the probability $p_n(t)$ of n customers *remaining* after t time units, we follow the arguments used with the pure birth model (Section 15.4.1). Thus,

$$p_N(t + h) = p_N(t)(1 - \mu h)$$

$$p_n(t + h) = p_n(t)(1 - \mu h) + p_{n+1}(t)\mu h, 0 < n < N$$

$$p_0(t + h) = p_0(t)(1) + p_1(t)\mu h$$

As $h \rightarrow 0$, we get

$$p'_N(t) = -\mu p_N(t)$$

$$p'_n(t) = -\mu p_n(t) + \mu p_{n+1}(t), 0 < n < N$$

$$p'_0(t) = \mu p_1(t)$$

The solution of these equations yields the following **truncated Poisson** distribution:

$$p_n(t) = \frac{(\mu t)^{N-n}e^{-\mu t}}{(N - n)!}, n = 1, 2, \ldots, N$$

$$p_0(t) = 1 - \sum_{n-1}^{N} p_n(t)$$

Example 15.4-2

The florist section in a grocery store stocks 18 dozen roses at the beginning of each week. On the average, the florist sells 3 dozens a day (one dozen at a time), but the actual demand follows a Poisson distribution. Whenever the stock level reaches 5 dozens, a new order of 18 new dozens is placed for delivery at the beginning of the following week. Because of the nature of the item, all roses left at the end of the week are disposed of. Determine the following:

(a) The probability of placing an order in any one day of the week.
(b) The average number of dozen roses that will be discarded at the end of the week.

Because purchases occur at the rate of $\mu = 3$ dozens per day, the probability of placing an order by the end of day t is given as

$$p_{n \le 5}(t) = p_0(t) + p_1(t) + \ldots + p_5(t)$$

$$= p_0(t) + \sum_{n=1}^{5} \frac{(3t)^{18-n}e^{-3t}}{(18 - n)!}, t = 1, 2, \ldots, 7$$

The calculations of $p_{n \le 5}(t)$ are best done using excelPoissonQ.xls or TORA. TORA's multiple scenarios may be more convenient in this case. The associated input data for the pure death model corresponding to $t = 1, 2, \ldots,$ and 7 are

Lambda $= 0$, Mu $= 3t$, $c = 1$, System Limit $= 18$, and Source Limit $= 18$

Note that t must be substituted out numerically as shown in file toraEx15.4-2.txt.

The output is summarized as follows:

t (days)	1	2	3	4	5	6	7
μt	3	6	9	12	15	18	21
$p_{n\le 5}(t)$	.0000	.0088	.1242	.4240	.7324	.9083	.9755

The average number of dozen roses discarded at the end of the week ($t = 7$) is $E(n|t = 7\}$. To calculate this value we need $p_n(7), n = 0, 1, 2, \ldots, 18$, which can be determined using provided software, which yields

$$E\{n|t = 7\} = \sum_{n=0}^{18} np_n(7) = .664 \approx 1 \text{ dozen}$$

PROBLEM SET 15.4B

1. In Example 15.4-2, use excelPoissonQ.xls or TORA to compute $p_n(7), n = 1, 2, \ldots, 18$, and then verify manually that these probabilities yield $E\{n|t = 7\} = .664$ dozen.
2. Consider Example 15.4-2. In each of the following cases, first write the answer algebraically, and then use excelPoissonQ.xls or TORA to provide numerical answers.
   *(a) The probability that the stock is depleted after 3 days.
   (b) The average number of dozen roses left at the end of the second day.
   *(c) The probability that at least one dozen is purchased by the end of the fourth day, given that the last dozen was bought at the end of the third day.
   (d) The probability that the time remaining until the next purchase is at most half a day given that the last purchase occurred a day earlier.
   (e) The probability that no purchases will occur during the first day.
   (f) The probability that no order will placed by the end of the week.
3. The Springdale High School band is performing a benefit jazz concert in its new 400-seat auditorium. Local businesses buy the tickets in blocks of 10 and donate them to youth organizations. Tickets go on sale to business entities for 4 hours only the day before the concert. The process of placing orders for tickets is Poisson with a mean 10 calls per hour. Any (blocks of) tickets remaining after the box office is closed are sold at a discount as "rush tickets" 1 hour before the concert starts. Determine
   (a) The probability that it will be possible to buy rush tickets.
   (b) The average number of rush tickets available.
4. Each morning, the refrigerator in a small machine shop is stocked with two cases (24 cans per case) of soft drinks for use by the shop's 10 employees. The employees can quench their thirst at any time during the 8-hour work day (8:00 A.M. to 4:00 P.M.), and each employee is known to consume approximately 4 cans a day, but the process is totally random (Poisson distribution). What is the probability that an employee will not find a drink at noon (the start of the lunch period)? Just before the shop closes?

*5. A freshman student receives a bank deposit of $100 a month from home to cover incidentals. Withdrawal checks of $20 each occur randomly during the month and are spaced according to an exponential distribution with a mean value of 1 week. Determine the probability that the student will run out of incidental money before the end of the fourth week.

6. Inventory is withdrawn from a stock of 80 items according to a Poisson distribution at the rate of 5 items per day. Determine the following:
   (a) The probability that 10 items are withdrawn during the first 2 days.
   (b) The probability that no items are left at the end of 4 days.
   (c) The average number of items withdrawn over a 4-day period.
7. A machine shop has just stocked 10 spare parts for the repair of a machine. Stock replenishment that brings the stock level back to 10 pieces occurs every 7 days. The time between breakdowns is exponential with mean 1 day. Determine the probability that the machine will remain broken for 2 days because no spare parts are available.
8. Demand for an item occurs according to a Poisson distribution with mean 3 per day. The maximum stock level is 25 items, which occurs on each Monday immediately after a new order is received. The order size depends on the number of units left at the end of the week on Saturday (business is closed on Sundays). Determine the following:
   ***(a)** The average weekly size of the order.
   ***(b)** The probability of incurring shortage when the business opens on Friday morning.
   (c) The probability that the weekly order size exceeds 10 units.
9. Prove that the distribution of the time between departures corresponding to the truncated Poisson in the pure death model is an exponential distribution with mean $\frac{1}{\mu}$ time units.
10. Derive the truncated Poisson distribution from the difference-differential equations of the pure death model using induction. [*Note*: See the hint in Problem 8, Set 15.4a.]

15.5 GENERALIZED POISSON QUEUING MODEL

This section develops a general queuing model that combines both arrivals and departures based on the Poisson assumptions—that is, the interarrival and the service times follow the exponential distribution. The model is the basis for the derivation of the specialized Poisson models in Section 15.6.

The development of the generalized model is based on the long-run or **steady-state** behavior of the queuing situation, which is achieved after the system has been in operation for a sufficiently long time. This type of analysis contrasts with the **transient** (or warm-up) behavior that prevails during the early operation of the system. One reason for not discussing the transient behavior in this chapter is its analytical complexity. Another reason is that the study of most queuing situations occurs under steady-state conditions.

The generalized model assumes that both the arrival and departure rates are **state dependent**, meaning that they depend on the number of customers in the service facility. For example, at a highway toll booth, attendants tend to speed up toll collection during rush hours. Another example occurs in a shop with a given number of machines where the rate of breakdown decreases as the number of broken machines increases (because only working machines are capable of generating new breakdowns).

Define

$$n = \text{Number of customers in the system (in-queue plus in-service)}$$

$$\lambda_n = \text{Arrival rate given } n \text{ customers in the system}$$

$$\mu_n = \text{Departure rate given } n \text{ customers in the system}$$

$$p_n = \text{Steady-state probability of } n \text{ customers in the system}$$

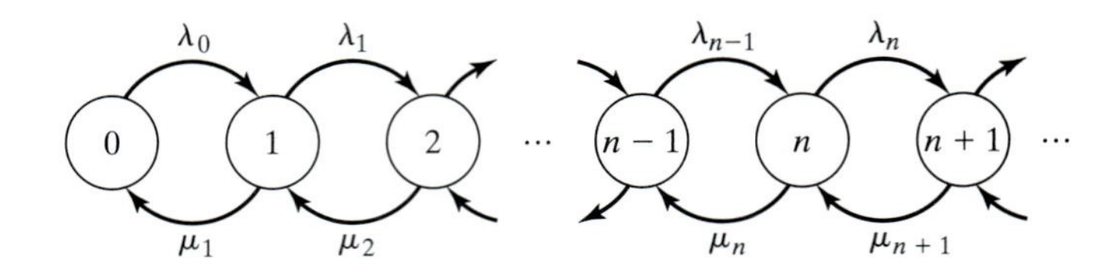

FIGURE 15.2
Poisson queues transition diagram

The generalized model derives p_n as a function of λ_n and μ_n. These probabilities are then used to determine the system's measures of performance, such as the average queue length, the average waiting time, and the average utilization of the facility.

The probabilities p_n are determined by using the **transition-rate diagram** in Figure 15.2. The queuing system is in state n when the number of customers in the system is n. As explained in Section 15.3, the probability of more than one event occurring during a small interval h tends to zero as $h \to 0$. This means that for $n > 0$, state n can change only to two possible states: $n - 1$ when a departure occurs at the rate μ_n, and $n + 1$ when an arrival occurs at the rate λ_n. State 0 can only change to state 1 when an arrival occurs at the rate λ_0. Notice that μ_0 is undefined because no departures can occur if the system is empty.

Under steady-state conditions, for $n > 0$, the *expected* rates of flow into and out of state n must be equal. Based on the fact that state n can be changed to states $n - 1$ and $n + 1$ only, we get

$$\begin{pmatrix}\text{Expected rate of} \\ \text{flow into state } n\end{pmatrix} = \lambda_{n-1}p_{n-1} + \mu_{n+1}p_{n+1}$$

Similarly,

$$\begin{pmatrix}\text{Expected rate of} \\ \text{flow out of state } n\end{pmatrix} = (\lambda_n + \mu_n)p_n$$

Equating the two rates, we get the following **balance equation**:

$$\lambda_{n-1}p_{n-1} + \mu_{n+1}p_{n+1} = (\lambda_n + \mu_n)p_n,\ n = 1, 2, \ldots$$

From Figure 15.2, the balance equation associated with $n = 0$, is

$$\lambda_0 p_0 = \mu_1 p_1$$

The balance equations are solved recursively in terms of p_0 as follows: For $n = 0$, we have

$$p_1 = \left(\frac{\lambda_0}{\mu_1}\right)p_0$$

Next, for $n = 1$, we have

$$\lambda_0 p_0 + \mu_2 p_2 = (\lambda_1 + \mu_1)p_1$$

Substituting $p_1 = \left(\frac{\lambda_0}{\mu_0}\right)p_0$ and simplifying, we get (verify!)

$$p_2 = \left(\frac{\lambda_1\lambda_0}{\mu_2\mu_1}\right)p_0$$

In general, we can show by induction that

$$p_n = \left(\frac{\lambda_{n-1}\lambda_{n-2}\cdots\lambda_0}{\mu_n\mu_{n-1}\cdots\mu_1}\right)p_0, \ n = 1, 2, \ldots$$

The value of p_0 is determined from the equation $\sum_{n=0}^{\infty} p_n = 1$

Example 15.5-1

B&K Groceries operates with three check-out counters. The manager uses the following schedule to determine the number of counters in operation, depending on the number of customers in store:

No. of customers in store	No. of counters in operation
1 to 3	1
4 to 6	2
More than 6	3

Customers arrive in the counters area according to a Poisson distribution with a mean rate of 10 customers per hour. The average check-out time per customer is exponential with mean 12 minutes. Determine the steady-state probability p_n of n customers in the check-out area.

From the information of the problem, we have

$$\lambda_n = \lambda = 10 \text{ customers per hour}, \qquad n = 0, 1, \ldots$$

$$\mu_n = \begin{cases} \frac{60}{12} = 5 \text{ customers per hour}, & n = 0, 1, 2, 3 \\ 2 \times 5 = 10 \text{ customers per hour}, & n = 4, 5, 6 \\ 3 \times 5 = 15 \text{ customers per hour}, & n = 7, 8, \ldots \end{cases}$$

Thus,

$$\begin{aligned}
p_1 &= \left(\tfrac{10}{5}\right)p_0 = 2p_0 \\
p_2 &= \left(\tfrac{10}{5}\right)^2 p_0 = 4p_0 \\
p_3 &= \left(\tfrac{10}{5}\right)^3 p_0 = 8p_0 \\
p_4 &= \left(\tfrac{10}{5}\right)^3\left(\tfrac{10}{10}\right)p_0 = 8p_0 \\
p_5 &= \left(\tfrac{10}{5}\right)^3\left(\tfrac{10}{10}\right)^2 p_0 = 8p_0 \\
p_6 &= \left(\tfrac{10}{5}\right)^3\left(\tfrac{10}{10}\right)^3 p_0 = 8p_0 \\
p_{n\geq 7} &= \left(\tfrac{10}{5}\right)^3\left(\tfrac{10}{10}\right)^3\left(\tfrac{10}{15}\right)^{n-6} p_0 = 8\left(\tfrac{2}{3}\right)^{n-6} p_0
\end{aligned}$$

The value of p_0 is determined from the equation

$$p_0 + p_0\{2 + 4 + 8 + 8 + 8 + 8 + 8(\tfrac{2}{3}) + 8(\tfrac{2}{3})^2 + 8(\tfrac{2}{3})^3 + \ldots\} = 1$$

or, equivalently

$$p_0\{31 + 8(1 + (\tfrac{2}{3}) + (\tfrac{2}{3})^2 + \ldots)\} = 1$$

Using the geometric sum series

$$\sum_{i=0}^{\infty} x^i = \frac{1}{1 - x}, |x| < 1$$

we get

$$p_0\left\{31 + 8\left(\frac{1}{1 - \frac{2}{3}}\right)\right\} = 1$$

Thus, $p_0 = \frac{1}{55}$.

Given p_0, we can now determine p_n for $n > 0$. For example, the probability that only one counter will be open is computed as the probability that there are at most three customers in the system:

$$p_1 + p_2 + p_3 = (2 + 4 + 8)(\tfrac{1}{55}) \approx .255$$

We can use p_n to determine measures of performance for the B&K situation. For example,

$$\begin{aligned}\begin{pmatrix}\text{Expected number}\\ \text{of idle counters}\end{pmatrix} &= 3p_0 + 2(p_1 + p_2 + p_3) + 1(p_4 + p_5 + p_6)\\ &\quad + 0(p_7 + p_8 + \ldots)\\ &= 1 \text{ counter}\end{aligned}$$

PROBLEM SET 15.5A

1. In Example 15.5-1, determine the following:
   - **(a)** The probability distribution of the number of open counters.
   - **(b)** The average number of busy counters.
2. In the B&K model of Example 15.5-1, suppose that the interarrival time at the checkout area is exponential with mean 5 minutes and that the checkout time per customer is also exponential with mean 10 minutes. Suppose further that B&K will add a fourth counter and that counters will open based on increments of two customers. Determine the following:
   - **(a)** The steady-state probabilities, p_n for all n.
   - **(b)** The probability that a fourth counter will be needed.
   - **(c)** The average number of idle counters.

*3. In the B&K model of Example 15.5-1, suppose that all three counters are always open and that the operation is set up such that the customer will go to the first empty counter. Determine the following:
   - **(a)** The probability that all three counters will be in use.
   - **(b)** The probability that an arriving customer will not wait.

4. First Bank of Springdale operates a one-lane drive-in ATM machine. Cars arrive according to a Poisson distribution at the rate of 12 cars per hour. The time per car needed to complete the ATM transaction is exponential with mean 6 minutes. The lane can accommodate a total of 10 cars. Once the lane is full, other arriving cars seek service in another branch. Determine the following:

(a) The probability that an arriving car will not be able to use the ATM machine because the lane is full.

(b) The probability that a car will not be able to use the ATM machine immediately on arrival.

(c) The average number of cars in the lane.

5. Have you ever heard someone repeat the contradictory statement, "The place is so crowded no one goes there any more"? This statement can be interpreted as saying that the opportunity for balking increases with the increase in the number of customers seeking service. A possible platform for modeling this situation is to say that the arrival rate at the system decreases as the number of customers in the system increases. More specifically, we consider the simplified case of M&M Pool Club, where customers usually arrive in pairs to "shoot pool." The normal arrival rate is 6 pairs (of people) per hour. However, once the number of pairs in the pool hall exceeds 8, the arrival rate drops to 5 pairs per hour. The arrival process is assumed to follow the Poisson distribution. Each pair shoots pool for an exponential time with mean 30 minutes. The pool hall has a total of 5 tables and can accommodate no more than 12 pairs at any one time. Determine the following:

(a) The probability that customers will balk.

(b) The probability that all tables are in use.

(c) The average number of tables in use.

(d) The average number of pairs waiting for a pool table to be available.

***6.** A barbershop serves one customer at a time and provides three seats for waiting customers. If the place is full, customers go elsewhere. Arrivals occur according to a Poisson distribution with mean four per hour. The time to get a haircut is exponential with mean 15 minutes. Determine the following:

(a) The steady-state probabilities.

(b) The expected number of customers in the shop.

(c) The probability that customers will go elsewhere because the shop is full.

7. Consider a one-server queuing situation in which the arrival and service rates are given by

$$\lambda_n = 10 - n, n = 0, 1, 2, 3$$

$$\mu_n = \frac{n}{2} + 5, n = 1, 2, 3, 4$$

This situation is equivalent to reducing the arrival rate and increasing the service rate as the number in the system, n, increases.

(a) Set up the transition diagram and determine the balance equation for the system.

(b) Determine the steady-state probabilities.

8. Consider the single queue model where only one customer is allowed in the system. Customers who arrive and find the facility busy never return. Assume that the arrivals distribution is Poisson with mean λ per unit time and that the service time is exponential with mean $\frac{1}{\mu}$ time units.

(a) Set up the transition diagram and determine the balance equations.

(b) Determine the steady-state probabilities.

(c) Determine the average number in the system.

9. The induction proof for deriving the general solution of the generalized model is applied as follows. Consider

$$p_k = \prod_{i=0}^{k-1}\left(\frac{\lambda_i}{\mu_{i+1}}\right)p_0,\ k = 0, 1, 2, \ldots$$

We substitute for p_{n-1} and p_{n-2} in the general difference equation involving p_n, p_{n-1}, and p_{n-2} to derive the desired expression for p_n. Verify this procedure.

15.6 SPECIALIZED POISSON QUEUES

Figure 15.3 depicts the specialized Poisson queuing situation with c parallel servers. A waiting customer is selected from the queue to start service with the first available server. The arrival rate at the system is λ customers per unit time. All parallel servers are identical, meaning that the service rate for any server is μ customers per unit time. The number of customers in the *system* is defined to include those *in service* and those waiting *in queue*.

A convenient notation for summarizing the characteristics of the queuing situation in Figure 15.3 is given by the following format:

$$(a/b/c):(d/e/f)$$

where

a = Arrivals distribution

b = Departures (service time) distribution

c = Number of parallel servers $(= 1, 2, \ldots, \infty)$

d = Queue discipline

e = Maximum number (finite or infinite) allowed in the system (in-queue plus in-service)

f = Size of the calling source (finite or infinite)

FIGURE 15.3

Schematic representation of a queuing system with c parallel servers

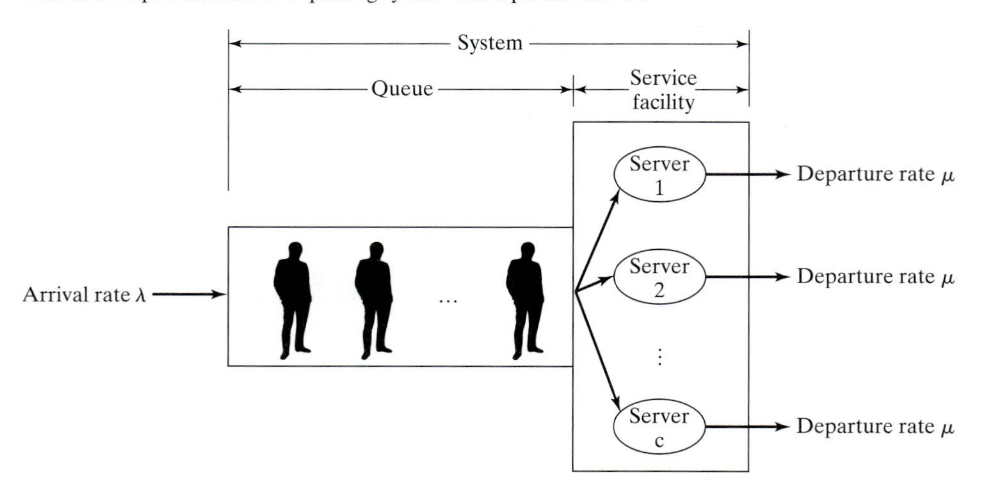

The standard notation for representing the arrivals and departures distributions (symbols a and b) is

$$M = \text{Markovian (or Poisson) arrivals or departures distribution (or equivalently exponential interarrival or service time distribution)}$$

$$D = \text{Constant (deterministic) time}$$

$$E_k = \text{Erlang or gamma distribution of time (or, equivalently, the sum of independent exponential distributions)}$$

$$GI = \text{General (generic) distribution of interarrival time}$$

$$G = \text{General (generic) distribution of service time}$$

The queue discipline notation (symbol d) includes

$$FCFS = \text{First come, first served}$$

$$LCFS = \text{Last come, first served}$$

$$SIRO = \text{Service in random order}$$

$$GD = \text{General discipline (i.e., any type of discipline)}$$

To illustrate the use of the notation, the model $(M/D/10)$:$(GD/20/\infty)$ uses Poisson arrivals (or exponential interarrival time), constant service time, and 10 parallel servers. The queue discipline is GD, and there is a limit of 20 customers on the entire system. The size of the source from which customers arrive is infinite.

As a historical note, the first three elements of the notation ($a/b/c$), were devised by D.G. Kendall in 1953 and are known in the literature as the **Kendall notation**. In 1966, A.M. Lee added the symbols d and e to the notation. This author added the last element, symbol f, in 1968.

Before presenting the details of the specialized Poisson queues, we show how the steady-state measures of performance of the generalized queuing situation can be derived from the steady-state probabilities p_n given in Section 15.5.

15.6.1 Steady-State Measures of Performance

The most commonly used measures of performance in a queuing situation are

$$L_s = \text{Expected number of customers in } \textit{system}$$

$$L_q = \text{Expected number of customers in } \textit{queue}$$

$$W_s = \text{Expected waiting time in } \textit{system}$$

$$W_q = \text{Expected waiting time in } \textit{queue}$$

$$\bar{c} = \text{Expected number of busy servers}$$

Recall that the *system* includes both the *queue* and the *service facility*.

We show now how these measures are derived (directly or indirectly) from the steady-state probability of n in the system, p_n as

$$L_s = \sum_{n=1}^{\infty} n p_n$$

$$L_q = \sum_{n=c+1}^{\infty} (n - c) p_n$$

The relationship between L_s and W_s (also L_q and W_q) is known as **Little's formula**, and is given as

$$L_s = \lambda_{\text{eff}} W_s$$

$$L_q = \lambda_{\text{eff}} W_q$$

These relationships are valid under rather general conditions. The parameter λ_{eff} is the *effective* arrival rate at the system. It equals the (nominal) arrival rate λ when all arriving customers can join the system. Otherwise, if some customers cannot join because the system is full (e.g., a parking lot), then $\lambda_{\text{eff}} < \lambda$. We will show later how λ_{eff} is determined.

A direct relationship also exists between W_s and W_q. By definition,

$$\begin{pmatrix}\text{Expected waiting}\\ \text{time in system}\end{pmatrix} = \begin{pmatrix}\text{Expected waiting}\\ \text{time in queue}\end{pmatrix} + \begin{pmatrix}\text{Expected service}\\ \text{time}\end{pmatrix}$$

This translates to

$$W_s = W_q + \frac{1}{\mu}$$

Next, we can relate L_s to L_q by multiplying both sides of the last formula by λ_{eff}, which together with Little's formula gives

$$L_s = L_q + \frac{\lambda_{\text{eff}}}{\mu}$$

By definition, the difference between the average number in the system, L_s, and the average number in the queue, L_q, must equal the average number of *busy* servers, $\bar{c}$. We thus have,

$$\bar{c} = L_s - L_q = \frac{\lambda_{\text{eff}}}{\mu}$$

It follows that

$$\begin{pmatrix}\text{Facility}\\ \text{utilization}\end{pmatrix} = \frac{\bar{c}}{c}$$

Example 15.6-1

Visitors' parking at Ozark College is limited to five spaces only. Cars making use of this space arrive according to a Poisson distribution at the rate of six cars per hour. Parking time is exponentially distributed with a mean of 30 minutes. Visitors who cannot find an empty space on arrival may temporarily wait inside the lot until a parked car leaves. That temporary space can hold only three cars. Other cars that cannot park or find a temporary waiting space must go elsewhere. Determine the following:

(a) The probability, p_n, of n cars in the system.
(b) The effective arrival rate for cars that actually use the lot.
(c) The average number of cars in the lot.
(d) The average time a car waits for a parking space inside the lot.
(e) The average number of *occupied* parking spaces.
(f) The average utilization of the parking lot.

We note first that a parking space acts as a server, so that the system has a total of $c = 5$ parallel servers. Also, the maximum capacity of the system is $5 + 3 = 8$ cars.

The probability p_n can be determined as a special case of the generalized model in Section 15.5 using

$$\lambda_n = 6 \text{ cars/hour}, n = 0, 1, 2, \ldots, 8$$

$$\mu_n = \begin{cases} n\left(\frac{60}{30}\right) = 2n \text{ cars/hour}, n = 1, 2, 3, 4, 5 \\ 5\left(\frac{60}{30}\right) = 10 \text{ cars/hour}, n = 6, 7, 8 \end{cases}$$

From Section 15.5, we get

$$p_n = \begin{cases} \frac{3^n}{n!}p_0, & n = 1, 2, 3, 4, 5 \\ \frac{3^n}{5!\,5^{n-5}}p_0, & n = 6, 7, 8 \end{cases}$$

The value of p_0 is computed by substituting p_n, $n = 1, 2, \ldots, 8$, in the following equation

$$p_0 + p_1 + \ldots + p_8 = 1$$

or

$$p_0 + p_0\left(\frac{3}{1!} + \frac{3^2}{2!} + \frac{3^3}{3!} + \frac{3^4}{4!} + \frac{3^5}{5!} + \frac{3^6}{5!\,5} + \frac{3^7}{5!\,5^2} + \frac{3^8}{5!\,5^3}\right) = 1$$

This yields $p_0 = .04812$ (verify!). From p_0, we can now compute p_1 through p_8 as

n	1	2	3	4	5	6	7	8
p_n	.14436	.21654	.21654	.16240	.09744	.05847	.03508	.02105

The effective arrival rate λ_{eff} can be computed by observing the schematic diagram in Figure 15.4, where customers arrive from the source at the rate λ cars per hour. An arriving car may enter the parking lot or go elsewhere with rates λ_{eff} or λ_{lost}, which means that $\lambda = \lambda_{\text{eff}} + \lambda_{\text{lost}}$. A car will not be able to enter the parking lot if 8 cars are already in. This means that the proportion of cars that will *not* be able to enter the lot is p_8. Thus,

$$\lambda_{\text{lost}} = \lambda p_8 = 6 \times .02105 = .1263 \text{ cars per hour}$$

$$\lambda_{\text{eff}} = \lambda - \lambda_{\text{lost}} = 6 - .1263 = 5.8737 \text{ cars per hour}$$

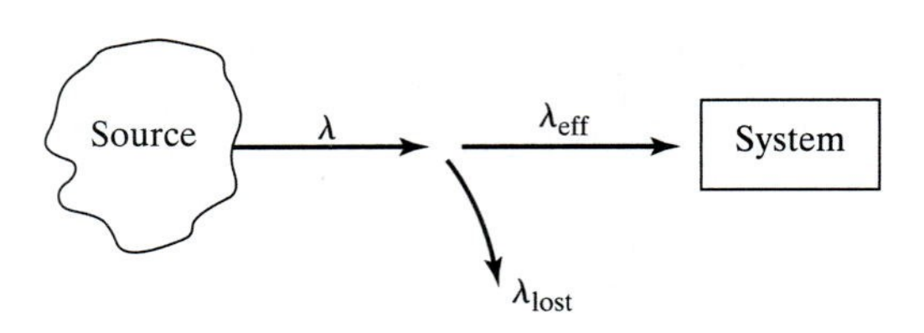

FIGURE 15.4
Relationship between λ, λ_{eff}, and λ_{lost}

The average number of cars in the lot (those waiting for or occupying a space) equals L_s, the average number in the system. We can compute L_s from p_n as

$$L_s = 0p_0 + 1p_1 + \ldots + 8p_8 = 3.1286 \text{ cars}$$

A car waiting in the temporary space is actually a car in queue. Thus, its waiting time until a space is found is W_q. To determine W_q we use

$$W_q = W_s - \frac{1}{\mu}$$

Thus,

$$W_s = \frac{L_s}{\lambda_{\text{eff}}} = \frac{3.1286}{5.8737} = .53265 \text{ hour}$$

$$W_q = .53265 - \frac{1}{2} = .03265 \text{ hour}$$

The average number of occupied parking spaces is the same as the average number of busy servers,

$$\bar{c} = L_s - L_q = \frac{\lambda_{\text{eff}}}{\mu} = \frac{5.8737}{2} = 2.9368 \text{ spaces}$$

From $\bar{c}$, we get

$$\text{Parking lot utilization} = \frac{\bar{c}}{c} = \frac{2.9368}{5} = .58736$$

PROBLEM SET 15.6A

1. In Example 15.6-1, do the following:
   *(a) Compute L_q directly using the formula $\sum_{n=c+1}^{\infty}(n - c)p_n$.
   (b) Compute W_s from L_q.
   *(c) Compute the average number of cars that will not be able to enter the parking lot during an 8-hour period.
   *(d) Show that $c - (L_s - L_q)$, the average number of empty spaces, equals $\sum_{n=0}^{c-1}(c - n)p_n$.
2. Solve Example 15.6-1 using the following data: number of parking spaces $= 6$, number of temporary spaces $= 4$, $\lambda = 10$ cars per hour, and average parking time $= 45$ minutes.

15.6.2 Single-Server Models

This section presents two models for the single server case ($c = 1$). The first model sets no limit on the maximum number in the system, and the second model assumes a finite system limit. Both models assume an infinite-capacity source. Arrivals occur at the rate λ customers per unit time and the service rate is μ customers per unit time.

The results of the two models (and indeed of all the remaining models in Section 15.6) are derived as special cases of the results of the generalized model of Section 15.5.

The Kendall notation will be used to summarize the characteristics of each situation. Because the derivations of p_n in Section 15.5 and of all the measures of performance in Section 15.6.1 are totally independent of a specific queue discipline, the symbol GD (general discipline) will be used with the notation.

$(M/M/1)$:$(GD/\infty/\infty)$. Using the notation of the generalized model, we have

$$\left.\begin{aligned}\lambda_n &= \lambda \\ \mu_n &= \mu\end{aligned}\right\}, n = 0, 1, 2, \ldots$$

Also, $\lambda_{\text{eff}} = \lambda$ and $\lambda_{\text{lost}} = 0$, because all arriving customers can join the system.

Letting $\rho = \frac{\lambda}{\mu}$, the expression for p_n in the generalized model then reduces to

$$p_n = \rho^n p_0, n = 0, 1, 2, \ldots$$

To determine the value of p_0, we use the identity

$$p_0(1 + \rho + \rho^2 + \ldots) = 1$$

Assuming $\rho < 1$, the geometric series will have the finite sum $\left(\frac{1}{1-\rho}\right)$, thus

$$p_0 = 1 - \rho, \text{ provided } \rho < 1.$$

The general formula for p_n is thus given by the following geometric distribution

$$p_n = (1 - \rho)\rho^n, n = 1, 2, \ldots (\rho < 1)$$

The mathematical derivation of p_n imposes the condition $\rho < 1$, or $\lambda < \mu$. If $\lambda \geq \mu$, the geometric series will not converge, and the steady-state probabilities p_n will not exist. This result makes intuitive sense, because unless the service rate is larger than the arrival rate, queue length will continually increase and no steady state can be reached.

The measure of performance L_q can be derived in the following manner:

$$\begin{aligned}L_s = \sum_{n=0}^{\infty} np_n &= \sum_{n=0}^{\infty} n(1 - \rho)\rho^n \\ &= (1 - \rho)\rho\frac{d}{d\rho}\sum_{n=0}^{\infty}\rho^n \\ &= (1 - \rho)\rho\frac{d}{d\rho}\left(\frac{1}{1-\rho}\right) = \frac{\rho}{1-\rho}\end{aligned}$$

Because $\lambda_{\text{eff}} = \lambda$ for the present situation, the remaining measures of performance are computed using the relationships in Section 15.6.1. Thus,

$$W_s = \frac{L_s}{\lambda} = \frac{1}{\mu(1-\rho)} = \frac{1}{\mu - \lambda}$$

$$W_q = W_s - \frac{1}{\mu} = \frac{\rho}{\mu(1-\rho)}$$

$$L_q = \lambda W_q = \frac{\rho^2}{1-\rho}$$

$$\bar{c} = L_s - L_q = \rho$$

Example 15.6-2

Automata car wash facility operates with only one bay. Cars arrive according to a Poisson distribution with a mean of 4 cars per hour, and may wait in the facility's parking lot if the bay is busy. The time for washing and cleaning a car is exponential, with a mean of 10 minutes. Cars that cannot park in the lot can wait in the street bordering the wash facility. This means that, for all practical purposes, there is no limit on the size of the system. The manager of the facility wants to determine the size of the parking lot.

For this situation, we have $\lambda = 4$ cars per hour, and $\mu = \frac{60}{10} = 6$ cars per hour. Because $\rho = \frac{\lambda}{\mu} < 1$, the system can operate under steady-state conditions.

The TORA or excelPoissonQ.xls input for this model is

Lambda	Mu	c	System limit	Source limit
4	6	1	infinity	infinity

The output of the model is shown in Figure 15.5. The average number of cars waiting in the queue, L_q, is 1.33 cars.

Generally, using L_q as the sole basis for the determination of the number of parking spaces is not advisable, because the design should, in some sense, account for the maximum possible length of the queue. For example, it may be more plausible to design the parking lot such that an arriving car will find a parking space at least 90% of the time. To do this, let S represent the number of parking spaces. Having S parking spaces is equivalent to having $S + 1$ spaces in the *system* (queue plus wash bay). An arriving car will find a space 90% of the time if there are *at most* S cars in the system. This condition is equivalent to the following probability statement:

$$p_0 + p_1 + \ldots + p_s \geq .9$$

From Figure 15.5, *cumulative* p_n for $n = 5$ is .91221. This means that the condition is satisfied for $S \geq 5$ parking spaces.

The number of spaces S can be determined also by using the mathematical definition of p_n—that is,

$$(1-\rho)(1 + \rho + \rho^2 + \ldots + \rho^S) \geq .9$$

```
Scenario1: (M/M/1):(GD/infinity/infinity)

Lambda =        4.00000                    Mu =          6.00000
Lambda eff = 4.00000                       Rho/c =       0.66667

Ls =          2.00000                      Lq =          1.33333
Ws =        0.50000                        Wq =          0.33333
```

n	Probability pn	Cumulative Pn	n	Probability pn	Cumulative Pn
0	0.33333	0.33333	13	0.00171	0.99657
1	0.22222	0.55556	14	0.00114	0.99772
2	0.14815	0.70370	15	0.00076	0.99848
3	0.09877	0.80247	16	0.00051	0.99899
4	0.06584	0.86831	17	0.00034	0.99932
5	0.04390	0.91221	18	0.00023	0.99955
6	0.02926	0.94147	19	0.00015	0.99970
7	0.01951	0.96098	20	0.00010	0.99980
8	0.01301	0.97399	21	0.00007	0.99987
9	0.00867	0.98266	22	0.00004	0.99991
10	0.00578	0.98844	23	0.00003	0.99994
11	0.00385	0.99229	24	0.00002	0.99996
12	0.00257	0.99486	25	0.00001	0.99997

FIGURE 15.5

TORA output of Example 15.6-2 (file toraEx15.6-2.txt)

The sum of the truncated geometric series equals $\frac{1 - \rho^{S+1}}{1 - \rho}$. Thus the condition reduces to

$$(1 - \rho^{S+1}) \geq .9$$

Simplification of the inequality yields

$$\rho^{S+1} \leq .1$$

Taking the logarithms on both sides (and noting that $\log(x) < 0$ for $0 < x < 1$, which reverses the direction of the inequality), we get

$$S \geq \frac{\ln(.1)}{\ln\left(\frac{4}{6}\right)} - 1 = 4.679 \approx 5$$

PROBLEM SET 15.6B

1. In Example 15.6-2, do the following.
   (a) Determine the percent utilization of the wash bay.
   (b) Determine the probability that an arriving car must wait in the parking lot prior to entering the wash bay.
   (c) If there are seven parking spaces, determine the probability that an arriving car will find an empty parking space.
   (d) How many parking spaces should be provided so that an arriving car may find a parking space 99% of the time?

*2. John Macko is a student at Ozark U. He does odd jobs to supplement his income. Job requests come every 5 days on the average, but the time between requests is exponential. The time for completing a job is also exponential with mean 4 days.

(a) What is the probability that John will be out of jobs?

(b) If John gets about $50 a job, what is his average monthly income?

(c) If at the end of the semester, John decides to subcontract on the outstanding jobs at $40 each. How much, on the average, should he expect to pay?

3. Over the years, Detective Columbo, of the Fayetteville Police Department, has had phenomenal success in solving every single crime case. It is only a matter of time before any case is solved. Columbo admits that the time per case is "totally random," but, on the average, each investigation will take about a week and half. Crimes in peaceful Fayetteville are not very common. They occur randomly at the rate of one crime per (four-week) month. Detective Columbo is asking for an assistant to share the heavy work load. Analyze Columbo's claim, particularly from the standpoint of the following points:

(a) The average number of cases awaiting investigation.

(b) The percentage of time the detective remains busy.

(c) The average time needed to solve a case.

4. Cars arrive at the Lincoln Tunnel toll gate according to a Poisson distribution, with a mean of 90 cars per hour. The time for passing the gate is exponential with mean 38 seconds. Drivers complain of the long waiting time, and authorities are willing to reduce the average passing time to 30 seconds by installing automatic toll collecting devices, provided two conditions are satisfied: (1) the average number of waiting cars in the present system exceeds 5, and (2) the percentage of the gate idle time with the new device installed does not exceed 10%. Can the new device be justified?

*5. A fast-food restaurant has one drive-in window. Cars arrive according to a Poisson distribution at the rate of 2 cars every 5 minutes. The space in front of the window can accommodate at most 10 cars, including the one being served. Other cars can wait outside this space if necessary. The service time per customer is exponential, with a mean of 1.5 minutes. Determine the following:

(a) The probability that the facility is idle.

(b) The expected number of customers waiting to be served.

(c) The expected waiting time until a customer reaches the window to place an order.

(d) The probability that the waiting line will exceed the 10-space capacity.

6. Customers arrive at a one-window drive-in bank according to a Poisson distribution, with a mean of 10 per hour. The service time per customer is exponential, with a mean of 5 minutes. There are three spaces in front of the window, including the car being served. Other arriving cars line up outside this 3-car space.

(a) What is the probability that an arriving car can enter one of the 3-car spaces?

(b) What is the probability that an arriving car will wait outside the designated 3-car space?

(c) How long is an arriving customer expected to wait before starting service?

*(d) How many car spaces should be provided in front of the window (including the car being served) so that an arriving car can find a space there at least 90% of the time?

7. In the $(M/M/1):(GD/\infty/\infty)$, give a plausible argument as to why L_s does not equal $L_q + 1$, in general. Under what condition will the equality hold?

8. For the $(M/M/1):(GD/\infty/\infty)$, derive the expression for L_q using the basic definition $\sum_{n=2}^{\infty}(n-1)p_n$.

9. For the $(M/M/1):(GD/\infty/\infty)$, show that

(a) The expected number in the queue given that the queue is not empty $= \frac{1}{(1-\rho)}$.

(b) The expected waiting time in the queue for those who must wait $= \left(\frac{1}{\mu - \lambda}\right)$.

Waiting Time Distribution for $(M/M/1)$:$(FCFS/\infty/\infty)$.[1] The derivation of p_n in the generalized model of Section 15.5 is *totally* independent of the queue discipline. This means that the average measures of performance (W_s, W_q, L_s, and L_q) apply to all queue disciplines.

Although the *average* waiting time is independent of the queue discipline, its probability density function is not. We illustrate this point by deriving the waiting-time distribution for the (*M/M/*1) model based on the FCFS discipline.

Let τ be the amount of time a person *just arriving* must be in the *system* (i.e., until the service is completed). Based on the FCFS discipline, if there are n customers in the system ahead of an arriving customer, then

$$\tau = t_1' + t_2 + \ldots + t_{n+1}$$

where t_1' is the time needed for the customer currently in service to complete service and $t_2, t_3, \ldots, t_n$ are the service times for the $n - 1$ customers in the queue. The time t_{n+1} represents the service time for the arriving customer.

Define $w(\tau|n+1)$ as the conditional density function of τ given n customers in the system ahead of the arriving customer. Because the distribution of the service time is exponential, the forgetfulness property (Section 15.3) tells us that t_1' is also exponential with the same distribution. Thus, τ is the sum of $n + 1$ identically distributed and independent exponential random variables. From probability theory, $w(\tau|n+1)$ follows a gamma distribution with parameters μ and $n + 1$. We thus have

$$\begin{aligned}
w(\tau) &= \sum_{n=0}^{\infty} w(\tau|n+1)p_n \\
&= \sum_{n=0}^{\infty} \frac{\mu(\mu\tau)^n e^{-\mu\tau}}{n!}(1-\rho)\rho^n \\
&= (1-\rho)\mu e^{-\mu\tau} \sum_{n=0}^{\infty} \frac{(\lambda\tau)^n}{n!} \\
&= (1-\rho)\mu e^{-\mu\tau} e^{\lambda\tau} \\
&= (\mu - \lambda)e^{-(\mu-\lambda)\tau}, \tau > 0
\end{aligned}$$

Thus, $w(\tau)$ is an exponential distribution with mean $W_s = \frac{1}{(\mu - \lambda)}$.

Example 15.6-3

In the car wash facility model of Example 15.6-2, it is reasonable to assume that this service is performed based on FCFS discipline. Assess the reliability of using W_s as an estimate of the waiting time in the system.

[1]This material may be skipped without loss of continuity.

One way of answering this question is to estimate the proportion of customers whose waiting time exceeds W_s. Noting that $W_s = \frac{1}{(\mu - \lambda)}$, we get

$$P\{\tau > W_s\} = 1 - \int_0^{W_s} w(\tau)\, d\tau$$
$$= e^{-(\mu-\lambda)W_s} = e^{-1} = .368$$

Thus, under FCFS discipline, about 37% of the customers will wait longer than W_s. This appears excessive, particularly since the current W_s for the car wash facility is already high (= .5 hour). We note that the computed probability ($= e^{-1} \approx .368$) is independent of the rates λ and μ for any $(M/M/1):(FCFS/\infty/\infty)$, which means that its value cannot be reduced. Thus, if we design the system based on the average W_s, then we should expect 36.8% of the customers to wait more than the average waiting time.

The situation can be improved in two ways: (1) we can increase the service rate μ to bring the value of W_s down to an acceptable level, or (2) we can select the service rate such that the probability that the waiting time exceeds a prespecified value (say, 10 minutes) remains under a reasonably small percentage (say, 10%). The first method is equivalent to finding μ such that $W_s < \overline{T}$, and the second method finds μ by solving the inequality $P\{\tau > \overline{T}\} < \alpha$, where $\overline{T}$ and α are specified by the analyst.

PROBLEM SET 15.6C

***1.** In Problem 3, Set 15.6b, determine the probability that detective Columbo will take more than 1 week to solve a crime case.

2. In Example 15.6-3, compute the following:
   - **(a)** The standard deviation of the waiting time τ in the system.
   - **(b)** The probability that the waiting time in the system will vary by half a standard deviation around the mean value.

3. In Example 15.6-3, determine the service rate μ that satisfies the condition $W_s < 10$ minutes.

4. In Example 15.6-3, determine the service rate μ that will satisfy the condition $P\{\tau > 10 \text{ minutes}\} < .1$.

***5.** Consider Problem 5, Set 15.6b. To attract more business, the owner of the restaurant will give free soft drinks to any customer who waits more than 5 minutes. Given that a drink costs 50 cents, how much will it cost daily to offer free drinks? Assume that the restaurant is open for 12 hours a day.

6. Show that for the $(M/M/1):(FCFS/\infty/\infty)$, the distribution of waiting time in the queue is

$$w_q(t) = \begin{cases} 1 - \rho, & t = 0 \\ \mu\rho(1 - \rho)e^{-(\mu-\lambda)t}, & t > 0 \end{cases}$$

Then find W_q from $w_q(t)$.

$(M/M/1)$:$(GD/N/\infty)$. This model differs from $(M/M/1):(GD/\infty/\infty)$ in that there is a limit N on the number in the system (maximum queue length $= N - 1$). Examples include manufacturing situations in which a machine may have a limited buffer area, and a one-lane drive-in window in a fast-food restaurant.

When the number of customers in the system reaches N, no more arrivals are allowed. Thus, we have

$$\lambda_n = \begin{cases} \lambda, & n = 0, 1, \ldots, N - 1 \\ 0, & n = N, N + 1 \end{cases}$$

$$\mu_n = \mu, \quad n = 0, 1, \ldots$$

Using $\rho = \frac{\lambda}{\mu}$, the generalized model in Section 15.5 yields

$$p_n = \begin{cases} \rho^n p_0 & n \leq N \\ 0, & n > N \end{cases}$$

The value of p_0 is determined from the equation $\sum_{n=0}^{\infty} p_n = 1$, which yields

$$p_0(1 + \rho + p^2 + \ldots + \rho^N) = 1$$

or

$$p_0 = \begin{cases} \dfrac{(1 - \rho)}{1 - \rho^{N+1}}, & \rho \neq 1 \\ \dfrac{1}{N + 1}, & \rho = 1 \end{cases}$$

Thus,

$$p_n = \left\{ \begin{array}{ll} \dfrac{(1 - \rho)\rho^n}{1 - \rho^{N+1}}, & \rho \neq 1 \\ \dfrac{1}{N + 1}, & \rho = 1 \end{array} \right\}, n = 0, 1, \ldots, N$$

The value of $\rho = \frac{\lambda}{\mu}$ need *not* be less than 1 in this model, because arrivals at the system are controlled by the system limit N. This means that λ_{eff}, rather than λ, is the rate that matters in this case. Because customers will be lost when there are N in the system, then, as shown in Figure 15.4,

$$\lambda_{\text{lost}} = \lambda p_N$$

$$\lambda_{\text{eff}} = \lambda - \lambda_{\text{lost}} = \lambda(1 - p_N)$$

In this case, $\lambda_{\text{eff}} < \mu$.

The expected number of customers in the system is computed as

$$\begin{aligned} L_s &= \sum_{n=1}^{N} n p_n \\ &= \frac{1 - \rho}{1 - \rho^{N+1}} \sum_{n=0}^{N} n \rho^n \\ &= \left(\frac{1 - \rho}{1 - \rho^{N+1}}\right) \rho \frac{d}{d\rho} \sum_{n=0}^{N} \rho^n \\ &= \frac{(1 - \rho)\rho}{1 - \rho^{N+1}} \frac{d}{d\rho} \left(\frac{1 - \rho^{N+1}}{1 - \rho}\right) \\ &= \frac{\rho[1 - (N + 1)\rho^N + N\rho^{N+1}]}{(1 - \rho)(1 - \rho^{N+1})}, \rho \neq 1 \end{aligned}$$

Scenario 1:(M/M/1):(GD/5/infinity)

Lambda =	4.00000	Mu =	6.00000
Lambda eff =	3.80752	Rho/c =	0.66667
Ls =	1.42256	Lq =	0.78797
Ws =	0.37362	Wq =	0.20695

n	Probability pn	Cumulative Pn	n	Probability pn	Cumulative Pn
0	0.36541	0.36541	3	0.10827	0.87970
1	0.24361	0.60902	4	0.07218	0.95188
2	0.16241	0.77143	5	0.04812	1.00000

FIGURE 15.6

TORA output of Example 15.6-4 (file toraEx15.6-4.txt)

When $\rho = 1$, $L_s = \frac{N}{2}$ (verify!). We can derive W_s, W_q, and L_q from L_s using λ_{eff} as shown in Section 15.6.1.

The use of a hand calculator to compute the queuing formulas is at best cumbersome (the formulas will get more complex in later models!). The use of TORA or template excelPoissonQ.xls to handle these computations is recommended.

Example 15.6-4

Consider the car wash facility of Example 15.6-2. Suppose that the facility has a total of four parking spaces. If the parking lot is full, newly arriving cars balk to other facilities. The owner wishes to determine the impact of the limited parking space on losing customers to the competition.

In terms of the notation of the model, the limit on the system is $N = 4 + 1 = 5$. The following input data provides the output in Figure 15.6.

Lambda	Mu	c	System limit	Source limit
4	6	1	5	infinity

Because the limit on the system is $N = 5$, the proportion of lost customers is $p_5 = .04812$, which, based on a 24-hour day, is equivalent to losing $(\lambda p_5) \times 24 = 4 \times .04812 \times 24 = 4.62$ cars a day. A decision regarding increasing the size of the parking lot should be based on the value of lost business.

Looking at the problem from a different angle, the expected total time in the system, W_s, is .3736 hour, or approximately 22 minutes, down from 30 minutes in Example 15.6-3 when all arriving cars are allowed to join the facility. This reduction of about 25% is secured at the expense of losing about 4.8% of all potential customers because of the limited parking space.

PROBLEM SET 15.6D

***1.** In Example 15.6-4, determine the following:

(a) Probability that an arriving car will go into the wash bay immediately on arrival.

(b) Expected waiting time until a service starts.

(c) Expected number of empty parking spaces.

(d) Probability that all parking spaces are occupied.

(e) Percent reduction in average service time that will limit the average time in the system to about 10 minutes. (*Hint*: Use trial and error with excelPoissonQ.xls or TORA.)

2. Consider the car wash facility of Example 15.6-4. Determine the number of parking spaces such that the percentage of cars that cannot find a space does not exceed 1%.

3. The time barber Joe takes to give a haircut is exponential with a mean of 12 minutes. Because of his popularity, customers usually arrive (according to a Poisson distribution) at a rate much higher than Joe can handle: six customers per hour. Joe really will feel comfortable if the arrival rate is effectively reduced to about four customers per hour. To accomplish this goal, he came up with the idea of providing limited seating in the waiting area so that newly arriving customers will go elsewhere when they discover that all the seats are taken. How many seats should Joe provide to accomplish his goal?

***4.** The final assembly of electric generators at Electro is produced at the Poisson rate of 10 generators per hour. The generators are then conveyed on a belt to the inspection department for final testing. The belt can hold a maximum of 7 generators. An electronic sensor will automatically stop the conveyor once it is full, preventing the final assembly department from assembling more units until a space becomes available. The time to inspect the generators is exponential, with a mean of 15 minutes.

(a) What is the probability that the final assembly department will stop production?

(b) What is the average number of generators on the conveyor belt?

(c) The production engineer claims that interruptions in the assembly department can be reduced by increasing the capacity of the belt. In fact, the engineer claims that the capacity can be increased to the point where the assembly department can operate 95% of the time without interruption. Is this claim justifiable?

5. A cafeteria can seat a maximum of 50 persons. Customers arrive in a Poisson stream at the rate of 10 per hour and are served (one at a time) at the rate of 12 per hour.

(a) What is the probability that an arriving customer will not eat in the cafeteria because it is full?

(b) Suppose that three customers (with random arrival times) would like to be seated together. What is the probability that their wish can be fulfilled? (Assume that arrangements can be made to seat them together as long as three seats are available.)

6. Patients arrive at a 1-doctor clinic according to a Poisson distribution at the rate of 20 patients per hour. The waiting room does not accommodate more than 14 patients. Examination time per patient is exponential, with a mean of 8 minutes.

(a) What is the probability that an arriving patient will not wait?

(b) What is the probability that an arriving patient will find a seat in the room?

(c) What is the expected total time a patient spends in the clinic?

7. The probabilities p_n of n customers in the system for an $(M/M/1):(GD/5/\infty)$ are given in the following table:

n	0	1	2	3	4	5
p_n	.399	.249	.156	.097	.061	.038

The arrival rate λ is five customers per hour. The service rate μ is eight customers per hour. Compute the following:

*(a) Probability that an arriving customer will be able to enter the system.

*(b) Rate at which the arriving customers will not be able to enter the system.

(c) Expected number in the system.

(d) Average waiting time in the queue.

8. Show that when $\rho = 1$ for the $(M/M/1):(GD/N/\infty)$ the expected number in the system, L_s, equals $\frac{N}{2}$. (*Hint*: $1 + 2 + \dots + i = \frac{i(i+1)}{2}$.)

9. Show that λ_{eff} for the $(M/M/1):(GD/N/\infty)$ can be computed from the formula

$$\lambda_{\text{eff}} = \mu(L_s - L_q)$$

15.6.3 Multiple-Server Models

This section considers three queuing models with multiple parallel servers. The first two models are the multiserver versions of the models in Section 15.6.2. The third model treats the self-service case, which is equivalent to having an infinite number of parallel servers.

Real-Life Application—Telephone Sales Manpower Planning at Qantas Airways

To reduce operating costs, Qantas Airways seeks to staff its main telephone sales reservation office efficiently while providing convenient service to its customers. Traditionally, staffing needs are estimated by forecasting future telephone calls based on historical increase in business. The increase in staff numbers is then calculated based on the projected average increase in telephone calls divided by the average number of calls an operator can handle. Because the calculations are based on averages, the additional number of hired staff does not take into account the fluctuations in demand during the day. In particular, long waiting time for service during peak business hours has resulted in customer complaints and lost business. The problem deals with the determination of a plan that strikes a balance between the number of hired operators and the customer needs. The solution uses $(M/M/c)$ queuing analysis imbedded into an integer programming model. Savings from the model in the Sydney office alone were around \$173,000 in fiscal year 1975–1976. The details of the study are given in Case 15, Chapter 24 on the CD.

$(M/M/c)$:$(GD/\infty/\infty)$. In this model, there are c parallel servers. The arrival rate is λ and the service rate per server is μ. Because there is no limit on the number in the system, $\lambda_{\text{eff}} = \lambda$.

The effect of using c parallel servers is a proportionate increase in the facility service rate. In terms of the generalized model (Section 15.5), λ_n and μ_n are thus defined as

$$\lambda_n = \lambda, \quad n \geq 0$$

$$\mu_n = \begin{cases} n\mu, & n < c \\ c\mu, & n \geq c \end{cases}$$

Thus,

$$p_n = \begin{cases} \dfrac{\lambda^n}{\mu(2\mu)(3\mu)\dots(n\mu)}p_0 = \dfrac{\lambda^n}{n!\,\mu^n}p_0 = \dfrac{\rho^n}{n!}p_0, & n < c \\ \dfrac{\lambda^n}{\left(\prod_{i=1}^{c} i\mu\right)(c\mu)^{n-c}}p_0 = \dfrac{\lambda^n}{c!\,c^{n-c}\mu^n}p_0 = \dfrac{\rho^n}{c!\,c^{n-c}}p_0, & n \geq c \end{cases}$$

Letting $\rho = \frac{\lambda}{\mu}$, and assuming $\frac{\rho}{c} < 1$, the value of p_0 is determined from $\sum_{n=0}^{\infty} p_n = 1$, which gives,

$$\begin{aligned} p_0 &= \left\{\sum_{n=0}^{c-1}\frac{\rho^n}{n!} + \frac{\rho^c}{c!}\sum_{n=c}^{\infty}\left(\frac{\rho}{c}\right)^{n-c}\right\}^{-1} \\ &= \left\{\sum_{n=0}^{c-1}\frac{\rho^n}{n!} + \frac{\rho^c}{c!}\left(\frac{1}{1-\frac{\rho}{c}}\right)\right\}^{-1}, \quad \frac{\rho}{c} < 1 \end{aligned}$$

The expression for L_q can be determined as follows:

$$\begin{aligned} L_q &= \sum_{n=c}^{\infty}(n-c)p_n \\ &= \sum_{k=0}^{\infty}kp_{k+c} \\ &= \sum_{k=0}^{\infty}k\frac{\rho^{k+c}}{c^k c!}p_0 \\ &= \frac{\rho^{c+1}}{c!\,c}p_0\sum_{k=0}^{\infty}k\left(\frac{\rho}{c}\right)^{k-1} \\ &= \frac{\rho^{c+1}}{c!\,c}p_0\frac{d}{d\left(\frac{\rho}{c}\right)}\sum_{k=0}^{\infty}\left(\frac{\rho}{c}\right)^k \\ &= \frac{\rho^{c+1}}{(c-1)!\,(c-\rho)^2}p_0 \end{aligned}$$

Because $\lambda_{\text{eff}} = \lambda$, $L_s = L_q + \rho$. The values of W_s and W_q can be determined by dividing L_s and L_q by λ.

Example 15.6-5

A community is served by two cab companies. Each company owns two cabs and both share the market equally, as evidenced by the fact that calls arrive at each company's dispatching office at the rate of eight per hour. The average time per ride is 12 minutes. Calls arrive according to a Poisson distribution, and the ride time is exponential. The two companies recently were bought by an investor who is interested in consolidating them into a single dispatching office to provide better service to customers. Analyze the new owner's proposal.

From the standpoint of queuing, the cabs are the servers, and the cab ride is the service. Each company can be represented by the model $(M/M/2){:}(GD/\infty/\infty)$ with $\lambda = 8$ calls per

```
Comparative analysis
```

c	Lambda	Mu	L'da eff	p0	Ls	Ws	Lq	Wq
2	8.000	5.000	8.00	0.110	4.444	0.556	2.844	0.356
4	16.000	5.000	16.00	0.027	5.586	0.349	2.386	0.149

FIGURE 15.7

TORA output for Example 15.6-5 (file toraEx15.6-5.txt)

hour and $\mu = \frac{60}{10} = 5$ rides *per cab* per hour. Consolidation will result in the model $(M/M/4):(GD/\infty/\infty)$ with $\lambda = 2 \times 8 = 16$ calls per hour and $\mu = 5$ rides *per cab* per hour.

A suitable measure for comparing the two models is the average time a customer waits for a ride, W_q. TORA comparative analysis input data are given as follows:

Scenario	Lambda	Mu	c	System limit	Source limit
1	8	5	2	infinity	infinity
2	16	5	4	infinity	infinity

Figure 15.7 provides the output for the two scenarios. The results show that the waiting time for a ride is .356 hour ($\approx$21 minutes) for the two-cab situation and .149 ($\approx$9 minutes) for the consolidated situation, a remarkable reduction of more than 50% and a clear evidence that the consolidation of the two companies is warranted.

Remark. The conclusion from the preceding analysis is that **pooling services** is *always* a more efficient mode of operation. This result is true even if the separate installations happen to be "very" busy (see Problems 2 and 10, Set 15.6e).

PROBLEM SET 15.6E

1. Consider Example 15.6-5.
  - **(a)** Show that the remarkable reduction in waiting time by more than 50% for the consolidated case is coupled with an increase in the percentage of time the servers remain busy.
  - **(b)** Determine the number of cabs that the consolidated company should have to limit the average waiting time for a ride to 5 minutes or less.

***2.** In the cab company example, suppose that the average time per ride is actually about 14.5 minutes, so that the utilization $\left(=\frac{\lambda}{\mu c}\right)$ for the 2- and 4-cab operations increases to more than 96%. Is it still worthwhile to consolidate the two companies into one? Use the average waiting time for a ride as the comparison measure.

3. Determine the minimum number of parallel servers needed in each of the following (Poisson arrival/departure) situations to guarantee that the operation of the queuing situation will be stable (i.e., the queue length will not grow indefinitely):
  - **(a)** Customers arrive every 5 minutes and are served at the rate of 10 customers per hour.
  - **(b)** The average interarrival time is 2 minutes, and the average service time is 6 minutes.
  - **(c)** The arrival rate is 30 customers per hour, and the service rate per server is 40 customers per hour.

4. Customers arrive at Thrift Bank according to a Poisson distribution, with a mean of 45 customers per hour. Transactions per customer last about 5 minutes and are exponentially distributed. The bank wants to use a single-line multiple-teller operation, similar to the ones used in airports and post offices. The manager is conscious of the fact that customers may switch to other banks if they perceive that their wait in line is "excessive." For this reason, the manager wants to limit the average waiting time in the queue to no more than 30 seconds. How many tellers should the bank provide?

***5.** McBurger fast food restaurant has 3 cashiers. Customers arrive according to a Poisson distribution every 3 minutes and form one line to be served by the first available cashier. The time to fill an order is exponentially distributed with a mean of 5 minutes. The waiting room inside the restaurant is limited. However, the food is good, and customers are willing to line up outside the restaurant, if necessary. Determine the size of the waiting room inside the restaurant (excluding those at the cashiers) such that the probability that an arriving customer does not wait outside the restaurant is at least .999.

6. A small post office has two open windows. Customers arrive according to a Poisson distribution at the rate of 1 every 3 minutes. However, only 80% of them seek service at the windows. The service time per customer is exponential, with a mean of 5 minutes. All arriving customers form one line and access available windows on an FCFS basis.

(a) What is the probability that an arriving customer will wait in line?
(b) What is the probability that both windows are idle?
(c) What is the average length of the waiting line?
(d) Would it be possible to offer reasonable service with only one window? Explain.

7. U of A computer center is equipped with four identical mainframe computers. The number of users at any time is 25. Each user is capable of submitting a job from a terminal every 15 minutes, on the average, but the actual time between submissions is exponential. Arriving jobs will automatically go to the first available computer. The execution time per submission is exponential with mean 2 minutes. Compute the following:

***(a)** The probability that a job is not executed immediately on submission.
(b) The average time until the output of a job is returned to the user.
(c) The average number of jobs awaiting execution.
(d) The percentage of time the entire computer center is idle.
***(e)** The average number of idle computers.

8. Drake Airport services rural, suburban, and transit passengers. The arrival distribution for each of the three groups is Poisson with mean rates of 15, 10, and 20 passengers per hour, respectively. The time to check in a passenger is exponential with mean 6 minutes. Determine the number of counters that should be provided at Drake under each of the following conditions:

(a) The total average time to check a customer in is less than 15 minutes.
(b) The percentage of idleness of the counters does not exceed 10%.
(c) The probability that all counters are idle does not exceed .01.

9. In the United States, the use of single-line, multiple-server queues is common in post offices and in passenger check-in counters at airports. However, both grocery stores and banks (especially in smaller communities) tend to favor single-line, single-server setups, despite the fact that single-line, multiple-server queues offer a more efficient operation. Comment on this observation.

10. For the $(M/M/c):(GD/\infty/\infty)$ model, Morse (1958, p. 103) shows that as $\frac{\rho}{c} \rightarrow 1$,

$$L_q = \frac{\rho}{c - \rho}$$

Noting that $\frac{\rho}{c} \to 1$ means that the servers are extremely busy, use this information to show that the ratio of the average waiting time in queue in the $(M/M/c):(GD/\infty/\infty)$ model to that in the $(M/M/1):(GD/\infty/\infty)$ model approaches $\frac{1}{c}$ as $\frac{\rho}{c} \to 1$. Thus, for $c = 2$, the average waiting time can be reduced by 50%. The conclusion from this exercise is that it is always advisable to pool services regardless of how "overloaded" the servers may be.

11. In the derivation of p_n for the $(M/M/c):(GD/\infty/\infty)$ model, indicate which part of the derivation requires the condition $\frac{\rho}{c} < 1$. Explain verbally the meaning of the condition. What will happen if the condition is not satisfied?

12. Prove that $L_s = L_q + \bar{c}$ starting with the definition $L_q = \sum_{n=c+1}^{\infty}(n - c)p_n$, where $\bar{c}$ is the average number of busy servers. Hence, show that $\bar{c} = \frac{\lambda_{\text{eff}}}{\mu}$.

13. Show that p_n for the $(M/M/1):(GD/\infty/\infty)$ model can be obtained from that of the $(M/M/c):(GD/\infty/\infty)$ by setting $c = 1$.

14. Show that for the $(M/M/c):(GD/\infty/\infty)$ that

$$L_q = \frac{c\rho}{(c - \rho)^2} p_c$$

15. For the $(M/M/c):(GD/\infty/\infty)$ model, show that

(a) The probability that a customer is waiting is $\frac{\rho}{(c - \rho)} p_c$.

(b) The average number in the queue given that it is not empty is $\frac{c}{(c - \rho)}$.

(c) The expected waiting time in the queue for customers who must wait is $\frac{1}{\mu(c - \rho)}$.

16. Prove that the probability density function of waiting time in the queue for the $(M/M/c):(GD/\infty/\infty)$ model is given as

$$w_q(T) = \begin{cases} 1 - \dfrac{\rho^c}{(c-1)!\,(c-\rho)} p_0, & T = 0 \\ \dfrac{\mu\rho^c e^{-\mu(c-\rho)T}}{(c-1)!} p_0, & T > 0 \end{cases}$$

(*Hint*: Convert the c-channel case into an *equivalent* single channel for which

$$P\{t > T\} = P\left\{\min_{1 \le i \le c} t_i > T\right\} = (e^{-\mu T})^c e^{-\mu c T}$$

where t is the service time in the equivalent single channel.)

17. Prove that for $w_q(T)$ in Problem 16

$$P\{T > y\} = P\{T > 0\} e^{-(c\mu - \lambda)y}$$

where $P\{T > 0\}$ is the probability that an arriving customer must wait.

18. Prove that the waiting time in the system for the $(M/M/c):(FCFS/\infty/\infty)$ model has the following probability density function:

$$w(\tau) = \mu e^{-\mu\tau} + \frac{\rho^c \mu e^{-\mu\tau}}{(c-1)!\,(c-\rho-1)} \left\{ \frac{1}{c-\rho} - e^{-\mu(c-\rho-1)\tau} \right\} p_0, \ \tau \ge 0$$

(*Hint:* τ is the convolution of the waiting time in queue, T [see Problem 16], and the service time distribution.)

$(M/M/c)$:$(GD/N/\infty)$, $c \leq N$. This model differs from that of the $(M/M/c)$:$(GD/\infty/\infty)$ model in that the system limit is finite and equal to N. This means that the maximum queue size is $N - c$. The arrival and service rates are λ and μ. The effective arrival rate λ_{eff} is less than λ because of the system limit, N.

In terms of the generalized model (Section 15.5), λ_n and μ_n for the current model are defined as

$$\lambda_n = \begin{cases} \lambda, & 0 \leq n \leq N \\ 0, & n > N \end{cases}$$

$$\mu_n = \begin{cases} n\mu, & 0 \leq n \leq c \\ c\mu, & c \leq n \leq N \end{cases}$$

Substituting λ_n and μ_n in the general expression in Section 15.5 and noting that $\rho = \frac{\lambda}{\mu}$, we get

$$p_n = \begin{cases} \dfrac{\rho^n}{n!} p_0, & 0 \leq n < c \\ \dfrac{\rho^n}{c!\, c^{n-c}} p_0, & c \leq n \leq N \end{cases}$$

where

$$p_0 = \begin{cases} \left(\displaystyle\sum_{n=0}^{c-1} \frac{\rho^n}{n!} + \frac{\rho^c\left(1 - \left(\frac{\rho}{c}\right)^{N-c+1}\right)}{c!\left(1 - \frac{\rho}{c}\right)}\right)^{-1}, & \dfrac{\rho}{c} \neq 1 \\ \left(\displaystyle\sum_{n=0}^{c-1} \frac{\rho^n}{n!} + \frac{\rho^c}{c!}(N - c + 1)\right)^{-1}, & \dfrac{\rho}{c} = 1 \end{cases}$$

Next, we compute L_q for the case where $\frac{\rho}{c} \neq 1$ as

$$\begin{aligned} L_q &= \sum_{n=c}^{N} (n - c) p_n \\ &= \sum_{j=0}^{N-c} j p_{j+c} \\ &= \frac{\rho^c \rho}{c!c} p_0 \sum_{j=0}^{N-c} j\left(\frac{\rho}{c}\right)^{j-1} \\ &= \frac{\rho^{c+1}}{cc!} p_0 \frac{d}{d\left(\frac{\rho}{c}\right)} \sum_{j=0}^{N-c} \left(\frac{\rho}{c}\right)^j \\ &= \frac{\rho^{c+1}}{(c-1)!(c-\rho)^2}\left\{1 - \left(\frac{\rho}{c}\right)^{N-c+1} - (N - c + 1)\left(1 - \frac{\rho}{c}\right)\left(\frac{\rho}{c}\right)^{N-c}\right\} p_0 \end{aligned}$$

It can be shown that for $\frac{\rho}{c} = 1$, L_q reduces to

$$L_q = \frac{\rho^c (N - c)(N - c + 1)}{2c!} p_0, \quad \frac{\rho}{c} = 1$$

To determine W_q and hence W_s and L_s, we compute the value of λ_{eff} as

$$\lambda_{\text{lost}} = \lambda p_N$$

$$\lambda_{\text{eff}} = \lambda - \lambda_{\text{lost}} = (1 - p_N)\lambda$$

```
Scenario1: (M/M/4):(GD/10/infinity)
```

Lambda =	16.00000	Mu =	5.00000
Lambda eff =	15.42815	Rho/c =	0.80000
Ls =	4.23984	Lq =	1.15421
Ws =	0.27481	Wq =	0.07481

n	Probability pn	Cumulative Pn	n	Probability pn	Cumulative Pn
0	0.03121	0.03121	6	0.08726	0.79393
1	0.09986	0.13106	7	0.06981	0.86374
2	0.15977	0.29084	8	0.05584	0.91958
3	0.17043	0.46126	9	0.04468	0.96426
4	0.13634	0.59760	10	0.03574	1.00000

FIGURE 15.8

TORA output of Example 15.6-6 (file toraEx15.6-6.txt)

Example 15.6-6

In the consolidated cab company problem of Example 15.6-5, suppose that new funds cannot be secured to purchase additional cabs. The owner was advised by a consultant that one way to reduce the waiting time is for the dispatching office to inform new customers of potential excessive delay once the waiting list reaches 6 customers. This move is certain to get new customers to seek service elsewhere, but will reduce the waiting time for those on the waiting list. Assess the friend's advice.

Limiting the waiting list to 6 customers is equivalent to setting $N = 6 + 4 = 10$ customers. We are thus investigating the model $(M/M/4):(GD/10/\infty)$, where $\lambda = 16$ customers per hour and $\mu = 5$ rides per hour. The following input data provide the results in Figure 15.8.

Lambda	Mu	c	System limit	Source limit
16	5	4	10	infinity

The average waiting time, W_q, before setting a limit on the capacity of the system is .149 hour ($\approx$9 minutes) (see Figure 15.7), which is about twice the new average of .075 hour ($\approx$4.5 minutes). This remarkable reduction is achieved at the expense of losing about 3.6% of potential customers ($p_{10} = .03574$). However, this result does not reflect the effect of possible loss of customer goodwill on the operation of the company.

PROBLEM SET 15.6F

1. In Example 15.6-6, determine the following:
   - **(a)** The expected number of idle cabs.
   - **(b)** The probability that a calling customer will be the last on the list.
   - **(c)** The limit on the waiting list if it is desired to keep the waiting time in the queue to below 3 minutes.

2. Eat & Gas convenience store operates a two-pump gas station. The lane leading to the pumps can house at most 3 cars, excluding those being serviced. Arriving cars go elsewhere if the lane is full. The distribution of arriving cars is Poisson with mean 20 per hour. The time to fill up and pay for the purchase is exponential with mean 6 minutes. Determine the following:

(a) Percentage of cars that will seek business elsewhere.

(b) Percentage of time one pump is in use.

*__(c)__ Percent utilization of the two pumps.

*__(d)__ Probability that an arriving car will not start service immediately but will find an empty space in the lane.

(e) Capacity of the lane that will ensure that, on the average, no more than 10% of the arriving cars are turned away.

(f) Capacity of the lane that will ensure that the probability that both pumps are idle is .05 or less.

3. A small engine repair shop is run by three mechanics. Early in March of each year, people bring in their tillers and lawn mowers for service and maintenance. The shop is willing to accept all the tillers and mowers that customers bring in. However, when new customers see the floor of the shop covered with waiting jobs, they go elsewhere for more prompt service. The floor shop can house a maximum of 15 mowers or tillers, excluding those being serviced. The customers arrive at the shop every 10 minutes on the average, and it takes a mechanic an average of 30 minutes to complete each job. Both the interarrival and the service times are exponential. Determine the following:

(a) Average number of idle mechanics:

(b) Amount of business lost to competition per 10-hour day because of the limited capacity of the shop.

(c) Probability that the next arriving customer will be serviced by the shop.

(d) Probability that at least one of the mechanics will be idle.

(e) Average number of tillers or mowers awaiting service.

(f) A measure of the overall productivity of the shop.

4. At U of A, newly enrolled freshmen students are notorious for wanting to drive their cars to class (even though most of them are required to live on campus and can conveniently make use of the university free transit system). During the first couple of weeks of the fall semester, traffic havoc prevails on campus as freshmen try desperately to find parking spaces. With unusual dedication, the students wait patiently in the lanes of the parking lot for someone to leave so they can park their cars. Let us consider a specific scenario: The parking lot has 30 parking spaces but can also accommodate 10 more cars in the lanes. These additional 10 cars cannot park in the lanes permanently and must await the availability of one of the 30 parking spaces. Freshman students arrive at the parking lot according to a Poisson distribution, with a mean of 20 cars per hour. The parking time per car averages about 60 minutes but actually follows an exponential distribution.

*__(a)__ What is the percentage of freshmen who are turned away because they cannot enter the lot?

*__(b)__ What is the probability that an arriving car will wait in the lanes?

(c) What is the probability that an arriving car will occupy the only remaining parking space on the lot?

*(d) Determine the average number of occupied parking spaces.

(e) Determine the average number of spaces that are occupied in the lanes.

*(f) Determine the number of freshmen who will not make it to class during an 8-hour period because the parking lot is totally full.

5. Verify the expression for p_0 for the $(M/M/c)$:$(GD/N/\infty)$ given that $\frac{\rho}{c} \neq 1$.
6. Prove the following equality for the $(M/M/c)$:$(GD/N/\infty)$

$$\lambda_{\text{eff}} = \mu\bar{c}$$

where $\bar{c}$ is the number of busy servers.

7. Verify the expression for p_0 and L_q for the $(M/M/c)$:$(GD/N/\infty)$ when $\frac{\rho}{c} = 1$.
8. For the $(M/M/c)$:$(GD/N/\infty)$ model in which $N = c$, define λ_n and μ_n in terms of the generalized model (Section 15.5), then show that the expression for p_n is given as

$$p_n = \frac{\rho^n}{n!} p_0, n = 1, 2, \ldots, c$$

where

$$p_0 = \left(1 + \sum_{n=1}^{c} \frac{\rho^n}{n!}\right)^{-1}$$

$(M/M/\infty)$:$(GD/\infty/\infty)$—Self-Service Model. In this model, the number of servers is unlimited because the customer is also the server. A typical example is taking the written part of a driver's license test. Self-service gas stations and 24-hour ATM banks do not fall under this model's description because the servers in these cases are actually the gas pumps and the ATM machines. The model assumes steady arrival and service rates, λ and μ, respectively.

In terms of the generalized model of Section 15.5, we have

$$\lambda_n = \lambda, \quad n = 0, 1, 2, \ldots$$

$$\mu_n = n\mu, n = 0, 1, 2, \ldots$$

Thus,

$$p_n = \frac{\lambda^n}{n!\,\mu^n} p_0 = \frac{\rho^n}{n!} p_0, n = 0, 1, 2, \ldots$$

Because $\sum_{n=0}^{\infty} p_n = 1$, it follows that

$$p_0 = \frac{1}{1 + \rho + \frac{\rho^2}{2!} + \ldots} = \frac{1}{e^{\rho}} = e^{-\rho}$$

As a result,

$$p_n = \frac{e^{-\rho}\rho^n}{n!}, n = 0, 1, 2, \ldots$$

which is Poisson with mean $L_s = \rho$. As should be expected, L_q and W_q are zero because it is a self-service model.

Example 15.6-7

An investor invests $1000 a month on average in one type of stock market security. Because the investor must wait for a good "buy" opportunity, the actual time of purchase is totally random. The investor usually keeps the securities for about 3 years on the average but will sell them at random times when a "sell" opportunity presents itself. Although the investor is generally recognized as a shrewd stock market player, past experience indicates that about 25% of the securities decline at about 20% a year. The remaining 75% appreciate at the rate of about 12% a year. Estimate the investor's (long-run) average equity in the stock market.

This situation can be treated as an $(M/M/\infty)$:$(GD/\infty/\infty)$ because, for all practical purposes, the investor does not have to wait in line to buy or to sell securities. The average time between order placements is 1 month, which yields $\lambda = 12$ securities per year. The rate of selling securities is $\mu = \frac{1}{3}$ security per year. You can secure the model output using the following input:

Lambda	Mu	c	System limit	Source limit
12	.3333333	infinity	infinity	infinity

Given the values of λ and μ, we obtain

$$L_s = \rho = \frac{\lambda}{\mu} = 36 \text{ securities}$$

The estimate of the (long-run) average *annual* net worth of the investor is

$$(.25L_s \times \$1000)(1 - .20) + (.75L_s \times \$1000)(1 + .12) = \$63{,}990$$

PROBLEM SET 15.6G

1. In Example 15.6-7, compute the following:

(a) The probability that the investor will sell out completely.

(b) The probability that the investor will own at least 10 securities.

(c) The probability that the investor will own between 30 and 40 securities, inclusive.

(d) The investor's net annual equity if only 10% of the securities depreciate by 30% a year, and the remaining 90% appreciate by 15% a year.

2. New drivers are required to pass written tests before they are given a road driving test. These tests are usually administered by the city police department. Records at the City of Springdale show that the average number of written tests is 100 per 8-hour day. The average time needed to complete the test is about 30 minutes. However, the actual arrival of test takers and the time each spends on the test are totally random. Determine the following:

***(a)** The average number of seats the police department should provide in the test hall.

***(b)** The probability that the number of test takers will exceed the average number of seats provided in the test hall.

(c) The probability that no tests will be administered in any one day.

3. Show (by using excelPoissonQ.xls or TORA) that for small $\rho = .1$, the values of L_s, L_q, W_s, W_q, and p_n for the $(M/M/c)$:$(GD/\infty/\infty)$ model can be estimated reliably using the less cumbersome formulas of the $(M/M/\infty)$:$(GD/\infty/\infty)$ model for c as small as 4 servers.

4. Repeat Problem 3 for large $\rho = 9$ and show that the same conclusion holds except that the value of c must be higher (at least 14). From the results of Problems 3 and 4, what general conclusion can be drawn regarding the use of the $(M/M/\infty)$:$(GD/\infty/\infty)$ to estimate the results of the $(M/M/c)$:$(GD/\infty/\infty)$ model?

15.6.4 Machine Servicing Model—$(M/M/R)$:$(GD/K/K)$, $R < K$

The setting for this model is a shop with K machines. When a machine breaks down, one of R available repairpersons is called upon to do the repair. The rate of breakdown *per machine* is λ breakdowns per unit time, and a repairperson will service broken machines at the rate of μ machines per unit time. All breakdowns and services are assumed to follow the Poisson distribution.

This model differs from all the preceding ones because it has a finite calling source. We can see this point by realizing that when all the machines in the shop are broken, no more calls for service can be generated. In essence, only machines in working order can break down and hence can generate calls for service.

Given the rate of breakdown per machine, λ, the rate of breakdown for the *entire shop* is proportional to the number of machines that are in working order. In terms of the queuing model, having n machines *in the system* signifies that n machines are broken. Thus, the rate of breakdown for the entire shop is

$$\lambda_n = (K - n)\lambda, 0 \leq n \leq K$$

In terms of the generalized model of Section 15.5, we have

$$\lambda_n = \begin{cases} (K - n)\lambda, & 0 \leq n \leq K \\ 0, & n \geq K \end{cases}$$

$$\mu_n = \begin{cases} n\mu, & 0 \leq n \leq R \\ R\mu, & R \leq n \leq K \end{cases}$$

From the generalized model, we can then obtain (verify!)

$$p_n = \begin{cases} C_n^K \rho^n p_0, & 0 \leq n \leq R \\ C_n^K \dfrac{n!\,\rho^n}{R!\,R^{n-R}} p_0, & R \leq n \leq K \end{cases}$$

$$p_0 = \left(\sum_{n=0}^{R} C_n^K \rho^n + \sum_{n=R+1}^{K} C_n^K \frac{n!\,\rho^n}{R!\,R^{n-R}} \right)^{-1}$$

There is no closed form expression for L_s, and hence it must be computed using the following basic definition:

$$L_s = \sum_{n=0}^{K} n p_n$$

The value of λ_{eff} is computed as

$$\lambda_{\text{eff}} = E\{\lambda(K - n)\} = \lambda(K - L_s)$$

Using the formulas in Section 15.6.1, we can compute the remaining measures of performance W_s, W_q, and L_q.

Example 15.6-8

Toolco operates a machine shop with a total of 22 machines. Each machine is known to break down once every 2 hours, on the average. It takes an average of 12 minutes to complete a repair. Both the time between breakdowns and the repair time follow the exponential distribution. Toolco is interested in determining the number of repairpersons needed to keep the shop running "smoothly."

The situation can be analyzed by investigating the productivity of the machines as a function of the number of repairpersons. Such productivity measure can be defined as

$$\begin{pmatrix}\text{Machines}\\ \text{productivity}\end{pmatrix} = \frac{\text{Available machines} - \text{Broken machines}}{\text{Available machines}} \times 100$$

$$= \frac{22 - L_s}{22} \times 100$$

The results for this situation can be obtained using the following input data: lambda = .5, mu = 5, R = 1, 2, 3, or 4, system limit = 22, and source limit = 22. Figure 15.9 provides the output. The associated productivity is computed as follows:

Repairperson, R	1	2	3	4
Machines productivity (100%)	45.44	80.15	88.79	90.45
Marginal increase (100%)	—	34.71	8.64	1.66

The results show that with one repairperson the productivity is low (= 45.44%). By increasing the number of repairpersons to two, the productivity jumps by 34.71% to 80.15%. When we employ three repairpersons, the productivity increases only by about 8.64% to 88.79%, whereas four repairpersons will increase the productivity by a meager 1.66% to 90.45%.

Judging from these results, the use of two repairpersons is justifiable. The case for three repairpersons is not as strong because it raises the productivity by only 8.64%. Perhaps a monetary comparison between the cost of hiring a third repairperson and the income attributed to the 8.64% increase in productivity can be used to settle this point (see Section 15.10 for discussion of cost models). As for hiring a fourth repairperson, the meager increase of 1.66% in productivity does not justify such an action.

FIGURE 15.9

TORA comparative analysis output for Example 15.6-8 (file toraEx15.6-8.txt)

Comparative Analysis

c	Lambda	Mu	L'da eff	p0	Ls	Lq	Ws	Wq
1	0.500	5.00	4.9980	0.0004	12.0040	11.0044	2.4018	2.2018
2	0.500	5.00	8.8161	0.0564	4.3677	2.6045	0.4954	0.2954
3	0.500	5.00	9.7670	0.1078	2.4660	0.5128	0.2525	0.0525
4	0.500	5.00	9.9500	0.1199	2.1001	0.1102	0.2111	0.0111

PROBLEM SET 15.6H

1. In Example 15.6-8, do the following:

(a) Verify the values of λ_{eff} given in Figure 15.9.

***(b)** Compute the expected number of idle repairpersons given $R = 4$.

(c) Compute the probability that all repairpersons are idle given $R = 3$.

***(d)** Compute the probability that the majority (more than half) of repairpersons are idle given $R = 3$.

2. In Example 15.6-8, define and compute the productivity of the repairpersons for $R = 1$, 2, 3, and 4. Use this information in conjunction with the measure of machine productivity to decide on the number of repairpersons Toolco should hire.

3. In the computations in Figure 15.9, it may appear confusing that the average rate of machine breakdown in the shop, λ_{eff}, increases with the increase in R. Explain why the increase in λ_{eff} should be expected.

***4.** An operator attends five automatic machines. After each machine completes a batch run, the operator must reset it before a new batch is started. The time to complete a batch run is exponential with mean 45 minutes. The setup time is also exponential with mean 8 minutes.

(a) Determine the average number of machines that are awaiting setup or are being set up.

(b) Compute the probability that all machines are working.

(c) Determine the average time a machine is down.

5. Kleen All is a service company that performs a variety of odd jobs, such as yard work, tree pruning, and house painting. The company's four employees leave the office with the first assignment of the day. After completing an assignment, the employee calls the office requesting instruction for the next job to be performed. The time to complete an assignment is exponential, with a mean of 45 minutes. The travel time between jobs is also exponential, with a mean of 20 minutes.

(a) Determine the average number of employees who are traveling between jobs.

(b) Compute the probability that no employee is on the road.

***6.** After a long wait, the Newborns were rewarded with quintuplets, two boys and three girls, thanks to the wonders of new medical advances. During the first 5 months, the babies' life consisted of two states: awake (and mostly crying) and asleep. According to the Newborns, the babies "awake-asleep" activities never coincide. Instead, the whole affair is totally random. In fact, Mrs. Newborn, a statistician by profession, believes that the length of time each baby cries is exponential, with a mean of 30 minutes. The amount of sleep each baby gets also happens to be exponential, with a mean of 2 hours. Determine the following:

(a) The average number of babies who are awake at any one time.

(b) The probability that all babies are asleep.

(c) The probability that the Newborns will not be happy because more babies are awake (and crying) than are asleep.

7. Verify the expression for p_n for the $(M/M/R)$:$(GD/K/K)$ model.

8. Show that the rate of breakdown in the shop can be computed from the formula

$$\lambda_{\text{eff}} = \mu \overline{R}$$

where $\overline{R}$ is the average number of busy repairpersons.

9. Verify the following results for the special case of one repairperson ($R = 1$):

$$p_n = \frac{K!\,\rho^n}{(K-n)!}p_0$$

$$p_0 = \left(1 + \sum_{n=1}^{R}\frac{K!\,\rho^n}{(K-n)}\right)^{-1}$$

$$L_s = K - \frac{(1-p_0)}{\rho}$$

15.7 $(M/G/1):(GD/\infty/\infty)$—POLLACZEK-KHINTCHINE (P-K) FORMULA

Queuing models in which the arrivals and departures do not follow the Poisson distribution are complex. In general, it is advisable to use simulation as an alternative tool for analyzing these situations (see Chapter 16).

This section presents one of the few non-Poisson queues for which analytic results are available. It deals with the case in which the service time, t, is represented by any probability distribution with mean $E\{t\}$ and variance $\text{var}\{t\}$. The results of the model include the basic measures of performance L_s, L_q, W_s, and W_q. The model does not provide a closed-form expression for p_n because of analytic intractability.

Let λ be the arrival rate at the single-server facility. Given $E\{t\}$ and $\text{var}\{t\}$ of the service time distribution and that $\lambda E\{t\} < 1$, it can be shown using sophisticated probability/Markov chain analysis that

$$L_s = \lambda E\{t\} + \frac{\lambda^2(E^2\{t\} + \text{var}\{t\})}{2(1-\lambda E\{t\})},\ \lambda E\{t\} < 1$$

The probability that the facility is empty (idle) is computed as

$$p_0 = 1 - \lambda E\{t\} = 1 - \rho$$

Because $\lambda_{\text{eff}} = \lambda$, the remaining measures of performance (L_q, W_s, and W_q) can be derived from L_s, as explained in Section 15.6.1.

Template excelPKFormula.xls automates the calculations of this model.

Example 15.7-1

In the Automata car wash facility of Example 15.6-2, suppose that a new system is installed so that the service time for all cars is constant and equal to 10 minutes. How does the new system affect the operation of the facility?

From Example 15.6-2, $\lambda_{\text{eff}} = \lambda = 4$ cars per hour. The service time is constant so that $E\{t\} = \frac{10}{60} = \frac{1}{6}$ hour and $\text{var}\{t\} = 0$. Thus,

$$L_s = 4\left(\tfrac{1}{6}\right) + \frac{4^2\left(\left(\tfrac{1}{6}\right)^2 + 0\right)}{2\left(1-\tfrac{4}{6}\right)} = 1.33 \text{ cars}$$

$$L_q = 1.333 - \left(\tfrac{4}{6}\right) = .667 \text{ cars}$$

$$W_s = \frac{1.333}{4} = .333 \text{ hour}$$

$$W_q = \frac{.667}{4} = .167 \text{ hour}$$

It is interesting that even though the arrival and departure rates are the same as in the Poisson case of Example 15.6-2 ($\lambda = 4$ cars per hour and $\mu = \frac{1}{E\{t\}} = 6$ cars per hour), the expected waiting time is lower in the current model because the service time is constant, as the following table shows.

	$(M/M/1):(GD/\infty/\infty)$	$(M/D/1):(GD/\infty/\infty)$
W_s (hr)	.500	.333
W_q (hr)	.333	.167

The results make sense because a constant service time indicates *more certainty* in the operation of the facility.

PROBLEM SET 15.7A

1. In Example 15.7-1, compute the percentage of time the facility is idle.

2. Solve Example 15.7-1 assuming that the service-time distribution is given as follows:

*__(a)__ Uniform between 8 and 20 minutes.

(b) Normal with $\mu = 12$ minutes and $\sigma = 3$ minutes.

(c) Discrete with values equal to 4, 8, and 15 minutes and probabilities .2, .6, and .2, respectively.

3. Layson Roofing Inc. installs shingle roofs on new and old residences in Arkansas. Prospective customers request the service randomly at the rate of nine jobs per 30-day month and are placed on a waiting list to be processed on a FCFS basis. Homes sizes vary, but it is fairly reasonable to assume that the roof areas are uniformly distributed between 150 and 300 squares. The work crew can usually complete 75 squares a day. Determine the following:

(a) Layson's average backlog of roofing jobs.

(b) The average time a customer waits until a roofing job is completed.

(c) If the work crew is increased to the point where they can complete 150 squares a day, how will this affect the average time until a job is completed?

*__4.__ Optica, Ltd., makes prescription glasses according to orders received from customers. Each worker is specialized in certain types of glasses. The company has been experiencing unusual delays in the processing of bifocal and trifocal prescriptions. The worker in charge receives 30 orders per 8-hour day. The time to complete a prescription is normally distributed, with a mean of 12 minutes and a standard deviation of 3 minutes. After spending between 2 and 4 minutes, uniformly distributed, to inspect the glasses, the worker can start on a new prescription. Determine the following:

(a) The percentage of time the worker is idle.

(b) The average backlog of bifocal and trifocal prescriptions in Optica.

(c) The average time until a prescription is filled.

5. A product arrives according to a Poisson distribution at the rate of one every 45 minutes. The product requires two tandem operations attended by one worker. The first operation uses a semiautomatic machine that completes its cycle in exactly 28 minutes. The second operation makes adjustments and minor changes, and its time depends on the condition of the product when it leaves operation 1. Specifically, the time of operation 2 is uniform between 3 and 6 minutes. Because each operation requires the complete attention of the worker, a new item cannot be loaded on the semiautomatic machine until the current item has cleared operation 2.
   (a) Determine the number of items that are awaiting processing on the semiautomatic machine.
   (b) What is the percentage of time the worker will be idle?
   (c) How much time is needed, on the average, for an arriving item to clear operation 2?
6. $(M/D/1):(GD/\infty/\infty)$. Show that for the case where the service time is constant, the P-K formula reduces to

$$L_s = \rho + \frac{\rho^2}{2(1-\rho)}$$

   where $\mu = \frac{1}{E\{t\}}$ and $\rho = \frac{\lambda}{\mu} = \lambda E\{t\}$.
7. $(M/E_m/1):(GD/\infty/\infty)$. Given that the service time is Erlang with parameters m and μ (i.e., $E\{t\} = \frac{m}{\mu}$ and $\text{var}\{t\} = \frac{m}{\mu^2}$), show that the P-K formula reduces to

$$L_s = m\rho + \frac{m(1+m)\rho^2}{2(1-m\rho)}$$

8. Show that the P-K formula reduces to L_s of the $(M/M/1):(GD/\infty/\infty)$ when the service time is exponential with a mean of $\frac{1}{\mu}$ time units.
9. In a service facility with c parallel servers, suppose that customers arrive according to a Poisson distribution, with a mean rate of λ. Arriving customers are assigned to servers (busy or free) on a strict rotational basis.
   (a) Determine the probability distribution of the interarrival time.
   (b) Suppose in part (a) that arriving customers are assigned randomly to the c servers with probabilities α_i, $\alpha_i \geq 0$, $i = 1, 2, \ldots, c$, and $\alpha_1 + \alpha_2 + \ldots + \alpha_c = 1$. Determine the probability distribution of the interarrival time.

15.8 OTHER QUEUING MODELS

The preceding sections have concentrated on the Poisson queuing models. Queuing literature is rich with other types of models. In particular, queues with priority for service, network queues, and non-Poisson queues form an important body of the queuing theory literature. These models can be found in most specialized books on queuing theory.

15.9 QUEUING DECISION MODELS

The *service level* in a queuing facility is a function of the service rate, μ, and the number of parallel servers, c. This section presents two decision models for determining "suitable" service levels for queuing systems: (1) a cost model, and (2) an aspiration-level model. Both models recognize that higher service levels reduce the waiting time in the system. Both models aim at striking a balance between the conflicting factors of service level and waiting.

15.9.1 Cost Models

Cost models attempt to balance two conflicting costs:

1. Cost of offering the service.
2. Cost of delay in offering the service (customer waiting time).

The two types of costs are in conflict because an increase in one automatically causes reduction in the other, as demonstrated earlier in Figure 15.1.

Letting x (= μ or c) represent the *service level*, the cost model can be expressed as

$$ETC(x) = EOC(x) + EWC(x)$$

where

$$ETC = \text{Expected total cost } \textit{per unit time}$$

$$EOC = \text{Expected cost of operating the facility } \textit{per unit time}$$

$$EWC = \text{Expected cost of waiting } \textit{per unit time}$$

The simplest forms for EOC and EWC are the following linear functions:

$$EOC(x) = C_1 x$$
$$EWC(x) = C_2 L_s$$

where

$$C_1 = \textit{Marginal} \text{ cost per unit of } x \text{ per unit time}$$

$$C_2 = \text{Cost of waiting per unit time per (waiting) customer}$$

The following two examples illustrate the use of the cost model. The first example assumes x to equal the service rate, μ, and the second assumes x to equal the number of parallel servers, c.

Example 15.9-1

KeenCo Publishing is in the process of purchasing a high-speed commercial copier. Four models whose specifications are summarized below have been proposed by vendors.

Copier model	Operating cost ($/hr)	Speed (sheets/min)
1	15	30
2	20	36
3	24	50
4	27	66

Jobs arrive at KeenCo according to a Poisson distribution with a mean of four jobs per 24-hour day. Job size is random but averages about 10,000 sheets per job. Contracts with the customers specify a penalty cost for late delivery of $80 per jobs per day. Which copier should KeenCo purchase?

Let the subscript i represent copier model i ($i = 1, 2, 3, 4$). The total expected cost *per day* associated with copier i is

$$\begin{aligned} ETC_i &= EOC_i + EWC_i \\ &= C_{1i} \times 24 + C_{2i}L_{si} \\ &= 24C_{1i} + 80L_{si}, i = 1, 2, 3, 4 \end{aligned}$$

The values of C_{1i} are given by the data of the problem. We determine L_{si} by recognizing that, for all practical purposes, each copier can be treated as an $(M/M/1):(GD/\infty/\infty)$ model. The arrival rate is $\lambda = 4$ jobs/day. The service rate μ_i associated with model i is computed as

Model i	Service rate μ_i (jobs/day)
1	4.32
2	5.18
3	7.20
4	9.50

Computation of the service rate is demonstrated for model 1.

$$\text{Average time per job} = \frac{10{,}000}{30} \times \frac{1}{60} = 5.56 \text{ hours}$$

Thus,

$$\mu_1 = \frac{24}{5.56} = 4.32 \text{ jobs/day}$$

The values of L_{si}, computed by TORA or excelPoissonQ.xls, are given in the following table:

Model i	λ_i (Jobs/day)	μ_i (Jobs/day)	L_{si} (Jobs)
1	4	4.32	12.50
2	4	5.18	3.39
3	4	7.20	1.25
4	4	9.50	0.73

The costs for the four models are computed as follows:

Model i	EOC_i (\$)	EWC_i (\$)	ETC_i (\$)
1	360.00	1000.00	1360.00
2	480.00	271.20	751.20
3	**576.00**	**100.00**	**676.00**
4	648.00	58.40	706.40

Model 3 produces the lowest cost.

PROBLEM SET 15.9A

1. In Example 15.9-1, do the following:

(a) Verify the values of μ_2, μ_3, and μ_4 given in the example.

(b) Suppose that the penalty of $80 per job per day is levied only on jobs that are *not* "in progress" at the end of the day. Which copier yields the lowest total cost per day?

***2.** Metalco is in the process of hiring a repairperson for a 10-machine shop. Two candidates are under consideration. The first candidate can carry out repairs at the rate of 5 machines per hour and earns $15 an hour. The second candidate, being more skilled, receives $20 an hour and can repair 8 machines per hour. Metalco estimates that each broken machine will incur a cost of $50 an hour because of lost production. Assuming that machines break down according to a Poisson distribution with a mean of 3 per hour and that repair time is exponential, which repairperson should be hired?

3. B&K Groceries is opening a new store boasting "state-of-the-art" check-out scanners. Mr. Bih, one of the owners of B&K, has limited the choices to two scanners: scanner A can process 10 items a minute, and the better-quality scanner B can scan 15 items a minute. The daily (10 hours) cost of operating and maintaining the scanners are $25 and $35 for models A and B, respectively. Customers who finish shopping arrive at the cashier according to a Poisson distribution at the rate of 10 customers per hour. Each customer's cart carries between 25 and 35 items, uniformly distributed. Mr. Bih estimates the average cost per waiting customer per minute to be about 20 cents. Which scanner should B&K acquire? (*Hint*: The service time per customer is not exponential. It is uniformly distributed.)

4. H&I Industry produces a special machine with different production rates (pieces per hour) to meet customer specifications. A shop owner is considering buying one of these machines and wants to decide on the most economical speed (in pieces per hour) to be ordered. From past experience, the owner estimates that orders from customers arrive at the shop according to a Poisson distribution at the rate of three orders per hour. Each order averages about 500 pieces. Contracts between the owner and the customers specify a penalty of $100 per late order per hour.

(a) Assuming that the actual production time per order is exponential, develop a general cost model as a function of the production rate, μ.

***(b)** From the cost model in (a), determine an expression for the optimal production rate.

***(c)** Using the data given in the problem, determine the optimal production rate the owner should request from H&I.

5. Jobs arrive at a machine shop according to a Poisson distribution at the rate of 80 jobs per week. An automatic machine represents the bottleneck in the shop. It is estimated that a unit increase in the production rate of the machine will cost $250 per week. Delayed jobs normally result in lost business, which is estimated to be $500 per job per week. Determine the optimum production rate for the automatic machine.

6. Pizza Unlimited sells two franchised restaurant models. Model A has a capacity of 20 groups of customers, and model B can seat 30 groups. The monthly cost of operating model A is $12,000 and that of model B is $16,000. An investor wants to set up a buffet-style pizza restaurant and estimates that groups of customers, each occupying one table, arrive according to a Poisson distribution at a rate of 25 groups per hour. If all the tables are occupied, customers will go elsewhere. Model A will serve 26 groups per hour, and model B will serve 29 groups per hour. Because of the variation in group sizes and in the types of orders, the service time is exponential. The investor estimates that the average cost of lost business per customer group per hour is \$15. A delay in serving waiting customers is estimated to cost an average of \$10 per customer group per hour.

(a) Develop an appropriate cost mode.

(b) Assuming that the restaurant will be open for business 10 hours a day, which model would you recommend for the investor?

7. Suppose in Problem 6 that the investor can choose any desired restaurant capacity based on a specific marginal cost for each additional capacity unit requested. Derive the associated general cost model, and define all its components and terms.

8. Second Time Around sells popular used items on consignment. Its operation can be viewed as an inventory problem in which the stock is replenished and depleted randomly according to Poisson distributions with rates λ and μ items per day. Every time unit the item is out of stock, Second Time loses $\$C_1$ because of lost opportunities, and every time unit an item is held in stock, a holding cost $\$C_2$ is incurred.

(a) Develop an expression for the expected total cost per unit time.

(b) Determine the optimal value of $\rho = \frac{\lambda}{\mu}$. What condition must be imposed on the relative values of C_1 and C_2 in order for the solution to be consistent with the assumptions of the $(M/M/1):(GD/\infty/\infty)$ model?

Example 15.9-2

In a multiclerk tool crib facility, requests for tool exchange occur according to a Poisson distribution at the rate of 17.5 requests per hour. Each clerk can handle an average of 10 requests per hour. The cost of hiring a new clerk in the facility is \$12 an hour. The cost of lost production per waiting machine per hour is approximately \$50. Determine the optimal number of clerks for the facility.

The situation corresponds to an $(M/M/c)$ model in which it is desired to determine the optimum value of c. Thus, in the general cost model presented at the start of this section, we put $x = c$, resulting in the following cost model:

$$ETC(c) = C_1c + C_2L_s(c)$$
$$= 12c + 50L_s(c)$$

Note that $L_s(c)$ is a function of the number of (parallel) clerks in the crib.

We use the $(M/M/c):(GD/\infty/\infty)$ model with $\lambda = 17.5$ requests per hour and $\mu = 10$ requests per hour. In this regard, the model will reach steady state only if $c > \frac{\lambda}{\mu}$—that is, for the present example, $c \geq 2$. The following table provides the necessary calculation for determining optimal c. The values of $L_s(c)$ (determined by excelPoissonQ.xls or TORA) given below show that the optimum number of clerks is 4.

c	$L_s(c)$ (requests)	ETC(c) (\$)
2	7.467	397.35
3	2.217	146.85
4	**1.842**	**140.10**
5	1.769	148.45
6	1.754	159.70

PROBLEM SET 15.9B

1. Solve Example 15.9-2, assuming that $C_1 = \$20$ and $C_2 = \$45$.

*2. Tasco Oil owns a pipeline booster unit that operates continuously. The time between breakdowns for each booster is exponential with a mean of 20 hours. The repair time is exponential with mean 3 hours. In a particular station, two repairpersons attend 10 boosters. The hourly wage for each repairperson is \$18. Pipeline losses are estimated to be \$30 per broken booster per hour. Tasco is studying the possibility of hiring an additional repairperson.
   - **(a)** Will there be any cost savings in hiring a third repairperson?
   - **(b)** What is the schedule loss in dollars per breakdown when the number of repairpersons on duty is two? Three?

3. A company leases a wide-area telecommunications service (WATS) telephone line for \$2000 a month. The office is open 200 working hours per month. At all other times, the WATS line service is used for other purposes and is not available for company business. Access to the WATS line during business hours is extended to 100 salespersons, each of whom may need the line at any time but averages twice per 8-hour day with exponential time between calls. A salesperson will always wait for the WATS line if it is busy at an estimated inconvenience of 1 cent per minute of waiting. It is assumed that no additional needs for calls will arise while the salesperson waits for a given call. The normal cost of calls (not using the WATS line) averages about 50 cents per minute, and the duration of each call is exponential, with a mean of 6 minutes. The company is considering leasing (at the same price) a second WATS line to improve service.
   - **(a)** Is the single WATS line saving the company money over a no-WATS system? How much is the company gaining or losing per month over the no-WATS system?
   - **(b)** Should the company lease a second WATS line? How much would it gain or lose over the single WATS case by leasing an additional line?

*4. A machine shop includes 20 machines and 3 repairpersons. A working machine breaks down randomly according to a Poisson distribution. The repair time per machine is exponential with a mean of 6 minutes. A queuing analysis of the situation shows an average of 57.8 calls for repair per 8-hour day for the entire shop. Suppose that the production rate per machine is 25 units per hour and that each produced unit generates \$2 in revenue. Further, assume that a repairperson is paid at the rate of \$20 an hour. Compare the cost of hiring the repairpersons against the cost of lost revenue when machines are broken.

5. The necessary conditions for $ETC(c)$ (defined earlier) to assume a minimum value at $c = c^*$ are

$$ETC(c^* - 1) \geq ETC(c^*) \text{ and } ETC(c^* + 1) \geq ETC(c^*)$$

   Show that these conditions reduce to

$$L_s(c^*) - L_s(c^* + 1) \leq \frac{C_1}{C_2} \leq L_s(c^* - 1) - L_s(c^*)$$

   Apply the result to Example 15.9-2 and show that it yields $c^* = 4$.

15.9.2 Aspiration Level Model

The viability of the cost model depends on how well we can estimate the cost parameters. Generally, these parameters are difficult to estimate, particularly the one associated with the waiting time of customers. The aspiration level model seeks to alleviate this difficulty by working directly with the measures of performance of the queuing situation. The idea is to determine an acceptable range for the service level (μ or c) by specifying reasonable limits on *conflicting* measures of performance. Such limits are the *aspiration levels* the decision maker wishes to reach.

We illustrate the procedure by applying it to the multiple-server model, where it is desired to determine an "acceptable" number of servers, c^*. We do so by considering the following two (conflicting) measures of performance:

1. The average time in the system, W_s.
2. The idleness percentage of the servers, X.

The idleness percentage can be computed as follows:

$$X = \frac{c - \bar{c}}{c} \times 100 = \frac{c - (L_s - L_q)}{c} \times 100 = \left(1 - \frac{\lambda_{\text{eff}}}{c\mu}\right) \times 100$$

(See Problem 12, Set 15.6e for the proof.)

The problem reduces to determining the number of servers c^* such that

$$W_s \leq \alpha \text{ and } X \leq \beta$$

where α and β are the levels of aspiration specified by the decision maker. For example, we may stipulate that $\alpha = 3$ minutes and $\beta = 10\%$.

The solution of the problem may be determined by plotting W_s and X as a function of c, as shown in Figure 15.10. By locating α and β on the graph, we can immediately determine an acceptable range for c^*. If the two conditions cannot be satisfied simultaneously, then one or both must be relaxed before a feasible range can be determined.

Example 15.9-3

In Example 15.9-2, suppose that it is desired to determine the number of clerks such that the expected waiting time until a tool is received stays below 5 minutes. Simultaneously, it is also required to keep the percentage of idleness below 20%.

Offhand, and before any calculations are made, an aspiration limit of 5 minutes on the waiting time until a tool is received (i.e., $W_s \leq 5$ minutes) is definitely unreachable because, accord-

FIGURE 15.10

Application of aspiration levels in queuing decision-making

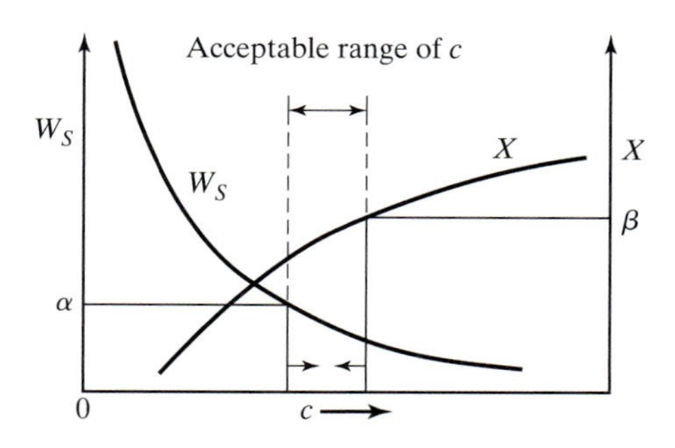

ing to the data of the problem, the average service time alone is 6 minutes. The following table summarizes W_s and X as a function of c:

c	2	3	4	5	6	7	8
W_s (min)	25.4	7.6	6.3	6.1	6.0	6.0	6.0
$X(\%)$	12.5	41.7	56.3	65.0	70.8	75.0	78.0

Based on these results, we should either reduce the service time or recognize that the source of the problem is that tools are being requested at a an unreasonably high rate ($\lambda = 17.5$ requests per hour). This, most likely, is the area that should be addressed. For example, we may want to investigate the reason for such high demand for tool replacement. Could it be that the design of the tool itself is faulty? Or could it be that the operators of the machines are purposely trying to disrupt production to express grievances?

PROBLEM SET 15.9C

***1.** A shop uses 10 identical machines. Each machine breaks down once every 7 hours on the average. It takes half an hour on the average to repair a broken machine. Both the breakdown and repair processes follow the Poisson distribution. Determine the following:

- **(a)** The number of repairpersons needed such that the average number of broken machines is less than 1.
- **(b)** The number of repairpersons needed so that the expected delay time until repair is started is less than 10 minutes.

2. In the cost model in Section 15.9.1, it is generally difficult to estimate the cost parameter C_2 (cost of waiting). As a result, it may be helpful to compute the cost C_2 implied by the aspiration levels. Using the aspiration level model to determine c^*, we can then estimate the implied C_2 by using the following inequality:

$$L_s(c^*) - L_s(c^* + 1) \leq \frac{C_1}{C_2} \leq L_s(c^* - 1) - L_s(c^*)$$

(See Problem 5, Set 15.9b, for the derivation.) Apply the procedure to the problem in Example 15.9-2, assuming $c^* = 3$ and $C_1 = \$12.00$.

REFERENCES

Bose, S., *An Introduction to Queuing Systems*, Kluwer Academic Publishers, Boston, 2001.

Hall, R., *Queuing Methods for Service and Manufacturing*, Prentice Hall, Upper Saddle River, NJ, 1991.

Lee, A., *Applied Queuing Theory*, St. Martin's Press, New York, 1966.

Lipsky, L., *Queuing Theory, A Linear Algebraic Approach*, Macmillan, New York, 1992.

Morse, P., *Queues, Inventories, and Maintenance*, Wiley, New York, 1958.

Parzen, E., *Stochastic Processes*, Holden-Day, San Francisco, 1962.

Saaty, T., *Elements of Queuing Theory with Applications*, Dover, New York, 1983.

Tanner, M., *Practical Queuing Analysis*, McGraw-Hill, New York, 1995.

Tijms, H. C., *Stochastic Models—An Algorithmic Approach*, Wiley, New York, 1994.

CHAPTER 16

Simulation Modeling

Chapter Guide. Simulation is the next best thing to observing a real system. It deals with a computerized imitation of the random behavior of a system for the purpose of estimating its measures of performance. Basically, simulation views an operational situation as a waiting line in a service facility. By literally following the movements of customers in the facility, pertinent statistics (e.g., waiting time and queue length) can be collected. The task of using simulation starts with the development of the logic of the computer model in a manner that will allow collecting needed data. A number of computer languages are available to facilitate these tedious computations.

A common misuse of simulation is to run the model for an arbitrary time period, and then view the results as the "true gospel." In fact, simulation output changes (sometimes drastically) with the length of the run. For this reason, simulation modeling deals with a statistical experiment whose output must be interpreted by appropriate statistical tests. As you study the material in this chapter, pay special attention to the peculiarities of the simulation experiment, including (1) the important role of (0,1) random numbers in sampling from probability distributions, and (2) the special methods used to collect observations to satisfy the underlying assumption of a true statistical experiment.

The prerequisite for this chapter is a basic knowledge of probability and statistics. A background in queuing theory is helpful.

This chapter includes 10 solved examples, 2 Excel templates, and 44 end-of-section problems. The AMPL/Excel/Solver/TORA programs are in folder ch16Files.

16.1 MONTE CARLO SIMULATION

A forerunner to present-day simulation is the Monte Carlo technique, a modeling scheme that estimates stochastic or deterministic parameters based on random sampling. Examples of Monte Carlo applications include evaluation of multiple integrals, estimation of the constant π ($\cong$ 3.14159), and matrix inversion.

This section uses an example to demonstrate the Monte Carlo technique. The objective of the example is to emphasize the statistical nature of the simulation experiment.

Example 16.1-1

We will use Monte Carlo sampling to estimate the area of a circle defined as

$$(x - 1)^2 + (y - 2)^2 = 25$$

The radius of the circle is $r = 5$ cm, and its center is $(x, y) = (1, 2)$.

The procedure for estimating the area requires enclosing the circle tightly in a square whose side equals the diameter of the circle, as shown in Figure 16.1. The corner points are determined from the geometry of the square.

The estimation of the area of the circle is based on the assumption that all the points in the square are equally likely to occur. Taking a random sample of n points in the square, if m of these points fall within the circle, then

$$\begin{pmatrix}\text{Estimate of the}\\ \text{area of the circle}\end{pmatrix} \cong \frac{m}{n}\begin{pmatrix}\text{Area of}\\ \text{the square}\end{pmatrix} = \frac{m}{n}(10 \times 10)$$

To ensure that all the points in the square occur with equal probabilities, we represent the coordinates x and y of a point in the square by the following *uniform* distributions:

$$f_1(x) = \frac{1}{10}, \ -4 \leq x \leq 6$$

$$f_2(y) = \frac{1}{10}, \ -3 \leq y \leq 7$$

A sampled point (x, y) based on the distribution $f_1(x)$ and $f_2(y)$ guarantees that all points in the square are equally likely to be selected.

The determination of a sample (x, y) is based on the use of independent and uniformly distributed random numbers in the range (0, 1). Table 16.1 provides a small list of such numbers which we will use in the example computations. For the purpose of general simulation, special arithmetic operations are used to generate the 0-1 random numbers, as will be shown in Section 16.4.

For a pair of 0-1 random numbers, R_1 and R_2, a random point (x, y) in the square is determined by mapping them on the x and y axes of Figure 16.1 using the following formulas:

$$x = -4 + [6 - (-4)]R_1 = -4 + 10R_1$$

$$y = -3 + [7 - (-3)]R_2 = -3 + 10R_2$$

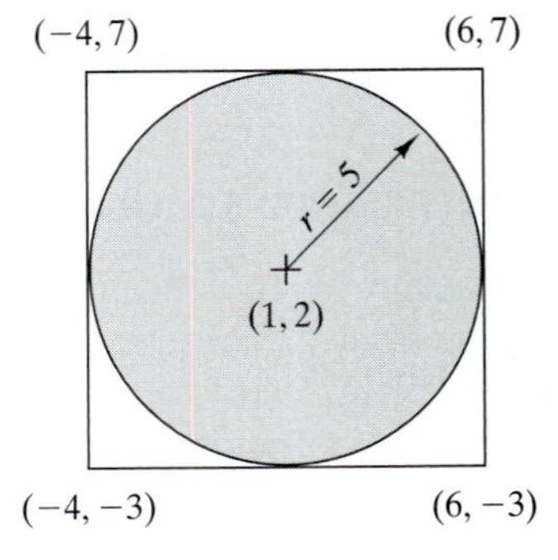

FIGURE 16.1
Monte Carlo estimation of the area of a circle

TABLE 16.1 A Short List of 0-1 Random Numbers

.0589	.3529	.5869	.3455	.7900	.6307
.6733	.3646	.1281	.4871	.7698	.2346
.4799	.7676	.2867	.8111	.2871	.4220
.9486	.8931	.8216	.8912	.9534	.6991
.6139	.3919	.8261	.4291	.1394	.9745
.5933	.7876	.3866	.2302	.9025	.3428
.9341	.5199	.7125	.5954	.1605	.6037
.1782	.6358	.2108	.5423	.3567	.2569
.3473	.7472	.3575	.4208	.3070	.0546
.5644	.8954	.2926	.6975	.5513	.0305

To demonstrate the application of the procedure, consider $R_1 = .0589$ and $R_2 = .6733$. Then

$$x = -4 + 10R_1 = -4 + 10 \times .0589 = -3.411$$

$$y = -3 + 10R_2 = -3 + 10 \times .6733 = 3.733$$

This point falls inside the circle because

$$(-3.411 - 1)^2 + (3.733 - 2)^2 = 22.46 < 25$$

The procedure is repeated n times, keeping track of the number of points m that fall within the circle. The estimate of the area is then computed as $100\frac{m}{n}$.

Remarks. To increase the reliability of estimating the area of the circle, we use the same procedures employed in ordinary statistical experiments:

1. Increase the sample size.
2. Use replications.

The discussion in Example 16.1-1 poses two questions regarding the simulation experiment:

1. How large should the sample size, n, be?
2. How many replications, N, are needed?

There are some formulas in statistical theory for determining n and N, and they depend on the nature of the simulation experiment as well as the desired confidence level. However, as in any statistical experiment, the golden rule is that higher values of n and N mean more reliable simulation results. In the end, the sample size will depend on the cost associated with conducting the simulation experiment. Generally speaking, however, a selected sample size is considered "adequate" if it produces a relatively "small" standard deviation.

Because of the random variation in the output of the experiment, it is necessary to express the results as a confidence interval. Letting $\overline{A}$ and s be the mean and variance of N replications, then, for a confidence level α, the confidence interval for the true area A is

$$\overline{A} - \frac{s}{\sqrt{N}} t_{\frac{\alpha}{2},N-1} \leq A \leq \overline{A} + \frac{s}{\sqrt{N}} t_{\frac{\alpha}{2},N-1}$$

The parameter $t_{\frac{\alpha}{2},N-1}$ is determined from the t-distribution tables given a confidence level α and $N - 1$ degrees of freedom (see the t-table in Appendix B or use excelStatTables.xls). Note that N equals the number of replications, which is distinct from n, the sample size.

Excel Moment

Because the computations associated with each sample in Example 16.1-1 are voluminous, Excel template excelCircle.xls (with VBA macros) is used to test the effect of sample size and number of replications on the accuracy of the area estimate. The input data include the circle radius, r, and its center, (cx, cy), sample size, n, and number of replications, N. The entry *Steps* in cell D4 allows executing several sample sizes in the same run. For example, if $n = 30{,}000$ and Steps $= 3$, the template will automatically produce output for $n = 30{,}000, 60{,}000, 90{,}000$. Each time the command button Press to Execute Monte Carlo is pressed, new estimates are realized, because Excel refreshes the random number generator to a different sequence.

Figure 16.2 summarizes the results for 5 replications and sample sizes of 30,000, 60,000, and 90,000. The exact area is 78.54 cm^2, and the Monte Carlo results show that

FIGURE 16.2

Excel output of Monte Carlo estimation of the area of a circle (file excelCircle.xls)

	B	C	D	E
1	**Monte Carlo Estimation of the Area of a Circle**			
2		**Input data**		
3	**Nbr. Replications, N =**	**5**		
4	**Sample size, n =**	**30,000**	**Steps =**	**3**
5	**Radius, r =**	**5**		
6	**Center, cx =**	**1**		
7	**Center, cy =**	**2**		
8		**Output results**		
9	Exact area =	78.540		
10	Press to Execute Monte Carlo			
11	Monte Carlo Calculations:			
12		n=30000	n=60000	n=90000
13	Replication 1	78.207	78.555	78.483
14	Replication 2	78.673	78.752	78.581
15	Replication 3	78.300	78.288	78.281
16	Replication 4	78.503	78.347	78.343
17	Replication 5	78.983	78.775	78.760
18				
19	Mean =	78.533	78.543	78.490
20	Std. Deviation =	0.308	0.225	0.191
21				
22	95% lower conf. limit =	78.151	78.263	78.253
23	95% upper conf. limit =	78.915	78.823	78.727

the mean estimated area for the three sample sizes varies from $\overline{A} = 78.533$ to $\overline{A} = 78.490$ cm^2. We note also that the standard deviation decreases from $s = .308$ for $n = 30{,}000$ to $s = .191$ for $n = 90{,}000$, an indication that accuracy increases with the increase in the sample size.

In terms of the present experiment, we are interested in establishing the confidence interval based on the largest sample size (i.e., $n = 90{,}000$). Given $N = 5$, $\overline{A} = 78.490$ cm^2, and $s = .191$ cm^2, $t_{.025,4} = 2.776$, and the resulting 95% confidence interval is $78.25 \le A \le 78.73$. In general, the value of N should be at least 5 to realize reasonable accuracy in the estimation of the confidence interval.

PROBLEM SET 16.1A

1. In Example 16.2-1, estimate the area of the circle using the first two columns of the (0, 1) random numbers in Table 16.1. (For convenience, go down each column, selecting R_1 first and then R_2.) How does this estimate compare with the ones given in Figure 16.2?

2. Suppose that the equation of a circle is

$$(x - 3)^2 + (y + 2)^2 = 16$$

(a) Define the corresponding distributions $f(x)$ and $f(y)$, and then show how a sample point (x, y) is determined using the (0, 1) random pair (R_1, R_2).

(b) Use excelCircle.xls to estimate the area and the associated 95% confidence interval given $n = 100{,}000$ and $N = 10$.

3. Use Monte Carlo sampling to estimate the area of the lake shown in Figure 16.3. Base the estimate on the first two columns of (0, 1) random numbers in Table 16.1.

4. Consider the game in which two players, Jan and Jim, take turns in tossing a fair coin. If the outcome is heads, Jim gets $10 from Jim. Otherwise, Jan gets $10 from Jan.

*(a) How is the game simulated as a Monte Carlo experiment?

(b) Run the experiment for 5 replications of 10 tosses each. Use the first five columns of the (0, 1) random numbers in Table 16.1, with each column corresponding to one replication.

(c) Establish a 95% confidence interval on Jan's winnings.

(d) Compare the confidence interval in (c) with Jan's expected theoretical winnings.

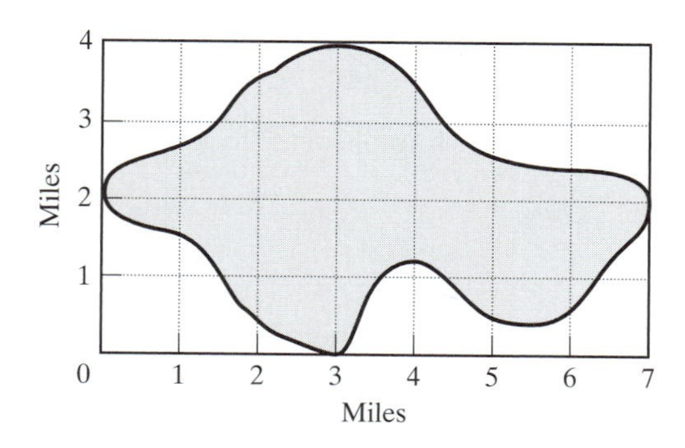

FIGURE 16.3

Lake map for Problem 3, Set 16.1a

5. Consider the following definite integral:

$$\int_0^1 x^2\, dx$$

(a) Develop the Monte Carlo experiment to estimate the integral.

(b) Use the first four columns in Table 16.1 to evaluate the integral based on 4 replications of size 5 each. Compute a 95% confidence interval, and compare it with the exact value of the integral.

6. Simulate five wins or losses of the following game of craps: The player rolls two fair dice. If the outcome sum is 7 or 11, the player wins \$10. Otherwise, the player records the resulting sum (called *point*) and keeps on rolling the dice until the outcome sum matches the recorded *point*, in which case the player wins \$10. If a 7 is obtained prior to matching the *point*, the player loses \$10.

***7.** The lead time for receiving an order can be 1 or 2 days, with equal probabilities. The demand *per day* assumes the values 0, 1, and 2 with the respective probabilities of .2, .7, and .1. Use the random numbers in Table 16.1 (starting with column 1) to estimate the joint distribution of the demand and lead time. From the joint distribution, estimate the pdf of demand during lead time. (*Hint*: The demand during lead time assumes discrete values from 0 to 4.)

8. Consider the Buffon needle experiment. A horizontal plane is ruled with parallel lines spaced D cm apart. A needle of length d cm ($d < D$) is dropped randomly on the plane. The objective of the experiment is to determine the probability that either end of the needle touches or crosses one of the lines. Define

h = Perpendicular distance from the needle center to a (parallel) line

θ = Inclination angle of the needle with a line

(a) Show that the needle will touch or cross a line only if

$$h \leq \frac{d}{2}\sin\theta,\ 0 \leq h \leq \frac{D}{2},\ 0 \leq \theta \leq \pi$$

(b) Design the Monte Carlo experiment, and provide an estimate of the desired probability.

(c) Use Excel to obtain 4 replications of size 10 each of the desired probability. Determine a 95% confidence interval for the estimate. Assume $D = 20$ cm and $d = 10$ cm.

(d) Prove that the theoretical probability is given by the formula

$$p = \frac{2d}{\pi D}$$

(e) Use the result in (c) together with the formula in (d) to estimate π.

16.2 TYPES OF SIMULATION

The execution of present-day simulation is based generally on the idea of sampling used with the Monte Carlo method. It differs in that it is concerned with the study of the behavior of real systems *as a function of time*. Two distinct types of simulation models exist.

1. Continuous models deal with systems whose behavior changes *continuously* with time. These models usually use difference-differential equations to describe the interactions among the different elements of the system. A typical example deals with the study of world population dynamics.

2. Discrete models deal primarily with the study of waiting lines, with the objective of determining such measures as the average waiting time and the length of the queue. These measures change only when a customer enters or leaves the system. The instants at which changes take place occur at specific discrete points in time (arrivals and departure events), giving rise to the name **discrete event simulation**.

This chapter presents the basics of discrete event simulation, including a description of the components of a simulation model, collection of simulation statistics, and the statistical aspect of the simulation experiment. The chapter also emphasizes the role of the computer and simulation languages in the execution of simulation models.

PROBLEM SET 16.2A

1. Categorize the following situations as either discrete or continuous (or a combination of both). In each case, specify the objective of developing the simulation model.
   *(a) Orders for an item arrive randomly at a warehouse. An order that cannot be filled immediately from available stock must await the arrival of new shipments.
   (b) World population is affected by the availability of natural resources, food production, environmental conditions, educational level, health care, and capital investments.
   (c) Goods arrive on pallets at a receiving bay of an automated warehouse. The pallets are loaded on a lower conveyor belt and lifted through an up-elevator to an upper conveyor that moves the pallets to corridors. The corridors are served by cranes that pick up the pallets from the conveyor and place them in storage bins.
2. Explain why you would agree or disagree with the following statement: "Most discrete event simulation models can be viewed in some form or another as queuing systems consisting of *sources* from which customers are generated, *queues* where customers may wait, and *facilities* where customers are served."

16.3 ELEMENTS OF DISCRETE-EVENT SIMULATION

This section introduces the concept of events in simulation and shows how the statistics of the simulated system are collected.

16.3.1 Generic Definition of Events

All discrete-event simulations describe, directly or indirectly, queuing situations in which customers arrive, wait in a queue if necessary, and then receive service before they depart the system. In general, any discrete-event model is composed of a network of interrelated queues.

Given that a discrete-event model is in reality a composite of queues, collection of simulation statistics (e.g., queue length and status of the service facility) take place only when a customer arrives or leaves the facility. This means that two principal events control the simulation model: arrivals and departures. These are the only two instants at which we need to examine the system. At all other instants, no changes affecting the statistics of the system take place.

Example 16.3-1

Metalco Jobshop receives two types of jobs: regular and rush. All jobs are processed on two consecutive machines with ample buffer areas. Rush jobs always assume nonpreemptive priority over regular jobs. Identify the events of the situation.

This situation consists of two tandem queues corresponding to the two machines. At first, one may be inclined to identify the events of the situation as follows:

$A11$: A regular job arrives at machine 1.

$A21$: A rush job arrives at machine 1.

$D11$: A regular job departs machine 1.

$D21$: A rush job departs machine 1.

$A12$: A regular job arrives at machine 2.

$A22$: A rush job arrives at machine 2.

$D12$: A regular job departs machine 2.

$D22$: A rush job departs machine 2.

In reality, we have only two events: an arrival of a (new) job at the shop and a departure of a (completed) job from a machine. First notice that events $D11$ and $A12$ are actually one and the same. The same applies to $D21$ and $A22$. Next, in discrete simulation we can use one event (arrival or departure) for both types of jobs and simply "tag" the event with an **attribute** that identifies the job type as either regular or rush. (We can think of the attribute in this case as a *personal identification number*, and, indeed, it is.) Given this reasoning, the events of the model reduce to (1) an arrival A (at the shop) and (2) a departure D (from a machine). The actions associated with the departure event will depend on the machine at which they occur.

Having defined the basic events of a simulation model, we show how the model is executed. Figure 16.4 gives a schematic representation of typical occurrences of events on the simulation time scale. After all the actions associated with a current event have been performed, the simulation advances by "jumping" to the next chronological event. In essence, the execution of the simulation occurs at the instants at which the events occur.

How does the simulation determine the occurrence time of the events? The arrival events are separated by the interarrival time (the interval between successive arrivals), and the departure events are a function of the service time in the facility. These times may be deterministic (e.g., a train arriving at a station every 5 minutes) or probabilistic (e.g., the random arrival of customers at a bank). If the time between events is deterministic, the determination of their occurrence

FIGURE 16.4

Example of the occurrence of simulation events on the time scale

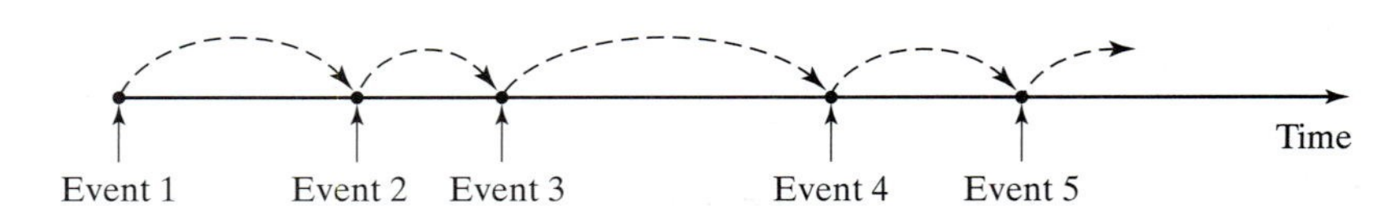

times is straightforward. If it is probabilistic, we use a special procedure to sample from the corresponding probability distribution. This point is discussed in the next section.

PROBLEM SET 16.3A

1. Identify the discrete events needed to simulate the following situation: Two types of jobs arrive from two different sources. Both types are processed on a single machine, with priority given to jobs from the first source.
2. Jobs arrive at a constant rate at a carousel conveyor system. Three service stations are spaced equally around the carousel. If the server is idle when a job arrives at the station, the job is removed from the conveyor for processing. Otherwise, the job continues to rotate on the carousel until a server becomes available. A processed job is stored in an adjacent shipping area. Identify the discrete events needed to simulate this situation.
3. Cars arrive at a two-lane, drive-in bank, where each lane can house a maximum of four cars. If the two lanes are full, arriving cars seek service elsewhere. If at any time one lane is at least two cars longer than the other, the last car in the longer lane will jockey to the last position in the shorter lane. The bank operates the drive-in facility from 8:00 A.M. to 3:00 P.M. each work day. Define the discrete events for the situation.

*4. The cafeteria at Elmdale Elementary provides a single-tray, fixed-menu lunch to all its pupils. Kids arrive at the dispensing window every 30 seconds. It takes 18 seconds to receive the lunch tray. Map the arrival-departure events on the time scale for the first five pupils.

16.3.2 Sampling from Probability Distributions

Randomness in simulation arises when the interval, t, between successive events is probabilistic. This section presents three methods for generating successive random samples ($t = t_1, t_2, \ldots$) from a probability distribution $f(t)$:

1. Inverse method.
2. Convolution method.
3. Acceptance-rejection method.

The inverse method is particularly suited for analytically tractable probability density functions, such as the exponential and the uniform. The remaining two methods deal with more complex cases, such as the normal and the Poisson. All three methods are rooted in the use of independent and identically distributed uniform (0, 1) random numbers.

Inverse Method. Suppose that it is desired to obtain a random sample x from the (continuous or discrete) probability density function $f(x)$. The inverse method first determines a closed-form expression of the cumulative density function $F(x) = P\{y \leq x\}$, where $0 \leq F(x) \leq 1$, for all defined values of y. Given that R is a random value obtained from a uniform (0, 1) distribution, and assuming that F^{-1} is the inverse of F, the steps of the method are as follows:

Step 1. Generate the (0, 1) random number, R.

Step 2. Compute the desired sample, $x = F^{-1}(R)$.

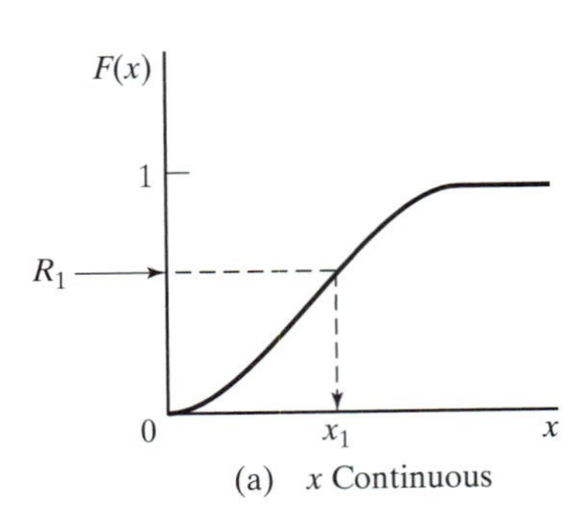

(a) x Continuous

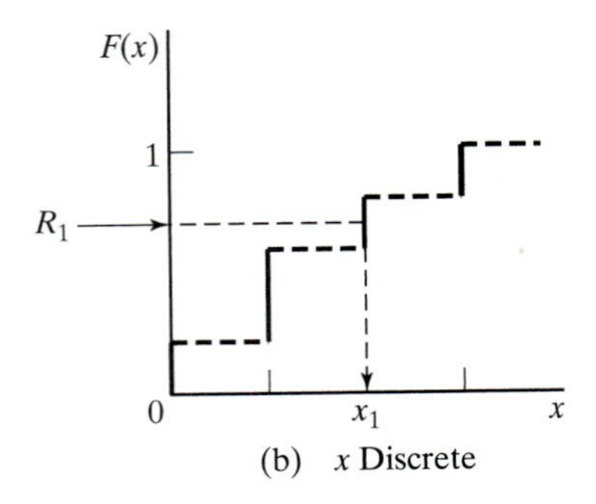

(b) x Discrete

FIGURE 16.5
Sampling from a probability distribution by the inverse method

Figure 16.5 illustrates the procedures for both a continuous and a discrete random distribution. The uniform (0, 1) random value R_1 is projected from the vertical $F(x)$-scale to yield the desired sample value x_1 on the horizontal scale.

The validity of the proposed procedure rests on showing that the random variable $z = F(x)$ is uniformly distributed in the interval $0 \leq z \leq 1$, as the following theorem proves.

Theorem 16.3-1. *Given the cumulative density function F(x) of the random variable* $x, -\infty < x < \infty$, *the random variable* $z = F(x), 0 \leq z \leq 1$, *has the following uniform 0-1 density function:*

$$f(z) = 1, 0 \leq z \leq 1$$

Proof. The random variable is uniformly distributed if, and only if,

$$P\{z \leq Z\} = Z, 0 \leq Z \leq 1$$

This result applies to $F(x)$ because

$$P\{z \leq Z\} = P\{F(x) \leq Z\} = P\{x \leq F^{-1}(Z)\} = F[F^{-1}(Z)] = Z$$

Additionally, $0 \leq Z \leq 1$ because $0 \leq P\{z \leq Z\} \leq 1$.

Example 16.3-2 (Exponential Distribution)

The exponential probability density function

$$f(t) = \lambda e^{-\lambda t}, t > 0$$

represents the interarrival time t of customers at a facility with a mean value of $\frac{1}{\lambda}$. Determine a random sample t from $f(t)$.

The cumulative density function is determined as

$$F(t) = \int_0^t \lambda e^{-\lambda x}\, dx = 1 - e^{-\lambda t}, t > 0$$

Setting $R = F(t)$, we can solve for t, which yields

$$t = -\left(\frac{1}{\lambda}\right)\ln(1 - R)$$

Because $1 - R$ is the complement of R, $\ln(1 - R)$ may be replaced with $\ln(R)$.

In terms of simulation, the result means that arrivals are spaced t time units apart. For example, for $\lambda = 4$ customers per hour and $R = .9$, the time period until the next arrival occurs is computed as

$$t_1 = -\left(\frac{1}{4}\right)\ln(1 - .9) = .577 \text{ hour} = 34.5 \text{ minutes}$$

The values of R used to obtain successive samples must be selected *randomly* from a uniform (0, 1) distribution. We will show later in Section 16.4 how these (0, 1) random values are generated during the course of the simulation.

PROBLEM SET 16.3B

***1.** In Example 16.3-2, suppose that the first customer arrives at time 0. Use the first three random numbers in column 1 of Table 16.1 to generate the arrival times of the next 3 customers and graph the resulting events on the time scale.

***2.** *Uniform Distribution.* Suppose that the time needed to manufacture a part on a machine is described by the following uniform distribution:

$$f(t) = \frac{1}{b - a}, a \le t \le b$$

Determine an expression for the sample t given the random number R.

3. Jobs are received randomly at a one-machine shop. The time between arrivals is exponential with mean 2 hours. The time needed to manufacture a job is uniform between 1.1 and 2 hours. Assuming that the first job arrives at time 0, determine the arrival and departure time for the first five jobs using the (0, 1) random numbers in column 1 of Table 16.1.

4. The demand for an expensive spare part of a passenger jet is 0, 1, 2, or 3 units per month with probabilities .2, .3, .4, and .1, respectively. The airline maintenance shop starts operation with a stock of 5 units, and will bring the stock level back to 5 units immediately after it drops below 2 units.

*__(a)__ Devise the procedure for sampling demand.

(b) How many months will elapse until the first replenishment occurs? Use successive values of R from the first column in Table 16.1.

5. In a simulation situation, TV units are inspected for possible defects. There is an 80% chance that a unit will pass inspection, in which case it is sent to packaging. Otherwise, the unit is repaired. We can represent the situation symbolically in one of two ways.

goto REPAIR/.2, PACKAGE/.8
goto PACKAGE/.8, REPAIR/.2

These two representations appear equivalent. Yet, when a given sequence of (0, 1) random numbers is applied to the two representations, different decisions (REPAIR or PACKAGE) may result. Explain why.

6. A player tosses a fair coin repeatedly until a head occurs. The associated payoff is 2^n, where n is the number of tosses until a head comes up.
   (a) Devise the sampling procedure of the game.
   (b) Use the random numbers in column 1 of Table 16.1 to determine the cumulative payoff after two heads occur.
7. *Triangular Distribution.* In simulation, the lack of data may make it impossible to determine the probability distribution associated with a simulation activity. In most of these situations, it may be easy to describe the desired variable by estimating its smallest, most likely, and largest values. These three values are sufficient to define a triangular distribution, which can then be used as a "rough cut" estimation of the real distribution.
   (a) Develop the formula for sampling from the following triangular distribution, whose respective parameters are a, b, and c:

$$f(x) = \begin{cases} \dfrac{2(x-a)}{(b-a)(c-a)}, & a \le x \le b \\ \dfrac{2(c-x)}{(c-b)(c-a)}, & b \le x \le c \end{cases}$$

   (b) Generate three samples from a triangular distribution with parameters $(1, 3, 7)$ using the first three random numbers in column 1 of Table 16.1.
8. Consider a probability distribution that consists of a rectangle flanked on the left and right sides by two symmetrical right triangles. The respective ranges for the triangle on the left, the rectangle, and the triangle on the right are $[a, b]$, $[b, c]$, and $[c, d]$, $a < b < c < d$. Both triangles have the same height as the rectangle.
   (a) Develop a sampling procedure
   (b) Determine five samples with $(a, b, c, d) = (1, 2, 4, 6)$ using the first five random numbers in column 1 of Table 16.1.

*9. *Geometric distribution*. Show how a random sample can be obtained from the following geometric distribution:

$$f(x) = p(1-p)^x, \quad x = 0, 1, 2, \ldots$$

   The parameter x is the number of (Bernoulli) failures until a success occurs, and p is the probability of a success, $0 < p < 1$. Generate five samples for $p = .6$ using the first five random numbers in column 1 of Table 16.1.

10. *Weibull distribution.* Show how a random sample can be obtained from the Weibull distribution with the following probability density function:

$$f(x) = \alpha\beta^{-\alpha}x^{\alpha-1}e^{-(x/\beta)^{\alpha}}, \ x > 0$$

   where $\alpha > 0$ is the shape parameter, and $\beta > 0$ is the scale parameter.

Convolution Method. The basic idea of the convolution method is to express the desired sample as the statistical sum of other easy-to-sample random variables. Typical among these distributions are the Erlang and the Poisson whose samples can be obtained from the exponential distribution samples.

Example 16.3-3 (Erlang Distribution)

The m-Erlang random variable is defined as the statistical sum (convolutions) of m independent and identically distributed exponential random variables. Let y represent the m-Erlang random variable; then

$$y = y_1 + y_2 + \cdots + y_m$$

where y_i, i $= 1, 2, \ldots, m$, are independent and identically distributed exponential random variables whose probability density function is defined as

$$f(y_i) = \lambda e^{-\lambda y_i},\ y_i > 0, i = 1, 2, \ldots, m$$

From Example 16.3-2, a sample from the ith exponential distribution is computed as

$$y_i = -\left(\frac{1}{\lambda}\right)\ln(R_i), i = 1, 2, \ldots, m$$

Thus, the m-Erlang sample is computed as

$$\begin{aligned} y &= -\left(\frac{1}{\lambda}\right)\{\ln(R_1) + \ln(R_2) + \cdots + \ln(R_m)\} \\ &= -\left(\frac{1}{\lambda}\right)\ln\left(\prod_{i=1}^{m} R_i\right) \end{aligned}$$

To illustrate the use of the formula, suppose that $m = 3$, and $\lambda = 4$ events per hour. The first 3 random numbers in column 1 of Table 16.1 yield $R_1R_2R_3 = (.0589)(.6733)(.4799) = .0190$, which yields

$$y = -\left(\tfrac{1}{4}\right)\ln(.019) = .991 \text{ hour}$$

Example 16.3-4 (Poisson Distributions)

Section 15.3.1 shows that if the distribution of the time between the occurrence of successive events is exponential, then the distribution of the number of events per unit time must be Poisson, and vice versa. We use this relationship to sample the Poisson distribution.

Assume that the Poisson distribution has a mean value of λ events per unit time. Then the time between events is exponential with mean $\frac{1}{\lambda}$ time units. This means that a Poisson sample, n, will occur during t time units if, and only if,

$$\text{Period till event } n \text{ occurs} \leq t < \text{Period till event } n + 1 \text{ occurs}$$

This condition translates to

$$\begin{aligned} t_1 + t_2 + \cdots + t_n \leq t < t_1 + t_2 + \cdots + t_{n+1}, n > 0 \\ 0 \leq t < t_1, n = 0 \end{aligned}$$

where $t_i, i = 1, 2 \ldots, n + 1$, is a sample from the exponential distribution with mean $\frac{1}{\lambda}$. From the result in Example 16.3-3, we have

$$-\left(\frac{1}{\lambda}\right) \ln\left(\prod_{i=1}^{n} R_i\right) \leq t < -\left(\frac{1}{\lambda}\right) \ln\left(\prod_{i=1}^{n+1} R_i\right), \; n > 0$$

$$0 \leq t < -\left(\frac{1}{\lambda}\right) \ln(R_1), n = 0$$

which reduces to

$$\prod_{i=1}^{n} R_i \geq e^{-\lambda t} > \prod_{i=1}^{n+1} R_i, n > 0$$

$$1 \geq e^{-\lambda t} > R_1, n = 0$$

To illustrate the implementation of the sampling process, suppose that $\lambda = 4$ events per hour and that we wish to obtain a sample for a period $t = .5$ hour. This gives $e^{-\lambda t} = .1353$. Using the random numbers in column 1 of Table 16.1, we note that $R_1 = .0589$ is less than $e^{-\lambda t} = .1353$. Hence, the corresponding sample is $n = 0$.

Example 16.3-5 (Normal Distribution)

The central limit theorem (see Section 12.4.4) states that the sum (convolution) of n independent and identically distributed random variables becomes asymptotically normal as n becomes sufficiently large. We use this result to generate samples from normal distribution with mean μ and standard deviation σ.

Define

$$x = R_1 + R_2 + \cdots + R_n$$

The random variable is asymptotically normal by the central limit theorem. Given that the uniform (0, 1) random number R has a mean of $\frac{1}{2}$ and a variance of $\frac{1}{12}$, it follows that x has a mean of $\frac{n}{2}$ and a variance of $\frac{n}{12}$. Thus, a random sample, y, from a normal distribution with mean μ and standard deviation σ, $N(\mu, \sigma)$, can be computed from x as

$$y = \mu + \sigma\left(\frac{x - \frac{n}{2}}{\sqrt{\frac{n}{12}}}\right)$$

In practice, we take $n = 12$ for convenience, which reduces the formula to

$$y = \mu + \sigma(x - 6)$$

To illustrate the use of this method, suppose that we wish to generate a sample from $N(10, 2)$ (mean $\mu = 10$ and standard deviation $\sigma = 2$). Taking the sum of the first 12 random numbers in columns 1 and 2 of Table 16.1, we get $x = 6.1094$. Thus, $y = 10 + 2(6.1094 - 6) = 10.2188$.

The disadvantage of this procedure is that it requires generating 12 random numbers for each normal sample, which is computationally inefficient. A more efficient procedure calls for using the transformation

$$x = \cos(2\pi R_2)\sqrt{-2\ln(R_1)}$$

Box and Muller (1958) prove that x is a standard $N(0, 1)$. Thus, $y = \mu + \sigma x$ will produce a sample from $N(\mu, \sigma)$. The new procedure is more efficient because it requires two (0, 1) random numbers only. Actually, this method is even more efficient than stated, because Box and Miller prove that the preceding formula will produce another $N(0, 1)$ sample if $\sin(2\pi R_2)$ replaces $\cos(2\pi R_2)$.

To illustrate the implementation of the Box-Muller procedure to the normal distribution $N(10, 2)$, the first two random numbers in column 1 of Table 16.1 yield the following $N(0, 1)$ samples:

$$x_1 = \cos(2\pi \times .6733)\sqrt{-2\ln(.0589)} \approx -1.103$$

$$x_2 = \sin(2\pi \times .6733)\sqrt{-2\ln(.0589)} \approx -2.109$$

Thus, the corresponding $N(10, 2)$ samples are

$$y_1 = 10 + 2(-1.103) = 7.794$$

$$y_2 = 10 + 2(-2.109) = 5.782$$

PROBLEM SET 16.3C[1]

*1. In Example 16.3-3, compute an Erlang sample, given $m = 4$ and $\lambda = 5$ events per hour.

2. In Example 16.3-4, generate three Poisson samples during a 2-hour period, given that the mean of the Poisson is 5 events per hour.

3. In Example 16.4-5, generate two samples from $N(8, 1)$ by using both the convolution method and the Box-Muller method.

4. Jobs arrive at Metalco jobshop according to a Poisson distribution, with a mean of six jobs per day. Received jobs are assigned to the five machining centers of the shop on a strict rotational basis. Determine one sample of the interval between the arrival of jobs at the first machine center.

5. The ACT scores for the 1994 senior class at Springdale High are normal, with a mean of 27 points and a standard deviation of 3 points. Suppose that we draw a random sample of six seniors from that class. Use the Box-Muller method to determine the mean and standard deviation of the sample.

*6. Psychology professor Yataha is conducting a learning experiment in which mice are trained to find their way around a maze. The base of the maze is square. A mouse enters the maze at one of the four corners and must find its way through the maze to exit at the same point where it entered. The design of the maze is such that the mouse must pass by each of the remaining three corner points exactly once before it exits. The multi-paths of the maze connect the four corners in a strict clockwise order. Professor Yataha estimates that the time the mouse takes to reach one corner point from another is uniformly

[1]For all the problems of this set, use the random numbers in Table 16.1 starting with column 1.

distributed between 10 and 20 seconds, depending on the path it takes. Develop a sampling procedure for the time a mouse spends in the maze.

7. In Problem 6, suppose that once a mouse makes an exit from the maze, another mouse instantly enters. Develop a sampling procedure for the number of mice that exit the maze in 5 minutes.

8. *Negative Binomial.* Show how a random sample can be determined from the negative binomial whose distribution is given as

$$f(x) = C_x^{r+x-1} p^r (1 - p)^x, x = 0, 1, 2, \text{K}$$

where x is the number of failures until the rth success occurs in a sequence of independent Bernoulli trials and p is the probability of success, $0 < p < 1$. (*Hint:* The negative binomial is the convolution of r independent geometric samples. See Problem 9, Set 16.3b.)

Acceptance-Rejection Method. The acceptance-rejection method is designed for complex pdfs that cannot be handled by the preceding methods. The general idea of the method is to replace the complex pdf $f(x)$ with a more analytically manageable "proxy" pdf $h(x)$. Samples from $h(x)$ can then be used to sample the original pdf $f(x)$.

Define the **majorizing function** $g(x)$ such that it dominates $f(x)$ in its entire range—that is,

$$g(x) \geq f(x), -\infty < x < \infty$$

Next, define the proxy pdf, $h(x)$, by normalizing $g(x)$ as

$$h(x) = \frac{g(x)}{\int_{-\infty}^{\infty} g(y)d(y)}, -\infty < x < \infty$$

The steps of the acceptance-rejection method are thus given as

Step 1. Obtain a sample $x = x_1$ from $h(x)$ using the inverse or the convolution method.

Step 2. Obtain a $(0, 1)$ random number R.

Step 3. If $R \leq \frac{f(x_1)}{g(x_1)}$, accept x_1 as a sample from $f(x)$. Otherwise, discard x_1 and return to step 1.

The validity of the method is based on the following equality:

$$P\{x \leq a | x = x_1 \text{ is accepted}, -\infty < x_1 < \infty\} = \int_{-\infty}^{a} f(y)\, dy, -\infty < a < \infty$$

This probability statement states that the sample $x = x_1$ that satisfies the condition of step 3 in reality is a sample from the original pdf $f(x)$, as desired.

The efficiency of the proposed method is enhanced by the decrease in the rejection probability of step 3. This probability depends on the specific choice of the majorizing function $g(x)$ and should decrease with the selection of a $g(x)$ that "majorizes" $f(x)$ more "snugly."

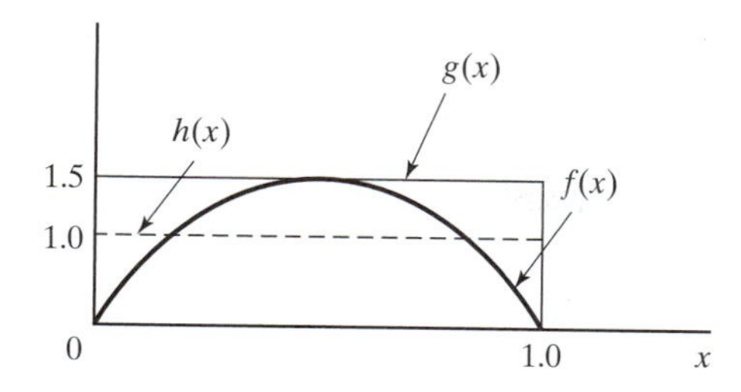

FIGURE 16.6
Majorizing function, $g(x)$, for the beta distribution, $f(x)$

Example 16.3-6 (Beta Distribution)

Apply the acceptance-rejection to the following beta distribution:

$$f(x) = 6x(1 - x), 0 \leq x \leq 1$$

Figure 16.6 depicts $f(x)$ and a majorizing function $g(x)$.

The height of the majorizing function $g(x)$ equals the maximum of $f(x)$, which occurs at $x = .5$. Thus, the height of the rectangle is $f(.5) = 1.5$. This means that

$$g(x) = 1.5, 0 \leq x \leq 1$$

The proxy pdf $h(x)$, also shown in Figure 16.6, is computed as

$$h(x) = \frac{g(x)}{\text{Area under } g(x)} = \frac{1.5}{1 \times 1.5} = 1, 0 \leq x \leq 1$$

The following steps demonstrate the procedure using the (0, 1) random sequence in Table 16.1.

Step 1. $R = .0589$ gives the sample $x = .0589$ from $h(x)$.

Step 2. $R = .6733$.

Step 3. Because $\frac{f(.0589)}{g(.0589)} = \frac{.3326}{1.5} = .2217$ is less than $R = .6733$, we accept the sample $x_1 = .0589$.

To obtain a second sample, we continue as follows:

Step 1. Using $R = .4799$, we get $x = .4799$ from $h(x)$.

Step 2. $R = .9486$.

Step 3. Because $\frac{f(.4799)}{g(.4799)} = .9984$ is larger than $R = .9486$, we reject $x = .4799$ as a valid beta sample. This means that the steps must be repeated again with "fresh" random numbers until the condition of step 3 is satisfied.

Remarks. The efficiency of the acceptance-rejection method is enhanced by selecting a majorizing function $g(x)$ that "jackets" $f(x)$ as tightly as possible while yielding an analytically tractable proxy $h(x)$. For example, the method will be more efficient if the rectangular majorizing function $g(x)$ in Figure 16.6 is replaced with a step-pyramid function (see Problem 2, Set 16.3d, for an illustration). The larger the number of steps, the more tightly will $g(x)$ majorize $f(x)$, and hence the higher is the probability of accepting a sample. However, a "tight" majorizing function generally entails additional computations which, if excessive, may offset the savings resulting from increasing the probability of acceptance.

PROBLEM SET 16.3D

1. In Example 16.3-5, continue the steps of the procedure until a valid sample is obtained. Use the (0, 1) random numbers in Table 16.1 in the same order in which they are used in the example.
2. Consider the beta pdf of Example 16.3-6. Determine a two-step pyramid majorizing function $g(x)$ with two equal jumps of height $\frac{1.5}{2} = .75$ each. Obtain one beta sample based on the new majorizing function using the same (0, 1) random sequence in Table 16.1 that was employed in Example 16.3-6. The conclusion, in general, is that a tighter majorizing function will increase the probability of acceptance. Observe, however, that the amount of the computations associated with the new function is larger.
3. Determine the functions $g(x)$ and $h(x)$ for applying the acceptance-rejection method to the following function:

$$f(x) = \frac{\sin(x) + \cos(x)}{2}, 0 \leq x \leq \frac{\pi}{2}.$$

   Use the (0, 1) random numbers from column 1 in Table 16.1 to generate two samples from $f(x)$. [*Hint*: For convenience, use a rectangular $g(x)$ over the defined range of $f(x)$.]
4. The interarrival time of customers at HairKare is described by the following distribution:

$$f_1(t) = \frac{k_1}{t}, 12 \leq t \leq 20$$

   The time to get a haircut is represented by the following distribution:

$$f_2(t) = \frac{k_2}{t^2}, 18 \leq t \leq 22$$

   The constant k_1 and k_2 are determined such that $f_1(t)$ and $f_2(t)$ are probability density functions. Use the acceptance-rejection method (and the random numbers in Table 16.1) to determine when the first customer will leave HairKare and when the next customer will arrive. Assume that the first customer arrives at $T = 0$.

16.4 GENERATION OF RANDOM NUMBERS

Uniform (0, 1) random numbers play a key role in sampling from distributions. True (0, 1) random numbers can only be generated by electronic devices. However, because simulation models are executed on the computer, the use of electronic devices to generate random numbers is much too slow for that purpose. Additionally, electronic devices are activated by laws of chance, and hence it will be impossible to duplicate the same sequence of random numbers at will. This point is important because debugging, verification, and validation of the simulation model often require duplicating the same sequence of random numbers.

The only plausible way for generating (0, 1) random numbers for use in simulation is based on arithmetic operations. Such numbers are not truly random because they can be generated in advance. It is thus more appropriate to refer to them as **pseudo-random numbers**.

The most common arithmetic operation for generating (0, 1) random numbers is the **multiplicative congruential method**. Given the parameters u_0, b, c, and m, a pseudo-random number R_n can be generated from the formulas:

$$u_n = (bu_{n-1} + c) \bmod (m), n = 1, 2, \ldots$$

$$R_n = \frac{u_n}{m}, n = 1, 2, \ldots$$

The initial value u_0 is usually referred to as the **seed** of the generator.

Variations of the multiplicative congruential method that improve the quality of the generator can be found in Law and Kelton (1991).

Example 16.4-1

Generate three random numbers based on the multiplicative congruential method using $b = 9, c = 5$, and $m = 12$. The seed is $u_0 = 11$.

$$u_1 = (9 \times 11 + 5) \bmod 12 = 8, R_1 = \frac{8}{12} = .6667$$

$$u_2 = (9 \times 8 + 5) \bmod 12 = 5, R_2 = \frac{5}{12} = .4167$$

$$u_3 = (9 \times 5 + 5) \bmod 12 = 2, R_3 = \frac{2}{12} = .1667$$

Excel Moment

Excel template excelRN.xls is designed to carry out the multiplicative congruential calculations. Figure 16.7 generates the sequence associated with the parameters of Example 16.4-1. Observe carefully that the cycle length is exactly 4, after which the sequence repeats itself. The conclusion here is that the choice of u_0, b, c, and m is critical in determining the (statistical) quality of the generator and its cycle length. Thus, "casual" implementation of the congruential formula is not advisable. Instead, one must use a reliable and tested generator. Practically all commercial computer programs are equipped with dependable random number generators.

PROBLEM SET 16.4A

*1. Use excelRN.xls with the following sets of parameters and compare the results with those in Example 16.4-1:

$$b = 17, c = 111, m = 103, \text{seed} = 7$$

2. Find a random number generator on your computer, and use it to generate 500 zero-one random numbers. Histogram the resulting values (using the Microsoft histogram tool, see Section 12.5) and visually convince yourself that the obtained numbers reasonably follow

	A	B
1	Multiplicative Congruential Method	
2	Input data(B7<=1000)	
3	b =	9
4	c =	5
5	u0 =	11
6	m =	12
7	How many numbers?	10
8	Output results	
9	Press to Generate Sequence	
10	Generated random numbers:	
11	1	0.66667
12	2	0.41667
13	3	0.16667
14	4	0.91667
15	5	0.66667
16	6	0.41667
17	7	0.16667
18	8	0.91667
19	9	0.66667
20	10	0.41667

FIGURE 16.7

Excel random numbers output for the data of Example 16.4-1 (file excelRN.xls)

the (0, 1) uniform distribution. Actually, to test the sequence properly, you would need to apply the following tests: chi-square goodness of fit (see Section 12.6), runs test for independence, and correlation test (see Law and Kelton [1991] for details).

16.5 MECHANICS OF DISCRETE SIMULATION

This section details how typical statistics are collected in a simulation model. The vehicle of explanation is a single-queue model. Section 16.5.1 uses a numeric example to detail the actions and computations that take place in a single-server queuing simulation model. Because of the tedious computations that typify the execution of a simulation model, Section 16.5.2 shows how the single-server model is modeled and executed using Excel spreadsheet.

16.5.1 Manual Simulation of a Single-Server Model

The interarrival time of customers at HairKare Barbershop is exponential with mean 15 minutes. The shop is operated by only one barber and it takes between 10 and 15 minutes, uniformly distributed, to do a haircut. Customers are served on a first-in, first-out (FIFO) basis. The objective of the simulation is to compute the following measures of performance:

1. The average utilization of the shop.
2. The average number of waiting customers.
3. The average time a customer waits in queue.

The logic of the simulation model can be described in terms of the actions associated with the arrival and departure events of the model.

Arrival Event

1. Generate and store chronologically the occurrence time of the next arrival event (= current simulation time + interarrival time).
2. If the facility (barber) is idle
   - **(a)** Start service and declare the facility busy. Update the facility utilization statistics.
   - **(b)** Generate and store chronologically the time of the departure event for the customer (= current simulation time + service time).
3. If the facility is busy, place the customer in the queue and update the queue statistics.

Departure Event

1. If the queue is empty, declare the facility idle. Update the facility utilization statistics.
2. If the queue is not empty
   - **(a)** Select a customer from the queue, and place it in the facility. Update the queue and facility utilization statistics.
   - **(b)** Generate and store chronologically the occurrence time of the departure event for the customer (= current simulation time + service time).

From the data of the problem, the interarrival time is exponential with mean 15 minutes, and the service time is uniform between 10 and 15 minutes. Letting p and q represent random samples of interarrival and service times, then, as explained in Section 16.3.2, we get

$$p = -15 \ln(R) \text{ minutes}, \quad 0 \leq R \leq 1$$

$$q = 10 + 5R \text{ minutes}, \quad 0 \leq R \leq 1$$

For the purpose of this example, we use R from Table 16.1, starting with column 1. We also use the symbol T to represent the simulation clock time. We further assume that the first customer arrives at $T = 0$ and that the facility starts empty.

Because the simulation computations are typically voluminous, the simulation is limited to the first 5 arrivals only. The example is designed to cover all possible situations that could arise in the course of the simulation. Later in the section we introduce the excelSingleServer.xls template that allows you to experiment with the model without the need to carry out the computations manually.

Arrival of Customer 1 at $T = 0$. Generate the arrival of customer 2 at

$$T = 0 + p_1 = 0 + [-15 \ln(.0589)] = 42.48 \text{ minutes}$$

Because the facility is idle at $T = 0$, customer 1 starts service immediately. The departure time is thus computed as

$$T = 0 + q_1 = 0 + (10 + 5 \times .6733) = 13.37 \text{ minutes}$$

The *chronological* list of future events is thus given as:

Time, T	Event
13.37	Departure of customer 1
42.48	Arrival of customer 2

Departure of Customer 1 at $T = 13.37$. Because the queue is empty, the facility is declared idle. At the same time, we record that the facility has been busy between $T = 0$ and $T = 13.37$ minutes. The updated list of future events becomes

Time, T	Event
42.48	Arrival of customer 2

Arrival of Customer 2 at $T = 42.48$. Customer 3 will arrive at

$$T = 42.48 + [-15 \ln(.4799)] = 53.49 \text{ minutes}$$

Because the facility is idle, customer 2 starts service and the facility is declared busy. The departure time is

$$T = 42.48 + (10 + 5 \times .9486) = 57.22 \text{ minutes}$$

The list of future events is updated as

Time, T	Event
53.49	Arrival of customer 3
57.22	Departure of customer 2

Arrival of Customer 3 at $T = 53.49$. Customer 4 will arrive at

$$T = 53.49 + [-15 \ln(.6139)] = 60.81 \text{ minutes}$$

Because the facility is currently busy (until $T = 57.22$), customer 3 is placed in queue at $T = 53.49$. The updated list of future events is

Time, T	Event
57.22	Departure of customer 2
60.81	Arrival of customer 4

Departure of Customer 2 at $T = 57.22$. Customer 3 is taken out of the queue to start service. The waiting time is

$$W_3 = 57.22 - 53.49 = 3.73 \text{ minutes}$$

The departure time is

$$T = 57.22 + (10 + 5 \times .5933) = 70.19 \text{ minutes}$$

The updated list of future events is

Time, T	Event
60.81	Arrival of customer 4
70.19	Departure of customer 3

Arrival of Customer 4 at $T = 60.81$. Customer 5 will arrive at

$$T = 60.81 + [-15\ln(.9341)] = 61.83 \text{ minutes}$$

Because the facility is busy until $T = 70.19$, customer 4 is placed in the queue. The updated list of future events is

Time, T	Event
61.83	Arrival of customer 5
70.19	Departure of customer 3

Arrival of Customer 5 at $T = 61.83$. The simulation is limited to 5 arrivals only, hence customer 6 arrival is not generated. The facility is still busy, hence the customer is placed in queue at $T = 61.83$. The updated list of events is

Time, T	Event
70.19	Departure of customer 3

Departure of Customer 3 at $T = 70.19$. Customer 4 is taken out of the queue to start service. The waiting time is

$$W_4 = 70.19 - 60.81 = 9.38 \text{ minutes}$$

The departure time is

$$T = 70.19 + [10 + 5 \times .1782] = 81.08 \text{ minutes}$$

The updated list of future events is

Time, T	Event
81.08	Departure of customer 4

Departure of Customer 4 at $T = 81.08$. Customer 5 is taken out of the queue to start service. The waiting time is

$$W_5 = 81.08 - 61.83 = 19.25 \text{ minutes}$$

The departure time is

$$T = 81.08 + (10 + 5 \times .3473) = 92.82 \text{ minutes}$$

The updated list of future events is

Time, T	Event
92.82	Departure of customer 5

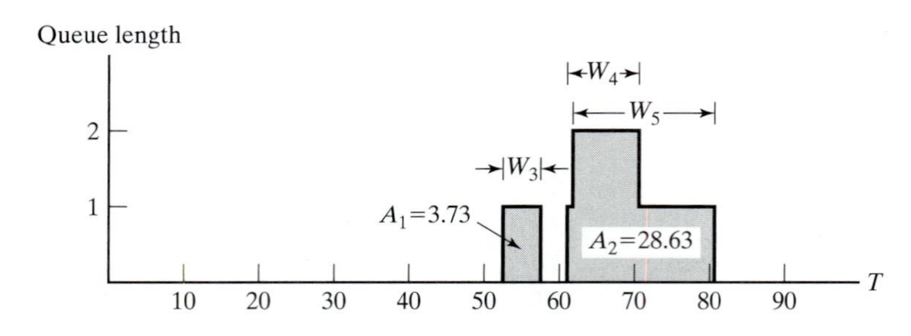

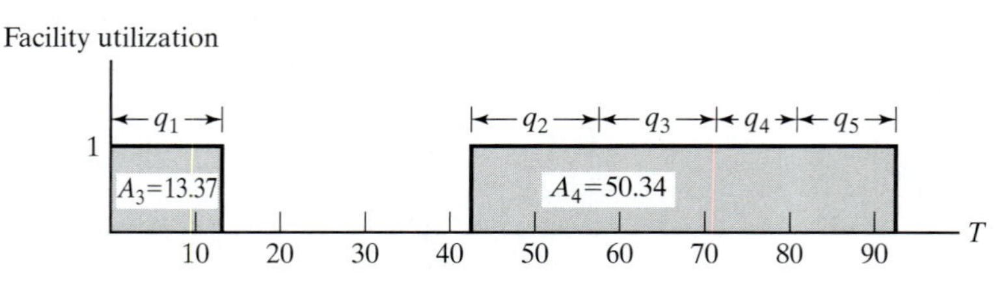

FIGURE 16.8
Changes in queue length and facility utilization as a function of simulation time, T

Departure of Customer 5 at T = 92.82. There are no more customers in the system (queue and facility) and the simulation ends.

Figure 16.8 summarizes the changes in the length of the queue and the utilization of the facility as a function of the simulation time.

The queue length and the facility utilization are known as **time-based** variables because their variation is a function of time. As result, their average values are computed as

$$\begin{pmatrix}\text{Average value of a}\\ \text{time-based variable}\end{pmatrix} = \frac{\text{Area under curve}}{\text{Simulated period}}$$

Implementing this formula for the data in Figure 16.8, we get

$$\begin{pmatrix}\text{Average queue}\\ \text{length}\end{pmatrix} = \frac{A_1 + A_2}{92.82} = \frac{32.36}{92.82} = .349 \text{ customer}$$

$$\begin{pmatrix}\text{Average facility}\\ \text{utilization}\end{pmatrix} = \frac{A_3 + A_4}{92.82} = \frac{63.71}{92.82} = .686 \text{ barber}$$

The average waiting time in the queue is an **observation-based** variable whose value is computed as

$$\begin{pmatrix}\text{Average value of an}\\ \text{observation-based variable}\end{pmatrix} = \frac{\text{Sum of observations}}{\text{Number of observations}}$$

Examination of Figure 16.8 reveals that the area under the queue-length curve actually equals the sum of the waiting time for the three customers who joined the queue; namely,

$$W_1 + W_2 + W_3 + W_4 + W_5 = 0 + 0 + 3.73 + 9.38 + 19.25 = 32.36 \text{ minutes}$$

The average waiting time in the queue for all customers is thus computed as

$$\overline{W_q} = \frac{32.36}{5} = 6.47 \text{ minutes}$$

PROBLEM SET 16.5A

1. Suppose that the barbershop of Section 16.5.1 is operated by two barbers, and customers are served on a FCFS basis. Suppose further that the time to get a haircut is uniformly distributed between 15 and 30 minutes. The interarrival time of customers is exponential, with a mean of 10 minutes. Simulate the system manually for 75 time units. From the results of the simulation, determine the average time a customer waits in queue, the average number of customers waiting, and the average utilization of the barbers. Use the random numbers in Table 16.1.

2. Classify the following variables as either *observation based* or *time based*:

*(a) Time-to-failure of an electronic component.

*(b) Inventory level of an item.

(c) Order quantity of an inventory item.

(d) Number of defective items in a lot.

(e) Time needed to grade test papers.

(f) Number of cars in the parking lot of a car-rental agency.

*3. The following table represents the variation in the number of waiting customers in a queue as a function of the simulation time.

Simulation time, T (hr)	No. of waiting customers
$0 \le T \le 3$	0
$3 < T \le 4$	1
$4 < T \le 6$	2
$6 < T \le 7$	1
$7 < T \le 10$	0
$10 < T \le 12$	2
$12 < T \le 18$	3
$18 < T \le 20$	2
$20 < T \le 25$	1

Compute the following measures of performance:

(a) The average length of the queue.

(b) The average waiting time in the queue for those who must wait.

4. Suppose that the barbershop of Example 16.5-1 is operated by three barbers. Assume further that the utilization of the servers (barbers) is summarized as given in the following table:

Simulation time, T (hr)	No. of busy servers
$0 < T \leq 10$	0
$10 < T \leq 20$	1
$20 < T \leq 30$	2
$30 < T \leq 40$	0
$40 < T \leq 60$	1
$60 < T \leq 70$	2
$70 < T \leq 80$	3
$80 < T \leq 90$	1
$90 < T \leq 100$	0

Determine the following measures of performance:

(a) The average utilization of the facility.

(b) The average busy time of the facility.

(c) The average idle time of the facility.

16.5.2 Spreadsheet-Based Simulation of the Single-Server Model

The presentation in Section 16.5.1 shows that simulation computations are typically tedious and voluminous. Thus, the use of the computer to execute simulation models is a must. This section develops a spreadsheet-based model for the single server model. The objective of the development is to reinforce the ideas introduced in Section 16.5.1. Of course, a single-server model is a simple situation, and for this reason can be modeled readily in a spreadsheet environment. Other situations require a more involved modeling effort, which is facilitated by available simulation packages (see Section 16.7).

The presentation in Section 16.5.1 shows that the simulation model of the single-server facility requires two basic elements:

1. A chronological list of the model's events.
2. A graph that keeps track of the changes in facility utilization and queue length.

These two elements remain essential in the development of the spreadsheet-based (indeed, any computer-based) simulation model. The difference is that the implementation is realized in a manner that is compatible with the use of the computer. As in Section 16.5.1, customers are served in order of arrival (FIFO).

Figure 16.9 provides the output of excelSingleServer.xls. The input data allow representing the interarrival and service time in one of four ways: constant, exponential, uniform, and triangular. The triangular distribution is useful in that it can be used as a rough initial estimate of any distribution, simply by providing three estimates a, b, and c that represent the smallest, the most likely, and the largest values of the interarrival or service time. The only other information needed to drive the simulation is the length of the simulation run, which in this model is specified by the number of arrivals that can be generated in the model.

	A	B	C	D	E	F	G	H	K	L	M	N	O	P	Q
1		**Simulation of a Single-Server Queueing Model**													
2		**Nbr of arrivals =**	**20**	<<Maximum 500											
3		**Enter x in column A to select interarrival pdf:**							Nbr	InterArvlT	ServiceT	ArrvlT	DepartT	Wq	Ws
4		**Constant =**							1	3.73	12.83	0.00	12.83	0.00	12.83
5	x	**Exponential:** λ =		**0.067**					2	5.37	14.71	3.73	27.55	9.10	23.82
6		**Uniform:**	a =		b =				3	3.86	12.21	9.09	39.75	18.45	30.66
7		**Triangular:**	a =		b =		c =		4	14.10	11.18	12.95	50.94	26.80	37.98
8		**Enter x in column A to select service time pdf:**							5	7.35	14.92	27.05	65.85	23.88	38.80
9		**Constant =**							6	35.70	14.22	34.41	80.07	31.45	45.67
10		**Exponential:** μ =							7	0.60	14.50	70.11	94.58	9.97	24.47
11	x	**Uniform:**	a =	10	b =	15			8	4.25	13.35	70.71	107.93	23.87	37.22
12		**Triangular:**	a =		b =		c =		9	4.85	12.45	74.96	120.38	32.97	45.41
13		**Output Summary**							10	7.43	11.57	79.81	131.94	40.56	52.13
14		Av. facility utilization =		0.98					11	8.99	14.65	87.24	146.59	44.70	59.34
15		Percent idleness (%) =		1.95					12	49.78	12.85	96.23	159.43	50.36	63.20
16					Press F9 to				13	0.42	14.12	146.01	173.55	13.43	27.54
17		Av. queue length, Lq =		1.57	trigger a				14	8.77	13.69	146.43	187.24	27.13	40.82
18		Av. nbr in system, Ls =		2.55	new simulation				15	11.19	10.50	155.20	197.75	32.05	42.55
19		Av. queue time, Wq =		21.24	run.				16	42.82	13.78	166.38	211.53	31.36	45.14
20		Av. system time, Ws =		34.47					17	19.87	12.29	209.20	223.82	2.33	14.62
21		Sum(ServiceTime) =		264.65					18	9.25	12.95	229.07	242.03	0.00	12.95
22		Sum(Wq) =		424.80					19	13.98	12.99	238.33	255.02	3.70	16.69
23		Sum(Ws) =		689.44					20	58.46	14.88	252.31	269.90	2.71	17.59

FIGURE 16.9

Excel output of a single-server simulation model (file excelSingleServer.xls)

The spreadsheet calculations reserve one row for each arrival. The interarrival and service times for each arrival are generated from the input data. The first arrival is assumed to occur at $T = 0$. Because the facility starts idle, the customer starts service immediately. Thus,

$$\begin{pmatrix}\text{Departure time}\\ \text{of customer 1}\end{pmatrix} = \begin{pmatrix}\text{Arrival time}\\ \text{of customer 1}\end{pmatrix} + \begin{pmatrix}\text{Service time}\\ \text{of customer 1}\end{pmatrix}$$

$$= 0 + 12.83 = 12.83$$

$$\begin{pmatrix}\text{Arrival time}\\ \text{of customer 2}\end{pmatrix} = \begin{pmatrix}\text{Arrival time}\\ \text{of customer 1}\end{pmatrix} + \begin{pmatrix}\text{Interarrival time}\\ \text{of customer 1}\end{pmatrix}$$

$$= 0 + 3.37 = 3.37$$

To determine the departure time of any customer i, we use the following formula

$$\begin{pmatrix}\text{Departure time}\\ \text{of customer } i\end{pmatrix}$$

$$= \max\left\{\begin{pmatrix}\text{Arrival time}\\ \text{of customer } i\end{pmatrix}, \begin{pmatrix}\text{Departure time}\\ \text{of customer } i-1\end{pmatrix}\right\} + \begin{pmatrix}\text{Service time}\\ \text{of customer } i\end{pmatrix}$$

The formula says that a customer cannot start service until the facility becomes available. To illustrate the use of this formula in Figure 16.9, we have

$$\text{Departure time of customer 3} = \max\{9.09, 27.55\} + 12.21 = 39.76$$

We now turn our attention to collecting the statistics of the model. First, note that for customer i, the waiting time in queue, $W_q(i)$, and in the entire system, $W_s(i)$, are computed as

$$W_q(i) = \begin{pmatrix}\text{Departure time}\\ \text{of customer } i\end{pmatrix} - \begin{pmatrix}\text{Arrival time}\\ \text{of customer } i\end{pmatrix} - \begin{pmatrix}\text{Service time}\\ \text{of customer } i\end{pmatrix}$$

$$W_s(i) = \begin{pmatrix}\text{Departure time}\\ \text{of customer } i\end{pmatrix} - \begin{pmatrix}\text{Arrival time}\\ \text{of customer } i\end{pmatrix}$$

Next, it may appear that computing the remaining statistics of the model necessitates tracking the changes in facility utilization and in queue length (as we did in Section 16.5.1). Fortunately, the calculations are simplified by two observations we made in Section 16.5.1 and explained in Figure 16.8:

1. Area under facility utilization curve = Sum of service times of all arrivals
2. Area under queue length curve = Sum of waiting times of all arrivals

To explain this point, Excel output in Figure 16.9 computes these two sums—namely,

$$\begin{aligned}
\text{Sum of service times} &= 264.65\\
\text{Sum of } W_q &= 424.8\\
\text{Sum of } W_s &= \text{Sum of } W_q + \text{Sum of service times}\\
&= 689.44\ (= 264.65 + 424.8)
\end{aligned}$$

Given that the last arrival (customer 20) departs at $T = 269.90$, it follows that

$$\begin{pmatrix}\text{Average facility}\\ \text{utilization}\end{pmatrix} = \frac{264.65}{269.90} = .9805$$

$$\begin{pmatrix}\text{Average queue}\\ \text{length}\end{pmatrix} = \frac{424.80}{269.90} = 1.57$$

Percent idleness of the facility is computed as $(1 - .98) \times 100 = 1.945\%$.

The remainder of the statistics are calculated in a straightforward manner; namely,

$$\begin{pmatrix}\text{Average waiting}\\ \text{time in queue}\end{pmatrix} = \frac{\text{Sum of } W_q}{\text{Number of arrivals}} = \frac{424.80}{20} = 21.24$$

$$\begin{pmatrix}\text{Average system}\\ \text{time}\end{pmatrix} = \frac{\text{Sum of } W_s}{\text{Number of arrivals}} = \frac{689.44}{20} = 34.47$$

Another spreadsheet was developed for simulating multiserver models (excelMultiServer.xls). The design of the template is based on the same ideas used in the single-server case. However, the determination of the departure time is not as straightforward and, hence, requires the use of VBA macros.

PROBLEM SET 16.5B

1. Using the input data in Section 16.5.1, run the Excel simulator for 10 arrivals and graph the changes in facility utilization and queue length as a function of the simulation time. Verify that the areas under the curves equal the sum of the service times and the sum of the waiting times, respectively.
2. Simulate the *M*/*M*/1 model for 500 arrivals given the arrival rate $\lambda = 4$ customers per hour and the service rate $\mu = 6$ departures per hour. Run 5 replications (by refreshing the spreadsheet—pressing F9) and determine a 95% confidence interval for all the measures of performance of the model. Compare the results with the steady-state theoretical values of the *M*/*M*/1 model.
3. Television units arrive on a conveyor belt every 15 minutes for inspection at a single-operator station. Detailed data for the inspection station are not available. However, the operator estimates that it takes 10 minutes "on the average" to inspect a unit. Under the worst conditions, the inspection time does not exceed 13 minutes, and for certain units inspection time may be as low as 9 minutes.
   (a) Use the Excel simulator to simulate the inspection of 200 TV units.
   (b) Based on 5 replications, estimate the average number of units awaiting inspection and the average utilization of the inspection station.

16.6 METHODS FOR GATHERING STATISTICAL OBSERVATIONS

Simulation is a statistical experiment and its output must be interpreted using proper statistical inference tools (e.g., confidence intervals and hypothesis testing). To accomplish this task, the observations of the simulation experiment must satisfy three conditions:

1. Observations are drawn from stationary (identical) distributions.
2. Observations are sampled from a normal population.
3. Observations are independent.

It so happens that, in the strict sense, the simulation experiment does not satisfy any of these conditions. Nevertheless, we can ensure that these conditions remain statistically viable by restricting the manner in which the simulation observations are gathered.

First, we consider the question of stationarity. Simulation output is a function of the length of the simulated period. The initial period produces erratic behavior and is usually referred to as the **transient** or **warm-up period**. When the output stabilizes, the system operates under **steady state**. Unfortunately, there is no way to predict the start point of steady state in advance. In general, a longer simulation run has better chance of reaching steady state. This point is demonstrated in Example 16.1-1, where the accuracy of estimating the area of a circle by Monte Carlo increases with the sample size. Thus, nonstationarity can be accounted for by using a sufficiently large sample size.

Next, we consider the requirement that simulation observations must be drawn from a normal population. This requirement is realized by using the *central limit theorem* (see Section 12.4.4), which states that the distribution of the average of a sample is asymptotically normal regardless of the parent population from which the sample is drawn. The central limit theorem is thus the main tool we use for satisfying the normal distribution assumption.

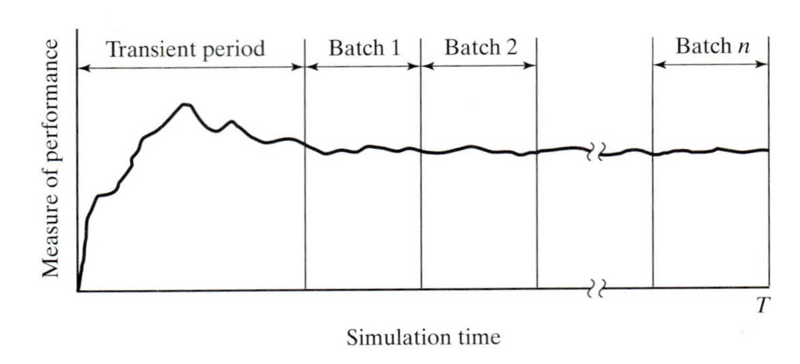

FIGURE 16.10

Collecting simulation data using the subinterval method

The third condition deals with the independence of the observations. The nature of the simulation experiment does not guarantee independence among successive simulation observations. However, by using sample average to represent a simulation observation, we can alleviate the problem of lack of independence. This is particularly true when we increase the time base used to compute the sample average.

Having discussed the peculiarities of the simulation experiment and ways to circumvent them, we present the three most common methods for collecting observations in simulation:

1. Subinterval method
2. Replication method
3. Regenerative (or cycles) method

16.6.1 Subinterval Method

Figure 16.10 illustrates the idea of the subinterval method. Suppose that the simulation is executed for T time units (i.e., run length $= T$) and that it is desired to collect n observations. The subinterval method first truncates an initial transient period, and then subdivides the remainder of the simulation run into n equal subintervals (or batches). The average of the desired measure of performance (e.g., queue length or waiting time in queue) within each subinterval is then used to represent a single observation. The truncation of the initial transient period implies that no statistical data are collected during the period.

The advantage of the subinterval method is that the effect of the transient (nonstationary) conditions is mitigated, particularly for those observations that are collected toward the end of the simulation run. The disadvantage of the method is that successive batches with common boundary conditions are necessarily correlated. The effect of correlation can be alleviated by increasing the time base for each batch.

Example 16.6-1

Figure 16.11 shows the change in queue length in a single-queue model as a function of the simulation time. The simulation run length is $T = 35$ hours, and the length of the transient period is

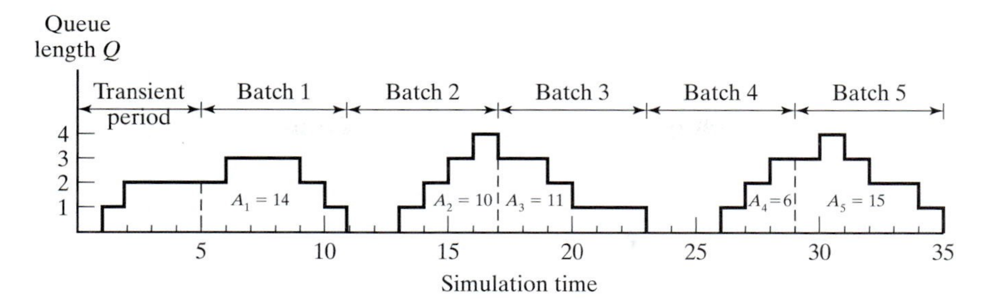

FIGURE 16.11

Change in queue length with simulation time in Example 16.6-1

estimated to equal 5 hours. It is desired to collect 5 observations—that is, $n = 5$. The corresponding time base for each batch thus equals $\frac{(35 - 5)}{5} = 6$ hours.

Let $\overline{Q}_i$ represent the average queue length in batch i. Because the queue length is a time-based variable, we have

$$\overline{Q}_i = \frac{A_i}{t}, i = 1, 2, \text{K}, 5$$

where A_i is the area under the queue length curve associated with batch (observation) i, and t is the time base per batch. In the present example, $t = 6$ hours.

The data in Figure 16.11 produce the following observations:

Observation i	1	2	3	4	5
A_i	14	10	11	6	15
$\overline{Q}_i$	2.33	1.67	1.83	1.00	2.5
Sample mean = 1.87		Sample standard deviation = .59			

The sample mean and variance can be used to compute a confidence interval, if desired. The computation of the sample variance in Example 16.6-1 is based on the following familiar formula:

$$s = \sqrt{\frac{\sum_{i=1}^{n} x_i^2 - n\overline{x}^2}{n - 1}}$$

This formula is only an approximation of the true variance because it ignores the effect of autocorrelation between the successive batches. The exact formula can be found in Law and Kelton (2000, pp. 249–253).

16.6.2 Replication Method

In the replication method, each observation is represented by an independent simulation run in which the transient period is truncated, as illustrated in Figure 16.12. The computation of the observation averages for each batch is the same as in the subinterval

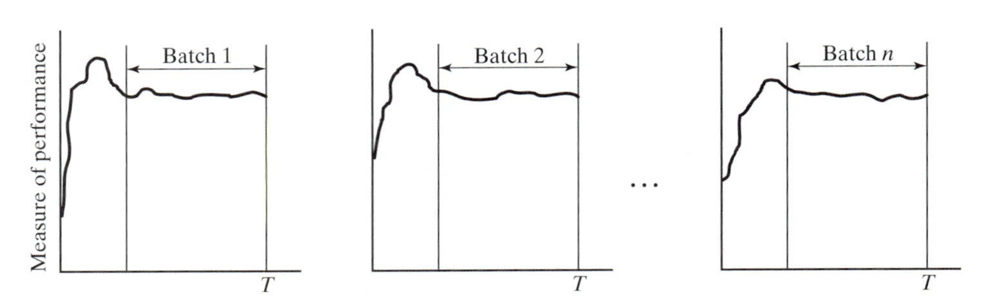

FIGURE 16.12
Collecting simulation data using the replication method

method. The only difference is that the standard variance formula is applicable because the batches are not correlated.

The advantage of the replication method is that each simulation run is driven by a distinct (0, 1) random number stream, which yields observations that are truly statistically independent. The disadvantage is that each observation may be biased by the initial effect of the transient conditions. Such a problem may be alleviated by making the run length sufficiently large.

16.6.3 Regenerative (Cycle) Method

The regenerative method may be regarded as an extended case of the subinterval method. The motivation behind the new method is that it attempts to reduce the effect of autocorrelation that characterizes the subinterval method by requiring similar starting conditions for each batch. For example, if the variable we are dealing with is the queue length, each batch would start at an instant where the queue length is zero. Unlike the subinterval method, the nature of the regenerative method may result in unequal time bases for the different batches.

Although the regenerative method may reduce autocorrelation, it has the disadvantage of yielding a smaller number of batches for a given run length. This follows because we cannot predict when a new batch will start or how long its time base will be. Under steady-state conditions, however, we should expect the starting points for the successive batches to be more or less evenly spaced.

The computation of the average for batch i in the regenerative method is generally defined as the ratio of two random variables a_i and b_i—that is, $x_i = \frac{a_i}{b_i}$. The definitions of a_i and b_i depend on the variable being computed. Specifically, if the variable is *time based*, then a_i would represent the area under the curve and b_i would equal the associated time base. If the variable is *observation based*, then a_i would be the total sum of the observations within batch i and b_i would be the associated number of observations.

Because x_i is the ratio of two random variables, an unbiased estimate of the sample average can be shown to be

$$\overline{y} = \frac{\sum_{i=1}^{n} y_i}{n}$$

where

$$y_i = \frac{n\bar{a}}{\bar{b}} - \frac{(n-1)(n\bar{a} - a_i)}{n\bar{b} - b_i}, i = 1, 2, \text{K}, n$$

$$\bar{a} = \frac{\sum_{i=1}^{n} a_i}{n}$$

$$\bar{b} = \frac{\sum_{i=1}^{n} b_i}{n}$$

In this case, a confidence interval is based on the mean and standard deviation of y_i.

Example 16.6-2

Figure 16.13 represents the number of busy servers in a single facility with three parallel servers. The length of the simulation run is 35 time units, and the length of the transient period is 4 time units. It is desired to estimate the average utilization of the facility based on the regenerative method.

After truncating the transient period, Figure 16.13 yields four batches with the common characteristic of starting with all three servers idle. The associated values of a_i and b_i are given in the following table:

Batch i	a_i	b_i
1	12	9
2	6	5
3	10	10
4	6	7
Averages	$\bar{a} = 8.50$	$\bar{b} = 7.75$

Based on these data, we have

$$y_i = \frac{4 \times 8.5}{7.75} - \frac{(4-1) \times (4 \times 8.5 - a_i)}{4 \times 7.75 - b_i} = 4.39 - \frac{102 - 3a_i}{31 - b_i}$$

These computations can be automated readily using Excel template excelRegenerative.xls.

FIGURE 16.13

Changes in the number of busy servers as a function of time in Example 16.6-2

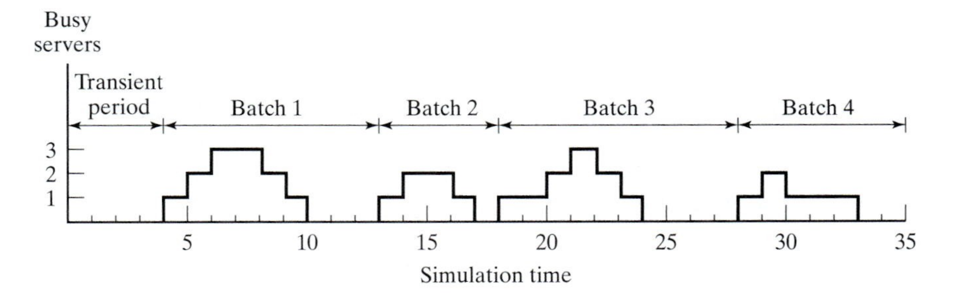

PROBLEM SET 16.6A

1. In Example 16.6-1, use the subinterval method to compute the average waiting time in the queue for those who must wait.

*2. In a simulation model, the subinterval method is used to compute batch averages. The transient period is estimated to be 100, and each batch has a time base of 100 time units as well. Using the following data, which provide the waiting times for customers as a function of the simulation time, estimate the 95% confidence interval for the mean waiting time.

Time interval	Waiting times
0–100	10, 20, 13, 14, 8, 15, 6, 8
100–200	12, 30, 10, 14, 16
200–300	15, 17, 20, 22
300–400	10, 20, 30, 15, 25, 31
400–500	15, 17, 20, 14, 13
500–600	25, 30, 15

3. In Example 16.6-2, suppose that the start point for each observation is the point in time where all the servers have just become idle. Thus, in Figure 16.13, these points correspond to $t = 10, 17, 24$, and 33. Compute the 95% confidence interval for the utilization of the servers based on the new definition of the regenerative points.

4. In a single-server queuing situation, the system is simulated for 100 hours. The results of the simulation show that the server was busy only during the following time intervals: (0, 10), (15, 20), (25, 30), (35, 60), (70, 80), and (90, 95). The length of the transient period is estimated to be 10 hours.
   - **(a)** Define the observation start point needed to implement the regenerative method.
   - **(b)** Compute the 95% confidence interval for the average utilization of the server based on the regenerative method.
   - **(c)** Apply the subinterval method to the same problem using a sample size $n = 5$. Compute the corresponding 95% confidence interval, and compare it with the one obtained from the regenerative method.

16.7 SIMULATION LANGUAGES

Execution of simulation models entails two distinct types of computations: (1) file manipulations that deal with the chronological storage and processing of model events, and (2) arithmetic and bookkeeping computations associated with generation of random samples and collection of model statistics. The first type of computation involves extensive logic development in list processing, and the second type entails tedious and time-consuming calculations. The nature of these computations makes the computer an essential tool for executing simulation models, and, in turn, prompts the development of special computer simulation languages for performing these computations conveniently and efficiently.

Available discrete simulation languages fall into two broad categories:

1. Event scheduling
2. Process oriented

In event scheduling languages, the user details the actions associated with the occurrence of each event, in much the same way they are given in Example 16.5-1. The main role of the language in this case is (1) automation of sampling from distributions, (2) storage and retrieval of events in chronological order, and (3) collection of model statistics.

Process-oriented languages use blocks or nodes that can be linked together to form a network that describes the movements of **transactions** or **entities** (i.e., customers) in the system. For example, the three most prominent blocks/nodes in any process-simulation language are a *source* from which transactions are created, a *queue* where they can wait if necessary, and a *facility* where service is performed. Each of these blocks/nodes is defined with all the information needed to drive the simulation automatically. For example, once the interarrival time for the source is specified, a process-oriented language automatically "knows" when arrival events will occur. In effect, each block/node of the model has standing instructions that define *how* and *when* transactions are moved in the simulation network.

Process-oriented languages are internally driven by the same actions used in event-scheduling languages. The difference is that these actions are automated to relieve the user of the tedious computational and logical details. In a way, we can regard process-oriented languages as being based on the input-output concept of the "black box" approach. This essentially means that process-oriented languages trade modeling flexibility for simplicity and ease of use.

Prominent event-scheduling languages include SIMSCRIPT, SLAM, and SIMAN. Over the years, these languages have evolved to include process-oriented capabilities. All three languages allow the user to write (a portion of) the model in higher-level language, such as FORTRAN or C. This capability is necessary to allow the user to model complex logic that otherwise cannot be achieved directly by the regular facilities of these languages. A major reason for this limitation is the restrictive and perhaps convoluted manner in which these languages move transactions (or entities) among the model's queues and facilities.

Several modern commercial packages currently dominate the simulation market, including Arena, AweSim, and GPSS/H, to mention only a few. These packages use extensive user interface to simplify the process of creating a simulation model. They also provide animation capabilities where changes in the system can be observed visually. However, to the experienced user, these interfaces may appear to reduce the development of a simulation model to a "slow-motion" pace. It is not surprising that some users prefer to write simulation models in such general programming languages as C, Basic, and FORTRAN.

PROBLEM SET 16.7A[2]

1. Patrons arrive randomly at a three-clerk post office. The interarrival time is exponential with mean 5 minutes. The time a clerk spends with a patron is exponential with a mean of

---

[2]Work these problems using a simulation language of your choice, or using BASIC, FORTRAN, or C.

10 minutes. All arriving patrons form one queue and wait for the first available free clerk. Run a simulation model of the system for 480 minutes to determine the following:

(a) The average number of patrons waiting in the queue.

(b) The average utilization of the clerks.

(c) Compare the simulation results with those of the *M/M/c* queuing model (Chapter 15) and with the spreadsheet MultiServerSimulator.xls.

2. Television units arrive for inspection on a conveyor belt at the constant rate of 5 units per hour. The inspection time takes between 10 and 15 minutes, uniformly distributed. Past experience shows that 20% of inspected units must be adjusted and then sent back for reinspection. The adjustment time is also uniformly distributed between 6 and 8 minutes. Run a simulation model for 480 minutes to compute the following:

(a) The average time a unit takes until it passes inspection.

(b) The average number of times a unit must be reinspected before it exits the system.

3. A mouse is trapped in a maze and desperately "wants out." After trying between 1 and 3 minutes, uniformly distributed, there is a 30% chance that it will find the right path. Otherwise, it will wander around aimlessly for between 2 and 3 minutes, uniformly distributed, and eventually end up where it started, only to try once again. The mouse can "try freedom" as many times as it pleases, but there is a limit to everything. With so much energy expended in trying and retrying, the mouse is certain to expire if it does not make it within a period that is normally distributed, with a mean of 10 minutes and a standard deviation of 2 minutes. Write a simulation model to estimate the probability that the mouse will be free. For the purpose of estimating the probability, assume that 100 mice will be processed by the model.

4. In the final stage of automobile manufacturing, a car moving on a transporter is situated between two parallel workstations to allow work to be done on both the left and right sides of the car simultaneously. The operation times for the left and right sides are uniform between 15 and 20 minutes and 18 and 22 minutes, respectively. The transporter arrives at the stations area every 20 minutes. Simulate the process for 480 minutes to determine the utilization of the left and right stations.

5. Cars arrive at a one-bay car wash facility where the interarrival time is exponential, with a mean of 10 minutes. Arriving cars line up in a single lane that can accommodate at most five waiting cars. If the lane is full, newly arriving cars will go elsewhere. It takes between 10 and 15 minutes, uniformly distributed, to wash a car. Simulate the system for 960 minutes, and estimate the time a car spends in the facility.

REFERENCES

Box, G., and M. Muller, "A Note on the Generation of Random Normal Deviates," *Annals of Mathematical Statistics*, Vol. 29, pp. 610–611, 1958.

Law, A., and W. Kelton, *Simulation Modeling & Analysis*, 3rd ed., McGraw-Hill, New York, 2000.

Ross, S., *A Course in Simulation*, Macmillan, New York, 1990.

Rubenstein, R., B. Melamed, and A. Shapiro, *Modern Simulation and Modeling*, Wiley, New York, 1998.

Taha, H., *Simulation Modeling and SIMNET*, Prentice Hall, Upper Saddle River, NJ, 1988.

CHAPTER 17

Markov Chains

Chapter Guide. This chapter provides a basic background about Markov chains and their use in practice, including cost-based models. Markov chain notation is "cumbersome" and its computations are tedious. To alleviate this problem, the more readable matrix notation is used where possible. With regard to the computations, two Excel templates are provided to handle the basic calculations for a Markov chain of any size, including n-step transition and absolute probabilities, steady-state probabilities, and first passage times in both ergodic and absorbing chains. Both spreadsheets should be helpful in solving end-of-section problems.

This chapter includes 17 solved examples, 42 end-of-section problems, and 2 Excel templates. The AMPL/Excel/Solver/TORA programs are in folder ch17Files.

17.1 DEFINITION OF A MARKOV CHAIN

Let X_t be a random variable that characterizes the state of the system at discrete points in time $t = 1, 2, \ldots$ The family of random variables $\{X_t\}$ forms a **stochastic process**. The number of states in a stochastic process may be finite or infinite, as the following two examples demonstrate:

Example 17.1-1 (Machine Maintenance)

The condition of a machine at the time of the monthly preventive maintenance is characterized as fair, good, or excellent. For month t, the stochastic process for this situation can be represented as:

$$X_t = \begin{Bmatrix} 0, & \text{if the condition is poor} \\ 1, & \text{if the condition is fair} \\ 2, & \text{if the condition is good} \end{Bmatrix}, t = 1, 2, \ldots$$

The random variable X_t is *finite* because it represents three states: poor (0), fair (1), and good (2).

Example 17.1-2 (Job Shop)

Jobs arrive randomly at a job-shop at the average rate of 5 jobs per hour. The arrival process follows a Poisson distribution which, theoretically, allows any number of jobs between zero and infinity to arrive at the shop during the time interval $(0, t)$. The inifinte-state process describing the number of arriving jobs is

$$X_t = 0, 1, 2, \ldots, t > 0$$

A stochastic process is a **Markov process** if the occurrence of a future state depends only on the immediately preceding state. This means that given the chronological times $t_0, t_1, \ldots, t_n$, the family of random variables $\{X_{t_n}\} = \{x_1, x_2, \ldots, x_n\}$ is said to be a Markov process if it possesses the following property:

$$P\{X_{t_n} = x_n | X_{t_{n-1}} = x_{n-1}, \ldots, X_{t_0} = x_0\} = P\{X_{t_n} = x_n | X_{t_{n-1}} = x_{n-1}\}$$

In a Markovian process with n exhaustive and mutually exclusive states (outcomes), the probabilities at a specific point in time $t = 0, 1, 2, \ldots$ is usually written as

$$p_{ij} = P\{X_t = j | X_{t-1} = i\}, (i, j) = 1, 2, \ldots, n, t = 0, 1, 2, \ldots, T$$

This is known as the **one-step transition probability** of moving from state i at $t - 1$ to state j at t. By definition, we have

$$\sum_j p_{ij} = 1, i = 1, 2, \ldots, n$$

$$p_{ij} \geq 0, (i, j) = 1, 2, \ldots, n$$

A convenient way for summarizing the one-step transition probabilities is to use the following matrix notation:

$$\mathbf{P} = \begin{pmatrix} p_{11} & p_{12} & p_{13} & \cdots & p_{1n} \\ p_{21} & p_{22} & p_{23} & \cdots & p_{2n} \\ \vdots & \vdots & \vdots & \vdots & \vdots \\ p_{n1} & p_{n2} & p_{n3} & \cdots & p_{nn} \end{pmatrix}$$

The matrix **P** defines the so-called **Markov chain**. It has the property that all its transition probabilities p_{ij} are fixed (stationary) and independent over time. Although a Markov chain may include an infinite number of states, the presentation in this chapter is limited to finite chains only, as this is the only type needed in the text.

Example 17.1-3 (The Gardener Problem)

Every year, at the beginning of the gardening season (March through September), a gardener uses a chemical test to check soil condition. Depending on the outcome of the test, productivity for the new season falls in one of three states: (1) good, (2) fair, and (3) poor. Over the years, the gardener has observed that last year's soil condition impacts current year's productivity and that the situation can be described by the following Markov chain:

$$\mathbf{P} = \text{State of the system this year} \begin{matrix} \\ 1 \\ 2 \\ 3 \end{matrix} \overset{\text{State of the system next year}}{\begin{matrix} 1 & 2 & 3 \\ \end{matrix}} \begin{pmatrix} .2 & .5 & .3 \\ 0 & .5 & .5 \\ 0 & 0 & 1 \end{pmatrix}$$

The transition probabilities show that the soil condition can either deteriotate or stay the same but never improve. If this year's soil is good (state 1), there is a 20% chance it will not change next year, a 50% chance it will become fair (state 2), and a 30% chance it will deteriorate to a poor condition (state 3). If this year's soil condition is fair (state 2), next year's productivity may remain fair with probability .5 or become poor (state 3), also with probability .5. Finally, a poor condition this year (state 3) can only lead to an equal condition next year (with probability 1).

The gardener can alter the transition probabilities **P** by using fertilizer to boost soil condition. In this case, the transition matrix becomes:

$$\mathbf{P}_1 = \begin{matrix} \\ 1 \\ 2 \\ 3 \end{matrix} \begin{matrix} 1 & 2 & 3 \\ \end{matrix} \begin{pmatrix} .30 & .60 & .10 \\ .10 & .60 & .30 \\ .05 & .40 & .55 \end{pmatrix}$$

The use of fertilizer now allows improvements in the deteriorating condition. There is a 10% chance that the soil condition will change from fair to good (state 2 to state 1), a 5% chance it will change from poor to good (state 3 to state 1), and a 40% chance that a poor condition will become fair (state 3 to state 2).

PROBLEM SET 17.1A

1. An engineering professor purchases a new computer every two years with preferences for three models: $M1$, $M2$, and $M3$. If the present model is $M1$, the next computer may be $M2$ with probability .2 or $M3$ with probability .15. If the present model is $M2$, the probabilities of switching to $M1$ and $M3$ are .6 and .25, respectively. And, if the present model is $M3$, then the probabilities of switching to $M1$ and $M2$ are .5 and .1, respectively. Represent the situation as a Markov chain.

***2.** A police car is on patrol in a neighborhood known for its gang activities. During a patrol, there is a 60% chance that the location where help is needed can be responded to in time, else the car will continue regular patrol. Upon receiving a call, there is a 10% chance for cancellation (in which case the car resumes its normal patrol) and a 30% chance that the car is already responding to a previous call. When the police car arrives at the scene, there is a 10% chance that the instigators will have fled (in which case the car returns back to patrol) and a 40% chance that apprehension is made immediately. Else, the officers will search the area. If apprehension occurs, there is a 60% chance of transporting the suspects to the police station, else they are released and the car returns to patrol. Express the probabilistic activities of the police patrol in the form of a transition matrix.

3. (Cyert and Associates, 1963) Bank1 offers loans which are either paid when due or are delayed. If the payment on a loan is delayed more than 4 quarters (1 year), Bank1 considers the loan a bad debt and writes it off. The following table provides a sample of Bank1's past experience with loans.

Loan amount	Quarters late	Payment history
\$10,000	0	\$2000 paid, \$3000 delayed by an extra quarter, \$3000 delayed by 2 extra quarters, and the rest delayed 3 extra quarters.
\$25,000	1	\$4000 paid, \$12,000 delayed by an extra quarter, \$6000 delayed by 2 extra quarters, and the rest delayed by 3 extra quarters.
\$50,000	2	\$7500 paid, \$15,000 delayed by an extra quarter, and the rest delayed by 2 extra quarters.
\$50,000	3	\$42,000 paid and the rest delayed by an extra quarter.
\$100,000	4	\$50,000 paid.

Express Bank1's loan situation as a Markov chain.

4. (Pliskin and Tell, 1981) Patients suffering from kidney failure can either get a transplant or undergo periodic dialysis. During any one year, 30% undergo cadaveric transplants and 10% receive living-donor kidneys. In the year following a transplant, 30% of the cadaveric transplants and 15% of living-donor recipients go back to dialysis. Death percentages among the two groups are 20% and 10%, respectively. Of those in the dialysis pool, 10% die and of the ones who survive more than one year after a transplant, 5% die and 5% go back to dialysis. Represent the situation as a Markov chain.

17.2 ABSOLUTE AND *n*-STEP TRANSITION PROBABILITIES

Given the initial probabilities $\mathbf{a}^{(0)} = \{a_j^{(0)}\}$ of starting in state j and the transition matrix $\mathbf{P}$ of a Markov chain, the absolute probabilities $\mathbf{a}^{(n)} = \{a_j^{(n)}\}$ of being in state j after n transitions $(n > 0)$ are computed as follows:

$$\mathbf{a}^{(1)} = \mathbf{a}^{(0)}\mathbf{P}$$
$$\mathbf{a}^{(2)} = \mathbf{a}^{(1)}\mathbf{P} = \mathbf{a}^{(0)}\mathbf{P}\mathbf{P} = \mathbf{a}^{(0)}\mathbf{P}^2$$
$$\mathbf{a}^{(3)} = \mathbf{a}^{(2)}\mathbf{P} = \mathbf{a}^{(0)}\mathbf{P}^2\mathbf{P} = \mathbf{a}^{(0)}\mathbf{P}^3$$

Continuing in the same manner, we get

$$\mathbf{a}^{(n)} = \mathbf{a}^{(0)}\mathbf{P}^n, n = 1, 2, \ldots$$

The matrix $\mathbf{P}^n$ is known as the ***n*-step transition matrix**. From these calculations we can see that

$$\mathbf{P}^n = \mathbf{P}^{n-1}\mathbf{P}$$

or

$$\mathbf{P}^n = \mathbf{P}^{n-m}\mathbf{P}^m, 0 < m < n$$

These are known as **Chapman-Kolomogorov** equations.

Example 17.2-1

The following transition matrix applies to the gardener problem with fertilizer (Example 17.1-3):

$$\mathbf{P} = \begin{matrix} \\ 1 \\ 2 \\ 3 \end{matrix}\begin{matrix} \begin{matrix} 1 & \quad 2 & \quad 3 \end{matrix} \\ \begin{pmatrix} .30 & .60 & .10 \\ .10 & .60 & .30 \\ .05 & .40 & .55 \end{pmatrix} \end{matrix}$$

The initial condition of the soil is good—that is $\mathbf{a}^{(0)} = (1, 0, 0)$. Determine the absolute probabilities of the three states of the system after 1, 8, and 16 gardening seasons.

$$\mathbf{P}^2 = \begin{pmatrix} .30 & .60 & .10 \\ .10 & .60 & .30 \\ .05 & .40 & .55 \end{pmatrix}\begin{pmatrix} .30 & .60 & .10 \\ .10 & .60 & .30 \\ .05 & .40 & .55 \end{pmatrix} = \begin{pmatrix} .1550 & .5800 & .2650 \\ .1050 & .5400 & .3550 \\ .0825 & .4900 & .4275 \end{pmatrix}$$

$$\mathbf{P}^4 = \begin{pmatrix} .1550 & .5800 & .2650 \\ .1050 & .5400 & .3550 \\ .0825 & .4900 & .4275 \end{pmatrix}\begin{pmatrix} .1550 & .5800 & .2650 \\ .1050 & .5400 & .3550 \\ .0825 & .4900 & .4275 \end{pmatrix}$$

$$= \begin{pmatrix} .10679 & .53295 & .36026 \\ .10226 & .52645 & .37129 \\ .09950 & .52193 & .37857 \end{pmatrix}$$

$$\mathbf{P}^8 = \begin{pmatrix} .10679 & .53295 & .36026 \\ .10226 & .52645 & .37129 \\ .09950 & .52193 & .37857 \end{pmatrix}\begin{pmatrix} .10679 & .53295 & .36026 \\ .10226 & .52645 & .37129 \\ .09950 & .52193 & .37857 \end{pmatrix}$$

$$= \begin{pmatrix} .101753 & .525514 & .372733 \\ .101702 & .525435 & .372863 \\ .101669 & .525384 & .372863 \end{pmatrix}$$

$$\mathbf{P}^{16} = \begin{pmatrix} .101753 & .525514 & .372733 \\ .101702 & .525435 & .372863 \\ .101669 & .525384 & .372863 \end{pmatrix}\begin{pmatrix} .101753 & .525514 & .372733 \\ .101702 & .525435 & .372863 \\ .101669 & .525384 & .372863 \end{pmatrix}$$

$$= \begin{pmatrix} .101659 & .52454 & .372881 \\ .101659 & .52454 & .372881 \\ .101659 & .52454 & .372881 \end{pmatrix}$$

Thus,

$$\mathbf{a}^{(1)} = (1 \quad 0 \quad 0)\begin{pmatrix} .30 & .60 & .10 \\ .10 & .60 & .30 \\ .05 & .40 & .55 \end{pmatrix} = (.30 \quad .60 \quad .1)$$

$$\mathbf{a}^{(8)} = (1 \quad 0 \quad 0)\begin{pmatrix} .101753 & .525514 & .372733 \\ .101702 & .525435 & .372863 \\ .101669 & .525384 & .372863 \end{pmatrix} = (.101753 \quad .525514 \quad .372733)$$

$$\mathbf{a}^{(16)} = (1 \quad 0 \quad 0)\begin{pmatrix} .101659 & .52454 & .372881 \\ .101659 & .52454 & .372881 \\ .101659 & .52454 & .372881 \end{pmatrix} = (.101659 \quad .52454 \quad .372881)$$

The rows of $\mathbf{P}^8$ and the vector of absolute probabilities $\mathbf{a}^{(8)}$ are almost identical. The result is more pronounced for $\mathbf{P}^{16}$. It demonstrates that, as the number of transitions increases, the absolute probabilities are independent of the initial $\mathbf{a}^{(0)}$. In this case the resulting probabilities are known as the **steady-state probabilities.**

Remarks. The computations associated with Markov chains are quite tedious. Template excelMarkovChains.xls provides a general easy-to-use spreadsheet for carrying out these calculations (see *Excel moment* following Example 17.4-1).

PROBLEM SET 17.2A

1. Consider Problem 1, Set 17.1a. Determine the probability that the professor will purchase the current model in four years.

***2.** Consider Problem 2, Set 17.1a. If the police car is currently at a call scene, determine the probability that an apprehension will take place in two patrols.

3. Consider Problem 3, Set 17.1a. Suppose that Bank1 currently has $500,000 worth of outstanding loans. Of these, $100,000 are new, $50,000 are one quarter late, $150,000 are two quarters late, $100,000 are three quarters late, and the rest are over four quarters late. What would the picture of these loans be like after two cycles of loans?

4. Consider Problem 4, Set 17.1a.

(a) For a patient who is currently on dialysis, what is the probability of receiving a transplant in two years?

(b) For a patient who is currently a more-than-one-year survivor, what is the probability of surviving four more years?

17.3 CLASSIFICATION OF THE STATES IN A MARKOV CHAIN

The states of a Markov chain can be classified based on the transition probability p_{ij} of $\mathbf{P}$.

1. A state j is **absorbing** if it returns to itself with certainty in one transition—that is $p_{jj} = 1$.
2. A state j is **transient** if it can reach another state but cannot itself be reached back from another state. Mathematically, this will happen if $\lim_{n\to\infty} p_{ij}^{(n)} = 0$, for all i.
3. A state j is **recurrent** if the probability of being revisited from other states is 1. This can happen if, and only if, the state is not transient.
4. A state j is **periodic** with period $t > 1$ if a return is possible only in $t, 2t, 3t, \ldots$ steps. This means that $p_{jj}^{(n)} = 0$ whenever n is not divisible by t.

Based on the given definitions, a *finite* Markov chain cannot consist of all transient states because, by definition, the transient property requires entering other "trapping" states, thus never revisiting the transient state. The "trapping" state need not be a single absorbing state. For example, in the chain

$$\mathbf{P} = \begin{pmatrix} 0 & 1 & 0 & 0 \\ 0 & 0 & 1 & 0 \\ 0 & 0 & .3 & .7 \\ 0 & 0 & .4 & .6 \end{pmatrix}$$

states 1 and 2 are transient because they cannot be reentered once the system is "trapped" in states 3 and 4. States 3 and 4, which, in a sense, play the role of an absorbing state, constitute a **closed set**. By definition, all the states of a *closed set* must **communicate**, which means that it is possible to go from any state to every other state in the set in one or more transitions—that is, $p_{ij}^{(n)} > 0$ for all $i \neq j$ and $n \geq 1$. Notice that states 3 and 4 can both be absorbing states if $p_{33} = p_{44} = 1$. In such a case, each state forms a closed set.

A *closed* Markov chain is said to be **ergodic** if all its states are *recurrent* and *aperiodic* (not periodic). In this case, the absolute probabilities after n transitions, $\mathbf{a}^{(n)} = \mathbf{a}^{(0)}\mathbf{P}^n$, always converge uniquely to a limiting (steady-state) distribution as $n \to \infty$ that is independent of the initial probabilities $\mathbf{a}^{(0)}$, as will be shown in Section 17.4.

Example 17.3-1 (Absorbing and Transient States)

Consider the gardener Markov chain with no fertilizer.

$$\mathbf{P} = \begin{pmatrix} .2 & .5 & .3 \\ 0 & .5 & .5 \\ 0 & 0 & 1 \end{pmatrix}$$

States 1 and 2 are transient because they reach state 3 but can never be reached back. State 3 is absorbing because $p_{33} = 1$. These classifications can also be seen when $\lim_{n \to \infty} p_{ij}^{(n)} = 0$ is computed. For example,

$$\mathbf{P}^{(100)} = \begin{pmatrix} 0 & 0 & 1 \\ 0 & 0 & 1 \\ 0 & 0 & 1 \end{pmatrix}$$

which shows that in the long run, the probability of ever reentering transient state 1 or 2 is zero, whereas the probability of being "trapped" in absorbing state 3 is certain.

Example 17.3-2 (Periodic States)

We can test the periodicity of a state by computing $\mathbf{P}^n$ and observing the values of $p_{ii}^{(n)}$ for $n = 2, 3, 4, \ldots$. These values will be positive only at the corresponding period of the state. For example, in the chain

$$\mathbf{P} = \begin{pmatrix} 0 & .6 & .4 \\ 0 & 1 & 0 \\ .6 & .4 & 0 \end{pmatrix}$$

we have

$$\mathbf{P}^2 = \begin{pmatrix} .24 & .76 & 0 \\ 0 & 1 & 0 \\ 0 & .76 & .24 \end{pmatrix}, \mathbf{P}^3 = \begin{pmatrix} 0 & .904 & .0960 \\ 0 & 1 & 0 \\ .144 & .856 & 0 \end{pmatrix}, \mathbf{P}^4 = \begin{pmatrix} .0576 & .9424 & 0 \\ 0 & 1 & 0 \\ 0 & .9424 & .0576 \end{pmatrix},$$

$$\mathbf{P}^5 = \begin{pmatrix} 0 & .97696 & .02304 \\ 0 & 1 & 0 \\ .03456 & .96544 & 0 \end{pmatrix}$$

Continuing with $n = 6, 7, \ldots, \mathbf{P}^n$ shows that p_{11} and p_{33} are positive for even values of n and zero otherwise. This means that the period for states 1 and 3 is 2.

PROBLEM SET 17.3A

1. Classify the states of the following Markov chains. If a state is periodic, determine its period:

*(a) $\begin{pmatrix} 0 & 1 & 0 \\ 0 & 0 & 1 \\ 1 & 0 & 0 \end{pmatrix}$

*(b) $\begin{pmatrix} \frac{1}{2} & \frac{1}{4} & \frac{1}{4} & 0 \\ 0 & 0 & 1 & 0 \\ \frac{1}{3} & 0 & \frac{1}{3} & \frac{1}{3} \\ 0 & 0 & 0 & 1 \end{pmatrix}$

(c) $\begin{pmatrix} 0 & 1 & 0 & 0 & 0 & 0 \\ 0 & .5 & .5 & 0 & 0 & 0 \\ 0 & .7 & .3 & 0 & 0 & 0 \\ 0 & 0 & 0 & 1 & 0 & 0 \\ 0 & 0 & 0 & 0 & .4 & .6 \\ 0 & 0 & 0 & 0 & .2 & 8 \end{pmatrix}$

(d) $\begin{pmatrix} .1 & 0 & .9 \\ .7 & .3 & 0 \\ .2 & .7 & .1 \end{pmatrix}$

17.4 STEADY-STATE PROBABILITIES AND MEAN RETURN TIMES OF ERGODIC CHAINS

In an ergodic Markov chain, the steady-state probabilities are defined as

$$\pi_j = \lim_{n \to \infty} a_j^{(n)}, \quad j = 0, 1, 2, \ldots$$

These probabilities, which are independent of $\{a_j^{(0)}\}$, can be determined from the equations

$$\boldsymbol{\pi} = \boldsymbol{\pi}\mathbf{P}$$

$$\sum_j \pi_j = 1$$

(One of the equations in $\boldsymbol{\pi} = \boldsymbol{\pi}\mathbf{P}$ is redundant.) What $\boldsymbol{\pi} = \boldsymbol{\pi}\mathbf{P}$ says is that the probabilities $\boldsymbol{\pi}$ remain unchanged after one transition, and for this reason they represent the steady-state distribution.

A direct by-product of the steady-state probabilities is the determination of the expected number of transitions before the systems returns to a state *j* for the first time. This is known as the **mean first return time** or the **mean recurrence time,** and is computed in an *n*-state Markov chain as

$$\mu_{jj} = \frac{1}{\pi_j}, j = 1, 2, \ldots, n$$

Example 17.4-1

To determine the steady-state probability distribution of the gardener problem with fertilizer (Example 17.1-3), we have

$$(\pi_1\ \pi_2\ \pi_3) = (\pi_1\ \pi_2\ \pi_3)\begin{pmatrix} .3 & .6 & .1 \\ .1 & .6 & .3 \\ .05 & .4 & .55 \end{pmatrix}$$

which yields the following set of equations:

$$\pi_1 = .3\pi_1 + .1\pi_2 + .05\pi_3$$

$$\pi_2 = .6\pi_1 + .6\pi_2 + .4\pi_3$$

$$\pi_3 = .1\pi_1 + .3\pi_2 + .55\pi_3$$

$$\pi_1 + \pi_2 + \pi_3 = 1$$

Recalling that one (any one) of the first three equations is redundant, the solution is $\pi_1 = 0.1017$, $\pi_2 = 0.5254$, and $\pi_3 = 0.3729$. What these probabilities say is that, in the long run, the soil condition approximately will be good 10% of the time, fair 52% of the time, and poor 37% of the time.

The mean first return times are computed as

$$\mu_{11} = \frac{1}{.1017} = 9.83, \mu_{22} = \frac{1}{.5254} = 1.9, \mu_{33} = \frac{1}{.3729} = 2.68$$

This means that, depending on the current state of the soil, it will take approximately 10 gardening seasons for the soil to return to a *good* state, 2 seasons to return to a *fair* state, and 3 seasons to return to a *poor* state. These results point to a more "bleak" than "promising" outlook for the soil condition under the proposed fertilizer program. A more aggressive program should improve the picture. For example, consider the following transition matrix in which the probabilities of moving to a good state are higher than in the previous matrix:

$$\mathbf{P} = \begin{pmatrix} .35 & .6 & .05 \\ .3 & .6 & .1 \\ .25 & .4 & .35 \end{pmatrix}$$

In this case, $\pi_1 = 0.31, \pi_2 = 0.58$, and $\pi_3 = 0.11$, which yields $\mu_{11} = 3.2, \mu_{22} = 1.7$, and $\mu_{33} = 8.9$, a reversal of the "bleak" outlook given previously.

Excel Moment

Figure 17.1 shows the output of the gardener example using the general Excel template excelMarkovChains.xls to compute n-step, absolute, and steady-state probabilities, and mean return time for a Markov chain of any size. The steps are self-explanatory. In step 2a, you may override the default state codes $(1, 2, 3, \ldots)$ by a code of your choice. These codes will be automatically updated everywhere else in the spreadsheet when you execute step 4.

FIGURE 17.1

Excel Spreadsheet for Markov chain computations

	A	B	C	D	E	F	G	H
1				Markov Chains				
2	Step 1:	Number of states =		3	Step 2a:	Initial probabilities:		
3	Step 2:	Click to enter Markov chain			Codes:	1	2	3
4						1	0	0
5	Step 3:	Number of transitions		8	Step 2b:	Input Markov chain		
6	Step 4:	Click to execute				1	2	3
7					1	0.3	0.6	0.1
8		Output Results			2	0.1	0.6	0.3
9		Absolute	Steady	Mean returr	3	0.05	0.4	0.55
10	State	(8-step)	state	time		Output (8-step) transition matrix		
11	1	0.10175	0.101695	9.8333254		1	2	3
12	2	0.52551	0.525424	1.9032248	1	0.10175	0.525514	0.372733
13	3	0.37273	0.372882	2.6818168	2	0.1017	0.525435	0.372864
14					3	0.10167	0.525384	0.372947

Example 17.4-2 (Cost Model)

Consider the gardener problem with fertilizer (Example 17.1-3). Suppose that the cost of the fertilizer is \$50 per bag and the garden needs two bags if the soil is good. The amount of fertilizer is increased by 25% if the soil is fair and 60% if the soil is poor. The gardener estimates the annual yield to be worth \$250 if no fertilizer is used and \$420 if fertilizer is applied. Is it worthwhile to use the fertilizer?

Using the steady state probabilities in Example 17.4-1, we get

$$\begin{aligned}\text{Expected annual cost of fertilizer} &= 2 \times \$50 \times \pi_1 + (1.25 \times 2) \times \$50 \times \pi_2 \\ &\quad + (1.60 \times 2) \times \$50 \times \pi_3 \\ &= 100 \times .1017 + 125 \times .5254 + 160 \times .3729 \\ &= \$135.51\end{aligned}$$

$$\text{Increase in the annual value of the yield} = \$420 - \$250 = \$170$$

The results show that, on the average, the use of fertilizer nets $170 - 135.51 = \$34.49$. Hence the use of fertilizer is recommended.

PROBLEM SET 17.4A

*1. On a sunny Spring day, MiniGolf can gross \$2000 in revenues. If the day is cloudy, revenues drop by 20%. A rainy day will reduce revenues by 80%. If today's weather is sunny, there is an 80% chance it will remain sunny tomorrow with no chance of rain. If it is cloudy, there is a 20% chance that tomorrow will be rainy and 30% chance it will be sunny. Rain will continue through the next day with a probability of .8, but there is a 10% chance it may be sunny.

   (a) Determine the expected daily revenues for MiniGolf.

   (b) Determine the average number of days the weather will not be sunny.

2. Joe loves to eat out in area restaurants. His favorite foods are Mexican, Italian, Chinese, and Thai. On the average, Joe pays \$10.00 for a Mexican meal, \$15.00 for an Italian meal, \$9.00 for a Chinese meal, and \$11.00 for a Thai meal. Joe's eating habits are predictable: There is a 70% chance that today's meal is a repeat of yesterday's, and equal probabilities of switching to one of the remaining three.

   (a) How much does Joe pay on the average for his daily dinner?

   (b) How often does Joe eat Mexican food?

3. Some ex-cons spend the rest of their lives in one four of states: free, on trial, in jail, or on probation. At the start of each year, statistics show that there is 50% chance that a free ex-con will commit a new crime and go on trial. The judge may send the ex-con to jail with probability .6 or grant probation with probability .4. Once in jail, 10% of ex-cons will be set free for good behavior. Of those who are on probation, 10% commit new crimes and are arraigned for new trials, 50% will go back to finish their sentence for violating probation orders, and 10% will be set free for lack of evidence. Taxpayers underwrite the costs associated with the punishment of the ex-felons. It is estimated that a trial will cost about \$5000, an average jail sentence will cost \$20,000, and an average probation period will cost \$2000.

(a) Determine the expected cost per ex-con.

(b) How often does an ex-con return to jail? Go on trial? Get set free?

4. A store sells a special item whose daily demand can described by the following pdf:

Daily demand, D	0	1	2	3
$P\{D\}$	.1	.3	.4	.2

The store is comparing two ordering policies: (1) Order up to 3 units every 3 days if the stock level is less than 2, else do not order. (2) Order 3 units every 3 days if the stock level is zero, else do not order. The fixed ordering cost per shipment is \$300 and the cost of holding excess units per unit per day is \$3. Immediate delivery is expected.

(a) Which policy should the store adopt to minimize the total expected daily cost of ordering and holding?

(b) For the two policies, compare the average number of days between successive inventory depletions.

***5.** There are three categories of income tax filers in the United States: those who never evade taxes, those who sometimes do it, and those who always do it. An examination of audited tax returns from one year to the next shows that of those who did not evade taxes last year, 95% continue in the same category this year, 4% move to the "sometimes" category, and the remainder move to the "always" category. For those who sometimes evade taxes, 6% move to "never," 90% stay the same, and 4% move to "always." As for the "always" evaders, the respective percentages are 0%, 10%, and 90%.

(a) Express the problem as a Markov chain.

(b) In the long run, what would be the percentages of "never," "sometimes," and "always" tax categories?

(c) Statistics show that a taxpayer in the "sometimes" category evades taxes on about \$5000 per return and in the "always" category on about \$12,000. Assuming an average income tax rate of 12% and a filers population of 70 million, determine the annual reduction in collected taxes due to evasion.

6. Warehouzer owns a renewable forest land for growing pine trees. Trees can fall into one of four categories depending on their age: baby (0–5 years), young (5–10 years), mature (11–15 years), and old (more than 15 years). Ten percent of baby and young trees die before reaching the next age group. For mature and old trees, 50% are harvested and only 5% die. Because of the renewal nature of the operation, all harvested and dead tree are replaced with new (baby) trees by the end of the next 5-year cycle.

(a) Express the forest dynamics as a Markov chain.

(b) If the forest land can hold a total of 500,000 trees, determine the long-run composition of the forest.

(c) If a new tree is planted at the cost of \$1 per tree and a harvested tree has a market value of \$20, determine the average annual income from the forest operation.

7. Population dynamics is impacted by the continual movement of people who are seeking better quality of life or better employment. The city of Mobile has an inner city population, a suburban population, and a surrounding rural population. The census taken in 10-year intervals shows that 10% of the rural population move to the suburbs and 5% to the inner city. For the suburban population, 30% move to rural areas and 15% to the inner city. Inner-city population would not move into suburbs, but 20% of them move to the quiet rural life.

(a) Express the population dynamics as a Markov chain.

(b) If the greater Mobile area currently includes 20,000 rural residents, 100,000 suburbanites, and 30,000 inner city inhabitants, what will the population distribution be in 10 years? In 20 years?

(c) Determine the long-run population picture of Mobile.

8. A car rental agency has offices in Phoenix, Denver, Chicago, and Atlanta. The agency allows one- and two-way rentals so that cars rented in one location may end up in another. Statistics show that at the end of each week 70% of all rentals are two-way. As for the one-way rentals: From Phoenix, 20% go to Denver, 60% to Chicago, and the rest goes to Atlanta; from Denver, 40% go to Atlanta and 60% to Chicago; from Chicago, 50% go to Atlanta and the rest to Denver; and from Atlanta, 80% go to Chicago, 10% to Denver, and 10% to Phoenix.

(a) Express the situation as a Markov chain.

(b) If the agency starts the week with 100 cars in each location, what will the distribution be like in two weeks?

(c) If each location is designed to handle a maximum of 110 cars, would there be a long-run space availability problem in any of the locations?

(d) Determine the average number of weeks that elapse before a car is returned to its originating location.

9. A bookstore keeps daily track of the inventory level of a popular book to restock it to a level of 100 copies at the start of each day. The data for the last 30 days provide the following end-of-day inventory position: 1, 2, 0, 3, 2, 1, 0, 0, 3, 0, 1, 1, 3, 2, 3, 3, 2, 1, 0, 2, 0, 1, 3, 0, 0, 3, 2, 1, 2, 2.

(a) Represent the daily inventory as a Markov chain.

(b) Determine the steady-state probability that the bookstore will run out of books in any one day.

(c) Determine the expected daily inventory.

(d) Determine the average number of days between successive zero inventories.

10. In Problem 9, suppose that the daily demand can exceed supply, which gives rise to shortage (negative inventory). The end-of-day inventory level for the past 30 days is given as: 1, 2, 0, −2, 2, 2, −1, −1, 3, 0, 0, 1, −1, −2, 3, 3, −2, −1, 0, 2, 0, −1, 3, 0, 0, 3, −1, 1, 2, −2.

(a) Express the situation as a Markov chain.

(b) Determine the long-term probability of a surplus inventory in any one day.

(c) Determine the long-term probability of a shortage inventory in any one day.

(d) Determine the long-term probability of the daily supply meeting the daily demand exactly.

(e) If the holding cost per (end-of-day) surplus book is $.15 per day and the penalty cost per shortage book is $4.00 per day, determine the expected inventory cost per day.

11. A store starts a week with at least 3 PCs. The demand per week is estimated at 0 with probability .15, 1 with probability .2, 2 with probability .35, 3 with probability .25, and 4 with probability .05. Unfilled demand is backlogged. The store's policy is to place an order for delivery at the start of the following week whenever the inventory level drops below 3 PCs. The new replenishment always brings the stock back to 5 PCs.

(a) Express the situation as a Markov chain.

(b) Suppose that the week starts with 4 PCs. Determine the probability that an order will be placed at the end of two weeks.

(c) Determine the long-run probability that no order will be placed in any week.

(d) If the fixed cost of placing an order is $200, the holding cost per PC per week is $5, and the penalty cost per shortage PC per week is $20, determine the expected inventory cost per week.

12. Solve Problem 11 assuming that the order size, when placed, is exactly 5 pieces.

13. In Problem 12, suppose that the demand for the PCs is 0, 1, 2, 3, 4, or 5 with equal probabilities. Further assume that the unfilled demand is not backlogged, but that the penalty cost is still incurred.

(a) Express the situation as a Markov chain.

(b) Determine the long-run probability that a shortage will take place.

(c) If the fixed cost of placing an order is $200, the holding cost per PC per week is $5, and the penalty cost per shortage PC per week is $20, determine the expected ordering and inventory cost per week.

*__14.__ The federal government tries to boost small business activities by awarding annual grants for projects. All bids are competitive, but the chance of receiving a grant is highest if the owner has not received any during the last three years and lowest if awards were given in each of the last three years. Specifically, the probability of getting a grant if none were awarded in the last three years is .9. It reduces to .8 if one grant was awarded, .7 if two grants were awarded, and only .5 if 3 were received.

(a) Express the situation as a Markov chain.

(b) Determine the expected number of awards per owner per year.

15. Jim Bob has a history of receiving many fines for driving violations. Unfortunately for Jim Bob, modern technology can keep track of his previous fines. As soon as he has accumulated 4 tickets, his driving license is revoked until he completes a new driver education class, in which case he starts with a clean slate. Jim Bob is most reckless immediately after completing the driver education class and he is invariably stopped by the police with a 50-50 chance of being fined. After each new fine, he tries to be more careful, which reduces the probability of a fine by .1.

(a) Express Jim Bob's problem as Markov chain.

(b) What is the average number of times Jim Bob is stopped by police before his license is revoked again?

(c) What is the probability that Jim Bob will lose his license?

(d) If each fine costs $100, how much, on the average, does Jim Bob pay between successive suspensions of his license?

17.5 FIRST PASSAGE TIME

In Section 17.4, we used the steady state probabilities to compute μ_{jj}, the *mean first return time* for state j. In this section, we are concerned with the determination of the **mean first passage time** μ_{ij}, the expected number of transitions needed to reach state j from state i for the first time. The calculations are rooted in the determination of the probability f_{ij} of *at least* one passage from state i to state j as $f_{ij} = \sum_{n=1}^{\infty} f_{ij}^{(n)}$, where $f_{ij}^{(n)}$ is the probability of a first passage from state i to state j in n transitions. An expression for $f_{ij}^{(n)}$ can be determined recursively from

$$p_{ij}^{(n)} = f_{ij}^{(n)} + \sum_{k=1}^{n-1} f_{ij}^{(k)} p_{jj}^{(n-k)}, \; n = 1, 2, \ldots$$

The transition matrix $\mathbf{P} = \|p_{ij}\|$ is assumed to have m states.

1. If $f_{ij} < 1$, it is not certain that the system will ever pass from state i to state j and $\mu_{ij} = \infty$.
2. If $f_{ij} = 1$, the Markov chain is ergodic and the *mean first passage time* from state i to state j is computed as

$$\mu_{ij} = \sum_{n=1}^{\infty} n f_{ij}^{(n)}$$

A simpler way to determine the mean first passage time for all the states in an m-transition matrix, $\mathbf{P}$, is to use the following matrix-based formula:

$$\|\mu_{ij}\| = (\mathbf{I} - \mathbf{N}_j)^{-1}\mathbf{1}, \; j \neq i$$

where

$\mathbf{I} = (m - 1)$-identity matrix

$\mathbf{N}_j =$ transition matrix $\mathbf{P}$ less its jth row and jth column of target state j

$\mathbf{1} = (m - 1)$ column vector with all elements equal to 1

The matrix operation $(\mathbf{I} - \mathbf{N}_j)^{-1}\mathbf{1}$ essentially sums the columns of $(\mathbf{I} - \mathbf{N}_j)^{-1}$.

Example 17.5-1

Consider the gardener Markov chain with fertilizers once again.

$$\mathbf{P} = \begin{pmatrix} .30 & .60 & .10 \\ .10 & .60 & .30 \\ .05 & .40 & .55 \end{pmatrix}$$

To demonstrate the computation of the first passage time to a specific state from all others, consider the passage from states 2 and 3 (fair and poor) to state 1 (good). Thus, $j = 1$ and

$$\mathbf{N}_1 = \begin{pmatrix} .60 & .30 \\ .40 & .55 \end{pmatrix}, (\mathbf{I} - \mathbf{N}_1)^{-1} = \begin{pmatrix} .4 & -.3 \\ -.4 & .45 \end{pmatrix}^{-1} = \begin{pmatrix} 7.50 & 5.00 \\ 6.67 & 6.67 \end{pmatrix}$$

Thus,

$$\begin{pmatrix} \mu_{21} \\ \mu_{31} \end{pmatrix} = \begin{pmatrix} 7.50 & 5.00 \\ 6.67 & 6.67 \end{pmatrix} \begin{pmatrix} 1 \\ 1 \end{pmatrix} = \begin{pmatrix} 12.50 \\ 13.34 \end{pmatrix}$$

This means that, on the average, it will take 12.5 seasons to pass from fair to good soil and 13.34 seasons to go from bad to good soil.

Similar calculations can be carried out to obtain μ_{12} and μ_{32} from $(\mathbf{I} - \mathbf{N}_2)$ and μ_{13} and μ_{23} from $(\mathbf{I} - \mathbf{N}_3)$, as shown below.

Excel Moment

The calculations of the mean first passage times can be carried out conveniently by Excel template excelFirstPassTime.xls. Figure 17.2 shows the calculations associated with Example 17.5-1. Step 2 of the spreadsheet automatically initializes the transition matrix **P** to zero values per the size given in step 1. In step 2a, you may override the

FIGURE 17.2

Excel spreadsheet calculations of first passage time of Example 17.5-1 (file excelFirstPassTime.xls)

	A	B	C	D	E	F	G	H
1		First Passage Times in Ergodic and Absorbing Markov Chains						
2	Step 1:	Number of states =		3	Step 2a:	You may override codes in ROW 6		
4	Step 2:	Click to enter Markov chain, P			Step 3:	Click to compute I- P		
5		Matrix P:(Blank cell may result in "Type mismatch" compiler error)						
6	Codes	1	2	3				
7	1	0.3	0.6	0.1				
8	2	0.1	0.6	0.3				
9	3	0.05	0.4	0.55				
10		Matrix I-P:						
11		1	2	3				
12	1	0.7	-0.6	-0.1				
13	2	-0.1	0.4	-0.3				
14	3	-0.05	-0.4	0.45				
15	Step 4: Perform first passage time calculations below:							
16		I-N			inv(I-N)			Mu
17	i=1	2	3		2	3		1
18	2	0.4	-0.3	2	7.5	5	2	12.5
19	3	-0.4	0.45	3	6.666667	6.6666667	3	13.33333
20								
21	i=2	1	3		1	3		2
22	1	0.7	-0.1	1	1.451613	0.3225806		1.774194
23	3	-0.05	0.45	3	0.16129	2.2580645		2.419355
24								
25	i=3	1	2		1	2		3
26	1	0.7	-0.6	1	1.818182	2.7272727		4.545455
27	2	-0.1	0.4	2	0.454545	3.1818182		3.636364

default state codes in row 6 with a code of your choice. The code will then be transferred automatically throughout the spreadsheet. After you enter the transition probabilities, step 3 creates the matrix $\mathbf{I} - \mathbf{P}$. Step 4 is carried out entirely by you using $\mathbf{I} - \mathbf{P}$ as the source for creating $\mathbf{I} - \mathbf{N}_j$ ($j = 1, 2$, and 3). You can do so by copying the entire $\mathbf{I} - \mathbf{P}$ and its state codes and pasting it in the target location, and then using appropriate Excel Cut and Paste operations to rid $\mathbf{I} - \mathbf{P}$ of row j and column j. For example, to create $\mathbf{I} - \mathbf{N}_2$, first copy $\mathbf{I} - \mathbf{P}$ and its state codes to the selected target location. Next, highlight column 3 of the copied matrix, cut it, and paste it in column 2, thus eliminating column 2. Similarly, highlight row 3 of the resulting matrix, cut it, and then paste it in row 2, thus eliminating row 2. The created $\mathbf{I} - \mathbf{N}_2$ automatically carries its correct state code.

Once $\mathbf{I} - \mathbf{N}_j$ is created, the inverse, $(\mathbf{I} - \mathbf{N}_j)^{-1}$, is computed in the target location. The associated operations are demonstrated by inverting $(\mathbf{I} - \mathbf{N}_1)$ in Figure 17.2:

1. Enter the formula =MINVERSE(B18:C19) in E18.
2. Highlight E18:F19, the area where the inverse will reside.
3. Press F2.
4. Press CTRL + SHIFT + ENTER.

The values of the first passage times from states 2 and 3 to state 1 are then computed by summing the rows of the inverse—that is, by entering =SUM(E18:F18) in H18 and then copying H18 into H19. After creating $\mathbf{I} - \mathbf{N}$ for $i = 2$ and $i = 3$, the remaining calculations are automated by copying E18:F19 into E22:F23 and E26:F27, and copying H18:H19 into H22:H23 and H26:H27.

PROBLEM SET 17.5A

*1. A mouse maze consists of the paths shown in Figure 17.3. Intersection 1 is the maze entrance and intersection 5 is the exit. At any intersection, the mouse has equal probabilities of selecting any of the available paths. When the mouse reaches intersection 5, it will be allowed to recirculate in the maze.
   - **(a)** Express the maze as a Markov chain.
   - **(b)** Determine the probability that, starting at intersection 1, the mouse will reach the exit after three trials.
   - **(c)** Determine the long-run probability that the mouse will locate the exit intersection.
   - **(d)** Determine the average number of trials needed to reach the exit point from intersection 1.

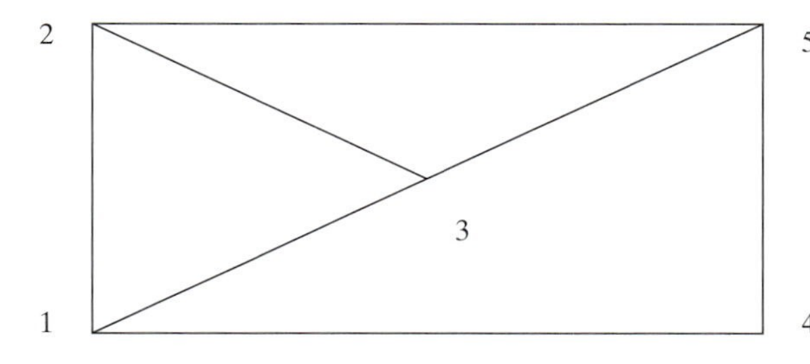

FIGURE 17.3
Mouse maze for Problem 1, Set 17.5a

2. In Problem 1, intuitively, if more options (routes) are added to the maze, will the average number of trials needed to reach the exit point increase or decrease? Demonstrate the answer by adding a route between intersections 3 and 4.

3. Jim and Joe start a game with five tokens, three for Jim and two for Joe. A coin is tossed and if the outcome is heads, Jim gives Joe a token, else Jim gets a token from Joe. The game ends when Jim or Joe has all the tokens. At this point, there is 30% chance that Jim and Joe will continue to play the game, again starting with three tokens for Jim and two for Joe.
   (a) Represent the game as a Markov chain.
   (b) Determine the probability that Joe will win in three coin tosses. That Jim will win in three coin tosses.
   (c) Determine the probability that a game will end in Jim's favor. Joe's favor.
   (d) Determine the average number of coin tosses needed before Jim wins. Joe wins.

4. An amateur gardener with training in botany is experimenting with scientific cross-pollination of pink irises with red, orange, and white irises. His annual experiments show that pink can produce 60% pink and 40% white, red can produce 40% red, 50% pink, and 10% orange, orange can produce 25% orange, 50% pink, and 25% white, and white can produce 50% pink and 50% white.
   (a) Express the gardener situation as a Markov chain.
   (b) If the gardener started the cross-pollination with equal numbers of each type iris, what would the distribution be like after 5 years? In the long run?
   (c) How many years on the average would a red iris take to produce a white bloom?

*5. Customers tend to exhibit loyalty to product brands but may be persuaded through clever marketing and advertising to switch brands. Consider the case of three brands: A, B, and C. Customer "unyielding" loyalty to a given brand is estimated at 75%, giving the competitors only a 25% margin to realize a switch. Competitors launch their advertising campaigns once a year. For brand A customers, the probabilities of switching to brands B and C are .1 and .15, respectively. Customers of brand B are likely to switch to A and C with probabilities .2 and .05, respectively. Brand C customers can switch to brands A and B with equal probabilities.
   (a) Express the situation as a Markov chain.
   (b) In the long run, how much market share will each brand command?
   (c) How long on the average will it take for a brand A customer to switch to brand B? To brand C?

17.6 ANALYSIS OF ABSORBING STATES

In the gardener problem without fertilizer the transition matrix is given as

$$\mathbf{P} = \begin{pmatrix} .2 & .5 & .3 \\ 0 & .5 & .5 \\ 0 & 0 & 1 \end{pmatrix}$$

States 1 and 2 (good and fair soil conditions) are *transient* and State 3 (poor soil condition) is *absorbing*, because once in that state the system will remain there indefinitely. A Markov chain may have more than one absorbing state. For example, an employee may remain employed with the same company until retirement or may quit a few years earlier (two absorbing states). In these types of chains, we are interested in determining the probability of reaching absorption and the expected number of transitions to absorption given that the system starts in a specific transient state. For example, in the gardener Markov chain given above, if the soil is currently good, we will be interested in determining the average number of gardening seasons till the soil becomes poor and also the probability associated with this transition.

The analysis of Markov chains with absorbing states can be carried out conveniently using matrices. First, the Markov chain is partitioned in the following manner:

$$\mathbf{P} = \left(\begin{array}{c|c} \mathbf{N} & \mathbf{A} \\ \hline \mathbf{0} & \mathbf{I} \end{array}\right)$$

The arrangement requires all the absorbing states to occupy the southeast corner of the new matrix. For example, consider the following transition matrix:

$$\mathbf{P} = \begin{array}{c} \\ 1 \\ 2 \\ 3 \\ 4 \end{array} \begin{array}{c} \begin{array}{cccc} 1 & 2 & 3 & 4 \end{array} \\ \begin{pmatrix} .2 & .3 & .4 & .1 \\ 0 & 1 & 0 & 0 \\ .5 & .3 & 0 & .2 \\ 0 & 0 & 0 & 1 \end{pmatrix} \end{array}$$

The matrix **P** can be rearranged and partitioned as

$$\mathbf{P}^* = \begin{array}{c} \\ 1 \\ 3 \\ 2 \\ 4 \end{array} \begin{array}{c} \begin{array}{cccc} 1 & 3 & 2 & 4 \end{array} \\ \begin{pmatrix} .2 & .4 & .3 & .1 \\ .5 & 0 & .3 & .2 \\ 0 & 0 & 1 & 0 \\ 0 & 0 & 0 & 1 \end{pmatrix} \end{array}$$

In this case, we have

$$\mathbf{N} = \begin{pmatrix} .2 & .4 \\ .5 & 0 \end{pmatrix}, \mathbf{A} = \begin{pmatrix} .3 & .1 \\ .3 & .2 \end{pmatrix}, \mathbf{I} = \begin{pmatrix} 1 & 0 \\ 0 & 1 \end{pmatrix}$$

Given the definition of **A** and **N** and the unit column vector **1** of all 1 elements, it can be shown that:

$$\text{Expected time in state } j \text{ starting in state } i = \text{element } (i, j) \text{ of } (\mathbf{I} - \mathbf{N})^{-1}$$

$$\text{Expected time to absorption} = (\mathbf{I} - \mathbf{N})^{-1}\mathbf{1}$$

$$\text{Probability of absorption} = (\mathbf{I} - \mathbf{N})^{-1}\mathbf{A}$$

Example 17.6-1[1]

A product is processed on two sequential machines, I and II. Inspection takes place after a product unit is completed on a machine. There is a 5% chance that the unit will be junked before inspection. After inspection, there is a 3% chance the unit will be junked and a 7% chance of its being returned to the same machine for reworking. Else, a unit passing inspection on both machines is good.

(a) For a part starting at machine I, determine the average number of visits to each station.
(b) If a batch of 1000 units is started on machine I, how many good units will be produced?

For the Markov chain, the production process has 6 states: start at I ($s1$), inspect after I ($i1$), start at II ($s2$), inspect after II ($i2$), junk after inspection I or II (J), and good after II (G). Units entering J and G are terminal and hence J and G are absorbing states. The transition matrix is given as

$$\mathbf{P} = \begin{array}{c} \\ s1 \\ i1 \\ s2 \\ i2 \\ J \\ G \end{array} \begin{array}{c} \begin{array}{cccccc} s1 & i1 & s2 & i2 & J & G \end{array} \\ \left(\begin{array}{cccc|cc} 0 & .95 & 0 & 0 & .05 & 0 \\ .07 & 0 & .9 & 0 & .03 & 0 \\ 0 & 0 & 0 & .95 & .05 & 0 \\ 0 & 0 & .07 & 0 & .03 & .9 \\ \hline 0 & 0 & 0 & 0 & 1 & 0 \\ 0 & 0 & 0 & 0 & 0 & 1 \end{array}\right) \end{array}$$

Thus,

$$\mathbf{N} = \begin{array}{c} \\ s1 \\ i1 \\ s2 \\ i2 \end{array} \begin{array}{c} \begin{array}{cccc} s1 & i1 & s2 & i2 \end{array} \\ \begin{pmatrix} 0 & .95 & 0 & 0 \\ .07 & 0 & .9 & 0 \\ 0 & 0 & 0 & .95 \\ 0 & 0 & .07 & 0 \end{pmatrix} \end{array}, \quad \mathbf{A} = \begin{array}{c} \begin{array}{cc} J & G \end{array} \\ \begin{pmatrix} .05 & 0 \\ .03 & 0 \\ .05 & 0 \\ .03 & .9 \end{pmatrix} \end{array}$$

Using the convenient spreadsheet calculations in excelEx17.6-1.xls (see *Excel moment* following Example 17.5-1), we get

$$(\mathbf{I} - \mathbf{N})^{-1} = \begin{pmatrix} 1 & -.95 & 0 & 0 \\ -.07 & 1 & -.9 & 0 \\ 0 & 0 & 0 & -.95 \\ 0 & 0 & -.07 & 1 \end{pmatrix}^{-1} = \begin{pmatrix} 1.07 & 1.02 & .98 & 0.93 \\ 0.07 & 1.07 & 1.03 & 0.98 \\ 0 & 0 & 1.07 & 1.02 \\ 0 & 0 & 0.07 & 1.07 \end{pmatrix}$$

$$(\mathbf{I} - \mathbf{N})^{-1}\mathbf{A} = \begin{pmatrix} 1.07 & 1.02 & .98 & 0.93 \\ 0.07 & 1.07 & 1.03 & 0.98 \\ 0 & 0 & 1.07 & 1.02 \\ 0 & 0 & 0.07 & 1.07 \end{pmatrix} \begin{pmatrix} .05 & 0 \\ .03 & 0 \\ .05 & 0 \\ .03 & .9 \end{pmatrix} = \begin{pmatrix} .16 & .84 \\ .12 & .88 \\ .08 & .92 \\ .04 & .96 \end{pmatrix}$$

[1]Adapted from J. Shamblin and G. Stevens, *Operations Research: A Fundamental Approach*, McGraw-Hill, New York, Chapter 4, 1974.

The top row of $(\mathbf{I} - \mathbf{N})^{-1}$ gives the average number of visits in each station for a part starting at machine I. Specifically, machine I is visited 1.07 times, inspection I is visited 1.02 times, machine II is visited .98 times, and inspection II is visited .93 times. The reason the number of visits in machine I and inspection I is greater than 1 is because of rework and re-inspection. On the other hand, the corresponding values for machine II are less than 1 because some parts are junked before reaching machine II. Indeed, under perfect conditions (no parts junked and no rework), the matrix $(\mathbf{I} - \mathbf{N})^{-1}$ will show that each station is visited exactly once (try it by assigning a transition probability of 1 for all the stations). Of course, the duration of stay at each station could differ. For example, if the processing times at machines I and II are 20 and 30 minutes and if the inspection times at I and II are 5 and 7 minutes, then a part starting at machine 1 will be processed (i.e., either junked or completed) in $1.07 \times 20 + 1.02 \times 5 + .98 \times 30 + .93 \times 7 = 62.41$ minutes.

To determine the number of completed parts in a starting batch of 1000 pieces, we can see from the top row of $(\mathbf{I} - \mathbf{N})^{-1}\mathbf{A}$ that

$$\text{Probability of a piece being junked} = .16$$

$$\text{Probability of a piece being completed} = .84$$

This means that $1000 \times .84 = 840$ pieces will be completed in a starting batch of 1000.

PROBLEM SET 17.6A

1. In Example 17.6-1, suppose that the labor cost for machines I and II is \$20 per hour and that for inspection is only \$18 per hour. Further assume that it takes 30 minutes and 20 minutes to process a piece on machines I and II, respectively. The inspection time at each of the two stations is 10 minutes. Determine the labor cost associated with a completed (good) piece.

***2.** When I borrow a book from the city library, I usually try to return it after one week. Depending on the length of the book and my free time, there is a 30% chance that I may keep it for another week. If I have had the book for two weeks, there is a 10% chance that I'll keep it for an additional week. Under no condition do I keep it for more than three weeks.

- **(a)** Express the situation as a Markov chain.
- **(b)** Determine the average number of weeks I keep a book before returning it to the library.

3. In Casino del Rio, a gambler can bet in whole dollars. Each bet will either gain \$1 with probability .4 or lose \$1 with probability .6. Starting with three dollars, the gambler will quit if all money is lost or the accumulation is doubled.

- **(a)** Express the problem as a Markov chain.
- **(b)** Determine the average number of bets until the game ends.
- **(c)** Determine the probability of ending the game with \$6. Of losing all \$3.

4. Jim must make five years worth of progress to complete his doctorate degree at ABC University. However, he enjoys the life of a student and is in no hurry to finish his degree. In any academic year there is a 50% chance he may take the year off and a 50% chance of his pursuing the degree full time. After completing three academic years, there is a 30% chance that Jim may "bail out" and simply get a master's degree, a 20% chance of his taking the next year off but continuing in the Ph.D. program, and 50% chance of his attending school full time toward his doctorate.

(a) Express Jim's situation as a Markov chain.

(b) Determine the expected number of academic years before Jim's student life comes to an end.

(c) Determine the probability that Jim will end his academic journey with only a master's degree.

(d) If Jim's fellowship pays an annual stipend of $15,000 (but only when he attends school), how much will he be paid before ending up with a degree?

5. An employee who is now 55 years old plans to retire at the age of 62 but does not rule out the possibility of quitting earlier. At the end of each year, he weighs his options (and state of mind regarding work). The probability of quitting after one year is only .1 but seems to increase by approximately .01 with each additional year.

(a) Express the problem as a Markov chain.

(b) What is the probability that the employee stay with the company until planned retirement at age 62?

(c) At age 57, what is the probability the employee will call it quits?

(d) At age 58, what is the expected number of years before the employee is off the payroll?

6. In Problem 3, Set 17.1a,

(a) Determine the expected number of quarters until a debt is either repaid or lost as bad debt.

(b) Determine the probability that a new loan will be written off as bad debt. Repaid in full.

(c) If a loan is six months old, determine the number of quarters until its status is settled.

7. In a men's singles tennis tournament, Andre and John are playing a match for the championship. The match is won when either player wins three out of five sets. Statistics show that there is 60% chance that Andre will win any one set.

(a) Express the match as a Markov chain.

(b) On the average, how long will the match last and what is the probability that Andre will win the championship?

(c) If the score is 1 set to 2, John's favor, what is the probability that Andre will win?

(d) In Part (c), determine the average number of sets till the match ends and interpret the result.

***8.** Students at U of A have expressed dissatisfaction with the fast pace at which the math department is teaching the one-semester Cal I. To cope with this problem, the math department is now offering Cal I in 4 modules. Students will set their individual pace for each module and, when ready, will take a test that will elevate them to the next module. The tests are given once every 4 weeks, so that a diligent student can complete all 4 modules in one semester. After a couple of years with this self-paced program, it is observed that for the first module 20% of the students do not complete it on time. The percentages for modules 2 through 4 are 22%, 25%, and 30%, respectively.

(a) Express the problem as a Markov chain.

(b) On the average, would a student starting with module 1 at the beginning of the current semester be able to take Cal II the next semester (Cal I is a prerequisite for Cal II)?

(c) Would a student who has completed only one module last semester be able to finish Cal I by the end of the current semester?

(d) Would you recommend that the use of the module idea be extended to other basic math classes? Explain.

9. At U of A, promotion from assistant to associate professor requires the equivalent of five years of seniority. Performance reviews are conducted once a year and the candidate is given either an average rating, a good rating, or an excellent rating. An average rating is the same as probation and the candidate gains no seniority toward promotion. A good rating is equivalent to gaining one year of seniority, and an excellent rating adds two years of seniority. Statistics show that in any year 10% of the candidates are rated average, 70% are rated good, and the rest are rated excellent.

(a) Express the problem as a Markov chain.

(b) Determine the average number of years until a new assistant professor is promoted.

***10.** (Pfifer and Carraway, 2000) A company targets its customers through direct mail advertising. During the first year, the probability that the customer will make a purchase is .5, which reduces to .4 in year 2, .3 in year 3, and .2 in year 4. If no purchases are made in four consecutive years, the customer is deleted from the mailing list. Making a purchase resets the count back to zero.

(a) Express the situation as a Markov chain.

(b) Determine the expected number of years a new customer will be on the mailing list.

(c) If a customer has not made a purchase in two years, determine the expected number of years on the mailing list.

11. An NC machine is designed to operate properly with power voltage setting between 108 and 112 volts. If the voltage falls outside this range, the machine will stop. The power regulator for the machine can detect variations in increments of one volt. Experience shows that change in voltage take place once every 15 minutes and that within the admissible range (118 to 112 volts), voltage can go up by one volt, stay the same, or go down by one volt, all with equal probabilities.

(a) Express the situation as a Markov chain.

(b) Determine the probability that the machine will stop because the voltage is low. High.

(c) What should be the ideal voltage setting that will render the longest working duration for the machine?

12. Consider Problem 4, Set 17.1a, dealing with patients suffering from kidney failure. Determine the following measures:

(a) The expected number of years a patient stays on dialysis.

(b) The longevity of a patient who starts on dialysis.

(c) The life expectancy of a patient who survives one year or longer after a transplant.

(d) The expected number of years before an at-least-one-year transplant survivor goes back to dialysis or dies.

(e) The quality of life for those who survive a year or more after a transplant (presumably, spending fewer years on dialysis signifies better quality of life).

REFERENCES

Derman, C., *Finite State Markovian Decision Processes,* Academic Press, New York, 1970.

Cyert, R., H. Davidson, and G. Thompson, "Estimation of the Allowance for Doubtful Accounts by Markov Chains," *Management Science,* Vol. 8, No. 4, pp. 287–303, 1963.

Pfifer, P., and R. Cassaway, "Modeling Customer Relations with Markov Chains," *Journal of Interactive Marketing,* Vol. 14, No. 2, pp. 43–55, 2000.

Grimmet, G., and D. Stirzaker, *Probability and Random Processes,* 2nd ed., Oxford University Press, Oxford, England, 1992.

Kallenberg, O., *Foundations of Modern Probability,* Springer-Verlag, New York, 1997.

Pliskin, J., and E. Tell, "Using Dialysis Need-Projection Model for Health Planning in Massachusetts," *Interfaces,* Vol. 11, No. 6, pp. 84–99, 1981.

Stewart, W., *Introduction to the Numerical Solution of Markov Chains,* Princeton University Press, Princeton, NJ, 1995.

CHAPTER 18

Classical Optimization Theory

Chapter Guide. Classical optimization theory uses differential calculus to determine points of maxima and minima (extrema) for unconstrained and constrained functions. The methods may not be suitable for efficient numerical computations, but the underlying theory provides the basis for most nonlinear programming algorithms (see Chapter 19). This chapter develops necessary and sufficient conditions for determining unconstrained extrema, the *Jacobian* and *Lagrangean* methods for problems with equality constraints, and the *Karush-Kuhn-Tucker* (KKT) conditions for problems with inequality constraints. The KKT conditions provide the most unifying theory for all nonlinear programming problems.

This chapter includes 10 solved examples, 1 Excel spreadsheet, and 23 end-of-chapter problems. The AMPL/Excel/Solver/TORA programs are in folder ch18Files.

18.1 UNCONSTRAINED PROBLEMS

An extreme point of a function $f(\mathbf{X})$ defines either a maximum or a minimum of the function. Mathematically, a point $\mathbf{X}_0 = (x_1^0, \ldots, x_j^0, \ldots, x_n^0)$ is a maximum if

$$f(\mathbf{X}_0 + \mathbf{h}) \leq f(\mathbf{X}_0)$$

for all $\mathbf{h} = (h_1, \ldots, h_j, \ldots, h_n)$ where $|h_j|$ is sufficiently small for all j. In other words, $\mathbf{X}_0$ is a maximum if the value of f at every point in the neighborhood of $\mathbf{X}_0$ does not exceed $f(\mathbf{X}_0)$. In a similar manner, $\mathbf{X}_0$ is a minimum if

$$f(\mathbf{X}_0 + \mathbf{h}) \geq f(\mathbf{X}_0)$$

Figure 18.1 illustrates the maxima and minima of a single-variable function $f(x)$ over the interval $[a,b]$. The points x_1, x_2, x_3, x_4, and x_6 are all extrema of $f(x)$, with x_1, x_3, and x_6 as maxima and x_2 and x_4 as minima. Because

$$f(x_6) = \max\{f(x_1), f(x_3), f(x_6)\}$$

$f(x_6)$ is a **global** or **absolute** maximum, and $f(x_1)$ and $f(x_3)$ are **local** or **relative** maxima. Similarly, $f(x_4)$ is a local minimum and $f(x_2)$ is a global minimum.

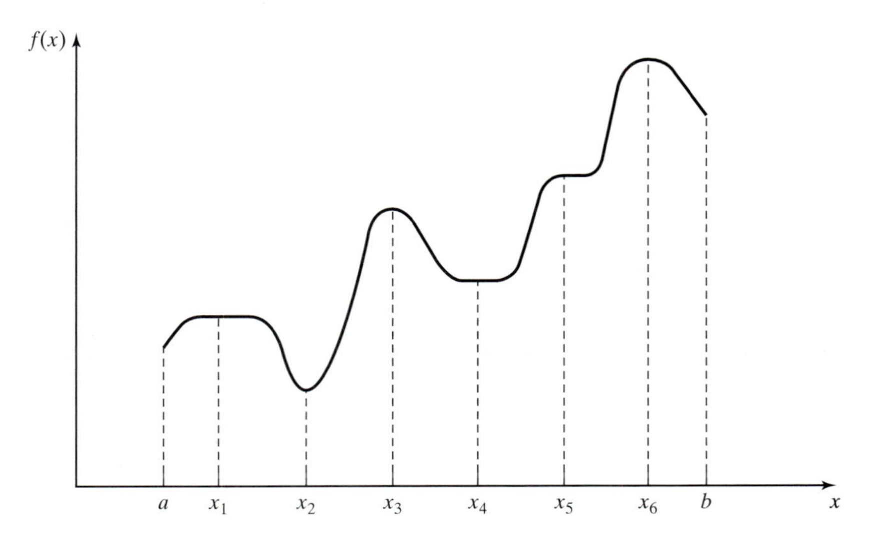

FIGURE 18.1

Examples of extreme points for a single-variable function

Although x_1 (in Figure 18.1) is a maximum point, it differs from remaining local maxima in that the value of f corresponding to at least one point in the neighborhood of x_1 equals $f(x_1)$. In this respect, x_1 is a **weak maximum**, whereas x_3 and x_6 are **strong maxima**. In general, for $\mathbf{h}$ as defined earlier, $\mathbf{X}_0$ is a weak maximum if $f(\mathbf{X}_0 + \mathbf{h}) \leq f(\mathbf{X}_0)$ and a strong maximum if $f(\mathbf{X}_0 + \mathbf{h}) < f(\mathbf{X}_0)$.

In Figure 18.1, the first derivative (slope) of f equals zero at all extrema. However, this property is also satisfied at **inflection** and **saddle** points, such as x_5. If a point with zero slope (gradient) is not an extremum (maximum or minimum), then it must be an inflection or a saddle point.

18.1.1 Necessary and Sufficient Conditions

This section develops the necessary and sufficient conditions for an n-variable function $f(\mathbf{X})$ to have extrema. It is assumed that the first and second partial derivatives of $f(\mathbf{X})$ are continuous for all $\mathbf{X}$.

Theorem 18.1-1. *A necessary condition for* $\mathbf{X}_0$ *to be an extreme point of* $f(\mathbf{X})$ *is that*

$$\nabla f(\mathbf{X}_0) = 0$$

Proof. By Taylor's theorem, for $0 < \theta < 1$,

$$f(\mathbf{X}_0 + \mathbf{h}) - f(\mathbf{X}_0) = \nabla f(\mathbf{X}_0)\mathbf{h} + \tfrac{1}{2}\mathbf{h}^T\mathbf{H}\mathbf{h}\big|_{\mathbf{X}_0+\theta\mathbf{h}}$$

For a sufficiently small $|h_j|$, the remainder term $\frac{1}{2}\mathbf{h}^T\mathbf{H}\mathbf{h}$ is of the order h_j^2 hence

$$f(\mathbf{X}_0 + \mathbf{h}) - f(\mathbf{X}_0) = \nabla f(\mathbf{X}_0)\mathbf{h} + 0(h_j^2) \approx \nabla f(\mathbf{X}_0)\mathbf{h}$$

We show by contradiction that $\nabla f(\mathbf{X}_0)$ must vanish at a minimum point $\mathbf{X}_0$. For suppose it does not, then for a specific j the following condition will hold.

$$\frac{\partial f(\mathbf{X}_0)}{\partial x_j} < 0 \text{ or } \frac{\partial f(\mathbf{X}_0)}{\partial x_j} > 0$$

By selecting h_j with appropriate sign, it is always possible to have

$$h_j \frac{\partial f(\mathbf{X}_0)}{\partial x_j} < 0$$

Setting all other h_j equal to zero, Taylor's expansion yields

$$f(\mathbf{X}_0 + \mathbf{h}) < f(\mathbf{X}_0)$$

This result contradicts the assumption that $\mathbf{X}_0$ is a minimum point. Thus, $\nabla f(\mathbf{X}_0)$ must equal zero. A similar proof can be established for the maximization case.

Because the necessary condition is also satisfied for inflection and saddle points, it is more appropriate to refer to the points obtained from the solution of $\nabla f(\mathbf{X}_0) = \mathbf{0}$ as **stationary** points. The next theorem establishes the sufficiency conditions for $\mathbf{X}_0$ to be an extreme point.

Theorem 18.1-2. *A sufficient condition for a stationary point $\mathbf{X}_0$ to be an extremum is that the Hessian matrix $\mathbf{H}$ evaluated at $\mathbf{X}_0$ satisfy the following conditions:*

(i) $\mathbf{H}$ *is positive definite if $\mathbf{X}_0$ is a minimum point.*
(ii) $\mathbf{H}$ *is negative definite if $\mathbf{X}_0$ is a maximum point.*

Proof. By Taylor's theorem, for $0 < \theta < 1$,

$$f(\mathbf{X}_0 + \mathbf{h}) - f(\mathbf{X}_0) = \nabla f(\mathbf{X}_0)\mathbf{h} + \tfrac{1}{2}\mathbf{h}^T\mathbf{H}\mathbf{h}|_{\mathbf{X}_0+\theta\mathbf{h}}$$

Given $\mathbf{X}_0$ is a stationary point, then $\nabla f(\mathbf{X}_0) = 0$ (Theorem 18.1-1). Thus,

$$f(\mathbf{X}_0 + \mathbf{h}) - f(\mathbf{X}_0) = \tfrac{1}{2}\mathbf{h}^T\mathbf{H}\mathbf{h}|_{\mathbf{X}_0+\theta\mathbf{h}}$$

If $\mathbf{X}_0$ is a minimum point, then

$$f(\mathbf{X}_0 + \mathbf{h}) > f(\mathbf{X}_0), \mathbf{h} \neq \mathbf{0}$$

Thus, for $\mathbf{X}_0$ to be a minimum point, it must be true that

$$\tfrac{1}{2}\mathbf{h}^T\mathbf{H}\mathbf{h}|_{\mathbf{X}_0+\theta\mathbf{h}} > 0$$

Given that the second partial derivative is continuous, the expression $\frac{1}{2}\mathbf{h}^T\mathbf{H}\mathbf{h}$ must have the same sign at both $\mathbf{X}_0$ and $\mathbf{X}_0 + \theta\mathbf{h}$. Because $\mathbf{h}^T\mathbf{H}\mathbf{h}|_{\mathbf{X}_0}$ defines a quadratic form (see Section D.3 on the CD), this expression (and hence $\mathbf{h}^T\mathbf{X}\mathbf{h}|_{\mathbf{X}_0+\theta\mathbf{h}}$) is positive if, and only if, $\mathbf{H}|_{\mathbf{X}_0}$ is positive-definite. This means that a sufficient condition for the stationary point $\mathbf{X}_0$ to be a minimum is that the Hessian matrix, $\mathbf{H}$, evaluated at the same point is positive-definite. A similar proof for the maximization case shows that the corresponding Hessian matrix must be negative-definite.

Example 18.1-1

Consider the function

$$f(x_1, x_2, x_3) = x_1 + 2x_3 + x_2x_3 - x_1^2 - x_2^2 - x_3^2$$

The necessary condition $\nabla f(\mathbf{X}_0) = 0$ gives

$$\frac{\partial f}{\partial x_1} = 1 - 2x_1 = 0$$

$$\frac{\partial f}{\partial x_2} = x_3 - 2x_2 = 0$$

$$\frac{\partial f}{\partial x_3} = 2 + x_2 - 2x_3 = 0$$

The solution of these simultaneous equations is

$$\mathbf{X}_0 = \left(\tfrac{1}{2}, \tfrac{2}{3}, \tfrac{4}{3}\right)$$

To determine the type of the stationary point, consider

$$\mathbf{H}|_{\mathbf{X}_0} = \begin{pmatrix} \frac{\partial^2 f}{\partial x_1^2} & \frac{\partial^2 f}{\partial x_1 \partial x_2} & \frac{\partial^2 f}{\partial x_1 \partial x_3} \\ \frac{\partial^2 f}{\partial x_2 \partial x_1} & \frac{\partial^2 f}{\partial x_2^2} & \frac{\partial^2 f}{\partial x_2 \partial x_3} \\ \frac{\partial^2 f}{\partial x_3 \partial x_1} & \frac{\partial^2 f}{\partial x_3 \partial x_2} & \frac{\partial^2 f}{\partial x_3^2} \end{pmatrix}_{\mathbf{X}_0} = \begin{pmatrix} -2 & 0 & 0 \\ 0 & -2 & 1 \\ 0 & 1 & -2 \end{pmatrix}$$

The principal minor determinants of $\mathbf{H}|_{\mathbf{X}_0}$ have the values -2, 4, and -6, respectively. Thus, as shown in Section D.3, $\mathbf{H}|_{\mathbf{X}_0}$ is negative-definite and $\mathbf{X}_0 = \left(\tfrac{1}{2}, \tfrac{2}{3}, \tfrac{4}{3}\right)$ represents a maximum point.

In general, if $\mathbf{H}|_{\mathbf{X}_0}$ is indefinite, $\mathbf{X}_0$ must be a saddle point. For nonconclusive cases, $\mathbf{X}_0$ may or may not be an extremum and the sufficiency condition becomes rather involved, because higher-order terms in Taylor's expansion must be considered.

The sufficiency condition established by Theorem 18.1-2 applies to single-variable functions as follows. Given that y_0 is a stationary point, then

(i) y_0 is a maximum if $f''(y_0) < 0$.
(ii) y_0 is a minimum if $f''(y_0) > 0$.

If $f''(y_0) = 0$, higher-order derivatives must be investigated as the following theorem requires.

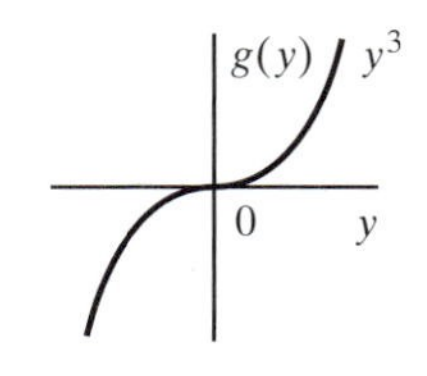

FIGURE 18.2

Extreme points of $f(y) = y^4$ and $g(y) = y^3$

Theorem 18.1-3. *Given y_0, a stationary point of $f(y)$, if the first $(n - 1)$ derivatives are zero and $f^{(n)}(y_0) \neq 0$, then*

(i) *If n is odd, y_0 is an inflection point.*
(ii) *If n is even then y_0 is a minimum if $f^{(n)}(y_0) > 0$ and a maximum if $f^{(n)}(y_0) < 0$.*

Example 18.1-2

Figure 18.2 graphs the following two functions:

$$f(y) = y^4$$
$$g(y) = y^3$$

For $f(y) = y^4, f'(y) = 4y^3 = 0$, which yields the stationary point $y_0 = 0$. Now

$$f'(0) = f''(0) = f^{(3)}(0) = 0, f^{(4)}(0) = 24 > 0$$

Hence, $y_0 = 0$ is a minimum point (see Figure 18.2).

For $g(y) = y^3, g'(y) = 3y^2 = 0$, which yields $y_0 = 0$ as a stationary point. Also

$$g'(0) = g''(0), g^{(3)}(0) = 6 \neq 0$$

Thus, $y_0 = 0$ is an inflection point.

PROBLEM SET 18.1A

1. *(a) $f(x) = x^3 + x$
   *(b) $f(x) = x^4 + x^2$
   (c) $f(x) = 4x^4 - x^2 + 5$
   (d) $f(x) = (3x - 2)^2(2x - 3)^2$
   *(e) $f(x) = 6x^5 - 4x^3 + 10$
2. Determine the extreme points of the following functions.
   (a) $f(\mathbf{X}) = x_1^3 + x_2^3 - 3x_1x_2$
   (b) $f(\mathbf{X}) = 2x_1^2 + x_2^2 + x_3^2 + 6(x_1 + x_2 + x_3) + 2x_1x_2x_3$

3. Verify that the function

$$f(x_1, x_2, x_3) = 2x_1x_2x_3 - 4x_1x_3 - 2x_2x_3 + x_1^2 + x_2^2 + x_3^2 - 2x_1 - 4x_2 + 4x_3$$

has the stationary points $(0, 3, 1)$, $(0, 1, -1)$, $(1, 2, 0)$, $(2, 1, 1)$, and $(2, 3, -1)$. Use the sufficiency condition to identify the extreme points.

***4.** Solve the following simultaneous equations by converting the system to a nonlinear objective function with no constraints.

$$x_2 - x_1^2 = 0$$
$$x_2 - x_1 = 2$$

[*Hint*: min $f^2(x_1, x_2)$ occurs at $f'(x_1, x_2) = 0$.]

5. Prove Theorem 18.1-3.

18.1.2 The Newton-Raphson Method

In general, the necessary condition equations, $\nabla f(\mathbf{X}) = \mathbf{0}$, may be difficult to solve numerically. The Newton-Raphson method is an iterative procedure for solving simultaneous nonlinear equations.

Consider the simultaneous equations

$$f_i(\mathbf{X}) = 0, \quad i = 1, 2, \ldots, m$$

Let $\mathbf{X}^k$ be a given point. Then by Taylor's expansion

$$f_i(\mathbf{X}) \approx f_i(\mathbf{X}_k) + \nabla f_i(\mathbf{X}_k)(\mathbf{X} - \mathbf{X}_k), i = 1, 2, \ldots, m$$

Thus, the original equations, $f_i(\mathbf{X}) = 0, \quad i = 1, 2, \ldots, m$, may be approximated as

$$f_i(\mathbf{X}_k) + \nabla f_i(\mathbf{X}_k)(\mathbf{X} - \mathbf{X}_k) = 0, i = 1, 2, \ldots, m$$

These equations may be written in matrix notation as

$$\mathbf{A}_k + \mathbf{B}_k(\mathbf{X} - \mathbf{X}_k) = \mathbf{0}$$

If $\mathbf{B}_k$ is nonsingular, then

$$\mathbf{X} = \mathbf{X}_k - \mathbf{B}_k^{-1}\mathbf{A}_k$$

The idea of the method is to start from an initial point $\mathbf{X}_0$ and then use the equation above to determine a new point. The process continues until two successive points, $\mathbf{X}_k$ and $\mathbf{X}_{k+1}$, are approximately equal.

A geometric interpretation of the method is illustrated by a single-variable function in Figure 18.3. The relationship between x_k and x_{k+1} for a single-variable function $f(x)$ reduces to

$$x_{k+1} = x_k - \frac{f(x_k)}{f'(x_k)}$$

or

$$f'(x_k) = \frac{f(x_k)}{x_k - x_{k+1}}$$

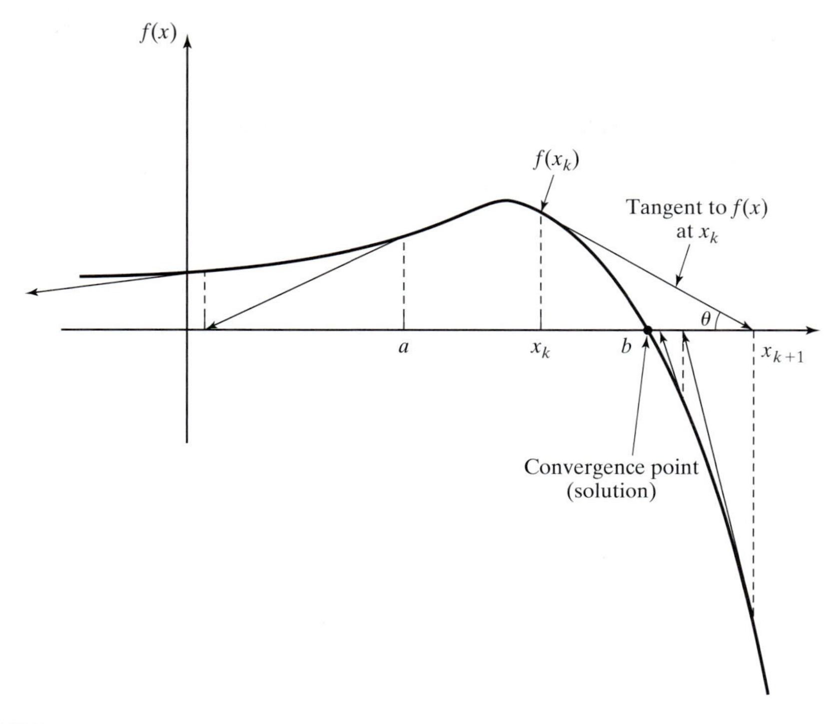

FIGURE 18.3
Illustration of the iterative process in the Newton-Raphson method

The figure shows that x_{k+1} is determined from the slope of $f(x)$ at x_k, where $\tan \theta = f'(x_k)$.

One difficulty with the method is that convergence is not always guaranteed unless the function f is well behaved. In Figure 18.3, if the initial point is a, the method will diverge. In general, trial and error is used to locate a "good" initial point.

Example 18.1-3

To demonstrate the use of the Newton-Raphson method, consider determining the stationary points of the function

$$g(x) = (3x - 2)^2(2x - 3)^2$$

To determine the stationary points, we need to solve

$$f(x) \equiv g'(x) = 72x^3 - 234x^2 + 241x - 78 = 0$$

Thus, for the Newton-Raphson method, we have

$$f'(x) = 216x^2 - 468x + 241$$

$$x_{k+1} = x_k - \frac{72x^3 - 234x^2 + 241x - 78}{216x^2 - 468x + 241}$$

Starting with $x_0 = 10$, the following table provides the successive iterations:

k	x_k	$\frac{f(x_k)}{f'(x_k)}$	x_{k+1}
0	10.000000	2.978923	7.032108
1	7.032108	1.976429	5.055679
2	5.055679	1.314367	3.741312
3	3.741312	0.871358	2.869995
4	2.869995	0.573547	2.296405
5	2.296405	0.371252	1.925154
6	1.925154	0.230702	1.694452
7	1.694452	0.128999	1.565453
8	1.565453	0.054156	1.511296
9	1.511296	.0108641	1.500432
10	1.500432	.00043131	1.500001

The method converges to $x = 1.5$. Actually, $f(x)$ has three stationary points at $x = \frac{2}{3}$, $x = \frac{13}{12}$, and $x = \frac{3}{2}$. The remaining two points can be found by selecting different values for initial x_0. In fact, $x_0 = .5$ and $x_0 = 1$ should yield the missing stationary points.

Excel Moment

Template excelNR.xls can be used to solve any single-variable equation. It requires entering $f(x)/f'(x)$ in cell C3. For Example 18.1-3, we enter

```
=(72*A3^3-234*A3^2+241*A3-78)/(216*A3^2-468*A3+241)
```

The variable x is replaced with A3. The template allows setting a tolerance limit Δ, which specifies the allowable difference between x_k and x_{k+1} that signals the termination of the iterations. You are encouraged to use different initial x_0 to get a feel of how the method works.

In general, the Newton-Raphson method requires making several attempts before "all" the solutions can be found. In Example 18.1-3, we know beforehand that the equation has three roots. This will not be the case with complex or multi-variable functions, however.

PROBLEM SET 18.1B

1. Use NewtonRaphson.xls to solve Problem 1(c), Set 18.1a.
2. Solve Problem 2(b), Set 18.1a by the Newton-Raphson method.

18.2 CONSTRAINED PROBLEMS

This section deals with the optimization of constrained continuous functions. Section 18.2.1 introduces the case of equality constraints and Section 18.2.2 deals with inequality constraints. The presentation in Section 18.2.1 is covered for the most part in Beightler and Associates (1979, pp. 45–55).

18.2.1 Equality Constraints

This section presents two methods: the **Jacobian** and the **Lagrangean**. The **Lagrangean** method can be developed logically from the Jacobian. This relationship provides an interesting economic interpretation of the Lagrangean method.

Constrained Derivatives (Jacobian) Method. Consider the problem

$$\text{Minimize } z = f(\mathbf{X})$$

subject to

$$\mathbf{g}(\mathbf{X}) = \mathbf{0}$$

where

$$\mathbf{X} = (x_1, x_2, \ldots, x_n)$$
$$\mathbf{g} = (g_1, g_2, \ldots, g_m)^T$$

The functions $f(\mathbf{X})$ and $g(\mathbf{X}), i = 1, 2, \ldots, m$, are twice continuously differentiable.

The idea of using constrained derivatives is to develop a closed-form expression for the first partial derivatives of $f(\mathbf{X})$ at all points that satisfy the constraints $\mathbf{g}(\mathbf{X}) = \mathbf{0}$. The corresponding stationary points are identified as the points at which these partial derivatives vanish. The sufficiency conditions introduced in Section 18.1 can then be used to check the identity of stationary points.

To clarify the proposed concept, consider $f(x_1, x_2)$ illustrated in Figure 18.4. This function is to be minimized subject to the constraint

$$g_1(x_1, x_2) = x_2 - b = 0$$

where b is a constant. From Figure 18.4, the curve designated by the three points A, B, and C represents the values of $f(x_1, x_2)$ for which the given constraint is always satisfied. The constrained derivatives method defines the gradient of $f(x_1, x_2)$ at any point on the curve ABC. Point B at which the constrained derivative vanishes is a stationary point for the constrained problem.

The method is now developed mathematically. By Taylor's theorem, for $\mathbf{X} + \Delta\mathbf{X}$ in the feasible neighborhood of $\mathbf{X}$, we have

$$f(\mathbf{X} + \Delta\mathbf{X}) - f(\mathbf{X}) = \nabla f(\mathbf{X})\,\Delta\mathbf{X} + O(\Delta x_j^2)$$

and

$$\mathbf{g}(\mathbf{X} + \Delta\mathbf{X}) - \mathbf{g}(\mathbf{X}) = \nabla \mathbf{g}(\mathbf{X})\,\Delta\mathbf{X} + O(\Delta x_j^2)$$

As $\Delta x_j \rightarrow 0$, the equations reduce to

$$\partial f(\mathbf{X}) = \nabla f(\mathbf{X})\,\partial\mathbf{X}$$

and

$$\partial \mathbf{g}(\mathbf{X}) = \nabla \mathbf{g}(\mathbf{X})\,\partial\mathbf{X}$$

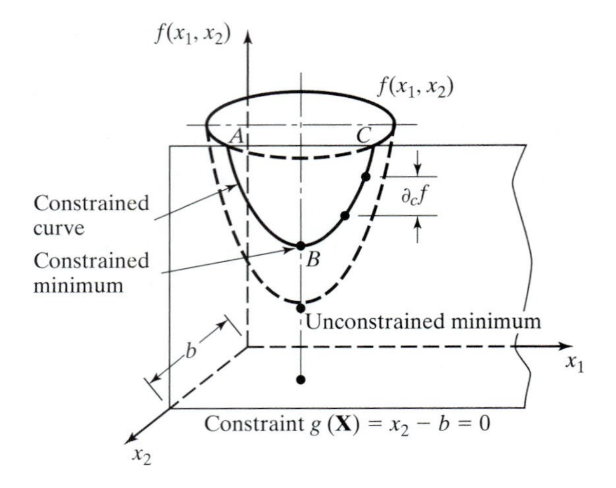

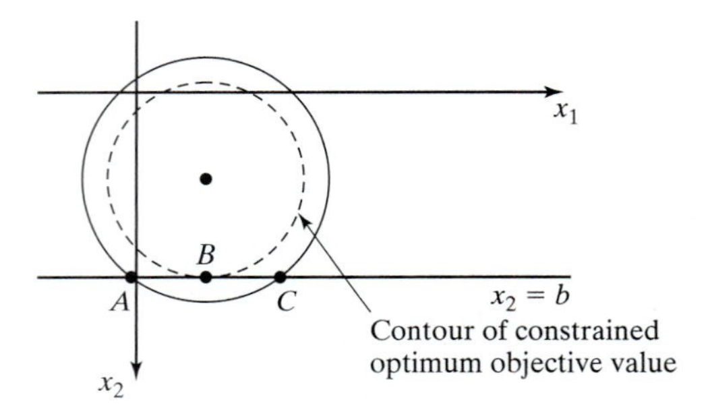

FIGURE 18.4
Demonstration of the idea of the Jacobian method

For feasibility, we must have $\mathbf{g}(\mathbf{X}) = \mathbf{0}$, $\partial\mathbf{g}(\mathbf{X}) = \mathbf{0}$, and it follows that

$$\partial f(\mathbf{X}) - \nabla f(\mathbf{X})\,\partial\mathbf{X} = 0$$
$$\nabla\mathbf{g}(\mathbf{X})\,\partial\mathbf{X} = \mathbf{0}$$

This gives $(m + 1)$ equations in $(n + 1)$ unknowns, $\partial f(\mathbf{X})$ and $\partial\mathbf{X}$. Note that $\partial f(\mathbf{X})$ is a dependent variable, and hence is determined as soon as $\partial\mathbf{X}$ is known. This means that, in effect, we have m equations in n unknowns.

If $m > n$, at least $(m - n)$ equations are redundant. Eliminating redundancy, the system reduces to $m \leq n$. If $m = n$, the solution is $\partial\mathbf{X} = \mathbf{0}$, and $\mathbf{X}$ has no feasible neighborhood, which means that the solution space consists of one point only. The remaining case, where $m < n$, requires further elaboration.

Define

$$\mathbf{X} = (\mathbf{Y}, \mathbf{Z})$$

such that

$$\mathbf{Y} = (y_1, y_2, \ldots, y_m), \mathbf{Z} = (z_1, z_2, \ldots, z_{n-m})$$

The vectors $\mathbf{Y}$ and $\mathbf{Z}$ are called the *dependent* and *independent* variables, respectively. Rewriting the gradient vectors of f and g in terms of $\mathbf{Y}$ and $\mathbf{Z}$, we get

$$\nabla f(\mathbf{Y}, \mathbf{Z}) = (\nabla_{\mathbf{Y}} f, \nabla_{\mathbf{Z}} f)$$

$$\nabla g(\mathbf{Y}, \mathbf{Z}) = (\nabla_{\mathbf{Y}} \mathbf{g}, \nabla_{\mathbf{Z}} \mathbf{g})$$

Define

$$\mathbf{J} = \nabla_{\mathbf{Y}} \mathbf{g} = \begin{pmatrix} \nabla_{\mathbf{Y}} g_1 \\ \vdots \\ \nabla_{\mathbf{Y}} g_m \end{pmatrix}$$

$$\mathbf{C} = \nabla_{\mathbf{Z}} \mathbf{g} = \begin{pmatrix} \nabla_{\mathbf{Z}} g_1 \\ \vdots \\ \nabla_{\mathbf{Z}} g_m \end{pmatrix}$$

$\mathbf{J}_{m \times m}$ is called the **Jacobian matrix** and $\mathbf{C}_{m \times n-m}$ the **control matrix**. The Jacobian $\mathbf{J}$ is assumed nonsingular. This is always possible because the given m equations are independent by definition. The components of the vector $\mathbf{Y}$ must thus be selected from among those of $\mathbf{X}$ such that $\mathbf{J}$ is nonsingular.

The original set of equations in $\partial f(\mathbf{X})$ and $\partial \mathbf{X}$ may be written as

$$\partial f(\mathbf{Y}, \mathbf{Z}) = \nabla_{\mathbf{Y}} f \partial \mathbf{Y} + \nabla_{\mathbf{Z}} f \partial \mathbf{Z}$$

and

$$\mathbf{J} \partial \mathbf{Y} = -\mathbf{C} \partial \mathbf{Z}$$

Because $\mathbf{J}$ is nonsingular, its inverse $\mathbf{J}^{-1}$ exists. Hence,

$$\partial \mathbf{Y} = -\mathbf{J}^{-1} \mathbf{C} \partial \mathbf{Z}$$

Substituting for $\partial \mathbf{Y}$ in the equation for $\partial f(\mathbf{X})$ gives ∂f as a function of $\partial \mathbf{Z}$—that is,

$$\partial f(\mathbf{Y}, \mathbf{Z}) = (\nabla_{\mathbf{Z}} f - \nabla_{\mathbf{Y}} f \mathbf{J}^{-1} \mathbf{C}) \partial \mathbf{Z}$$

From this equation, the constrained derivative with respect to the independent vector $\mathbf{Z}$ is given by

$$\nabla_c f = \frac{\partial_c f(\mathbf{Y}, \mathbf{Z})}{\partial_c \mathbf{Z}} = \nabla_z f - \nabla_{\mathbf{Y}} f \mathbf{J}^{-1} \mathbf{C}$$

where $\nabla_c f$ is the **constrained gradient** vector of f with respect to $\mathbf{Z}$. Thus, $\nabla_c f(\mathbf{Y}, \mathbf{Z})$ must be null at the stationary points.

The sufficiency conditions are similar to those developed in Section 18.1. The Hessian matrix will correspond to the independent vector $\mathbf{Z}$, and the elements of the Hessian matrix must be the *constrained* second derivatives. To show how this is obtained, let

$$\nabla_c f = \nabla_z f - \mathbf{WC}$$

It thus follows that the ith row of the (constrained) Hessian matrix is $\frac{\partial \nabla_c f}{\partial z_i}$. Notice that **W** is a function of **Y** and **Y** is a function of **Z**. Thus, the partial derivative of $\nabla_c f$ with respect to z_i is based on the following chain rule:

$$\frac{\partial w_j}{\partial z_i} = \frac{\partial w_j}{\partial y_j}\frac{\partial y_j}{\partial z_i}$$

Example 18.2-1

Consider the following problem:

$$f(\mathbf{X}) = x_1^2 + 3x_2^2 + 5x_1x_3^2$$

$$g_1(\mathbf{X}) = x_1x_3 + 2x_2 + x_2^2 - 11 = 0$$

$$g_2(\mathbf{X}) = x_1^2 + 2x_1x_2 + x_3^2 - 14 = 0$$

Given the feasible point $\mathbf{X}^0 = (1, 2, 3)$, we wish to study the variation in $f(= \partial_c f)$ in the feasible neighborhood of $\mathbf{X}^0$.

Let

$$\mathbf{Y} = (x_1, x_3) \quad \text{and} \quad \mathbf{Z} = x_2$$

Thus,

$$\nabla_{\mathbf{Y}} f = \left(\frac{\partial f}{\partial x_1}, \frac{\partial f}{\partial x_3}\right) = \left(2x_1 + 5x_3^2, 10x_1x_3\right)$$

$$\nabla_{\mathbf{Z}} f = \frac{\partial f}{\partial x_2} = 6x_2$$

$$\mathbf{J} = \begin{pmatrix} \frac{\partial g_1}{\partial x_1} & \frac{\partial g_1}{\partial x_3} \\ \frac{\partial g_2}{\partial x_1} & \frac{\partial g_2}{\partial x_3} \end{pmatrix} = \begin{pmatrix} x_3 & x_1 \\ 2x_1 + 2x_2 & 2x_3 \end{pmatrix}$$

$$\mathbf{C} = \begin{pmatrix} \frac{\partial g_1}{\partial x_2} \\ \frac{\partial g_2}{\partial x_2} \end{pmatrix} = \begin{pmatrix} 2x_2 + 2 \\ 2x_1 \end{pmatrix}$$

Suppose that we need to estimate $\partial_c f$ in the feasible neighborhood of the feasible point $\mathbf{X}_0 = (1, 2, 3)$ given a small change $\partial x_2 = .01$ in the independent variable x_2. We have

$$\mathbf{J}^{-1}\mathbf{C} = \begin{pmatrix} 3 & 1 \\ 6 & 6 \end{pmatrix}^{-1}\begin{pmatrix} 6 \\ 2 \end{pmatrix} = \begin{pmatrix} \frac{6}{12} & -\frac{1}{12} \\ -\frac{6}{12} & \frac{3}{12} \end{pmatrix}\begin{pmatrix} 6 \\ 2 \end{pmatrix} \approx \begin{pmatrix} 2.83 \\ -2.50 \end{pmatrix}$$

Hence, the incremental value of constrained f is given as

$$\partial_c f = (\nabla_{\mathbf{Z}} f - \nabla_{\mathbf{Y}} f \mathbf{J}^{-1}\mathbf{C})\partial \mathbf{Z} = \left(6(2) - (47, 30)\begin{pmatrix} 2.83 \\ -2.50 \end{pmatrix}\right)\partial x_2 = -46.01 \partial x_2$$

By specifying the value of ∂x_2 for the *independent* variable x_2, feasible values of ∂x_1 and ∂x_2 are determined for the dependent variables x_1 and x_3 using the formula

$$\partial \mathbf{Y} = -\mathbf{J}^{-1}\mathbf{C}\partial \mathbf{Z}$$

Thus, for $\partial x_2 = .01$,

$$\begin{pmatrix} \partial x_1 \\ \partial x_3 \end{pmatrix} = -\mathbf{J}^{-1}\mathbf{C}\partial x_2 = \begin{pmatrix} -.0283 \\ .0250 \end{pmatrix}$$

We now compare the value of $\partial_c f$ as computed above with the difference $f(\mathbf{X}_0 + \partial \mathbf{X}) - f(\mathbf{X}_0)$, given $\partial x_2 = .01$.

$$\mathbf{X}_0 + \partial \mathbf{X} = (1 - .0283, 2 + .01, 3 + .025) = (.9717, 2.01, 3.025)$$

This yields

$$f(\mathbf{X}_0) = 58, f(\mathbf{X}_0 + \partial \mathbf{X}) = 57.523$$

or

$$f(\mathbf{X}_0 + \partial \mathbf{X}) - f(\mathbf{X}_0) = -.477$$

The amount $-.477$ compares favorably with $\partial_c f = -46.01 \partial x_2 = -.4601$. The difference between the two values is the result of the linear approximation in computing $\partial_c f$ at $\mathbf{X}_0$.

PROBLEM SET 18.2A

1. Consider Example 18.2-1.

(a) Compute $\partial_c f$ by the two methods presented in the example, using $\partial x_2 = .001$ instead of $\partial x_2 = .01$. Does the effect of linear approximation become more negligible with the decrease in the value of ∂x_2?

***(b)** Specify a relationship among the elements of $\partial \mathbf{X} = (\partial x_1, \partial x_2, \partial x_3)$ at the feasible point $\mathbf{X}_0 = (1, 2, 3,)$ that will keep the point $\mathbf{X}_0 + \partial \mathbf{X}$ feasible.

(c) If $\mathbf{Y} = (x_2, x_3)$ and $\mathbf{Z} = x_1$, what is the value of ∂x_1 that will produce the same value of $\partial_c f$ given in the example?

Example 18.2-2

This example illustrates the use of constrained derivatives. Consider the problem

$$\text{Minimize } f(\mathbf{X}) = x_1^2 + x_2^2 + x_3^2$$

subject to

$$g_1(\mathbf{X}) = x_1 + x_2 + 3x_3 - 2 = 0$$

$$g_2(\mathbf{X}) = 5x_1 + 2x_2 + x_3 - 5 = 0$$

We determine the constrained extreme points as follows. Let

$$\mathbf{Y} = (x_1, x_2) \text{ and } \mathbf{Z} = x_3$$

Thus,

$$\nabla_{\mathbf{Y}} f = \left(\frac{\partial f}{\partial x_1}, \frac{\partial f}{\partial x_2}\right) = (2x_1, 2x_2), \nabla_{\mathbf{Z}} f = \frac{\partial f}{\partial x_3} = 2x_3$$

$$\mathbf{J} = \begin{pmatrix} 1 & 1 \\ 5 & 2 \end{pmatrix}, \mathbf{J}^{-1} = \begin{pmatrix} -\frac{2}{3} & \frac{1}{3} \\ \frac{5}{3} & -\frac{1}{3} \end{pmatrix}, \mathbf{C} = \begin{pmatrix} 3 \\ 1 \end{pmatrix}$$

Hence,

$$\begin{aligned} \nabla_c f = \frac{\partial_c f}{\partial_c x_3} &= 2x_3 - (2x_1, 2x_2)\begin{pmatrix} -\frac{2}{3} & \frac{1}{3} \\ \frac{5}{3} & -\frac{1}{3} \end{pmatrix}\begin{pmatrix} 3 \\ 1 \end{pmatrix} \\ &= \tfrac{10}{3}x_1 - \tfrac{28}{3}x_2 + 2x_3 \end{aligned}$$

The equations for determining the stationary points are thus given as

$$\begin{aligned} \nabla_c f &= 0 \\ g_1(\mathbf{X}) &= 0 \\ g_2(\mathbf{X}) &= 0 \end{aligned}$$

or

$$\begin{pmatrix} 10 & -28 & 6 \\ 1 & 1 & 3 \\ 5 & 2 & 1 \end{pmatrix}\begin{pmatrix} x_1 \\ x_2 \\ x_3 \end{pmatrix} = \begin{pmatrix} 0 \\ 2 \\ 5 \end{pmatrix}$$

The solution is

$$\mathbf{X}_0 \approx (.81, .35, .28)$$

The identity of this stationary point is checked using the sufficiency condition. Given that x_3 is the independent variable, it follows from $\nabla_c f$ that

$$\frac{\partial_c^2 f}{\partial_c x_3^2} = \tfrac{10}{3}\left(\frac{dx_1}{dx_3}\right) - \tfrac{28}{3}\left(\frac{dx_2}{dx_3}\right) + 2 = \left(\tfrac{10}{3}, -\tfrac{28}{3}\right)\begin{pmatrix} \frac{dx_1}{dx_3} \\ \frac{dx_2}{dx_3} \end{pmatrix} + 2$$

From the Jacobian method,

$$\begin{pmatrix} \frac{dx_1}{dx_3} \\ \frac{dx_2}{dx_3} \end{pmatrix} = -\mathbf{J}^{-1}\mathbf{C} = \begin{pmatrix} \frac{5}{3} \\ -\frac{14}{3} \end{pmatrix}$$

Substitution gives $\frac{\partial_c^2 f}{\partial_c x_3^2} = \frac{460}{9} > 0$. Hence, $\mathbf{X}_0$ is the minimum point.

Sensitivity Analysis in the Jacobian Method. The Jacobian method can be used to study the effect of small changes in the right-hand side of the constraints on the optimal value of f. Specifically, what is the effect of changing $g_i(\mathbf{X}) = 0$ to $g_i(\mathbf{X}) = \partial g_i$ on the optimal value of f? This type of investigation is called **sensitivity analysis** and is similar to that carried out in linear programming (see Chapters 3 and 4). However, sensitivity analysis in nonlinear programming is valid only in the small neighborhood of the extreme point. The development will be helpful in studying the Lagrangean method.

We have shown previously that

$$\partial f(\mathbf{Y}, \mathbf{Z}) = \nabla_{\mathbf{Y}} f \partial \mathbf{Y} + \nabla_{\mathbf{Z}} f \partial \mathbf{Z}$$

$$\partial \mathbf{g} = \mathbf{J} \partial \mathbf{Y} + \mathbf{C} \partial \mathbf{Z}$$

Given $\partial \mathbf{g} \neq 0$, then

$$\partial \mathbf{Y} = \mathbf{J}^{-1} \partial \mathbf{g} - \mathbf{J}^{-1} \mathbf{C} \partial \mathbf{Z}$$

Substituting in the equation for $\partial f(\mathbf{Y}, \mathbf{Z})$ gives

$$\partial f(\mathbf{Y}, \mathbf{Z}) = \nabla_{\mathbf{Y}} f \mathbf{J}^{-1} \partial \mathbf{g} + \nabla_c f \partial \mathbf{Z}$$

where

$$\nabla_c f = \nabla_{\mathbf{Z}} f - \nabla_{\mathbf{Y}} f \mathbf{J}^{-1} \mathbf{C}$$

as defined previously. The expression for $\partial f(\mathbf{Y}, \mathbf{Z})$ can be used to study variation in f in the feasible neighborhood of a feasible point $\mathbf{X}_0$ resulting from small changes $\partial \mathbf{g}$ and $\partial \mathbf{Z}$.

At the extreme (indeed, any stationary) point $\mathbf{X}_0 = (\mathbf{Y}_0, \mathbf{Z}_0)$ the constrained gradient $\nabla_c f$ must vanish. Thus

$$\partial f(\mathbf{Y}_0, \mathbf{Z}_0) = \nabla_{\mathbf{Y}_0} f \mathbf{J}^{-1} \partial \mathbf{g}(\mathbf{Y}_0, \mathbf{Z}_0)$$

or

$$\frac{\partial f}{\partial \mathbf{g}} = \nabla_{\mathbf{Y}_0} f \mathbf{J}^{-1}$$

The effect of the small change $\partial \mathbf{g}$ on the *optimum* value of f can be studied by evaluating the rate of change of f with respect to $\mathbf{g}$. These rates are usually referred to as **sensitivity coefficients**.

Example 18.2-3

Consider the same problem of Example 18.2-2. The optimum point is given by $\mathbf{X}_0 = (x_{01}, x_{02}, x_{03}) = (.81, .35, .28)$. Given $\mathbf{Y}_0 = (x_{01}, x_{02})$, then

$$\nabla_{\mathbf{Y}_0} f = \left(\frac{\partial f}{\partial x_1}, \frac{\partial f}{\partial x_2}\right) = (2x_{01}, 2x_{02}) = (1.62, .70)$$

Consequently,

$$\left(\frac{\partial f}{\partial g_1}, \frac{\partial f}{\partial g_2}\right) = \nabla_{\mathbf{Y}_0} f \mathbf{J}^{-1} = (1.62, .7)\begin{pmatrix} -\frac{2}{3} & \frac{1}{3} \\ \frac{5}{3} & -\frac{1}{3} \end{pmatrix} = (.0876, .3067)$$

This means that for $\partial g_1 = 1$, f will increase *approximately* by .0867. Similarly, for $\partial g_2 = 1$, f will increase *approximately* by .3067.

Application of the Jacobian Method to an LP Problem. Consider the linear program

$$\text{Maximize } z = 2x_1 + 3x_2$$

subject to

$$\begin{aligned} x_1 + x_2 + x_3 \quad &= 5 \\ x_1 - x_2 \quad + x_4 &= 3 \\ x_1, x_2 x_3, x_4 &\geq 0 \end{aligned}$$

To account for the nonnegativity constraints $x_j \geq 0$, substitute $x_j = w_j^2$. With this substitution, the nonnegativity conditions become implicit and the original problem becomes

$$\text{Maximize } z = 2w_1^2 + 3w_2^2$$

subject to

$$\begin{aligned} w_1^2 + w_2^2 + w_3^2 &= 5 \\ w_1^2 - w_2^2 + w_4^2 &= 3 \end{aligned}$$

To apply the Jacobian method, let

$$\mathbf{Y} = (w_1, w_2) \quad \text{and} \quad \mathbf{Z} = (w_3, w_4)$$

(In the terminology of linear programming, $\mathbf{Y}$ and $\mathbf{Z}$ correspond to the basic and nonbasic variables, respectively.) Thus

$$\mathbf{J} = \begin{pmatrix} 2w_1 & 2w_2 \\ 2w_1 & -2w_2 \end{pmatrix}, \mathbf{J}^{-1} = \begin{pmatrix} \frac{1}{4w_1} & \frac{1}{4w_1} \\ \frac{1}{4w_2} & \frac{-1}{4w_2} \end{pmatrix}, w_1 \text{ and } w_2 \neq 0$$

$$\mathbf{C} = \begin{pmatrix} 2w_3 & 0 \\ 0 & 2w_4 \end{pmatrix}, \nabla_{\mathbf{Y}} f = (4w_1, 6w_2), \nabla_{\mathbf{Z}} f = (0, 0)$$

so that

$$\nabla_c f = (0,0) - (4w_1, 6w_2) \begin{pmatrix} \frac{1}{4w_1} & \frac{1}{4w_1} \\ \frac{1}{4w_2} & \frac{-1}{4w_2} \end{pmatrix} \begin{pmatrix} 2w_3 & 0 \\ 0 & 2w_4 \end{pmatrix} = (-5w_3, w_4)$$

The solution of the equations comprised of $\nabla_c f = \mathbf{0}$ and the constraints of the problem yield the stationary point $(w_1 = 2, w_2 = 1, w_3 = 0, w_4 = 0)$. The Hessian is given by

$$\mathbf{H}_c = \begin{pmatrix} \frac{\partial^2_c f}{\partial_c w_3^2} & \frac{\partial^2_c f}{\partial_c w_3 \partial_c w_4} \\ \frac{\partial^3_c f}{\partial_c w_3 \partial_c w_4} & \frac{\partial^2_c f}{\partial_c w_4^2} \end{pmatrix} = \begin{pmatrix} -5 & 0 \\ 0 & 1 \end{pmatrix}$$

Because $\mathbf{H}_c$ is indefinite, the stationary point does not yield a maximum.

The reason the solution above does not yield the optimum solution is that the specific choices of **Y** and **Z** are not optimum. In fact, to find the optimum, we need to keep on altering our choices of **Y** and **Z** until the sufficiency condition is satisfied. This will be equivalent to locating the optimum extreme point of the linear programming solution space. For example, consider $\mathbf{Y} = (w_2, w_4)$ and $\mathbf{Z} = (w_1, w_3)$. The corresponding constrained gradient vector becomes

$$\nabla_c f = (4w_1, 0) - (6w_2, 0)\begin{pmatrix} \frac{1}{2w_2} & 0 \\ \frac{1}{2w_4} & \frac{1}{2w_4} \end{pmatrix}\begin{pmatrix} 2w_1 & 2w_3 \\ 2w_1 & 0 \end{pmatrix} = (-2w_1, 6w_3)$$

The corresponding stationary point is given by $w_1 = 0, w_2 = \sqrt{5}, w_3 = 0, w_4 = \sqrt{8}$. Because

$$\mathbf{H}_c = \begin{pmatrix} -2 & 0 \\ 0 & -6 \end{pmatrix}$$

is negative-definite, the solution is a maximum point.

The result is verified graphically in Figure 18.5. The first solution $(x_1 = 4, x_2 = 1)$ is not optimal, and the second $(x_1 = 0, x_2 = 5)$ is. You can verify that the remaining two (feasible) extreme points of the solution space are not optimal. In fact, the extreme point $(x_1 = 0, x_2 = 0)$ can be shown by the sufficiency condition to yield a minimum point.

The sensitivity coefficients $\nabla_{\mathbf{Y}_0} f\mathbf{J}^{-1}$ when applied to linear programming yield the dual values. To illustrate this point for the given numerical example, let u_1 and u_2 be the corresponding dual variables. At the optimum point $\left(w_1 = 0, w_2 = \sqrt{5}, w_3 = 0, w_4 = \sqrt{8}\right)$, these dual variables are given by

$$(u_1, u_2) = \nabla_{\mathbf{Y}_0}\mathbf{J}^{-1} = (6w_2, 0)\begin{pmatrix} \frac{1}{2w_2} & 0 \\ \frac{1}{2w_4} & \frac{1}{2w_4} \end{pmatrix} = (3, 0)$$

The corresponding dual objective value is $5u_1 + 3u_2 = 15$, which equals the optimal primal objective value. The given solution also satisfies the dual constraints and hence is optimal and feasible. This shows that the sensitivity coefficients are the same as the dual variables. In fact, both have the same interpretation.

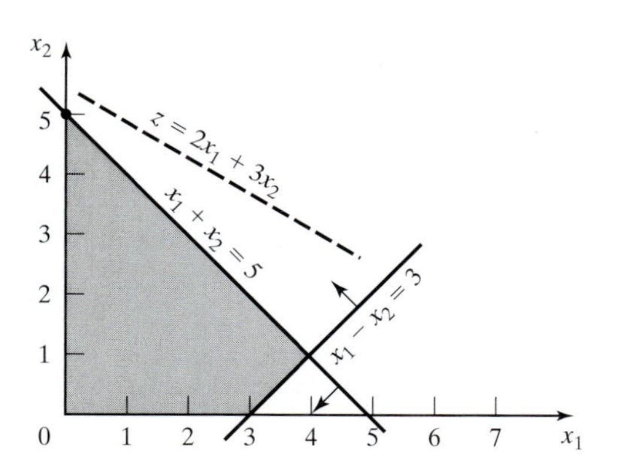

FIGURE 18.5

Extreme points of the solution space of the linear program

We can draw some general conclusions from the application of the Jacobian method to the linear programming problem. From the numerical example, the necessary conditions require the independent variables to equal zero. Also, the sufficiency conditions indicate that the Hessian is a diagonal matrix. Thus, all its diagonal elements must be positive for a minimum and negative for a maximum. The observations demonstrate that the necessary condition is equivalent to specifying that only basic (feasible) solutions are needed to locate the optimum solution. In this case the independent variables are equivalent to the nonbasic variables in the linear programming problem. Also, the sufficiency condition demonstrates the strong relationship between the diagonal elements of the Hessian matrix and the optimality indicator $z_j - c_j$ (see Section 7.2) in the simplex method.[1]

PROBLEM SET 18.2B

1. Suppose that Example 18.2-2 is solved in the following manner. First, use the constraints to express x_1 and x_2 in terms of x_3; then use the resulting equations to express the objective function in terms of x_3 only. By taking the derivative of the new objective function with respect to x_3, we can determine the points of maxima and minima.

(a) Would the derivative of the new objective function (expressed in terms of x_3) be different from that obtained by the Jacobian method?

(b) How does the suggested procedure differ from the Jacobian method?

2. Apply the Jacobian method to Example 18.2-1 by selecting $\mathbf{Y} = (x_2, x_3)$ and $\mathbf{Z} = (x_1)$.

***3.** Solve by the Jacobian method:

$$\text{Minimize } f(\mathbf{X}) = \sum_{i=1}^{n} x_i^2$$

subject to

$$\prod_{i=1}^{n} x_i = C$$

where C is a positive constant. Suppose that the right-hand side of the constraint is changed to $C + \delta$, where δ is a small positive quantity. Find the corresponding change in the optimal value of f.

4. Solve by the Jacobian method:

$$\text{Minimize } f(\mathbf{X}) = 5x_1^2 + x_2^2 + 2x_1x_2$$

subject to

$$g(\mathbf{X}) = x_1x_2 - 10 = 0$$

(a) Find the change in the optimal value of $f(\mathbf{X})$ if the constraint is replaced by $x_1x_2 - 9.99 = 0$.

(b) Find the change in value of $f(\mathbf{X})$ if the neighborhood of the feasible point (2, 5) given that $x_1x_2 = 9.99$ and $\partial x_1 = .01$.

---

[1]For a formal proof of the validity of these results for the general linear programming problem, see H. Taha and G. Curry, "Classical Derivation of the Necessary and Sufficient Conditions for Optimal Linear Programs," *Operations Research*, Vol. 19, pp.1045–1049, 1971. The paper shows that the key ideas of the simplex method can be derived by the Jacobian method.

5. Consider the problem:

$$\text{Maximize } f(\mathbf{X}) = x_1^2 + 2x_2^2 + 10x_3^2 + 5x_1x_2$$

subject to

$$g_1(\mathbf{X}) = x_1 + x_2^2 + 3x_2x_3 - 5 = 0$$

$$g_2(\mathbf{X}) = x_1^2 + 5x_1x_2 + x_3^2 - 7 = 0$$

Apply the Jacobian method to find $\partial f(\mathbf{X})$ in the neighborhood of the feasible point $(1, 1, 1)$. Assume that this neighborhood is specified by $\partial g_1 = -.01$, $\partial g_2 = .02$, and $\partial x_1 = .01$

6. Consider the problem

$$\text{Minimize } f(\mathbf{X}) = x_1^2 + x_2^2 + x_3^2 + x_4^2$$

subject to

$$g_1(\mathbf{X}) = x_1 + 2x_2 + 3x_3 + 5x_4 - 10 = 0$$

$$g_2(\mathbf{X}) = x_1 + 2x_2 + 5x_3 + 6x_4 - 15 = 0$$

(a) Show that by selecting x_3 and x_4 as independent variables, the Jacobian method fails to provide a solution and state the reason.

***(b)** Solve the problem using x_1 and x_3 as independent variables and apply the sufficiency condition to determine the type of the resulting stationary point.

(c) Determine the sensitivity coefficients given the solution in (b).

7. Consider the linear programming problem.

$$\text{Maximize } f(\mathbf{X}) = \sum_{j=1}^{n} c_j x_j$$

subject to

$$g_i(\mathbf{X}) = \sum_{j=1}^{n} a_{ij}x_j - b_i = 0 \quad i = 1, 2, \ldots, m$$

$$x_j \geq 0, \quad j = 1, 2, \ldots, n$$

Neglecting the nonnegativity constraint, show that the constrained derivatives $\nabla_c f(\mathbf{X})$ for this problem yield the same expression for $\{z_j - c_j\}$ defined by the optimality condition of the linear programming problem (Section 7.2)—that is,

$$\{z_j - c_j\} = \{\mathbf{C}_B\mathbf{B}^{-1}\mathbf{P}_j - c_j\}, \text{ for all } j$$

Can the constrained-derivative method be applied directly to the linear programming problem? Why or why not?

Lagrangean Method. In the Jacobian method, let the vector $\boldsymbol{\lambda}$ represent the sensitivity coefficients—that is

$$\boldsymbol{\lambda} = \nabla \mathbf{Y}_0 \mathbf{J}^{-1} = \frac{\partial f}{\partial \mathbf{g}}$$

Thus,

$$\partial f - \boldsymbol{\lambda}\,\partial \mathbf{g} = 0$$

This equation satisfies the necessary conditions for stationary points because $\frac{\partial f}{\partial \mathbf{g}}$ is computed such that $\nabla_c f = \mathbf{0}$. A more convenient form for presenting these equations is to take their partial derivatives with respect to all x_j. This yields

$$\frac{\partial}{\partial x_j}(f - \boldsymbol{\lambda}\mathbf{g}) = 0, \quad j = 1, 2, \ldots, n$$

The resulting equations together with the constraint equations $\mathbf{g}(\mathbf{X}) = \mathbf{0}$ yield the feasible values of $\mathbf{X}$ and $\boldsymbol{\lambda}$ that satisfy the *necessary* conditions for stationary points.

The given procedure defines the *Lagrangean method* for identifying the stationary points of optimization problems with *equality* constraints. Let

$$L(\mathbf{X}, \lambda) = f(\mathbf{X}) - \lambda\mathbf{g}(\mathbf{X})$$

The function L is called the **Lagrangean function** and the parameters λ the **Lagrange multipliers**. By definition, these multipliers have the same interpretation as the sensitivity coefficients of the Jacobian method.

The equations

$$\frac{\partial L}{\partial \boldsymbol{\lambda}} = 0, \frac{\partial L}{\partial \mathbf{X}} = 0$$

give the necessary conditions for determining stationary points of $f(\mathbf{X})$ subject to $\mathbf{g}(\mathbf{X}) = \mathbf{0}$. Sufficiency conditions for the Lagrangean method exist but they are generally computationally intractable.

Example 18.2-4

Consider the problem of Example 18.2-2. The Lagrangean function is

$$L(\mathbf{X}, \boldsymbol{\lambda}) = x_1^2 + x_2^2 + x_3^2 - \lambda_1(x_1 + x_2 + 3x_3 - 2) - \lambda_2(5x_1 + 2x_2 + x_3 - 5)$$

This yields the following necessary conditions:

$$\frac{\partial L}{\partial x_1} = 2x_1 - \lambda_1 - 5\lambda_2 = 0$$

$$\frac{\partial L}{\partial x_2} = 2x_2 - \lambda_1 - 2\lambda_2 = 0$$

$$\frac{\partial L}{\partial x_3} = 2x_3 - 3\lambda_1 - \lambda_2 = 0$$

$$\frac{\partial L}{\partial \lambda_1} = -(x_1 + x_2 + 3x_3 - 2) = 0$$

$$\frac{\partial L}{\partial \lambda_2} = -(5x_1 + 2x_2 + x_3 - 5) = 0$$

The solution to these simultaneous equations yields

$$\mathbf{X}_0 = (x_1, x_2, x_3) = (.8043, .3478, .2826)$$
$$\boldsymbol{\lambda} = (\lambda_1, \lambda_2) = (.0870, .3043)$$

This solution combines the results of Examples 18.2-2 and 18.2-3. The values of the Lagrange multipliers $\boldsymbol{\lambda}$ equal the sensitivity coefficients obtained in Example 18.2-3. The result shows that these coefficients are independent of the specific choice of the dependent vector $\mathbf{Y}$ in the Jacobian method.

PROBLEM SET 18.2C

1. Solve the following linear programming problem by both the Jacobian and the Lagrangean methods:

$$\text{Maximize } f(\mathbf{X}) = 5x_1 + 3x_2$$

subject to

$$g_1(\mathbf{X}) = x_1 + 2x_2 + x_3 \qquad - 6 = 0$$
$$g_2(\mathbf{X}) = 3x_1 + x_2 \qquad + x_4 - 9 = 0$$
$$x_1, x_2, x_3, x_4 \geq 0$$

***2.** Find the optimal solution to the problem

$$\text{Minimize } f(\mathbf{X}) = x_1^2 + 2x_2^2 + 10x_3^2$$

subject to

$$g_1(\mathbf{X}) = x_1 + x_2^2 + x_3 - 5 = 0$$
$$g_2(\mathbf{X}) = x_1 + 5x_2 + x_3 - 7 = 0$$

Suppose that $g_1(\mathbf{X}) = .01$ and $g_2(\mathbf{X}) = .02$. Find the corresponding change in the optimal value of $f(\mathbf{X})$.

3. Solve Problem 6, Set 18.2b, by the Lagrangean method and verify that the values of the Lagrange multipliers are the same as the sensitivity coefficients obtained in Problem 6, Set 18.2b.

18.2.2 Inequality Constraints—Karush-Kuhn-Tucker (KKT) Conditions[2]

This section extends the Lagrangean method to problems with inequality constraints. The main contribution of the section is the development of the general Karush-Kuhn-Tucker (KKT) *necessary* conditions for determining the stationary points. These conditions are also sufficient under certain rules that will be stated later.

Consider the problem

$$\text{Maximize } z = f(\mathbf{X})$$

[2]Historically, W. Karush was the first to develop the KKT conditions in 1939 as part of his M.S. thesis at the University of Chicago. The same conditions were developed independently in 1951 by W. Kuhn and A. Tucker.

subject to

$$\mathbf{g}(\mathbf{X}) \leq \mathbf{0}$$

The inequality constraints may be converted into equations by using *nonnegative* slack variables. Let S_i^2 (≥ 0) be the slack quantity added to the ith constraint $g_i(\mathbf{X}) \leq 0$ and define

$$\mathbf{S} = (S_1, S_2, \ldots, S_m)^T, \mathbf{S}^2 = (S_1^2, S_2^2, \ldots, S_m^2)^T$$

where m is the total number of inequality constraints. The Lagrangean function is thus given by

$$L(\mathbf{X}, \mathbf{S}, \boldsymbol{\lambda}) = f(\mathbf{X}) - \boldsymbol{\lambda}[\mathbf{g}(\mathbf{X}) + \mathbf{S}^2]$$

Given the constraints

$$\mathbf{g}(\mathbf{X}) \leq \mathbf{0}$$

a necessary condition for optimality is that $\boldsymbol{\lambda}$ be nonnegative (nonpositive) for maximization (minimization) problems. This result is justified by noting that the vector $\boldsymbol{\lambda}$ measures the rate of variation of f with respect to $\mathbf{g}$—that is,

$$\boldsymbol{\lambda} = \frac{\partial f}{\partial \mathbf{g}}$$

In the maximization case, as the right-hand side of the constraint $\mathbf{g}(\mathbf{X}) \leq \mathbf{0}$ increases from $\mathbf{0}$ to the vector $\partial \mathbf{g}$, the solution space becomes less constrained and hence f cannot decrease, meaning that $\boldsymbol{\lambda} \geq \mathbf{0}$. Similarly for minimization, as the right-hand side of the constraints increases, f cannot increase, which implies that $\boldsymbol{\lambda} \leq \mathbf{0}$. If the constraints are equalities, that is, $\mathbf{g}(\mathbf{X}) = \mathbf{0}$, then $\boldsymbol{\lambda}$ becomes unrestricted in sign (see Problem 2, Set 18.2d).

The restrictions on $\boldsymbol{\lambda}$ holds as part of the KKT necessary conditions. The remaining conditions will now be developed.

Taking the partial derivatives of L with respect to **X, S**, and $\boldsymbol{\lambda}$, we obtain

$$\frac{\partial L}{\partial \mathbf{X}} = \nabla f(\mathbf{X}) - \boldsymbol{\lambda}\nabla \mathbf{g}(\mathbf{X}) = \mathbf{0}$$

$$\frac{\partial L}{\partial S_i} = -2\lambda_i S_i = 0, i = 1, 2, \ldots, m$$

$$\frac{\partial L}{\partial \boldsymbol{\lambda}} = -(\mathbf{g}(\mathbf{X}) + \mathbf{S}^2) = \mathbf{0}$$

The second set of equations reveals the following results:

1. If $\lambda_i \neq 0$, then $S_i^2 = 0$, which means that the corresponding resource is scarce, and, hence, it is consumed completely (equality constraint).
2. If $S_i^2 > 0$, then $\lambda_i = 0$. This means resource i is not scarce and, consequently, it has no affect on the value of f (i.e., $\lambda_i = \frac{\partial f}{\partial g_i} = 0$).

From the second and third sets of equations, we obtain

$$\lambda_i g_i(\mathbf{X}) = 0, i = 1, 2, \ldots, m$$

This new condition essentially repeats the foregoing argument, because if $\lambda_i > 0$, $g_i(\mathbf{X}) = 0$ or $S_i^2 = 0$; and if $g_i(\mathbf{X}) < 0$, $S_i^2 > 0$, and $\lambda_i = 0$.

The KKT necessary conditions for maximization problem are summarized as:

$$\begin{aligned} \boldsymbol{\lambda} &\geq \mathbf{0} \\ \nabla f(\mathbf{X}) - \boldsymbol{\lambda}\nabla \mathbf{g}(\mathbf{X}) &= \mathbf{0} \\ \lambda_i g_i(\mathbf{X}) &= 0, \quad i = 1, 2, \ldots, m \\ \mathbf{g}(\mathbf{X}) &\leq \mathbf{0} \end{aligned}$$

These conditions apply to the minimization case as well, except that $\boldsymbol{\lambda}$ must be non-positive (verify!). In both maximization and minimization, the Lagrange multipliers corresponding to equality constraints are unrestricted in sign.

Sufficiency of the KKT Conditions. The Kuhn-Tucker necessary conditions are also sufficient if the objective function and the solution space satisfy specific conditions. These conditions are summarized in Table 18.1.

It is simpler to verify that a function is convex or concave than to prove that a solution space is a convex set. For this reason, we provide a list of conditions that are easier to apply in practice in the sense that the convexity of the solution space can be established by checking the convexity or concavity of the constraint functions. To provide these conditions, we define the generalized nonlinear problems as

$$\text{Maximize or minimize } z = f(\mathbf{X})$$

subject to

$$\begin{aligned} g_i(\mathbf{X}) &\leq 0, \quad i = 1, 2, \ldots, r \\ g_i(\mathbf{X}) &\geq 0, \quad i = r + 1, \ldots, p \\ g_i(\mathbf{X}) &= 0, \quad i = p + 1, \ldots, m \end{aligned}$$

$$L(\mathbf{X}, \mathbf{S}, \boldsymbol{\lambda}) = f(\mathbf{X}) - \sum_{i=1}^{r} \lambda_i[g_i(\mathbf{X}) + S_i^2] - \sum_{i=r+1}^{p} \lambda_i[g_i(\mathbf{X}) - S_i^2] - \sum_{i=p+1}^{m} \lambda_i g_i(\mathbf{X})$$

where λ_i is the Lagrangean multiplier associated with constraint i. The conditions for establishing the sufficiency of the KKT conditions are summarized in Table 18.2.

The conditions in Table 18.2 represent only a subset of the conditions in Table 18.1 because a solution space may be convex without satisfying the conditions in Table 18.2.

TABLE 18.1

Sense of optimization	Required conditions	
	Objective function	*Solution space*
Maximization	Concave	Convex set
Minimization	Convex	Convex set

TABLE 18.2

Sense of optimization	Required conditions			
	$f(\mathbf{X})$	$g_i(\mathbf{X})$	λ_i	
Maximization	Concave	Convex	≥ 0	$(1 \leq i \leq r)$
		Concave	≤ 0	$(r+1 \leq i \leq p)$
		Linear	Unrestricted	$(p+1 \leq i \leq m)$
Minimization	Convex	Convex	≤ 0	$(1 \leq i \leq r)$
		Concave	≥ 0	$(r+1 \leq i \leq p)$
		Linear	Unrestricted	$(p+1 \leq i \leq m)$

Table 18.2 is valid because the given conditions yield a concave Lagrangean function $L(\mathbf{X}, \mathbf{S}, \boldsymbol{\lambda})$ in case of maximization and a convex $L(\mathbf{X}, \mathbf{S}, \boldsymbol{\lambda})$ in case of minimization. This result is verified by noticing that if $g_i(x)$ is convex, then $\lambda_i g_i(x)$ is convex if $\lambda_i \geq 0$ and concave if $\lambda_i \leq 0$. Similar interpretations can be established for all the remaining conditions. Observe that a linear function is both convex and concave. Also, if a function f is concave, then $(-f)$ is convex, and vice versa.

Example 18.2-5

Consider the following minimization problem:

$$\text{Minimize } f(\mathbf{X}) = x_1^2 + x_2^2 + x_3^2$$

subject to

$$\begin{aligned} g_1(\mathbf{X}) &= 2x_1 + x_2 - 5 \leq 0 \\ g_2(\mathbf{X}) &= x_1 + x_3 - 2 \leq 0 \\ g_3(\mathbf{X}) &= 1 - x_1 \leq 0 \\ g_4(\mathbf{X}) &= 2 - x_2 \leq 0 \\ g_5(\mathbf{X}) &= -x_3 \leq 0 \end{aligned}$$

This is a minimization problem, hence $\boldsymbol{\lambda} \leq \mathbf{0}$. The KKT conditions are thus given as

$$(\lambda_1, \lambda_2, \lambda_3, \lambda_4, \lambda_5) \leq \mathbf{0}$$

$$(2x_1, 2x_2, 2x_3) - (\lambda_1, \lambda_2, \lambda_3, \lambda_4, \lambda_5)\begin{pmatrix} 2 & 1 & 0 \\ 1 & 0 & 1 \\ -1 & 0 & 0 \\ 0 & -1 & 0 \\ 0 & 0 & -1 \end{pmatrix} = \mathbf{0}$$

$$\lambda_1 g_1 = \lambda_2 g_2 = \dots = \lambda_5 g_5 = 0$$

$$\mathbf{g}(\mathbf{X}) \leq \mathbf{0}$$

These conditions reduce to

$$\begin{aligned}
\lambda_1, \lambda_2, \lambda_3, \lambda_4, \lambda_5 &\leq 0 \\
2x_1 - 2\lambda_1 - \lambda_2 + \lambda_3 &= 0 \\
2x_2 - \lambda_1 + \lambda_4 &= 0 \\
2x_3 - \lambda_2 + \lambda_5 &= 0 \\
\lambda_1(2x_1 + x_2 - 5) &= 0 \\
\lambda_2(x_1 + x_3 - 2) &= 0 \\
\lambda_3(1 - x_1) &= 0 \\
\lambda_4(2 - x_2) &= 0 \\
\lambda_5 x_3 &= 0 \\
2x_1 + x_2 &\leq 5 \\
x_1 + x_3 &\leq 2 \\
x_1 \geq 1, x_2 \geq 2, x_3 &\geq 0
\end{aligned}$$

The solution is $x_1 = 1, x_2 = 2, x_3 = 0, \lambda_1 = \lambda_2 = \lambda_5 = 0, \lambda_3 = -2, \lambda_4 = -4$. Because both $f(\mathbf{X})$ and the solution space $\mathbf{g}(\mathbf{X}) \leq \mathbf{0}$ are convex, $L(\mathbf{X}, \mathbf{S}, \boldsymbol{\lambda})$ must be convex and the resulting stationary point yields a global constrained minimum. The KKT conditions are central to the development of the nonlinear programming algorithms in Chapter 19.

PROBLEM SET 18.2D

1. Consider the problem:

$$\text{Maximize } f(\mathbf{X})$$

subject to

$$\mathbf{g}(\mathbf{X}) \geq \mathbf{0}$$

Show that the KKT conditions are the same as in Section 18.2.2, except that the Lagrange multipliers $\boldsymbol{\lambda}$ are nonpositive.

2. Consider the following problem:

$$\text{Maximize } f(\mathbf{X})$$

subject to

$$\mathbf{g}(\mathbf{X}) = 0$$

Show that the KKT conditions are

$$\begin{aligned}
\nabla f(\mathbf{X}) - \boldsymbol{\lambda}\nabla \mathbf{g}(\mathbf{X}) &= \mathbf{0} \\
\mathbf{g}(\mathbf{X}) &= \mathbf{0}
\end{aligned}$$

$$\boldsymbol{\lambda} \text{ unrestricted in sign}$$

3. Write the KKT necessary conditions for the following problems.

(a) Maximize $f(\mathbf{X}) = x_1^3 - x_2^2 + x_1x_3^2$
subject to

$$x_1 + x_2^2 + x_3 = 5$$
$$5x_1^2 - x_2^2 - x_3 \geq 2$$
$$x_1, x_2, x_3 \geq 0$$

(b) Minimize $f(\mathbf{X}) = x_1^4 + x_2^2 + 5x_1x_2x_3$
subject to

$$x_1^2 - x_2^2 + x_3^3 \leq 10$$
$$x_1^3 + x_2^2 + 4x_3^2 \geq 20$$

4. Consider the problem

$$\text{Maximize } f(\mathbf{X})$$

subject to

$$\mathbf{g}(\mathbf{X}) = \mathbf{0}$$

Given $f(\mathbf{X})$ is concave and $g_i(\mathbf{X})(i = 1, 2, \ldots, m)$ is a *linear* function, show that the KKT necessary conditions are also sufficient. Is this result true if $g_i(\mathbf{X})$ is a convex *non*linear function for all i? Why?

5. Consider the problem

$$\text{Maximize } f(\mathbf{X})$$

subject to

$$g_1(\mathbf{X}) \geq 0,\ g_2(\mathbf{X}) = 0,\ g_3(\mathbf{X}) \leq 0$$

Develop the KKT conditions and give the stipulations under which the conditions are sufficient.

REFERENCES

Bazarra, M., H. Shrali, and C. Shetty, *Nonlinear Programming Theory and Algorithms*, 2nd ed., Wiley, New York, 1993.

Beightler, C., D. Phillips, and D. Wilde, *Foundations of Optimization*, 2nd ed., Prentice Hall, NJ, 1979.

Rardin, R., *Optimization in Operations Research*, Prentice Hall, NJ, 1998.

CHAPTER 19

Nonlinear Programming Algorithms

Chapter Guide. The solution methods of nonlinear programming generally can be classified as either *direct* or *indirect* algorithms. Examples of the direct methods are the gradient algorithms, where the maximum (minimum) of a problem is found by following the fastest rate of increase (decrease) of the objective function. In the indirect methods, the original problem is replaced by an auxiliary one from which the optimum is determined. Examples of these situations include quadratic programming, separable programming, and stochastic programming.

This chapter includes 9 solved examples. 1 AMPL model, 2 Solver models, and 24 end-of-section problems. The AMPL/Excel/Solver/TORA programs are in folder ch19Files.

19.1 UNCONSTRAINED ALGORITHMS

This section presents two algorithms for the unconstrained problem: the *direct search* algorithm and the *gradient* algorithm.

19.1.1 Direct Search Method

Direct search methods apply primarily to strictly unimodal single-variable functions. Although the case may appear trivial, Section 19.1.2 shows that optimization of single-variable functions is key in the development of the more general multivariable algorithms.

The idea of direct search methods is to identify the **interval of uncertainty** that is known to include the optimum solution point. The procedure locates the optimum by iteratively narrowing the interval of uncertainty to any desired level of accuracy.

Two closely related algorithms are presented in this section: **dichotomous** and **golden section** search methods. Both algorithms seek the maximization of a unimodal function $f(x)$ over the interval $a \leq x \leq b$, which is known to include the optimum point x^*. The two methods start with $I_0 = (a, b)$ representing the initial interval of uncertainty.

General Step *i*. Let $I_{i-1} = (x_L, x_R)$ be the current interval of uncertainty (at iteration 0, $x_L = a$ and $x_R = b$). Next, identify x_1 and x_2 in the following manner:

Dichotomous method	Golden section method
$x_1 = \frac{1}{2}(x_R + x_L - \Delta)$	$x_1 = x_R - (\frac{\sqrt{5}-1}{2})(x_R - x_L)$
$x_2 = \frac{1}{2}(x_R + x_L + \Delta)$	$x_2 = x_L + (\frac{\sqrt{5}-1}{2})(x_R - x_L)$

The selection of x_1 and x_2 guarantees that

$$x_L < x_1 < x_2 < x_R$$

The next interval of uncertainty, I_i, is determined in the following manner:

1. If $f(x_1) > f(x_2)$, then $x_L < x^* < x_2$. Let $x_R = x_2$ and set $I_i = (x_L, x_2)$ [see Figure 19.1(a)].
2. If $f(x_1) < f(x_2)$, then $x_1 < x^* < x_R$. Let $x_L = x_1$ and set $I_i = (x_1, x_R)$ [see Figure 19.1(b)].
3. If $f(x_1) = f(x_2)$, then $x_1 < x^* < x_2$. Let $x_L = x_1$ and $x_R = x_2$; set $I_i = (x_1, x_2)$.

The manner in which x_1 and x_2 are determined guarantees that $I_{i+1} < I_i$, as will be shown shortly. The algorithm terminates at iteration k if $I_k \leq \Delta$, where Δ is a user-specified level of accuracy.

FIGURE 19.1

Illustration of the general step of the dichotomous/golden section search methods

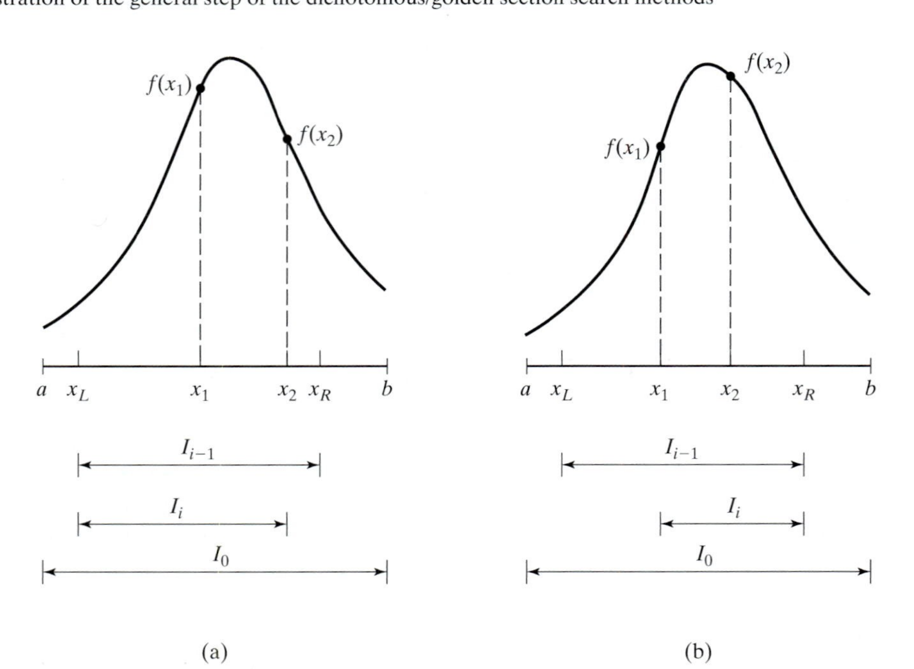

In the dichotomous method, the values x_1 and x_2 sit symmetrically around the midpoint of the current interval of uncertainty. This means that

$$I_{i+1} = .5(I_i + \Delta)$$

Repeated application of the algorithm guarantees that the length of the interval of uncertainty will approach the desired accuracy, Δ.

In the golden section method, the idea is more involved. We notice that each iteration of the dichotomous method requires calculating the two values $f(x_1)$ and $f(x_2)$, but ends up discarding one of them. What the golden section proposes is to save computations by reusing the discarded value in the immediately succeeding iteration.

Define for $0 < \alpha < 1$,

$$x_1 = x_R - \alpha(x_R - x_L)$$

$$x_2 = x_L + \alpha(x_R - x_L)$$

Then the interval of uncertainty I_i at iteration i equals (x_L, x_2) or (x_1, x_R). Consider the case $I_i = (x_L, x_2)$, which means that x_1 is included in I_i. In iteration $i + 1$, we select x_2 equal to x_1 in iteration i, which leads to the following equation:

$$x_2(\text{iteration } i + 1) = x_1(\text{iteration } i)$$

Substitution yields

$$x_L + \alpha[x_2(\text{iteration } i) - x_L] = x_R - \alpha(x_R - x_L)$$

or

$$x_L + \alpha[x_L + \alpha(x_R - x_L) - x_L] = x_R - \alpha(x_R - x_L)$$

which finally simplifies to

$$\alpha^2 + \alpha - 1 = 0$$

This equation yields $\alpha = \frac{-1 \pm \sqrt{5}}{2}$. Because $0 \le \alpha \le 1$, we select the positive root $\alpha = \frac{-1 + \sqrt{5}}{2} \approx .681$.

The design of the golden section computations guarantees an α-reduction in successive intervals of uncertainty—that is

$$I_{i+1} = \alpha I_i$$

The golden section method converges more rapidly than the dichotomous method because, in the dichotomous method, the narrowing of the interval of uncertainty slows down appreciably as $I \to \Delta$. In addition, each iteration in the golden section method requires half the computations because the method always recycles one set of computations from the immediately preceding iteration.

Example 19.1-1

$$\text{Maximize } f(x) = \begin{cases} 3x, & 0 \le x \le 2 \\ \frac{1}{3}(-x + 20), & 2 \le x \le 3 \end{cases}$$

The maximum value of $f(x)$ occurs at $x = 2$. The following table demonstrates the calculations for iterations 1 and 2 using the dichotomous and the golden section methods. We will assume $\Delta = .1$.

Dichotomous method	Golden section method
Iteration 1	*Iteration 1*
$I_0 = (0, 3) \equiv (x_L, x_R)$	$I_0 = (0, 3) \equiv (x_L, x_R)$
$x_1 = 0 + .5(3 - 0 - .1) = 1.45, f(x_1) = 4.35$	$x_1 = 3 - .618(3 - 0) = 1.146, f(x_1) = 3.438$
$x_2 = 0 + .5(3 - 0 + .1) = 1.55, f(x_2) = 4.65$	$x_2 = 0 + .618(3 - 0) = 1.854, f(x_2) = 5.562$
$f(x_2) > f(x_1) \Rightarrow x_L = 1.45, I_1 = (1.45, 3)$	$f(x_2) > f(x_1) \Rightarrow x_L = 1.146, I_1 = (1.146, 3)$
Iteration 2	*Iteration 2*
$I_1 = (1.45, 3) \equiv (x_L, x_R)$	$I_1 = (1.146, 3) \equiv (x_L, x_R)$
$x_1 = 1.45 + .5(3 - 1.45 - .1) = 2.175,$ $f(x_1) = 5.942$	$x_1 = x_2$ in iteration $0 = 1.854,$ $f(x_1) = 5.562$
$x_2 = \frac{3 + 1.45 + .1}{2} = 2.275, f(x_2) = 5.908$	$x_2 = 1.146 + .618(3 - 1.146) = 2.292, f(x_2) = 5.903$
$f(x_1) > f(x_2) \Rightarrow x_R = 2.275, I_2 = (1.45, 2.275)$	$f(x_2) > f(x_1) \Rightarrow x_L = 1.854, I_1 = (1.854, 3)$

Continuing in the same manner, the interval of uncertainty will eventually narrow down to the desired Δ-tolerance.

Excel Moment

Template excelDiGold.xls handles either method. The input data include $f(x)$, a, b, and Δ. The function $f(x)$ is entered in cell E3 as

```
=IF(C3<=2,3*C3,(-C3+20)/3)
```

Cell C3 plays the role of x in $f(x)$. Limits a and b are entered in cells B4 and D4 to represent the admissible search range for $f(x)$. Also, the tolerance limit, Δ, is entered in cell B3. The search method is selected by entering the letter x in either D5 (dichotomous) or F5 (golden section).

Figure 19.2 compares the two methods. The golden section method requires less than half as many iterations as the dichotomous method. Additionally, each iteration requires half the calculations, as explained previously.

PROBLEM SET 19.1A

1. Use Excel template excelDiGold.xls to solve Example 19.1-1 assuming that $\Delta = .01$. Compare the amount of computations and the accuracy of the results with those in Figure 19.2.
2. Find the maximum of each of the following functions by dichotomous search. Assume that $\Delta = .05$.

(a) $f(x) = \dfrac{1}{|(x-3)^3|}, \quad 2 \le x \le 4$

(b) $f(x) = x \cos x, \quad 0 \le x \le \pi$

***(c)** $f(x) = x \sin \pi x, \quad 1.5 \le x \le 2.5$

	A	B	C	D	E	F
1	Dichotomous/Golden Section Search					
2	**Input data:** Type f(C3) in E3, where C3 represents x in f(x)					
3	Δ =	0.1	C3		#VALUE!	Clear Calculations
4	**Minimum x =**	0	**Maximum x =**	3		
5	**Solution:**	Enter x to select>	Dichotomous:	x	GoldenSection:	
6	x* =	2.04001	f(x*) =	5.97002		
7	**Clalculations:**				Perform calculation	
8	xL	xR	x1	x2	f(x1)	f(x2)
9	0.000000	3.000000	1.450000	1.550000	4.350000	4.650000
10	1.450000	3.000000	2.175000	2.275000	5.941667	5.908333
11	1.450000	2.275000	1.812500	1.912500	5.437500	5.737500
12	1.812500	2.275000	1.993750	2.093750	5.981250	5.968750
13	1.812500	2.093750	1.903125	2.003125	5.709375	5.998958
14	1.903125	2.093750	1.948438	2.048438	5.845313	5.983854
15	1.948438	2.093750	1.971094	2.071094	5.913281	5.976302
16	1.971094	2.093750	1.982422	2.082422	5.947266	5.972526
17	1.982422	2.093750	1.988086	2.088086	5.964258	5.970638
18	1.988086	2.093750	1.990918	2.090918	5.972754	5.969694
19	1.988086	2.090918	1.989502	2.089502	5.968506	5.970166
20	1.989502	2.090918	1.990210	2.090210	5.970630	5.969930
21	1.989502	2.090210	1.989856	2.089856	5.969568	5.970048
22	1.989856	2.090210	1.990033	2.090033	5.970099	5.969989
23	1.989856	2.090033	1.989944	2.089944	5.969833	5.970019
24	1.989944	2.090033	1.989989	2.089989	5.969966	5.970004
25	1.989989	2.090033	1.990011	2.090011	5.970033	5.969996
26	1.989989	2.090011	1.990000	2.090000	5.969999	5.970000
27	1.990000	2.090011	1.990005	2.090005	5.970016	5.969998
28	1.990000	2.090005	1.990003	2.090003	5.970008	5.969999

	A	B	C	D	E	F
1	Dichotomous/Golden Section Search					
2	**Input data:** Type f(C3) in E3, where C3 represents x in f(x)					
3	Δ =	0.1	C3		#VALUE!	Clear Calculations
4	**Minimum x =**	0	**Maximum x =**	3		
5	**Solution:**	Enter x to select>	Dichotomous:		GoldenSection:	x
6	x* =	2.00909	f(x*) =	5.99290		
7	**Clalculations:**				Perform calculation	
8	xL	xR	x1	x2	f(x1)	f(x2)
9	0.000000	3.000000	1.145898	1.854102	3.437694	5.562306
10	1.145898	3.000000	1.854102	2.291796	5.562306	5.902735
11	1.854102	3.000000	2.291796	2.562306	5.902735	5.812565
12	1.854102	2.562306	2.124612	2.291796	5.958463	5.902735
13	1.854102	2.291796	2.021286	2.124612	5.992905	5.958463
14	1.854102	2.124612	1.957428	2.021286	5.872283	5.992905
15	1.957428	2.124612	2.021286	2.060753	5.992905	5.979749
16	1.957428	2.060753	1.996894	2.021286	5.990683	5.992905
17	1.996894	2.060753	2.021286	2.036361	5.992905	5.987880

FIGURE 19.2

Excel output of the dichotomous and golden section methods applied to Example 19.1-1 (file excelDiGold.xls)

(d) $f(x) = -(x - 3)^2, \quad 2 \le x \le 4$

***(e)** $f(x) = \begin{cases} 4x, & 0 \le x \le 2 \\ 4 - x, & 2 \le x \le 4 \end{cases}$

19.1.2 Gradient Method

This section develops a method for optimizing functions that are twice continuously differentiable. The idea is to generate successive points in the direction of the gradient of the function.

The Newton-Raphson method presented in Section 18.1.2 is a gradient method for solving simultaneous equations. This section presents another technique, called the **steepest ascent** method.

Termination of the gradient method occurs at the point where the gradient vector becomes null. This is only a necessary condition for optimality. Optimality cannot be verified unless it is known a priori that $f(\mathbf{X})$ is concave or convex.

Suppose that $f(\mathbf{X})$ is maximized. Let $\mathbf{X}_0$ be the initial point from which the procedure starts and define $\nabla f(\mathbf{X}_k)$ as the gradient of f at point $\mathbf{X}_k$. The idea is to determine a particular path p along which $\frac{\partial f}{\partial p}$ is maximized at a given point. This result is achieved if successive points $\mathbf{X}_k$ and $\mathbf{X}_{k+1}$ are selected such that

$$\mathbf{X}_{k+1} = \mathbf{X}_k + r_k \nabla f(\mathbf{X}_k)$$

where r_k is the optimal **step size** at $\mathbf{X}_k$.

The step size r_k is determined such that the next point, $\mathbf{X}_{k+1}$, leads to the largest improvement in f. This is equivalent to determining $r = r_k$ that maximizes the function

$$h(r) = f[\mathbf{X}_k + r \nabla f(\mathbf{X}_k)]$$

Because $h(r)$ is a single-variable function, the search method in Section 19.1.1 may be used to find the optimum, provided that $h(r)$ is strictly unimodal.

The proposed procedure terminates when two successive trial points $\mathbf{X}_k$ and $\mathbf{X}_{k+1}$ are approximately equal. This is equivalent to having

$$r_k \nabla f(\mathbf{X}_k) \approx \mathbf{0}$$

Because $r_k \neq 0$, the necessary condition $\nabla f(\mathbf{X}_k) = \mathbf{0}$ is satisfied at $\mathbf{X}_k$.

Example 19.1-2

Consider the following problem:

$$\text{Maximize } f(x_1, x_2) = 4x_1 + 6x_2 - 2x_1^2 - 2x_1x_2 - 2x_2^2$$

The exact optimum occurs at $(x_1^*, x_2^*) = \left(\frac{1}{3}, \frac{4}{3}\right)$.

We show how the problem is solved by the steepest ascent method. The gradient of f is given as

$$\nabla f(\mathbf{X}) = (4 - 4x_1 - 2x_2, 6 - 2x_1 - 4x_2)$$

The quadratic nature of the function dictates that the gradients at any two successive points are orthogonal (perpendicular to one another).

Suppose that we start at the initial point $\mathbf{X}_0 = (1, 1)$. Figure 19.3 shows the successive solution points.

Iteration 1

$$\nabla f(\mathbf{X}_0) = (-2, 0)$$

The next point $\mathbf{X}_1$ is obtained by considering

$$\mathbf{X} = (1, 1) + r(-2, 0) = (1 - 2r, 1)$$

Thus,

$$h(r) = f(1 - 2r, 1) = -2(1 - 2r)^2 + 2(1 - 2r) + 4$$

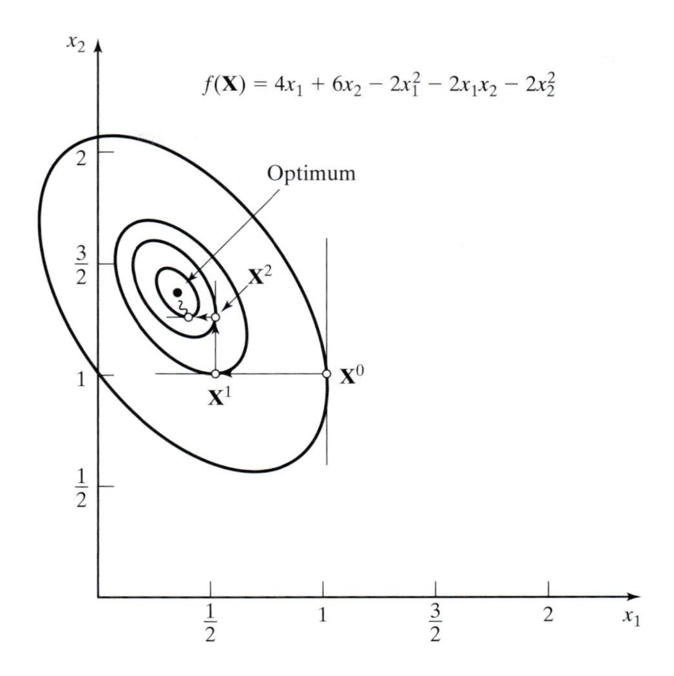

FIGURE 19.3
Maximization of $f(x_1, x_2) = 4x_1 + 6x_2 - 2x_1^2 - 2x_1x_2 - 2x_2^2$ by the steepest-ascent method

The optimal step size is obtained using the classical necessary conditions in Chapter 18 (you may also use the search algorithms in Section 19.1.1 to determine the optimum). The maximum value of $h(r)$ is $r_1 = \frac{1}{4}$, which yields the next solution point as $\mathbf{X}_1 = \left(\frac{1}{2}, 1\right)$.

Iteration 2

$$\nabla f(\mathbf{X}_1) = (0, 1)$$
$$\mathbf{X} = \left(\tfrac{1}{2}, 1\right) + r(0, 1) = \left(\tfrac{1}{2}, 1 + r\right)$$
$$h(r) = -2(1 + r)^2 + 5(1 + r) + \tfrac{3}{2}$$

This gives $r_2 = \frac{1}{4}$ and $\mathbf{X}_2 = \left(\frac{1}{2}, \frac{5}{4}\right)$.

Iteration 3

$$\nabla f(\mathbf{X}_2) = \left(-\tfrac{1}{2}, 0\right)$$
$$\mathbf{X} = \left(\tfrac{1}{2}, \tfrac{5}{4}\right) + r\left(-\tfrac{1}{2}, 0\right) = \left(\tfrac{1 - r}{2}, \tfrac{5}{4}\right)$$
$$h(r) = -\tfrac{1}{2}(1 - r)^2 + \tfrac{3}{4}(1 - r) + \tfrac{35}{8}$$

Hence, $r_3 = \frac{1}{4}$ and $\mathbf{X}_3 = \left(\frac{3}{8}, \frac{5}{4}\right)$.

Iteration 4

$$\nabla f(\mathbf{X}_3) = \left(0, \tfrac{1}{4}\right)$$

$$\mathbf{X} = \left(\tfrac{3}{8}, \tfrac{5}{4}\right) + r\left(0, \tfrac{1}{4}\right) = \left(\tfrac{3}{8}, \tfrac{5+r}{4}\right)$$

$$h(r) = -\tfrac{1}{8}(5+r)^2 + \tfrac{21}{16}(5+r) + \tfrac{39}{32}$$

Thus, $r_4 = \frac{1}{4}$ and $\mathbf{X}_4 = \left(\frac{3}{8}, \frac{21}{16}\right)$.

Iteration 5

$$\nabla f(\mathbf{X}_4) = \left(-\tfrac{1}{8}, 0\right)$$

$$\mathbf{X} = \left(\tfrac{3}{8}, \tfrac{21}{16}\right) + r\left(-\tfrac{1}{8}, 0\right) = \left(\tfrac{3-r}{8}, \tfrac{21}{16}\right)$$

$$h(r) = -\tfrac{1}{32}(3-r)^2 + \tfrac{11}{64}(3-r) + \tfrac{567}{128}$$

This gives $r_5 = \frac{1}{4}$ and $\mathbf{X}_5 = \left(\frac{11}{32}, \frac{21}{16}\right)$.

Iteration 6

$$\nabla f(\mathbf{X}_5) = \left(0, \tfrac{1}{16}\right)$$

Because $\nabla f(\mathbf{X}_5) \approx \mathbf{0}$, the process can be terminated at this point. The *approximate* maximum point is given by $\mathbf{X}_5 = (.3438, 1.3125)$. The exact optimum is $\mathbf{X}^* = (.3333, 1.3333)$.

PROBLEM SET 19.1B

*1. Show that, in general, the Newton-Raphson method (Section 18.1.2) when applied to a strictly concave quadratic function will converge in exactly one step. Apply the method to the maximization of

$$f(\mathbf{X}) = 4x_1 + 6x_2 - 2x_1^2 - 2x_1x_2 - 2x_2^2$$

2. Carry out at most five iterations for each of the following problems using the method of steepest descent (ascent). Assume that $\mathbf{X}^0 = 0$ in each case.

   (a) $\min f(\mathbf{X}) = \min f(\mathbf{X}) = (x_2 - x_1^2)^2 + (1 - x_1)$

   (b) $\max f(\mathbf{X}) = \mathbf{cX} + \mathbf{X}^T\mathbf{AX}$

   where

$$\mathbf{c} = (1, 3, 5)$$

$$\mathbf{A} = \begin{pmatrix} -5 & -3 & -\frac{1}{2} \\ -3 & -2 & 0 \\ -\frac{1}{2} & 0 & -\frac{1}{2} \end{pmatrix}$$

   (c) $\min f(\mathbf{X}) = x_1 - x_2 + x_1^2 - x_1x_2$

19.2 CONSTRAINED ALGORITHMS

The general constrained nonlinear programming problem is defined as

$$\text{Maximize (or minimize) } z = f(\mathbf{X})$$

subject to

$$\mathbf{g}(\mathbf{X}) \leq \mathbf{0}$$

The nonnegativity conditions, $\mathbf{X} \geq \mathbf{0}$, are part of the constraints. Also, at least one of the functions $f(\mathbf{X})$ and $\mathbf{g}(\mathbf{X})$ is nonlinear and all the functions are continuously differentiable.

No single algorithm exists for handling the general nonlinear models, because of the erratic behavior of the nonlinear functions. Perhaps the most general result applicable to the problem is the KKT conditions (Section 18.2.2). Table 18.2 shows that unless $f(\mathbf{X})$ and $\mathbf{g}(\mathbf{X})$ are well behaved (based on the convexity and concavity conditions), the KKT conditions are only necessary for realizing optimality.

This section presents a number of algorithms that may be classified generally as *indirect* and *direct* methods. Indirect methods solve the nonlinear problem by dealing with one or more *linear* programs derived from the original program. Direct methods deal with the original problem.

The indirect algorithms presented in this section include separable, quadratic, and chance-constrained programming. The direct algorithms include the method of linear combinations and a brief discussion of SUMT, the sequential unconstrained maximization technique. Other important nonlinear techniques can be found in the list of references at the end of the chapter.

19.2.1 Separable Programming

A function $f(x_1, x_2, \ldots, x_n)$ is **separable** if it can be expressed as the sum of n single-variable functions $f_1(x_1), f_2(x_2), \ldots, f_n(x_n)$—that is,

$$f(x_1, x_1, \ldots, x_n) = f_1(x_1) + f_2(x_2) + \cdots + f_n(x_n)$$

For example, the linear function

$$h(x_1, x_2, \ldots, x_n) = a_1x_1 + a_2x_2 + \cdots + a_nx_n$$

is separable (the parameters a_i, $i = 1, 2, \ldots, n$, are constants). Conversely, the function

$$h(x_1, x_2, x_3) = x_1^2 + x_1 \sin(x_2 + x_3) + x_2e^{x3}$$

is not separable.

Some nonlinear functions are not directly separable but can be made so by appropriate substitutions. Consider, for example, the case of maximizing $z = x_1x_2$. Letting $y = x_1x_2$, then $\ln y = \ln x_1 + \ln x_2$ and the problem becomes

$$\text{Maximize } z = y$$

subject to

$$\ln y = \ln x_1 + \ln x_2$$

which is separable. The substitution assumes that x_1 and x_2 are *positive* variables because the logarithmic function is undefined for nonpositive values.

The case where x_1 and x_2 assume zero values (i.e., $x_1, x_2 \geq 0$) may be handled in the following manner. Let δ_1 and δ_2 be positive constants and define

$$w_1 = x_1 + \delta_1$$

$$w_2 = x_2 + \delta_2$$

In the substitution

$$x_1 x_2 = w_1 w_2 - \delta_2 w_1 - \delta_1 w_2 + \delta_1 \delta_2$$

the new variables, w_1 and w_2, are strictly positive. Letting $y = w_1 w_2$, the problem is expressed as

$$\text{Maximize } z = y - \delta_2 w_1 - \delta_1 w_2 + \delta_1 \delta_2$$

subject to

$$\ln y = \ln w_1 + \ln w_2$$

$$w_1 \geq \delta_1, w_2 \geq \delta_2$$

which is separable.

This section shows how an approximate solution can be obtained for *any* separable problem by linear approximation and the simplex method of linear programming. The single-variable function $f(x)$ can be approximated by a piecewise linear function using mixed integer programming (Chapter 9). Suppose that $f(x)$ is to be approximated over an interval $[a, b]$. Define a_k, $k = 1, 2, \ldots, K$, as the kth breakpoint on the x-axis such that $a_1 < a_2 < \cdots < a_K$. The points a_1 and a_K coincide with end points a and b of the designated interval. Thus, $f(x)$ is approximated as follows:

$$f(x) \approx \sum_{k=1}^{K} f(a_k) w_k$$

$$x = \sum_{k=1}^{K} a_k w_k$$

where w_k is a nonnegative weight associated with the kth breakpoint such that

$$\sum_{k=1}^{K} w_k = 1, w_k \geq 0, k = 1, 2, \ldots, K$$

Mixed integer programming ensures the validity of the approximation by imposing two conditions:

1. At most two w_k are positive.
2. If w_k is positive, then only an adjacent w_{k+1} or w_{k-1} can assume a positive value. To show how these conditions are satisfied, consider the separable problem

$$\text{Maximize (or minimize) } z = \sum_{j=1}^{n} f_j(x_j)$$

subject to

$$\sum_{j=1}^{n} g_{ij}(x_j) \leq b_i, i = 1, 2, \ldots, m$$

This problem can be approximated by a mixed integer program as follows. Let[1]

$$\left.\begin{aligned} a_{jk} &= \text{breakpoint } k \text{ for variable } x_j \\ w_{jk} &= \text{weight with breakpoint } k \text{ of variable } x_j \end{aligned}\right\} k = 1, 2, \ldots, K_j, j = 1, 2, \ldots, n$$

Then the equivalent mixed problem is

$$\text{Maximize (or minimize) } z = \sum_{j=1}^{n}\sum_{k=1}^{K_j} f_j(a_{jk})w_{jk}$$

subject to

$$\sum_{j=1}^{n}\sum_{k=1}^{K_j} g_{jk}(a_{jk})w_{jk} \leq b_i, \quad i = 1, 2, \ldots, m$$

$$0 \leq w_{j1} \leq y_{j1}, \quad j = 1, 2, \ldots, n$$

$$0 \leq w_{jk} \leq y_{j,k-1} + y_{jk}, \quad k = 2, 3, \ldots, K_j - 1, \; j = 1, 2, \ldots, n$$

$$0 \leq w_{jK_j} \leq y_{j,K_j-1}, \quad j = 1, 2, \ldots, n$$

$$\sum_{k=1}^{K_j-1} y_{jk} = 1, \quad j = 1, 2, \ldots, n$$

$$\sum_{k=1}^{K_j} w_{jk} = 1, \quad j = 1, 2, \ldots, n$$

$$y_{jk} = (0, 1), k = 1, 2, \ldots, K_j, j = 1, 2, \ldots, n$$

The variables for the approximating problem are w_{jk} and y_{jk}.

This formulation shows how any separable problem can be solved, at least in principle, by mixed integer programming. The difficulty is that the number of constraints increases rather rapidly with the number of breakpoints. In particular, the computational feasibility of the procedure is questionable because there are no consistently reliable computer codes for solving large mixed integer programming problems.

Another method for solving the approximate model is the regular simplex method (Chapter 3) using **restricted basis**. In this case the additional constraints involving y_{jk} are dropped. The restricted basis modifies the simplex method optimality condition by selecting the entering variable w_j with the *best* $(z_{jk} - c_{jk})$ such that two w-variables can be positive only if they are adjacent. The process is repeated until the optimality condition is satisfied or until it is impossible to introduce new w_{jk} without violating the restricted basis condition, whichever occurs first. The last tableau gives the *approximate* optimal solution to the problem.

---

[1]It is more accurate to replace the index k with k_j to correspond uniquely to variable j. However, we will not do so to simplify the notation.

The mixed integer programming method yields a global optimum to the approximate problem, but the restricted basis method can only guarantee a local optimum. Additionally, in the two methods, the approximate solution may not be feasible for the original problem. In fact, the approximate model may give rise to additional points that are not part of the solution space of the original problem.

Example 19.2-1

Consider the problem

$$\text{Maximize } z = x_1 + x_2^4$$

subject to

$$3x_1 + 2x_2^2 \leq 9$$

$$x_1, x_2 \geq 0$$

The exact optimum solution to this problem, obtained by AMPL or Solver, is $x_1 = 0$, $x_2 = 2.12132$, and $z^* = 20.25$. To show how the approximating method is used, consider the separable functions

$$\begin{aligned} f_1(x_1) &= x_1 \\ f_2(x_2) &= x_2^4 \\ g_1(x_1) &= 3x_1 \\ g_2(x_2) &= 2x_2^2 \end{aligned}$$

The functions $f_1(x_1)$ and $g_1(x_1)$ remain the same because they are already linear. In this case, x_1 is treated as one of the variables. Considering $f_2(x_2)$ and $g_2(x_2)$, we assume four breakpoints: $a_{2k} = 0, 1, 2,$ and 3 for $k = 1, 2, 3,$ and 4, respectively. Because the value of x_2 cannot exceed 3, it follows that

k	a_{2k}	$f_2(a_{2k}) = a_{2k}^4$	$g_2(a_{2k}) = 2a_{2k}^2$
1	0	0	0
2	1	1	2
3	2	16	8
4	3	81	18

This yields

$$\begin{aligned} f_2(x_2) &\approx w_{21}f_2(a_{21}) + w_{22}f_2(a_{22}) + w_{23}f_2(a_{23}) + w_{24}f_2(a_{24}) \\ &\approx 0w_{21} + 1w_{22} + 16w_{23} + 81w_{24} = w_{22} + 16w_{23} + 81w_{24} \end{aligned}$$

Similarly,

$$g_2(x_2) \approx 2w_{22} + 8w_{23} + 18w_{24}$$

The approximating problem thus becomes

$$\text{Maximize } z = x_1 + w_{22} + 16w_{23} + 81w_{24}$$

subject to

$$3x_1 + 2w_{22} + 8w_{23} + 18w_{24} \leq 9$$

$$w_{21} + w_{22} + w_{23} + w_{24} = 1$$

$$x_1 \geq 0, w_{2k} \geq 0, k = 1, 2, 3, 4$$

The values of w_{2k}, $k = 1, 2, 3, 4$, must satisfy the restricted basis condition.

The initial simplex tableau (with rearranged columns to give a starting solution) is given by

Basic	x_1	w_{22}	w_{23}	w_{24}	s_1	w_{21}	Solution
z	-1	-1	-16	-81	0	0	0
s_1	3	2	8	18	1	0	9
w_{21}	0	1	1	1	0	1	1

The variable s_1 (≥ 0) is a slack. (This problem happened to have an obvious starting solution. In general, one may have to use artificial variables, Section 3.4.)

From the z-row coefficients, w_{24} is the entering variable. Because w_{21} is currently basic and positive, the restricted basis condition dictates that it must leave before w_{24} can enter the solution. By the feasibility condition, s_1 must be the leaving variable, which means that w_{24} cannot enter the solution. The next-best entering variable, w_{23}, requires w_{21} to leave the basic solution, a condition that happens to be satisfied by the feasibility condition. The new tableau thus becomes

Basic	x_1	w_{22}	w_{23}	w_{24}	s_1	w_{21}	Solution
z	-1	15	0	-65	0	16	16
s_1	3	-6	0	10	1	-8	1
w_{23}	0	1	1	1	0	1	1

Next, w_{24} is the entering variable, which is admissible because w_{23} is positive. The simplex method shows that s_1 will leave. Thus,

Basic	x_1	w_{22}	w_{23}	w_{24}	s_1	w_{21}	Solution
z	$\frac{37}{2}$	-24	0	0	$\frac{13}{2}$	-36	$22\frac{1}{2}$
w_{24}	$\frac{3}{10}$	$-\frac{6}{10}$	0	1	$\frac{1}{10}$	$-\frac{8}{10}$	$\frac{1}{10}$
w_{23}	$-\frac{3}{10}$	$\frac{16}{10}$	1	0	$-\frac{1}{10}$	$\frac{18}{10}$	$\frac{9}{10}$

The tableau shows that w_{21} and w_{22} are candidates for the entering variable. Because w_{21} is not adjacent to basic w_{23} or w_{24}, it cannot enter. Similarly, w_{22} cannot enter because w_{24} cannot leave. The last tableau thus is the best restricted-basis solution for the approximate problem.

The optimum solution to the original problem is

$$x_1 = 0$$

$$x_2 \approx 2w_{23} + 3w_{24} = 2\left(\tfrac{9}{10}\right) + 3\left(\tfrac{1}{10}\right) = 2.1$$

$$z = 0 + 2.1^4 = 19.45$$

The value $x_2 = 2.1$ approximately equals the true optimum value ($= 2.12132$).

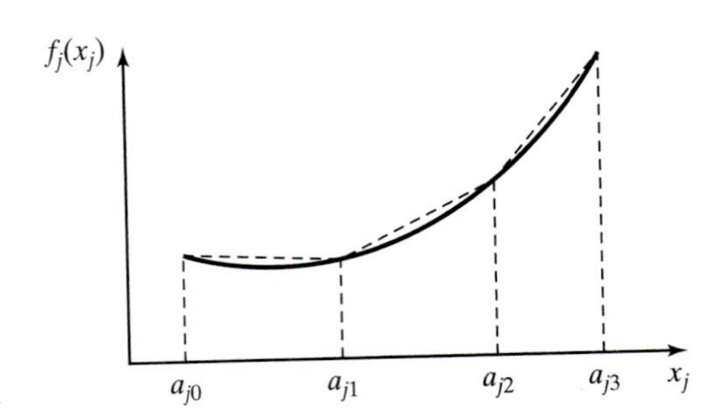

FIGURE 19.4
Piecewise-linear approximation of a convex function

Separable Convex Programming. A special case of separable programming occurs when $g_{ij}(x_j)$ is convex for all i and j, which ensures a convex solution space. Additionally, if $f_j(x_j)$ is convex (minimization) or concave (maximization) for all j, then the problem has a global optimum (see Table 18.2, Section 18.2.2). Under such conditions, the following simplified approximation can be used.

Consider a minimization problem and let $f_j(x_j)$ be as shown in Figure 19.4. The breakpoints of the function $f_j(x_j)$ are $x_j = a_{jk}$, $k = 0, 1, \ldots, K_j$. Let x_{jk} define the increment of the variable x_j in the range $(a_{j,k-1}, a_{jk})$, $k = 1, 2, \ldots, K_j$, and let r_{jk} be the corresponding rate of change (slope of the line segment) in the same range. Then

$$f_j(x_j) \approx \sum_{k=1}^{K_j} r_{jk}x_{jk} + f_j(a_{j0})$$

$$x_j = \sum_{k=1}^{K_j} x_{jk}$$

$$0 \leq x_{jk} \leq a_{jk} - a_{j,k-1}, k = 1, 2, \ldots, K_j$$

The fact that $f_j(x_j)$ is convex ensures that $r_{j1} < r_{j2} < \cdots < r_{jK_j}$. This means that in the minimization problem, for $p < q$, the variable x_{jp} is more attractive than x_{jq} Consequently, x_{jp} will always reach its maximum limit before x_{jq} can assume a positive value.

The convex constraint functions $g_{ij}(x_j)$ are approximated in essentially the same way. Let r_{ijk} be the slope of the kth line segment corresponding to $g_{ij}(x_j)$. It follows that

$$g_{ij}(x_j) \approx \sum_{k=1}^{K_j} r_{ijk}x_{jk} + g_{ij}(a_{j0})$$

The complete problem is thus given by

$$\text{Minimize } z = \sum_{j=1}^{n}\left(\sum_{k=1}^{K_j} r_{jk}x_{jk} + f_j(a_{j0})\right)$$

subject to

$$\sum_{j=1}^{n}\left(\sum_{k=1}^{K_j} r_{ijk}x_{jk} + g_{ij}(a_{j0})\right) \leq b_i, i = 1, 2, \ldots, m$$

$$0 \leq x_{jk} \leq a_{jk} - a_{j,k-1}, k = 1, 2, \ldots, K_j, \quad j = 1, 2, \ldots, n$$

where

$$r_{jk} = \frac{f_j(a_{jk}) - f_j(a_{j,k-1})}{a_{jk} - a_{j,k-1}}$$

$$r_{ijk} = \frac{g_{ij}(a_{jk}) - g_{ij}(a_{j,k-1})}{a_{jk} - a_{j,k-1}}$$

The maximization problem is treated essentially the same way. In this case, $r_{j1} > r_{j2} > \cdots > r_{jK_j}$, which means that, for $p < q$, the variable x_{jp} will always reach its maximum value before x_{jq} is allowed to assume a positive value (see Problem 7, Set 19.2a, for proof).

The new problem can be solved by the simplex method with upper-bounded variables (Section 7.3). The restricted basis concept is not needed because the convexity (concavity) of the functions guarantees correct selection of basic variables.

Example 19.2-2

Consider the problem

$$\text{Maximize } z = x_1 - x_2$$

subject to

$$3x_1^4 + x_2 \le 243$$

$$x_1 + 2x_2^2 \le 32$$

$$x_1 \ge 2.1$$

$$x_2 \ge 3.5$$

The separable functions of this problem are

$$f_1(x_1) = x_1, \quad f_2(x_2) = -x_2$$

$$g_{11}(x_1) = 3x_1^4, \quad g_{12}(x_2) = x_2$$

$$g_{21}(x_1) = x_1, \quad g_{22}(x_2) = 2x_2^2$$

These functions satisfy the convexity condition required for the minimization problems. The functions $f_1(x_1)$, $f_2(x_2)$, $g_{12}(x_2)$, and $g_{21}(x_1)$ are already linear and need not be "approximated."

The ranges of the variables x_1 and x_2 (estimated from the constraints) are $0 \le x_1 \le 3$ and $0 \le x_2 \le 4$. Let $K_1 = 3$ and $K_2 = 4$. The slopes corresponding to the separable functions are determined as follows.

For $j = 1$,

k	a_{1k}	$g_{11}(a_{1k}) = 3a_{1k}^4$	r_{11k}	x_{1k}
0	0	0	—	—
1	1	3	3	x_{11}
2	2	48	45	x_{12}
3	3	243	195	x_{13}

For $j = 2$,

k	a_{2k}	$g_{22}(a_{2k}) = 2a_{2k}^2$	r_{22k}	x_{2k}
0	0	0	—	—
1	1	2	2	x_{21}
2	2	8	6	x_{22}
3	3	18	10	x_{23}
4	4	32	14	x_{24}

The complete problem then becomes

$$\text{Maximize } z = x_1 - x_2$$

subject to

$$3x_{11} + 45x_{12} + 195x_{13} + x_2 \leq 243 \quad (19.1)$$

$$x_1 + 2x_{21} + 6x_{22} + 10x_{23} + 14x_{24} \leq 32 \quad (19.2)$$

$$x_1 \geq 2.1 \quad (19.3)$$

$$x_2 \geq 3.5 \quad (19.4)$$

$$x_{11} + x_{12} + x_{13} - x_1 = 0 \quad (19.5)$$

$$x_{21} + x_{22} + x_{23} + x_{24} - x_2 = 0 \quad (19.6)$$

$$0 \leq x_{1k} \leq 1, \quad k = 1, 2, 3 \quad (19.7)$$

$$0 \leq x_{2k} \leq 1, \quad k = 1, 2, 3, 4 \quad (19.8)$$

$$x_1, x_2 \geq 0$$

Constraints 5 and 6 are needed to maintain the relationship between the original and new variables. The optimum solution is

$$z = -.52, x_1 = 2.98, x_2 = 3.5, x_{11} = x_{12} = 1, x_{13} = .98, x_{21} = x_{22} = x_{23} = 1, x_{24} = .5$$

AMPL Moment

The modeling of nonlinear problems in AMPL is very much the same as in linear problems. Obtaining the solution is an entirely different matter, however, because of the "unpredictable" behavior of the nonlinear functions. Figure 19.5 gives the AMPL model of the original problem of Example 19.2-2 (file amplEx19.2-2.txt). The only deviation from LP (other than the nonlinearity, of course) is that you may need to specify "appropriate" initial values for the variables to get the solution iteration to converge. In Figure 19.5, the arbitrary initial values $x_1 = 10$ and $x_2 = 10$ are specified by appending `:=10` to the

```
var x1>=0  :=10;  #inital value = 10
var x2>=0  :=10;  #initial value = 10

maximize z: x1-x2;
subject to c1: 3*x1^4+x2<=243;
subject to c2: x1+2*x2^2<=32;
subject to c3: x1>=2.1;
subject to c4: x2>=3.5;

solve;
display z,x1,x2;
```

FIGURE 19.5

AMPL model of Example 19.2-2

definition of the two variables. If you do not specify initial values at all, AMPL will not reach the optimum solution and will print the message "too many major iterations." Although a solution is given in this case, it usually is not correct. In essence, the most logical way to deal with a nonlinear problem is to specify different initial values for the variables and then decide if a consensus can be reached regarding the optimum solution.

Remark. AMPL provides a special syntax for handling convex separable programs. This representation appears to work more reliably if the nonlinearity occurs in the objective function only. Else, the behavior is quite erratic and, indeed, AMPL may claim that the constraints are infeasible when in fact they are not.

PROBLEM SET 19.2A

1. Approximate the following problem as a mixed integer program.

$$\text{Maximize } z = e^{-x_1} + x_1 + (x_2 + 1)^2$$

subject to

$$x_1^2 + x_2 \leq 3$$
$$x_1, x_2 \geq 0$$

***2.** Repeat Problem 1 using the restricted basis method. Then find the optimal solution.

3. Consider the problem

$$\text{Maximize } z = x_1x_2x_3$$

subject to

$$x_1^2 + x_2 + x_3 \leq 4$$
$$x_1, x_2, x_3 \geq 0$$

Approximate the problem as a linear program for use with the restricted basis method.

***4.** Show how the following problem can be made separable.

$$\text{Maximize } z = x_1x_2 + x_3 + x_1x_3$$

subject to

$$x_1x_2 + x_2 + x_1x_3 \leq 10$$
$$x_1, x_2x_3 \geq 0$$

5. Show how the following problem can be made separable.

$$\text{Minimize } z = e^{2x_1+x_2^2} + (x_3 - 2)^2$$

subject to

$$x_1 + x_2 + x_3 \leq 6$$
$$x_1, x_2 x_3 \geq 0$$

6. Show how the following problem can be made separable.

$$\text{Maximize } z = e^{x_1 x_2} + x_2^2 x_3 + x_4$$

subject to

$$x_1 + x_2 x_3 + x_3 \leq 10$$
$$x_1, x_2, x_3 \geq 0$$
$$x_4 \text{ unrestricted in sign}$$

7. Show that in separable convex programming, it is never optimal to have $x_{ki} > 0$ when $x_{k-1,i}$ is not at its upper bound.

8. Solve as a separable convex programming problem.

$$\text{Minimize } z = x_1^4 + x_2 + x_3^2$$

subject to

$$x_1^2 + x_2 + x_3^2 \leq 4$$
$$|x_1 + x_2| \leq 0$$
$$x_1, x_3 \geq 0$$
$$x_2 \text{ unrestricted in sign}$$

9. Solve the following as a separate convex programming problem.

$$\text{Minimize } z = (x_1 - 2)^2 + 4(x_2 - 6)^2$$

subject to

$$6x_1 + 3(x_2 + 1)^2 \leq 12$$
$$x_1, x_2 \geq 0$$

19.2.2 Quadratic Programming

A quadratic programming model is defined as

$$\text{Maximize } z = \mathbf{CX} + \mathbf{X}^T\mathbf{DX}$$

subject to

$$\mathbf{AX} \leq \mathbf{b}, \mathbf{X} \geq \mathbf{0}$$

where

$$\mathbf{X} = (x_1, x_2, \ldots, x_n)^T$$
$$\mathbf{C} = (c_1, c_2, \ldots, c_n)$$
$$\mathbf{b} = (b_1, b_2, \ldots, b_m)^T$$
$$\mathbf{A} = \begin{pmatrix} a_{11} & \cdots & a_{1n} \\ \vdots & \vdots & \vdots \\ a_{m1} & \cdots & a_{mn} \end{pmatrix}$$
$$\mathbf{D} = \begin{pmatrix} d_{11} & \cdots & d_{1n} \\ \vdots & \vdots & \vdots \\ d_{n1} & \cdots & d_{nn} \end{pmatrix}$$

The function $\mathbf{X^TDX}$ defines a quadratic from (Section D.3 on the CD). The matrix $\mathbf{D}$ is assumed symmetric and negative definite. This means that z is strictly concave. The constraints are linear, which guarantees a convex solution space.

The solution to this problem is based on the KKT necessary conditions. Because z is strictly concave and the solution space is a convex set, these conditions (as shown in Table 18.2, Section 18.2.2) are also sufficient for a global optimum.

The quadratic programming problem will be treated for the maximization case. Conversion to minimization is straightforward. The problem may be written as

$$\text{Maximize } z = \mathbf{CX} + \mathbf{X}^T\mathbf{DX}$$

subject to

$$\mathbf{G}(\mathbf{X}) = \begin{pmatrix} \mathbf{A} \\ -\mathbf{I} \end{pmatrix}\mathbf{X} - \begin{pmatrix} \mathbf{b} \\ \mathbf{0} \end{pmatrix} \leq \mathbf{0}$$

Let

$$\boldsymbol{\lambda} = (\lambda_1, \lambda_2, \ldots, \lambda_m)^T$$
$$\mathbf{U} = (\mu_1, \mu_2, \ldots, \mu_n)^T$$

be the Lagrange multipliers corresponding to constraints $\mathbf{AX} - \mathbf{b} \leq \mathbf{0}$ and $-\mathbf{X} \leq \mathbf{0}$, respectively. Application of the KKT conditions yields

$$\boldsymbol{\lambda} \geq \mathbf{0}, \mathbf{U} \geq \mathbf{0}$$
$$\nabla z - (\boldsymbol{\lambda}^T, \mathbf{U}^T)\, \nabla\mathbf{G}(\mathbf{X}) = \mathbf{0}$$
$$\lambda_i\left(b_i - \sum_{j=1}^{n} a_{ij}x_j\right) = 0, i = 1, 2, \ldots, m$$
$$\mu_j x_j = 0, \quad j = 1, 2, \ldots, n$$
$$\mathbf{AX} \leq \mathbf{b}$$
$$-\mathbf{X} \leq \mathbf{0}$$

Now

$$\nabla z = \mathbf{C} + 2\mathbf{X}^T\mathbf{D}$$

$$\nabla \mathbf{G}(\mathbf{X}) = \begin{pmatrix} \mathbf{A} \\ -\mathbf{I} \end{pmatrix}$$

Let $\mathbf{S} = \mathbf{b} - \mathbf{AX} \geq \mathbf{0}$ be the slack variables of the constraints. The conditions reduce to

$$-2\mathbf{X}^T\mathbf{D} + \boldsymbol{\lambda}^T\mathbf{A} - \mathbf{U}^T = \mathbf{C}$$

$$\mathbf{AX} + \mathbf{S} = \mathbf{b}$$

$$\mu_j x_j = 0 = \lambda_i S_i \quad \text{for all } i \text{ and } j$$

$$\boldsymbol{\lambda}, \mathbf{U}, \mathbf{X}, \mathbf{S} \geq \mathbf{0}$$

Because $\mathbf{D}^T = \mathbf{D}$, the transpose of the first set of equations can be written as

$$-2\mathbf{DX} + \mathbf{A}^T\boldsymbol{\lambda} - \mathbf{U} = \mathbf{C}^T$$

Hence, the necessary conditions may be combined as

$$\begin{pmatrix} -2\mathbf{D} & \mathbf{A}^T & -\mathbf{I} & \mathbf{0} \\ \mathbf{A} & \mathbf{0} & \mathbf{0} & \mathbf{I} \end{pmatrix} \begin{pmatrix} \mathbf{X} \\ \boldsymbol{\lambda} \\ \mathbf{U} \\ \mathbf{S} \end{pmatrix} = \begin{pmatrix} \mathbf{C}^T \\ \mathbf{b} \end{pmatrix}$$

$$\mu_j x_j = 0 = \lambda_i S_i, \quad \text{for all } i \text{ and } j$$

$$\boldsymbol{\lambda}, \mathbf{U}, \mathbf{X}, \mathbf{S} \geq \mathbf{0}$$

Except for the conditions $\mu_j x_j = 0 = \lambda_i S_i$, the remaining equations are linear functions in $\mathbf{X}$, $\boldsymbol{\lambda}$, $\mathbf{U}$, and $\mathbf{S}$. Thus, the problem is equivalent to solving a set of linear equations with the additional conditions $\mu_j x_j = 0 = \lambda_i S_i$. Because z is strictly concave and the solution space is convex, the *feasible* solution satisfying all these conditions must yield a unique optimum solution.

The solution of the system is obtained by using phase I of the two-phase method (Section 3.4.2). The only restriction is to satisfy the conditions $\lambda_i S_i = 0 = \mu_j x_j$. This means that λ_i and s_i cannot be positive simultaneously, and neither can μ_j and x_j. This is the same idea of the **restricted basis** used in Section 19.2.1.

Phase I will render all the artificial variables equal to zero provided the problem has a feasible space.

Example 19.2-3

Consider the problem

$$\text{Maximize } z = 4x_1 + 6x_2 - 2x_1^2 - 2x_1x_2 - 2x_2^2$$

subject to

$$x_1 + 2x_2 \leq 2$$

$$x_1, x_2 \geq 0$$

This problem can be put in matrix form as follows:

$$\text{Maximize } z = (4, 6)\begin{pmatrix} x_1 \\ x_2 \end{pmatrix} + (x_1, x_2)\begin{pmatrix} -2 & -1 \\ -1 & -2 \end{pmatrix}\begin{pmatrix} x_1 \\ x_2 \end{pmatrix}$$

subject to

$$(1, 2)\begin{pmatrix} x_1 \\ x_2 \end{pmatrix} \leq 2$$

$$x_1, x_2 \geq 0$$

The KKT conditions are given as

$$\begin{pmatrix} 4 & 2 & 1 & -1 & 0 & 0 \\ 2 & 4 & 2 & 0 & -1 & 0 \\ 1 & 2 & 0 & 0 & 0 & 1 \end{pmatrix}\begin{pmatrix} x_1 \\ x_2 \\ \lambda_1 \\ \mu_1 \\ \mu_2 \\ s_1 \end{pmatrix} = \begin{pmatrix} 4 \\ 6 \\ 2 \end{pmatrix}, \mu_1 x_1 = \mu_2 x_2 = \lambda_1 s_1 = 0$$

The initial tableau for phase 1 is obtained by introducing the artificial variables R_1 and R_2 and updating the objective row. Thus

Basic	x_1	x_2	λ_1	μ_1	μ_2	R_1	R_2	s_1	Solution
r	6	6	3	−1	−1	0	0	0	10
R_1	4	2	1	−1	0	1	0	0	4
R_2	2	4	2	0	−1	0	1	0	6
s_1	1	2	0	0	0	0	0	1	2

Iteration 1 Because $\mu_1 = 0$, the most promising entering variable x_1 can be made basic with R_1 as the leaving variable. This yields the following tableau:

Basic	x_1	x_2	λ_1	μ_1	μ_2	R_1	R_2	s_1	Solution
R	0	3	$\frac{3}{2}$	$\frac{1}{2}$	−1	$-\frac{3}{2}$	0	0	4
x_1	1	$\frac{1}{2}$	$\frac{1}{4}$	$-\frac{1}{4}$	0	$\frac{1}{4}$	0	0	1
R_2	0	3	$\frac{3}{2}$	$\frac{1}{2}$	−1	$-\frac{1}{2}$	1	0	4
s_1	0	$\frac{3}{2}$	$-\frac{1}{4}$	$\frac{1}{4}$	0	$-\frac{1}{4}$	0	1	1

Iteration 2 The most promising variable x_2 can be made basic because $\mu_2 = 0$. This gives

Basic	x_1	x_2	λ_1	μ_1	μ_2	R_1	R_2	s_1	Solution
r	0	0	2	0	−1	−1	0	−2	2
x_1	1	0	$\frac{1}{3}$	$-\frac{1}{3}$	0	$\frac{1}{3}$	0	$-\frac{1}{3}$	$\frac{2}{3}$
R_1	0	0	2	0	−1	0	1	−2	2
x_1	0	1	$-\frac{1}{6}$	$\frac{1}{6}$	0	$-\frac{1}{6}$	0	$\frac{2}{3}$	$\frac{2}{3}$

Iteration 3 Because $s_1 = 0$, λ_1 can be introduced into the solution. This yields

Basic	x_1	x_2	λ_1	μ_1	μ_2	R_1	R_2	s_1	Solution
r	0	0	0	0	0	-1	-1	0	0
x_1	1	0	0	$-\frac{1}{3}$	$\frac{1}{6}$	$\frac{1}{3}$	$-\frac{1}{6}$	0	$\frac{1}{3}$
λ_1	0	0	1	0	$-\frac{1}{2}$	0	$\frac{1}{2}$	-1	1
x_2	0	1	0	$\frac{1}{6}$	$-\frac{1}{12}$	$-\frac{1}{6}$	$\frac{1}{12}$	$\frac{1}{2}$	$\frac{5}{6}$

The last tableau gives the optimal solution for phase I. Because $r = 0$, the solution, $x_1^* = \frac{1}{3}$, $x_2^* = \frac{5}{6}$, is feasible. The optimal value of z, computed from the original problem, is 4.16.

Solver Moment

Figure 19.6 provides the solution for Example 19.2-3 using Solver (file excelQP.xls). The data are entered in a manner similar to the one used in linear programming (see Section 2.4.2). The main difference occurs in the way the nonlinear functions are entered. Specifically, in Example 19.2-3, the nonlinear objective function

$$z = 4x_1 + 6x_2 - 2x_1^2 - 2x_1x_2 - 2x_2^2$$

FIGURE 19.6

Solver solution of the quadratic program of Example 19.2-3 (file excelQP.xls)

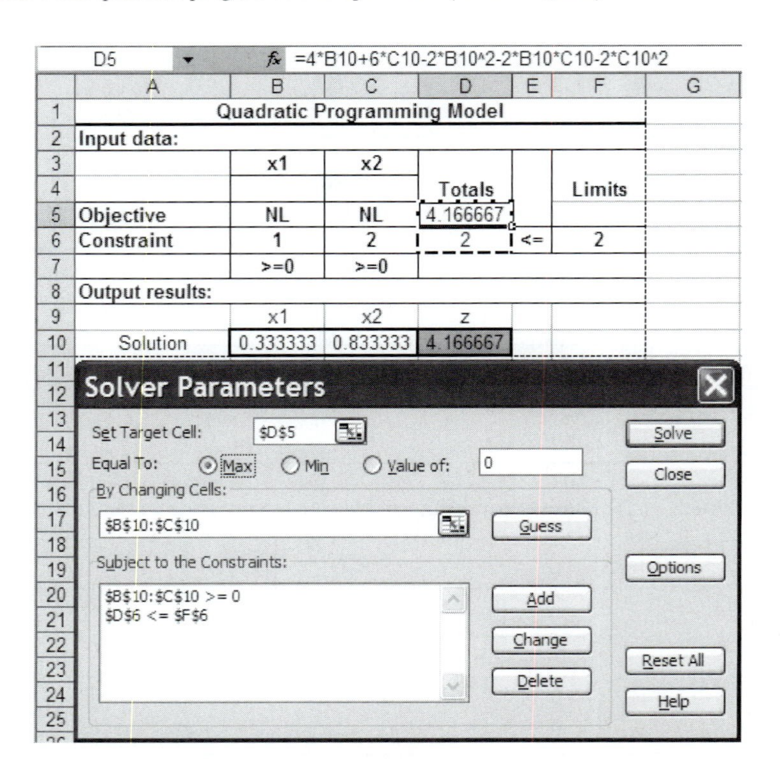

is entered in target cell D5 as

```
=4*B10+6*c10-2*B10^2-2*B10*C10-2*C10^2
```

Here, the changing cells are B10:C10 [$\equiv (x_1, x_2)$]. Notice that cells B5:C5 are not used at all in the model. For readability, we entered the symbol NL to indicate that the associated constraint is nonlinear. Also, you can specify the nonnegativity of the variables either in the Options dialogue box or by adding explicit nonnegativity constraints.

PROBLEM SET 19.2B

*__1.__ Consider the problem

$$\text{Maximize } z = 6x_1 + 3x_2 - 4x_1x_2 - 2x_1^2 - 3x_2^2$$

subject to

$$x_1 + x_2 \leq 1$$
$$2x_1 + 3x_2 \leq 4$$
$$x_1, x_2 \geq 0$$

Show that z is strictly concave and then solve the problem using the quadratic programming algorithm.

*__2.__ Consider the problem:

$$\text{Minimize } z = 2x_1^2 + 2x_2^2 + 3x_3^2 + 2x_1x_2 + 2x_2x_3 + x_1 - 3x_2 - 5x_3$$

subject to

$$x_1 + x_2 + x_3 \geq 1$$
$$3x_1 + 2x_2 + x_3 \leq 6$$
$$x_1, x_2, x_3 \geq 0$$

Show that z is strictly convex and then solve by the quadratic programming algorithm.

19.2.3 Chance-Constrained Programming

Chance-constrained programming deals with situations in which the parameters of the constraints are random variables and the constraints are realized with a minimum probability. Mathematically, the problem is defined as

$$\text{Maximize } z = \sum_{j=1}^{n} c_j x_j$$

subject to

$$P\left\{\sum_{j=1}^{n} a_{ij}x_j \leq b_i\right\} \geq 1 - \alpha_i, i = 1, 2, \ldots, m, x_j \geq 0, \text{for all } j$$

The parameters a_{ij} and b_i are random variables and constraint i is realized with a minimum probability of $1 - \alpha_i$, $0 < \alpha_i < 1$.

Three cases are considered:

1. Only a_{ij} is random for all i and j.
2. Only b_i is random for all i.
3. Both a_{ij} and b_i are random for all i and j.

In all three cases, it is assumed that the parameters are normally distributed with known means and variances.

Case 1. Each a_{ij} is normally distributed with mean $E\{a_{ij}\}$, variance $\text{var}\{a_{ij}\}$, and $\text{cov}\{a_{ij}, a_{i'j'}\}$ of a_{ij} and $a_{i'j'}$.

Consider

$$P\left\{\sum_{j=1}^{n} a_{ij}x_j \leq b_i\right\} \geq 1 - \alpha_i$$

and define

$$h_i = \sum_{j=1}^{n} a_{ij}x_j$$

Then h_i is normally distributed with

$$E\{h_i\} = \sum_{j=1}^{n} E\{a_{ij}\}x_j$$

$$\text{var}\{h_i\} = \mathbf{X}^T\mathbf{D}_i\mathbf{X}$$

where

$$\mathbf{X} = (x_1, \ldots, x_n)^T$$

$$\mathbf{D}_i = i\text{th covariance matrix}$$

$$= \begin{pmatrix} \text{var}\{a_{i1}\} & \ldots & \text{cov}\{a_{i1}, a_{in}\} \\ \vdots & \vdots & \vdots \\ \text{cov}\{a_{in}, a_{i1}\} & \ldots & \text{var}\{a_{in}\} \end{pmatrix}$$

Now

$$P\{h_i \leq b_i\} = P\left\{\frac{h_i - E\{h_i\}}{\sqrt{\text{var}\{h_i\}}} \leq \frac{b_i - E\{h_i\}}{\sqrt{\text{var}\{h_i\}}}\right\} \geq 1 - \alpha_i$$

where $\frac{h_i - E\{h_i\}}{\sqrt{\text{var}\{h_i\}}}$ is standard normal with mean zero and variance one. This means that

$$P\{h_i \leq b_i\} = F\left(\frac{b_i - E\{h_i\}}{\sqrt{\text{var}\{h_i\}}}\right)$$

where F represents the CDF of the standard normal distribution.

Let K_{α_i} be the standard normal value such that

$$F(K_{\alpha_i}) = 1 - \alpha_i$$

Then the statement $P\{h_i \le b_i\} \ge 1 - \alpha_i$ is realized if and only if

$$\frac{b_i - E\{h_i\}}{\sqrt{\text{var}\{h_i\}}} \ge K_{\alpha_i}$$

This yields the following nonlinear deterministic constraint:

$$\sum_{j=1}^{n} E\{a_{ij}\}x_j + K_{\alpha_i}\sqrt{\mathbf{X}^T\mathbf{D}_i\mathbf{X}} \le b_i$$

For the special case where the parameters a_{ij} are independent,

$$\text{cov}\{a_{ij}, a_{i'j'}\} = 0$$

and the last constraint reduces to

$$\sum_{j=1}^{n} E\{a_{ij}\}x_j + K_{\alpha_i}\sqrt{\sum_{j=1}^{n}\text{var}\{a_{ij}\}x_j^2} \le b_i$$

This constraint can be put in the separable programming form (Section 19.2.1) by using the substitution

$$y_i = \sqrt{\sum_{j=1}^{n}\text{var}\{a_{ij}\}x_j^2}, \text{ for all } i$$

Thus, the original constraint is equivalent to

$$\sum_{j=1}^{n} E\{a_{ij}\}x_j + K_{\alpha_i}y_i \le b_i$$

and

$$\sum_{j=1}^{n}\text{var}\{a_{ij}\}x_j^2 - y_i^2 = 0$$

Case 2. Only b_i is normal with mean $E\{b_i\}$ and variance $\text{var}\{b_i\}$. The analysis is similar to that of case 1. Consider the stochastic constraint

$$P\left\{b_i \ge \sum_{j=1}^{n} a_{ij}x_j\right\} \ge \alpha_i$$

As in case 1,

$$P\left\{\frac{b_i - E\{b_i\}}{\sqrt{\text{var}\{b_i\}}} \ge \frac{\sum_{j=1}^{n} a_{ij}x_j - E\{b_i\}}{\sqrt{\text{var}\{b_i\}}}\right\} \ge \alpha_i$$

This can hold only if

$$\frac{\sum_{j=1}^{n} a_{ij}x_j - E\{b_i\}}{\sqrt{\text{var}\{b_i\}}} \le K_{\alpha_i}$$

Thus, the stochastic constraint is equivalent to the deterministic linear constraint

$$\sum_{j=1}^{n} a_{ij}x_j \le E\{b_i\} + K_{\alpha_i}\sqrt{\text{var}\{b_i\}}$$

Case 3. In this case all a_{ij} and b_i are normal random variables. Consider the constraint

$$\sum_{j=1}^{n} a_{ij}x_j \le b_i$$

This may be written

$$\sum_{j=1}^{n} a_{ij}x_j - b_i \le 0$$

Because all a_{ij} and b_i are normal, $\sum_{j=1}^{n} a_{ij}x_j - b_i$ is also normal. This shows that the chance constraint reduces to the situation in case 1 and is treated in a similar manner.

Example 19.2-4

Consider the chance-constrained problem

$$\text{Maximize } z = 5x_1 + 6x_2 + 3x_3$$

subject to

$$P\{a_{11}x_1 + a_{12}x_2 + a_{13}x_3 \le 8\} \ge .95$$
$$P\{5x_1 + x_2 + 6x_3 \le b_2\} \ge .10$$
$$x_1, x_2, x_3 \ge 0$$

Assume that the parameters a_{1j}, $j = 1, 2, 3$, are independent and normally distributed random variables with the following means and variances:

$$E\{a_{11}\} = 1, E\{a_{12}\} = 3, E\{a_{13}\} = 9$$
$$\text{var}\{a_{11}\} = 25, \text{var}\{a_{12}\} = 16, \text{var}\{a_{13}\} = 4$$

The parameter b_2 is normally distributed with mean 7 and variance 9.

From standard normal tables in Appendix B (or excelStatTables.xls),

$$K_{\alpha_1} = K_{.05} \approx 1.645, \quad K_{\alpha_2} = K_{.10} \approx 1.285$$

For the first constraint, the equivalent deterministic constraint is

$$x_1 + 3x_2 + 9x_3 + 1.645\sqrt{25x_1^2 + 16x_2^2 + 4x_3^2} \le 8$$

and for the second constraint

$$5x_1 + x_2 + 6x_3 \le 7 + 1.285(3) = 10.855$$

The resulting problem can be solved as a nonlinear program using AMPL or Solver, or it can be converted to a separable program as follows:

$$y^2 = 25x_1^2 + 16x_2^2 + 4x_3^2$$

The problem becomes

$$\text{Maximize } z = 5x_1 + 6x_2 + 3x_3$$

subject to

$$x_1 + 3x_2 + 9x_3 + 1.645y \leq 8$$
$$25x_1^2 + 16x_2^2 + 4x_3^2 - y^2 = 0$$
$$5x_1 + x_2 + 6x_3 \leq 10.855$$
$$x_1, x_2, x_3, y \geq 0$$

which can be solved by separable programming.

Solver Moment

Excel optimum solution of the nonlinear problem of Example 19.2-4 is given in Figure 19.7 (file excelCCP.xls). Only the left-hand side of constraint 2 is nonlinear, and it is entered in cell F7 as

```
=25*B12^2+16*c12^2+4*D12^2-E12^2
```

PROBLEM SET 19.2C

*1. Convert the following stochastic problem into an equivalent deterministic model.

$$\text{Maximize } z = x_1 + 2x_2 + 5x_3$$

subject to

$$P\{a_1x_1 + 3x_2 + a_3x_3 \leq 10\} \geq 0.9$$
$$P\{7x_1 + 5x_2 + x_3 \leq b_2\} \geq 0.1$$
$$x_1, x_2, x_3 \geq 0$$

Assume that a_1 and a_3 are independent and normally distributed random variables with means $E\{a_1\} = 2$ and $E\{a_3\} = 5$ and variances $\text{var}\{a_1\} = 9$ and $\text{var}\{a_3\} = 16$, and that b_2 is normally distributed with mean 15 and variance 25.

2. Consider the following stochastic programming model:

$$\text{Maximize } z = x_1 + x_2^2 + x_3$$

subject to

$$P\{x_1^2 + a_2x_2^3 + a_3\sqrt{x_3} \leq 10\} \geq 0.9$$
$$x_1, x_2, x_3 \geq 0$$

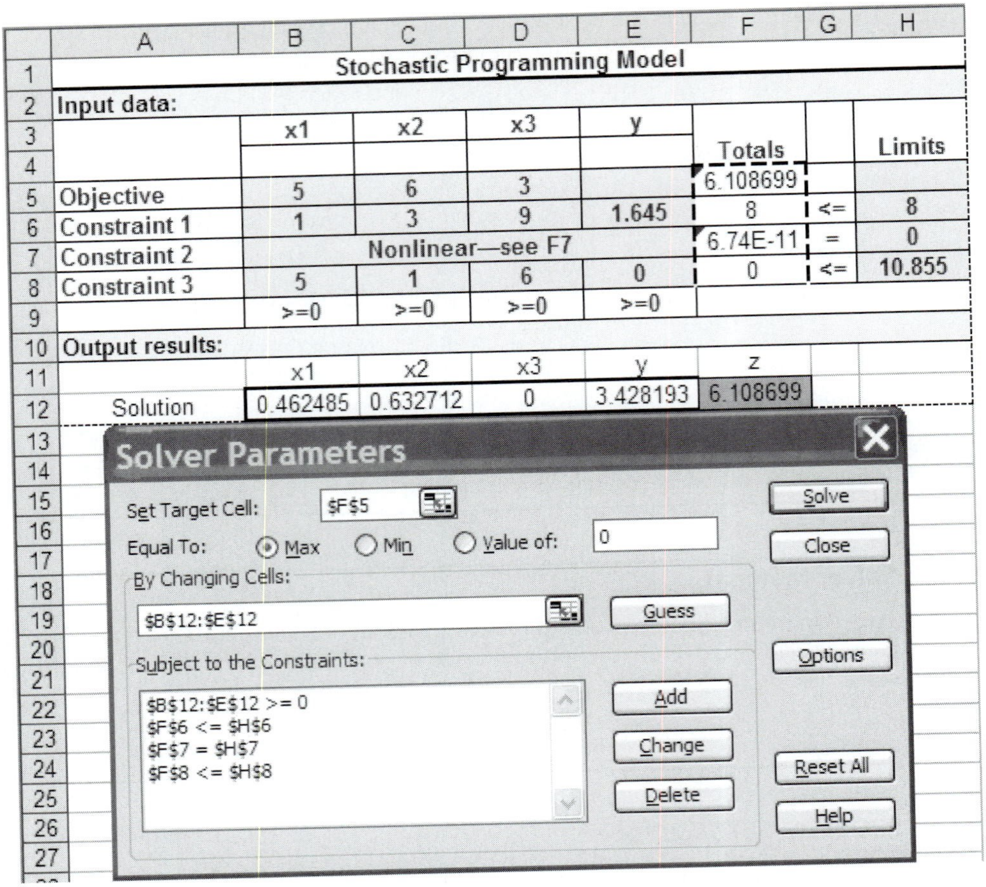

	A	B	C	D	E	F	G	H
1			Stochastic Programming Model					
2	**Input data:**							
3		x1	x2	x3	y			
4						Totals		Limits
5	Objective	5	6	3		6.108699		
6	Constraint 1	1	3	9	1.645	8	<=	8
7	Constraint 2		Nonlinear—see F7			6.74E-11	=	0
8	Constraint 3	5	1	6	0	0	<=	10.855
9		>=0	>=0	>=0	>=0			
10	**Output results:**							
11		x1	x2	x3	y	z		
12	Solution	0.462485	0.632712	0	3.428193	6.108699		

Solver Parameters

Set Target Cell: $F$5

Equal To: (•) Max ( ) Min ( ) Value of: 0

By Changing Cells:

$B$12:$E$12

Subject to the Constraints:

$B$12:$E$12 >= 0
$F$6 <= $H$6
$F$7 = $H$7
$F$8 <= $H$8

Solve | Close | Guess | Options | Add | Change | Delete | Reset All | Help

FIGURE 19.7
Solver solution of the chance-constrained program of Example 19.2-4 (file excelCCP.xls)

The parameters a_2 and a_3 are independent and normally distributed random variables with means 5 and 2, and variance 16 and 25, respectively. Convert the problem into a (deterministic) separable programming form.

19.2.5 Linear Combinations Method

This method deals with the following problem in which all constraints are linear:

$$\text{Maximize } z = f(\mathbf{X})$$

subject to

$$\mathbf{AX} \leq \mathbf{b}, \mathbf{X} \geq \mathbf{0}$$

The procedure is based on the steepest-ascent (gradient) method (Section 19.1.2). However, the direction specified by the gradient vector may not yield a feasible solution for the constrained problem. Also, the gradient vector will not necessarily be null at the optimum (constrained) point. The steepest ascent method thus must be modified to handle the constrained case.

Let $\mathbf{X}_k$ be the *feasible* trial point at iteration k. The objective function $f(\mathbf{X})$ can be expanded in the neighborhood of $\mathbf{X}_k$ using Taylor's series. This gives

$$f(\mathbf{X}) \approx f(\mathbf{X}_k) + \nabla f(\mathbf{X}_k)(\mathbf{X} - \mathbf{X}_k) = (f(\mathbf{X}_k) - \nabla f(\mathbf{X}_k)\mathbf{X}_k) + \nabla f(\mathbf{X}_k)\mathbf{X}$$

The procedure calls for determining a feasible point $\mathbf{X} = \mathbf{X}^*$ such that $f(\mathbf{X})$ is maximized subject to the (linear) constraints of the problem. Because $f(\mathbf{X}_k) - \nabla f(\mathbf{X}_k)\mathbf{X}_k$ is a constant, the problem for determining $\mathbf{X}^*$ reduces to solving the following linear program:

$$\text{Maximize } w_k(\mathbf{X}) = \nabla f(\mathbf{X}_k)\mathbf{X}$$

subject to

$$\mathbf{AX} \leq \mathbf{b}, \mathbf{X} \geq \mathbf{0}$$

Given that w_k is constructed from the gradient of $f(\mathbf{X})$ at $\mathbf{X}_k$, an improved solution point can be secured if and only if $w_k(\mathbf{X}^*) > w_k(\mathbf{X}_k)$. From Taylor's expansion, the condition does not guarantee that $f(\mathbf{X}^*) > f(\mathbf{X}_k)$ unless $\mathbf{X}^*$ is in the neighborhood of $\mathbf{X}_k$. However, given $w_k(\mathbf{X}^*) > w_k(\mathbf{X}_k)$, there must exist a point $\mathbf{X}_{k+1}$ on the line segment $(\mathbf{X}_k, \mathbf{X}^*)$ such that $f(\mathbf{X}_{k+1}) > f(\mathbf{X}_k)$. The objective is to determine $\mathbf{X}_{k+1}$. Define

$$\mathbf{X}_{k+1} = (1 - r)\mathbf{X}_k + r\mathbf{X}^* = \mathbf{X}^k + r(\mathbf{X}^* - \mathbf{X}_k), 0 < r \leq 1$$

This means that $\mathbf{X}_{k+1}$ is a **linear combination** of $\mathbf{X}_k$ and $\mathbf{X}^*$. Because $\mathbf{X}_k$ and $\mathbf{X}^*$ are two feasible points in a *convex* solution space, $\mathbf{X}_{k+1}$ is also feasible. In terms of the steepest-ascent method (Section 19.1.2), the parameter r represents step size.

The point $\mathbf{X}_{k+1}$ is determined such that $f(\mathbf{X})$ is maximized. Because $\mathbf{X}_{k+1}$ is a function of r only, $\mathbf{X}_{k+1}$ is determined by maximizing

$$h(r) = f(\mathbf{X}_k + r(\mathbf{X}^* - \mathbf{X}_k))$$

The procedure is repeated until, at the kth iteration, $w_k(\mathbf{X}^*) \leq w_k(\mathbf{X}_k)$. At this point, no further improvements are possible, and the process terminates with $\mathbf{X}_k$ as the best solution point.

The linear programming problems generated at the successive iterations differ only in the coefficients of the objective function. Post-optimal analysis procedures presented in Section 4.5 thus may be used to carry out calculations efficiently.

Example 19.2-5

Consider the quadratic programming of Example 19.2-3.

$$\text{Maximize } f(\mathbf{X}) = 4x_1 + 6x_2 - 2x_1^2 - 2x_1x_2 - 2x_2^2$$

subject to

$$x_1 + 2x_2 \leq 2$$

$$x_1, x_2 \geq 0$$

Let the initial trial point be $\mathbf{X}_0 = \left(\frac{1}{2}, \frac{1}{2}\right)$, which is feasible. Now

$$\nabla f(\mathbf{X}) = (4 - 4x_1 - 2x_2, 6 - 2x_1 - 4x_2)$$

Iteration 1

$$\nabla f(\mathbf{X}_0) = (1, 3)$$

The associated linear program maximizes $w_1 = x_1 + 3x_2$ subject to the constraints of the original problem. This gives the optimal solution $\mathbf{X}^* = (0, 1)$. The values of w_1 at $\mathbf{X}_0$ and $\mathbf{X}^*$ equal 2 and 3, respectively. Hence, a new trial point is determined as

$$\mathbf{X}_1 = \left(\frac{1}{2}, \frac{1}{2}\right) + r\left[(0, 1) - \left(\frac{1}{2}, \frac{1}{2}\right)\right] = \left(\frac{1 - r}{2}, \frac{1 + r}{2}\right)$$

The maximization of

$$h(r) = f\left(\frac{1 - r}{2}, \frac{1 + r}{2}\right)$$

yields $r_1 = 1$. Thus $\mathbf{X}_1 = (0, 1)$ with $f(\mathbf{X}_1) = 4$.

Iteration 2

$$\nabla f(\mathbf{X}_1) = (2, 2)$$

The objective function of the new linear programming problem is $w_2 = 2x_1 + 2x_2$. The optimum solution to this problem yields $\mathbf{X}^* = (2, 0)$. Because the values of w_2 at $\mathbf{X}_1$ and $\mathbf{X}^*$ are 2 and 4, a new trial point must be determined. Thus

$$\mathbf{X}_2 = (0, 1) + r[(2, 0) - (0, 1)] = (2r, 1 - r)$$

The maximization of

$$h(r) = f(2r, 1 - r)$$

yields $r_2 = \frac{1}{6}$. Thus $\mathbf{X}_2 = \left(\frac{1}{3}, \frac{5}{6}\right)$ with $f(\mathbf{X}_2) \approx 4.16$.

Iteration 3

$$\nabla f(\mathbf{X}_2) = (1, 2)$$

The corresponding objective function is $w_3 = x_1 + 2x_2$. The optimum solution of this problem yields the alternative solutions $\mathbf{X}^* = (0, 1)$ and $\mathbf{X}^* = (2, 0)$. The value of w_3 for both points equals its value at $\mathbf{X}_2$. Consequently, no further improvements are possible. The *approximate* optimum solution is $\mathbf{X}_2 = \left(\frac{1}{3}, \frac{5}{6}\right)$ with $f(\mathbf{X}_2) \approx 4.16$. This happens to be the exact optimum.

PROBLEM SET 19.2D

1. Solve the following problem by the linear combinations method.

$$\text{Minimize } f(\mathbf{X}) = x_1^3 + x_2^3 - 3x_1x_2$$

subject to

$$3x_1 + x_2 \le 3$$
$$5x_1 - 3x_2 \le 5$$
$$x_1, x_2 \ge 0$$

19.2.6 SUMT Algorithm

In this section, a more general gradient method is presented. It is assumed that the objective function $f(\mathbf{X})$ is concave and each constraint function $g_i(\mathbf{X})$ is convex. Moreover, the solution space must have an interior. This rules out both implicit and explicit use of *equality* constraints.

The SUMT (Sequential Unconstrained Maximization Technique) algorithm is based on transforming the constrained problem into an equivalent *un*constrained problem. The procedure is more or less similar to the Lagrange multipliers method. The transformed problem can then be solved using the steepest-ascent method (Section 19.1.2).

To clarify the concept, consider the new function

$$p(\mathbf{X},t) = f(\mathbf{X}) + t\left(\sum_{i=1}^{m} \frac{1}{g_i(\mathbf{X})} - \sum_{j=1}^{n} \frac{1}{x_j}\right)$$

where t is a nonnegative parameter. The second summation sign accounts for the nonnegativity constraints, which must be put in the form $-x_j \le 0$ to be consistent with the original constraints. Because $g_i(\mathbf{X})$ is convex, $\frac{1}{g_i(\mathbf{X})}$ is concave. This means that $p(\mathbf{X},t)$ is concave in $\mathbf{X}$. Consequently, $p(\mathbf{X},t)$ possesses a unique maximum. Optimization of the original constrained problem is equivalent to optimization of $p(\mathbf{X},t)$.

The algorithm is initiated by arbitrarily selecting an initial *nonnegative* value for t. An initial point $\mathbf{X}_0$ is selected as the first trial solution. This point must be an interior point—that is, it must not lie on the boundaries of the solution space. Given the value of t, the steepest-ascent method is used to determine the corresponding optimal solution (maximum) of $p(\mathbf{X},t)$.

The new solution point will always be an interior point, because if the solution point is close to the boundaries, at least one of the functions $\frac{1}{g_i(\mathbf{X})}$ or $-\frac{1}{x_i}$ will acquire a very large negative value. Because the objective is to maximize $p(\mathbf{X},t)$, such solution points are automatically excluded. The main result is that successive solution points will always be interior points. Consequently, the problem can always be treated as an unconstrained case.

Once the optimum solution corresponding to a given value of t is obtained, a new value of t is generated and the optimization process (using the steepest-ascent method) is repeated. If t' is the current value of t, the next value, t'', must be selected such that $0 < t'' < t'$.

The SUMT algorithm ends when, for two successive values of t, the corresponding *optimum* values of $\mathbf{X}$ obtained by maximizing $p(\mathbf{X},t)$ are approximately the same. At this point further trials will produce little improvement.

Actual implementation of SUMT involves more details than have been presented here. Specifically, the selection of an initial value of t is an important factor that can affect the speed of convergence. Further, the determination of an initial interior point may require special techniques. These details can be found in Fiacco and McCormick (1968).

REFERENCES

Bazaraa, M., H. Sherall, and C. Shetty, *Nonlinear Programming, Theory and Algorithms*, 2nd ed., Wiley, New York, 1993.

Beightler, C., D. Phillips, and D. Wilde, *Foundations of Optimization*, 2nd ed., Prentice Hall, Upper Saddle River, NJ, 1979

Fiacco, A., and G. McCormick, *Nonlinear Programming: Sequential Unconstrained Minimization Techniques*, Wiley, New York, 1968.

Luenberger, D., *Linear and Nonlinear Programming*, Kluwer Academic Publishers, Boston, 2003.

Rardin, D., *Optimization in Operations Research*, Prentice Hall, Upper Saddle River, NJ, 1998.

A P P E N D I X A

AMPL Modeling Language[1]

This appendix presents the principal syntactic rules of AMPL needed for the development and solution of complex mathematical programming models. For additional details, you may consult the basic language reference given at the end of this appendix (Fourer and Associates, 2003). You may also consult the Web site *www.ampl.com* for additional resources as well as the latest news and updates.

A.1 RUDIMENTARY AMPL MODEL

AMPL provides a facility for modeling mathematical programs (linear, integer, and nonlinear) in a long hand format. Figure A.1 gives the (self-explanatory) LP code for the Reddy Mikks model (file RM1.txt). All reserved words, other than the special operators (+ − *, ; : >< =) are in bold. The remaining symbols are generated by the user.

AMPL uses command lines and operates in a DOS environment. A recent beta version of a Windows interface can be found in *www.OptiRisk-Systems.com.*

You can execute a model by clicking on **ampl.exe** in the AMPL folder and, at the **ampl** prompt, typing the following command followed by Return:

```
ampl: model RM1.txt;
```

```
var x1 >=0;
var x2 >=0;
maximize z: 5*x1+4*x2;
subject to
  c1: 6*x1+4*x2<=24;
  c2: x1+2*x2<=6;
  c3: -x1+x2<=1;
  c4: x2<=2;
solve;
display z,x1,x2;
```

FIGURE A.1

Rudimentary AMPL model (file RM1.txt)

[1]Folder AppenAFiles on the CD includes all the files for this appendix.

The output will be displayed on the screen as[2]

```
MINOS 5.5: Optimal solution found.
2 iterations
z = 21
x1 = 3
x2 = 1.5
```

The rudimentary long hand format given here is not recommended for solving practical problems because it is problem-specific. The remainder of this appendix provides the details of how AMPL is used in practice.

A.2 COMPONENTS OF AMPL MODEL

Figure A.2 specifies the general structure of an AMPL model. The model is comprised of two basic segments: The top segment (elements 1 through 4) is the algebraic representation of the model, and the bottom segment (elements 5 through 7) supplies the data that drive the algebraic model. Thus, in LP, the algebraic representation in AMPL exactly parallels the following mathematical model:

$$\text{Maximize } z = \sum_{j=1}^{n} c_j x_j$$

subject to

$$\sum_{j=1}^{n} a_{ij} x_j \leq b_i, i = 1, 2, \ldots, m$$

FIGURE A.2

Basic structure of an AMPL model

Algebraic representation	1. Sets definitions. 2. Parameters definitions. 3. Variables definitions. 4. Model representation (objective and constraints)
Model implementation	5. Input data. 6. Solution of the model. 7. Output results.

[2]Every version of AMPL has a default *solver* that carries out the computations needed to optimize the AMPL model. In the student version, MINOS is the default solver and it can handle linear and nonlinear problems. The CD includes other solvers: CPLEX, KNITRO, LPSOLVE, and LOQO. CPLEX handles linear, integer, and quadratic problems. LPSOLVE handles linear and integer problems. KNITRO and LOQO handle linear and nonlinear problems.

The advantage of this arrangement is that the same algebraic model can be used to solve an LP problem of any size simply by changing the input data: m, n, c_j, a_{ij}, and b_i.

A number of syntax rules apply to the development of an AMPL model:

1. AMPL files must be plain text (Windows Notepad editor creates plain text).
2. Commented text may appear anywhere in the model preceded with #.
3. Each AMPL statement, comments excluded, must terminate with a semicolon (;).
4. An AMPL statement may occupy more than one line. Breakpoints occur at a proper separator, such as a blank space, colon, comma, parenthesis, brace, bracket, or mathematical operator. An exception to this rule occurs with strings (enclosed in quotes `' '` or `" "`) where a breakpoint is designated by adding a backslash (\).
5. All keywords (with few exceptions) are in lower case.
6. User-generated names are case sensitive. A name must be alphanumeric, interspersed with underscores, if desired. No other special characters are allowed.

We will use the Reddy Mikks problem of Section 2.1 to show how AMPL works. Figure A.3 gives the corresponding model (file RM2.txt). For convenience, key (or reserved) words are emphasized in bold.

The algebraic model starts with the *sets* that define the indices of the general LP model. The user-generated names `resource` and `paint` each preceded by the keyword `set` correspond to the sets $\{i\}$ and $\{j\}$ in the general LP model. The specific elements of the sets `resource` and `paint` that define the Reddy Mikks model are given in the input data section of the model.

The *parameters* are user-generated names preceded by the keyword `param` that define the coefficients of the objective function and the constraints as a function of the variable and constraint sets. The parameters `unitprofit{paint}`, `aij{resource, paint}`, and `rhs{resource}` correspond, respectively, to the mathematical symbols c_j, a_{ij}, and b_i in the general LP model. The subscripts i and j are represented by AMPL sets `resource` and `paint`, respectively. The input data provide specific values of the parameters.

The *variables* of the model, x_j, are given the name `product` preceded by the keyword `var`. Again, `product` is a function of the set `paint`. We can add the nonnegativity condition (>=0) in the same statement. Else, the default is that the variables are unrestricted in sign.

Having defined the sets, parameters, and variables of the model, the next step is to express the optimization problem in terms of these elements. The objective-function statement specifies the sense of optimization using the keyword `maximize` or `minimize`. The objective value z is given the user name `profit` followed by a colon (:) and its AMPL statement is a direct translation of the mathematical expression $\sum_j c_j x_j$:

```
sum{j in paint} unitprofit[j]*product[j];
```

The index `j` is user specified. Note the use of braces in `{j in paint}` to indicate that `j` is a member of the set `paint`, and the use of brackets in `[j]` to represent a subscript.

```
#*******************ALGEBRAIC MODEL******************
#----------------------------------------------------sets
set paint;
set resource;
#------------------------------------------------parameters
param  unitprofit{paint};
param  rhs {resource};
param  aij {resource,paint};
#------------------------------------------------variables
var product{paint} >= 0;
#---------------------------------------------------model
maximize profit: sum{j in paint} unitprofit[j]*product[j];
subject to limit{i in resource}:
        sum{j in paint} aij[i,j]*product[j] <= rhs[i];
#**********************DATA*************************
data;
set paint := exterior interior;
set resource := m1 m2 demand market;
param unitprofit :=
              exterior 5
              interior 4;
param rhs:=
              m1      24
              m2      6
              demand  1
              market  2;

param aij:  exterior  interior :=
       m1      6       4
       m2      1       2
       demand  -1      1
       market  0       1;

#*********************SOLUTION**********************
solve;
#-----------------------------------------output results
display profit, product, limit.dual, product.rc;
```

FIGURE A.3

AMPL model for the Reddy Mikks problem (file RM2.txt)

A model may include one or more constraint statements, and each such statement can be preceded by the keywords `subject to` or simply `s.t.` Actually, `s.t.` and `subject to` are optional, and AMPL assumes that any statement that does not start with a keyword is a constraint. The ReddyMikks model has only one set of constraints named `limit` and indexed over the set `resource`:

```
limit{i in resource}:
     sum{j in paint} aij[i,j]*product[j] <= rhs[i];
```

The statement is a direct translation of constraint i, $\sum_j a_{ij}x_{ij} \leq b_i$.

The idea of declaring variables as nonnegative can be generalized to allow establishing upper and lower bounds on the variables, thus eliminating the need to declare these bounds as explicit constraints. First, the two bounds are declared with the user-generated names `lowerbound` and `upperbound` as

```
param lowerbound{paint};
param upperbound{paint};
```

Next, the variables are defined as

```
var product{j in paint}>=lowerbound[j],<=upperbound[j];
```

Notice that the syntax does not allow comparing "vectors". Thus, an error is generated if we use

```
var product{paint}>=lowerbound{paint},<=upperbound{paint};
```

We can use the same syntax to set conditions on *parameters* as well. For example, the statement

```
param upperbound{j in paint}>=lowerbound[j];
```

will guarantee that `upperbound` is never less than `lowerbound`. Else AMPL will issue an error. The main purpose of using bounds on parameters is to prevent entering conflicting data inadvertently. Another instance where such checks may be used is when a parameter is required to assume nonnegative values only.

The algebraic model in Figure A.3 is general in the sense that it applies to any number of variables and constraints. It can be tailored to the Reddy Mikks situation by specifying the data of the problem. Following the statement `data`; we first define the members of the sets and then use these definitions to assign numeric values to the different parameters.

The set `paint` includes the names of two variables, which we suggestively call `exterior` and `interior`. Members of the set `resource` are given the names `m1, m2, demand,` and `market`. The associated statements in the `data` section are thus given as

```
set paint := exterior interior;
set resource := m1 m2 demand  market;
```

Members of each set appear to the right of the reserved operator `:=` separated by a blank space (or a comma). String indices must be enclosed in double quotes when used *outside* the data segment—that is, `paint["exterior"], paint["interior"], limit["m1"], limit["m2"], limit["demand"]`, and `limit["market"]`. Otherwise, the string index will be incorrectly interpreted as a (numeric) parameter.

We could have *defined* the sets at the start of the algebraic model (instead of in the data segment) as

```
set resource ={"m1","m2","demand","market"};
set paint = {"exterior","interior"};
```

(Note the mandatory use of the double quotes " ", the separating commas, and the braces.) This convention is not advisable in general because it is problem specific, which may limit tailoring the model to different input data scenarios. When this convention is used, AMPL will not allow modifying the set members in the `data` segment.

The use of alphanumeric names for the members of the sets `resource` and `paint` can be cumbersome in large problems. For this reason, AMPL allows the use of purely numeric sets—that is, we can use

```
set paint:= 1 2;
set resource:= 1..4;
```

The range `1..4` replaces the explicit `1 2 3 4` representation and is useful for sets with a large number of members. For example, `1..1000` is a set with 1000 members starting with 1 and ending with 1000 in increments of 1.

The range representation can be made more general by first defining `m` and `n` as parameters

```
param m;
param n;
```

In this case, the sets `1..m` and `1..n` can be used directly throughout the entire model as shown in Figure A.4 (file RM2a.txt), eliminating altogether the need to use the set names `resource` and `paint`.

FIGURE A.4

AMPL model for the Reddy Mikks problem (file RM2a.txt)

```
param m;
param n;
param  unitprofit{1..n};
param  rhs{1..m};
param  aij{1..m,1..n};
#------------------------------------------variables
var product{1..n}>= 0;
#---------------------------------------------model
maximize profit:sum{j in 1..n}unitprofit[j]*product[j];
subject to limit{i in 1..m}:
     sum{j in 1..n}aij[i,j]*product[j]<=rhs[i];
data;
param m:=4;
param n:=2;
param unitprofit := 1 5  2 4;
param rhs:= 1 24  2 6  3 1  4 2;
param aij: 1    2:=
     1     6    4
     2     1    2
     3    -1    1
     4     0    1;
solve;
display profit, product, limit.dual, product.rc;
```

Actually, the syntax `1..m` (or `1..n`) has the general format

```
start..end by step
```

where `start`, `end`, and `step` are defined AMPL parameters whose values are specified under `data`. If `start < end` and `step > 0`, then members of the set begin with `start` and advance by the amount `step` to the highest value less than or equal to `end`. The opposite occurs if `start > end` and `step < 0`. For example, `3..10 by 2` produces the members 3, 5, 7, and 9, and `10..3 by -2` produces the members 10, 8, 6, and 4. The default for `step` is 1, which means that `start..end by 1` is the same as `start..end`.

Actually, the parameters `start`, `end`, and `step` can be any legitimate AMPL mathematical expressions computed during execution. For example, given the parameters `m` and `n`, the set `j in 2*n..m+n^2 by n/2` is perfectly legal. Note, however, that a fractional `step` is used directly to create the members of the set. For example, for `m = 5`, `n = 13`, the members of the set `m..n step m/2` are 5, 7.5, 10, and 12.5.

The Reddy Mikks model includes single- and two-dimensional parameters. The parameters `unitprofit` and `rhs` fall in the first category and the parameter `aij` in the second. In the first category, data are specified by listing each set member followed by a numeric value, as the following statements show:

```
param unitprofit :=
     exterior 5
     interior 4;
param rhs:=
     m1          24
     m2          6
     demand      1
     market      2;
```

The elements of the list may be "strung" into one line, if desired. The only requirement is a separation of at least one blank space. The format given here promotes better readability.

Input data for the two-dimensional parameter `aij` are prepared similar to that of the one-dimensional case, except that the order of the columns must be specified after `aij:` to eliminate ambiguity, as the following statement shows:

```
param aij:          exterior   interior :=
          m1              6          4
          m2              1          2
          demand         -1          1
          market          0          1;
```

Again, the list is totally free-formatted so long as the logical sequential order is preserved and the elements are separated by blank spaces.

AMPL allows assigning default values to all the elements of a parameter.

For example, suppose that for a parameter c, $c_1 = 11$ and $c_8 = 22$ with $c_i = 0$ for $i = 2, 3, \ldots, 7$. We can use the following statements to specify the data for c:

```
param c{1..8};
.
.
.
data;
param c:=1 11 2 0 3 0 4 0 5 0 6 0 7 0 8 22;
```

A more compact way of achieving the same result is to use the following statements:

```
param c{1..8} default 0;
.
.
.
data;
param c:=1 11 8 22;
```

Initially, `c[1]` through `c[8]` assume the default value 0, with `c[1]` and `c[8]` changed to 11 and 22 in the `data` segment. In general, `default` may be followed by any mathematical expression. This expression is evaluated only once at the start of the execution.

The final segment of the AMPL model deals with obtaining the solution and the presentation of the output. The command `solve` is all that is needed to solve the model. Once completed, specific output results may be requested. The command `display` followed by an output list is but one way to view the results. In the Reddy Mikks model, the statement

```
display profit, product, limit.dual, product.rc;
```

requests the optimal values of the objective function and the variables, `profit` and `product`, the dual values of the constraints, `limit.dual`, and the reduced costs of the variables, `product.rc`. The keywords `dual` and `rc` are suffixed to the names of the constraints `limit` and variables `product` separated by a period. They may not be used as stand-alone keywords. The output defaults to the screen. It may be directed to an external file by inserting >*filename* immediately before the semicolon. Section A.5 provides more details about how output is directed to files and spreadsheets.

The execution command in DOS is

```
ampl: model RM2.txt;
```

The associated output is displayed on the screen, as the snapshot in Figure A.5 shows.

The layout of the output in Figure A.5 is a bit "cluttered" because it mixes the indices of the constraints and the variables. We can streamline the output by placing its elements in groups of the same dimension using the following two `display` statements:

```
display profit, product, product.rc;
display limit.dual;
```

In a typical AMPL model, such as the one in Figure A.3, the segment associated with the logic of the model preferably should remain static. The data and output segments

```
MINOS 5.5: optimal solution found.
2 iterations, objective 21

profit = 21

:           product limit.dual   product.rc      :=
demand         .          0             .
exterior     3            .       6.66134e-16
interior     1.5          .       0
m1             .          0.75          .
m2             .          0.5           .
market         .          0             .
;
```

FIGURE A.5

AMPL output using `display profit, product, limit.dual. product.rc;` in the Reddy Mikks model

are changed as needed to match specific LP scenarios. For this purpose, the AMPL model is represented by two separate files: RM2b.txt providing the logic of the model and RM2b.dat accounting for the input data and the output results.[3] In this case, the DOS line commands are entered sequentially as:

```
ampl: model RM2b.txt;
ampl: data RM2b.dat;
```

We will see in Section A.7 how commands such as `solve` and `display` can be issued interactively rather than being hard-coded in the model.

The Reddy Mikks model provides only a "glimpse' of the capabilities of AMPL. We will show later how input data may be read from external files and spreadsheet tables. We will also show how tailored (formatted) output can be sent to these media. Also, AMPL interactive commands are important debugging and execution tools, as will be explained in Sectiton A.7.

PROBLEM SET A.2A

1. Modify the Reddy Mikks AMPL model of Figure A.3 (file RM2.txt) to account for a third type of paint named "marine." Requirements per ton of raw materials `m1` and `m2` are .5 and .75 ton, respectively. The daily demand for the new paint lies between .5 ton and 1.5 tons. The revenue per ton is $3.5 (thousand). No other restrictions apply to this product.
2. In the ReddyMikks model of Figure A.3 (file RM2.txt), rewrite the AMPL code using the following set definitions:
   (a) `paint` and `{1..m}`.
   (b) `{1..n}` and `resource`.
   (c) `{1..m}` and `{1..n}`.

[3]Actually, the output command may be processed separately instead of being included in the .dat file, as will be explained in Section A.7

3. Modify the definition of the variables in the Reddy Mikks model of Figure A.3 (file RM2.txt) to include a minimum demand of 1 ton of exterior paint and maximum demands of 2 and 2.5 tons of exterior and interior paints, respectively.
4. In the Reddy Mikks model of Figure A.3, the command

```
display profit;
```

provides the value of the objective function. We can use the same command to display the contribution of each variable to the total profit as follows:

```
display profit, {j in paint} unitprofit[j]*product[j];
```

Another convenient way to accomplish the same result is to use **defined variable** statements as follows:

```
var extProfit=unitprofit["exterior"]*product["exterior"]
var intProfit=unitprofit["interior"]*product["interior"]
```

In this case, the objective function and display statements may be written in a less complicated form as

```
maximize profit: extProfit + intProfit;
display profit, extProfit, intProfit;
```

In fact, defined variables can be in indexed form as:

```
var varProfit{j in paint} = unitprofit[j]*product[j];
```

The resulting objective function and display statement will then read as

```
maximize profit: sum {j in paint} varProfit[j];
display profit, varProfit;
```

Use *defined variables* with the Reddy Mikks model to allow displaying each variable's profit contribution and resource consumption of raw materials `m1` and `m2`.
5. Develop and solve an AMPL model for the diet problem of Example 2.2-2 and find the optimum solution. Determine and interpret the associated dual values and the reduced costs.

A.3 MATHEMATICAL EXPRESSIONS AND COMPUTED PARAMETERS

We have seen that AMPL allows placing upper and lower bounds on parameters. Actually, the language affords more flexibility in defining parameters as complex mathematical expressions, modified conditionally, if desired.

To illustrate the use of computed parameters, consider the case of a bank offering n types of loans that charges an interest rate r_i for loan i, $0 < r_i < 1, i = 1, 2, \ldots, n$. Unrecoverable bad debt, both principal and interest, for loan i equals v_i of the amount of loan i. The objective is to determine the amount x_i the bank allocates to loan i to maximize the total return subject to a set of restrictions.

The use of computed parameters will be demonstrated by concentrating on the objective function. Algebraically, the objective function is expressed as

$$\text{Maximize } z = \sum_{i=1}^{n} r_i(1 - v_i)x_i - \sum_{i=1}^{n} v_i x_i = \sum_{i=1}^{n} [r_i - v_i(r_i + 1)]x_i$$

A direct translation of z into AMPL is the following:

```
param r{1..n}>0, <1;
param v{1..n}>0, <1;
var x{1..n}>=0;

maximize z: sum{i in 1..n}(r[i]-v[i]*(r[i]+1))*x[i];

(constraints)
```

Another way to handle the bank situation is to use a computed parameter to represent the objective function coefficients in the following manner:

```
param r{1..n}>0, <1;
param v{1..n}>0, <1;
param c{i in 1..n}=(r[i]-v[i]*(r[i]+1));
var x{1..n}>=0;

maximize z: sum{i in 1..n}c[i]*x[i];

(constraints)
```

AMPL will compute the parameter `c[i]` and use its value in the objective statement `z`. The new formulation enhances readability. But in some cases the use of computed parameters may be essential.

In general, the expression defining the value of a computed parameter can be of any complexity and may include any of the built-in arithmetic functions familiar to any programming language (e.g., `sin`, `max`, `log`, `sqrt`, `exp`). An important requirement is that the expression evaluate to a numeric value.[4]

Computed parameters may also be evaluated conditionally using the construct

parameter = `if` *condition* `then` *expression*`1` `else` *expression*`2;`

The *condition* compares arithmetic quantities and strings using the familiar operators =, <, >, <=, >=, and <> (together with `and/or`). Note that nonlinearity will result if *condition* is a function of the model variables. As in other programming languages, the construct may be used without `else` *expression2*. Nested `if` is also allowed following `then` and `else`.

The `if-then-else` construct gives the computed parameters the numeric value of either *expression*1 or *expression*2. This is the reason the `if-then-else` presented here is an *expression* and not a *statement*. (Section A.7 introduces the `if-then-else` *statement* together with the loop statements `for{}`, `repeat while{}`, and `repeat until{}`. These statements are used mainly for automating solution scenarios and formatting output.)

We will use a simple case to demonstrate the use of the `if` expression. In a multi-period manufacturing situation, units of a certain item are produced to meet variable demand. Unit production cost is estimated at p dollars for the first m periods, and increases by 10% for the next m periods and by 20% for the following m periods.

---

[4]AMPL manual provides an "exception" when a parameter is declared `binary`, in which case it can also be treated as *logical*. This distinction is artificial, because treating such a parameter as numeric still produces the same result.

The constraints of this model deal with capacity restrictions for each period and the balance equations that relate inventory, production, and demand. To demonstrate the use of the `if` expression, we will concentrate on the objective function. Let

$$x_j = \text{units produced in period } j, j = 1, 2, \ldots, 3m$$

The objective function is given as

Minimize

$$z = p(x_1 + x_2 + \cdots + x_m) + 1.1p(x_{m+1} + x_{m+2} + \cdots + x_{2m}) + 1.2p(x_{2m+1} + x_{2m+2} + \cdots + x_{3m})$$

We can model this function in AMPL as

```
param p;
var x{1..3*m}>=0;

minimize cost: p*(sum{j in 1..m}x[j]+1.1*sum{j in m+1..2*m}x[j]+
      1.2*sum{j in 2*m+1..3*m}x[j]);

(constraints)
```

A more compact way that also enhances readability is to use `if-then-else` to represent the objective-function parameter `c[j]`:

```
param m;
param n=3*m;
param p;
param c{j in 1..n}= if j<=m then p else
                    (if j>m and j<=2*m then 1.1*p else 1.2*p);

var x{j in 1..n};

minimize z: sum{j in 1..n}c[j]*x[j];

(constraints)
```

Note the nesting of the conditions. The parentheses () enclosing the second `if` are not necessary and are used to enhance readability. Observe that `then` and `else` are always followed by what must evaluate to numeric values. Note also that `c` can be defined as

```
param c{j in 1..n}=p*( if j<=m then 1 else
                   (if j>m and j<=2*m then 1.1 else 1.2));
```

A particularly useful implementation of `if-then-else` occurs in the situation where parameters or variables are defined *recursively*. A typical example of such a parameter occurs in determining the inventory level I_t in period $t, t = 1, 2, \ldots, n$, with initial zero inventory. The production amount and demand in period t are p_t and d_t, respectively. Thus, the inventory level is

$$I_0 = 0$$

$$I_t = I_{t-1} + p_t - d_t, t = 1, 2, \ldots, n$$

The amount I_t can be computed recursively in AMPL as follows:

```
param p{1..n};
param d{1..n};
var I{t in 1..n}= if i=1 then 0 else I[t-1])+p[t]-d[t];
```

Notice that it would be somewhat cumbersome to compute I_t were it not for the use of the `if-then-else` expression (see also Set A.3a).

PROBLEM SET A.3A

***1.** Consider the following set of constraints:

$$x_i + x_{i+1} \geq c_i, i = 1, 2, \ldots, n - 1$$

$$x_1 + x_n \geq c_n$$

Use `if-then-else` to develop a single set of constraints that represents all n inequalities.

2. In a multiperiod production-inventory problem, let x_t, z_t, and d_t be, respectively, the amount of entering inventory, production quantity, and demand for period $t, t = 1, 2, \ldots, T$. The balance equation associated with period t is $x_t + z_t - d_t - x_{t+1} = 0$. In a specific situation, $x_1 = c\ (>0)$ and $x_{T+1} = 0$. Write the AMPL constraints corresponding to the balance equations using `if-then-else` to account for $x_1 = c$ and $x_{T+1} = 0$.

A.4 SUBSETS AND INDEXED SETS

Subsets. Suppose that we have the following constraint:

$$x_1 + x_2 + x_5 + x_6 + x_7 \leq 15$$

There are 7 variables in the model, and this particular constraint does not include the variables x_3 and x_4.

We can model this constraint by using subsets in a number of ways (all *new* keywords are in bold):

```
#----------------------- method 1 ----------------------------------
var x{1..7}>=0;
subject to lim: sum{j in 1..7: j<=2 or j>=5}x[j]<=15;
#----------------------- method 2 ----------------------------------
var x{1..7}>=0;
subject to lim: sum{j in 1..2 union 5..7}x[j]<=15;
#----------------------- method 3 ----------------------------------
var x{1..7}>=0;
subject to lim: sum{j in 1..7 diff 3..4}x[j]]<=15;
#----------------------- method 4 ----------------------------------
var x{1..7}>=0;
subject to lim: sum{j in 1..7 diff (1..4 inter 3..7)}x[j]]<=15;
#------------------------------------------------------------------
```

In method 1, the set `{j in 1..7}` deletes the elements 3 and 4 by imposing restrictions on `j`. A colon separates the modified set from the condition(s). Keywords `union`, `diff`, and `inter` play the roles of $A \cup B$, $A - B$, and $A \cap B$, respectively. Method 4 is a convoluted set representation. Nevertheless, it serves to represent the use of the operator `inter`.

Indexed sets. A powerful feature of AMPL allows indexing sets over the elements of a regular set. Suppose that two components A and B are used to produce products 1, 2, 3, 4, and 5. Component A is used in products 1, 3, and 5, and component B is used in products 1, 2, 4, and 5. Each product requires one unit of the specified components. The maximum availabilities of components A and B are 200 and 300 units, respectively. The problem deals with determining the number of assembly units of each product. Other pertinent data will be needed to complete the description of the problem, but we will concentrate only on the constraints dealing with the components availability.

Let x_i be the production quantity of product $i, i = 1, 2, \ldots, 5$. Then the constraints for components A and B can be expressed mathematically as

$$\text{Component } A\text{: } x_1 + x_3 + x_5 \leq 200$$

$$\text{Component } B\text{: } x_1 + x_2 + x_4 + x_5 \leq 300$$

The AMPL representation of the constraints can be achieved using indexed sets as follows:

```
set comp;
set prod{comp};
param d{comp};
var x{1..5}>=0;
#------objective function here
subject to
     C{i in comp}:sum{j in prod[i]}x[j]]<=d[i];
#------other constraints here
data;
set comp:= A B;
set prod[A]:=1 3 5;
set prod[B]:=1 2 4 5;
param d:= A 200 B 300;
```

The indices of set `prod` are the elements `A` and `B` of set `comp`, thus defining the two indexed sets `prod[A]` and `prod[B]`. Next, the data of the problem define the elements of `prod[A]` and `prod[B]`. With these data, the constraints of the components (regardless of how many there are) are defined by the single statement:

```
C{i in comp}:sum{j in prod[i]}x[j]]<=d[i];
```

The applications of indexed sets are demonstrated aptly in the *AMPL moments* following Examples 6.5-4 and 9.1-2.

PROBLEM SET A.4A

1. Use subsets to express the left-hand side by means of a single `sum{}` function:

   (a) $\sum_{j=1}^{m} x_j + \sum_{j=m+k}^{n} x_j + \sum_{j=n+p}^{q} x_j \geq c$

   (b) $\sum_{i=m}^{n} x_i + \sum_{i=n+k}^{2n+k} x_i \leq c, k > 1$

*2. Suppose that 5 components (one unit per product unit) are used in the production of 10 products according to the following schedule:

Component	Products that use the component	Minimum availability
1	1, 2, 5, 10	500
2	3, 6, 7, 8, 9	400
3	1, 2, 3, 5, 6, 7, 9	900
4	2, 4, 6, 8, 10	700
5	1, 3 4, 5, 6, 7, 9, 10	100

The unit assembly cost of each product is a function of the component used: $9, $4, $6, $5, and $8 for components 1 through 5, respectively. The maximum demand for any of the products is 300 units. Use AMPL *indexed sets* to determine the optimal product mix that minimizes the installation cost. (*Hint*: Let x_{ij} be the number of units of product *i* that use component *j*.)

3. Repeat Problem 2 assuming that the unit installation cost of the components is a function of the assembled product: $1, $3, $2, $6, $4, $9, $2, $5, $10, and $7 for products 1 through 10, respectively.

A.5 ACCESSING EXTERNAL FILES

So far, we have used "hard-coded" data to drive AMPL models. Actually, AMPL data may be accessed from external files, spreadsheets, and/or databases. The same is true for retrieving output results. This section deals with reading data from or writing output to

1. External files, including screen and keyboard.
2. Spreadsheets.

More details can be found in Fourer and Associates, 2003, Chapter 10.

A.5.1 Simple Read Files

The statement for reading data from an unformatted external file is

```
read item-list <filename;
```

The *item-list* is a comma-separated list of nonindexed or indexed parameters. In the indexed case, the syntax is $\{indexing\}paramName[index]$. The list can include parameters only. This means that any set members must be accounted for under data *prior to* invoking the `read` statement. (We will see in Sections A.5.3 and A.5.4 how set members are read from formatted files and spreadsheets.)

To illustrate the use of `read`, consider the Reddy Mikks model where all the data for the parameters `unitprofit`, `rhs`, and `aij` are read from a file named RM3.dat per the model in file RM3.txt. The associated `read` statement is:

```
read {j in paint}unitprofit[j],
     {i in resource}rhs[i],
     {i in resource, j in paint}aij[i,j]<RM3.dat;
```

File RM3.dat lists the data in the exact order in which the items appear in the read list—that is,

```
5       4
24      6       1       2
6       4
1       2
-1      1
0       1
```

The multiple-row organization of the data enhances readability, in the sense that we could have had all the elements on one line (separated by blank spaces).[5] Note that this file happens to be all numeric. For convenience, nonnumeric data (such as parameter names) can appear in the data file provided that they are declared `symbolic` (for details, see Sections 7.8 and 9.5 in Fourer and Associates, 2003).

The `read` statement allows accessing data from the keyboard. In this case, the *filename* is replaced with a minus sign—that is, using `<-`. The execution of `read` in this case will produce the DOS prompt `ampl?,` and will be repeated until all the data requested by `read` have been accounted for.

PROBLEM SET A.5A

1. Prepare the input file RM3x.dat for the Reddy Mikks model (file RM3.txt), assuming the that the `read` statement is given as

```
read {j in paint}
     {i in resource}
     (
     rhs[i],
     {j in paint}aij[i,j]
     )<RM3x.dat;
```

***2.** For the Reddy Mikks model, explain why the following `read` statement is cumbersome:

```
read {i in resource}
     (
     rhs[i],
     {j in paint}(unitprofit[j],aij[i,j])
     )<RM3xx.dat;
```

[5]Hidden codes in `.dat` files (and in `.tab` files which will be presented later in this section) can trigger AMPL errors such as "too few elements in line xx" or "unexpected end of file" (xx stands for a numeric value) even though the text file may appear perfectly legal. To get rid of these hidden codes, click immediately to the right of the last data element in the file, then press the following keys in succession: Return, Backspace, and Return.

A.5.2 Using Print or Printf to Retrieve Output

A simple way to retrieve output data in AMPL is to use pre-formatted `print` or formatted `printf`. As an illustration, in the Reddy Mikks model we can use the following statements to send output data to a file we name `file.out` (output defaults to the screen if a file is not designated):

```
printf "Objective value is %6.2f\n",profit >file.out;
printf {j in paint}:
     "%8s%8.2f%8.3f\n",j,product[j],product[j].rc >file.out;
```

The output format always precedes the output list and must be enclosed in double quotes. The same statement can be used with `print` simply by removing the format code.

In the first `printf` statement, the format includes the optional descriptive text `Objective value is` and mandatory specifications of how the output list is printed. The code `%6.2f` says that the value of `profit` is printed in a field of length 6 with two decimal points. The code `\n` moves printing to the next line in the file. These format codes are the same as in C programming.

In the second print statement, the output list includes `j, product[j], product[j].rc`, where `j` is one of the members (`exterior, interior`) in the AMPL set `paint`. The code `%8s` reserves the first eight fields for printing the name `exterior` or `interior`. If `j` were numeric (e.g., `{j in 1..2}`), then the format specification would have to be integer; e.g., `%3i`.

The format specifications in this section are limited to `%s, %i, %f,` and `\n`. AMPL provides other specifications (see Table A-10 in Fourer and Associates, 2003).

PROBLEM SET A.5B

1. Use `printf` statements to present the optimal solution of the Reddy Mikks model (file RM2.txt) in the following format where the suffixes `.slack` and `.dual` are used to retrieve slack amount and the dual price:

```
Objective value =
-----------------------------------------
Product        Quantity        Profit($)
-----------------------------------------
   .              .               .
   .              .               .
   .              .               .
Constraint    Slack amount   Dual price
-----------------------------------------
   .              .               .
   .              .               .
   .              .               .
-----------------------------------------
```

A.5.3 Input Table Files

The `read` statement in Section A.5.1 does not allow reading set members. This situation is accounted for using `table` statements.

In `table` files, the data are presented as tables with properly-labeled rows and columns using the members of the defining sets. Access to `table` files requires a companion `read` statement. The `table` statement formats the data, and the `read` statement makes the data available to the model.

The syntax of `table` and `read` statements is as follows:

```
table tableName IN "fileName": SetName<-[SetColHdng], parameters~ParamColHdng;
read table tableName;
```

This syntax allows reading *both* the members of AMPL sets and the parameters from *tableName* in *fileName.*

The default *fileName* where the text table is stored is *tableName* `.tab`. It may be overridden by explicitly specifying *fileName* (in double quotes) with mandatory `.tab` extension following the keyword `IN`. `IN` (in caps) means INput (as contrasted with `OUT` which, as shown later, is used to OUTput data to a table file). *SetColHdng* may be an arbitrary heading name in the table which is cross-referenced to the elements of *SetName* using `<-`. Similarly, AMPL parameters are cross-referenced to the arbitrary names *ParamColHdng* using ~. If *SetColHdng* happens to be the same as *SetName*, the syntax *SetName*`<-`[*SetColHdng*] may be replaced with [*SetName*] IN. In the case of parameters, ~*ParamColHdng* is deleted from the statement.

To illustrate the use of tables, Figure A.6 gives the contents of the files named RM4profit.tab, RM4rhs.tab, and RM4aij.tab for inputting the parameters `unitprofit`, `rhs`, and `aij` of the Reddy Mikks model. The first line in each file must always follow the format

```
ampl.tab    nbr_indexing_sets    nbr_read_parameters
```

The first element, `ampl.tab`, identifies the table as a `.tab` file, with the succeeding two elements providing the number of indexing sets of the parameters that will be read from the table. In `RMprofit.tab` and `RMrhs.tab`, only *one* set is needed to define the

File RM4profit.tab:

```
ampl.tab 1 1
COL1            COL2
exterior        5
interior        4
```

File RM4rhs.tab:

```
ampl.tab 1 1
resource rhs
m1       24
m2       6
demand   1
market   2
```

File RM4aij.tab:

```
ampl.tab 2 1
resource paint          aij
m1       exterior       6
m1       interior       4
m2       exterior       1
m2       interior       2
demand   exterior       -1
demand   interior       1
market   exterior       0
market   interior       1
```

FIGURE A.6

Contents of the table files for inputting the parameters `unitprofit`, `rhs`, and `aij` of the Reddy Mikks model parameters unitprofit and rhs, and for this reason ampl.tab 1 1 is used as the header line in these two files. For the parameter aij, *two* sets are needed, which requires the use of the header line ampl.tab 2 1.

The header line is followed by a list of the *exact* or *substitute* names of the sets and the parameters. The succeeding rows in the respective file list the values of the input parameter as an explicit function of its indexing set(s) using blank space(s) as separators. For unitprofit and rhs, the listing is straightforward. For the double-indexed parameter aij, each parameter list is identified by two *explicit* indices, even at the expense of redundancy.

For the ReddyMikks model, the associated tables are defined as follows:

```
table RM4profit IN: paint <- [COL1], unitprofit~COL2;
table RM4rhs IN: [resource] IN, rhs;
table RM4aij IN: [resource,paint], aij;
```

Following the declaration of the table statements, we can use the following statements to read in the data:

```
read table RMprofit;
read table RMrhs;
read table RMaij;
```

For readability, it is recommended that the table declaration statements follow the constraints segment. The read statements are then placed immediately below the table declarations (see file RM4.txt).

The table statements above illustrate four syntactic rules:

1. In all three tables, the default file name is the table name with .tab extension (else a file name enclosed in "" must be given immediately before the colon).
2. In the profit statement, the syntax paint<-[COL1] tells AMPL that the entries in the COL1-column in file RM4profit.tab define the members of the set paint.
3. In the profit statement, the syntax unitprofit~COL2 cross-references the entries in COL2 with the parameter unitprofit.
4. In the rhs statement, [resource] IN automatically defines the members of the set resource because the name resource is used as a column heading in the table.
5. In the aij statement, aij has (at least) two dimensions, hence the statement *cannot* be used to read the members of the associated sets. Instead, these sets must be read from the single-dimensional tables RM4profit and RM4rhs. Thus, [resource,paint] in the aij statement are used only to define the indices of the parameter aij. In general, if a model has no indexed parameters, a table can be declared for the sole purpose of reading in the members of a set from a file. In this case, the header line in the .tab file must read ampl.tab 1 0, indicating that the file includes one column for the set members and no parameters. For example, the following statement declares the table for reading the elements of the set paint from file paintSet.tab:

```
table paintSet IN: [paint] IN;
```

In this case, the contents of `paintSet.tab` will be

```
ampl.tab 1 0
paint
exterior
interior
```

In some cases it may be convenient to read the data of a two-dimensional parameter as an array in place of two indexed single elements, as given above for `aij`. AMPL allows this by changing the definition of the table to:

```
table RM4arrayAij IN:[i~resource],{j in paint}<aij[i,j]~(j)>;
```

(The new table definition is somewhat "overcoded" in the sense that `~(j)` appears redundant. Nevertheless, it gets the job done.) In this case the file RM4arrayAij.tab must appear as

```
ampl.tab 1 2
resource   exterior  interior
m1                6         4
m2                1         2
demand           -1         1
market            0         1
```

Note that the header `ampl.tab 1 2` indicates that table `RM4arrayAij` has one key index (namely, `[i~resource]`) and two data columns with the headings `exterior` and `interior`. The new table, `RM4arrayAij`, does not permit reading the members of the sets `resource` and `paint`, the same restriction table `RM4aij` has. (See file RM4.txt.)

A.5.4 Output Table Files

Table files may also receive output from AMPL *after* the `solve` command has been executed. The syntax is similar to that of the input files, except that in the table declaration, `IN` is replaced with `OUT`. For example, in the Reddy Mikks model, suppose that we are interested in retrieving the following information:

1. Values of the variables and their reduced costs.
2. Slack and dual values associated with the constraints.

This information requires the use of *two* tables because the two item are functions of distinct sets: `paint` and `resource`:

```
table varData OUT:[paint],product,product.rc;
table conData OUT:
      [resource],limit.slack~slack,limit.dual~Dual;
```

The OUT-table declaration statements should be placed after the constraints segment to ensure that names of variables and constraints used in these statements have already been defined (see file RM4.txt). The syntax `limit.slack~slack` and `limit.dual~Dual` assigns the descriptive header names `slack` and `Dual` to the columns where the corresponding data are written in the file. Otherwise, the default header names will be `limit.slack` and `limit.dual`.

To retrieve the output, we need to issue the command `solve` and then follow it with the following `write` statements:

```
write table varData;
write table conData;
```

The output will be sent to files `varData.tab` and `conData.tab`, respectively. As with the input case, we can override the default file name by entering (in double quotes) a specific name (with `.tab` extension) following the keyword `OUT` and immediately before the colon.

Output tables can also be used to send two-dimensional arrays to a file. For example, either one of the following two definitions can be used to send the array `aij` to a `.tab` file:

```
table AijMatrix OUT: [resource,paint], aij;

table Aijout OUT:{i in resource}->[RESOURCES],{j in paint}<aij[i,j]~(j)>;
```

In the first definition, file AijMatrix.tab lists each element of `aij` with its two indices on the same row. In the second, file Aijout.tab lists `aij` in an array format, with the user-specified name `RESOURCES` being the heading of the first (key) column.

PROBLEM SET A.5C

*1. In RM4.txt, suppose the statements

```
read table RM4profit;
read table RM4rhs;
read table RM4aij;
```

are replaced with

```
read table RM4aij;
data;
param unitprofit:= exterior 5 interior 4;
param rhs:=m1 24 m2 6 demand 1 market 2;
```

Explain why AMPL will not execute properly with the proposed change. (*`Hint`*: The best way to find out the answer is to experiment with the model.)

2. Suppose that the contents of file RM4rhs.tab read as

```
ampl.tab 1 1
constrName Availability
m1      24
m2      6
demand 1
market 2
```

Make the necessary changes in RM4.txt and execute the model.

A.5.5 Spreadsheet Input/Output Tables

Accessing data from and sending data to a spreadsheet uses a syntax similar to that of the table files presented in Section A.5.4. The following statements show how the input data of the Reddy Mikks model can be accessed from an Excel spreadsheet file named RM5.xls:

```
table profitVector IN "ODBC" "RM5.xls":paint<-[COL1], unitprofit~COL2;
table rhsVector IN "ODBC" "RM5.xls":[resource] IN,rhs;
table aijMatrix IN "ODBC" "RM5.xls":[resource,paint], aij;
```

The user-generated names `profitVector, rhsVector,` and `aijMatrix` are those of the tables within the spreadsheet RM5.xls. These names define the ranges in the spreadsheet that correspond to the respective data tables.[6] `"ODBC"` is the standard data-handling interface for the spreadsheet. A `read table` then inputs the data to the model (see file RM5.txt). Note the use of COL1 and COL2 in table `profitVector,` which correspond to the (arbitrary) column names in the spreadsheet. The syntax is the same as in input tables (Section A.5.3). Each data table of the model may be stored in a separate sheet, if desired.

As in Section A.5.3, two-dimensional data can be read in an array format using the following table definition:

```
table aijArray IN "ODBC" "RM5.xls":[i~resource],{j in paint}<aij[i,j]~(j)>;
```

In this case, the array `aij` appears in the range `aijArray` of RM5.xls and must include the proper row and column headings. *It is also important to remember that numeric column headings when used in the table must be converted to strings by using Excel* `TEXT` *function, else AMPL will issue some undecipherable error messages.*

The same table declaration can be used to export output data to a spreadsheet. The only difference is to replace `IN` with `OUT,` exactly as in the case of table files. In this case, a `write table` command (following the `solve` command) will send the output to the spreadsheet. The following examples demonstrate the use of `OUT` tables:

```
table variables OUT "ODBC" "RM5a.xls":
        [paint],product~solution,product.rc~reducedCost;
table constraints OUT "ODBC" "RM5a.xls":
        [resource],limit.slack~slack,limit.dual~dual;
```

The output tables `variables` and `constraints` will go to Excel file `RM5a.xls` following the execution of the `write table` command, each appearing automatically in the northwest corner of a separate sheet.

A.6 INTERACTIVE COMMANDS

AMPL allows the user to solve the model interactively and to check/modify data and retrieve output to the screen or to a file. The following is a partial list of a number of useful commands:

`delete` *comma-separated names of objective function and constraints;*

`drop` *comma-separated names of objective function and constraints;*

---

[6]To name a range, highlight it and type its name in the "name box" to the left of the Excel formula bar then click Enter, or use Excel's Insert/Names/Define.

`restore` *comma-separated names of objective function and constraints;*

`display` *comma-separated item_list;*

`print/printf` *unformatted/formatted item_list;*

`expand` *comma-separated names of objective function and constraints;*

`let` *parameter or variable (indexed or nonindexed)* `:=`*value;*

`fix` *variable (indexed or nonindexed)* `:=`*value;*

`unfix` *variable (indexed or noindexed);*

`reset;`

`reset data;`

`solve;`

Such commands are entered interactively at the `ampl:` prompt. Some, such as `display` and `print`, may appropriately be hard-coded in the model, if desired.

The `delete` command completely removes the listed objective function and/or constraints, whereas `drop` temporarily yanks them out of the model. The `drop` command may be annulled by the `restore` command. A new objective function or constraint may be added to the model by entering it from the keyboard, exactly as we do in a hard-coded model. (See Example 9.2-1 for an application to the B&B algorithm.)

We have used `display` with the Reddy Mikks model. The output may be directed to an external file using >*filename* immediately before the terminating semicolon. Else, the output defaults to the screen.

The `print/printf` command has been discussed earlier in Section A.5.2. The output defaults to the screen or it may be directed to an output file as in `display`.

The `expand` command provides a long hand representation of the objective function and the constraints. For example, in the Reddy Mikks model, the command

```
expand profit;
```

prints out the objective function as

```
maximize profit:5*product["exterior"]+4*product["interior"];
```

In this manner, the user can see if the model has retrieved the input data correctly. In a similar manner, the command

```
expand limit;
```

will expand all the constraints of the model. If you are interested in a specific constraint, then `limit` must be properly indexed. For example,

```
expand limit["m1"];
```

will display the first constraint of the model.

The `let` command allows entering new values of parameters and variables (using: = as assignment operator). The right-hand side may be a simple numeric value or a mathematical expression. It is used to test different solution scenarios as we will show in Section A.7.

The `fix` command is used to assign a specific value to a variable prior to solving the model. For example, suppose that the following statements are issued interactively prior to solving the ReddyMikks model

```
ampl: fix product["exterior"]:=1.5;
ampl: solve;
```

With these commands, AMPL solves the problem with the added restriction `product["exterior"]=1.5`. The change caused by `fix` can be undone by issuing the `unfix` command as

```
ampl: unfix product["exterior"];
```

The `fix/unfix` commands can be useful in experimenting with the model when some of the variables are either eliminated $(=0)$ or held constant. (See *AMPL Moment* following Example 9.3-4 on page 392 for an application to the traveling salesperson problem.)

The command `reset` removes all reference to the current model from AMPL. A fresh `model` command will thus be necessary to restart the model. Also, the command `reset data;` will delete all the data, while `reset data a b c`; will delete the values of parameters `a, b,` and `c.` Also, AMPL requires the use of `reset;` between successive `model` commands. Else, undecipherable errors will result.

There is a large number of useful interactive commands in AMP, but their detailed presentation is beyond the scope of this abridged presentation.

A.7 ITERATIVE AND CONDITIONAL EXECUTION OF AMPL COMMANDS

Suppose in the Reddy Mikks model we are interested in studying the sensitivity of the optimal solution to changes in specific parameters. For example, in file RM2.txt, how is the optimal solution affected when the availability of raw material m1 `(=rhs["m1"])` is changed from its current value of 24 tons to the new values of 27 and 30 tons? After executing RM2.txt and getting the solution for `rhs["m1"]=24,` we can enter the following statements interactively:

```
ampl: let rhs["m1"]:= 27;
ampl: solve;
ampl: display profit, product;
```

The output will be displayed on the screen (it can also be sent to a file, if desired, as we explained earlier). To secure results for `rhs["m1"]=30,` the same statements are repeated with the `let` statement specifying the new value. This, however, is not the most efficient way to do the task.

AMPL allows building convenient `commands` files that will eliminate the unnecessary chore of retyping commands. Specifically, for the present example, a command file (which we arbitrarily name cmd.txt) may have the following statements:

```
for (i in 1..2}
   {
   let rhs["m1"]:=rhs["m1"]+3;
```

```
solve;
display profit, product;
}
```

Following the execution of the model (with `rhs["m1"]=24`), we can execute the remaining two cases by entering

```
ampl: commands cmd.txt;
```

Of course, we can modify cmd.txt to include `rhs["m1"]=24` as well. See Problem 1, Set A.7a.

We can use the statement `repeat while` *`condition`*`{...};` or `repeat until` *`condition`*`{...};` to replace `for{...}` as follows:

```
repeat while rhs["m1"]<=30
{
let rhs["m1"]:=rhs["m1"] +3;
solve;
display profit, product;
};
```

Alternatively, we may use

```
repeat until rhs["m1"]<30
{
let rhs["m1"]:=rhs["m1"]+3;
solve;
display profit,rhs["m1"], product;
};
```

Note that `repeat while` will loop so long the condition is true, whereas `repeat until` will loop so long as the condition is false.

Another useful statement in commands file is `if-then-else`. In this case, `if` may be followed by any legitimate condition, whereas `then` and `else` can be followed only by command statements. With the `if` statement, AMPL commands `continue;` and `break;` may be used within the loop construct to either skip to the next index of the loop or exit the loop altogether.

Example 9.3-5 (Figure 9.14) provides a good illustration of the use of the loop and conditional statements to print formatted output.

PROBLEM SET A.7A

1. Modify RM2.txt so that `rhs["m1"]` will assume the values 20 to 35 tons in steps of 5 tons. All `solve` commands must be executed from within the command file cmd.txt in the following manner:

```
ampl: model RM2.txt;
ampl: commands cmd.txt;
```

The command file cmd.txt is developed using the three different versions to construct the loop:

(a) `for{}.`

(b) `repeat while{};.`

(c) `repeat until{};.`

A.8 SENSITIVITY ANALYSIS USING AMPL

We have seen previously how the dual values and the reduced costs can be determined in an AMPL LP model by using the *ConstraintName*`.dual` and *VariableName*`.rc` in the `display` command. To complete the standard LP sensitivity analysis report, AMPL additionally provides facilities for the determination of the optimality ranges for the objective-function coefficients and the feasibility ranges for the (constant) right-hand sides of the constraints. We will use file RM2.txt (see Figure A.3) to demonstrate how AMPL generates the sensitivity analysis report.

In the model in Figure A.3, replace the `solve` and `display` statements with

```
option solver cplex;
option cplex_options 'sensitivity';
solve;
display limit.down,limit.current,limit.up,limit.dual;
display product.down,product.current,product.up,product.rc;
```

The output can be directed to a file if desired (see file RM6.txt). The two `option` statements must precede the `solve` command. The first `display` command provides the feasibility ranges for all the constraints (named `limit` in the model). The suffixes `.down`, `.current`, and `.up` give the lower, current, and upper values for the right-hand side of each member constraint. In a similar manner, the second `display` command provides the optimality ranges for the objective-function coefficients. The following output is self-explanatory.

```
profit = 21
        product.down  product.current  product.up  product.rc
exterior    2                5              6          0
interior    3.33333          4             10          0

        limit.down  limit    limit.up
demand      -1.5       0       1e+20
m1          20         0.75    36
m2           4         0.5      6.66667
market       0         0        0
```

REFERENCE

Fourer, R., D. Gay, and B. Kernighan, *AMPL, A Modeling Language for Mathematical Programming,* 2nd ed., Brooks/Cole-Thomson, Pacific Grove, CA, 2003.

APPENDIX B

Statistical Tables[1]

TABLE B.1 Normal Distribution Function

$$F(z) = \frac{1}{\sqrt{2\pi}} \int_{-\infty}^{z} e^{-(\frac{1}{2})t^2} dt$$

z	0.00	0.01	0.02	0.03	0.04	0.05	0.06	0.07	0.08	0.09
0.0	0.5000	0.5040	0.5080	0.5120	0.5160	0.5199	0.5239	0.5279	0.5319	0.5359
0.1	0.5398	0.5438	0.5478	0.5517	0.5557	0.5596	0.5636	0.5675	0.5714	0.5753
0.2	0.5793	0.5832	0.5871	0.5910	0.5948	0.5987	0.6026	0.6064	0.6103	0.6141
0.3	0.6179	0.6217	0.6255	0.6293	0.6331	0.6368	0.6406	0.6443	0.6480	0.6517
0.4	0.6554	0.6591	0.6628	0.6664	0.6700	0.6736	0.6772	0.6808	0.6844	0.6879
0.5	0.6915	0.6950	0.6985	0.7019	0.7054	0.7088	0.7123	0.7157	0.7190	0.7224
0.6	0.7257	0.7291	0.7324	0.7357	0.7389	0.7422	0.7454	0.7486	0.7517	0.7549
0.7	0.7580	0.7611	0.7642	0.7673	0.7704	0.7734	0.7764	0.7794	0.7823	0.7852
0.8	0.7881	0.7910	0.7939	0.7967	0.7995	0.8023	0.8051	0.8078	0.8106	0.8133
0.9	0.8159	0.8186	0.8212	0.8238	0.8264	0.8289	0.8315	0.8340	0.8365	0.8389
1.0	0.8413	0.8438	0.8461	0.8485	0.8508	0.8531	0.8554	0.8577	0.8599	0.8621
1.1	0.8643	0.8665	0.8686	0.8708	0.8729	0.8749	0.8770	0.8790	0.8810	0.8830
1.2	0.8849	0.8869	0.8888	0.8907	0.8925	0.8944	0.8962	0.8980	0.8997	0.9015
1.3	0.9032	0.9049	0.9066	0.9082	0.9099	0.9115	0.9131	0.9147	0.9162	0.9177
1.4	0.9192	0.9207	0.9222	0.9236	0.9251	0.9265	0.9279	0.9292	0.9306	0.9319
1.5	0.9332	0.9345	0.9357	0.9370	0.9382	0.9394	0.9406	0.9418	0.9429	0.9441
1.6	0.9452	0.9463	0.9474	0.9484	0.9495	0.9505	0.9515	0.9525	0.9535	0.9545
1.7	0.9554	0.9564	0.9573	0.9582	0.9591	0.9599	0.9608	0.9616	0.9625	0.9633
1.8	0.9641	0.9649	0.9656	0.9664	0.9671	0.9678	0.9686	0.9693	0.9699	0.9706
1.9	0.9713	0.9719	0.9726	0.9732	0.9738	0.9744	0.9750	0.9756	0.9761	0.9767
2.0	0.9772	0.9778	0.9783	0.9788	0.9793	0.9798	0.9803	0.9808	0.9812	0.9817
2.1	0.9821	0.9826	0.9830	0.9834	0.9838	0.9842	0.9846	0.9850	0.9854	0.9857

[1]Spreadsheet excelStatTable.xls replaces the (hard-copy) statistical tables of 12 common distributions, including the ones presented in this appendix.

TABLE B.1 Continued

z	0.00	0.01	0.02	0.03	0.04	0.05	0.06	0.07	0.08	0.09
2.2	0.9861	0.9864	0.9868	0.9871	0.9875	0.9878	0.9881	0.9884	0.9887	0.9890
2.3	0.9893	0.9896	0.9898	0.9901	0.9904	0.9906	0.9909	0.9911	0.9913	0.9916
2.4	0.9918	0.9920	0.9922	0.9925	0.9927	0.9929	0.9931	0.9932	0.9934	0.9936
2.5	0.9938	0.9940	0.9941	0.9943	0.9945	0.9946	0.9948	0.9949	0.9951	0.9952
2.6	0.9953	0.9955	0.9956	0.9957	0.9959	0.9960	0.9961	0.9962	0.9963	0.9964
2.7	0.9965	0.9966	0.9967	0.9968	0.9969	0.9970	0.9971	0.9972	0.9973	0.9974
2.8	0.9974	0.9975	0.9976	0.9977	0.9977	0.9978	0.9979	0.9979	0.9980	0.9981
2.9	0.9981	0.9982	0.9982	0.9983	0.9984	0.9984	0.9985	0.9985	0.9986	0.9986
3.0	0.9987	0.9987	0.9987	0.9988	0.9988	0.9989	0.9989	0.9989	0.9990	0.9990
3.1	0.9990	0.9991	0.9991	0.9991	0.9992	0.9992	0.9992	0.9992	0.9993	0.9993
3.2	0.9993	0.9993	0.9994	0.9994	0.9994	0.9994	0.9994	0.9995	0.9995	0.9995
3.3	0.9995	0.9995	0.9995	0.9996	0.9996	0.9996	0.9996	0.9996	0.9996	0.9997
3.4	0.9997	0.9997	0.9997	0.9997	0.9997	0.9997	0.9997	0.9997	0.9997	0.9998
3.5	0.9998									
4.0	0.99997									
5.0	0.9999997									
6.0	0.999999999									

Source: MILLER, I., and J. FREUND, *Probability and Statistics for Engineers,* Prentice Hall, Upper Saddle River, NJ, 1985.

TABLE B.2 $t_{\alpha,\nu}$ (Student t) Values*

ν	$\alpha = 0.10$	$\alpha = 0.05$	$\alpha = 0.025$	$\alpha = 0.01$	$\alpha = 0.005$	ν
1	3.078	6.314	12.706	31.821	63.657	1
2	1.886	2.920	4.303	6.965	9.925	2
3	1.638	2.353	3.182	4.541	5.841	3
4	1.533	2.132	2.776	3.747	4.604	4
5	1.476	2.015	2.571	3.365	4.032	5
6	1.440	1.943	2.447	3.143	3.707	6
7	1.415	1.895	2.365	2.998	3.499	7
8	1.397	1.860	2.306	2.896	3.355	8
9	1.383	1.833	2.262	2.821	3.250	9
10	1.372	1.812	2.228	2.764	3.169	10
11	1.363	1.796	2.201	2.718	3.106	11
12	1.356	1.782	2.179	2.681	3.055	12
13	1.350	1.771	2.160	2.650	3.012	13
14	1.345	1.761	2.145	2.624	2.977	14
15	1.341	1.753	2.131	2.602	2.947	15
16	1.337	1.746	2.120	2.583	2.921	16
17	1.333	1.740	2.110	2.567	2.898	17
18	1.330	1.734	2.101	2.552	2.878	18
19	1.328	1.729	2.093	2.539	2.861	19
20	1.325	1.725	2.086	2.528	2.845	20
21	1.323	1.721	2.080	2.518	2.831	21
22	1.321	1.717	2.074	2.508	2.819	22

TABLE B.2 Continued

v	$\alpha = 0.10$	$\alpha = 0.05$	$\alpha = 0.025$	$\alpha = 0.01$	$\alpha = 0.005$	v
23	1.319	1.714	2.069	2.500	2.807	23
24	1.318	1.711	2.064	2.492	2.797	24
25	1.316	1.708	2.060	2.485	2.787	25
26	1.315	1.706	2.056	2.479	2.779	26
27	1.314	1.703	2.052	2.473	2.771	27
28	1.313	1.701	2.048	2.467	2.763	28
29	1.311	1.699	2.045	2.462	2.756	29
Inf.	1.282	1.645	1.960	2.326	2.576	inf.

*Abridged by permission of Macmillan Publishing Co., Inc., from *Statistical Methods for Research Workers,* 14th ed, by R. A. Fisher. Copyright © 1970 University of Adelaide.

TABLE B.3 $\chi^2_{\alpha,\nu}$ (Chi-square) Values*

v	$\alpha = 0.995$	$\alpha = 0.99$	$\alpha = 0.975$	$\alpha = 0.95$	$\alpha = 0.05$	$\alpha = 0.025$	$\alpha = 0.01$	$\alpha = 0.005$	v
1	0.0000393	0.000157	0.000982	0.00393	3.841	5.024	6.635	7.879	1
2	0.0100	0.0201	0.0506	0.103	5.991	7.378	9.210	10.597	2
3	0.0717	0.115	0.216	0.352	7.815	9.348	11.345	12.838	3
4	0.207	0.297	0.484	0.711	9.488	11.143	13.277	14.860	4
5	0.412	0.554	0.831	1.145	11.070	12.832	15.056	16.750	5
6	0.676	0.872	1.237	1.635	12.592	14.449	16.812	18.548	6
7	0.989	1.239	1.690	2.167	14.067	16.013	18.475	20.278	7
8	1.344	1.646	2.180	2.733	15.507	17.535	20.090	21.955	8
9	1.735	2.088	2.700	3.325	16.919	19.023	21.666	23.589	9
10	2.156	2.558	3.247	3.940	18.307	20.483	23.209	25.188	10
11	2.603	3.053	3.816	4.575	19.675	21.920	24.725	26.757	11
12	3.074	3.571	4.404	5.226	21.026	23.337	26.217	28.300	12
13	3.565	4.107	5.009	5.892	22.362	24.736	27.688	29.819	13
14	4.075	4.660	5.629	6.571	23.685	26.119	29.141	31.319	14
15	4.601	5.229	6.262	7.261	24.996	27.488	30.578	32.801	15
16	5.142	5.812	6.908	7.962	26.296	28.845	32.000	34.267	16
17	5.697	6.408	7.564	8.672	27.587	30.191	33.409	35.718	17
18	6.265	7.015	8.231	9.390	28.869	31.526	34.805	37.156	18
19	6.844	7.633	8.907	10.117	30.144	32.852	36.191	38.582	19
20	7.434	8.260	9.591	10.851	31.410	34.170	37.566	39.997	20
21	8.034	8.897	10.283	11.591	32.671	35.479	38.932	41.401	21
22	8.643	9.542	10.982	12.338	33.924	36.781	40.289	42.796	22
23	9.260	10.196	11.689	13.091	35.172	38.076	41.638	44.181	23
24	9.886	10.856	12.401	13.484	36.415	39.364	42.980	45.558	24
25	10.520	11.524	13.120	14.611	37.652	40.646	44.314	46.928	25
26	11.160	12.198	13.844	15.379	38.885	41.923	45.642	48.290	26
27	11.808	12.879	14.573	16.151	40.113	43.194	46.963	49.645	27
28	12.461	13.565	15.308	16.928	41.337	44.461	48.278	50.993	28
29	13.121	14.256	16.047	17.708	42.557	45.772	49.588	52.336	29
30	13.787	14.953	16.791	18.493	43.773	46.979	50.892	53.672	30

*This table is based on Table 8 of *Biometrika Tables for Statisticians,* Vol. 1, by permission of the Biometrika trustees.

A P P E N D I X C

Partial Answers to Selected Problems[1]

CHAPTER 1

Set 1.1a

4. 17 minutes

5. (a) Jim's alternatives: Throw curve or fast ball.
Joe's alternatives: Prepare for curve or fast ball.
(b) Joe wants to increase his batting average.
Jim wants to reduce Joe's batting average.

CHAPTER 2

Set 2.1a

1. (a) $-x_1 + x_2 \geq 1$
(c) $x_1 - x_2 \leq 0$
(e) $.5x_1 - .5x_2 \geq 0$

3. Unused $M1 = 4$ tons/day

Set 2.2a

1. (a and e) See Figure C.1.

2. (a and d) See Figure C.2.

[1]Solved problems in this appendix are designated by * in the text.

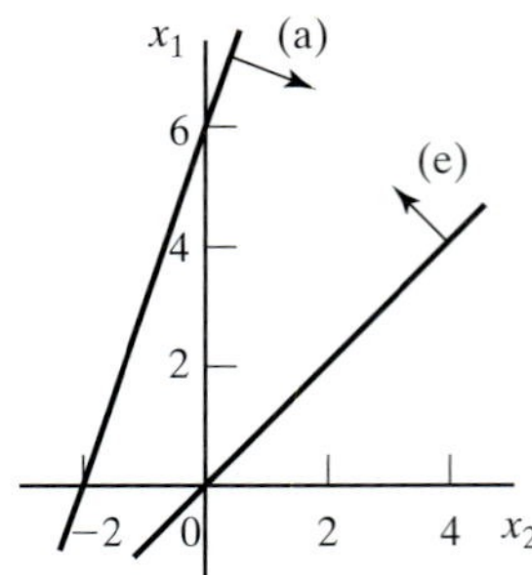

FIGURE C.1

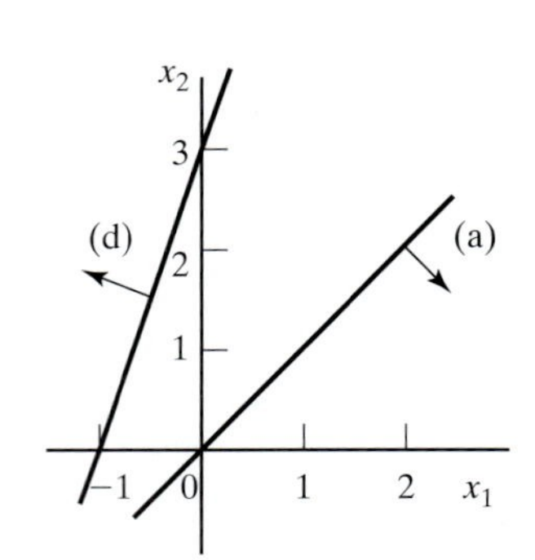

FIGURE C.2

5. Let

x_1 = Number of units of A

x_2 = Number of units of B

Maximize $z = 20x_1 + 50x_2$ subject to

$$-.2x_1 + .8x_2 \leq 0, 2x_1 + 4x_2 \leq 240$$
$$x_1 \leq 100, x_1, x_2 \geq 0$$

Optimum: $(x_1, x_2) = (80, 20), z = \$2{,}600$

7. Let

x_1 = Dollars invested in A

x_2 = Dollars invested in B

Maximize $z = .05x_1 + .08x_2$ subject to

$$.75x_1 - .25x_2 \geq 0, .5x_1 - .5x_2 \geq 0,$$
$$x_1 - .5x_2 \geq 0, x_1 + x_2 \leq 5000, x_1, x_2 \geq 0$$

Optimum: $(x_1, x_2) = (2500, 2500), z = \$325$

11. Let

x_1 = Play hours per day

x_2 = Work hours per day

Maximize $z = 2x_1 + x_2$ subject to

$$x_1 + x_2 \leq 10, x_1 - x_2 \leq 0$$
$$x_1 \leq 4, x_1, x_2 \geq 0$$

Optimum: $(x_1, x_2) = (4, 6), z = 14$

14. Let

x_1 = Tons of $C1$ per hour

x_2 = Tons of $C2$ per hour

Maximize $z = 12000x_1 + 9000x_2$ subject to

$$-200x_1 + 100x_2 \leq 0, 2.1x_1 + .9x_2 \leq 20, x_1, x_2 \geq 0$$

Optimum: $(x_1, x_2) = (5.13, 10.26)$, $z = 153{,}846$ lb

(a) Optimum ratio $C1:C2 = .5$.

(b) Optimum ratio is the same, but steam generation will increase by 7692 lb/hr.

18. Let

x_1 = Number of HiFi1 units

x_2 = Number of HiFi2 units

Minimize $z = 1267.2 - (15x_1 + 15x_2)$ subject to

$$6x_1 + 4x_2 \leq 432, 5x_1 + 5x_2 \leq 412.8$$

$$4x_1 + 6x_2 \leq 422.4, x_1, x_2 \geq 0$$

Optimum: $(x_1, x_2) = (50.88, 31, 68)$, $z = 28.8$ idle min.

Set 2.2b

1. (a) See Figure C.3

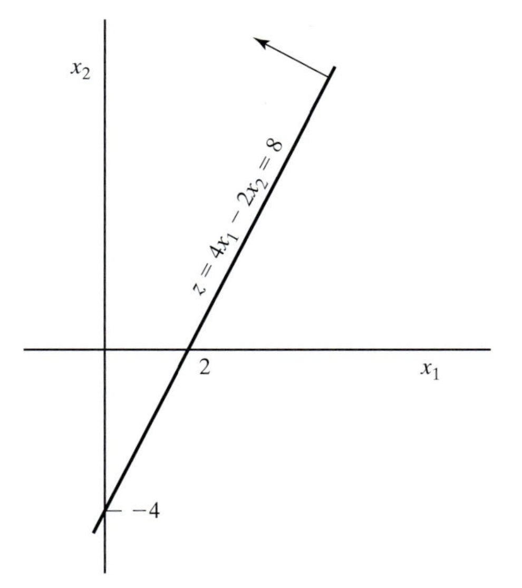

FIGURE C.3

5. Let

x_1 = Thousand bbl/day from Iran

x_2 = Thousand bbl/day from Dubai

Minimize $z = x_1 + x_2$ subject to

$$-.6x_1 + .4x_2 \leq 0, .2x_1 + .1x_2 \geq 14$$
$$.25x_1 + .6x_2 \geq 30, .1x_1 + .15x_2 \geq 10$$
$$.15x_1 + .1x_2 \geq 8, x_1, x_2 \geq 0$$

Optimum: $x_1 = 55, x_2 = 30, z = 85$

7. Let

x_1 = Ratio of scrap A alloy

x_2 = Ratio of scrap B alloy

Minimize $z = 100x_1 + 80x_2$ subject to

$$.03 \leq .06x_1 + .03x_2 \leq .06, .03 \leq .03x_1 + .06x_2 \leq .05$$
$$.03 \leq .04x_1 + .03x_2 \leq .07, x_1 + x_2 = 1, x_1, x_2 \geq 0$$

Optimum: $x_1 = .33, x_2 = .67, z = \$86,667$

Set 2.3a

3. Let

x_{ij} = Portion of project i completed in year j

$$\text{Maximize } z = .05(4x_{11} + 3x_{12} + 2x_{13}) + .07(3x_{22} + 2x_{23} + x_{24})$$
$$+ .15(4x_{31} + 3x_{32} + 2x_{33} + x_{34}) + .02(2x_{43} + x_{44})$$

subject to

$$x_{11} + x_{12} + x_{13} = 1, x_{43} + x_{44} = 1$$
$$.25 \leq x_{22} + x_{23} + x_{24} + x_{25} \leq 1$$
$$.25 \leq x_{31} + x_{32} + x_{33} + x_{34} + x_{35} \leq 1$$
$$5x_{11} + 15x_{31} \leq 3, 5x_{12} + 8x_{22} + 15x_{32} \leq 6$$
$$5x_{13} + 8x_{23} + 15x_{33} + 1.2x_{43} \leq 7$$
$$8x_{24} + 15x_{34} + 1.2x_{44} \leq 7, 8x_{25} + 15x_{35} \leq 7$$
$$\text{all } x_{ij} \geq 0$$

Optimum: $x_{11} = .6, x_{12} = .4, x_{24} = .255, x_{25} = .025$

$x_{32} = .267, x_{33} = .387, x_{34} = .346, x_{43} = 1, z = \$523,750$

Set 2.3b

2. The model can be generalized to account for any input currency p and any output currency q. Define x_{ij} as in Example 2.3-2 and r_{ij} as the exchange rate from currency i to currency j. The associated model is

Maximize $z = y$ subject to

$$\text{capacity: } x_{ij} \leq c_i, \text{ for all } i \neq j$$

$$\text{Input currency } p\text{: } I + \sum_{j \neq p} r_{jp} x_{jp} = \sum_{j \neq p} x_{pj}$$

$$\text{Output currency } q\text{: } y + \sum_{j \neq q} x_{qj} = \sum_{j \neq q} r_{jq} x_{jq}$$

$$\text{Currency } i \neq p \text{ or } q\text{: } \sum_{j \neq i} r_{ji} x_{ji} = \sum_{j \neq i} x_{ij}$$

$$\text{all } x_{ij} \geq 0$$

Rate of return: 1.8064% for \$ → \$, 1.7966% for \$ → €, 1.8287% for \$ → £, 2.8515% for \$ → ¥, and 1.0471% for \$ → KD. Wide discrepancies in ¥ and KD currencies may be attributed to the fact that the given exchange rates may not be totally consistent with the other rates. Nevertheless, the problem demonstrates the advantage of targeting accumulation in different currencies.

[*Note*: Interactive AMPL (file ampl2.3b-2.txt) or Solver (file solver2.3b-2.xls) is ideal for solving this problem. See Section 2.4.]

Set 2.3c

2. Let

x_i = Dollars invested in project i, $i = 1, 2, 3, 4$

y_j = Dollars invested in bank in year j, $j = 1, 2, 3, 4$

Maximize $z = y_5$ subject to

$$x_1 + x_2 + x_4 + y_1 \leq 10{,}000$$

$$.5x_1 + .6x_2 - x_3 + .4x_4 + 1.065y_1 - y_2 = 0$$

$$.3x_1 + .2x_2 + .8x_3 + .6x_4 + 1.065y_2 - y_3 = 0$$

$$1.8x_1 + 1.5x_2 + 1.9x_3 + 1.8x_4 + 1.065y_3 - y_4 = 0$$

$$1.2x_1 + 1.3x_2 + .8x_3 + .95x_4 + 1.065y_4 - y_5 = 0$$

$$x_1, x_2, x_3, x_4, y_1, y_2, y_3, y_4, y_5 \geq 0$$

Optimum solution:

$$x_1 = 0, x_2 = \$10{,}000, x_3 = \$6000, x_4 = 0$$

$$y_1 = 0, y_2 = 0, y_3 = \$6800, y_4 = \$33{,}642$$

$$z = \$53{,}628.73 \text{ at the start of year 5}$$

5. Let x_{iA} = amount invested in year i using plan A, $i = 1, 2, 3$
x_{iB} = amount invested in year i using plan B, $i = 1, 2, 3$
Maximize $z = 3x_{2B} + 1.7x_{3A}$ subject to

$$x_{1A} + x_{1B} \leq 100 \text{ (start of year 1)}$$

$$-1.7x_{1A} + x_{2A} + x_{2B} = 0 \quad \text{(start of year 2)}$$

$$-3x_{1B} - 1.7x_{2A} + x_{3A} = 0 \quad \text{(start of year 3)}$$

$$x_{iA}, x_{iB} \geq 0, i = 1, 2, 3$$

Optimum solution: Invest \$100,000 in plan A in year 1 and \$170,000 in plan B in year 2. Problem has alternative optima.

Set 2.3d

3. Let x_j = number of units of product j, $j = 1, 2, 3$
Maximize $z = 30x_1 + 20x_2 + 50x_3$ subject to

$$2x_1 + 3x_2 + 5x_3 \leq 4000$$

$$4x_1 + 2x_2 + 7x_3 \leq 6000$$

$$x_1 + .5x_2 + .33x_3 \leq 1500$$

$$2x_1 - 3x_2 = 0$$

$$5x_2 - 2x_3 = 0$$

$$x_1 \geq 200, x_2 \geq 200, x_3 \geq 150$$

$$x_1, x_2, x_3 \geq 0$$

Optimum solution: $x_1 = 324.32$, $x_2 = 216.22$, $x_3 = 540.54$, $z = \$41{,}081.08$

7. Let x_{ij} = Quantity produced by operation i in month j, $i = 1, 2$, $j = 1, 2, 3$
I_{ij} = Entering inventory of operation i in month j, $i = 1, 2$, $j = 1, 2, 3$

Minimize $z = \sum_{j=1}^{3}(c_{1j}x_{1j} + c_{2j}x_{2j} + .2I_{1j} + .4I_{2j})$ subject to

$$.6x_{11} \leq 800, .6x_{12} \leq 700, .6x_{13} \leq 550$$

$$.8x_{21} \leq 1000, .8x_{22} \leq 850, .8x_{23} \leq 700$$

$$x_{1j} + I_{1,j-1} = x_{2j} + I_{1j}, x_{2j} + I_{2,j-1} = d_j + I_{2j}, j = 1, 2, 3$$

$$I_{1,0} = I_{2,0} = 0, \text{ all variables} \geq 0$$

$$d_j = 500, 450, 600 \text{ for } j = 1, 2, 3$$

$$c_{1j} = 10, 12, 11 \text{ for } j = 1, 2, 3$$

$$c_{2j} = 15, 18, 16 \text{ for } j = 1, 2, 3$$

Optimum: $x_{11} = 1333.33$ units, $x_{13} = 216.67$, $x_{21} = 1250$ units, $x_{23} = 300$ units, $z = \$39{,}720$.

Set 2.3e

2. Let x_s = lb of screws/package, x_b = lb of bolts/package, x_n = lb of nuts/package, x_w = lb of washers/package

Minimize $z = 1.1x_s + 1.5x_b + \left(\frac{70}{80}\right)x_n + \left(\frac{20}{30}\right)x_w$ subject to

$$y = x_s + x_b + x_n + x_w$$

$$y \geq 1, x_s \geq .1y, x_b \geq .25y, x_n \leq .15y, x_w \leq .1y$$

$$\left(\tfrac{1}{10}\right)x_b \leq x_n, \left(\tfrac{1}{50}\right)x_b \leq x_w$$

All variables are nonnegative

Solution: $z = \$1.12, y = 1, x_s = .5, x_b = .25, x_n = .15, x_w = .1$

5. Let x_A = bbl of crude A/day, x_B = bbl of crude B/day, x_r = bbl of regular/day x_p = bbl of premium/day, x_j = bbl of jet fuel/day

$$\begin{aligned}\text{Maximize } z = {} & 50(x_r - s_r^+) + 70(x_p - s_p^+) + 120(x_j - s_j^+) \\ & - (10s_r^- + 15s_p^- + 20s_j^- + 2s_r^+ + 3s_p^+ + 4s_j^+) \\ & - (30x_A + 40x_B) \text{ subject to}\end{aligned}$$

$$x_A \leq 2500, x_B \leq 3000, x_r = .2x_A + .25x_B, x_p = .1x_A + .3x_B, x_j = .25x_A + .1x_B$$

$$x_r + s_r^- - s_r^+ = 500, x_p + s_p^- - s_p^+ = 700, x_j + s_j^- - s_j^+ = 400, \text{ All variables} \geq 0$$

Solution:

$$z = \$21{,}852.94, x_A = 1176.47 \text{ bbl/day}, x_B = 1058.82, x_r = 500 \text{ bbl/day}$$

$$x_p = 435.29 \text{ bbl/day}, x_j = 400 \text{ bbl/day}, s_p^- = 264.71$$

Set 2.3f

1. Let $x_i(y_i)$ = Number of 8-hr (12-hr) buses starting in period i

Minimize $z = 2\sum_{i=1}^{6} x_i + 3.5\sum_{i=1}^{6} y_i$ subject to

$$x_1 + x_6 + y_1 + y_5 + y_6 \geq 4, x_1 + x_2 + y_1 + y_2 + y_6 \geq 8,$$

$$x_2 + x_3 + y_1 + y_2 + y_3 \geq 10, x_3 + x_4 + y_2 + y_3 + y_4 \geq 7,$$

$$x_4 + x_5 + y_3 + y_4 + y_5 \geq 12, x_5 + x_6 + y_4 + y_5 + y_6 \geq 4$$

All variables are nonnegative

Solution: $x_1 = 4, x_2 = 4, x_4 = 2, x_5 = 4, y_3 = 6$, all others $= 0, z = 49$.

Total number of buses $= 20$. For the case of 8-hr shift, number of buses $= 26$ and comparable $z = 2 \times 26 = 52$. Thus, (8-hr + 12-hr) shift is better.

5. Let $x_i =$ Number of students starting in period i ($i = 1$ for 8:01 A.M., $i = 9$ for 4:01 P.M.)

Minimize $z = x_1 + x_2 + x_3 + x_4 + x_6 + x_7 + x_8 + x_9$ subject to

$$x_1 \geq 2, x_1 + x_2 \geq 2, x_1 + x_2 + x_3 \geq 3,$$
$$x_2 + x_3 + x_4 \geq 4, x_3 + x_4 \geq 4, x_4 + x_6 \geq 3,$$
$$x_6 + x_7 \geq 3, x_6 + x_7 + x_8 \geq 3, x_7 + x_8 + x_9 \geq 3$$
$$x_5 = 0, \text{ all other variables are nonnegative}$$

Solution: Hire 2 at 8:01, 1 at 10:01, 3 at 11:01, and 3 at 2:01. Total $= 9$ students

Set 2.3g

1. (a) $1150L$ ft^2

(b) (3,0,0), (1,1,0), (1,0,1), and (0,2,0) with respective 0, 3, 1, and 1 trim loss per foot.

(c) Number of standard 20′-rolls decreased by 30.

(d) Number of standard 20′-rolls increased by 50.

6.

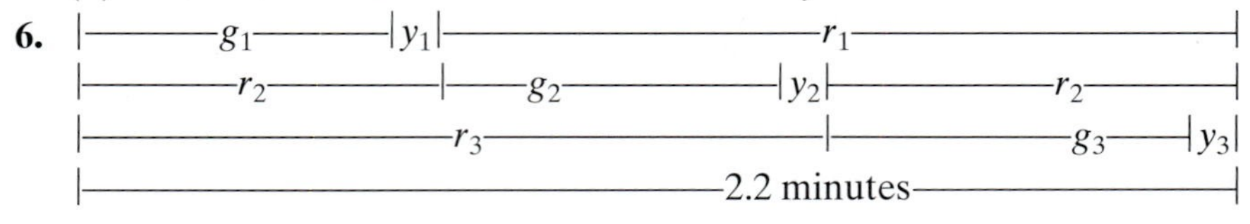

Let g_i, y_i, and r_i be the durations of green, yellow, and red lights for cars exiting highway i. All time units are in seconds. No cars move on yellow.

maximize $z = 3(500/3600)g_1 + 4(600/3600)g_2 + 5(400/3600)g_3$ subject to

$$(500/3600)g_1 + (600/3600)g_2 + (400/3600)g_3 \leq (510/3600)(2.2 \times 60 - 3 \times 10)$$
$$g_1 + g_2 + g_3 + 3 \times 10 \leq 2.2 \times 60, g_1 \geq 25, g_2 \geq 25, g_3 \geq 25$$

Solution: $g_1 = 25$ sec., $g_2 = 43.6$ sec., $g_3 = 33.4$ sec. Booth income $=$ \$58.04/hr

Set 2.4a

2. (d) See file solver2.4a-2(d).xls in folder AppenCFiles.

Set 2.4b

2. (c) See file ampl2.4b-2(c).txt in folder AppenCFiles.

(f) See file ampl2.4b-2(f).txt in folder AppenCFiles.

CHAPTER 3

Set 3.1a

1. 2 tons/day and 1 ton/day for raw materials $M1$ and $M2$, respectively.

4. Let x_{ij} = units of product i produced on machine j.
Maximize $z = 10(x_{11} + x_{12}) + 15(x_{21} + x_{22})$ subject to

$$x_{11} + x_{21} - x_{12} - x_{22} + s_1 = 5$$
$$-x_{11} - x_{21} + x_{12} + x_{22} + s_2 = 5$$
$$x_{11} + x_{21} + s_3 = 200$$
$$x_{12} + x_{22} + s_4 = 250$$
$$s_i, x_{ij} \geq 0, \text{ for all } i \text{ and } j$$

Set 3.1b

3. Let x_j = units of product j, $j = 1, 2, 3$.
Maximize $z = 2x_1 + 5x_2 + 3x_3 - 15x_4^+ - 10x_5^+$
subject to

$$2x_1 + x_2 + 2x_3 + x_4^- - x_4^+ = 80$$
$$x_1 + x_2 + 2x_3 + x_5^- - x_5^+ = 65$$
$$x_1, x_2, x_3, x_4^-, x_4^+, x_5^-, x_5^+ \geq 0$$

Optimum solution: $x_2 = 65$ units, $x_4^- = 15$ units, all others $= 0$, $z = \$325$.

Set 3.2a

1. (c) $x_1 = \frac{6}{7}, x_2 = \frac{12}{7}, z = \frac{48}{7}$.
(e) Corner points $(x_1 = 0, x_2 = 3)$ and $(x_1 = 6, x_2 = 0)$ are infeasible.

3. Infeasible basic solutions are:

$$(x_1, x_2) = \left(\tfrac{26}{3}, -\tfrac{4}{3}\right),\ (x_1, x_3) = (8, -2)$$
$$(x_1, x_4) = (6, -4),\ (x_2, x_3) = (16, -26)$$
$$(x_2, x_4) = (3, -13), (x_3, x_4) = (6, -16)$$

Set 3.3a

3. (a) Only (A, B) represents successive simplex iterations because corner point A and B are adjacent. In all the remaining pairs the associated corner points are not adjacent.
(b) (i) Yes. (ii) No, C and I are not adjacent. (iii) No, path returns to a previous corner point, A.

5. (a) x_3 enters at value 1, $z = 3$ at corner point D.

Set 3.3b

3.

New basic variable	x_1	x_2	x_3	x_4
Value	1.5	1	0	.8
Leaving variable	x_7	x_7	x_8	x_5

6. (b) x_2, x_5, and x_6 can increase value of z. If x_2 enters, x_8 leaves and $\Delta z = 5 \times 4 = 20$. If x_5 enters, x_1 leaves and $\Delta z = 0$ because x_5 equals 0 in the new solution. If x_6 enters, no variable leaves because all the constraint coefficients of x_6 are less than or equal to zero. $\Delta z = \infty$ because x_6 can be increased to infinity without causing infeasibility.

9. Second best value of $z = 20$ occurs when s_2 is made basic.

Set 3.4a

3. (a) Minimize $z = (8M - 4)x_1 + (6M - 1)x_2 - Ms_2 - Ms_3 = 10M$
(b) Minimize $z = (3M - 4)x_1 + (M - 1)x_2 = 3M$

6. The starting tableau is

Basic	x_1	x_2	x_3	x_4	Solution
z	−1	−12	0	0	−8
x_3	1	1	1	0	4
x_4	1	4	0	1	8

Set 3.4b

1. Always minimize the sum of artificial variables because the sum represents the amount of infeasibility in the problem.

7. Any nonbasic variable having nonzero objective coefficients at end of Phase I cannot become positive in Phase II because it will mean that the optimal objective value in Phase I will be positive; that is, infeasible Phase I solution.

Set 3.5a

1. (a) $A \rightarrow B \rightarrow C \rightarrow D$.
(b) 1 at A, 1 at B, $C_2^4 = 6$ at C, and 1 at D.

Set 3.5b

1. Alternative basic optima: $\left(0, 0, \frac{10}{3}\right)$, $(0, 5, 0)$, $\left(1, 4, \frac{1}{3}\right)$. Nonbasic alternative optima: $\left(\alpha_3, 5\alpha_2 + 4\alpha_3, \frac{10}{3}\alpha_1 + \frac{1}{3}\alpha_3\right)$, $\alpha_1 + \alpha_2 + \alpha_3 = 1, 0 \leq \alpha_i \leq 1, i = 1, 2, 3$.

Set 3.5c

2. (a) Solution space is unbounded in the direction of x_2.
   (b) Objective value is unbounded because each unit increase in x_2 increases z by 10.

Set 3.5d

1. The most that can be produced is 275 units.

Set 3.6a

2. Let

$$x_1 = \text{number of Type 1 hats per day}$$
$$x_2 = \text{number of Type 2 hats per day}$$

Maximize $z = 8x_1 + 5x_2$ subject to

$$2x_1 + x_2 \leq 400$$
$$x_1 \leq 150, x_2 \leq 200$$
$$x_1, x_2 \geq 0$$

   (a) See Figure C.4: $x_1 = 100$, $x_2 = 200$, $z = \$1800$ at point B.
   (b) \$4 per Type 2 hat in the range (200, 500).
   (c) No change because the dual price is \$0 per unit in the range $(100, \infty)$.
   (d) \$1 worth per unit in the range (100, 400). Maximum increase = 200 Type 2.

Set 3.6b

3. (a) $0 \leq \frac{c_1}{c_2} \leq 2$.
   (b) New $\frac{c_1}{c_2} = 1$. Solution remains unchanged.

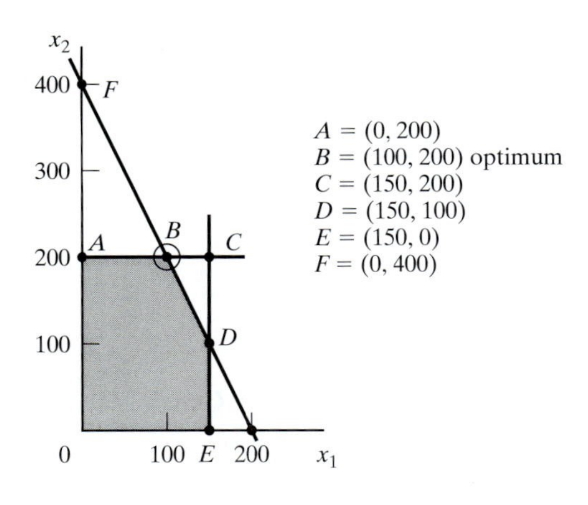

FIGURE C.4

Set 3.6c

2. (a) Yes, because additional revenue per min = \$1 (for up to 10 min of overtime) exceeds additional cost of \$.83/min.

(b) Additional revenue is \$2/min (for up to 400 min of overtime) = \$240 for 2 hr. Additional cost for 2 hr = \$110. Net revenue = \$130.

(c) No, its dual price is zero because the resource is already abundant.

(d) $D_1 = 10$ min. Dual price = \$1/min for $D_1 \leq 10$. $x_1 = 0, x_2 = 105, x_3 = 230$, net revenue $= (\$1350 + \$1 \times 10 \text{ min}) - \left(\frac{\$40}{60} \times 10 \text{ min}\right) = \$1353.33$.

(e) $D_2 = -15$. Dual price = \$2/min for $D_2 \geq -20$. Decrease in revenue = \$30. Decrease in cost = \$7.50. Not recommended.

6. Let

$$x_1 = \text{radio minutes}, x_2 = \text{TV minutes}, x_3 = \text{newspaper ads}$$

Maximize $z = x_1 + 50x_2 + 10x_3$ subject to

$$15x_1 + 300x_2 + 50x_3 + s_1 = 10{,}000,\ x_3 - S_2 = 5,$$

$$x_1 + s_3 = 400,\ -x_1 + 2x_2 + s_4 = 0,\ x_1, x_2, x_3 \geq 0,$$

$$s_1, s_2, s_3, s_4 \geq 0$$

(a) $x_1 = 59.09$ min, $x_2 = 29.55$ min, $x_3 = 5$ ads, $z = 1561.36$

(b) From TORA, $z + .158s_1 + 2.879S_2 + 0s_3 + 1.364s_4 = 156.364$. Dual prices for the respective constraints are .158, −2.879, 0, and 1.36. Lower limit set on newspaper ads can be decreased because its dual price is negative (= −2.879). There is no advantage in increasing the upper limit on radio minutes because its dual price is zero (the present limit is already abundant).

(c) From TORA, $x_1 = 59.9091 + .00606D_1 \geq 0$, $x_3 = 5$, $s_3 = 340.90909 + .00606D_1 \geq 0$, $x_2 = 29.54545 + .00303D_1 \geq 0$. Thus, dual price = .158 for the range $-9750 \leq D_1 \leq 56{,}250$. A 50% increase in budget ($D_1 = \$5000$) is recommended because the dual price is positive.

11. (a) Scarce: resistor and capacitor resource; abundant: chip resource.

(b) Worths per unit of resistor, capacitor, and chips are \$1.25, \$.25, and \$0.

(e) Change $D_3 = 350 - 800 = -450$ falls outside the feasibility range $D_3 \geq -400$. Hence problem must be solved anew.

13. (b) Solution $x_1 = x_2 = 2 + \frac{\Delta}{3}$ is feasible for all $\Delta > 0$. For $0 < \Delta \leq 3$, $r_1 + r_2 = \frac{\Delta}{3} \leq 1 \Rightarrow$ feasibility confirmed. For $3 \leq \Delta < 6$, $r_1 + r_2 = \frac{\Delta}{3} > 1 \Rightarrow$ feasibility not confirmed. For $\Delta > 6$, the change falls outside the ranges for D_1 and D_2.

Set 3.6d

2. (a) x_1 = Cans of A1, x_2 = Cans of A_2, x_3 = Cans of BK.

Maximize $z = 80x_1 + 70x_2 + 60x_3$ subject to

$$x_1 + x_2 + x_3 \le 500,\ x_1 \ge 100,\ 4x_1 - 2x_2 - 2x_3 \le 0$$

Optimum: $x_1 = 166.67$, $x_2 = 333.33$, $x_3 = 0$, $z = 36666.67$.

(b) From TORA, reduced cost per can of BK $= 10$. Price should be increased by more than 10 cents.

(c) $d_1 = d_2 = d_3 = -5$ cents. From TORA, the reduced costs for the nonbasic variables are

$$x_3\text{: } 10 + d_2 - d_3 \ge 0, \text{ satisfied}$$

$$s_1\text{: } 73.33 + .67d_2 + .33d_1 \ge 0, \text{ satisfied}$$

$$s_3\text{: } 1.67 - .17d_2 + .17d_1 \ge 0, \text{ satisfied}$$

Solution remains the same.

5. (a) $x_i =$ Number of units of motor i, $i = 1, 2, 3, 4$.
Maximize $z = 60x_1 + 40x_2 + 25x_3 + 30x_4$ subject to

$$8x_1 + 5x_2 + 4x_3 + 6x_4 \le 8000,\ x_1 \le 500,\ x_2 \le 500,$$

$$x_3 \le 800,\ x_4 \le 750,\ x_1, x_2, x_3, x_4 \ge 0$$

Optimum: $x_1 = 500$, $x_2 = 500$, $x_3 = 375$, $x_4 = 0$, $z = \$59{,}375$

(b) From TORA, $8.75 + d_2 \ge 0$. Type 2 motor price can be reduced by up to \$8.75.

(c) $d_1 = -\$15$, $d_2 = -\$10$, $d_3 = -\$6.25$, $d_4 = -\$7.50$. From TORA,

$$x_4\text{: } 7.5 + 1.5d_3 - d_4 \ge 0, \text{ satisfied}$$

$$s_1\text{: } 6.25 + .25d_3 \ge 0, \text{ satisfied}$$

$$s_2\text{: } 10 - 2d_3 + d_1 \ge 0, \text{ satisfied}$$

$$s_3\text{: } 8.75 - 1.25d_3 + d_2 \ge 0, \text{ satisfied}$$

Solution remains the same, but z will be reduced by 25%.

(d) Reduced cost of $x_4 = 7.5$. Increase price by more than \$7.50.

Set 3.6e

5. The dual price for the investment constraint $x_{1A} + x_{1B} \le 100$ is \$5.10 per dollar invested for *any* amount of investment.

9. (a) Dual price for raw material A is \$10.27. The cost of \$12.00 per lb exceeds the expected revenue. Hence, purchase of additional raw material A is not recommended.

(b) Dual price for raw material B is \$0. Resource is already abundant and no additional purchase is warranted.

CHAPTER 4

Set 4.1a

2. Let y_1, y_2, and y_3 be the dual variables.
Maximize $w = 3y_1 + 5y_2 + 4y_3$ subject to

$$y_1 + 2y_2 + 3y_3 \leq 15, 2y_1 - 4y_2 + y_3 \leq 12$$

$$y_1 \geq 0, y_2 \leq 0, y_3 \text{ unrestricted}$$

4. (c) Let y_1 and y_2 be the dual variables.
Minimize $z = 5y_1 + 6y_2$ subject to

$$2y_1 + 3y_2 = 1, y_1 - y_2 = 1$$

$$y_1, y_2 \text{ unrestricted}$$

5. Dual constraint associated with the artificial variables is $y_2 \geq -M$.
Mathematically, $M \rightarrow \infty \Rightarrow y \geq -\infty$, which is the same as y_2 being unrestricted.

Set 4.2a

1. (a) $\mathbf{AV}_1$ is undefined.
(e) $\mathbf{V}_2\mathbf{A} = (-14 \quad -32)$

Set 4.2b

1. (a) $\text{Inverse} = \begin{pmatrix} \frac{1}{4} & -\frac{1}{2} & 0 & 0 \\ -\frac{1}{8} & \frac{3}{4} & 0 & 0 \\ \frac{3}{8} & -\frac{5}{4} & 1 & 0 \\ \frac{1}{8} & -\frac{3}{4} & 0 & 1 \end{pmatrix}$

Set 4.2c

3. Let y_1 and y_2 be the dual variables.
Minimize $w = 30y_1 + 40y_2$ subject to

$$y_1 + y_2 \geq 5, 5y_1 - 5y_2 \geq 2, 2y_1 - 6y_2 \geq 3$$

$$y_1 \geq -M(\Rightarrow y_1 \text{ unrestricted}), y_2 \geq 0$$

Solution: $y_1 = 5, y_2 = 0, w = 150$.

6. Let y_1 and y_2 be the dual variables.
Minimize $w = 3y_1 + 4y_2$ subject to

$$y_1 + 2y_2 \geq 1, 2y_1 - y_2 \geq 5, y_1 \geq 3$$

$$y_2 \text{ unrestricted}$$

Solution: $y_1 = 3, y_2 = -1, w = 5$

8. (a) $(x_1, x_2) = (3, 0), z = 15, (y_1, y_2) = (3, 1), w = 14$. Range $= (14, 15)$

9. (a) Dual solution is infeasible, hence cannot be optimal even though $z = w = 17$.

Set 4.2d

2. (a) Feasibility: $(x_2, x_4) = (3, 15) \Rightarrow$ feasible.
Optimality: Reduced costs of $(x_1, x_3) = (0, 2) \Rightarrow$ optimal.

4.

Basic	x_1	x_2	x_3	x_4	x_5	Solution
z	0	0	$-\frac{2}{5}$	$-\frac{1}{5}$	0	$\frac{12}{5}$
x_1	1	0	$-\frac{3}{5}$	$\frac{1}{5}$	0	$\frac{3}{5}$
x_2	0	1	$\frac{4}{5}$	$-\frac{3}{5}$	0	$\frac{6}{5}$
x_5	0	0	-1	1	1	0

Solution is optimal and feasible.

7. Objective value: From primal, $z = c_1x_1 + c_2x_2$, and from dual, $w = b_1y_1 + b_2y_2 + b_3y_3$. $b_1 = 4, b_2 = 6, b_3 = 8, c_1 = 2, c_2 = 5 \Rightarrow z = w = 34$.

Set 4.3a

2. (a) Let $(x_1, x_2, x_3, x_4) =$ daily units of SC320, SC325, SC340, and SC370
Maximize $z = 9.4x_1 + 10.8x_2 + 8.75x_3 + 7.8x_4$ subject to

$$10.5x_1 + 9.3x_2 + 11.6x_3 + 8.2x_4 \leq 4800$$
$$20.4x_1 + 24.6x_2 + 17.7x_3 + 26.5x_4 \leq 9600$$
$$3.2x_1 + 2.5x_2 + 3.6x_3 + 5.5x_4 \leq 4700$$
$$5x_1 + 5x_2 + 5x_3 + 5x_4 \leq 4500$$
$$x_1 \geq 100,\ x_2 \geq 100,\ x_3 \geq 100,\ x_4 \geq 100$$

(b) Only soldering capacity can be increased because it has a positive dual price (= .4944).

(c) Dual prices for lower bounds are ≤ 0 ($-.6847, -1.361, 0$, and -5.3003), which means that the bounds have an adverse effect on profitability.

(d) Dual price for soldering is \$.4944/min valid in the range (8920, 10201.72), which corresponds to a maximum capacity increase of 6.26% only.

Set 4.3b

2. New fire truck toy is profitable because its reduced cost $= -2$.

3. Parts PP3 and PP4 are not part of the optimum solution. Current reduced costs are .1429 and 1.1429. Thus, rate of deterioration in revenue per unit is \$.1429 for PP3 and \$1.1429 for PP4.

Set 4.4a

1. (b) No, because point E is feasible and the dual simplex must stay infeasible until optimum is reached.

4. (c) Add the artificial constraint $x_1 \leq M$. Problem has no feasible solution.

Set 4.5a

4. Let Q be the weekly feed in lb (= 5200, 9600, 15000, 20000, 26000, 32000, 38000, 42000, for weeks 1, 2, ..., and 8). Optimum solution: Limestone = .028Q, corn = .649Q, and soybean meal = .323Q. Cost = .81221Q.

Set 4.5b

1. (a) Additional constraint is redundant.

Set 4.5c

2. (a) New dual values $= \left(\frac{1}{2}, 0, 0, 0\right)$. Current solution remains optimal.

(c) New dual values $= \left(-\frac{1}{8}, \frac{11}{4}, 0, 0\right)$. $z - .125s_1 + 2.75s_2 = 13.5$. New solution: $x_1 = 2, x_2 = 2, x_3 = 4, z = 14$.

Set 4.5d

1. $\frac{p}{100}(y_1 + 3y_2 + y_3) - 3 \geq 0$. For $y_1 = 1, y_2 = 2$, and $y_3 = 0, p \geq 42.86\%$.

3. (a) Reduced cost for fire engines $= 3y_1 + 2y_2 + 4y_3 - 5 = 2 > 0$. Fire engines are not profitable.

CHAPTER 5

Set 5.1a

4. Assign a very high cost, M, to the route from Detroit to dummy destination.

6. (a and b) Use $M = 10{,}000$. Solution is shown in bold. Total cost = \$49,710.

	1	2	3	Supply
Plant 1	600	700	400; **25**	**25**
Plant 2	320; **23**	300; **17**	350	**40**
Plant 3	500	480; **25**	450; **5**	**30**
Excess Plant 4	1000; **13**	1000	M	**13**
Demand	**36**	**42**	**30**	

(c) City 1 excess cost = \$13,000.

9. Solution (in million gallons) is shown in bold. Area 2 will be 2 million gallons short. Total cost = \$304,000.

	A1	A2	A3	Supply
Refinery 1	12 **4**	18 **2**	M	**6**
Refinery 2	30	10 **4**	8 **1**	**5**
Refinery 3	20	25	12 **6**	**6**
Dummy	M	50 **2**	50	**2**
Demand	**4**	**8**	**7**	

Set 5.2a

2. Total cost = \$804. Problem has alternative optima.

		Sharpening service			
Day	*New*	*Overnight*	*2-day*	*3-day*	Disposal
Monday	24	0	6	18	0
Tuesday	12	12	0	0	0
Wednesday	2	14	0	0	0
Thursday	0	0	20	0	0
Friday	0	14	0	0	4
Saturday	0	2	0	0	12
Sunday	0	0	0	0	22

5. Total cost = \$190,040. Problem has alternative optima.

Period	Capacity	Produced amount	Delivery
1	500	500	400 for (period) 1 and 100 for 2
2	600	600	200 for 2, 220 for 3, and 180 for 4
3	200	200	200 for 3
4	300	200	200 for 4

Set 5.3a

1. (a) Northwest: cost = \$42. Least-cost: cost = \$37. Vogel: cost = \$37.

Set 5.3b

5. (a) Cost = \$1475.

(b) $c_{12} \geq 3, c_{13} \geq 8, c_{23} \geq 13, c_{31} \geq 7$.

Set 5.4a

5. Use the code (city, date) to define the rows and columns of the assignment problem. Example: The assignment (D, 3)-(A,7) means leaving Dallas on Jun 3 and returning from Atlanta June 7 at a cost of $400. Solution is shown in bold. Cost = $1180. Problem has alternative optima.

	(A, 7)	(A, 12)	(A, 21)	(A, 28)
(D, 3)	400	300	300	**280**
(D, 10)	**300**	400	300	300
(D, 17)	300	**300**	400	300
(D, 25)	300	300	**300**	400

6. Optimum assignment: I-*d*, II-*c*, III-*a*, IV-*b*.

Set 5.5a

4. Total cost = $1550. Optimum solution summarized below. Problem has alternative optima.

	Store 1	Store 2	Store 3
Factory 1	50	0	0
Factory 2	50	200	50

CHAPTER 6

Set 6.1a

1. For network (i): (a) 1-3-4-2. (b) 1-5-4-3-1. (c and d) See Figure C.5.
4. Each square is a node. Adjacent squares are connected by arcs. Each of nodes 1 and 8 has the largest number of emanating arcs, and hence must appear in the center. Problem has more than one solution. See Figure C.6.

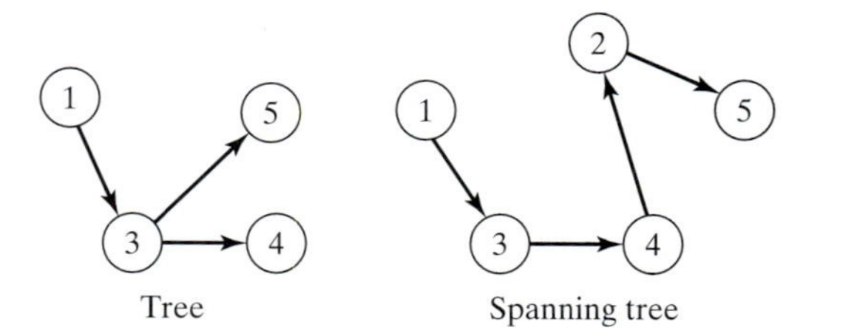

FIGURE C.5

FIGURE C.6

	3	5	
7	1	8	2
	4	6	

Set 6.2a

2. (a) 1-2, 2-5, 5-6, 6-4, 4-3. Total length = 14 miles.

5. High pressure: 1-2-3-4-6. Low pressure: 1-5-7 and 5-9-8.

Set 6.3a

1. Buy new car in years 1 and 4. Total cost = \$8900. See Figure C.7.

4. For arc (i, v_i)-$(i + 1, v_{i+1})$, define $p(q)$ = value(number of item i). Solution: Select one unit of each of items 1 and 2. Total value = \$80. See Figure C.8.

Set 6.3b

1. (c) Delete all nodes but 4, 5, 6, 7, and 8. Shortest distance = 8 associated with routes 4-5-6-8 and 4-6-8.

Set 6.3c

1. (a) 5-4-2-1, distance = 12.

4. Figure C.9 summarizes the solution. Each arc has unit length. Arrows show one-way routes. Example solution: Bob to Joe: Bob-Kay-Rae-Kim-Joe. Largest number of contacts = 4.

Set 6.3d

1. (a) Right-hand side of equations for nodes 1 and 5 are 1 and −1, respectively, all others = 0. Optimum solution: 1-3-5 or 1-3-4-5, distance = 90.

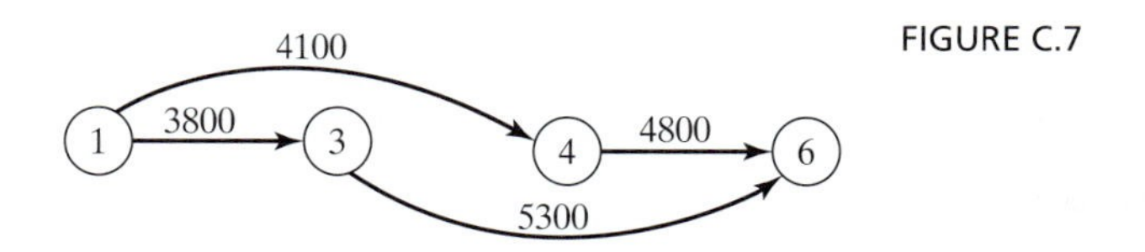

FIGURE C.7

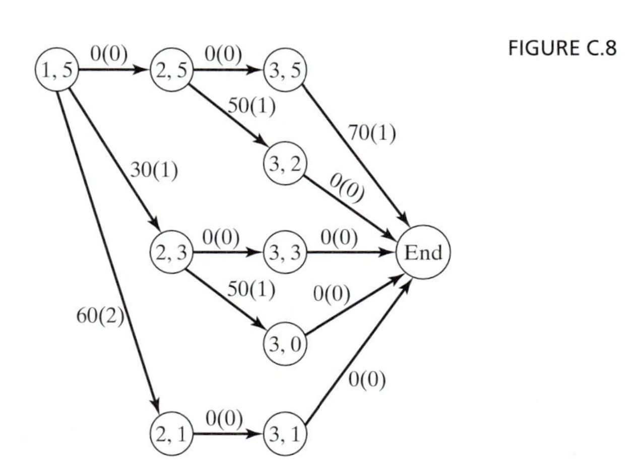

FIGURE C.8

FIGURE C.9

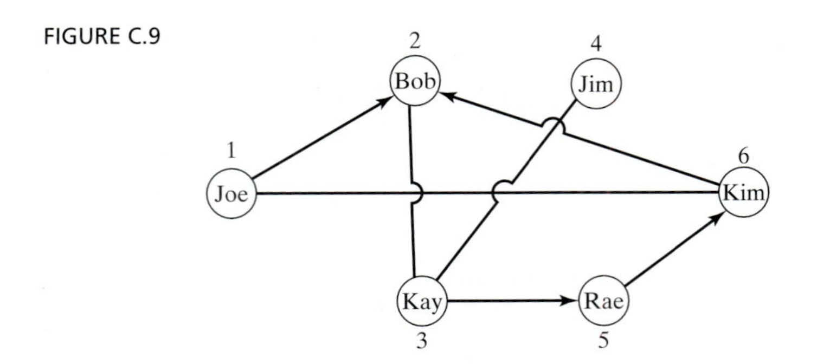

Set 6.4a

1. Cut 1: 1-2, 1-4, 3-4, 3-5, capacity = 60.

Set 6.4b

1. (a) Surplus capacities: arc (2-3) = 40, arc (2-5) = 10, arc (4-3) = 5.
   (b) Node 2: 20 units, node 3: 30 units, node 4: 20 units.
   (c) No, because there is no surplus capacity out of node 1.
7. Maximum number of chores is 4. Rif-3, Mai-1, Ben-2, Kim-5. Ken has no chore.

Set 6.5a

3. See Figure C.10.

Set 6.5b

1. Critical path: 1-3-4-5-6-7. Duration = 19.

Set 6.5c

3. (a) 10. (b) 5. (c) 0.
5. (a) Critical path: 1-3-6, duration = 45 days.
   (b) A, D, and E.

FIGURE C.10

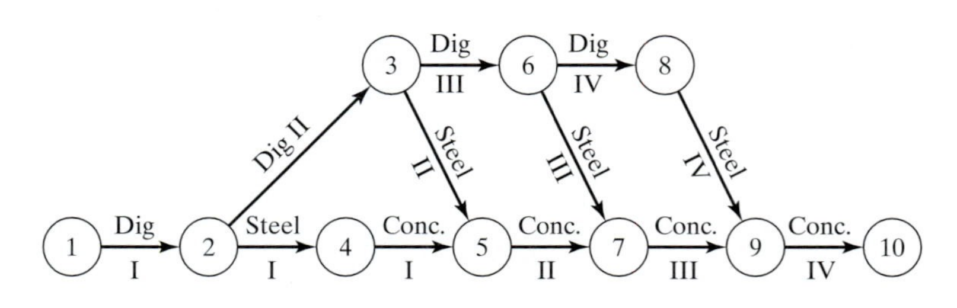

(c) Each of C, D, and G will be delayed by 5 days. E will not be affected.
(d) Minimum equipment = 2 units.

CHAPTER 7

Set 7.1a

2. (1, 0) and (0, 2) are in Q, but $\lambda(1, 0) + (1 - \lambda)(0, 2) = (\lambda, 2 - 2\lambda)$ does not lie in Q for $0 < \lambda < 1$.

Set 7.1b

2. (b) Unique solution with $x_1 > 1$ and $0 < x_2 < 1$. See Figure C.11.
(d) An infinite number of solutions.
(f) No solution.

3. (a) Basis because $\det \mathbf{B} = -4$.
(d) Not a basis because a basis must include exactly 3 independent vectors.

Set 7.1c

1.

$$\mathbf{B}^{-1} = \begin{pmatrix} .3 & -.2 \\ .1 & .1 \end{pmatrix}$$

Basic	x_1	x_2	x_3	x_4	Solution
z	1.5	−.5	0	0	21.5
x_3	0	.5	1	0	2
x_4	.5	0	0	1	1.5

Solution is feasible but nonoptimal.

4. Optimal $z = 34$.
Maximize $z = 2x_1 + 5x_2$ subject to $x_1 \leq 4$, $x_2 \leq 6$, $x_1 + x_2 \leq 8$, $x_1, x_2 \geq 0$

FIGURE C.11

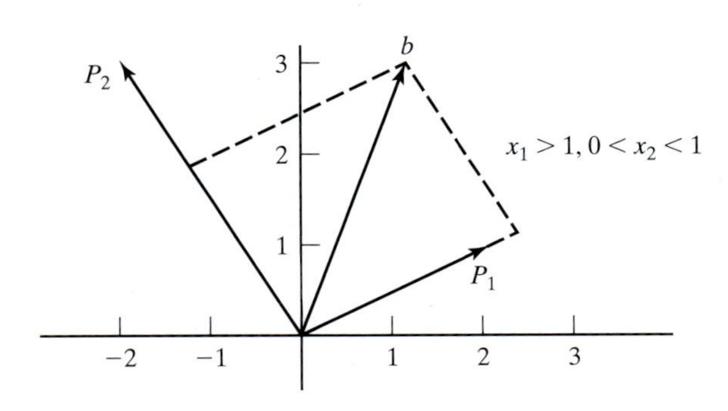

Set 7.2a

1. (a) $\mathbf{P}_1$ must leave.
(b) $\mathbf{B} = (\mathbf{P}_2, \mathbf{P}_4)$ is a feasible basis.

2. For the basic vector $\mathbf{X}_B$, we have

$$\{z_j - c_j\} = \mathbf{c}_B\mathbf{B}^{-1}\mathbf{B} - \mathbf{c}_B = \mathbf{c}_B\mathbf{I} - \mathbf{c}_B = \mathbf{c}_B - \mathbf{c}_B = \mathbf{0}$$

7. Number of adjacent extreme points is $n - m$, assuming nondegeneracy.

10. In case of degeneracy, number of extreme points is less than the number of basic solutions, else they are equal.

11. (a) new $x_j = \frac{1}{\alpha}$ old x_j.
(b) new $x_j = \frac{\beta}{\alpha}$ old x_j.

Set 7.2b

2. (b) $(x_1, x_2, x_3) = (1.5, 2, 0), z = 5$.

Set 7.3a

2. (b) $(x_1, x_2, x_3, x_4, x_5, x_6) = (0, 1, .75, 1, 0, 1), z = 22$.

Set 7.4a

2. Maximize $w = \mathbf{Yb}$ subject to $\mathbf{YA} \leq \mathbf{c}, \mathbf{Y} \geq \mathbf{0}$.

Set 7.4b

5. Method 1: $(b_1, b_2, b_3) = (4, 6, 8) \Rightarrow$ dual objective value $= 34$.
Method 2: $(c_1, c_2) = (2, 5) \Rightarrow$ primal objective value $= 34$.

7. Minimize $w = \mathbf{Yb}$ subject to $\mathbf{YA} = \mathbf{C}, \mathbf{Y}$ unrestricted.

Set 7.5a

1. $-\frac{2}{7} \leq t \leq 1$

2. (a)

Basic solution	Applicable range of t
$(x_2, x_3, x_6) = (5, 30, 10)$	$0 \leq t \leq \frac{1}{3}$
$(x_2, x_3, x_1) = (\frac{25}{4}, \frac{90}{4}, 5)$	$\frac{1}{3} \leq t \leq \frac{5}{2}$
$(x_2, x_4, x_1) = (\frac{5}{2}, 15, 20)$	$\frac{5}{2} \leq t \leq \infty$

5. $\{z_j - c_j\}_{j=1,4,5} = \left(4 - \frac{3t}{2} - \frac{3t^2}{2}, 1 - t^2, 2 - \frac{t}{2} + \frac{t^2}{2}\right)$. Basis remains optimal for $0 \leq t \leq 1$.

Set 7.5b

1. (a) $t_1 = 10$, $\mathbf{B}_1 = (\mathbf{P}_2, \mathbf{P}_3, \mathbf{P}_4)$
2. At $t = 0$, $(x_1, x_2, x_4) = (.4, 1.8, 1)$. It remains basic for $0 \le t \le 1.5$. No feasible solution for $t > 1.5$.

CHAPTER 8

Set 8.1a

1. G_5: Minimize s_5^+, $55x_p + 3.5x_f + 5.5x_s - .0675x_g + s_5^- - s_5^+ = 0$.
3. Let x_1 = No. of in-state freshmen, x_2 = No. of out-of-state freshmen, x_3 = No. of international freshmen.

   G_i: Minimize s_i^-, $i = 1, 2, \ldots, 5$, subject to $x_1 + x_2 + x_3 + s_1^- - s_1^+ = 1200$,

   $$2x_1 + x_2 - 2x_3 + s_2^- - s_2^+ = 0, -.1x_1 - .1x_2 + .9x_3 + s_3^- - s_3^+ = 0,$$

   $$.125x_1 - .05x_2 - .556x_3 + s_4^- - s_4^+ = 0, -.2x_1 + .8x_2 - .2x_3 + s_5^- - s_5^+ = 0$$

   All variables are nonnegative
5. Let x_j = No. of production runs in shift j, $j = 1, 2, 3$.
   Minimize $z = s_1^- + s_1^+$, subject to $-100x_1 + 40x_2 - 80x_3 + s_1^- - s_1^+ = 0$,

   $$4 \le x_1 \le 5, 10 \le x_2 \le 20, 3 \le x_3 \le 20$$

Set 8.2a

1. Objective function: Minimize $z = s_1^- + s_2^- + s_3^- + s_4^+ + s_5^+$
   Solution: $x_p = .0201$, $x_f = .0457$, $x_s = .0582$, $x_g = 2$ cents, $s_5^+ = 1.45$
   Gasoline tax is \$1.45 million short of goal.
4. x_1 = lb of limestone/day, x_2 = lb of corn/day, x_3 = lb of soybean meal/day.
   Objective function: Minimize $z = s_1^- + s_2^+ + s_3^- + s_4^- + s_5^+$
   Solution: $x_1 = 166.08$ lb, $x_2 = 2778.56$ lb, $x_3 = 3055.36$ lb, $z = 0$. Problem has alternative optima. All goals are satisfied but goals 3 and 4 are overachieved.
7. x_j = No. of units of product j, $j = 1, 2$.
   Assign a relatively high weight to the quota constraints.
   Objective function: Minimize $z = 100s_1^- + 100s_2^- + s_3^+ + s_4^+$
   Solution: $x_1 = 80$, $x_2 = 60$, $s_3^+ = 100$ minutes, $s_4^+ = 120$ minutes.
   Production quota can be met with 100 minutes of overtime for machine 1 and 120 minutes of overtime for machine 2.

Set 8.2b

2. G_1 solution: $x_p = .01745$, $x_f = .0457$, $x_s = .0582$, $x_g = 21.33$, $s_4^+ = 19.33$, all others $= 0$. Goals G_1, G_2, and G_3 are satisfied. G_4 is not.

G_4 problem: Same constraints as G_1 plus $s_1^- = 0, s_2^- = 0, s_3^- = 0$.

G_4 solution: $x_p = .0201, x_f = .0457, x_s = .0582, x_g = 2, s_5^+ = 1.45$. All other variables $= 0$. Goal G_5 is not satisfied.

G_5 problem: Same as G_4 plus $s_4^+ = 0$.

$G5$ solution: Same as G_4, which means that goal 5 cannot be satisfied ($s_5^+ = 1.45$).

CHAPTER 9

Set 9.1a

3. x_{ij} = No. of bottles of type i assigned to individual j, where $i = 1$ (full), 2 (half full), 3 (empty).

Constraints:

$$x_{11} + x_{12} + x_{13} = 7, x_{21} + x_{22} + x_{23} = 7, x_{31} + x_{32} + x_{33} = 7$$

$$x_{11} + .5x_{21} = 3.5, x_{12} + .5x_{22} = 3.5, x_{13} + .5x_{23} = 3.5$$

$$x_{11} + x_{21} + x_{31} = 7, x_{12} + x_{22} + x_{32} = 7, x_{13} + x_{23} + x_{33} = 7$$

All x_{ij} are nonnegative integers

Solution: Use a dummy objective function.

	No. bottles assigned to individual		
Status	*1*	*2*	*3*
Full	1	3	3
Half full	5	1	1
Empty	1	3	3

6. y = Original sum of money. x_j = Amount taken on night $j, j = 1, 2, 3$.

x_4 = Amount given to each mariner by first officer.

Minimize $z = y$ subject to $3x_1 - y = 2, x_1 + 3x_2 - y = 2, x_1 + x_2 + 3x_3 - y = 2, y - x_1 - x_2 - x_3 - 3x_4 = 1$. All variables are nonnegative integers.

Solution: $y = 79 + 81n, n = 0, 1, 2, \ldots$

10. Side 1: 5, 6, and 8 (27 minutes). Side 2: 1, 2, 3, 4, and 7 (28 minutes). Problem has alternative optima.

12. $x_{ij} = 1$ if student i selects course j, and zero otherwise, c_{ij} = associated preference score, C_j = course j capacity. Maximize $z = \sum_{i=1}^{10}\sum_{j=1}^{6} c_{ij}x_{ij}$ subject to

$$\sum_{j=1}^{6} x_{ij} = 2, i = 1, 2, \ldots, 10, \sum_{i=1}^{10} x_{ij} \leq C_j, j = 1, 2, \ldots, 6$$

Solution: Course 1: students (2, 4, 9), 2: (2, 8), 3:(5, 6, 7, 9), 4:(4, 5, 7, 10), 5: (1,3, 8, 10), 6: (1,3). Total score = 1775.

Set 9.1b

1. Let $x_j = 1$ if route j is selected and 0 otherwise. Total distance of route (ABC, 1, 2, 3, 4, ABC) $= 10 + 32 + 4 + 15 + 9 = 80$ miles.

Minimize $z = 80x_1 + 50x_2 + 70x_3 + 52x_4 + 60x_5 + 44x_6$ subject to

$$x_1 + x_3 + x_5 + x_6 \geq 1, x_1 + x_3 + x_4 + x_5 \geq 1, x_1 + x_2 + x_4 + x_6 \geq 1,$$

$$x_1 + x_2 + x_5 \geq 1, x_2 + x_3 + x_4 + x_6 \geq 1, x_j = (0, 1), \text{ for all } j.$$

Solution: Select routes (1, 4, 2) and (1, 3, 5), $z = 104$. Customer 1 should be skipped in one of the two routes.

2. Solution: Committee is formed of individuals a, d, and f. Problem has alternative optima.

7. $x_t = 1$ if transmitter t is selected, 0 otherwise. $x_c = 1$ if community c is covered, 0 otherwise. c_t = cost of transmitter t. S_c = set of transmitters covering community c. P_j = population of community j.

Maximize $z = \sum_{c=1}^{15} P_c x_c$ subject to

$$\sum_{t \in S_c} x_t \geq x_c, c = 1, 2, \ldots, 15, \sum_{t=1}^{7} c_t x_t \leq 15$$

Solution: Build transmitters 2,4,5,6, and 7. All but community 1 are covered.

Set 9.1c

2. Let x_j = Number of widgets produced on machine j, $j = 1, 2, 3$. $y_j = 1$ if machine j is used and 0 otherwise. Minimize $z = 2x_1 + 10x_2 + 5x_3 + 300y_1 + 100y_2 + 200y_3$ subject to $x_1 + x_2 + x_3 \geq 2000$, $x_1 - 600y_1 \leq 0$, $x_2 - 800y_2 \leq 0$, $x_3 - 1200y_3 \leq 0$, $x_1, x_2, x_3 \geq 500$ and integer, $y_1, y_2, y_3 = (0, 1)$.

Solution: $x_1 = 600$, $x_2 = 500$, $x_3 = 900$, $z = \$11{,}300$.

3. Solution: Site 1 is assigned to targets 1 and 2, and site 2 is assigned to targets 3 and 4. $z = 18$.

10. x_e = Number of Eastern (one-way) tickets, x_u = Number of US Air tickets, x_c = Number of Continental tickets. e_1, and e_2 binary variables. u and c nonnegative integers. Maximize $z = 1000(x_e + 1.5x_u + 1.8x_c + 5e_1 + 5e_2 + 10u + 7c)$ subject to $e_1 \leq x_e/2$, $e_2 \leq x_e/6$, $u \leq x_u/6$, and $c \leq x_c/5$, $x_e + x_u + x_c = 12$.

Solution: Buy 2 tickets on Eastern and 10 tickets on Continental. Bonus = 39000 miles.

Set 9.1d

1. Let x_{ij} = Integer amount assigned to square (i, j). Use a dummy objective function with all zero coefficients.

Constraints:

$$\sum_{j=1}^{3} x_{ij} = 15, i = 1, 2, 3, \sum_{i=1}^{3} x_{ij} = 15, j = 1, 2, 3,$$

$$x_{11} + x_{22} + x_{33} = 15, x_{31} + x_{22} + x_{13} = 15,$$

$$(x_{11} \geq x_{12} + 1 \text{ or } x_{11} \leq x_{12} - 1), (x_{11} \geq x_{13} + 1 \text{ or } x_{11} \leq x_{13} - 1),$$

$$(x_{12} \geq x_{13} + 1 \text{ or } x_{12} \leq x_{13} - 1), (x_{11} \geq x_{21} + 1 \text{ or } x_{11} \leq x_{21} - 1),$$

$$(x_{11} \geq x_{31} + 1 \text{ or } x_{11} \leq x_{31} - 1), (x_{21} \geq x_{31} + 1 \text{ or } x_{21} \leq x_{31} - 1),$$

$$x_{ij} = 1, 2, \ldots, 9, \text{ for all i and } j$$

Solution:

2	9	4
7	5	3
6	1	8

Alternative solutions are direct permutations of rows and/or columns.

3. x_j = Daily number of units of product j.
Maximize $z = 25x_1 + 30x_2 + 22x_3$ subject to

$$\begin{pmatrix} 3x_1 + 4x_2 + 5x_3 \leq 100 \\ 4x_1 + 3x_2 + 6x_3 \leq 100 \end{pmatrix} \text{ or } \begin{pmatrix} 3x_1 + 4x_2 + 5x_3 \leq 90 \\ 4x_1 + 3x_2 + 6x_3 \leq 120 \end{pmatrix}$$

$$x_1, x_2, x_3 \geq 0 \text{ and integer}$$

Solution: Produce 26 units of product 1, 3 of product 2, and none of product 3, and use location 2.

Set 9.2a[2]

2. (a) $z = 6, x_1 = 2, x_2 = 0.$
(d) $z = 12, x_1 = 0, x_2 = 3.$

3. (a) $z = 7.25, x_1 = 1.75, x_2 = 1.$
(d) $z = 10.5, x_1 = .5, x_2 = 2.$

9. Equivalent 0-1 ILP:
Maximize $z = 18y_{11} + 36y_{12} + 14y_{21} + 28y_{22} + 8y_{31} + 16y_{32} + 32y_{33}$
subject to $15y_{11} + 30y_{12} + 12y_{21} + 24y_{22} + 7y_{31} + 14y_{32} + 28y_{33} \leq 43$
All variables are binary.
Solution: $z = 50, y_{12} = 1, y_{21} = 1,$ all others $= 0$. Equivalently, $x_1 = 2, x_2 = 1$.
The 0-1 version required 41 nodes. The original requires 29.

[2]Use TORA integer programming module to generate the B&B tree.

Set 9.2b

1. (a) Legitimate cut because it passes through an integer point and does not eliminate any feasible integer point. You can verify this result by plotting the cut on the LP solution space.

6. (a) Optimum integer solution: $(x_1, x_2, x_3) = (2, 1, 6)$, $z = 26$.
   Rounded solution: $(x_1, x_2, x_3) = (3, 1, 6)$, which is infeasible.

Set 9.3a

1. The table below gives the number of distinct employees who enter/leave the manager's office when we switch from project i to project j. The objective is to find a "tour" through all projects that will minimize the total traffic.

	1	2	3	4	5	6
1	—	4	4	6	6	5
2	4	—	6	4	6	3
3	4	6	—	4	8	7
4	6	4	4	—	6	5
5	6	6	8	6	—	5
6	5	3	7	5	5	—

Set 9.3c

2. See Figure C.12.

FIGURE C.12

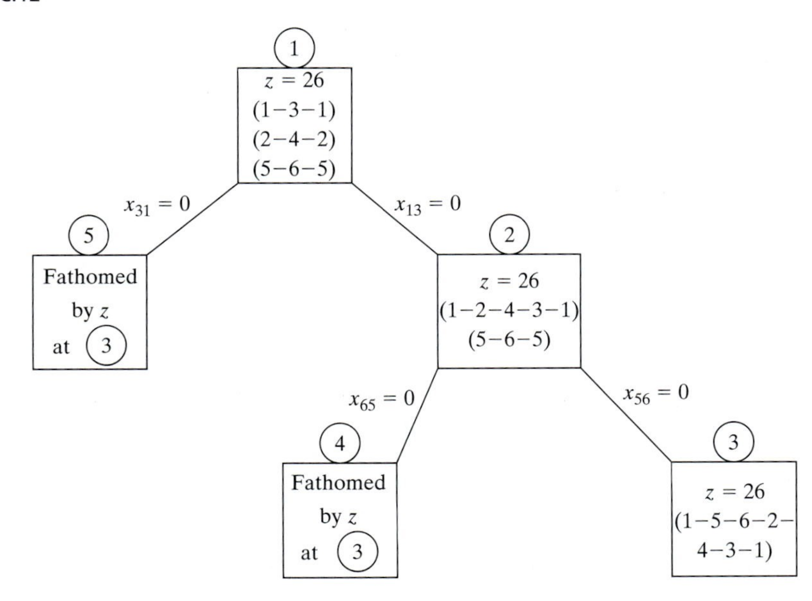

CHAPTER 10

Set 10.1a

1. Solution: Shortest distance = 21 miles. Route: 1-3-5-7.

Set 10.2a

3. Solution: Shortest distance = 17. Route: 1-2-3-5-7.

Set 10.3a

2. (a) Solution: Value = 120. $(m_1, m_2, m_3) = (0, 0, 3), (0, 4, 1), (0, 2, 2)$, or $(0, 6, 0)$.
5. Solution: Total points = 250. Select 2 courses from I, 3 from II, 4 from III, and 1 from IV,
7. Let $x_j = 1$ if application j is accepted, and 0 otherwise. Equivalent knapsack model is
   Maximize $z = 78x_1 + 64x_2 + 68x_3 + 62x_4 + 85x_5$ subject to

   $$7x_1 + 4x_2 + 6x_3 + 5x_4 + 8x_5 \leq 23, x_j = (0, 1), j = 1, 2, \ldots, 5$$

   Solution: Accept all but the first application. Value = 279.

Set 10.3b

1. (a) Solution: Hire 6 for week 1, fire 1 for week 2, fire 2 for week 3, hire 3 for week 4, and hire 2 for week 5.
3. Solution: Rent 7 cars for week 1, return 3 for week 2, rent 4 for week 3, and no action for week 4.

Set 10.3c

2. Decisions for next 4 years: Keep, Keep, Replace, Keep. Total cost = \$458.

Set 10.3d

3. (a) Let x_i and y_i be the number of sheep kept and sold at the end of period i and define $z_i = x_i + y_i$.

   $$f_n(z_n) = \max_{y_n = z_n}\{p_n y_n\}$$

   $$f_i(z_i) = \max_{y_i \leq z_i}\{p_i y_i + f_{i+1}(2z_i - 2y_i)\}, i = 1, 2, \ldots, n - 1$$

CHAPTER 11

Set 11.3a

2. (a) Total cost per week = \$51.50.
   (b) Total cost per week = \$50.20, $y^* = 239.05$ lb.

4. (a) Choose policy 1 because its cost per day is \$2.17 as opposed to \$2.50 for policy 2.
   (b) Optimal policy: Order 100 units whenever the inventory level drops to 10 units.

Set 11.3b

2. Optimal policy: Order 500 units whenever level drops to 130 units. Cost per day = \$258.50.
4. No advantage if $\text{TCU}_1(y_m) \le \text{TCU}_2(q)$, which translates to no advantage if the discount factor does not exceed .9344%.

Set 11.3c

1. AMPL/Solver solution: $(y_1, y_2, y_3, y_4, y_5) = (4.42, 6.87, 4.12, 7.2, 5.8)$, cost = \$568.12,
4. Constraint: $\sum_{i=1}^{4} \frac{365D_i}{y_i} \le 150.$
   Solver/AMPL solution: $(y_1, y_2, y_3, y_4) = (155.3, 118.82, 74.36, 90.09)$, cost = \$54.71.

Set 11.4a

1. (a) 500 units required at the start of periods 1, 4, 7, and 10.

Set 11.4b

3. Produce 173 units in period 1, 180 in period 2, 240 in period 3, 110 in period 4, and 203 in period 5.

Set 11.4c

1. (a) No, because inventory should not be held needlessly at end of horizon.
   (b) (i) $0 \le z_1 \le 5, 1 \le z_2 \le 5, 0 \le z_3 \le 4; x_1 = 4, 1 \le x_2 \le 6, 0 \le x_3 \le 4.$
   (ii) $5 \le z_1 \le 14, 0 \le z_2 \le 9, 0 \le z_3 \le 5; x_1 = 0, 0 \le x_2 \le 9, 0 \le x_3 \le 5.$
2. (a) $z_1 = 7, z_2 = 0, z_3 = 6, z_4 = 0$. Total cost = \$33.

Set 11.4d

1. Use initial inventory to satisfy the entire demand of period 1 and 4 units of period 2, thus reducing demand for the four periods to 0, 22, 90, and 67, respectively. Optimal solution: Order 112 units in period 2 and 67 units in period 4. Total cost = \$632.

Set 11.4e

1. Solution: Produce 210 units in January, 255 in April, 210 in July, and 165 in October.

CHAPTER 12

Set 12.1a

1. (a) .15 and .25, respectively. (b) .571. (c) .821.
2. $n \geq 23$.
3. $n > 253$.

Set 12.1b

3. $\frac{5}{32}$.
4. Let p = probability Liz wins. Probability John wins is $3p$, which equals the probability Jim will win. Probability Ann wins is $6p$. Because one of the four wins, $p + 3p + 3p + 3p + 6p = 1$.

(a) $\frac{3}{13}$.

(b) $\frac{7}{13}$.

(c) $\frac{6}{13}$.

Set 12.1c

3. (a) .375. (b) .6.
7. .9545.

Set 12.2a

2. (a) $K = 20$.
3. $P\{\text{Demand} \geq 1100\} = .3$.

Set 12.3a

3. (a) $P\{50 \leq \text{copies sold} \leq 70\} = .6667$.
(b) Expected number of unsold copies = 2.67
(c) Expected net profit = \$22.33

Set 12.3b

1. Mean = 3.667, variance = 1.556.

Set 12.3c

1. (a) $P(x_1 = 1) = P(x_2 = 1) = .4$, $P(x_1 = 2) = P(x_2 = 2) = .2$, $P(x_1 = 3) = P(x_2 = 3) = .4$.
(b) No, because $P(x_1, x_2) \neq P(x_1)P(x_2)$.

Set 12.4a

1. $\left(\frac{1}{2}\right)^{10}$.
3. .0547.

Set 12.4b

1. .8646.
3. (a) $P\{n = 0\} = 0$.
(b) $P\{n \geq 3\}$; 1.

Set 12.4c

1. $\lambda = 12$ arrivals/min. $P\{t \leq 5 \text{ sec}\} = .63$.

Set 12.4d

2. .001435.

CHAPTER 13

Set 13.1a

1. Weights for A, B, and C = (.44214, .25184, .30602).

Set 13.1b

2. CR > .1 for all matrices except **A**. $(w_S, w_J, w_M) = (.331, .292, .377)$. Select Maisa.
4. All matrices are consistent. $(w_H, w_P) = (.502, .498)$. Select H.

Set 13.2a

2. (a) See Figure C.13.
(b) EV(corn) = −\$8250, EV(soybeans) = \$250. Select soybeans.
6. (a) See Figure C.14.
(b) EV(game) = −\$.025. Do not play the game.

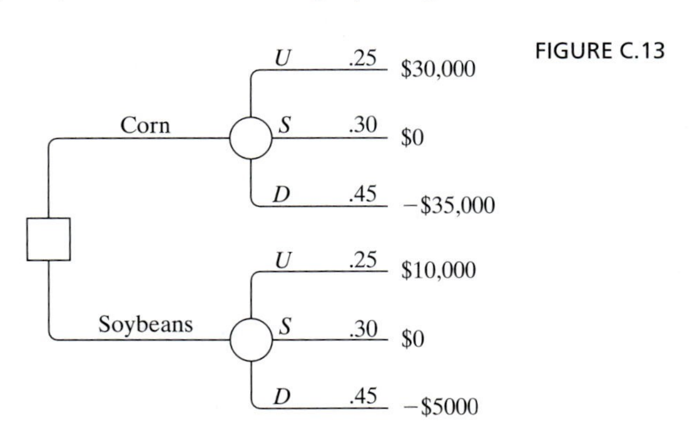

FIGURE C.13

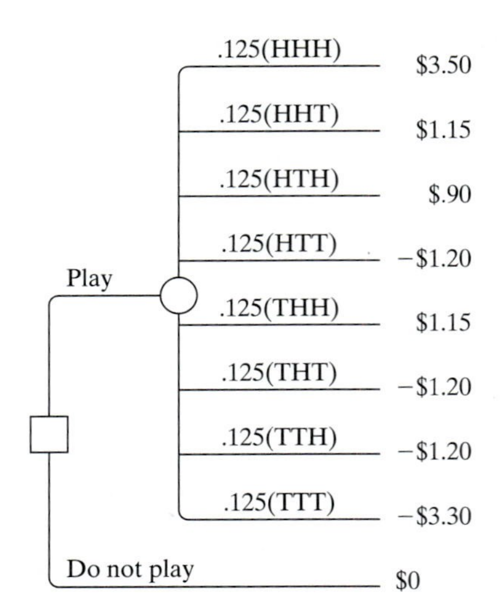

FIGURE C.14

12. Optimum maintenance cycle = 8 years. Cost per year = \$397.50.
15. Optimum production rate = 49 pieces per day.
19. Level must be between 99 and 151 gallons.

Set 13.2b

2. Let z be the event of having one defective item in a sample of size 5.
Answer: $P\{A|z\} = .6097$, $P\{B|z\} = .3903$.
4. (a) Expected revenue if you self-publish = \$196,000.
Expected revenue if you use a publisher = \$163,000.
(b) If survey predicts success, self-publish, else use a publisher.
7. (b) Ship lot to B if both items are bad, else ship lot to A.

Set 13.2c

1. (a) Expected value = \$5, hence there is no advantage.
(b) For $0 \le x < 10, U(x) = 0$, and for $x = 10, U(x) = 100$.
(c) Play the game.
2. Lottery: $U(x) = 100 - 100p$, with $U(-\$1{,}250{,}000) = 0$ and $U(\$900{,}000) = 100$.

Set 13.3a

1. (a) All methods: Study all night (action a_1).
(b) All methods: Select actions a_2 or a_3.

Set 13.4a

2. (a) Saddle-point solution at $(2, 3)$. Value of game = 4.
3. (a) $2 < v < 4$.

Set 13.4b

1. Each player should mix strategies 50-50. Value of game = 0.
2. Police payoff matrix:

	100%A	50%A-50%B	100%B
A	100	50	0
B	0	30	100

Strategy for Police: Mix 50-50 strategies 100%A and 100%B.
Strategy for Robin: Mix 50-50 strategies A and B. Value of game = \$50 (= expected fine paid by Robin).

Set 13.4c

1. (a) Payoff matrix for team 1:

	AB	AC	AD	BC	BD	CD
AB	1	0	0	0	0	−1
AC	0	1	0	0	−1	0
AD	0	0	1	−1	0	0
BC	0	0	−1	1	0	0
BD	0	−1	0	0	1	0
CD	−1	0	0	0	0	1

Optimal strategy for both teams: Mix AB and CD 50-50. Value of the game = 0.

3. (a) (m, n) = (Number of regiments at location 1, No. of regiments at locations 2). Each location has a payoff of 1 if won and −1 if lost. For example, Botto's strategy (1, 1) against the enemy's (0, 3) will win location 1 and lose location 2, with anet payoff of $1 + (-1) = 0$. Payoff matrix for Colonel Blotto:

	3, 0	2, 1	1, 2	0, 3
2, 0	−1	−1	0	0
1, 1	0	−1	−1	0
0, 2	0	0	−1	−1

Optimal strategy for Blotto: Blotto mixes 50-50 strategies (2-0) and (0-2), and the enemy mixes 50-50 strategies (3-0) and (1-2). Value of the game = $-.5$, and Blotto loses. Problem has alternative optima.

CHAPTER 14

Set 14.1a

1. (a) Order 1000 units whenever inventory level drops to 537 units.

Set 14.1b

2. Solution: $y^* = 317.82$ gallons, $R^* = 46.82$ gallons.

3. Solution: $y^* = 316.85$ gallons, $R^* = 58.73$ gallons. In Example 14.1-2, $y^* = 319.44$ gallons, $R^* = 93.61$ gallons. Order quantity remains about the same as in Example 14.1-2, but R^* is smaller because the demand pdf has a smaller variance.

Set 14.2a

3. $.43 \leq p \leq .82$

6. 32 coats.

Set 14.2b

1. Order $9-x$ if $x < 4.53$, else do not order.

Set 14.3a

2. Order $4.61-x$ if $x < 4.61$, else do not order.

CHAPTER 15

Set 15.1a

1. (a) Productivity = 71%.

(b) The two requirements cannot be met simultaneously.

Set 15.2a

1.

Situation	Customer	Server
(a)	Plane	Runway
(b)	Passenger	Taxi
(h)	Car	Parking space

Set 15.3a

1. (b) (i) $\lambda = 6$ arrivals per hour, average interarrival time $= \frac{1}{6}$hour.
(c) (i) $\mu = 5$ services per hour, average service time $= .2$ hour.

3. (a) $f(t) = 20e^{-20t}, t > 0$.
(b) $P\left\{t > \frac{15}{60}\right\} = .00674$.

7. Jim's payoff is 2 cents with probability $P\{t \leq 1\} = .4866$ and -2 cents with probability $P\{t \geq 1\} = .5134$. In 8 hours, Jim pays Ann $= 17.15$ cents.

10. (a) $P\{t \leq 4 \text{ minutes}\} = .4866$.
(b) Average discount percentage $= 6.208$.

Set 15.4a

1. $p_{n\geq 5}(1 \text{ hour}) = .55951$.

4. (a) $p_2(t = 7) = .24167$.

6. (a) Combined $\lambda = \frac{1}{10} + \frac{1}{7}$, $p_2(t = 5) = .219$.

Set 15.4b

2. (a) $p_0(t = 3) = .00532$.
(c) $p_{n\leq 17}(t = 1) = .9502$.

5. $p_0(4) = .37116$.

8. (a) Average order size $= 25 - 7.11 = 17.89$ items.
(b) $p_0(t = 4) = .00069$.

Set 15.5a

3. (a) $p_{n\geq 3} = .4445$.
(b) $p_{n\leq 2} = .5555$.

6. (a) $p_j = .2, j = 0, 1, 2, 3, 4$.
(b) Expected number in shop $= 2$ customers.
(c) $p_4 = .2$.

Set 15.6a

1. (a) $L_q = 1p_6 + 2p_7 + 3p_8 = .1917$ car.
(c) $\lambda_{\text{lost}} = .1263$ car per hour. Average number lost in 8 hr $= 1.01$ cars.
(d) No. of empty spaces $= c - (L_s - L_q) = c - \sum_{n=0}^{8} np_n + \sum_{n=c+1}^{8} (n - c)p_n$.

Set 15.6b

2. (a) $p_0 = .2$.
(b) Average monthly income $= \$50 \times \mu t = \$375$.
(c) Expected payment $= \$40 \times L_q = \$128$.

5. (a) $p_0 = .4$.
(b) $L_q = .9$ car.
(c) $W_q = 2.25$ min.
(d) $p_{n \geq 11} = .0036$.

6. (d) No. of spaces is at least 13.

Set 15.6c

1. $P\{\tau > 1\} = .659$.

5. \$37.95 per 12-hour day.

Set 15.6d

1. (a) $p_0 = .3654$.
(b) $W_q = .207$ hour.
(c) Expected number of empty spaces $= 4 - L_q = 3.212$.
(d) $p_5 = .04812$.
(e) 40% reduction lowers W_s to about 9.6 min ($\mu = 10$ cars/hr).

4. (a) $p_8 = .6$.
(b) $L_q = 6.34$ generators.
(c) Probability of finding an empty space cannot exceed .4 regardless of belt capacity. This means that the best utilization of the assembly department is 60%.

7. (a) $1 - p_5 = .962$.
(b) $\lambda_{\text{lost}} = \lambda p_5 = .19$ customer per hour.

Set 15.6e

2. For $c = 2$, $W_q = 3.446$ hour and for $c = 4$, $W_q = 1.681$ hour, an improvement of over 51%.

5. Let K be the number of waiting-room spaces. Using TORA, $p_0 + p_1 + \cdots + p_{K+2} \geq .999$ yields $K \geq 10$.

7. (a) $p_{n \geq 4} = .65772$.
(e) Average number of idle computers $= .667$ computer.

Set 15.6f

2. (c) Utilization $= 81.8\%$.
(d) $p_2 + p_3 + p_4 = .545$.

4. (a) $p_{40} = .00014$.
(b) $p_{30} + p_{31} + \text{L} + p_{39} = .02453$.
(d) Expected number of occupied spaces $= L_s - L_q = 20.043 - .046 \approx 20$.
(f) Probability of not finding a parking space $= 1 - p_{n \leq 29} = .02467$. Number of students who cannot park in an 8-hour period is approximately 4.

Set 15.6g

2. (a) Approximately 7 seats.
(b) $p_{n \geq 8} = .2911$.

Set 15.6h

1. (b) Average number of idle repairpersons = 2.01.
(d) $P\{2 \text{ or } 3 \text{ idle servers}\} = p_0 + p_1 = .34492$.

4. (a) $L_s = 1.25$ machines.
(b) $p_0 = .33342$.
(c) $W_s = .25$ hour.

6. $\lambda = 2$ calls per hour per baby, $\mu = .5$ baby per hour, $R = 5$, $K = 5$.
(a) Number of awake babies $= 5 - L_s = 1$ baby.
(b) $p_5 = .32768$.
(c) $p_{n \leq 2} = .05792$.

Set 15.7a

2. (a) $E\{t\} = 14$ minutes and $\text{var}\{t\} = 12$ minutes^2. $L_s = 7.8672$ cars.

4. $\lambda = .0625$ prescriptions per minute, $E\{t\} = 15$ minutes, $\text{var}\{t\} = 9.33$ minutes^2.
(a) $p_0 = .0625$.
(b) $L_q = 7.3$ prescriptions
(c) $W_s = 132.17$ minutes.

Set 15.9a

2. Use $(M/M/1):(GD/10/10)$. Cost per hour is \$431.50 for repairperson 1 and \$386.50 for repairperson 2.

4. (b) $\mu = \lambda + \sqrt{\frac{c_2\lambda}{c_1}}$
(c) Optimum production rate = 2725 pieces per hour.

Set 15.9b

2. (a) Hourly cost per hour is \$86.4 for two repairpersons and \$94.80 for three.
(b) Schedule loss per breakdown $= \$30 \times W_s = \$121.11$ for two repairpersons and \$94.62 for three.

4. Rate of breakdowns per machine, $\lambda = .36125$ per hour, $\mu = 10$ per hour. Model $(M/M/3):(GD/20/20)$ yields $L_s = .70529$ machine. Lost revenue = \$36.60 and cost of three repairpersons = \$60.

Set 15.9c

1. (a) Number of repairpersons ≥ 5.
(b) Number of repairpersons ≥ 4.

CHAPTER 16

Set 16.1a

4. (a) $P\{H\} = P\{T\} = .5$. If $0 \le R \le .5$, Jim gets \$10.00. If $.5 < R \le 1$, Jan gets \$10.00.

7. Lead time sampling: If $0 \le R \le .5$, $L = 1$ day. If $.5 < R \le 1$, $L = 2$ days. Demand per day sampling: If $0 \le R \le .2$, demand = 0 unit. If $.2 < R \le .9$, demand = 1 unit. If $.9 < R \le 1$, demand = 2 units. Use one R to sample L. If $L = 1$, use another R to sample demand for one day, else if $L = 2$, use one R to generate demand for day 1 and then another R to generate demand for day 2.

Set 16.2a

1. (a) Discrete.

Set 16.3a

4. See Figure C.15.

Set 16.3b

1. $t = -\frac{1}{\lambda}\ln(1 - R)$, $\lambda = 4$ customers per hour.

Customer	R	t (hr)	Arrival time
1	—	—	0
2	0.0589	0.015176	0.015176
3	0.6733	0.279678	0.294855
4	0.4799	0.163434	0.458288

2. $t = a + (b - a)R$.

4. (a) $0 \le R < .2: d = 0$, $.2 \le R < .5: d = 1$, $.5 \le R < .9: d = 2$, $.9 \le R \le 1$: $d = 3$.

9. If $0 \le R \le p$, then $x = 0$, else $x = \left(\text{largest integer} \le \frac{\ln(1 - R)}{\ln q}\right)$.

FIGURE C.15

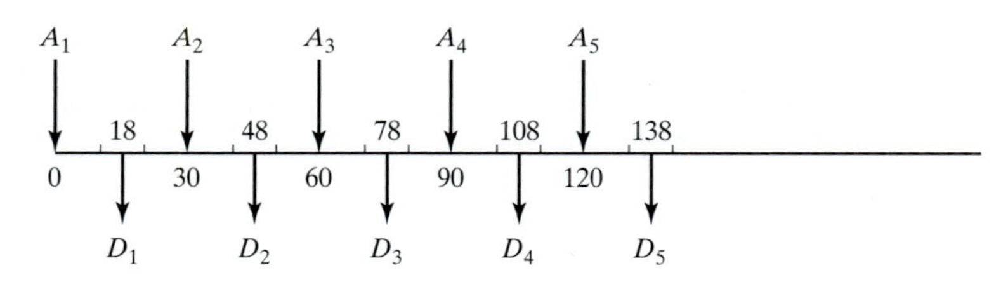

Set 16.3c

1. $y = -\frac{1}{5}\ln(.0589 \times .6733 \times .4799 \times .9486) = .803$ hour.
6. $t = x_1 + x_2 + x_3 + x_4$, where $x_i = 10 + 10R_i$, $i = 1, 2, 3, 4$.

Set 16.4a

1. In Example 16.4-1, cycle length = 4. With the new parameters, cycling was not evident after 50 random numbers were generated. The conclusion is that judicious selection of the parameters is important.

Set 16.5a

2. (a) Observation-based.
(b) Time-based.
3. (a) 1.48 customers.
(b) 7.4 hours.

Set 16.6a

2. Confidence interval: $15.07 \leq \mu \leq 23.27$.

CHAPTER 17

Set 17.1a

2. S1: Car on patrol
S2: Car responding to a call
S3: Car at call scene
S4: Apprehension made.
S5: Transport to police station

	S1	S2	S3	S4	S5
S1	0.4	0.6	0	0	0
S2	0.1	0.3	0.6	0	0
S3	0.1	0	0.5	0.4	0
S4	0.4	0	0	0	0.6
S5	1	0	0	0	0

Set 17.2a

2. Initial probabilities:

S1	S2	S3	S4	S5
0	0	1	0	0

Input Markov chain:

	S1	S2	S3	S4	S5
S1	0.4	0.6	0	0	0
S2	0.1	0.3	0.6	0	0
S3	0.1	0	0.5	0.4	0
S4	0.4	0	0	0	0.6
S5	1	0	0	0	0

Output (2-step or 2 patrols) transition matrix ($\mathbf{P}^2$)

	S1	S2	S3	S4	S5
S1	0.22	0.42	0.36	0	0
S2	0.13	0.15	0.48	0.24	0
S3	0.25	0.06	0.25	0.2	0.24
S4	0.76	0.24	0	0	0
S5	0.4	0.6	0	0	0

Absolute 2-step probabilities $= (0\,0\,1\,0\,0)\mathbf{P}^2$

State	Absolute (2-step)
S1	0.25
S2	0.06
S3	0.25
S4	0.2
S5	0.24

$P\{\text{apprehension, S4, in 2 patrols}\} = .2$

Set 17.3a

1. (a) Using excelMarkovChains.xls, the chain is periodic with period 3.
(b) States 1, 2, and 3 are transient, State 4 is absorbing.

Set 17.4a

1. (a)

Input Markov chain:

	S	C	R
S	0.8	0.2	0
C	0.3	0.5	0.2
R	0.1	0.1	0.8

Steady state probabilities:

$(\pi_1, \pi_2, \pi_3) = (\pi_1, \pi_2, \pi_3)\mathbf{P}$

$\pi_1 + \pi_2 + \pi_3 = 1$

Output Results:

State	*Steady state*	*Mean return time*
S	0.50	2.0
C	0.25	4.0
R	0.25	4.0

Expected revenues $= 2 \times .5 + 1.6 \times .25 + .4 \times .25 =$ \$1,500

(b) Sunny days will return every $\mu_{SS} = 2$ days—meaning two days on no sunshine.

5. (a)

Input Markov chain:

	never	some	always
never	0.95	0.04	0.01
some	0.06	0.9	0.04
always	0	0.1	0.9

(b)

Output Results

State	*Steady state*	*Mean return time*
never	0.441175	2.2666728
some	0.367646	2.7200089
always	0.191176	5.2307892

44.12% never, 36.76% sometimes, 19.11% always

(c) Expected uncollected taxes/year $= .12(\$5000 \times .3676 + 12{,}000 \times .1911) \times 70{,}000{,}000 = \$34{,}711{,}641{,}097.07$

14. (a) State = (i, j, k) = (No. in year -2, No. in year -1, No. in current year), i, j, k =(0 or 1)

Example: (1-0-0) this year links to (0-0-1) if a contract is secured next yr.

	0-0-0	1-0-0	0-1-0	0-0-1	1-1-0	1-0-1	0-1-1	1-1-1
0-0-0	0.1	0	0	0.9	0	0	0	0
1-0-0	0.2	0	0	0.8	0	0	0	0
0-1-0	0	0.2	0	0	0	0.8	0	0
0-0-1	0	0	0.2	0	0	0	0.8	0
1-1-0	0	0.3	0	0	0	0.7	0	0
1-0-1	0	0	0.3	0	0	0	0.7	0
0-1-1	0	0	0	0	0.3	0	0	0.7
1-1-1	0	0	0	0	0.5	0	0	0.5

(b)

State	Steady state
0-0-0	0.014859
1-0-0	0.066865
0-1-0	0.066865
0-0-1	0.066865
1-1-0	0.178306
1-0-1	0.178306
0-1-1	0.178306
1-1-1	0.249629

$$\begin{aligned}\text{Expected nbr. of contracts in 3 yrs} &= 1(0.066865 + 0.066865 + 0.066865)\\ &\quad + 2(0.178306 + 0.178306 + 0.178306)\\ &\quad + 3(0.249629) = 2.01932\end{aligned}$$

$$\text{Expected nbr. of contracts/yr} = 2.01932/3 = 0.67311$$

Set 17.5a

1. (a) Initial probabilities:

1	2	3	4	5
1	0	0	0	0

Input Markov chain:

1	2	3	4	5
0	.3333	.3333	.3333	0
.3333	0	.3333	0	.3333
.3333	.3333	0	0	.3333
.5	0	0	0	.5
0	.3333	.3333	.3333	0

State	Absolute (3-step)	Steady state
1	.07407	.214286
2	.2963	.214286
3	.2963	.214286
4	.25926	.142857
5	**.07407**	**.214286**

(b) $a_5 = .07407$
(c) $\pi_5 = .214286$
(d) $\mu_{15} = 4.6666$.

	$(\mathbf{I} - \mathbf{N})^{-1}$				Mu
	1	2	3	5	
1	2	1	1	.6667	4.6666
2	1	1.625	.875	.3333	3.8333
3	1	.875	1.625	.3333	3.8333
4	1	.5	.5	1.3333	3.3333

5. (a) Input Markov chain:

	A	B	C
A	.75	.1	.15
B	.2	.75	.05
C	.125	.125	.75

(b)

State	Steady state
A	.394737
B	.307018
C	.298246

A: 39.5%, B: 30.7%, C: 29.8%

(c)

$(\mathbf{I}-\mathbf{N})^{-1}$ and Mu

	A	C		B
A	5.71429	3.42857	A	**9.14286**
C	2.85714	5.71429	C	8.57143

	1	2		C
A	5.88235	2.35294	A	**8.23529**
B	4.70588	5.88235	B	1.5882

A → B: 9.14 years
A → C: 8.23 years

Set 17.6a

2. (a) States: 1wk, 2wk, 3wk, Library

Matrix **P**:

	1	2	3	lib
1	0	0.3	0	0.7
2	0	0	0.1	0.9
3	0	0	0	1
lib	0	0	0	1

(b)

$(\mathbf{I}-\mathbf{N})^{-1}$ and Mu

	1	2	3		lib
1	1	0.3	.03	1	1.33
2	0	1	.01	2	1.1
3	0	0	1	3	1

I keep the book 1.33 wks on the average.

8. (a)

Matrix P:

	1	2	3	4	F
1	0.2	0.8	0	0	0
2	0	0.22	0.78	0	0
3	0	0	0.25	0.75	0
4	0	0	0	0.3	0.7
F	0	0	0	0	1

(b)

	$(\mathbf{I}-\mathbf{N})^{-1}$ 1	2	3	4		Mu F
1	1.25	1.282	1.333	1.429	1	5.29
2	0	1.282	1.333	1.429	2	4.04
3	0	0	1.333	1.429	3	2.76
4	0	0	0	1.429	4	1.43

(c) To be able to take Cal II, the student must finish in 16 weeks (4 transitions) or less. Average number of transitions needed = 5.29. Hence, an average student will not be able to finish Cal I on time.

(d) No, per answer in (c).

10. (a) states:0, 1, 2, 3, D (delete)

Matrix **P**:

	0	1	2	3	D
0	0.5	0.5	0	0	0
1	0.4	0	0.6	0	0
2	0.3	0	0	0.7	0
3	0.2	0	0	0	0.8
D	0	0	0	0	1

(b) A new customer stays 12 years on the list.

	$(\mathbf{I}-\mathbf{N})^{-1}$ 0	1	2	3		Mu D
0	5.952	2.976	1.786	1.25	0	12
1	3.952	2.976	1.786	1.25	1	9.96
2	2.619	1.31	1.786	1.25	2	6.96
3	1.19	0.595	0.357	1.25	3	3.39

(c) 6.96 years.

CHAPTER 18

Set 18.1a

1. (a) No stationary points.

(b) Minimum at $x = 0$.

(e) Inflection point at $x = 0$, minimum at $x = .63$, and maximum at $x = -.63$.

4. $(x_1, x_2) = (-1, 1)$ or $(2,4)$.

Set 18.2a

1. (b) $(\partial x_1, \partial x_2) = (2.83, -2.5)\,\partial x_2$

Set 18.2b

3. Necessary conditions: $2\left(x_i - \frac{x_n^2}{x_i}\right) = 0, i = 1, 2, \ldots, n - 1$. Solution is $x_i = \sqrt[n]{C}$, $i = 1, 2, \ldots, n$. $\partial f = 2\delta\sqrt[n]{C^{2-n}}$.

6. (b) Solution $(x_1, x_2, x_3, x_4) = \left(-\frac{5}{74}, -\frac{10}{74}, \frac{155}{74}, \frac{60}{74}\right)$, which is a minimum point.

Set 18.2c

2. Minima points: $(x_1, x_2, x_3) = (-14.4, 4.56, -1.44)$ and $(4.4, .44, .44)$.

CHAPTER 19

Set 19.1a

2. (c) $x = 2.5$, achieved with $\triangle = .000001$.
(e) $x = 2$, achieved with $\triangle = .000001$.

Set 19.1b

1. By Taylor's expansion, $\nabla f(\mathbf{X}) = \nabla f(\mathbf{X}^0) + \mathbf{H}(\mathbf{X} - \mathbf{X}^0)$. The Hessian $\mathbf{H}$ is independent of $\mathbf{X}$ because $f(\mathbf{X})$ is quadratic. Also, the given expansion is exact because higher-order derivatives are zero. Thus, $\nabla f(\mathbf{X}) = \mathbf{0}$ yields $\mathbf{X} = \mathbf{X}^0 - \mathbf{H}^{-1}\nabla f(\mathbf{X}^0)$. Because $\mathbf{X}$ satisfies $\nabla f(\mathbf{X}) = \mathbf{0}$, $\mathbf{X}$ must be optimum regardless of the choice of initial $\mathbf{X}^0$.

Set 19.2a

2. Optimal solution: $x_1 = 0, x_2 = 3, z = 17$.

4. Let $w_j = x_j + 1, j = 1, 2, 3, v_1 = w_1 w_2, v_2 = w_1 w_3$. Then,
Maximize $z = v_1 + v_2 - 2w_1 - w_2 + 1$
subject to $v_1 + v_2 - 2w_1 - w_2 \leq 9$, $\ln v_1 - \ln w_1 - \ln w_2 = 0$,

$$\ln v_2 - \ln w_1 - \ln w_3 = 0, \text{ all variables are nonnegative.}$$

Set 19.2b

1. Solution: $x_1 = 1, x_2 = 0, z = 4$.

2. Solution: $x_1 = 0, x_2 = .4, x_3 = .7, z = -2.35$.

Set 19.2c

1. Maximize $z = x_1 + 2x_2 + 5x_3$
   subject to $2x_1 + 3x_2 + 5x_3 + 1.28y \leq 10$
   $9x_1^2 + 16x_3^2 - y^2 = 0$
   $7x_1 + 5x_2 + x_3 \leq 12.4, x_1, x_2, x_3, y \geq 0$

CHAPTER 20

Set 20.1a

1. See Figure C.16.

Set 20.1b

1. Case 1: Lower bound is not substituted out.

	x_{12}	x_{13}	x_{24}	x_{32}	x_{34}	
Minimize z	1	5	3	4	6	
Node 1	1	1				= 50
Node 2	−1		1	−1		= −40
Node 3		−1		1	1	= 20
Node 4			−1		−1	= −30
Lower bound	0	30	10	10	0	
Upper bound	∞	40	∞	∞	∞	

Case 2: Lower bound is substituted out.

	x'_{12}	x'_{13}	x'_{24}	x'_{32}	x'_{34}	
Minimize z	1	5	3	4	6	
Node 1	1	1				= 20
Node 2	−1		1	−1		= −40
Node 3		−1		1	1	= 40
Node 4			−1		−1	= −20
Upper bound	∞	10	∞	∞	∞	

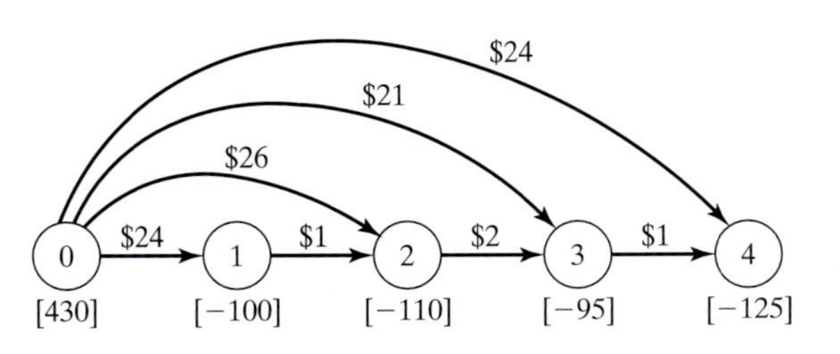

FIGURE C.16

Set 20.1c

1. Optimum cost = \$9895. Produce 210 units in period 1 and 220 units in period 3.

5. Optimal solution: Total student miles = 24,300. Problem has alternative optima.

	Number of students	
	School 1	*School 2*
Minority area 1	0	500
Minority area 2	450	0
Minority area 3	0	300
Nonminority area 2	1000	0
Nonminority area 2	0	1000

Set 20.2a

1. (c) Add the artificial constraint $x_2 \leq M$. Then

$$(x_1, x_2) = \alpha_1(0,0) + \alpha_2(10,0) + \alpha_3(20,10) + \alpha_4(20,M) + \alpha_5(0,M)$$

$$\alpha_1 + \alpha_2 + \alpha_3 + \alpha_4 + \alpha_5 = 1, \alpha_j \geq 0, j = 1, 2, \ldots, 5$$

2. Subproblem 1: $(x_1, x_2) = \alpha_1(0,0) + \alpha_2\left(\frac{12}{5}, 0\right) + \alpha_3(0,12)$
Subproblem 2: $(x_4, x_5) = \beta_1(5,0) + \beta_2(50,0) + \beta_3(0,10) + \beta_4(0,5)$
Optimal solution: $\alpha_1 = \alpha_2 = 0, \alpha_3 = 1 \Rightarrow x_1 = 0, x_2 = 12$

$$\beta_1 = .4889, \beta_2 = .5111, \beta_3 = \beta_4 = 0 \Rightarrow x_4 = 28, x_5 = 0.$$

6. Since the original problem is minimization, we must maximize each subproblem. Optimal solution: $(x_1, x_2, x_3, x_4) = \left(\frac{5}{3}, \frac{15}{3}, 0, 20\right), z = 195$.

CHAPTER 22

Set 22.1a

2. Solution: Day 1: Accept if offer is high. Day 2: Accept if offer is medium or high. Day 3: Accept any offer.

Set 22.2a

1. Solution: Year 1: Invest \$10,000. Year 2: Invest all. Year 3: Do not invest. Year 4: Invest all. Expected accumulation = \$35,520.

4. Allocate 2 bikes to center 1, 3 to center 2, and 3 to center 3.

Set 22.3a

3. Solution: First game: Bet \$1. Second game: Bet \$1. Third game: Bet \$1 or none. Maximum probability = .109375.

CHAPTER 23

Set 23.1a

2. Do not fertilize, fertilize when in state 1, fertilize when in state 2, fertilize when in state 3, fertilize when in state 1 or 2, fertilize when in state 1 or 3, fertilize when in state 2 or 3, or fertilize regardless of state.

Set 23.2a

1. Years 1 and 2: Don't advertise if product is successful; otherwise, advertise. Year 3: Don't advertise.
3. If stock level at the start of month is zero, order 2 refrigerators; otherwise, do not order.

Set 23.3a

1. Advertise whenever in state 1.

APPENDIX A

Set A.3a

1. `rest{i in 1..n}:(if i<=n-1 then x[i]+x[i+1] else x[1]+x[n])>=c[i];`

Set A.4a

2. See file A.4a-2.txt

Set A.5a

2. Data for `unitprofit` must be re-read four times with convoluted ordering of data elements.

```
24 5  6 4 4
 6 5  1 4 2
 1 5 -1 4 1
 2 5  0 4 1
```

Set A.5c

1. Error will result because members of sets `paint` and `resource` cannot be read from the double-subscripted table `RMaij`.

Index[1]

[1] *IBC* in page numbers refers to the software summary listed in the Inside Back Cover. The prefix CD refers to chapter/appendix material on the accompanying CD.

B

C

F

G

H

I

J

K

L

M

N

O

P

Q

R

S

T

U

V

W

Z